ASHGATE
RESEARCH
COMPANION

THE ASHGATE RESEARCH COMPANION TO WAR

ASHGATE
RESEARCH
COMPANION

The *Ashgate Research Companions* are designed to offer scholars and graduate students a comprehensive and authoritative state-of-the-art review of current research in a particular area. The companions' editors bring together a team of respected and experienced experts to write chapters on the key issues in their speciality, providing a comprehensive reference to the field.

Other Research Companions available in Politics and International Relations:

The Ashgate Research Companion to Non-State Actors
Edited by Bob Reinalda
ISBN 978-0-7546-7906-6

The Ashgate Research Companion to New Public Management
Edited by Tom Christensen and Per Lægreid
ISBN 978-0-7546-7806-9

The Ashgate Research Companion to Modern Warfare
Edited by George Kassimeris and John Buckley
ISBN 978-0-7546-7410-8

The Ashgate Research Companion to US Foreign Policy
Edited by Robert J. Pauly, Jr.
ISBN 978-0-7546-4862-8

The Ashgate Research Companion to Political Leadership
Edited by Joseph Masciulli, Mikhail A. Molchanov and W. Andy Knight
ISBN 978-0-7546-7182-4

The Ashgate Research Companion to Ethics and International Relations
Edited by Patrick Hayden
ISBN 978-0-7546-7101-5

The Ashgate Research Companion to Federalism
Edited by Ann Ward and Lee Ward
ISBN 978-0-7546-7131-2

The Ashgate Research Companion to the Politics of Democratization in Europe
Edited by Kari Palonen, Tuija Pulkkinen and José María Rosales
ISBN 978-0-7546-7250-0

The Ashgate Research Companion to War
Origins and Prevention

Edited by

HALL GARDNER and OLEG KOBTZEFF
The American University of Paris, France

Routledge
Taylor & Francis Group
LONDON AND NEW YORK

First published 2012 by Ashgate Publishing

2 Park Square, Milton Park, Abingdon, Oxon OX14 4RN
711 Third Avenue, New York, NY 10017, USA

Routledge is an imprint of the Taylor & Francis Group, an informa business

First issued in paperback 2017

British Library Cataloguing in Publication Data
The Ashgate research companion to war : origins and
prevention.
1. War--Causes. 2. Military history. 3. Cold War. 4. Cold
War--Influence. 5. War and society.
I. Gardner, Hall. II. Kobtzeff, Oleg.
355'.02-dc22

Library of Congress Cataloging-in-Publication Data
Gardner, Hall.
The Ashgate research companion to war : origins and prevention / by Hall Gardner and Oleg Kobtzeff.
p. cm.
Includes bibliographical references and index.
ISBN 978-0-7546-7826-7 (hbk)
1. War--Causes. 2. War--Prevention. 3. War--History. 4. International relations--History. I. Kobtzeff, Oleg. II. Title.
JZ6385.G35 2012
355.02--dc23

2011030408

ISBN 978-0-7546-7826-7 (hbk)
ISBN 978-1-138-11109-7 (pbk)

Contents

PART II MAJOR WARS IN HISTORY

PART III COLD WAR AND BEYOND ...

List of Figures

List of Tables

Notes on Contributors

Jean-Paul Charnay is the director of research at CNRS (the National Center for Scientific Research in France), as well as founder and president of the Center for Philosophy of Strategy at the Sorbonne in Paris. Publications include: *Principes de stratégie arabe* [*Principles of Arab Strategy*] (Paris, France: L'Herne, 2003), *Regards sur l'Islam, Freud, Marx, Ibn Khaldun* [*Perspectives on Islam, Freud, Marx, Ibn Khaldun*] (Paris, France: L'Herne, 2003) and *La Charîa et l'Occident* [*Sharia and the Occident*] (Paris, France: L'Herne, 2001).

Ben Cramer is a Paris-based researcher of the political dimensions of the arms race and a journalist specializing in ecological security. He co-authored the UNESCO-sponsored award-winning interactive encyclopedia on water, *L'Or Bleu* (*Blue Gold*), and served as a consultant to Green Cross for their 'Legacy of the Cold War' program. He is co-author with Camille Saïsset of *La Descente aux Enfers Nucléaires* [*Descent into Nuclear Hell*] (Paris, France: L'Esprit Frappeur, 2004). Vice-president of the French Association des Journalistes de l'Environnement, he is a lecturer at the Geneva International Peace and Research Institute (GIPRI) and wrote *Nuclear Weapons: At What Cost?* (Geneva: International Peace Bureau, 2009).

Anthony D'Agostino is a professor in the department of history at San Francisco State University. Publications include: *The Russian Revolution, 1917–1945* (forthcoming), *Gorbachev's Revolution, 1985–1991* (New York, NY: NYU Press, 1998), *Soviet Succession Struggles: Kremlinology and the Russian Question from Lenin to Gorbachev* (London, UK: Routledge, 1988) and *Marxism and the Russian Anarchists* (Cambridge, UK: Germinal Press, 1977).

Troy Davis is the president of the World Citizen Foundation, as well as vice-president of Business Development at Ecozeo and an administrator at ECO-Conseil, the European Institute of Environmental Counseling. He has written extensively on global politics, reflected in his book *Appel pour une Démocratie Mondiale* [*Call for a Global Democracy*] (Paris, France: Desclée de Brouwer, 1998).

Debra L. DeLaet is a professor at Drake University specializing in human rights, international law, the United Nations, gender and world politics. Publications include *The Global Struggle for Human Rights: Universal Principles in World Politics* (Belmont, CA: Wadsworth Publishing, 2006) and *U.S. Immigration Policy in an Age of Rights* (Westport, CT: Praeger Publishers, 2000).

Steven Ekovich is an associate professor at the American University of Paris in the International and Comparative Politics Department, and a former associate professor at the *École Polytechnique* (1984–2000) as well as the *Institut d'Études Politiques de Paris*. He has also been a lecturer at the *Institut Supérieur d'Armement et Défense* at the University of Paris II, and a tutor at the French Ministry of Foreign Affairs (1995–2001). He graduated from the University of California, Irvine, with a BA in Philosophy and History (1975), an MA in History (1979) and a PhD in History (1984). Prior to studying at the University of California, he served with the United States Air Force in Vietnam from 1969–1972. He has lectured and consulted abroad in more than 30 countries in Europe and Africa.

Peter Emerson is the director of the de Borda Institute in Belfast, Northern Ireland. Selected publications include *Designing an All-Inclusive Democracy* (editor) (Berlin, Germany: Springer, 2007), *Defining Democracy* (Belfast, Northern Ireland: The de Borda Institute, 2002) and *Beyond the Tyranny of the Majority* (Belfast, Northern Ireland: The de Borda Institute, 1998).

Hall Gardner is Professor and Chair of the Department of International and Comparative Politics, The American University of Paris. He is the author of *Averting Global War* (New York, NY: Palgrave Macmillan, 2007; 2010 paperback); *American Global Strategy and the 'War on Terrorism'* (Burlington, VT: Ashgate, 2007); *Dangerous Crossroads* (Westport, CT: Praeger Publishers, 1997); *Surviving the Millennium* (Westport, CT: Praeger Publishers, 1994), among many other edited books and articles, including *NATO and the European Union: New World, New Europe, New Threats* (Burlington, VT: Ashgate, 2004). He was a member of the Group of Experts which produced the EastWest Institute report *Euro-Atlantic Security: One Vision, Three Paths* (New York, NY: EWI, June 2009) that was presented at the OSCE Summit in Corfu in June 2009. His website is: http://www.epsilen.com/hgardner.

Azar Gat is the Ezer Weitzman Professor for National Security at Tel Aviv University. His publications include *Victorious and Vulnerable: Why Democracy Won in the 20th Century and How it is Still Imperiled* (New York, NY: Rowman & Littlefield Publishers, Inc., 2009); *War in Human Civilization* (Oxford, UK: Oxford University Press, 2006); and *A History of Military Thought: From Enlightenment to the Cold War* (Oxford, UK: Oxford University Press, 2001).

François Géré has published extensively on nuclear proliferation, counter terrorism, psychological warfare, and military strategy. *The Strategic Role of Information and Disinformation* is forthcoming. He is author of *L'Iran et le nucléaire, les tourments perses* [*Iran and Nuclear Power: Persian Torments*] (Paris, France: Lignes de Repères, 2006); *Dictionnaire de la désinformation* [*Dictionary of Disinformation*] (Paris, France: Armand Colin, 2011); *La pensée militaire allemande* [*German Military Thought*] (with Eugène Carrias and Beatrice Heuser) (Paris, France: Economica, 2010); *Iran, l'état de crise* [*Iran: State of Crisis*] (Paris, France: Karthala, 2010); *La guerre moderne* [*Modern War*] (with Roger Trinquier) (Paris, France: Economica, 2008); *Pourquoi le terrorisme?*

[*Why Terrorism?*] (Paris, France: Larousse, 2006); *Les Volontaires de la mort: L'Arme du suicide* [*Volunteers of Death: Suicide as a Weapon*] (Paris, France: Bayard, 2003); *Pourquoi les guerres?* [*Why Wars?*] (Paris, France: Larousse, 2003).

Athina Karatzogianni is a lecturer in Media, Culture and Society at the University of Hull, UK. She is the author of *The Politics of Cyberconflict* (London, UK: Routledge, 2006), *Power, Conflict and Resistance: Social Movements, Networks and Hierarchies* (with Andrew Robinson) (London, UK: Routledge, 2010), and editor of *Cyber Conflict and Global Politics* (London, UK: Routledge, 2009). Athina has theorized and written extensively on various incidents of cyberconflict, while other work has explored world-system theory and global networks of resistance. She is also contributing to work theorizing ultraviolent subjectivities in cyberspace, examining conflict analysis and war coverage of crises in global hotspots and exploring the potential of information communication technologies and network forms of organization for social movements, resistance and open knowledge production. She can be contacted at: athina.k@gmail.com.

Oleg Kobtzeff, Assistant Professor at The American University of Paris, is a Fellow of the Royal Geographical Society. He studied at the Sorbonne primarily under Pierre Chaunu, leading figure of the second generation of the *Annales*, then turned to Slavic and Arctic studies, working on Stalinism and colonization in the Asia-Pacific region under François-Xavier Coquin. His publications often compare the great European powers and the US, and range over topics such as violence, colonialism, environmental politics and identity. He is also a regular commentator for major radio and TV stations on three continents. A full list of publications and broadcasts is available at: http://olegkobtzeff.blogspot.com.

Carol Mann, sociologist, anthropologist and historian, author and lecturer specializing in Gender and Conflict, is Research Associate at SOAS (School of Oriental and African Studies, University of London), and teacher of Gender and Conflict at the American Graduate School in Paris. She is Director and Founder of FemAid and Women in War, both charities registered in France working on gender problems in conflict zones in the developing world. Her most recent books include *Femmes dans la Guerre (1914–1945)* (Paris, France: Pygmalion/Flammarion, 2010) and *Femmes Afghanes dans la Guerre* (Paris, France: Le Croquant, 2010).

George Modelski is Professor Emeritus, Political Science, at the University of Washington, Seattle. His most recent books are *World Cities: -3000 to 2000* (Washington, DC: FAROS 2000, 2003); *Leading Sectors and World Powers: The Coevolution of Global Economics and Politics* (with William R. Thompson) (Columbia, SC: University of South Carolina Press, 1995); *Documenting Global Leadership* (with Sylvia Modelski) (Humanity Press/Prometheus Bk, 1988); *Globalization as Evolutionary Process: Modeling Global Change* (with Tessaleno Devezas and William R. Thompson) (London, UK: Routledge, 2008).

Susan Hitchcock Perry, a sinologist and specialist in international human rights law, holds degrees from Brown, Yale, and Oxford universities and the École des Hautes Études en Sciences Sociales (University of Paris). She is the current Director of the Division of International Politics, Economics, and Public Policy (IPEPP) at The American University of Paris. Dr Perry recently served in Chad as a member of the US Delegation to the drafting of the N'djamena Declaration to End the Recruitment and Use of Child Soldiers, a regional best practices document designed to push for an accelerated demobilization and rehabilitation of the thousands of child soldiers in Central Africa.

Ilya Platov is an associate professor and researcher in the Russian department of INALCO, the National Institute of Languages and Oriental Civilizations in Paris. He is the author of a doctoral thesis *The Slavic Crusade: War, Culture and Memory in Russia and in the Balkans, 1876–1914* [in French]. He is a specialist of Russian cultural history in the nineteenth and early twentieth centuries, with a specific focus on wartime culture, and sacred and symbolic sources of national identity.

Marco Rimanelli is Professor in European Affairs and International Security, and Director of International Studies at Saint Leo University, Tampa, Florida, USA. He served twice in the US government in Washington DC as Scholar-in-Residence/Senior Analyst on NATO–European Security and US/NATO–USSR/Russia nuclear arms control (1999–2001, 1990–92), and in 2004–05 was US Fellow at the Institut des Hautes Études de Défense Nationale in France's École Militaire (War College), Paris. He has published seven books, including *Historical Dictionary of NATO and Other International Security Organizations* (Lanham MD: Scarecrow/Rowman, 2009).

Andrew Robinson is a political theorist and contributor to *Ceasefire*, an independent magazine based in the United Kingdom. Publications include *Power, Resistance and Conflict in the Contemporary World* (with Athina Karatzogianni) (London, UK: Routledge, 2010).

William R. Thompson is the Rogers Professor of Political Science at Indiana University. His publications include *Causes of War* (with Jack S. Levy) (Malden, MA: Blackwell, 2010), *Limits to Globalization: North-South Divergence* (with Rafael Reuveny) (London, UK: Routledge, 2009) and *Leading Sectors and World Powers: The Coevolution of Global Economics and Politics* (with George Modelski) (Columbia, SC: University of South Carolina Press, 1995).

Preface

In many ways this book came about through sheer serendipity. My previous edited books were the result of planned conferences.

Putting together scholarship written in both French and English, *The Ashgate Research Companion to War: Origins and Prevention* publishes scholars who already possess international reputations, plus a number of authors who are well known in France, but who have never been published in English, while also introducing new voices. My sole regret is that my now retired mentor at *The Johns Hopkins Paul H. Nitze School of Advanced International Studies*, George (Jiri) Liska, who in many ways inspired me to examine the world from a long term geohistorical perspective, engaging both theory and history, and whose later work on empires and systemic conflicts has generally not received the attention it deserves, could not participate.

I thank the editors at Ashgate for making this book possible. And I thank my family once again for coping with yet another one of my projects. My assistants, Elizabeth Austin, Molly Fee and Alexandre Schmitt helped review texts, endnotes and translations. Co-editor Oleg Kobtzeff helped translate the complex and brilliantly original texts of Jean-Paul Charnay from French, while contributing his own unique and deeply researched perspectives. I am convinced that a number of these chapters will prove to be major contributions to the field of conflict studies—and to the conceptual development of *polemology*.

Hall Gardner
Paris, November 11, 2011

Perhaps wars weren't won any more. Maybe they went on forever.

Ernest Hemingway, *A Farewell to Arms*
(Simon & Schuster, 1929; 1995), 118

The chain reaction of evil—hate begetting hate, wars producing more wars—must be broken, or we shall be plunged into the dark abyss of annihilation.

Martin Luther King, Jr, *Loving Your Enemies*
(Sermon, Dexter Avenue Baptist Church,
Montgomery, Alabama, Christmas, 1957)

General Introduction: Polemology

Hall Gardner and Oleg Kobtzeff[1]

One can dispute the fragments of Heraclitus that "war is the father of all things." One can attack the gender bias of such a statement raising the question "and who is the mother?" And one can even develop a theory of spontaneous generation; but one cannot deny the fact that major wars are often at the origins of, or originate from, profound transformations in the nature of state and societal interaction. One also cannot deny that major—and even some so-called "minor"—conflicts can possess significant and long term socio-historical, if not globally systemic, ramifications.

Throughout history, questions concerning the origins of both civil conflict and international/ interstate wars have challenged scholars from differing disciplines (historians, social psychologists, sociologists, economists, political scientists, anthropologists, cultural studies specialists, philosophers, military strategists, international relations theorists and many others). The causes of conflict, the psychology and motivations of the belligerents, the strategies and tactics, the mechanisms leading to the breakdown of peaceful relations have all been debated (whether those conflicts are caused on purpose, by mishap or by accidents "waiting to happen," or even by actions taken "accidentally on purpose"). Scholars have likewise debated the social and historical consequences and general ramifications of war; they have additionally speculated on the question as to whether many of these conflicts were inevitable or could have been avoided, and what, if any, alternative policies or actions might have prevented the outbreak of conflict, or else transformed the outcomes.

Polemology

From the 1920s to the early 1990s, the "New Historians" (as their younger generation liked to be identified)—the "Annales school" as they are known in the English-speaking world—discredited the practice of history as a mere literary

1 Hall Gardner wrote the General Introduction, with the exception of the section on polemology, written by Oleg Kobtzeff.

exercise in story-telling and reconstruction of chains of events. The history of wars, along with the never-ending narratives of diplomatic intrigues and accounts of battles, symbolized what they loathed in traditional historical literature. Neither could the New Historians, although some remained Marxist, be content with fitting events into drawers conveniently marked "class struggle" or "dialectical materialism." Nothing could be so simple after the cataclysms of the twentieth century. The Years 1914–1918, which wiped out a large portion of Europe's history students and teachers, also swept away the survivors' certitudes, including historiographic traditions, particularly the positivist assumption that historians need only to identify logical linear cause-and-effect mechanisms in order to explain past events and readjust humanity's normally unstoppable advance on the road of peace and progress. Unfortunately, even in the twenty-first century, many students learning history must still suffer reading the industrially-produced textbooks of the last of the historians who, like soap-opera script-writers, spin a thread made of incidents rythmed by adverbs ("consequently," "subsequently," "accordingly," "therefore," "typically," "incidentally," "inadvertently," "not surprisingly," etc.). New historians challenged this near-sightedness by rejecting *events* and exploring large-scale *situations* involving multi-layered environmental, economic, social, or cultural phenomena evolving over long periods, with the tools of all available social sciences and even natural sciences.

Several research undertakings should have given hope that the Annales school was finally going to produce a new kind of literature on the history of war. Such examples of research were Pierre Chaunu's team's work on violence and criminal justice, Jean Delumeau's famous studies of collective fears in the Middle Ages, or Alain Besançon's psychoanalytical, sociological and cultural analysis of centuries of ideological construction and power in Russia. Unfortunately, such projects would remain isolated. By the time postmodernism became very influential in France as well as in the world, around the 1980s and 1990s, marking the decline of the Annales school (which had itself reached the peak of its influence in the French intelligentsia and public in the 1970s), war was still not a subject that the New Historians had been able to interpret. The original prejudice against war studies in particular and political science in general still prevailed after three generations.[2]

This probably explains the isolation of French sociologist Gaston Bouthoul (1896–1980) who devoted his career to fusing the knowledge and methodologies of all social sciences into a new interdisciplinary field of research and reflection for which he invented a new word: *polemology*. Originally published in 1951 as *Les guerres* [*Wars*], and renamed *Traité de Polémologie* [*Treatise on Polemology*] in 1970 (published in Paris by Payot), Bouthoul's landmark monograph and his other writings are known to only a few specialists. The word "polemology" remains

2 As demonstrated by Annales dissident Hervé Coutau-Bégarie in his *Le phénomène "Nouvelle Histoire": Stratégie et idéologie des nouveaux historiens* [*The Phenomenom of New History: Strategy and Ideology of the New Historians*] (Paris: Economica, 1983), petty campus politics in Paris prevented the junction of research teams close to the Annales and those, mainly in sociology departments, who were working on collective violence.

either almost unheard of or is mistakenly associated with the kind of practical academic programs delivered by military institutions. In the aftermath of World War I, and later, again, following World War II, it had been too difficult to explain to a battle-fatigued public the need for a large-scale effort to create a science of wars. The tens of millions of casualties made it difficult to accept that war can be quietly studied as an object of science, an "ordinary" social phenomenon. Bouthoul had to defend his project and explain that it was not just a laboratory for the amusement of cold-blooded scientists oblivious to the sufferings of the victims. Also, when Bouthoul and his team of researchers founded, in 1945, a modest research center named the French Institute of Polemology (Institut Français de Polémologie) they were met with suspicion which was typical of the late forties to mid-to-late 1960s. On one hand, the now dominating Annales school suspected too much political comment over war; on the other, an increasingly radical left-wing intelligentsia rejected any activity in the arts or social sciences that could not justify itself as politically and socially useful for the Revolution (peace studies, like hippies, risked demobilizing the working masses). The fact that Bouthoul's institute cooperated with the French military academic system made him even more suspicious. Nevertheless, Bouthoul, a socialist who loathed war, strongly argued that his scientific activities were pacifist in nature. Although he conceived *polemology* as a discipline based on scientific methods, academic objectivity was not meant to abolish the scientist's ability to function as a thinking and responsible citizen striving for peace. Fortunately, other scholars initiated comparable efforts in other parts of the world. At a time when nuclear holocaust appeared to be a real possibility, it became evident that conflict prevention or resolution was hampered by the dramatic lack of research structures that could offer the necessary broad array of conflict analysis tools. The urgency of the need for more research on war and peace became evident after such dramatic moments as the Cuban missile crisis, or under the growing pressure from peace movements in the 1960s. The emerging disciplines or sub-disciplines of sociology or political science became better known as "Peace Studies" with the creation of such groups and institutions as PRIO, the Peace Research Institute Oslo (associated with Johan Galtung); or the Malmö group that established the Peace Research Society (which initially included Walter Isard, Kenneth Boulding and Anatol Rapoport, and which became in 1973 the Peace Science Society), or many teaching and research institutions in a variety of departments housed by prestigious universities in Anglo-Saxon academia in Scandinavia, Italy and the Netherlands.

It is with this in mind that *The Ashgate Research Companion to War: Origins and Prevention* seeks to advance a new interdisciplinary field of research and reflection for which Bouthoul coined the term "polemology," stemming from the ancient Greek word for war, *pólemos*. As professors at the American University of Paris, the editors have gained closer contact with the many perspectives on historiography of their French and European colleagues than they would have had they taught on a more typical American campus in North America. The editors thus consider it essential to break the linguistic barrier of the French frontier and, for the first time, bring together North American and European authors—*polemologists*—who have developed and analyzed the origins and strategies of war in different fields and from different theoretical and cultural perspectives. As a tribute, at least, to the many efforts of Gaston Bouthoul, the neologism "polemology" deserves to enter the English language and be propagated as the best designation for the inter-disciplinary field presented in the following pages. The scale of this presentation is, however, considerably more modest than Bouthoul's monumental enterprise and it can only aim at presenting some aspects of how theories of war causation and histories of warfare could be written from new perspectives.[3]

Toward an Interdisciplinary and Historical Approach to War

The *polemological* approach to war causation represents an essentially *interdisciplinary* and *historical* approach to the study of war. It seeks to explore the interacting geostrategic, military technological, political-economic, legal (including domestic and international laws and norms), socio-cultural (including religion and values), bio-political (including ethnic identity, age, gender and demography), ideological (including conceptions of justice and peace), dialogical (including diplomacy, media and propaganda), natural-environmental and psychological factors (including alienation) that influence and impact upon the causes of conflict and war. The *polemological* approach argues that there is rarely a *mono-causal* factor for conflict and war, but interrelated causes, of which some may be more significant or relevant than others *depending upon the confluence of forces in specific situations and historical periods*.

From a polemological perspective, war cannot be considered merely as "organized violence carried on by political units against each other" in Hedley Bull's formulation, as there are many other factors, in addition to politics, at play as well, while the preparation for war, if not the act of war itself, helps to organize (or possibly undermine) the entire social, economic and political system.[4] Likewise, the Marxist emphasis on class struggle is not sufficient to explicate the full range of factors that help generate conflict. While Karl Marx analyzed conflict through an essentially socio-economic concept of alienation, *an expanded concept of alienation that reveals its multidimensional features* helps to better explain the complex

3 On the contribution of French social scientists and the Annales school to polemology, see Oleg Kobtzeff, Chapters 4, 10, and 15. On the Annales school, see also D'Agostino, Chapter 14. The conceptual development of "long cycle theory" is explained by George Modelski in Chapter 28. Ben Cramer discusses the interdisciplinary approach to environment and conflict and critiques Bouthoul's ecological argument of "carrying capacity" in Chapter 22. Azar Gat discusses his interdisciplinary approach to war in Chapter 2. See discussion of need for an interdisciplinary approach to cyber conflict in Athina Karatzogianni, Chapter 23.

4 Hedley Bull, *The Anarchical Society* (London: Macmillan, 1977). See also Charles Tilly, *Coercion, Capital and European States* (Cambridge, MA: Blackwell, 1992) on war as a force organizing the state and the society.

interaction between the powerful and powerless, between the decision-making process and the larger society in interaction with often contending domestic and international interests, values and beliefs, not to overlook other geostrategic, military-technological, political-economic, socio-cultural and ideological forces and factors that can lead to inter-state and inter-societal disputes—if not to conflict and war. (See Gardner, Chapter 1.)

A *polemological* or interdisciplinary *historical* approach, as articulated in this book, seeks to enlarge the perspective of Kenneth Waltz in his work, *Man, State and War*, in which Waltz recognized that any full understanding of the origins of conflict must show the interrelationship and interaction between the first image (human psychology), the second image (socio-political formation and state governance) and the third image (international structure and system). In his later, essentially *ahistorical* work, *A Theory of International Politics*, however, Waltz focuses primarily on the third image (international structure and system) alone, by arguing, in essence, that international anarchy sets the conditions for perpetual conflict, although in Waltz's view "bipolar" systems are generally more stable than so-called "multipolar" systems.[5] In effect, if the international state system cannot transcend the lack of international governance and conditions of "anarchy," which are largely seen by Waltz as permanent features of the international system—regardless of the fact that differing systems of socio-political governance behave very differently in diverse historical epochs—then state leaderships are basically helpless to prevent conflict.[6]

Waltz's later emphasis on the third image of international anarchy, downplaying the first and second images, opened his theory to attack by Alexander Wendt who

5 See Gardner, Chapter 1, this book. Despite the obvious elements of US-Soviet rivalry, the bicentric Cold War was also characterized by elements of tacit cooperation against emerging powers that might threaten the interests of the both the US/NATO and Soviet Union/Warsaw Pact, which were able to maintain a rough military-technological parity (despite Soviet weaknesses) against one another and hence remained stable until Soviet break-up. The post-Cold War period, however, is characterized by the emergence of polycentric system in which states (and anti-state movements) possess highly uneven capabilities of force and influence that can threaten US and Russian interests, and which have not yet congealed into rival alliance formations as was the case for World Wars I or II, for example. See Hall Gardner, *American Global Strategy and the "War on Terrorism"* (Ashgate: 2010), 24–26.

6 In *Man, the State, and War: A Theoretical Analysis* (New York: Columbia University Press, 1959), Kenneth Waltz outlined the "three images" (structure of the international system, nature of the domestic society and psychology of the leadership). He argued that the anarchical structure of the international system largely predominated over the nature of the domestic system and the psychology of elites—even if all three elements were interrelated. At the same time, he gave very little attention to actual strategy and tactics (at the state level), state concerns at the sub-systemic levels with neighbors and regions and the psychology of leadership (at the individual level) as leadership interacts with differing and often clashing domestic socio-political-economic actors and forces. See also Kenneth Waltz, *A Theory of International Politics* (New York: McGraw Hill, 1979).

gave primacy to the second image, arguing that "anarchy is what states make of it."[7] Wendt's point is that there is an interaction between international system and structure and the policy decisions of individual states. In the post-World War II period, for example, the majority of European states have sought to transcend historically generated conditions of international "anarchy" at least *within that specific region*. This has led some to claim that Europe is now "post-historical"—although savage wars along the European periphery, global financial crisis, international migration, among other factors, have certainly been straining inter-state and inter-societal European cooperation in the post-Cold War era.

At the same time, Wendt's emphasis on the second image appears to overlook the fact that states are made up of often contending individual leaderships with differing socio-psychological perspectives and socio-cultural values and ideologies; various political factions or parties; rivalries among state bureaucracies; and differing, often conflicting, societal groups with differing interests, values and beliefs that may or may not support state policies, while, at the same time, state leaderships are not only concerned with international systemic concerns, as Waltz indicates, but with regional actors and issues as well. In effect, each state leadership must make key *strategic choices* that are based, at least in part, on Waltz's first image, involving the social-psychology of the leadership's decision-making process and that leadership's method of analyzing domestic, regional *and* international disputes, issues and goals. States must then act upon those assessments, often with that leadership's own interests in mind. In effect, the question of conflict among states—and particularly that of systemic conflict (or global war)—is to a large extent dependent upon the *strategic choices* of state elites who can choose to go, or not to go, to war, taking the risks and consequences involved in those decisions. Such choices are *relatively autonomous* and are generally implemented in reaction to the perceived nature of the global system and related regional sub-systems—as well as to the perceived nature, interests, values and beliefs of domestic political factions, groups with common identities or values and socio-economic classes.

In essence, this book seeks to address the *systemic* nature of conflict, which includes both international/interstate and domestic/civil wars—with an emphasis upon major and regional power warfare *over the long term*. It is often stated that interstate wars appear to be waning in number, in large part due to the fact that peace is more profitable (see Azar Gat, Chapter 2), or that war is "obsolete" (see critique by Modelski, Chapter 28). A number of studies have indicated that interstate wars appear to be decreasing in intensity and in number of deaths, and that in the post-Cold War era there have been more internal conflicts than inter-state conflicts, at least since 1991.[8]

7 Alexander Wendt, "Anarchy is What States Make of It," *International Organization* Vol. 46, No. 2 (Spring 1992).

8 Human Security Report Project, *Human Security Report (2009–2010): The Causes of Peace and the Shrinking Causes of War* (Vancouver: HSRP, 2010), http://www.hsrgroup.org/human-security-reports/20092010/text.aspx. See also debates in Raimo Väyrynen (ed.), *The Waning of Major War: Theories and Debates* (London and New York: Routledge, 2006).

At the same time, however, it should be pointed out that many of these so-called "internal" conflicts, which can involve genocides and democides, can possess devastating effects upon the populations involved, creating both internal and external refugees, and making economic growth more difficult, if not impossible, to achieve, with significant impact upon the levels of poverty, the educational systems of the inhabitants, as well as upon human health (war-related diseases) and the immediate environment. More to the point, one can question whether these ostensibly "internal" wars necessarily originate "within" states—as many civil wars have been secretly armed or financed by outside powers. Moreover, a number of these so-called internal conflicts have resulted in outside military interventions, whether by unilateral or multilateral state actions, in which some, but not all, interventions have been legitimized by the UN Security Council in the post-Cold War era.

Post-Cold War Liberal Interventionism and the "Responsibility to Protect"

A number of post-Cold War military interventions have, at least in part, been rationalized through the development of a "new" legal and essentially cosmopolitan moral/ethical principle: the "responsibility to protect," in which state sovereignty is no longer regarded as possessing absolute impunity within a specific territory under Article 2(7) of the UN Charter. Rather, state sovereignty is increasingly seen as a "right" granted or legitimized by the "international community" based upon mutual respect between and among states and societies with mutual acceptance of international law and norms. In other words, sovereignty is no longer granted by a mere monopoly of force over territory alone, in Max Weber's views, but represents a more complex interrelationship and accord among states and their civil societies—a new form of global "social contract". (See Troy Davis, Chapter 7.) State leaderships must accordingly act "responsibly" towards other states and populations, as well as toward their own citizens, including migrants. If, however, state leaderships do not or cannot act responsibly toward their own citizens, and if they engage in extreme human rights abuses, then the states of the "international community" can claim the "responsibility to protect" those civilian populations which are not being properly protected by their own authorities.[9]

The decision to intervene militarily under the "responsibility to protect" doctrine depends on an appropriate analysis of what is "large scale" loss of life, actual or "apprehended." Yet "apprehension" of a *future* large scale loss of life raises questions as to how the event is perceived, interpreted and then *mediatized*

9 International Commission on Intervention and State Sovereignty, *The Responsibility to Protect* (December 2001), http://www.iciss.ca/pdf/Commission-Report.pdf. See also UN Resolution 1674 (2006), http://www.un.org/News/Press/docs/2006/sc8710.doc.htm.

(often with political purposes in mind involving mis- and/or dis-information), so as to justify or not justify an intervention based on predictions that *may* or *may not* prove accurate. "Responsibility to protect" doctrine also depends upon "right intention." Yet the "right intention" of states is questionable in that states generally hide ulterior motives, even if working multilaterally.

Moreover, state intervention is always *selective* (and not universal): Why intervene militarily in Afghanistan, Iraq, Pakistan, Bosnia/Kosovo/Serbia, Libya or elsewhere while ignoring other states that are also engaging in egregious human rights violations? Furthermore, the process of choosing which domestic political groups and organizations to support against the "oppressive" government in power (or which political factions to oppose) can also be selective. Will those political factions supported by external powers necessarily be perceived as *legitimate* by the local population? What kind of policies will these factions enact once, and if, they achieve power? How will they treat their former opponents?

Using force only as a "last resort" is likewise crucial to the responsibility to protect doctrine, yet it is not at all clear that states always engage in all possible measures to prevent conflict, even in situations where the outbreak of conflict has been predicted long before it actually erupts. It is generally only once the possibility of conflict appears *imminent* (however defined) that states engage in last minute efforts to mediate, and hence attempt to avert conflict. The dilemma is that they often fail, doing far too little, too late. A major concern raised here is that state leaderships generally do not give sufficient long term attention to transforming or reforming the *deeper* systemic and structural causes of conflict and are consequently unable to prevent conflict in the first place. Even if the UN or OSCE possessed peacekeeping forces at their immediate disposal, military intervention and peacemaking/peacekeeping really represent palliatives that can only attempt to ameliorate the nature of specific conflicts in the short term—as neither the domestic causes, nor the contextual causes (involving regional and international concerns), are generally addressed in full. Military intervention can additionally lead to long term peacekeeping, which can, in turn, lead to a perceptions of an "occupation" despite denials to the contrary—a situation which is ironically not the goal set by the "responsibility to protect" doctrine itself.

One possibility to strengthen the doctrine would be the deployment of peacekeepers *prior to* the outbreak of conflict, assuming that conflict appears "imminent." Demands for UN preventive war forces to be deployed in Bosnia in 1991 and in Rwanda in 1994 before the majority of massacres took place, however, were not implemented as the UN Security Council was not prepared or willing to act; neither was a proposal to deploy a joint NATO-Russia inter-positionary force in Kosovo implemented in 1999.[10] The United Nations Preventive Deployment

10 In 1991, Bosnian Vice President Ejup Ganic asked for at least 1,000 UN Blue Helmets as a preventive war force. He was told by an assistant to UN negotiator Cyrus Vance that his demand was not a function of the UN. In effect, to obtain UN protection, Bosnia had to become a state recognized by the UN and then wait until the war spread to Bosnian territory. *Tribune de Genève* (4 January 1993). In 1994, Major General Romeo Dallaire

Force (UNPREDEP) deployed in 1995 in Macedonia continues to represent a possible model for future UN, OSCE, or even NATO Partnership for Peace, deployments in other countries—in order to limit the chances that conflict might break out. But UNPREDEP (which was disbanded in 1999 due to Chinese opposition in the UN, purportedly in retaliation for Macedonian recognition of Taiwan) was nevertheless a limited and specific action taken to prevent conflict from spreading to the Macedonian border. The point raised here is that war prevention can be much less costly in human and material terms than conflict management and/or military intervention; yet to be truly effective, *war prevention must take place as part of a larger, more integrated strategy that seeks to implement systems of security and development throughout an entire region of states, backed by the major powers*. (See Gardner, Chapter 29.)

Assuming, however, that a particular war cannot be prevented, and once the decision has been made to intervene by military means, actions are expected to be "proportional." Yet in an age of asymmetrical warfare, how can proportionality be defined? (See Géré, Chapter 24.) Even assuming that the use of force is actually considered *as a last resort* (and, here, it is still possible that strong economic sanctions short of engaging in war can cause the death of innocent individuals), another moral ambiguity arises. No matter how a military intervention might be legitimized or justified, whether by the UN Security Council or not, the effort to stop others from killing involves killing. Military intervention furthermore requires sending soldiers to risk their lives for a cause that they might not consider their own. Even if the target is strictly "military" and hit by ostensibly "surgical strikes," state leaderships who are engaged in military interventions are nevertheless conscious of the significant risk of killing civilians who become "collateral damage"—not to overlook those killed on one's own side. Such interventions, which, in the post-Cold War era, have tended to involve aerial bombardment by the US, Europeans and NATO so as to minimize the risk to one's own forces ("zero death"), consequently result in the ostensibly "accidental" killing of civilians (so-called "collateral damage") in order to prevent the further *deliberate* killing of civilians by the opponent.

The liberal interventionist doctrine furthermore expects a "reasonable" chance for the "success" of military action. But "success" according to whose standards, values and perspectives? By whose "reason"? And by what means? There may be other approaches besides direct military intervention involving diplomacy, sanctions, deployment of peacekeepers, *coup d'état*, even assassination, among other possibilities, that could, at least to a certain extent, help reduce the extent of the slaughter—assuming the goal is truly to protect the population and not

called for a mechanized brigade of 5,000 soldiers to prevent the slaughter of several hundred thousand people in Rwanda; the UN Security Council (with US opposition) failed to act, in part due to Rwandan governmental opposition to expanding the original UN mandate. On proposals for a joint NATO-Russia inter-positionary force to be deployed in Kosovo in 1991, see "Update: June 4, 1999—World Policy Institute—Research Project," http://www.worldpolicy.org/projects/arms/updates/june4.html.

engage in a politically motivated "regime change."[11] This is not to rule out the "responsibility to protect" doctrine altogether, as it may be the right approach in certain circumstances, but to raise the question as to how to implement the doctrine in practical terms *without making the situation even worse than it already had become*—in terms of the external military intervention's overall impact upon the domestic society, the economy and the environment, and not to overlook the international political, economic and human "costs" of that intervention for those who have opted for intervention.[12]

From this perspective, not only are international military interventions exceedingly costly,[13] but there is a real prospect that US and European military interventions—in Afghanistan, Iraq, Pakistan, as well as Libya—will not necessarily resolve the deeper systemic and structural roots of these conflicts, leaving these regions to fester in social/political/economic instability without prospects for real human development and "good governance" that could seek to blend some elements of US or European models of democracy with local traditions of decision-making and conflict resolution. The most pessimistic scenario points to the possible creation of an even wider shatterbelt of instability in the "greater Middle East" (or region of the former Ottoman empire)—involving gray and black market activities and near permanent factional, sectarian and socio-economic conflict both within and between collapsed and collapsing states—in which nomadic anti-state "terrorists" can migrate almost at will, and in which some states could adopt some form of authoritarian rule. A less pessimistic view suggests that there may be pockets of stability and development in certain regions of the "greater Middle East." But without steps toward a resolution of a number of conflicts, the possibility of a series of Balkan-like wars that preceded World War I remains highly plausible. (For scenarios, see Gardner, Chapter 29.)

And finally remains the fundamental question of "right authority" that legitimizes military action in the first place. Without a radical reform of the UN system, it is not clear that the present United Nations truly represents the post-Cold War "world community," and yet it is the only, if limited, global governance

11 Military intervention in 2011 against the Qaddafi regime in Libya was justified by the Gulf Cooperation Council and the Arab League and then the UN Security Council on the basis of the "responsibility to protect" civilians, but the goal transformed in the process of military intervention toward "regime change." See Gardner, Chapter 29.

12 Andrew Enterline and J. Michael Greig, "The History of Imposed Democracy and the Future of Iraq and Afghanistan," *Foreign Policy Analysis* Vol. 4, No. 4 (October 2008). In an examination of forty-three cases of imposed democratic regimes between 1800 and 1994, it was found that 63 percent failed, although there is some evidence that supportive intervention by UN blue helmets can help democratize target states. Jeffrey Pickering and Mark Peceny, "Forging Democracy at Gunpoint," *International Studies Quarterly* Vol. 50, No. 3 (September 2006).

13 For overall human and financial costs of post-September 11, 2001 conflicts, see Costs of War: http://costs ofwar.org. For military costs, see Amy Belasco, "The Costs of Iraq, Afghanistan and Other Global War on Terror Operations Since 9/11," Congressional Research Service (March 29, 2011). See also, Gardner, Chapter 29.

authority that exists at present. This fact raises questions for Just War Theory as to the "right authority" to engage in warfare. (On Just War Theory, see Gardner, Chapter 1; Davis, Chapter 7.) The lack of a global authority with fully perceived legitimacy makes it more difficult, but not impossible, to implement appropriate measures that can limit extreme violence and raises questions as to what kinds of steps should be taken both during and after the conflict.

While the goals of the "responsibility to protect" doctrine may appear to be noble and just, and could work in certain situations, the actual implementation of those goals may not be feasible or "successful" in all cases, and may represent only partial or stop gap measures. A more ambitious option is for the major powers to work in concert to prevent conflict altogether, where possible, through the step-by-step implementation of internationalized, yet relatively autonomous, "regional security and development communities" that would seek to secure and develop unstable states and regions and localities.[14] Such a project, leading to the establishment of a number of interlinking regional confederations, may prove to be necessary, if not *obligatory*—if global war is to be averted.

The Relevance of Soviet Collapse

Ironically, it was in large part the collapse of a geo-strategically contained and de-capitalized Soviet Union as a countervailing power that has supplemented the US and European temptation to intervene militarily in the post-Cold War era, whether those interventions were based upon the elimination of "weapons of mass destruction," the "responsibility to protect" populations, or support for democratic political movements, etc. Prior to Soviet collapse, the so-called superpowers generally refused to engage in direct military interventions *within each other's mutually recognized spheres of influence and security*. During the Cold War, for example, NATO did not intervene in numerous conflicts inside eastern Europe involving human rights abuses (Hungary in 1956; the Czech Republic in 1968;

14 See my argument in Hall Gardner, *Averting Global War: Regional Challenges, Overextension and Options for American Strategy* (New York: Palgrave Macmillan, 2010). See also Chester Crocker, "The Place of Grand Strategy, Statecraft, and Power in Conflict Management" in Crocker, Hampton, Aall (eds), *Leasing the Dogs of War* (Washington, DC: US Institute of Peace Press, 2007). These regions could include: ex-Yugoslavia; the Black Sea/Caucasus; India-Pakistan-Afghanistan; the Korean peninsula; China, Taiwan and the South China sea; North Africa (Libya, Egypt, Tunisia and Algeria); Israel-Syria-Palestine-Jordan; Central Asian states; the Arab Gulf states plus Iran; the Horn of Africa; the West Africa conflict zone (Sierra Leone, Guinea, Liberia, Ivory Coast); the Central African Great Lakes region, plus the conflict zone surrounding Sudan, south Sudan, and Chad, among others. As one cannot achieve human development without security, and as the emphasis on security alone implies perpetual tension, conflict, if not repression, I have expanded the concept to "regional security and development communities."

Poland in 1980, among others) as the Soviet Union could effectively check outside interference within its "sphere of influence and security."

From this perspective, it was only after the collapse of the Soviet empire, and the subsequent breakdown of clearly defined spheres of influence and security, that NATO intervened in Bosnia and ex-Yugoslavia, a formerly neutral Communist state. NATO then acted without obtaining a UN Security Council mandate in its war "over" Kosovo in 1999, in what it subsequently called an "exceptional" military intervention. (See Rimanelli, Chapter 18; Kobtzeff, Chapter 19. See also Emerson, Chapter 6, fns 18 and 19.) The US did, however, obtain a general UN Security Council mandate to intervene in Iraq in 1990–91 and then in Afghanistan in 2001 (with NATO peacekeeping and "peacemaking" supports by 2003). The US and its coalition partners once again intervened in Iraq in 2003 ostensibly to eliminate weapons of mass destruction (which were not discovered) but, this time, without UN Security Council backing, and against the views of democratic France and Germany, in addition to Russian and Chinese opposition. The US has subsequently not been able to check inter-communal Sunni, Shi'a, Kurdish conflicts, nor Al-Qaeda activities in Iraq; Washington nevertheless announced the withdrawal of US military forces (but not necessarily civilian experts and security services) by the end of 2011. (See Chapter 29.)

In early 2011, backed by the Gulf Cooperation Council, the Arab League and with a UN mandate, France, Britain and the US, then NATO, intervened in Libya following Colonel Qaddafi's violent crackdown on protesters, which had been depicted by the media as part of the region-wide Arab "democracy" movement. These military actions—justified under the "responsibility to protect"—completely reversed policy toward the Qaddafi regime in which the US and Europeans had sought to engage diplomatically with Libya from 1999 to 2010 in order to assure that Qaddafi would keep his promises not to develop weapons of mass destruction, that he would stop his support for "terrorist" organizations abroad, and that he would permit European and US access to Libya's significant energy and mineral resources and other markets. In 2011, European and US policy shifted overnight from "regime recognition" to "responsibility to protect," and then to "regime change," by support of insurgents opposed to Qaddafi's rule. Having previously sought to contain, if not overthrow Qaddafi since his seizure of power in 1969, US and European steps (from 1999 to 2010) to recognize the Qaddafi dictatorship did not work to significantly "reform" that regime from the viewpoint of the majority of the domestic population, nor from the perspective of the Europeans and United States.[15]

15 On US-European steps to recognize the Libyan regime before 2011, see Chester Crocker, "The Place of Grand Strategy, Statecraft, and Power in Conflict Management" in Crocker, Hampton, Aall (eds), *Leasing the Dogs of War* (Washington, DC: US Institute of Peace Press, 2007). One can question whether France, Britain and the US would have intervened against the Qaddafi dictatorship in 2011—had Qaddafi opened the country more overtly to European and American interests.

Not only was military intervention in Libya protested (but not adamantly opposed) by two permanent UN Security Council members, Russia and China, but the Libyan intervention was also disputed, at least initially, by democratic states such as Turkey, Germany, India, Brazil and South Africa. South Africa and the African Union, along with Russia, had hoped to mediate the crisis between Qaddafi and the insurgents led by the National Transition Council. The war could consequently have profound global and regional ramifications as states interpret its meaning relative to their own interests and security concerns. Here, for instance, US and European military actions in Libya could justify an attempt to acquire nuclear weapons by Iran or other states, as the promises of US and Europe not to engage in forceful "regime change" after a state such a Libya promises to give up its nuclear weapons capability cannot be trusted. Likewise, US and European military intervention in Libya (following interventions in Kosovo and Iraq) could be used to rationalize an eventual Russian and/or Chinese backlash as these wars could be perceived as asserting European and US interests and potentially excluding the interests of Russia, China and other states. Other African countries, presently supported by the US or Europeans, may begin to question the ultimate intent of the US and European policy, particularly if they likewise represent dictatorships with socially and politically divided populations. Concurrently, the Libyan and other post-Cold War military interventions have raised profound questions as to whether democratic states can necessarily forge common strategies among themselves with respect to the use of force and the goals of military intervention.

In effect, the point is that what has *permitted* international military intervention, whether under a UN mandate or through "coalitions of the willing" is not so much the rise of a "new" cosmopolitan moral/ethical climate, but the breakdown of Soviet countervailing power over regional spheres of influence and security, and the general willingness, thus far, of Russian policy not to veto US or European military actions (despite some protest by Russian leaders and elites). At the same time, however, Russian defense policy has continued to claim a nationally defined version of the "responsibility to protect" doctrine—in that Moscow has proclaimed its willingness to protect Russian nationals or allied ethnic groups abroad, at least since 1992 under Boris Yeltsin.[16] In carrying out its proclaimed doctrine, Russia opted, for example, to intervene unilaterally to "protect" South Ossetians and Abkhazians in Georgia in the Georgia-Russia war in August 2008.

Later in 2008, Moscow recognized the independence of South Ossetia and Abkhazia in an ostensible *quid pro quo* following US support for unilaterally declared Kosovar independence from Serbia—an action that had been opposed by both Moscow and Belgrade (among other states, such as democratic Spain which has similarly opposed Basque secessionism). US support for Kosovar independence in 2008 was ostensibly intended as a means to protect Albanian Kosovars from Serb pressures, following the war "over" Kosovo in 1999, but was regarded by

16 James F. Holcomb and Michael M. Boll, *Russia's New Doctrine: Two Views*, Strategic Studies Institute (July 20, 1994), http://www.fas.org/nuke/guide/russia/doctrine/rusnudoc.pdf.

Russia as an affront to the sovereignty of its Serbian ally. (See Kobtzeff, Chapter 19.) The point raised here is that there is still a danger that Russia (or other powers) could eventually decide to intervene at cross purposes against the interests of the Europeans or Americans in support of opposing partisan groups or states in a given conflict—and under the grounds of the "responsibility to protect" broadly defined.

A significant US-European confrontation with Russia (or China or other states) could take place in a situation in which major and regional powers act unilaterally and not in true concert—in the belief that each state possesses the legitimate right or responsibility to protect a chosen side—potentially setting off major power warfare. A number of major confrontations with Russia could have hypothetically taken place had Russia opted to strongly oppose US-led military interventions in Iraq in 1990 and 2003, or in the Balkans in the 1990s, or with regard to Georgia in 2008, assuming the US/NATO had opted to support Tbilisi against Moscow. In past eras, a number of leaders asserted with confidence that major power conflict would not repeat itself—ironically just before such conflicts, including the French Revolutionary Wars and 1870–71 Franco-Prussian war, broke out.[17] The possibility of major power warfare—or that of wider regional wars with interwoven theaters—cannot simply be ruled out in accord with ideological preferences and must still be considered a real possibility, *precisely in order to prevent such a scenario*. (See George Modelski, Chapter 28.)

Three Rationales for the Waning of Major Power War

There are essentially three mainstream theories as to why the Cold War and immediate post-Cold War eras did not, or will not, result in either major power warfare or else in wider conflicts involving interlinking regional theaters. The first is "democratic peace theory"[18] which claims that the peaceful spread of liberal democracy will ultimately help prevent major power wars. As previously

17 In February 1792, British Prime Minister William Pitt stated his conviction that the situation in Europe could expect 15 years of peace; yet two months later, the European continent was at war, and less than one year later Great Britain was involved. Likewise, once he became British Foreign Secretary in 1870, Lord Granville had been informed by the Permanent Undersecretary that there were no major issues that Britain needed to contend with. Three weeks later the Franco-Prussian war broke out. See Hans Morgenthau, *Politics Among Nations* (New York: McGraw Hill, 1993), 23.

18 On democratic peace theory, see Bruce Russett and John Oneal, *Triangulating Peace: Democracy, Interdependence, and International Organizations* (New York: W.W. Norton, 2001); Errol A. Henderson, *Democracy and War* (Boulder, CO: Lynne Rienner Publishers, 2002); Michael E. Brown, Sean M. Lynn-Jones and Steven E. Miller, eds. *Debating the Democratic Peace* (Cambridge: MIT Press, 1999); James Lee Ray, *Democracy and International Conflict: An Evaluation of the Democratic Peace Proposition* (Columbia, SC: University of South Carolina Press, 1995), among many other books on a subject which has become a cottage industry.

discussed, a variant of liberal interventionism argues that military force may be necessary (ostensibly *as a last resort*) to protect civilians and support pro-democracy movements, in accord with the tenets of Just War Theory, but such interventions are to be concerted and limited, and not intended to provoke conflict with third states. The second is neo-realist theory which, in essence, claimed that "bipolarity plus nuclear weaponry" helped sustain global peace during the Cold War, but that the *gradual* spread of nuclear weaponry to politically and economically stable states can help preserve global peace, or at least keep regional wars "limited" in the post-Cold War period.[19] The third, hegemonic regime theory, in association with "neo-conservatism,"[20] has argued that American "leadership" (coupled with NATO enlargement to counter any potential threat) must sustain a position of hegemonic pre-eminence in order to help sustain global peace against all potential threats—even if this means overtly engaging in "limited" wars or "regime change" against certain emergent or "rogue" states, or taking action against anti-state "terrorist" groups, in the "global war on terrorism." These three claims have represented the dominant paradigms or rationales as to why major power war has not yet broken out in the post-Cold War period; yet each paradigm presents a number of problems and contradictions.

Democratic peace

Democratic peace theory has been challenged by a number of the authors in this book, but supported by others (see Gat, Chapter 2). Democratic states do not necessarily set the conditions for peace, and may well be aggressive. It is possible for a majority of individuals, even in a democratic government, to support aggressive leaderships; the latter in turn can repress minority groups or demonize "other" societies regardless of the actual nature of the latter's socio-political system (see Oleg Kobtzeff, chapters 4 and 15). It can furthermore be argued that in so-called "fledgling" democracies, a number of domestic political disputes, conflicts and civil wars—in ex-Yugoslavia, Chechnya, Northern Ireland, Lebanon, Rwanda, Sudan, Ukraine, Kenya, Ivory Coast, among others—have actually been exacerbated by the very rules of democratic majoritarian "winner take all" voting and/or binary "for or against" plebiscites. The latter systems of democratic majority rule can actually work to cause or exacerbate disputes and conflict by excluding minority views and alternative options that might otherwise

19 Kenneth Waltz, *Theory of International Politics* (New York: McGraw Hill, 1979); John Lewis Gaddis, *The United States and the End of the Cold War: Reconsideration, Implications, Provocations* (Oxford: Oxford University Press, 1992); John J. Mearsheimer, *The Tragedy of Great Power Politics* (New York: W.W. Norton 2003), among others, have essentially argued this neo-realist perspective. For an alternative to the neo-realist perspective on the Cold War, see Hall Gardner, *Surviving the Millennium* (Westport, CT: Praeger, 1994), Chapters 1 and 2. See also Hall Gardner, *Dangerous Crossroads* (Westport, CT: Praeger, 1997), and Hall Gardner, Chapter 29 of this book.

20 On the rise of neo-conservatism and US military intervention in Iraq, see Hall Gardner, *American Global Strategy and the "War on Terrorism"* (Farnham: Ashgate, 2007), Chapter 2.

help sustain peaceful relations among differing socio-political identity groups and factions. Rather than helping to implement systems of *consensual* democracy involving *power sharing*, the international community has tended to export the right of national self-determination and the practice of binary plebiscites to a number of regions of ethnic or tribal diversity—with potentially explosive consequences. (See Peter Emerson, Chapter 6.)

As proved the case with the democracy movement in China in June 1989, "fledgling" democratic movements can be suppressed. They can also be hijacked by extremists. The latter was the case for the anti-Shah movement in Iran in 1979, just prior to the Ayatollah Khomeini's seizure of absolute power, and for the January 2006 Palestinian parliamentary elections which led to a Hamas victory. Extremists (however defined) could also hijack the Arab movements against dictatorship in Tunisia and Egypt in 2011, if not in Syria, if the minoritarian regime of Bashar al-Assad eventually collapses. Lebanon and Israel represent a clear example of two "democratic" states that have fought a number of wars against one another since 1978 due to an inability or reluctance to resolve geopolitical disputes through diplomacy—even if Israel claimed to focus its attacks mainly on the Palestine Liberation Organization in 1978 and in 1982, and then against Hizb'allah in 2006—actions which alienated many in Lebanon, even if they were not Hizb'allah supporters.[21]

While the majority of democratic states do live side-by-side each other in peace, other examples (depending upon how one defines "war" and "democracy") where the basic principle of democratic peace theory—that democratic states do not fight one another—does *not* appear to hold include the American Civil War (1861–65), the British–South African "Boer War" (1899–1902), and US backing for the September 11, 1973 *coup d'état* against the democratically elected Socialist government in Chile—that is, if one defines a *coup d'état* as a form of warfare.

Another major concern, from an anarchist perspective, is that the liberal state itself no longer appears "democratic"—and that the "democratic peace" itself represents a potential cause of domestic societal and class conflict *across* borders. From this perspective, the liberal view of the modern democratic state as representing the interests of a society as a whole has become increasingly untenable due to the globalization of the political economy and fact that even developed states have increasingly tended to act on behalf of international financial concerns and multinational corporations, resulting in highly uneven levels of development in core, semi-peripheral and peripheral countries. Highly uneven levels of socio-

21 Against this position, one could argue that the PLO and Hizb'allah had both forged a "state within a state" inside Lebanon, or that the Lebanese confessional system (a form of consociationalism) is not a "true" democracy. But this does not address the question of the failure/refusal of democratic states, such as Israel, to resolve disputes through diplomacy. For a critique of the Israeli decision to use force against Hizb'allah in 2006, rather than choose a range of other options, see Raphael Cohen-Almagor and Sharon Haleva-Amir, "The Israel–Hezbollah War and the Winograd Committee," in *Violence and War in Culture and the Media*, Athina Karatzogianni (ed.) (Routledge, forthcoming).

economic development have often impelled domestic socio-political groups and anarchist factions (among others) to engage in transnational, and sometimes violent, protest against governance by the Group of Eight (G-8) and Group of 20 (G-20) developed countries and against their own democratically elected governments for not providing sufficient jobs (the elementary basis for survival) and for imposing austerity measures, for example. (See Robinson, Chapter 5.)

Additional concerns with democratic peace theory include the fact that the theory does not appear to explain why democratic India, for example, tended to side more with the Soviet Union during the Cold War, than with the US, while the US itself tended to back military regimes in Pakistan. Or why the US tended to back a number of Arab dictatorships prior to the Arab "democracy" movements in 2011, and then why Washington has generally been reluctant to support the latter movements. (See Gardner, Chapter 29.) Or, as previously argued, why democratic states do not necessarily forge common policies with regard to the use of force toward non-democratic governments.

This book takes an additional approach toward the subject. Few theorists argue that democracies *by themselves* are inherently more peaceful, as democracies have historically fought against what were regarded as authoritarian regimes of differing forms (not to overlook wars against indigenous peoples with non-European, and ostensibly non-democratic, forms of governance).[22] The key issue raised in this book is that the focus of much democratic peace theory research has primarily been upon individual *dyads*, and thus upon bilateral conflicts between two ostensibly democratic states—and *not* upon more *systemic* conflicts, involving a number of complex interacting dyads that could possibly pit democratic states against each other in differing alliance formations. Democratic peace theory does not, for example, explain why, Britain aligned with Tsarist Russia when Imperial Germany could have been considered more "democratic" at the time—and why Britain and Tsarist Russia both sought to repress the Persian constitutional movement at the turn of the nineteenth century. Democratic peace theory also does not help to explain why the United States aligned with the Soviet Union (rather than oppose all totalitarian regimes) during World War II.

It is furthermore not at all certain that democratic forms of governance will necessarily be able to prevent the next major power war. For democracies to make macro-decisions, and to act decisively in favor of implementing democratic forms of governance that would preclude the possibility of global conflict, it might be necessary for the world system "to reach a condition of overwhelming majority for democracies, accounting for, say, 90 percent of the world population." (See Modelski, Chapter 28.) This, of course, appears to represent a highly demanding condition that would be very difficult to fulfill so as to guarantee global peace in the long term. One could speculate that a peaceful democratic transformation in China might ultimately help bring about such a precondition. Yet, in the meantime, in awaiting China's hoped-for metamorphosis toward democratic governance,

22 For a critique of democratic peace theory, see Errol Anthony Henderson, *Democracy and War: the End of an Illusion?* (Boulder, CO: Lynne Rienner Publishers, 2002).

democratic peace theory says very little as to how democratic states should deal with non-democratic states *so as to prevent war*. (On China, see Thompson, Chapter 27.) In fact, after the crackdown on the student-led democracy movement in June 1989 on Tiananmen Square and throughout China, one could argue that Chinese authorities have been willing to use force, censorship and repression in order to stifle democratic transformation, and that international demands for democratic change in China have thus far been considered as threatening the peace from the perspective of Communist Party elites (despite some promises to engage in social, political and economic reforms relevant to Chinese conditions).

Nuclear peace

With respect to claims of nuclear peace, one can question the belief that nuclear weaponry necessarily deters or dissuades violent conflicts, whether domestic or interstate. Nuclear weapons did not prevent a number of revolutionary wars against states that actually possessed nuclear weaponry, including those struggles waged against the US (Vietnam), United Kingdom (Zimbabwe), the Soviet Union (Afghanistan) and Israel (Palestinian *intifada*), for example. Nuclear weaponry furthermore did not prevent major border clashes between Russia and China in the 1960s. Nor did the "existential" Israeli nuclear deterrent prevent war with Egypt in 1973 (see Rimanelli, Chapter 9), nor conflicts with Lebanon previously mentioned. Nor has nuclear weaponry prevented major border clashes and acts of terrorism between India and Pakistan (the 1999 Kargil crisis, for example, in which the threat to use nuclear weaponry appeared very plausible). The UK–Argentine high tech war over the resource-rich Falklands/Malvinas islands in 1982, in which British nuclear weaponry did not deter Argentina from attacking, could represent a possible model for future "resource wars"—assuming such wars cannot be prevented by concerted diplomacy. (See Cramer, Chapter 22.) The Falklands/Malvinas model also implies that major inter-state wars of the future will be waged more by professional high tech and computerized armies and by special forces than by mass or popular armies—more along the lines of eighteenth-century warfare than mass warfare since Napoleon.

It had been argued (prior to 2001) that the US and Europe did not procure and field more advanced weaponry following the high-tech 1973 Arab–Israeli wars and 1982 Falklands/Malvinas Wars (wars which could represent models for future conflicts between major and rising regional powers) as might have been expected if major powers were actually preparing for major power warfare.[23] Yet it nevertheless appears that what can be called "cruise missile diplomacy" has been on the rise after the major military interventions in Iraq (in 1990–91, 1998–2003 and then, in full scale intervention, in 2003). Cruise (and now drone) missile diplomacy,

23 Prior to the September 11, 2001 attacks, van Crevald argued that the possibility of major power war was waning. Martin van Crevald, *The Rise and the Decline of the State* (Cambridge: Cambridge University Press, 1999).

which has taken the place of nineteenth century American "gunboat diplomacy," has been put into effect during the wars over Bosnia in 1995 and Kosovo in 1999, over Afghanistan since 2001, on the Pakistani border with Afghanistan since at least 2008, and now in the war against Libya in 2011. These wars have been waged against lesser states, but they nevertheless put major and regional powers such as Russia, China and Iran, among others, on warning. Here, the use of such high-technology, plus stealth weaponry, provides optimism that one can defeat the "enemy" with a minimum of deaths on one's own side.

There are, furthermore, numerous signs that potentially significant conflict could take place in the near future if deterrence fails—if one looks at the signals sent by US naval maneuvers. Although not sparking war, tensions between Taiwan and China impelled the US to deploy two aircraft carriers in the region after China fired unarmed missiles into Taiwanese waters during Taiwanese elections in 1996; the US also deployed aircraft carriers close to Taiwan during elections in March 2008. Likewise, in April 2008, the US threatened the temporary deployment of a second aircraft carrier in the Gulf region, although the Pentagon downplayed a potential escalation in tensions between Iran, the US and Israel. In February 2008, a Russian strategic bomber was said to "buzz" the US Aircraft Carrier USS Nimitz. In November 2010, the US sent an aircraft carrier for military exercises off the west coast of the Korean Peninsula just after Pyongyang shelled South Korean territory. Each of these examples points to the possibility of conventional, if not nuclear, escalation in the not so long term—in that such military maneuvers respond to immediate inter-state and inter-societal tensions, but do not necessarily address *by themselves* the deeper systemic and contextual roots of geostrategic, political-economic and socio-cultural disputes that could eventually cause regional, if not global wars.

While strategic nuclear weapons might impact decision-making in the midst of conflict, it is not clear that such weaponry *by itself* will always deter or dissuade (in the French formula) the onset of conflict; other diplomatic factors must also be involved in preventing the outbreak of war. Moreover, the threat to use ostensibly lower-yield tactical nuclear weaponry in conventional conflicts potentially lowers the nuclear threshold, creating the illusion that state leaderships can deploy "less destructive" weapons and take the risk that they can get away with it. (See Géré, Chapter 24; for a historical approach to technologies of warfare, see also Charnay, Chapter 25.)

Despite the pleas of US President Ronald Reagan and Soviet President Mikhail Gorbachev to abolish nuclear weaponry altogether (see Modelski, Chapter 28), nuclear weapons continued to spread to a number of states which may or may not prove to be politically or financially stable in the long term. The issue raised here is that the neo-realist viewpoint, which argues that the spread of nuclear weaponry to certain politically or economically "stable" states can help sustain regional and global peace, works to delegitimize the 1970 Nuclear Non-Proliferation Treaty (NPT). The NPT permits only the US, Soviet Union (now Russia), Britain, France and China the legal right to possess nuclear weaponry.[24] Ironically, these are the permanent members of the UN Security Council.

Political support for the gradual spread of nuclear weaponry raises charges of "double standards" by those states that are not necessarily considered "stable," such as North Korea and Iran. Pakistan and India, which have not yet signed the NPT, have likewise denounced nuclear "double standards," while Israel, which has likewise refused to sign the NPT, has wanted to keep the world guessing as to the nature of its nuclear weapons capacity. The possession of nuclear weaponry by one state generally results in demands by rival states to possess nuclear arms or other forms of weapons of mass destruction; it can also lead states to demand that they be protected by a major power that possesses a nuclear capability. This has been the case with NATO enlargement since 1999, which promises Article V security guarantees to its members. Despite its efforts to reach out to Moscow after the September 11 attacks, NATO policy has been countered by Russian nuclear weapons policy, which, in mirroring NATO planning, does not rule out the "first use" of nuclear weapons. Similarly, Iranian threats to acquire nuclear weaponry could lead other regional powers, such as Saudi Arabia or Turkey, to do the same. This dynamic works to cause a destabilizing "insecurity–security dialectic" (Hall Gardner, Chapter 1) or the "Red Queen effect" (Azar Gat, Chapter 2) on a regional, if not international, scale—assuming states cannot find ways to limit, if not abolish, nuclear weapons and other weapons of mass destruction altogether.[25]

While nuclear weapons *might* make states think twice about engaging in direct conflict, they do not necessarily prevent secret state engagement in unconventional or asymmetrical wars involving acts of "terrorism" by surrogate actors or by special forces; nor do nuclear weapons prevent attacks by electronic viruses, for example, to destroy communications and computer systems, or the use of other high-tech defense systems. Likewise, Missile Defense (MD) systems can be used for both defensive and offensive purposes, so as to protect states after one side

24 See the early arguments of John J. Mearsheimer, "Back to the Future: Instability in Europe After the Cold War," *International Security* Vol. 15, No. 4 (Summer 1990), 5–56; John J. Mearsheimer, "The Case for a Ukrainian Nuclear Deterrent," *Foreign Affairs* (Summer 1993), http://mearsheimer.uchicago.edu/pdfs/A0020.pdf. Mearsheimer argued for the spread of nuclear weaponry to Germany and Ukraine; yet the US and Russia worked more or less in tandem to pressure Ukraine, Belarus and Kazakhstan to give up their nuclear weaponry. See my early argument *contra* Mearsheimer, in Hall Gardner, *Surviving the Millennium: American Global Strategy, the Collapse of the Soviet Empire and the Question of Peace* (Westport, CT: Praeger, 1994), 17, 186–195. At the same time, the fear that the US might reduce its military presence in Europe, plus the need to cut defense costs, appears to be drawing France and Britain into closer nuclear collaboration, in tacit defense of Germany and the European Union, http://www.natowatch.org/sites/default/files/NATO_Watch_Briefing_Paper_No.16.pdf.

25 In Europe and western Russia, steps should be taken to abolish tactical nuclear weapons. Steps in the greater Middle East should be taken to forge a "no weapons of mass destruction zone" perhaps preceded by a pledge of "no first use" of nuclear and other weapons of mass destruction.

has launched a first strike. The real danger is that a number of governments throughout the world have been preparing for differing kinds of war scenarios, ranging from asymmetrical, to symmetrical, to the use of tactical and strategic nuclear weaponry, or other weapons of mass destruction, while differing anti-state partisan organizations have been preparing new surprise attacks in what forewarns to be a very long-term struggle. (See Chapters 18, 23, 24, 25 and 26.)

Hegemonic stability (plus neo-conservatism)

The hegemonic stability argument (associated with "neo-conservatism") depends largely upon continuing American military and financial/economic predominance well beyond the Cold War era. The not-so-gradual shift from an essentially bicentric global system[26] during the Cold War toward a highly uneven polycentric global system of states and anti-state actors with significantly different force capabilities, however, raises the probability of wider regional conflicts, if not wars between major and regional powers. In effect, by destabilizing a number of states and regions, Soviet disaggregation can be compared and contrasted to the collapse of four empires in the aftermath of World War I. These include the Tsarist empire, Imperial Germany, the Austro-Hungarian empire, as well as the disaggregation of the Ottoman empire. (See Gardner, Chapter 29.) Additionally, Soviet collapse has indirectly helped to interlink regions with differing degrees of socio-political and economic instability within the "greater Middle East" (or former Ottoman empire) which can now be said to range from the Sub-Saharan/Sahel and North Africa to the Gulf region to former Soviet Central Asia and South and Southwest Asia. Soviet collapse has, in effect, put the US in the posture of the predominant power as a "world policeman" that can intervene militarily in the affairs of other states multilaterally or else unilaterally, if deemed necessary, but largely overstretching US capabilities to manage such a vast area unilaterally (even with significant multilateral assistance).

A number of the chapters in this book argue that the US may have already begun to lose its position of *relative* predominance or global leadership. (See Chapters 27, 28, 29.) While the US will probably retain its overall lead in military capabilities for at least two decades, the dilemma is that emerging powers could seek to challenge the United States and its allies inside key *regional* "theaters" in the not-too-distant future. This possibility could provoke major and regional power conflict, while the US or other major powers concurrently seek to contain, preclude or preempt those challengers (see Gardner, Chapter 29, for scenarios.) The issue raised here is that the hegemonic stability (and "neo-conservative") argument appears to rest upon a presumption of long-term American *military* and *political economic* predominance. Yet, in addition to the advent of the global financial crisis, American engagement in

26 One could argue, however, that the US and Soviet Union were never at true parity, that the Soviet Union was in a position of relative inferiority throughout much of the Cold War, except perhaps in terms of nuclear "throw-weight" by the 1980s.

a number of regional conflicts (Afghanistan, Iraq, Pakistan, if not Libya) may have actually helped accelerate the political-economic (including financial) aspects of a *relative* decline in American global leadership. A further irony is that these costly conflicts have represented wars of strategic choice—rather than wars of existential necessity and survival.[27] (See also Géré, Chapter 24.)

Although the US should be able to sustain its status as the predominant *military* power until about 2030 (see Thompson and Modelski, Chapters 27 and 28), this does not necessarily signify that the American leadership and general population will be willing to engage in direct military interventions involving troops on the ground as has been the case for Afghanistan and Iraq. Instead, Washington may increasingly opt for a back seat in opting for "cruise (and drone) missile diplomacy" as has been the case in Libya, in which the US has backed up Britain, France and NATO in 2011, but in which Washington did not want to appear to be taking the lead in this particular conflict. In addition, the significant US financial crisis (coupled with the apparently burgeoning crisis along the Mexican-American border and in Central America) could result in a much faster shift in American geostrategic interests than many might expect. Such a geostrategic shift could augment domestic American demands for a return to a more traditional, pre-World War II, form of geostrategic "isolationism." This scenario could lead to a greater US focus on North and South America than on the emerging powers and a number of threats throughout the rest of the world, possibly making a number of regional conflicts more difficult to manage due to a lack of diplomatic support from an isolationist US.

From this perspective, not one of the three paradigms—democratic peace, gradual nuclear spread and hegemonic stability—necessarily helps to explain why major power warfare will not take place in the not-so-distant future. Nor do these paradigms provide much guidance for conflict resolution or the transformation of conflicts in a more positive direction, toward mutual cooperation, or, at a minimum, toward a *modus vivendi* (or an agreement to disagree) among conflicting states and socio-political movements. Democratic peace theory, for example, says little about how democratic states should deal with non-democratic states that are not governed by rule of law, or that severely oppress their populations, so as to prevent the very possibility of conflict, while hegemonic stability theory (coupled with "neo-conservatism") has largely failed to anticipate (and hence prepare appropriate policies for) the possible consequences of a relative American decline from global leadership, followed by the emergence of a polycentric global system characterized by highly uneven centers of power, influence and development.

The purpose of this book is to first explicate differing and interacting theories as to the origins of conflict and war under the general concept of *polemology* in Part I. Part II then analyzes the historical origins and global ramifications of a number of significant individual conflicts within the longer historical context of conflicts, from

27 See Chester Crocker, "The Place of Grand Strategy, Statecraft, and Power in Conflict Management" in Crocker, Hampton, Aall (eds), *Leasing the Dogs of War* (Washington, DC: US Institute of Peace Press, 2007). On costs of wars after September 11, see fn 13, this chapter. See also Gardner, Chapter 29, this book.

the Peloponnesian Wars to World War II. Part III examines more closely the nature of conflict and its ramifications during the Cold War and post-Cold War periods within a larger geohistorical framework, likewise choosing two key individual conflicts, Vietnam and Kosovo, which appear representative of those two eras. And finally, Part IV looks at long cycles of systemic conflict, and speculates, in part, whether another global war is theoretically possible—and if so, whether it can be averted.

Although "history" can be misused, abused and manipulated (see the introduction on the Annales school, and D'Agostino, Chapter 14), the primary intent of this book is to bring history back into what has become a largely *ahistorical* debate as to what direction the post-Cold War global system is evolving—in an ideological climate that has tended to advance the largely metaphysical conception that the US, and Europe in particular, have somehow reached an "end of history" or have attained a "post-historical" phase, transcending the "large scale" conflicts of previous eras. As it is not at all certain that major power conflict is necessarily "passing from the scene,"[28] the purpose of this book is to show that an historically-oriented *polemological* approach can possess relevance for understanding the nature of the contemporary crisis and conflicts—that is, before the so-called post-historical realm is tossed, perhaps even more violently and destructively, back into the future of a vengeful past.

Chapter Outline

The Ashgate Research Companion to War: Origins and Prevention is divided into four sections. "Part I: Alienation, Legitimacy and the Roots of War" examines a number of interacting theoretical concerns that explain the origins of war from a *polemological* perspective: alienation, socio-biological influences (including ethnic identity, age, gender and demography), socio-political formation (including democratization, majority rule and anarchist theories), not to overlook the key question of *legitimacy* that can justify or rationalize the decision to go to war. The theories presented are not necessarily comprehensive, and some of the scholars present contending perspectives, based on differing theoretical assumptions and data, but each points to a differing piece of the much larger multidimensional and multidisciplinary puzzle.

28 Even though "terrorism and wars of national liberation will continue to be an important item on the international agenda … large-scale conflict must involve large states still caught in the grip of history, and they are what appear to be passing from the scene." See Francis Fukuyama, "The End of History," *The National Interest* (Summer, 1989); Francis Fukuyama, *The End of History and the Last Man* (New York: Macmillan, 1992). See also Francis Fukuyama, "The History at the End of History," *The Guardian* (3 April 2007), http://www.guardian.co.uk/commentisfree/2007/apr/03/thehistoryattheendofhist.

A few of the chapters propose differing prognoses which could help prevent, or at least ameliorate conditions that can lead to conflict, but only if all sides accept. The irony is that the very effort to propose alternative strategies and policies can actually work to provoke disputes and potential conflict, if opponents of those policies believe their possible implementation must be blocked at all costs. The editors nevertheless insist that it is necessary to present a number of alternative viewpoints, particularly when status quo approaches appear to be failing so dramatically. The intent is to spark a debate and an ongoing dialectic between the pros and cons of differing strategies and policies in an effort to prevent, manage, resolve or "transform" conflict in the long term, with the goal of establishing a modicum of local, regional and global peace and justice.

In Chapter 1, as a general theoretical introduction to the origins of disputes that could lead to conflict and war, **Hall Gardner** analyzes the multiple interacting dimensions of alienation with respect to interrelationships between decision-making power and lack of power—not to overlook alienation from the natural environment as a whole. Among the many interacting geostrategic, military technological, political-economic, socio-cultural, ideological, dialogical, natural-environmental, bio-political and psychological factors that influence and impact upon the causes of conflict and war, the chapter places primary emphasis upon historically shifting forms of alienation characterized by highly uneven socio-political-economic inequities and territorial divisions among separate, relatively autonomous, decision-making centers of state power. These inequities and territorial divisions are furthermore conditioned by the nature of the equilibrium (or disequilibrium) of the global system, in which there is a lack of cooperative or concerted geostrategic and political-economic decision-making processes linking the local, national, regional and global levels.

After arguing that the views of Hobbes are generally more correct than those of Rousseau in light of anthropological evidence, **Azar Gat** examines a number of socio-biological and political-economic reasons as to why war has occurred in the past. In contending that violence under the state has actually produced *fewer* casualties than pre-state violence and that the state has likewise offered more protection, Gat then argues that interstate war in the affluent and liberal democratic world generally appears to be receding, in that peace is more profitable than war, but that war with other non-liberal states and societies will nevertheless remain with us. Gat argues that this appears true in that most countries are still far behind on the road to modernization, and some, such as China and Russia, are developing in directions which are not liberal and democratic. Concurrently, weapons of mass destruction appear to be trickling down to *below* the state level where they cannot be controlled. While Gat looks primarily at the masculine gender issues as impacting upon the causes and tactics of war, **Debra L. DeLaet** takes a socially constructed gender perspective in discussing the recruitment of soldiers, war propaganda and battlefield tactics. She argues how the nature of gender (not biological sex) tends to influence the causes, tactics and consequences of war, in that gender "not only helps to explain the underlying causes of war but is also critical to understanding *how* states and non-state actors fight wars."

The next three chapters look directly or indirectly at the question of "democratic peace" and whether democratic states are necessarily peaceful. **Oleg Kobtzeff** examines the rise of mass warfare, genocide and democide since the Middle Ages, in which the "transfer of the regalian right to use force to the people and the development of large popular armed forces" could be qualified as the "democratization of warfare," which has otherwise been dubbed as the *Tocqueville paradox*. In part by way of critiquing Hobbes and realism, **Andrew Robinson** looks at the history of anarchist theories in the view that the state does not necessarily protect its citizens but *produces* conditions of atomization and mutual rivalry, both domestic and external. From this perspective, the Westphalian state system, coupled with the highly uneven nature of state capitalist development and historical processes of global marketization, tends to generate both domestic conflict and interstate wars. In reference to the "global war on terror" and the "war on drugs," anarchist theorists tend to argue that such wars can only be perpetual; that these so-called wars represent a "continuous and uninterrupted exercise of power" that is indistinguishable from policing, but in a situation in which claims to justice and democracy, that generally attempt to legitimize such police actions, are abandoned. In essence, anarchists argue that minoritarian socio-political groups and factions need to *decelerate*—or counter what is regarded as the increasing speed of war—in order to resist the formation of permanent war machines and national security bureaucracies, by way of gaining greater public support for their cause. **Peter Emerson** then questions the contemporary, almost sacrosanct, belief in majority voting in an effort to establish more inclusive, informed and *consensual* democratic procedures. By contrast with the majoritarian approach, *consensual* democracy seeks to prevent the disputes and conflicts that can result from decisions in which the majority overrules the minority, often by only a very small percentage of votes. Emerson argues that the international community tends to promote a majority vote interpretation of democracy in a number of developing countries, and only when such a policy goes wrong does the argument swing in favor of that which, in some ways, represents a complete opposite—*power-sharing*.

In concluding Part I, **Troy Davis** critiques Just War Theory and raises the question as to who actually possesses the *legitimacy* to declare war, particularly given the fact that the UN system no longer appears to provide sufficient legitimacy for the entire world community. His argument for "preemptive democracy"—as opposed to "preemptive war" as was the case for the 2003 US military intervention in Iraq—is not quite as utopian as it might sound. In many ways, his argument that opposition leadership (which can include former military leaders) can seek to delegitimize a dictatorial regime and call for its overthrow represents, in effect, an early twenty-first century globally mediatized version of a Spanish or Mexican *pronunciamento*, in which a group of military officers *publicly* declare their opposition to the current government and wait to see if other officers and the population will support them.[29]

29 Such an approach can be used by any faction. Ironically, the attempt to overthrow Soviet leader Mikhail Gorbachev in August 1991 represented a Latin-style *pronunciamento* which failed to achieve its ultimate goals, in part because it failed to obtain significant

The question remains, however, whether and what form of democracy, if any, might replace those dictatorial regimes that are challenged by "preemptive democracy."

"Part II: Major Conflicts in History" analyzes the origins and processes of wars from a closer perspective, in part for the purposes of comparing *and* contrasting past and present conflicts, and for the purpose of revealing the nature of key individual conflicts within a larger and long-term systemic context. Part II thus examines a number of significant historical conflicts, including the classical Peloponnesian conflict between Athens and Sparta, long-term Arab–Israeli conflicts arising in the "greater Middle East," and conflicts in Europe from the Hundred Years War to the Thirty Years War.[30] The section then looks more closely at the origins of more "modern" conflicts after the Thirty Years War: the French Revolutionary/Napoleonic Wars, nineteenth-century wars between the Ottoman empire and Russia, followed by World Wars I and II. Part II likewise looks at how theories of the causes of World War I (the World War I analogy) have impacted upon policy decisions before World War II, among other Cold War conflicts. The last chapter focuses its attention on the deeper origins of World War II in which national socialist, fascist, communist and other militarist and totalitarian socio-political movements took advantage of economic depression and politically weak regimes in order to seize power.

In opening Part II, **Athina Karatzogianni** takes a fresh look at ancient Athens as a model of hegemonic decline in the context of the Peloponnesian wars. She focuses on the less popularized aspects of Athenian hegemony, exploring the role of political innovation through the establishment of "knowledge networks" in ancient Athens, both as an enabling force to capture hegemony, but also as a factor for inciting fear and suspicion in Athens' own allies. The Peloponnesian wars represented a series of systemic wars *within the same civilization* (despite some distinct socio-cultural differences between the peoples of Athens and Sparta). These ancient wars—involving the formation of alliances and counter-alliances between sea and land powers and the fears generated by political, social, economic and technological *innovation*—retain their relevance for modern wars, if not for future systemic wars. The most important Athenian innovations included "the innovation-promoting and learning-based context of democratic institutions and culture."

Marco Rimanelli then traces the long-term origins of Arab–Israeli conflicts in the contemporary "greater Middle East" up until the 1973 Yom Kippur War, which, in many ways, set the background conditions for the ongoing disputes between the Israelis and Palestinians today. As a land bridge linking Europe, Asia

public support. It did remove Gorbachev, but resulted in Boris Yeltsin, who was not initially supported by the putschists, coming into power. On *pronunciamento,* see Edward Luttwak, *Coup d'État: A Practical Handbook* (Greenwich, CT: Fawcett, 1969).

30 While this study examines the Peloponnesian wars, it does not cover the rise of the Roman empire and the Punic Wars with Carthage, as an example of pre-modern major power warfare. Yet, due to the fact that the Peloponnesian wars were fought among peoples who generally shared a common language, culture and civilization, these ancient Greek wars appear to represent an even more problematic case than wars fought between states and societies which do *not* share a common language, culture or civilization.

and North Africa, as well as strategic/trade sea- and air-routes from Europe to Africa and Central Asia, and as the birthplace of the world's three rival monotheist religions, conflicts in the "greater Middle East" have come very close to sparking major power warfare during the Cold War.[31] After a discussion of the geohistorical development of the "greater Middle East," **Oleg Kobtzeff** reminds us of when Europe itself represented a shatterbelt of conflicting feudal socio-political entities: the "age of planetary upheaval" from the end of the Middle Ages to the Treaty of Westphalia in 1648. This period introduced new technologies of violence, and ultimately established the modern territorial state and the still contentious concept of "balance of power."

Following the Thirty Years War, Part II traces conflicts from Louis XIV (the Sun King) until the rise of Napoleon and the advent of modern mass warfare, before it examines factors that led to mass warfare in the late nineteenth and early twentieth centuries. After briefly tracing the geohistorical roots of the French Revolutionary/Napoleonic Wars, **Marco Rimanelli** details Napoleon's military strategy (and its influence on future military strategy, mass propaganda and "national democracy") as Napoleon engaged in both symmetrical and asymmetrical warfare (against the Spanish guerrillas, for example) and in what would later be called lightning wars or *blitzkrieg*—and with continuing relevance for contemporary American military strategy in Afghanistan and Iraq.

In reference primarily to Russian and Ottoman conflict in the late nineteenth century, a conflict which set the geopolitical conditions for conflict in the Balkans prior to World War I in the aftermath of the 1878 Treaty of Berlin, **Ilya Platov** explores the relationship between war and the sacred in arguing that when war "is infused with religious values, it becomes a crusade; it is even truer for modern warfare, which tends to become even more total and to mobilize the masses in an unprecedented way." Platov's argument additionally indicates some foreboding parallels between Tsarist Russia in the late nineteenth century and Russia today, with implications for the contemporary "global war on terrorism" as well. **Hall Gardner** then takes a very close look at the diplomatic failure to prevent World War I, in arguing that Britain, France and Imperial Germany failed to forge an entente or alliance that could have either prevented the formation of the 1894 Franco-Russian Dual Alliance, or else have broken the tightening alliance links between France and Russia—which Berlin eventually decried as "encirclement"—six years before Imperial Germany exploded into a two-front war that was intended to last only six months. Gardner likewise points out that civil society groups formed "democratic internationalist" factions that nevertheless failed to prevent their pan-nationalist rivals from provoking war. **Anthony D'Agostino** examines how the differing interpretations of the causes of the Great War and the question of "war guilt" not only became part of the historical and diplomatic debate as to the causes of World

31 The "greater Middle East" continues to risk drawing the major and regional powers into confrontation, given the triangular nature of geostrategic conflict between the major regional actors, Israel, Iran and Saudi Arabia (with Turkey in the background). See Gardner, Chapter 29.

War II, but also how that horrific conflict influenced actual policy toward both Weimar and Nazi Germany in the interwar period. Moreover, the World War I analogy also entered into the debate as to how to deal with the Soviet Union during the Cuban missile crisis, among other conflicts. From this perspective, the World War I analogy appeared to provide a false basis for American foreign policy during the Cold War—in that Moscow, unlike Imperial Germany, collapsed in a whimper and not a bang. In the last chapter of Part II, **Oleg Kobtzeff** takes a systemic look at the socio-political origins of World War II in the aftermath of economic crisis: the rise of Japanese militarism; the collapse of Austro-Hungarian "cosmopolitanism"; Nazi German *revanche* (from Versailles to Munich); and the onset of communism and fascism in its different incarnations. World War II can be considered the planet's first *truly* "global" war among major powers, linking war "theaters" in Asia, Europe, the "greater Middle East" and Africa, although only touching the United States directly at Pearl Harbor.

"Part III: Cold War and Beyond" looks at the complex nature of US–Soviet co-existence and NATO–Warsaw Pact military-strategic confrontation in the aftermath of World War II. It then studies the nature of conflict and military intervention in the post-Cold War era following the Soviet collapse. In the opening chapter, **Marco Rimanelli** details the role of NATO in preventing US–Soviet conflict during the Cold War, at the risk of World War III. With the US and Soviet Union unable to forge a diplomatic compromise, Cold War conflict escalated in large part due to disputes over the fate of Germany and eastern Europe in the aftermath of World War II. **Steven Ekovich** then looks at the Vietnam war—"the first TV war"—from the angle of media use and its manipulation by official, corporate and anti-war sources. Vietnam represented one of the major conflicts during the Cold War era—intended to prevent other countries from following to Communist revolutions like "dominos" (the "domino effect") and to draw the line against worldwide Communist expansionism. Ekovich argues that "it was not the mainstream media alone that turned the public against the war" but that the lack of consensus over the prosecution of the war laid "the groundwork for an ever-growing anti-war movement—fortified by veterans of the conflict who spoke out against the war." **Marco Rimanelli** then looks at NATO's role as a peacemaking and peacekeeping organization in the post-Cold War period, while simultaneously enlarging its membership and engaging in wars in Bosnia and "over" Kosovo, in support of its purpose in defending liberal democracies, despite strong criticism of its actions. **Oleg Kobtzeff** follows with a critical outlook on NATO's war "over" Kosovo, as a conflict that appears representative of post-Cold War "humanitarian" interventionism based on the "responsibility to protect." Kobtzeff shows how the conflict fueled Serb pan-nationalism, while creating ambivalent feelings in Moscow which upset the hopes of many Serb pan-nationalists who sought greater Russian support. Yet Russian policy was generally against the actions of Slobodan Milošević, contrary to Western perceptions. This observation raises questions as to whether a more refined US-NATO-European diplomacy toward Russia and Serbia could have been pursued—without alienating Russian elites.

In reference to conflicts involving a number of collapsed and collapsing states in Africa, with a focus on Chad and the Sudan in particular, **Susan Hitchcock Perry** discusses contemporary legal and human rights concerns and questions of "peace" versus "justice" involved in the not entirely new phenomenon of child soldiers, their role and post-conflict fate. Perry likewise raises the question as to whether children should be considered victims or perpetrators of organized warfare. **Carol Mann** looks at the role of women in war, examining, in part, rape as a weapon of war in a number of conflicts. With respect to the ongoing US and NATO war with the Taliban in Afghanistan, she argues, in reference to the origins of war, that women "remain one of the major excuses to start or pursue a war, often to justify colonialist expansion." Mann accordingly examines how gender roles have been instrumentalized in wartime propaganda.

With respect to actual conflicts related to resources and climate change in general, and not entirely futuristic concerns dealing with possible socio-political conflict resulting from resource depletion and climate change, **Ben Cramer** provides a critical overview that explains how global climate changes could make wars more likely, if diplomacy, combined with environmentally sustainable technological change, cannot help resolve "green" disputes. Cramer argues that the structural causes of conflict, such as the weakening of the state, the emergence of markets of violence, the exclusion or extermination of certain population groups, can be reinforced and accelerated by ecological problems and the loss of resources such as water and soil.

The final two chapters in Part III examine new forms of asymmetrical and cyber warfare. **Athina Karatzogianni** provides a close study of the new forms of cyber warfare, looking, in part, as to how cyber warfare links death and entertainment, in the process of examining how cyber attacks and quest for information have become part of the new technology of geopolitical rivalries. In effect, the new information and computerized media no longer only represents a means to influence the politics of war; it has also become an actual tool of warfare. In arguing that "the [11 September 2001] aggression has generated a new relationship between security and defense, between law and justice operations, and the use of military forces," **François Géré** concludes Part III by studying the historical development of military strategy (including the Napoleonic ability to strike the enemy from all directions) as it impacts upon the so-called "global war on terrorism." While the costs of interstate warfare and military interventions are mounting, states are continuing to prepare new technologies (including cyber systems, stealth technology and robotics); states are also developing new tactics for contingencies involving *both* symmetrical *and* asymmetrical warfare. Concurrently, anti-state partisan actors can cause considerable damage using relatively inexpensive means and tactics that can often only be countered at excessive expense.

While the previous two sections closely analyzed a number of key wars in history, "Part IV: Long Cycles and Major Power Conflict" presents the dynamics of global conflicts *in a long term perspective*. Not representing a summary, this section seeks to indirectly incorporate previous *polemological* discussions in larger, more globally systemic, theoretical and historical perspectives. From essentially geo-sociological (Charnay), political-economic (Thompson and Modelski) and geostrategic (Gardner) perspectives, these chapters challenge the theories of democratic peace and hegemonic stability as well as the theory that a gradual spread of nuclear weaponry can help prevent war and conflict (as previously discussed in the General Introduction), and tend to emphasize *differing forms of innovation* as a major factor in provoking warfare.

In the first chapter of Part IV, **Jean-Paul Charnay** examines differing systems, forms and principles of warfare from feudal times up until the era of nuclear dissuasion/deterrence. His geo-sociological methodology does not base itself upon a traditional history of wars, *but upon mutations of the art of war through the major periods of Western civilization*. In effect, he shows how *innovations* in warfare have helped to generate conflict in different phases of history since the fall of Rome. In his second contribution to this final section, **Jean-Paul Charnay** provides a magisterial study of Islamic conceptions of *jihâd* as *both* war *and* spiritual renewal over the long term—providing the ideological basis for asymmetrical warfare on the one hand, but, on the other, also potentially opening the door to greater mutual understanding and the possibility of finding an eventual *modus vivendi* between the western and non-Islamic worlds and those worlds perceived by differing Islamic movements. In the next chapter, **William Thompson** looks at major power wars since the late 15th century in accord with "leadership long cycle analysis"—with a focus on the insular roles of Portugal, Holland, Great Britain, Great Britain again, and then the United States as key *leaders* and technological/economic *innovators* in the global system in rivalry with major powers in each epoch—with global wars recurring after periods of hegemonic decline followed by coalition building. Thompson argues that constraints on a possible global conflict in the not-so-long term include the rising costs of warfare, the increased benefits of peace, and the Kantian cluster of variables (democratization, international organizations and economic interdependence). Likewise looking at long cycles of leadership, in part with a focus on the global implications of China's "peaceful rise," **George Modelski** examines long cycles involving five global wars, and provides a critical look at possible "substitutions" for global warfare, including democratic peace theory, abolition of nuclear weaponry, and planetary defense (against asteroids)—in addition to examining the effectiveness of multilateral state engagement (the "G-20") and global civil society movements (or unions of democracies), as possible means to avert global conflict.

In the final chapter, by way of addressing key points raised by the previous chapters, **Hall Gardner** discusses the rise of Anglo-French geostrategic conflict from the Seven Years War to the Napoleonic wars and Anglo-German conflict from World War I to World War II. He argues that the latter global wars represent a sequence involving wars of *initial challenge* followed by geopolitical collapse and regional and/or global *revenge*, that could possibly be followed by American and/or European conflict with *revanchist* Russia and/or China in alliance with other states whose geostrategic and political economic interests and force capabilities had been generally restrained before and immediately after the Cold War period. In this scenario, an alliance of neo-Kantian states, emphasizing socio-political and technological *innovation*, could confront an alliance of neo-Hegelian states, emphasizing state sovereignty, but which could nevertheless include a number of democratic states. Another possible scenario would involve wider regional wars involving collapsed or collapsing states among interwoven theaters—possibly resulting in wars between rival regional and major powers. The latter scenario would represent a possible sequel to the "global war on terrorism" in which the attacks on the World Trade Center and Pentagon on September 11, 2001 were far more devastating than the Japanese attack on Pearl Harbor, but which were waged by a nomadic anti-state actor with *suspected* links to foreign states.

In reflecting upon the *polemological* arguments of previous chapters, it is argued that so-called "post-historical" Europe could eventually be dragged into major or regional power war—in that other states, including the United States, have not yet "transcended" history—but neither has all of Europe, particularly in relationship to its former colonies and with respect to the overall European relationship with Russia itself. In discussing US strategy toward a rising China, Gardner proposes the formation of a US-European-Russian-Japanese-Indian entente or alliance that would seek to break the apparent long cycles of global conflict while concurrently working to dampen apparently widening socio-political conflict throughout the "greater Middle East" and the Far East, among other regions, but without alienating Beijing. It is argued that building global peace will require a step-by-step, region by region approach that could be achieved by means of the creation of a number of relatively autonomous, yet interlinked, regional "security and development communities" backed by major and regional power security guarantees, financial assistance and international political-economic cooperation.

PART I
Alienation, Legitimacy and the Roots of War

Alienation and the Origins and Prevention of War

Hall Gardner

Introduction

The concept of alienation has multiple meanings, which can possess a number of social and political ramifications with respect to separation or isolation from the power to make key decisions that directly affect an individual's life, if not an individual's very survival within a larger historical context in the broadest sense.[1] As the nature of power is at its roots a *socio-psychological interrelationship*, the separation or isolation from the processes and sources of decision-making can possess profound consequences that can impact issues involving conflict and war.

In this view, alienation is not the separation from a higher being or from a presumed divine unity in a neo-Platonist sense, nor something generated by the "loss" of innocence in a religious sense, as in the myth of the Garden of Eden, nor does it refer to the "loss" of primitive forms of solidarity in a socialist or anarchist sense, as argued by Rousseau, for example. Nor does it refer to a presumed harmony of classical Greece disrupted by the advent of Christianity in the Hegelian sense.[2] Instead, alienation refers to actual socio-political interactions in differing kinds of societies and governments *in differing historical epochs.* The concept seeks

1 Karl Marx recognized the dual aspects of man's alienation from man and from nature, but focused his analysis primarily upon socio-economic aspects of alienation. Marx gave very limited attention to man's alienation from the natural environment and to the artificial division of the earth by the territorial divisions between states. See Karl Marx, "Economic and Philosophical Manuscripts of 1844," in Robert C. Tucker (ed.), *The Marx-Engels Reader* (New York: Norton, 1978), 75.

2 On Rousseau and the myth of a non-violent communal past, see Azar Gat, Chapter 2 of this book. At the same time, much as Andrew Robinson points out in Chapter 5 of this book, one needs to examine both the external environment and the nature of the norms of the society itself to help explain why particular socio-political groups, whether limited in numbers or not, are characterized by violence. For a concise genealogy of the changing concepts of alienation, see James Der Derian, *Critical Practices in International Theory* (Routledge, 2009), chapter 1. Der Derian calls for an application

to explain interrelations and interactions in both public and private spheres, as well as between and among differing states and socio-political communities, between those who are in power (who are most *directly* involved in the decision-making and executive processes) and those who are without power (and who often suffer the consequences of key decisions made beyond their control and without their direct participation or indirect influence through representation or lobbying, for example).

The multidimensional conditions of alienation interact with differing levels of the male/female relationship, within the family, the immediate community, the workplace, the locality, the larger socio-political collective, the complex of state bureaucracies, the region of differing states, as well as with the global system and structure of states which include overlapping self-identified socio-political communities or differing "societies"—if not "civilizations" within "humanity" as a species being as a whole. These multidimensional conditions of alienation often result in socio-political *disequilibrium* caused by *exclusion* (for differing reasons) and by lack of meaningful participation in all levels of power relationships and decision-making processes. Various minority groups, for example, are often excluded from power for differing reasons; women are, for the most part, alienated from an essentially male constructed and dominated socio-political and economic systems of governance. (See DeLaet, Chapter 3.) In addition to a more general condition of human alienation from the entire Cosmos, multidimensional conditions of alienation likewise imply the general lack of careful interaction with, and intervention in, the natural world.

By contrast, the often long-term effort to overcome or transcend alienation (the process of *disalienation*) requires courage and willingness to assert individual and collective interests, identity and values, but in dialogue with those of others, generally in an effort to achieve some form of *power sharing* arrangement involving mutually acceptable rights and responsibilities between states or within a particular society. *Disalienation* is generally a long-term process between individuals and socio-political collectives that can potentially take place at different levels of socio-political interaction, but is never a foregone or inevitable conclusion, in that steps toward *disalienation* can be thrown back by countervailing forces and influences that tend to undermine the effort to achieve a not altogether utopian goal of inter-state, inter-regional, inter-community and inter-individual cooperation, involving power sharing and mutual respect in careful interaction with the natural environment.

In other words, since alienation is a multidimensional concept, the major dilemma is that the effort to mediate one dimension of alienation (to find a geopolitical compromise through some form of power sharing agreement) could be undermined by another dimension such as a financial crisis or the rise of inter-communal disputes or lack of public support for that compromise, among other possibilities. Due to the multidimensional nature of the concept, the focus of this chapter will primarily be upon alienation between territorial states, whose centers

---

of the multidimensional concept of alienation to diplomacy defined as the *mediation* of estrangement.

of decision-making power are *relatively* autonomous from societal influence, with respect to the difficulties involved in establishing modes of inter-state and inter-community power sharing and mutual respect.

Multiple Dimensions of Alienation

On a psychological level, alienation can imply some form of "imbalance" generally due to mental illness or the inability to properly control one's own mind and body. Psychological "imbalance" (a Macbeth complex, for example) can accordingly distort a leader's decision-making capacities. Such a psychological imbalance need not be due to mental illness, but to a general inability to fully comprehend a changing and complex reality. There is often a fixation on past understandings ("rear-view mirrorism") even when the situation has changed radically; socio-political pressures can produce "groupthink" which checks innovative thinking. (See the discussion of "dialectics of insecurity and security" in this chapter.)

In a more philosophical context, alienation stems from the questioning and analysis of the complex origins, evolution and purpose of the Cosmos, in the sense that neither religion nor scientific analysis can explain the precise reasons for creation and why the Cosmos (and human evolution) has taken the *particular* forms and direction that it has taken.[3] Differing interpretations of the Cosmos and human evolution can directly or indirectly (if not sub-rationally) influence the nature, norms and goals of differing socio-political ideologies and forms of governance, both secular and religious; such interpretations can provide rationalizations for war and conflict. Disputes among differing secular and religious factions, and, sometimes more crucially, schisms *within* the same orthodoxy, can likewise heighten social and political tensions resulting in violence.

In a socio-psychological context, alienation means some form of separation, isolation or estrangement from the historically evolving norms, values or actions and decision-making processes of a larger society, or else *disenchantment* (in the sense of Max Weber or even Franz Kafka) with respect to modern bureaucratic inflexibility, for example. These factors can check or limit thinking and options that are "outside the (bureaucratic) box" and can result in socio-psychological expressions of isolation which could potentially instigate dissent and/or conflict—acts of "terrorism"—if it does not result in *anomie*. Policy-driven state bureaucracies can accordingly impose policies, standards, values or laws that individuals and minoritarian socio-political collectives cannot accept or identify with, thereby resulting in a lack of recognition, clash of values, and the exclusion of differing individuals and socio-political collectives from full participation in the decision-making processes of the larger society and state (the refusal to vote, for example).

3 As a concrete example of alienation from the Cosmos, see Modelski, Chapter 28, on the international cooperation needed to defend the planet from stray asteroids.

In addition to psychological distortions, and often conflicting socio-cultural-religious differences in values and practices that can impact upon interpersonal and socio-political relations and policy decisions, alienation can stem from a socio-psychological interrelationship between a leadership's goals and decisions and the opposition or resistance to such goals by divergent factions within the population as well as due to potential opposition by states and other self-defined groups outside the socio-political collective. Conflicting socio-cultural/ideological beliefs accordingly play *in the background* in a struggle for direction and purpose, as well as in a struggle for a state leadership's or political faction's right and legitimacy to make key *political* and *strategic choices* with regard to war or other important issues. The proclaimed conviction that a particular socio-political collective's beliefs and values are legitimate, if not universal or sacrosanct, makes possible compromise over specific issues and interests deemed "vital" much more difficult and provides the fuel for conflict and war.[4]

Moreover, as conflict and war risk the ultimate alienation—the risk of an individual dying for an abstract cause—socio-cultural, ideological or religious beliefs must provide grounds for renouncing rationality based on self-preservation and for taking the ultimate risk of death by means of a leap toward belief. (On religion and the sacralization of war, see Platov, Chapter 12.) The dilemma posed here is that the socio-psychological expectations raised by the grandiose goals of war as often promulgated by state leaderships tend to conflict with the values and moral relationship initially established between an individual and his society in the assumption that the latter reacts negatively to brutal and inhuman actions of both individuals and the military during that conflict. The fact that one is generally forbidden to kill *within* the state, but that one can be given license to kill *outside* state borders, troubles the individual's moral conscience; governmental demands that one kill others and/or risk one's life for an uncertain cause can likewise alienate the individual from the leadership and society which imposes those demands. At the same time, socio-psychological aspects of alienation or *anomie* are not exclusively a result of war and conflict, but can result from other state or social actions which collide with individual expectations or personal values that have often been nurtured within that very society. These factors can cause dissent among elites who may disagree with such policy, if not socio-political conflict within the state or society, assuming leaderships cannot find ways to repress, compromise, co-opt or *channel* that dissent in new directions.[5]

---

4 Without necessarily adopting Reinhold Niebuhr's religious (Protestant) approach, his statement in support of "religious humility" (and humility in general) applies to both non-religious and anti-religious movements as well: "Every absolute devotion to relative political ends (and all political ends are relative) is a threat to communal peace." Reinhold Niebuhr, *The Children of Light and the Children of Darkness* (New York: Scribner, 1960).

5 In the 1960s and 1970s, the Weathermen faction of the Students for Democratic Society were willing to use violence against the American government, which they did not see as exemplifying "true" democracy; somewhat similarly, pan-Islamic groups see a number of Islamic regimes as corrupt and not practicing "pure" Islam. Here, there is

In an ecological context, alienation stems from difficulties involved in analyzing and interpreting the natural world (irrespective of differing interpretations of nature's origins and meaning) and can further result from inappropriate intervention and interaction with the natural environment and ineffective management of the earth's resources (in terms of human productivity and behavior). Not only are certain resources in the natural environment potentially scarce, if over-used and unprotected, but scarcity is also artificially induced by the highly uneven and inequitable socio-economic distribution of those often scarce resources. The actual exploitation of natural resources, potable water and energy cannot be linked to demography alone, but to socio-cultural habits, given excessive US and European demand as compared to much of the developing world, including more populous countries such as India and China—although resource demand of a number of the developing countries is beginning to expand significantly in potential conflict with the resource demand of other states and societies. In addition, wasteful production can create harmful externalities whose costs are not always borne by the producer/exploiter itself, but are often placed upon consumers or other producers (oil spills on beaches affecting tourism and fishing industries, for example).

Alienation likewise refers to the transformation of the natural world into things or commodities that are separated from nature by means of human intervention; the latter objects are, in turn, socialized by man in ways that may transform/overturn the delicate ecological equilibrium. The development of "ever-revolutionizing" technologies (from fire to atomic energy) has greatly augmented the human capacity to intervene in the natural world and radically transform its physical and biological potential and capabilities. Yet technology itself is not entirely neutral.[6] On a social level, certain forms of technology may require formation of a highly skilled elite to manage and control that technology (as well as its potentially negative human and environmental impact), while other forms of technology can work to humanize the workplace and more carefully and safely interact with nature. The choice (or lack of choice) of technology can accordingly prove dehumanizing and/or environmentally destructive. Alienation accordingly stems from the general inability to sustain a positive interplay with the natural world with and within which humans must intervene and interact in order to survive.

In terms of war and conflict, some technologies are more destructive to human health and the natural world than others. The displacement of genes and organisms from their social and ecological context (transforming genes, for example, into commodities for private use that are separate from the organism that produced them) represents a controversial contemporary form of alienation with profound implications for both man as a species being and for the natural world in general.[7]

---

a gap between differing ideal views of ideology/religion and the way that ideology/religion is practiced. See Charnay, Chapter 26, on schisms *within* Islam.

6 Herbert Marcuse, *One Dimensional Man* (Boston: Beacon, 1964). See also Robinson, Chapter 5.

7 For a discussion of the legal ramifications of privatization in terms of alienation, see Scott Prudham, "The Fictions of Autonomous Invention" in Becky Mansfield

On the one hand, bioengineering raises prospects for new cures of genetic disorders; on the other, genetic manipulations could be utilized in new forms of biological warfare, if not risk the partial or incontrollable contamination of the biosphere. Global warming, mostly likely due to anthropogenic causes, such as the emission of greenhouse gases, can potentially uproot the agricultural base of whole societies or flood entire regions, potentially causing an outflux of environmental refugees, situations that could exacerbate the possibility of domestic or international political conflict. (See Cramer, Chapter 22.)

The so-called "peaceful" nuclear atom, which has generally helped subsidize the overall costs of the "war" atom, requires formation of a technocratic elite to control and manage its complex processes, the decommissioning of power plants and safeguarding of radioactive waste product, and to prevent such plants from meltdown (as was the case for Chernobyl in 1986 and now Fukushima in 2011). The export of nuclear technology has helped to create geopolitical and economic rivalry for control and expertise over the fission process and for the capacity to make nuclear weaponry. The use (and even testing) of nuclear weaponry can result in genetic genocide once differing radioactive isotopes climb up the food chain through the "biological amplifier." It is furthermore not certain that possession of nuclear weaponry necessarily deters war; instead nuclear weaponry forms part of a panoply of possible weaponry for use in conflict. (See Introduction, Géré, Chapter 24, and Gardner, Chapter 29.)

In a legal context, alienation refers to an act of transferring property or title to another, for either state control or private interests. Here, for example, the privatization of common lands (enclosure) in Great Britain resulted in the inequitable ownership and division of land and territory, as well as population displacement, if not urbanized (and often violent) squalor. By the same token, the nationalization or expropriation of land or property can alienate property owners as well. The state can use differing means to expropriate property, including extremely high levels of inflation or else "excessive" taxation, resulting in property disputes, if not revolution, due to "taxation without representation," as was the case for the American and French revolutions, which in turn resulted in major interstate wars. The question is not to abolish private property, nor is it to abolish local, regional or global markets, but to find ways to make property distribution more equitable, balanced or proportional and to better *channel* market forces.

Likewise in a legal context, the term "alien" can refer to a legal (or illegal) migrant who has not been naturalized by the society in which that individual now lives. Even former aliens who acquire citizenship are often stigmatized and initially prevented from access to power and influence, and hence find themselves "alienated." As compared with the past, when Europeans tended to emigrate abroad from the seventeenth to early twentieth centuries (in many cases, imposing European socio-political-cultural systems by colonizing indigenous populations), population flows since World War II have tended to move from south and east

(ed.), *Privatization, Property and the Remaking of Nature-Society Relations* (Oxford: Blackwell, 2008).

to the United States and Europe. Certain waves of immigrants have caused socio-political tensions due to difficulties of integration in their host societies on account of differing languages, religions, ethnic backgrounds and social customs, among other factors, including competition for jobs. Since the downswing in the global economy in the late twentieth century, the US and European countries have generally, but not very effectively, sought to tighten legislation in an effort to better control the free movement of population, thereby exacerbating both the global political economic and territorial dimensions of socio-political alienation, and giving rise to differing xenophobic movements with differing ideological formations of the "alien" or "other." At the same time, in addition to east to west and south to north migration, south to south migration has been raising socio-political-economic tensions as well, magnifying inequities among peripheral and semi-peripheral countries.

Alienation can be observed in how profit-seeking corporate bureaucracy imposes its own forms of rationality upon an unwilling society or even upon the natural environment. In a corporate context, alienation can refer to lack of control or influence over the key decision-making processes that affect the entire workplace as to what to produce and how to produce and distribute that product or service. On the one hand, such alienation may express itself in the form of strikes, work stoppages, if not acts of sabotage or violence. On the other, the production process, often involving planned obsolescence, re-impacts upon both the natural environment and the psychological and socio-economic well-being of differing societies in new and differing ways. Alienation can also be a consequence of disproportionate incomes and benefits granted to the executive class by contrast with salary/wages of employees of differing ranks coupled with the general lack of *equitable* or *proportional* profit sharing among *all* employees—without employee stock ownership or other forms of employee participation in corporate decision-making, for example. Differing forms of alienation, depending in part upon socio-economic class and status, can likewise develop from the highly differentiated nature of *risks* that both employees and executives take as the "price" of doing business.

In a macroeconomic context, alienation stems from the lack of control and influence (once again resulting in highly differentiated *risks* for financiers and executives as opposed to lower-level employees) over the erratic and "unbalanced" nature of ever-revolutionizing and innovating means of production, and increasingly global markets and financial forces, in which Adam Smith's "invisible hand" does not always operate within a pure textbook Pareto optimum of a "dynamic equilibrium." In the contemporary context, new financial "innovations" have proven highly destabilizing: The buying and selling of differing forms of "toxic assets" and "liar's loans" (often known to be fraudulent) indicates the over-commodification of goods and services and indicates the tendency of the "free market" to self-destruct[8]—that is, without *effective* national and international regulation that can be achieved

8 James K. Galbraith, "The Final Death (and Next Life) of Keynes." Fifth annual "Dijon" conference on Post Keynesian economics, Roskilde University (May 13, 2011).

by either quasi-governmental or non-governmental organizations or regimes. In this respect, global finance and transnational investments and transactions appear to have increasingly abstracted or "disembedded" (in Karl Polanyi's terms)[9] themselves from actual productive processes, while concurrently leading to highly uneven levels of development in localities, regions and countries around the globe.

The key dilemma is that the dynamic and systemic nature of global economy coupled with the liberal democratization process itself—which generally involves multiparty parliamentarianism, monetary standardization, budgetary controls, economic liberalization and constitutional rule of law, plus respect for human and minority rights—can prove destabilizing for both authoritarian and non-authoritarian regimes alike—as it may be difficult for differing political systems and societies to implement appropriate political-economic reforms that meet these criteria due to a lack of resiliency.[10] At the same time, however, due to the fact that increasingly international or global financial and economic crises are not entirely caused by state institutions themselves, and may, in fact, be derived from international economic and financial forces *outside* the state and society, the pressure to transform political-economic systems in order to reduce state budgets, for example, tends to make individual societies liable for flaws of the international financial system that were not necessarily (or entirely) created by those states or societies. In effect, both transnational financial organizations and profligate state governments themselves can pass the buck onto unwilling consumers on an international scale.[11]

9 Karl Polanyi, *The Great Transformation* (Boston, MA: Beacon Press, 2001).

10 I have extrapolated from Karl Polanyi, op. cit.

11 As James K. Galbraith, op. cit. points out, the crisis in Europe stems to a large extent from the American housing crisis (liar's loans), among other factors. Greece's weak tax system and big civil service, coupled with Ireland's unsustainable housing boom, were not new issues: "The initial shock to Europe didn't come from the discovery of these facts, it came from American mortgage markets. When European banks and other investors realized the extent of their losses, beginning in late 2008, they looked for ways to protect themselves … by selling the weak assets and buying strong ones: German, French bonds and above all United States treasuries. That is why yields rose on all the small peripheral countries and fell on the big ones despite the very different circumstances in the countries that were badly affected."

The Greek crisis has led to socio-political protest and violent conflict in Greece. England, Ireland, France, Germany, Spain, Italy, Iceland (among other European countries) have all seen significant socio-economic protest as well, as has the United States. As of this writing, the crisis is now implicating the value of the Euro itself, if not the political-economic role of the European Union as a whole, with France and Germany at odds as to what to do. For a brief outline of the roots of the global financial crisis, see Barry Ritholtz, "What caused the financial crisis? The Big Lie goes viral" *Washington Post* (November 5, 2011). For a study of the close relationship between austerity and social conflict, see Ponticelli, Jacopo and Voth, Hans-Joachim, "Austerity and Anarchy: Budget Cuts and Social Unrest in Europe, 1919–2009" CEPR Discussion Paper No. DP8513 (August 2011). See also, "Unrest in peace," *The Economist*, 00130613, 10/22/2011, Vol. 400, Issue 8756.

While global financial and economic innovations, coupled with potential economic crises, do not, by themselves, necessarily generate wars, they can nevertheless exacerbate significant domestic and international disputes among differing factions and socio-political collectives. The possibility of wider inter-state and inter-societal conflict will consequently depend upon the depth and breadth of the contemporary financial crisis and whether or not states, transnational corporations and societies can coordinate domestic, regional and international policies to sufficiently to transcend that crisis.

Alienation and Majoritarian Democracy

Even indirect systems of majoritarian democracy, which have accompanied late twentieth century transnational capitalist development, can be alienating. One major problem is that indirect systems of democratic governance can generate alienation precisely because they tend to create promises and illusions of direct democracy and participation in the decision-making processes which are not always fulfilled. "Winner take all" systems of democracy can cut all potential rivals out of power, even if the latter received a significant percentage (as much as 49 percent) of the vote. Populist practices, such as over-simplified "yes or no" referendums, do not spell out all the options. In this sense, majoritarian systems of governance can become "tyrannies of the majority," in de Tocqueville's view. But even then, it is not clear that the majority necessarily agrees with all policy decisions due to the ways complex strategic and political choices are often oversimplified, and in which outcomes do not meet expectations. (See Kobtzeff, Chapter 4; Emerson, Chapter 6.)

The dilemma is that "democratic" states (based upon indirect majoritarian democracy, "checks and balances" and specified procedures of "majority rule") have not yet been able to overcome the alienation of individuals and various minoritarian socio-political groups who do not necessarily agree with the *presumed* majority opinion. Nor do those minorities (which can be quite substantial in number) necessarily accept those values, norms and goals that might challenge their *perceived* identity, which remains largely a question of personal and social construction, in that even groups with the same ethnic identity can still splinter into clans or political factions, as has largely been the case for the Kurds and ethnically homogeneous Somalia—even if genetic or biological bases for ethnic identity can be established (as argued by Gat, Chapter 2).

Likewise, individuals with multiple identities may find themselves caught between rival socio-political worlds, and can be impelled, against their will, to pledge *allegiance* or *loyalty* to one side or another in times of conflict. In times of peace, it is easier to hold onto conflicting identities such as dual nationality, for example. In case of conflict, however, does one identify with a larger civilization or religion? With a particular state? With a national or ethnic group? With a transnational corporation which tends to interlink societies and regions that do not necessarily possess strong political or cultural connections? With a gender group?

With a clan or family? Or with one's individual conscience or personal interests? With whom do orphans, the homeless, the state-less, or others generally rejected by society, identify?

Opposition to individually-held beliefs and self- or socially-affirmed identities can result in tensions, socio-political conflict, lack of "solidarity" or even accusations of sedition and treason, if not apostasy in political-religious communities. Furthermore, external states and socio-political collectives can possibly support domestic dissidents against the state or socio-political collective, whether democratic or not. Such external support for domestic dissent may or may not be exaggerated, but can, in either case, provide an excuse for a repression of the socio-political demands of various individuals, minority groups or factions.

Conditions of perceived alienation can be somewhat ameliorated (but never thoroughly alleviated) if individuals and groups *willingly* accept the procedure or nature of the decision-making process—even if participation is limited and they are not necessarily active participants in those decisions. The essential irony (and the very dilemma with respect to the process of *disalienation*) is that individuals can identify with the ideology or general norms and values of a society—whether decisions in that system of governance are made by differing forms of majority rule, consensus building, differing forms of power sharing, or even by authoritarian means and dictatorship. (Here, authoritarian rule which ostensibly abides by principles of good governance may be considered less alienating than dictatorship or tyranny which rules by arbitrary or oppressive decree.) By contrast, the possible refusal to accept the nature and *legitimacy* of a particular form of socio-political structure, or else the rejection of the decision-making processes and procedures, coupled with the *lack of veto power* in the making of specific decisions (even if the decision-making process appears *legitimate*), can exacerbate conditions of *alienation* for specific political factions and socio-political collectives as well as for individuals, thus exacerbating the possibility of disputes leading to conflict.[12]

The issue raised here—and which is not generally raised in the "democratic peace" debate (see Introduction)—is that there are differing forms of direct and indirect democracy which may or may not prove compatible. Liberal majoritarian democracy, social democracy, socialist democracy, national or "sovereign democracy,"[13] libertarian democracy, consociational (or confessional) democracy,

12 Effective veto power is perhaps key to continued participation in the United Nations by members of the UN Security Council, for example, despite the fact that the UNSC veto can render the organization ineffective. The threat to use the veto ironically helps to keep the UN together; states then may have to opt to act outside the UN Security Council through the 1950 Uniting for Peace Resolution, for example, or through the OSCE, or through "coalitions of the willing" in order to engage in force if they cannot build consensus for the use of force. For the importance of the veto, see argument DG54 of John C. Calhoun, *A Disquisition on Government* (1849), in support of "negative power" and a "concurrent majority" involving a dynamic interaction between both majority and minority groups.

13 Sovereign democracy means, in the Russian sense, that each state be permitted to choose its own path to "democracy" with mutual respect and geostrategic parity, but

consensual and anarchist democracy (where it exists in localities) and "workplace democracy" involving employee decision making and degrees of employee ownership—not to overlook "democratic internationalism" (also known as "neo-conservatism") which contrasts with "world democracy"[14]—are all very different.

All these differing forms of "democracy" can be conceived with differing degrees of state intervention in civil society and enterprise and with differing degrees of individual or employee participation in the state, community and the workplace, if not within transnational corporations and international organizations. Here, for example, forms of democracy and federalism within the European Union differ greatly from more centralized forms of democracy and federalism in the United States.[15] The point is that differing forms of "democracy" involving differing hierarchies of values and different practices in leadership and decision-making can lead to schisms and alienation among "democratic" countries themselves—unless new forms of consensual democracy, confederation and power sharing can be established and generally legitimized.

Quest for Legitimacy

In general, political factions and differing leaderships attempt to seek out the lowest common denominator of ideology, interests or identity in order to draw as many supporters as possible, and thus claim *legitimacy* as leaders, in order to be in a position to make executive decisions (whether or not those decisions are necessarily carried out effectively). Such a lowest common denominator may exclude certain groups or socio-political collectives, or represent a significant segment of the population, whether a majority or plurality. The issue raised here is that these differing ideologies, interests and/or identities are not only multiple, and hence overlapping, if not interchangeable, *but are also evolving historically*. For these reasons, it can never be certain *which* overlapping ideologies, interests and/or identities might predominate and become engaged in a particular social or political project. In other words, one can identify with a locality, a state, a corporate brand, a larger region (Europe, for example), a partisan group, a religious denomination, a culture, a civilization, etc. Political factions and leaderships must consequently seek to manipulate these differing and ever-changing ideologies, interests, and/or

that some democratic states, i.e. the United States, are more "sovereign" than others.

14 World democracy seeks to re-legitimize international regimes, such as the UN and World Trade Organization, by permitting greater civil society participation in those organizations; it has also demanded the formation of a global assembly within the UN. Neo-conservatism or "democratic internationalism" advocates the imposition of democratic structures of government against dictatorship by external force of arms, if necessary, as in the cases of Afghanistan (2001), Iraq (2003) and Libya (2011).

15 See my argument in Hall Gardner, *American Global Strategy and the "War on Terrorism"* (Farnham: Ashgate, 2007).

identities for specific goals and purposes, in order to obtain supporters, and engage in proposed socio-political projects.

From this perspective, the quest for *legitimacy* is self-propagated by elites who seek to formulate the general contours of common interests, ideology and identity where possible in order to win over followers in the hope that the latter will acquiesce to their proposed policies and actions (either willingly or unwillingly). Moreover, as the quest to achieve and sustain legitimacy needs to be reinvented or reconstructed over time, elites must often appeal to the collective history and common (mis)understandings of those presently living, or engage in reforms when previous understandings are not necessarily practiced as preached. Yet even the path of reform may require the use of both persuasion and force. In this regard, legitimacy is not inherent or god-given, nor is it necessarily based upon "legal" or "moral" governance.[16] The South African system of Apartheid and the American system of segregation were "legal," for example, but they did not represent forms of governance that could be considered either "moral" or "legitimate" particularly once predominant values began to change worldwide. At the same time, concepts and practices of "morality" differ from society to society, and from state to state, hence making claims to legitimacy on the basis of "morality" appear tenuous, without compromise among differing values and traditions.

Even the legitimacy of majoritarian "winner take all" democracy can be achieved by the use of force and political pressure, and not by persuasion alone. In the evolution of the American form of democratic governance, for example, after the American Revolution had achieved power against Great Britain by force, the more centralized Federalist approach to governance gained precedence over the more decentralized anti-Federalist approach (yet only following a peacefully achieved mutual agreement to forge the Bill of Rights). Both of these approaches could be considered "legitimate"; yet the anti-Federalist approach would have led the United States down a very different path involving more decentralized rule, granting individual states much more power. The Federalist approach gained even greater influence and ostensibly more legitimacy *by force* after northern and southern states failed to reach a compromise over slavery, taxation and veto rights, among other issues, resulting in the American Civil War (otherwise called the "war of secession"). A stronger federal government was granted even greater standing and legitimacy after the Great Depression and World War II. Whether the US government can sustain that self-achieved legitimacy in the long term remains to be seen as the US now confronts massive national, state and urban debt and deficits, corporate bankruptcy and imperial overstretch.

Ideally, legitimacy should involve practicing what a state or society preaches, and a willingness to engage in peaceful reform—particularly if formerly accepted beliefs and practices fail to remain relevant in new dynamic circumstances. In effect, an *ideal* conception of legitimacy should stem from clearly understood and "transparent" decision-making processes, which are open for all members of the

16 Hans Morgenthau, *Politics Among Nations: The Struggle for Power and Peace* (New York: McGraw Hill, 1993), 32.

society to participate in some fashion, and in which procedures for implementation are generally accepted by all, but which can be openly challenged by individuals or minoritarian groups who contest those decisions. At the same time, however, the goal of pure "transparency" is difficult, if not impossible, to achieve. One example that contradicts the ideal of transparency is that voting procedures and controversial issues often require secret ballots so that individuals are not put under pressure to vote against their will. Likewise, complex negotiations often require a degree of secrecy so as to discuss options that may prove to be contentious.

As alienation largely results from imbalance or *disequilibrium*, the problem then is to achieve a *degree* of "balance" and "proportionality." Decisions reached in total secrecy could be opposed by the general public and may appear to represent a means to hide illicit or self-serving activities. Ways to test proposals by giving them a public airing, and ways to bring in differing points of view from all perspectives must be found, even if it is impossible for all sides to actually participate. In effect, decisions made in total secrecy can be alienating, but so too can decisions made in a totally open forum, particularly when issues are contentious and perceived "vital" interests and identities are at stake.[17] Then again, failure to find the proper balance between what should be kept secret, and what should not, can generate socio-political tensions, while, ironically, the very effort to achieve "balance" can alienate those factions who oppose compromise.

Territorial State Alienation

Territorial state behavior should be analyzed in the context of global systemic interrelationships and interactions, both *within* and *without* individually demarcated territorial units, even if these relations are not always precisely delineated. State behavior thus takes place within a *dynamic interaction* of multiple countervailing or contending geostrategic, military-technological, political-economic, bio-political (including ethnic identity, gender, age and demography), natural-environmental, inter-bureaucratic, legal (including international norms and law), dialogical (including diplomacy as well as media and propaganda), socio-cultural/ideological and psychological factors, forces and influences, including ethics/values (such as humanitarian concerns).[18] These influences can thus include international

17 Throughout 2010, Wikileaks provoked significant debate as to what information should be made public and what should not, and who should make the decision to expose what governments consider secret, with jobs, reputations and lives—and legitimacy—on the line following exposure of previously "secret" activities. http://wikileaks.ch.nyud.net/.

18 Postmodernists are not wrong to critique *one-dimensional* meta-narratives. Yet this approach, as advocated in this chapter, emphasizes long-term *multi-dimensional* and *interacting* forces and influences, and does not overlook unexpected and accidental confluences that can occur on account of a clash between unexpected interactions and transactions. Hence, the *alternative realist* theory (which emphasizes *strategic choices*

and domestic corporate and private interests, various forms of pressure from differing socio-political factions and groups, the nature of diplomatic initiatives and endeavors, as well as internal bureaucratic policy recommendations or other international or domestic interests and issues that impact upon the quality of state decision-making processes in which decision-makers generally consider and weigh those various options from a position of *relative* autonomy.

Each of these countervailing forces possesses its own internal and external dynamics of action, reaction and counter-reaction which result in consensus, conflict, dissent or *anomie* in response to differing decision-making processes and that interact with the evolution, revolution or devolution and other intrinsic and extrinsic forces and factors. The international system is consequently in a condition of constant tension between equilibration, dissipation and *possible* steps toward re-equilibration (in terms of "wholes" and "not wholes" in the views of Heraclitus) as leaderships seek to adjust and re-adjust policies (successfully or unsuccessfully in accord with their initial intent and goals) and according to changing domestic and international/interstate circumstances.

The dilemma remains that the dynamic nature of contending internal and external forces and innovations (not to overlook haphazard occurrences and fortuitous confluences of interacting forces) in the international system, combined with the highly uneven distribution of force capabilities and influence in decision-making processes, risks undermining any preconceived *stable* local, domestic, regional and global equilibrium. This dynamic interaction of contending forces impacts very differently upon historically changing forms of socio-political collectives (from rival city states, principalities or monarchies, unions, confederations, multinational empires, up to the modern territorial state, not to overlook what can be called collapsed and collapsing states which threaten a "neo-medievalism" in some parts of the world) and makes it even more difficult to sustain a general *equilibrium* over time. The dynamic nature of these intersecting, yet often contending, processes on differing local, domestic, regional and global levels can consequently exacerbate a *multidimensional* sense of alienation, which is all the more magnified if re-adaption and re-equilibration cannot be achieved.

Here, a *presumed equilibrium* based upon a rough *estimation of contending forces* functions more effectively to help manage, minimize or attempt to mediate disputes when there is a general recognition and acceptance by all the significant actors as to the nature of the *status quo*, and with a general acceptance of norms, including legal or ethical restraints and sanctions.[19] If, however, the status quo and socio-political identities and norms are not generally accepted or legitimized, then conflicts over specific interests (over territory deemed strategically or economically "vital," for example, or over regions in which socio-political collectives possess irredentist

---

and *diplomatic options*) developed here is not a case where "one size fits all." See also footnotes 21 and 37, this chapter.

19 The 1648 Treaty of Westphalia helped reduce, but not eliminate, the general condition of interstate alienation by establishing general rules for interstate behavior in Europe, but not necessarily overseas. See Gardner, Chapter 29.

claims) may prove more difficult to resolve over time, or to *disalienate* in a more peaceful and mutually cooperative direction in the not-so-long term.

A rough, yet dynamic *estimation of contending forces* can break down or dissipate in large part due to the inability to accurately assess the force capabilities and ultimate *intent* of potentially rival states or socio-political collectives. Along with the general inability, reluctance or unwillingness to mediate disputes or conflicts, the inability to accurately assess and estimate the force capabilities and the ultimate political and strategic *intent* of potentially rival state leaderships and socio-political collectives lies at the root of inter-societal or inter-state *alienation*. This is true in that the lack of trust and a general reluctance, inability or unwillingness to engage in cooperation and *power sharing* between states and other socio-political entities (such as anti-state "revolutionary" organizations) exacerbates mutual suspicions and imprecations, possibly leading to conflict and war.

This dynamic thus implies a continual shaping and re-shaping of the global, regional and internal domestic aspects of the international territorial state system and global economy in which the centers of power and influence—those who possess the capability to decide and then attempt to carry out major political and strategic decisions—are generally unevenly distributed and unbalanced. Not only are centers of power, influence and forms of governance unevenly distributed, but so too are the very capabilities that enhance power and influence, including technological and military capabilities, as well as natural resources, industrial capacities and access to finance. This is not to overlook the differing capacities of highly uneven centers of power to utilize political discourse, religion, ideology, propaganda, and media to help shape the general contours of laws, values and/or norms (with or without clearly defining specific content) in order to better influence and legitimize domestic and international policy and activities. In addition, differing centers of power (even those that vie for power and influence within the same state or socio-political collective) can operate at cross purposes, whether wittingly or unwittingly, with varying degrees of domestic and international impact.

Decision-Making Power and Alienation

In observing centers of power, one should make clear distinctions between the *power to decide*, the *ability to execute* and the *ability to influence* those decisions. Once the decision-making process is completed by those in power (a process which, in itself, may be characterized by improper procedures, perverted by self-interest and lacking in "transparency"), decisions may or may not be executed as initially envisioned. In addition, such decisions may not necessarily take into sufficient account the interests and concerns of those who had no influence upon those decisions. Concurrently, different factors of influence, coupled with socio-political opposition, may likewise impact upon both the decision-making process as well as the implementation of policy.

On the one hand, the vast majorities of individuals are alienated from the power to make and execute decisions, but must often bear the consequences of those decisions. On the other hand, those generally privileged few involved in executive decision-making and in the implementation processes might be acting in accord with mis- or dis-information, and might accordingly take a differing degree and kind of risk in making their decisions than those who are directly impacted by those decisions. Decisions and their execution might additionally be influenced by other extraneous interests or factors that can possibly lead to inappropriate or unexpected outcomes, thus exacerbating the chances for continued disputes and conflict. In many ways, the problematic related to the power to make decisions, the ability to execute and influence decisions, and the impact of those decisions upon those who did not make or influence those decisions, represents the roots of socio-political forms of alienation, and of potential disputes—if not conflict and war.

The essence of geopolitical alienation among and within centers of power can most clearly be regarded in debates over the level at which the governmental decision-making process should take place. Should geostrategic and/or political-economic decisions (among others) be made and implemented at the local level, the state level, the regional level or else at the international level, through some form of global governance? There is a major irony in the fact that the local level represents the least alienating and possesses the most potential for individual participation. (See Robinson's discussion of decentralized variants of anarchism, Chapter 5.) By contrast, the international level is potentially the most alienating—but perhaps the most necessary in the contemporary international system—due to the need to better coordinate global interstate and regional strategies and political-economic policies toward various kinds of disputes, conflicts and issues.

Here it appears necessary to regulate political-economic policy (finance and trade policy, for example) at the regional and international levels. Yet how much regional and international cooperation and coordination is actually needed? How much political-economic, if not military, "integration" is really needed? Is greater decentralization truly possible (what the European Union, for example, calls "subsidiarity")—if financing tends to come from central authorities and cannot be raised locally? What if states, localities or corporate interests contest national or international mandates, regulations or political decisions? What if individual states cannot balance their budgets or become excessively indebted? Is it possible for any leadership or independent organization to play the arbitrator between and among differing states with differing financial and power capabilities, if not between and among different levels of governance *within* individual states?[20] Is it possible to permit greater civil society and minoritarian participation in government decision-making processes, and hence reduce degrees of alienation from government and

20 The World Trade Organization has attempted to adjudicate tensions between California tuna fishing laws to protect dolphins which conflicted with Mexican fishing interests, for example, placing the US Federal government in an uncomfortable position in-between California and Mexico. On the WTO and environmental issues, see http://www.wto.org/english/thewto_e/whatis_e/tif_e/bey2_e.htm#dolphin.

corporate decision-making, *but without creating new forms of alienation due to the very complexity involved in coordinating decisions at differing levels*?

Alienation and Estimations of Contending Force Capabilities

In general, diplomatic skills, socio-cultural values and ideological influence, demographic and natural-environmental forces, as well as political-economic and financial capabilities, are rarely mentioned in neo-realist "balance of power" accounts, but are nevertheless key to sustaining a dynamic equilibrium and in reducing or preventing conflict. Neo-realist accounts have tended to focus primarily on military-technological factors and offensive/defensive capabilities as the primary issues affecting the force potential of differing states and socio-political collectives. Nevertheless, classical realists included a number of additional factors in analyzing the so-called "balance of power"—or what really should be considered as *estimations of contending force capabilities*.[21] These factors include geographic positioning (geostrategy), natural resources, industrial capacity and finance, military preparedness, population, national character and morale (which can include mediatized images of state "prestige") and perhaps most overlooked—the quality of dialogue and diplomacy.

From this perspective, so-called "balance of power" or what is really "estimations of contending force capabilities, strategic intent and international norm" is much closer in conception to the dynamics of a Calder mobile than to a traditional balance or scale. On the one hand, the general equilibrium (which is often dissipating and in need of re-equilibration) can collapse rapidly if a single cord is cut due to lack of finance, for example; on the other, the "balance of power" can lose its presumed flexibility when states solidify into rival alliance formations. The fact that the global, regional, national and local actors are in constant flux makes a stable equilibrium (which is not the same as the neo-realist conception of "anarchy") difficult to estimate and achieve. The condition of constant flux leads to potential misperceptions and misinterpretations of the strategic or political intent of

21 A closer reading of both Hans Morgenthau and Hedley Bull indicates that both non-military and non-economic factors also need to be factored into calculations of "balance of power," which can be better described as estimations of "contending force capabilities, strategic intent, and international norm." In essence, by oversimplification, neo-realists have done a disservice to some of the more complex arguments of traditional realists, as have liberals, cosmopolitans, constructionists and post-modernists. See Hans Morgenthau, *Politics Among Nations*, parts 3 to 6; Hedley Bull, *The Anarchical Society: A Study of Order in World Politics* (New York: Columbia University Press, 1995), Chapter 5. Neo-realists have additionally overlooked the traditional realist emphasis on *strategic choices* and *diplomatic options* aimed at conflict resolution (as explicated by Morgenthau, ibid., parts 8 to 10) and which helped form the basis of what I call "alternative realism." See Hall Gardner, *American Global Strategy and the "War on Terrorism"* (Farnham: Ashgate, 2007).

potential rivals, thus exacerbating *pre-existing* conditions of *alienation* and opening the door to potential disputes, physical violence, if not war. In this regard, due the fact that states and socio-political collectives do not always accept the status quo, the act of imperialism—which should be defined as a policy that seeks to overthrow an initial "equilibrium" of inter-state relationships by force and then to sustain that new uneven relationship—likewise contains multiple political-economic and socio-cultural dimensions (such as controls over economic investment and trade, or imposition of a predominant religion or language) in addition to geostrategic—as a means to obtain and sustain regional, if not global, predominance.

In addition to the concept of "balance," the neo-realist concept of "power" itself likewise requires closer critical examination, as do the concepts of "unipolarity" and "multipolarity," if not that of "anarchy." Power is often confused with force and influence, but is better regarded as the center of the decision-making process. Moreover, when one speaks of a "balance of power," one is really speaking of *estimations of contending force capabilities*. It is the center of power that then determines what should be done with those force capabilities, through strategic decisions and executive and legislative policy making. In addition, influence should be not confused with power: Al-Qaeda has a *negative* global influence well beyond its actual means and force capabilities. Al-Qaeda's "anti-power" (using loosely linked networks) lies in its ability to influence those Muslims (and others) alienated from the societies in which they live and to take advantage of *weaknesses* and flaws in the defense mechanisms of differing states and societies. (See Géré, Chapter 24; Charnay, Chapter 26.)

There is furthermore no such thing as "multipolarity" as it relates to different centers of power. The concept of "polarity" is not particularly helpful here, being biased toward the anticipation of antagonism and "polarization." A better term is *centers of power* where significant decisions are made and executed. Such centers of power can be unicentric (with one major power center), bicentric (with two major centers of power) or polycentric (with three or more centers of power). These centers of decision-making power, which can possess highly uneven geostrategic, military-technological, political-economic and socio-cultural capabilities and influences, are not necessarily at odds with one another as is implied in the case of the term "poles"; they can link together in different combinations, and can cooperate or not cooperate on different decision-making levels. There can also be differing centers of power *within* a single state which may cooperate or conflict with one another in the decision-making process, often resulting in a system of "checks and imbalances" that can counteract one another or even impact upon decisions of outside centers of power.

Dialectics of Insecurity–Security

On a day-to-day level, alienation among differing centers of power manifests itself in the difficulties arising from analyzing complex behavior (obtaining complete and accurate information), making the "right" diagnosis (in mediating different interpretations of that information) and then making decisions that may or may not "correctly" resolve the problem *from the perspective of those who made and then tried to implement the decisions*. Alienation can accordingly result from the failure to obtain correct and accurate information, and in the interaction between what is perceived and how it is interpreted. After the interpretation of information, the next step is the actual decision as to whether to intervene and how to intervene. If the decision to intervene is then made, the consequences of that intervention must be predicted in terms of the nature of that intervention, and how that intervention then impacts upon the domestic society, actors outside that society, as well as the natural environment.

Alienation can also stem from socio-psychological blockages—the stereotyping of complex reality, the inability to change or transform preconceived notions, or the misreading of data due to social pressures or "groupthink"—which block or check innovative perspectives and conceptions. The very nature of compromise can result in alienation if the various sides are not able to reconcile themselves to the outcome produced by a general compromise which may be regarded as watering down cherished values or "vital" interests, and thus does not really satisfy all parties. Alienation is magnified when information is manipulated, distorted or invented for self-interested or political purposes, or if accidental occurrences undermine mutually agreed understandings. Due to the fact that problems are generally to be managed or resolved in accord with the perspective of those who make and attempt to execute decisions, alienation also stems from the gap between expectations raised by political leaderships and the actual implementation and impact of those decisions.

The inability to accurately assess the political and strategic *intent* of presumed rivals generally due to a general lack of understanding or "trust" (often accompanied by the development of innovative social, political or technological capabilities of differing forms) creates a *dialectic of security and insecurity*.[22] The latter dialectic provides a deeper understanding for the potential outbreak of conflict and retribution than does the generally acknowledged "security dilemma." (What Azar Gat has more interestingly called the "Red Queen Effect." See Gat, Chapter 2.)

In the conceptualization of the "security dilemma," states take retaliatory actions to secure themselves against perceived threats, but can end up making the situation even more insecure, provoking a counter-reaction (an arms race, for example) by one or more actors. Yet the general inadequacy of the concept stems from the fact that it implies a relatively simple action–reaction process. The overly simplistic concept of the security dilemma does not fully analyze the question of *anticipation* and how the domestic *processes* of perception, analysis, interpretation, and policy formulation can actually help generate social and political tensions even before the policy is implemented—and before some form of peaceful diplomatic or violent military intervention even takes place. The concept also does not examine how state leaderships, factions and socio-political

22 Hall Gardner, *Averting Global War*.

collectives can attempt to manipulate the policies of their rivals from *inside* the state, by supporting dissident opinions and financing surrogate groups, for example. As opposed to the essentially one-dimensional action–reaction nature of the "security dilemma," the concept of an "insecurity–security dialectic" seeks to examine how the quest for "security" impacts upon *both* the internal domestic *and* external foreign policy actions and the reactions of state leaderships or socio-political collectives, in which the quest to assert identity or difference can lead to misperceptions, disputes and subsequent conflict.

In effect, perceived threats can create *anticipatory* or preemptive responses that may not have any real basis in reality, but that are based upon presumed or projected future capabilities of the rival side—if not *threat-obsessed hyperrationality*. The roots of the "insecurity–security dialectic" accordingly take place in the perception or observation of an "activity" or statement of *intent* by a state leadership or socio-political collective that is generally perceived as an actual or potential rival or enemy. That statement or activity can then be interpreted (or misinterpreted) as an actual or potential "threat" (however defined) to state or societal interests, or to the interests of a particular party or faction in power. The problem then arises in the process of how the perceived "threat" is analyzed, the interaction between perception and exactly what is perceived, and how that perception is interpreted and then acted upon (assuming it is deemed "necessary" to act). On a deeper level, alienation also arises when original perceptions or proposals that are accurate or could possibly be "effective" (but from *whose* perspective?) are somehow compromised or corrupted in discussion and/or application.

Policy formulation often goes through a process involving discussions within differing governmental structures (at least in more "open" societies). This process first involves internal, and generally secretive, debate within the executive branch and related bureaucracies. Assuming the issues are discussed thoroughly (and not prematurely leaked to the media and internet on purpose by either supporters or opponents), the issue then goes out for external parliamentary debate and public discussion. Differing factions in civil society, business and labor interests, among others, with both national and international concerns and interests, become involved, as do differing media perspectives which impact on both the general population and elites.[23] The point is that the very process of responding to perceptions of threat may totally transform and politicize or somehow compromise an issue that need not, nor should not, be politicized or compromised, mixing accurate and inaccurate information, and ultimately magnifying misperceptions and misinterpretations.

This process issues forth a *dialectic of security and insecurity* among two or more states (and among differing societies or socio-political collectives) on at least two levels: 1) the upper level of state leadership in which bureaucratic elites clash over the appropriate policy response to perceived threats from a rival state or states or

23 The American and Soviet/Russian governments have often forged "executive agreements" intended to prevent a nuclear arms rivalry, for example, but, in effect, by-passing democratic procedures, precisely because the leaderships could not obtain either US congressional or Russian parliamentary support for such accords.

socio-political collectives; and 2) the domestic or local level in which representatives or leaders of various political factions interact with the general population, often attempting to expose, if not heavily exaggerate, those threats, for their own political purposes and initiatives. This exaggeration of threats helps explain the potential obsession with the "over-securitization" of domestic affairs and why military spending tends to be excessive, often without strict controls or oversight, even if a certain degree of defense spending can be justified.

All this takes place in a time frame in which the rival or opposition actor or actors can attempt to seize the moment (in a Machiavellian sense) as speed or chronopolitics may be crucial in the effort to assert interests, either by initiating a new action or acting in response to a previous one. While rival authoritarian states can engage in somewhat similar processes, such governments generally possess less developed systems of checks and balances and lack substantial popular or significantly broad-based participation in governance. By contrast, a more "open" society generally permits greater public and media debate (although it can still act secretly before the debate "escapes" to the public). Yet even competing bureaucracies within the same government can become involved in *dialectics of insecurity and security*. By contrast, smaller, more unified factions, such as nomadic anti-state "terrorist" organizations, which can move from state to state, can possibly engage in more direct action to take advantage of perceived political divisions in a government or among governments whose "checks and imbalances" stall government or inter-governmental reactions.

The interactive process on differing levels thus opens a great deal of space for misperception, misdiagnosis and botched interventions, thus exacerbating the general condition of inter- and intra- state and societal alienation. Major threats can be downplayed or ignored; or, by contrast, cause overreaction and over-securitization that can repress political freedoms, if not basic human rights. Non-threats can become major threats hyped by media; or, by contrast, conflicts with significant human rights abuses and loss of life can be downplayed as "non-wars." (See Robinson, Chapter 5.) Or wars can be manipulated as a means "to engage a disengaged citizenry." (See Karatzogianni, Chapter 23.)

Moreover, third parties may interpret those actions/interventions very differently, thus taking new countermeasures due to their differing perceptions of the need for "security," however defined and implemented, intensifying the over-securitization process in which individuals and minoritarian oppositional groups, whose political demands are not necessarily related to a specific "threat," could nevertheless become suspect. A degree of security is essential for societies to develop and prosper; yet too great an emphasis on security results in paranoia that can place shackles on original thought and creativity, with critical commentators potentially regarded as enemies. Even actors who perceive roughly similar "threats" may interpret those threats differently and be unable to coordinate policy due to differing criteria, interests, values and the nature of their domestic political leadership and structures. This is true even if the values, norms and interests of allies, for example, appear somewhat similar, as differing state leaderships in an alliance may not agree to the same *order* or *hierarchy* of concerns.[24]

The general condition of alienation is further exacerbated when state leaderships, political factions or socio-political collectives begin to use forceful bargaining and the threat to use violence (discriminate and/or indiscriminate) as a form of strategic leveraging so as to assert their interests. Somewhat ironically, in addition to pursuing specific interests, state leaderships, political factions and socio-political collectives can all choose to use force in the effort to press for recognition, respect, if not degrees of power sharing, from rival states or collectives. Here, even the use of extreme or indiscriminate violence by "terrorist" groups can represent a means to weaken the resolve of the government through wars of attrition, to threaten those who support the government, and to achieve limited aims (such as the freeing of political prisoners), but it can also ironically represent a demonstration of force intended to gain recognition for the "cause" (however portrayed) or else seek out international recognition for the group itself. Such use of violence and force, however, may or may not ultimately gain the other side's respect and recognition, particularly if respect is not mutual or if suspicions that one side or the other possesses malevolent intent continue to fester.

The Decision to Go to War

The decision to go to war possesses both rational and irrational aspects: rational due to the *appearance* of cool calculation; irrational due to the risks and uncertainties involved. Wars can be predatory/imperialist, preclusive, preemptive, defensive or revolutionary; wars can be of *initial challenge*, wars of *revanche*, or wars of attrition. Predatory/imperialist wars seek to gain territory, resources, possessions, plus greater influence and control over weak states or opposition groups. Preclusive wars seek to prevent an opponent from gaining an advantage in the long term. By contrast, preemptive wars take place when war appears *imminent* and attempt to prevent the opponent from striking in the very near term. Defensive wars are those made in self defense, after being attacked, but can become predatory or imperialist if the opponent who attacked first is defeated, and then occupied. Revolutionary wars can be fought for "independence" against an external power or against the leadership of a particular state; secessionist wars are fought to break away from a state and to control a particular territory. Revolutionary movements can possess broad socio-political support involving a wide range of classes and socio-political collectives; or they can be more narrowly based, involving demands for the "national self-determination" of a particular socio-political collective, such as an ethnic or religious group, which may rationalize its demands on the basis of "persecution" and/or the right to control the wealth of a particular territory. Differing forms of revolutionary wars may appear to be fought only *inside* a particular state, but

24 See Hall Gardner, *American Global Strategy*, Chapter 7.

revolutionary partisans generally obtain diplomatic and financial support from *outside* that state, if not through self-financing by black and grey market activities, for example.

Wars of initial challenge can represent a series of wars in which each side tests the capabilities of the other. By contrast, wars of regional or global *revanche* take place after defeat and can be driven by the collective memory of elites and socio-political interest groups who seek the return of territory, lost possessions, influence and control—even if that memory is often distorted for political purposes. (See Kobtzeff, Chapter 15.) Wars of attrition involve the effort to undermine or disrupt for as long as possible the capabilities of the enemy and to continue to jam the actions and operations of the opposing state and/or socio-political group. Revolutionary wars raise the question of the political economic viability of the new state once it breaks away from the controls of an external imperial power or once the new regime seeks to establish itself, while wars of secessionism raise the prospects of state disaggregation in which certain secessionist regions may prove to be more viable in political-economic terms than others. In general, wars may result from the refusal, reluctance or inability to engage in differing forms of power sharing, autonomy or confederal arrangements with other rival powers or socio-political collectives for fear of losing (or of failing to gain) a significant advantage relative to the adversary. At the same time, as argued above, conflicts among states and socio-political collectives are multidimensional and often involve clashes in geostrategic, political-economic, military technological, socio-cultural-ideological interests and values.

An optimistic belief that one's cause is "just" or *legitimate* and that a war will be "successful" (due to presumed military-technological advantage, for example) represents an additional stimulus for initiating conflict against one's adversaries, in addition to considerations that the war will be "short" in length (but often with very little thought for post-war consequences and *jus post bellum*). By contrast, pessimism as to the odds for "victory" (however defined) and fears that a conflict could turn into a long-term and costly struggle possibly resulting in defeat could discourage the decision to go to war—unless there appears to be no other option. *Decisive* victories tend to set conditions for peace (with or without a modicum of justice for the losing side) in the long term, while *indecisive* victories can result in renewed wars of attrition, initial challenge or regional, if not global, revenge[25]—but only if necessary compromises or irenic policies cannot appropriately be implemented. If there can be no general agreement among disputing leaderships and their supporters, then recurrent conflict and wars of attrition can be expected; at the same time, decisive defeat does not necessarily prevent various minoritarian partisan factions from keeping the flame of revenge alive.

Decisions to go to war are often accompanied by legal arguments in accord with international norms, even if a military intervention does not necessarily possess clear legal grounds. Obtaining international legitimacy (in abiding by international norms and organization procedures, and in finding a legal justification) represents

25 Geoffrey Blainey, *Causes of War* (New York: The Free Press, 1988).

icing on the cake, but it does not necessarily guarantee that a decision to go to war is necessarily appropriate, nor does it guarantee a politically "successful" war, nor does it guarantee a suitable outcome that can ultimately "reconcile" the differing sides. In addition, questions as to whether *all* measures and options to resolve the crisis before opting for war were truly tried with conviction *as is the basic requirement of just war theory* continue to linger for a very long time after any conflict. (See Gardner, Introduction; Davis, Chapter 7.)

The so-called right of "humanitarian intervention" or "responsibility to protect" which has been regarded by some as a form of "just war" may or may not be justified in that the decision to intervene is always a question of *geopolitical selection*: Why intervene in the domestic or sovereign affairs of one particular state as opposed to others when a number of states are engaged in egregious violations of human rights? Why does one case deserve more international attention than another? Moreover, while the concept of sovereignty *should* involve the respect for international norms and values, and hence respect for citizens *within* that state (due to the fact that sovereignty can be seen as derived in part from international diplomatic recognition), it is not clear that forceful international intervention will necessarily "teach" that respect. The use of force could, instead, set up a different hierarchy of leadership and new, yet differing, forms of alienation.

More promising might be multilateral pressures and incentives to change a government's repressive policies, that is, if a change in human rights policy is truly the *only* goal and that it is not confounded with other possible goals and interests. Also at question in "just war" theory is the issue of possessing a reasonable chance for "success" in the post-intervention phase. "Success" on what grounds and on whose terms? "Success" in terms of political defeat of the rival leadership? "Success" in terms of rebuilding and reconciling the defeated society? But in what form should that state and society be rebuilt? How should "success" be defined? (See also, Introduction, this book.)

Even the right to go to war in ostensible "self-defense" can be questioned as illegitimate by factions that oppose the policies of a particular leadership and that believe alternative options for dealing with the rival or rivals were possible to implement prior to engaging in a conflict. From these perspectives, war can never be "just" as someone always suffers the consequences of the decision to go to war. War can never be "just" because those who suffer the most are not necessarily the initiators, nor the original causes, of the conflict. The decision to go to war can only be "justified" or rationalized by various leaders or political factions for their own purposes based upon the failure, refusal or inability to seek out all options short of war.

Role of Diplomacy

Alienation among states is further extended (in a socio-psychological sense) by situations and conditions in which socio-political, political-economic and/or military-technological *innovations* of differing forms take place, often making it even more difficult to properly assess force capabilities. (On innovation in the modern world system, see Thompson, Chapter 27.) Changing socio-political circumstances make it more difficult to determine precisely which political faction might be in command and thus what is the true *intent* of the leadership. Alienation is likewise exacerbated if political factions (which may or may not be able to obtain strong domestic or external supports) use extra-legal means of violence, strikes or acts of "terrorism" in order to assert their interests or gain recognition for their cause. If deterrence or dissuasion fails, then conflict can accordingly break out. Here, diplomacy (as an integral aspect of the rough *estimation of contending forces*) needs to be willing and skilled enough to find a way to mediate, if not reconcile, disputes while attempting to *reconcile* or *transform* perceptions of "vital" interests (and the social identification with those interests)—with the realization that compromise by itself could create further divisions and alienation, particularly if the various sides and partisan groups cannot reconcile themselves to that compromise. Yet to truly minimize perceptions of alienation, various forms or degrees of power sharing generally need to be implemented, or new participatory systems of governance, including consensual voting procedures, need to be developed.

The key dilemma is consequently to determine whether states, political factions or socio-political collectives are engaging in political activities (including armed struggle) in an effort to obtain recognition in terms of *parity* or *superiority*: Are states or socio-political collectives are driven by *isothymia* (the desire to be recognized as an equal) or else *megalothymia* (the desire to be recognized as a superior)? In many ways, states, political factions and socio-political collectives may be interested in socio-political *recognition* as much as power, particularly in confronting those who already possess a superior status. Not all states and socio-political groups seek conquest and superiority. The question, then, is to decide when and how to negotiate on the basis of mutual respect despite differences in values, norms and interests, and whether or not to accept some degree of mutual compromise or even unilateral concessions, or else to continue to engage in differing forms of strategic leveraging in an effort to pressure, if not *compel*, the other side to make more concessions. Effective diplomacy is key in helping to mediate disputes and in mitigating conditions of alienation, but it cannot repair conditions of alienation altogether.

The role of diplomacy in achieving possible transformation or reconciliation between conflicting socio-political collectives is even more difficult in conditions in which one or more sides is convinced that its demands are just. The lack of flexibility in intentions and goals (and lack of trust leading to uncertainty as to how the other side might act) is inherent to the *insecurity–security dialectic* and leads to claims and counter-claims, against other states and/or against factions or socio-political collectives both within and outside that state and society. This lack of flexibility is particularly evident in situations in which each side is attempting to build, if not invent, a case to justify or legitimize a possible decision to go to war. Even in authoritarian societies, the decision to go to war must generally obtain some form of justification or ideological rationalization. The process may be more complex in societies in which affairs of state are more openly debated; nevertheless, the executive branch generally seeks to bring the differing factions and interests into some form of consensus. If the leadership believes it cannot obtain consensus, then it can engage unilaterally, without consultations with parliamentary bodies or even allies, in the name of "national security."

Disalienation of States and Populations

The fact that humanity is divided into separate territorial states with largely artificial borders (that have generally been acquired or imposed by force and which have shifted over time) further exacerbates the general condition of the individual's alienation and self-defined socio-political collectives from other states/societies and from the natural world as a whole. In effect, mankind has alienated the natural world as a physical and organic whole by artificially dividing it into both private property and state boundaries for political-economic purposes and interests. (Though the cosmos can be alienating in itself, the natural world does not recognize either private property or state boundaries!) This fact makes it more difficult for states and societies to effectively coordinate actions among themselves while sustaining complex ecological equilibriums that protect the natural environment through careful intervention and interaction with the natural world.

Alienation between territorially divided states (which Marx and Engels presumed would magically "wither away" under the rule of socialist government) is the result of the historical, yet largely arbitrary and uneven division of the earth's geography and resources under presumed monopolies of force, much as Max Weber argued. Yet monopolies of force can be challenged by social collectives and political factions from both *inside* and *outside* that territory. As argued previously, effective diplomacy is key in helping to resolve or transform disputes and in mitigating conditions of alienation, if possible, but it cannot transcend conditions of alienation altogether. The major dilemma raised here is that diplomacy can be two-faced. On the one hand, force-based diplomacy (as diplomacy "by other means" in the sense of Clausewitz) may use violent aspects of strategic leveraging to pressure a rival state or socio-political grouping (the "other") to come to terms. On the other hand, some diplomatic activities may be insincere and may be purposely used to divert attention from other pressing issues, or even to stall for time, for example. Instead of fully engaging in sincere efforts to mediate disputes, what can be called "incomplete diplomacy" may be unwilling to fully resolve all aspects of the dispute and might actually seek to avoid processes of conflict resolution or transformation in more troublesome regions.[26] Here, the refusal to engage in open and more or less "transparent" diplomacy could be directed at fooling an external "enemy," but it

26 See James Der Derian, in Michael W. Doyle and G. John Ikenberry (eds), *New Thinking in International Relations Theory* (Boulder, CO: Westview, 1997).

might also be directed at outmaneuvering domestic political rivals who might seek to block such initiatives.

The refusal or reluctance or inability to fully engage in conflict resolution or transformation can be carried out in the name of managing, co-opting or "appeasing" powerful interests (either domestic or international); it can also be rationalized upon grounds of "national security" or "human rights violations." Certain issues may be appropriately adjudicated while others may be ignored or left untouched in the effort to establish "peace and order" in the long term without stepping on too many toes. The problem raised here is that one or more sides may well possess ulterior motives in pursuing negotiations. As negotiations, in effect, reside on a plane "above" the level of geostrategic, military-technological, political-economic, and socio-cultural conflict, only those negotiated settlements which truly pertain to the mutual interests, values and norms of territorial states or political collectives, and which have established a rough "parity" in relationships, will begin to change the dynamics of inter-territorial state and inter-communal relations.

If truly mutual interests, values and norms are not fully taken into account, there will be no guarantee that even signed accords will be executed in the long term as initially agreed. Even in cases of sincere diplomacy, disputes can arise over key questions with respect to who has legitimacy to act (and to what extent to let disputing actors resolve the conflict on their own); whether to pursue a peace settlement (ignoring crimes against humanity, for example) or else attempt to implement differing conceptions of justice and reconciliation; whether or not to use force (and how much force should be used); whether or not to engage significant resources to try to *prevent* a potential crisis or else let the conflict play itself out.[27] These dilemmas and complications generally make the conflict resolution/ transformation a long-term affair.

The point raised here is that mutual interests (in the realist sense) are *not* always sufficient: differing states and socio-political collectives generally need to establish some form of *mutual recognition* leading to *mutual respect* for each other's identity and values if the conflict is to dissipate in the long term. If successful, mutual recognition can then lead to a *transformation* in how two (or more) sides perceive one another, hence ultimately transforming the nature of perceived intent or threat. If successful, the process, which is gradual (and frequently interrupted), not only changes perceptions but identities as well, in that the former "enemy" and the society, as well as the leadership (assuming it does not step down from power), is understood in new ways. The key and initial problem in preventing a further decaying of interrelationships in the first place is how to build trust and confidence; otherwise, any apparent changes will prove superficial and limited to a political—and not a deeper—social level.

27 On key dilemmas of conflict resolution, see I. William Zartman (ed.), *Peacekeeping in International Conflict* (US Institute for Peace, 2007), Introduction. See also, Chester A. Crocker, Fen Osler Hampson, and Pamela Aall, *Taming Intractable Conflicts* (US Institute of Peace: Washington, DC, 2004).

Another rationale for which a state leadership might be unwilling to thoroughly mediate disputes through differing forms of diplomacy could be due to the fact that certain elites might fear that they cannot remain in power without instigating perpetual tension, if not war. The manipulation of perpetual tensions and conflict represents a form of strategic leveraging intended to *legitimize* the position of power and influence of a particular leadership or specific factions in a society (such as national arms producers) while concurrently pressing and exhorting its domestic population to accept the necessity of conflict.[28] In effect, leadership might actually seek to *identify itself* with the conflict itself in its rivalry with other state leaderships, while domestic factions and the society in general may or may not question the policies of perpetual tension and war.[29]

In this view, any compromise with a particular faction's ideological convictions and larger goals or mission could result in the demise of a political party or leadership, and thus lead to the growth of dissent within a socio-political group or the society in general. Here, political ideology itself often plays a major role in the effort to patch over or hide the general conditions of alienation, in that such ideologies often attempt to explain the inexplicable, or else set the general goals and values for the society as a whole. The risk is that changing political tactics may result in the need for compromise (particularly when compromise impacts self-proclaimed "vital" interests); the option to compromise may appear to violate the proclaimed longer term goals and values of that leadership, thus undermining its sense of legitimacy.

Depending on the circumstances, a strategy of avoiding negotiation and of sustaining perpetual tension can ultimately result in a degradation of the economy and long term socio-political crisis.[30] The failure to engage in a process of deep and sincere conflict resolution or in engaged transformation among conflicting states and socio-political movements can consequently result in the fact that apparently

28 Jean Jacques Rousseau recognized that elites whose power and status are threatened may be in "perpetual need of war" (even if it ruins the state) "as a price of keeping their own office ... as a means of oppressing the people on the plea of national necessity ... [and] of rigging the market and setting up a thousand odious monopolies." Rousseau, "Saint Pierre's Project for Peace," in Stanley Hoffman and David P. Fidler, *Rousseau on International Relations* (Oxford: Clarendon Press, 1991), 93. See also Robinson, Chapter 5 of this book.

29 From a Hegelian perspective (contra Rousseau, Kant and Marx, plus anarchists in general), society *as a whole* and its state leadership can identify with war—particularly those wars that are either "defensive" or "victorious"—as a means to regenerate the state and its social values. Here, however, it is not clear that war will necessarily "regenerate" social values. A "lost" war is more likely to pervert or degenerate social values, as was the case for Germany in the aftermath of World War I. But even a revolutionary or victorious war can lead to perverse behavior and wars of extermination, which Hegel himself opposes. See Hegel, *Philosophy of Right*, sections 324, 333, 334.

30 Anti-state actors and surrogates financed by outside powers may attempt to use wars of attrition against established states to their advantage assuming they can sustain pressure in the long term.

inert socio-political disputes could eventually return to life in the form of renewed violence and conflict, even if states appear to be engaging in mutual compromise. Such renewed conflict generally takes place in new forms and in very different historical circumstances, and may or may not result in conflict resolution or conflict "transformation."[31]

In short, "regime recognition" may or may not lead to domestic "regime reform." US-European efforts to engage diplomatically with the Qaddafi regime in Libya from 1999 to 2010, for example, in order to assure that Qaddafi would keep to his promises not to develop weapons of mass destruction, not to support international terrorism, and open the country's market to multinational investment in Libya's vast energy and mineral reserves, did not work toward "transforming" or significantly "reforming" that regime from the perspective of the majority of the indigenous population, nor from the external perspectives of the US and Europeans. The Europeans, with US and Arab League backing, then altered course in 2011 under a general UN mandate, shifting overnight from "regime recognition" to "responsibility to protect," and then to "regime change," by support of insurgents opposed to Qaddafi's rule. What kind of social and political regime will arise next remains to be seen. (See Gardner, Introduction and Chapter 29.)

Relevance of the Peloponnesian Wars

The question of socio-cultural-ideological, political-economic and military-technological *innovation* leading to alienation among major states and societies likewise represents a tremendous barrier to reconciliation. In this respect, much as was the case for World War I, among other major power conflicts, the alienation of Sparta (seen as a classic and archetypical example) represents the deeper meaning of Thucydides' argument as to the cause of the Peloponnesian wars. From this perspective,[32] it was the fear of differing dimensions of Athenian

31 Der Derian argues in Nietzschean terms of "eternal return" as to how promises for the establishment of an independent Armenia and Kurdistan vanished under the 1923 Treaty of Lausanne, for example, and yet these issues and conflicts have, in effect, returned to haunt the present. But by contrast with the Nietzschean view, such renewed conflict takes place in very *different* geohistorical conditions and is not necessarily "eternal." See James Der Derian, in Michael W. Doyle and G. John Ikenberry (eds), *New Thinking in International Relations Theory* (Boulder, CO: Westview, 1997). See also James Der Derian, *On Diplomacy: A Genealogy of Western Estrangement* (Oxford: Basil Blackwell, 1997).

32 Thucydides, *History of the Peloponnesian War*, Book I. See also Richard Ned Lebow, *The Tragic Vision of Politics* (Cambridge: Cambridge University Press, 2003), 88 and passim. Lebow argues that war could have been avoided had Athens revoked the Megarian Decree, made some symbolic gestures toward Aegina and terminated the Potidaean siege, thus placating Spartan hardliners, while simultaneously isolating Corinth which sought Spartan support against Corycea. Justin Rosenberg points out that "Thucydides'

innovation (geostrategic positioning and control of trade, such as corn and other cereals, development of naval and other military capabilities, as well as the newly democratic Athenian social-political and ideological orientation, in addition to Athenian alliances with Argos and then Megara which controlled the trade route to southern Greece) that led to Spartan alienation and war with Athens. (See Athina Karatzogianni, Chapter 8.) Sparta's alienation from Athens can furthermore be understood with respect to the fact that Athens and Sparta had been allies versus Persia prior to the outbreak of the Peloponnesian wars, and thus had previously engaged in power sharing for a period of time. Once again, at the roots of the crisis was Spartan alienation: The fact that Sparta no longer possessed its previous power and ability to influence the policy of Athens (which in turn had tried to dictate Spartan policies) helped to generate long term conflict and war. The breakdown in Athenian-Spartan cooperation, which was interlinked with a domestic power struggle in Athens, can be seen in the unsuccessful efforts of Cimon, as commander of the Delian League, and as a voluntary *Proxenos* for Sparta within Athens, to sustain cooperation between the two city-states. In 462 BC, Cimon obtained Athenian support to assist Sparta after the latter appealed to the Delian League for help in forcibly repressing Helot and Messenian insurrections after an earthquake. But Cimon and the 4,000 Athenian hoplites were then accused by Sparta itself of "designing innovations" in the words of Plutarch; Cimon's forces alone, unlike those of other Greek states, were expelled from Sparta, rupturing Athenian-Spartan trust.[33]

In addition to advocating closer ties with Sparta (and angering Athenian citizens through his overt praise for Spartan values), Cimons was seen as supporting the aristocratic Areopagite constitution. By contrast, the radical democrat Ephialites opposed alliance with Sparta in the hope that the Helot insurrection would weaken the Spartan rival; Ephialites took steps to strengthen the power of the popular assembly (*ekklesia*) and engaged in other radical democratic reforms that sought to limit the power of the Areopagus council—at the same time that Cimons was engaged overseas, in an intervention intended to assist Sparta repress the Helots. Upon his forced return from Sparta, after being accused of "designing innovations," Cimons was humiliated in Athens and ostracized for ten years, while Ephialites was assassinated in 461 BC. This led the populist Pericles, who sought to consolidate majoritarian democratic gains so as to strengthen Athenian power, to take command as military Archon or *strategos*. With the complete rupture of trust between Athens and Sparta, the first Peloponnesian war broke out in 460 BC.

famous one-liner is emphatically *not* an instance of a substantive realist explanation"; nor were the causes of the war limited to "balance of power" politics. Correct on the surface, this argument is based on stereotyped conceptions of realism and balance of power. See Justin Rosenberg, *The Empire of Civil Society: A Critique of Realist Theory of International Relations* (London: Verso, 1994).

33 Plutarch's Parallel Lives http://ancienthistory.about.com/library/bl/bl_text_plutarch_cimon.htm.

In contemporary circumstances, the expansion of NATO into former Soviet space, coupled with inter-allied disputes over "power" and "burden" sharing, appear somewhat reminiscent of Athenian disputes with both Sparta and its Delian League allies. Despite the fact that NATO has claimed to revamp itself as a new form of "defensive" alliance with no nuclear weapons and no substantial combat forces to be deployed on the territory of new NATO members, Moscow has remained suspicious.[34] One could argue that NATO enlargement, coupled the building of missile defense systems that could be used either defensively or offensively, and opposed by Russia (and China), could parallel in strategic and symbolic terms the Athenian decision to build long walls of defense, an action opposed by Sparta, causing the latter "secret annoyance" in the words in Thucydides.[35] Likewise, the Athenian decision to forge a new "defensive" alliance with Corycea against Corinth, which was aligned with Sparta, helped to spark the first Peloponnesian war following forceful Athenian actions against Potidaea and Megara. Over the long term, the breakdown of US-Soviet wartime alliance led to the Cold War in which the Soviet Union still sustained a rough parity with the United States—even if both empires did attempt to outmaneuver the other. Yet Soviet disaggregation in 1991, followed by NATO enlargement and US intervention abroad, has further destabilized the bi-centric "equilibrium" of the Cold War US-Soviet relationship. In this respect, NATO's new "defensive" yet expansionary stance (combined with both US and NATO "out of area" military interventions) appear to possess some parallels with the geostrategy and alliances of ancient Athens, once the Athenian-Spartan wartime alliance against Persia began to collapse. This appears plausible in that overlapping US, NATO and European defense ties with significant third powers could possibly draw the US and NATO into conflict with a number of states that might look back to either Russia or eventually, China, for military protection. Whether the US, NATO and Russia can eventually establish a relationship of trust and mutual respect—particularly in the aftermath of the August 2008 Georgia-Russia war—remains to be seen. From this perspective, it seems increasingly necessary to *transcend* the Athens-Sparta analogy. (See Gardner, Chapter 29.)

Mitigating Conditions of Alienation

Effective state leadership can seek to overcome or prevent *certain elements* of the so-called "anarchical" international system from degenerating into conflict through the establishment of entente and alliance relationships. The issue raised here is that so-called "anarchy" is not necessarily as universal or as extensive as neo-realists (and post-structuralists) claim, and can theoretically be transcended

34 See my original argument, "NATO Enlargement and Geohistory" in *NATO for a New Century: Enlargement and Intervention in the Atlantic Alliance* (ed.) Carl Hodge (Westport: Praeger, 2002).

35 Thucydides, *History of the Peloponnesian War*, Book I.

region-by-region by effective multilateral leadership leading to greater degrees of socio-economic and security cooperation in the formation of internationalized, yet relatively autonomous, "regional security and development communities."[36]

These communities seek to bring neighboring states and local socio-political communities into cooperation with each other through differing security and economic accords and confederal arrangements where possible. Mutually-agreed territorial adjustments may represent one means to achieve a larger regional accord that helps to alleviate the alienation of opposing states and socio-political communities ultimately leading toward peaceful political, social and economic interaction. If such territorial adjustments are not sufficiently internationalized by means of cooperation with major and regional powers and international regimes, however, they could result in new tensions and conflicts. Territorial adjustments, coupled with agreements that can lead to new forms of trade diversion relative to states not included in agreements, and the implementation of conflicting legal and regulatory frameworks among differing regions, can lead to new disputes in which a newfound confederation may not remain stable. If international and regional security guarantees fail, a weak confederation may be subject to imperialist action and seizures of territory by aggressive neighbors, which may attempt to draw lesser states into a tighter military bloc or alliance. For these reasons, steps to forge regional security and development communities must be taken through concerted measures involving both regional and major powers, as well as local community actors, in part in the effort to limit the number of accidental or unanticipated actions that might provoke conflict.[37]

In this respect, positive aspects of engaged diplomacy seek to *mitigate* perceptions of alienation through efforts to negotiate specific issues and interests as a step toward mutual understanding leading to *mutual trust* and *recognition*, while gradually moving from *regime recognition* toward *regime reform* and *transformation*. Steps to transform relations between France and a unified Germany toward mutual trust, understanding, recognition and power sharing, for example, took place only after the Napoleonic wars, the Franco-Prussian War, World War I and World War II. Despite the pooling of sovereignty in the European Community and European Union as a form of power sharing, coupled with German integration into NATO,

36 On "security communities," see Karl Deutsch, *Political Community at the International Level* (Garden City, NY: Doubleday, 1954). See also Chester A. Crocker, "The Place of Grand Strategy, Statecraft and Power in Conflict Management" in Chester A. Crocker, Fen Osler Hampson, and Pamela Aall (eds), *Leashing the Dogs of War* (Washington, DC: US Institute of Peace Press, 2007), 362. See also Hall Gardner, *Averting Global War*. As one cannot achieve development and disalienation without security, and as the emphasis on security alone implies perpetual tension, conflict, if not repression, I have expanded the concept to "regional security and development communities." See Introduction and Chapter 29.

37 By expecting the accidental, and by seeking to *re-structure* the international system through proposed interlinked security and development communities, such an approach seeks to address post-modernist concerns and attempt to minimize both hazard and the hazardous wherever possible. See footnotes 21 and 24, this chapter.

full Franco-German state and social reconciliation really did not take place until German troops (as part of Eurocorp) symbolically participated in the July 14 Bastille Day celebration in Paris in 1994, and when it was announced in 2009 that German forces would be based in Alsace. Here, France and Germany were assisted in the process of reconciliation by the formation of interlinking "regional security and development communities." These steps included the formation of NATO in 1949, the European Coal and Steel Community in 1950, the membership of West Germany in NATO and, ultimately, the formation of the European Union. These represented the major international steps toward the *disalienation* of interstate Franco-German relations and the transformation of their inter-societal ties, a process which, so far, appears successful but which could be challenged by the Euro financial crisis.

From this perspective, it is clear that efforts to mitigate the deep-seated alienation among conflicting states and societies can take place, but only in a long-term geohistorical process of *disalienation*. Just as efforts to reconcile France and Germany took place in a larger systemic context, it is possible that other forms of internationalized, yet relatively autonomous, "regional security and development communities" could be implemented in other conflict zones throughout the globe. Sincere policies of engagement need to be followed to resolve or transform disputes between "democracies" and "non-democracies," as well as among democracies themselves. Here, for example, one could argue that Indian-Pakistani conflict over Kashmir and Afghanistan plays a role somewhat similar to German-French conflict over Alsace, among other possible analogies. A multilateral and concerted approach involving the US, Russia, Europeans, China, India, Pakistan and central Asian states to the ongoing crisis in south and southwest Asia, for example, could ultimately help wind down that seemingly intractable conflict.

The differences in values between democratic and authoritarian states may make conflict resolution appear more difficult, but this fact did not prevent Great Britain and Tsarist Russia, for example, from coming to terms (largely due to common geostrategic interests) before World War I. Confronted with the rise of Imperial Germany, Britain and Tsarist Russia came to terms in the 1907 Anglo-Russian Conventions over Tibet, Afghanistan, Baluchistan and Persia. Steps toward alliance took place despite the fact that Imperial Germany and Great Britain possessed even closer ties between states (through monarchical relations) and between societies (through Anglo-Saxon "civilizational" ties) than did Britain and Russia. In contemporary circumstances, however, a great power US-Russian condominium will not work in quite the same way as did the early twentieth-century Anglo-Russian entente; both the US and Russia will need to work together, but in developing concerted relationships with other major regional and local actors as well. (See Gardner, Chapters 13 and 29.)

Geohistorical efforts to *disalienate* relations among differing conflicting states and socio-political collectives generally go through two stages: from *antipathy* and *conflict* to *empathy* and *détente*, ultimately leading to *mutual sympathy* and *entente* or *alliance*. The move from *antipathy* to *empathy* can take place by taking steps toward mutual understanding and by finding common interests; the move from *empathy* (mutual recognition) to *sympathy* can take place by taking steps toward policy cooperation and possibly toward power sharing (*entente*). These steps, in effect, do transcend the original positions of the differing sides, but do not necessarily achieve a greater sense "universality"—in that perspectives are only enlarged to incorporate those states and societies implicated in those power sharing accords. Power sharing can possibly lead to closer defense coordination (*alliance*) with some acceptance of shared values likewise involving a step-by-step transformation of inter-societal relations.

Conclusions

Multidimensional elements of alienation stem from lack of power sharing or lack of real influence in decision-making processes between the sexes, the family, the workplace, the immediate local community, the larger state, the regions surrounding states, as well as within the global—and not yet realized—"community" of humankind.

Alienation continues to be illustrated by a refusal to recognize "the other" and to engage in degrees of cooperation and power sharing with differing individuals and self-defined socio-political collectives, as well as with specific states. Contrary to Marxian interpretations, it is unfortunately not inevitable that either states or socio-political collectives can eventually transcend historically conditioned dimensions of alienation; there is no guarantee that states and socio-political collectives with divergent factions will necessarily progress in moving beyond *antipathy* toward *empathy* and then *sympathy*, at both the inter-state and inter-societal level. Nor is there a guarantee that steps toward *disalienation* taken by two states and societies will necessarily lead those states toward a more universal understanding that would incorporate other states and societies, although such steps can provide a model to emulate and follow.

On the negative side, it is also possible for state leaderships to reach mutual executive accords with other states; such accords may appear to bring peace, but could be deceptive in that they may, or may not, possess legislative or popular backing, and thus would not represent true steps toward *disalienation*. It is also possible for significant numbers of a population to support the domestic or foreign policies of a rival state, but one which is not necessarily on positive terms with that population's own leadership or government. Such situations, involving mixed, if not contradictory, perceptions and interests within a state and society, can augment social and political tensions to the point of initiating domestic civil war or international conflict.

In addition, the effort to reconcile differing norms, values and rules through compromise by international mediation or though international organizations represents yet another way to ameliorate conditions of inter-state and inter-societal alienation on a global scale. The dilemma, however, is that this process could alienate those who do not want to water down norms and cultural values too greatly, resulting in too significant a degree of homogenization. Steps taken by international mediation must be able to engage both states and the larger societies if they are to successfully achieve long term goals of *disalienation*.

If the processes of inter-state and inter-societal *disalienation* are to succeed, the first step is to forge mutual agreements of common interest, followed by verification of those agreements, in the case of arms control and reductions, for example, or compromises over property rights, debt repayments, reparations, territorial trade-offs, etc. Most importantly, steps toward *disalienation* must ultimately establish conditions of mutual recognition and trust that can eventually lead to a transformation of inter-societal values and identities in addition to establishing shared interests, often through the implementation of some form of power sharing accords. At the same time, the long term nature of the process can nevertheless involve the use, or threat to use, force and violence of differing forms and circumstances as part of the bargaining process.

From this perspective, the formation of larger overlapping confederations of states, combined with positive incentives (as well as negative pressures) from both *inside* and *outside*, can theoretically help transcend conditions of inter-state and inter-societal alienation on a regional basis. Here, both major and regional powers need to take decisive steps toward *détente* and then *entente* or *alliance*. Cooperation among major and regional powers can then provide lesser states and societies with the clear prospects of socio-economic development plus greater incentives of *power sharing* in order to help resolve or transcend conflicts. The formation of internationalized yet interlinking "regional security and development communities"—*involving differing and relatively autonomous confederal arrangements that can be built from the ground up and backed by major and regional powers with security assurances/guarantees and that are dedicated to conflict mediation and socio-economic development*—could represent a more decisive step toward the *disalienation* of conflicting states and socio-political collectives in differing regions throughout the world.

The Roots and Evolution of Conflict: From Cain to the Present

Azar Gat

My book, *War in Human Civilization*, set out to unravel the 'riddle of war', to find the answers to the most fundamental and long-elusive questions concerning war.[1] Why do people fight? Is war rooted in human nature, or did people start to engage in it only with the advent of agriculture, the state and civilization? How has war affected and how has it been affected by modernity and by the spread of democracy? What is the future of war?

These questions are not new, and they have resisted conclusive answers to the point that both questions and answers appear almost as clichés. But they have very rarely been subjected to a rigorous comprehensive investigation. In pursing the answers, the book took a global view, and ranged from pre-history to the twenty-first century and analyzed interdisciplinary approaches, combining anthropology, evolutionary theory, archaeology, history, sociology, economics, and political science.

The first part of the book – entitled only half-whimsically the 'The First Two Million Years' – addresses what I call the human state of nature, when humans lived as hunter-gatherers. This part aims at resolving such fundamental questions as why people fight and since when they have been doing so.

Let us begin with the question of when humans started to fight. Was warfare always there – is it as old as the species? Or was it a relatively new phenomenon, only emerging with humankind's cultural evolution? For instance, did it begin with the advent of agriculture (that occurred in the most pioneering groups of people some ten thousand years ago), or the emergence of the state (again, some five thousand years ago in some parts of the world and much later in other parts: in the now advanced societies of northern Europe and Japan, for example, only about fifteen hundred years ago)?

What disciplines deal with this question? History does not deal with it at all. It only begins with literacy, that is, 5,000 years ago in the most advanced societies. However, the genus *Homo* has existed for about two million years, and our own

1 Azar Gat, *War in Human Civilization* (Oxford: Oxford University Press, 2006).

species *Homo sapiens*, people who are biologically similar to us, has existed for over 150,000 years. Thus, human prehistory comprises 99.5 and 95 percent of our time on this planet. Compared with these time spans, historical times are just the tip of the iceberg. This only underlines the question of whether fighting is a new cultural invention or has always been around.

It was in the field of political philosophy that the question was first addressed systematically. Here the two classical and conflicting answers were formulated – by Thomas Hobbes in the seventeenth century and by Jean-Jacques Rousseau in the eighteenth. For Hobbes, the pre-state condition was characterized by a war of every man against every man, when, in the absence of a peace-enforcing authority, life was 'poore, nasty, brutish, and short'. For Rousseau, things were quite the opposite: the aboriginal condition of humans was fundamentally peaceful and innocent. In the absence of property before agriculture, there was little to fight over. War, according to Rousseau, is a late development, one of the ills of civilization. So convincing were each of these positions that they have remained with us until today.

But who was right: Hobbes or Rousseau? For political philosophers the question barely exists at all. They usually claim that Hobbes and Rousseau postulated the state of nature as hypothetical, and leave it at that. But this is not an entirely accurate description of Hobbes's and Rousseau's views, and, irrespective of how they saw it, the question is unquestionably historical and empirical. Living in the age of exploration and of European contact with a variety of aboriginal societies, Hobbes and Rousseau were well aware of this. And since then, our insight into the distant past of humankind has increased dramatically, and with it also the realization of how recent in human history the development of agriculture and the state was.

This leads to archaeology. What light can it shed on our question? Unfortunately, not that much, and the main reason for this is that weapons for fighting before the introduction of metals are practically indistinguishable from hunting implements: stone axes, spears, and arrows. Were they used for hunting only, or were they dual-purposed and used also for fighting? We cannot tell with certainty. Even when missile-heads are found embedded in human bones, this can be attributed to hunting accidents. As for specialized fighting equipment such as shields that are made of perishable material such as wood or leather, these have not survived. Also, people did not live in sedentary dwellings until the transition to agriculture during the Neolithic Era (and in a few cases as early as the Upper Paleolithic or Mesolithic), and therefore evidence of fortifications and destruction that we find from the Neolithic on does not exist. The same applies to evidence from cemeteries, which also appear only when people settled down in sedentary settlements.

Biology has gone through an interesting development which serves as a sobering lesson regarding the evolution of ideas, including scientific ideas. During the 1960s, the founder of ethology, the science of animal behaviour, Noble laureate Conrad Lorenz, popularized a set of controversial ideas regarding violence in nature. Lorenz claimed that fighting between animals of the same species is mostly 'ritualistic' and mainly involves display. The loser retreats or submits, while the victor refrains from pressing its advantage to the finish. According to Lorenz, the reason for this pattern of behaviour was the need to preserve the species.

It thus appeared that humans, who fight to kill their own kind, are a deviation from the normal pattern in nature. This notion of a murderous human perversion – a killer ape – sat well with the Rousseauite doctrines of the 1960s regarding nature's purity, the corruption of civilization, and man as a blank slate, wholly moulded by culture. The view that fighting was a late by-product of human civilization became widely held.

Since then, however, evolutionary theory has undergone a sweeping change, known as the neo-Darwinian revival. It became clear that natural selection mostly takes place within, rather than between, species. As Darwin himself argued, the struggle among individuals from the same species is the most intense, because they compete for the same sorts of food and for the same mates, in the very same ecological niches.

At the same time, there have been major new empirical findings regarding animal fighting in nature. It has been revealed in innumerable studies carried out since the 1970s that within species in nature perpetual violent and lethal competition takes place. This also includes our closest cousins, the chimpanzees, studied in their natural habitats by Jane Goodall and others, and documented to engage in murder, as well as in group fighting and killing. It is true that adult animal males usually avoid a fight to the finish among themselves for reasons of self-preservation – any serious injury might render an animal incapable of getting food and result in starvation. There is no social security in nature. Yet when deterrence by the display and demonstration of force fails, serious fighting, injuries and death often follow. Furthermore, most killing within species in nature is carried out against the young and the weak, including eggs, chicks and cubs. Such asymmetrical killing is performed with relative safety for the killer. No benevolence towards one's kind exists. Lions killing the cubs of the former monarch of the pride; chicks in a nest pecking their siblings to death when there is a food shortage; birds throwing the eggs of other birds of their kind from nests; and so on.

So as biology has completed a full circle since the 1960s, as it turns out, humans are no longer unique in nature in extensively killing their kind, and do not call for some special explanation. Widespread killing within the species is actually the norm in nature. This still leaves the empirical question: what are the actual findings regarding humans in the state of nature? Is there concrete evidence that people fought before agriculture and the state?

The discipline that is the richest in relevant information for answering this question is anthropology, which studies extant and recently extinct pre-state and pre-agricultural societies – not that access to, and the interpretation of, that information is easy. The main problem is the so-called 'contact paradox'. These societies have no written records of their own, and documentation therefore requires contact with literate societies that necessarily affect the former. As in quantum mechanics, the very activity of observation changes the object under observation. For example, literate societies have goods – such as agricultural products, livestock and manufactured tools – which hunter-gatherers might want to steal. How can it be determined that a war-like behaviour on their part did not originate only with contact, but had existed before? How can one observe pure hunter-gatherer societies that are free from contact with agriculturalists and states? This is like the light in a refrigerator: does it really turn off when the door is closed?

Because of this built-in ambiguity, anthropologists to this day continue to debate who was right, Hobbes or Rousseau, with their answers changing with the *Zeitgeist*: the nineteenth century was dominated by the Hobbesian image of the savage and the brute, whereas the twentieth century, with its critique of civilization, was predominantly Rousseauite.

So the challenge is how to observe pure hunter-gatherer societies to determine whether they fought or not. The most significant test case is Australia, an entire continent of Aboriginal hunter-gatherers, with no agriculturalists and pastoralists, whose isolation came to an end only two hundred years ago, in 1788. This is the closest to a pure laboratory on a continental scale that we are ever going to get. Australia incorporated about 300 regional groups, or tribes, when the Europeans arrived. The evidence shows that the Aboriginals fought incessantly among themselves, including the material evidence of shields, which were not of course used for hunting kangaroos. Almost as good a laboratory is the American North-West, from Oregon to Alaska, a huge space of nearly isolated complex hunter-gatherers who also are revealed to have fought incessantly. Even the Kalahari Bushmen which were the focus of study in the 1960s, and were celebrated as peaceful by their chief researcher, the Harvard Marxist anthropologist Richard Lee, among others, were soon recognized to have had four times the 1990 US homicide rate, which was itself by far the highest in the developed world.

To summarize the findings with respect to the many pre-agricultural and pre-state societies studied around the globe, findings which might be vague in any particular case but which consistently repeat themselves in one separate anthropological case study after the other, thereby becoming an unmistakable pattern: around 25 percent of the adult males in these societies found a violent death, while all the rest were covered with scars. Lawrence Keeley's book, *War before Civilization*, systematically demolished Rousseauite anthropology, providing extensive and varied anthropological and archeological evidence. This point is reinforced by the fact that mostly different case studies were used, as Professor Keeley concentrated on the Neolithic and horticulturalists, whereas I mainly focus on hunter-gatherers, because Rousseau's original claim concerned the effect of the transition to agriculture. In any case, the findings are very similar. Contrary to prevailing views, pre-agriculture and pre-state societies suffered from a much *higher* violent death rate than that incurred in modern societies, with only the world wars coming close. An endemic state of insecurity and fear prevailed in these societies which shaped all aspects of their lives. Hobbes was much closer to the truth than Rousseau in describing the human state of nature.

I shall now pass on very briefly to the question of why people fight. People, like all organisms, fight for the attainment or defence of the very same objects of desire that underlie their lives in general, and which are originally shaped by the calculus of survival and reproduction in the evolutionary state of nature. People can cooperate, compete, or use violence in order to achieve these objects, depending on what they believe can serve them best in any given circumstances. Violence is not a primary drive which requires release, like the desire for food or sex, as some 1960s theorists and popular writers believed, following Freud's latter-day theorizing about *thanatos*, which was controversial even among his followers. Think about the Swedes or the Swiss, who have not fought for centuries, and yet exhibit no particular distress on that account. But try to imagine their reaction if deprived of food or sex for any substantial period. So if violence is 'hardwired' in us, so to speak, this is not as a primary drive, but as a means, a tactic for achieving desired aims. And it is a very dangerous means, which is therefore mostly activated if other, more peaceful, means fail, or are too costly, and if the chances of success are judged good. Violence is the hammer in our evolution-shaped behavioural tool kit, which also contains a variety of more delicate instruments, to be selected according to the circumstances at hand.

So what did people in the state of nature fight over? Resources such as hunting territories were fiercely competed and fought over – because scarcity often meant the difference between life and death; the stakes were very high. There was no primordial state of abundance and free common land of the Rousseauite anthropological imagination. Women were hotly fought over, again because reproduction is a tremendous selection force, and is therefore central to our system of motivation. Polygamy for the more successful men was the rule, as was fighting caused by the abduction of women, rape, and extra-marital relations.

Status in society has always been a means for getting one's way and reaping material and reproductive benefits, and has been hotly pursued and fought over as such. People staunchly defend their honour, for if they do not they may be trampled over and encourage further encroachments, that is, create a process of victimization. To paraphrase Winston Churchill: those who prefer shame to war, may beget shame and then war. People retaliate to deter injuries inflicted by others, because as a famous computer game has demonstrated, tit-for-tat is the most effective mechanism in human relations: respond with good will to acts of good will, but pay back on injures. However, tit-for-tat often results in a vicious cycle of retribution, when people find themselves locked into conflict in a sort of 'Prisoner's Dilemma', irrespective of the original causes that generated that conflict in the first place.

Indeed, under conditions of competition and potential conflict, the other's very existence constitutes a threat, which requires vigilance, increase in strength and even preemption, all of which, in turn, only intensify the sense of mutual insecurity. The result is yet more vicious circles, such as the 'Security Dilemma' variant of the 'Prisoner's Dilemma', when fear and suspicion of the other, in themselves, even in the absence of offensive intent on his part, pushes the sided to arm themselves, fight, and even expand, all for defensive reasons. One result is arms races, which exist all over nature; they are the reason why trees have trunks: they take the enormous expenditure involved in growing trunks only in order to outdo one another in getting to the sun. But arms races often result in a 'Red Queen Effect', called so after one of the paradoxes in Lewis Carroll's *Through the Looking Glass*, where the sides run as fast as they can, spending energy and resources, only to remain in the same place relative to one another. So there are real enough reasons for conflict and war, which, however, may then get inflated because of mutual hostility, fear and suspicion, with conflict seemingly assuming a life of its own – escalating and perpetuating itself – causing heavy losses to all sides almost irrespective of their original motives.

Finally, people prefer kin to non-kin – which is the root cause of tribalism, ethnocentrism and nationalism – and they may support kin even by force. Also, they have cultures and comprehensive outlooks – religious and secular ideologies – which they regard as crucially important for ordering life in this world, and sometimes the afterlife, and which again they may defend and promote by force.

Part Two of my book, *War and Civilization*, traces the interrelationship between warfare and such landmark transformations of human history as the transition to agriculture and the rise of the state and civilization. Just a few of the general themes will be discussed here.

Increasingly dense and sedentary populations, stationary means of production and accumulated property now made possible a differential appropriation and concentration of surpluses. In a process first outlined by Rousseau, existing natural differences between people were enormously magnified and objectified by accumulated resources. On this at least Rousseau was right – relying on concentrated wealth, and the social power that emanated from wealth, the rich and powerful dominated social life. State emergence was the culmination of this process, as a single power nucleus won against all others in an often-violent competition. It established command over a population, institutionalizing power, driving other social power nuclei into subordination, and introducing hitherto unprecedented levels of hierarchic organization, coercion, systematic resource extraction and force mobilization, while competing with neighbouring state structures.

This highly violent process was the main driving force for the growth of civilization. As political societies grew in size through forceful means, they created economies of scale, aggregated and purposefully directed resources and human activity, and gave rise to monumental building, literacy, and high culture.

Another major consequence of state emergence was the gulf that opened up between internal and external – 'in-group' and 'out-group' – violence, which had been far less distinct from one another in both scale and methods in pre-state societies. To be sure, the prospect of violent conflict dominated state existence not only abroad but also at home, as in struggles for the throne and civil wars. Still, in 'normal' times statehood created a sharp difference between small-scale murder and feud, now outlawed, on the one hand, and large-scale organized violence, construed as war, on the other, a distinction which we take for granted but which is as recent as the state itself.

State wars, with their coercive mobilization of people and resources and growing scale, resulted in a continuous increase in the size of the fighting hosts and in a wholly new level of regimentation: participation became obligatory rather than voluntary as in pre-state societies; armed forces grew from scores of men-warriors into thousands, tens of thousands and hundreds of thousands; warriors became soldiers; disorganized kin-based hosts gave way to orderly fighting formations; stricter hierarchic command replaced leadership by example. Fortifications, much denser sedentary settlement, and greater distances spelled a decline in the significance of the raid – the predominant and most lethal form of warfare in pre-state societies – because wholesale surprise of the enemy community became more difficult to achieve. The siege and the battle became almost synonymous with war. Large-scale and long-distance campaigning required complex logistics, and the state's bureaucracy was called upon to support and sustain larger and more permanent armies, securing finance and supervising over the acquisition and requisition of provisions. Indeed, it was the state apparatus that also made possible permanent conquest and direct rule over foreign people – that crucial upgrade in the activity of war which was responsible for the state's continuous increase in size.

It should be realized, however, that although growing scale was an underlying trend of state warfare, overall violent mortality rates actually *decreased* with the growth of the state. The state's success in imposing internal peace – limited and fragile as it was – was probably the major reason for this decrease. But there was another factor involved, largely at variance with commonly-held intuitions. As states grew in size, their civilian populations became less exposed to fighting, and adult male participation rates in their armed forces *declined* in comparison to tribal societies, simply because of the constraints imposed by far greater distances. Despite the state's coercive powers, in large-scale polities it was possible to call into military service only a limited part of a country's manpower, because the rest of the men had to be left behind to produce food. Warfare was no longer close to home as in small scale societies, when people could leave their productive occupations for a day's, or a few days', campaigning. For these reasons, whereas armies, wars and killing all grew conspicuously larger in *absolute* terms, they actually *decreased* in proportion to population.

In the Second Punic War (218–202 BC), ancient Rome's most devastating conflict, of which we have relatively good census and other demographic statistics, Rome (and Italy) lost, according to one minimalist estimate, at least 17, if not more than 20, percent of its adult male population.[2] But a calamity of such magnitude was exceptional. Some parts of Germany are estimated to have suffered even greater demographic losses during the Thirty Years War. In relation to the general mortality, death in war in France, one of the most warlike nations in Europe, is estimated by one source at 1.1 percent in the seventeenth century, 2.7 percent in the eighteenth century, 3 percent in the nineteenth century and 6.3 percent in the first three decades of the twentieth century.[3] In the American Civil War 1.3 percent of the population were either killed or wounded. In World War I about 3 percent of both the French and German populations died, representing roughly 15 percent of the adult males. In World War II over 15 percent of the Soviet Union's population perished, and around 5 percent in Germany. However, when averaged over time, even the dreadful figures from these cataclysmic events fall short of those for

2 Peter Brunt, *Italian Manpower, 225 B.C.–A.D. 14* (Oxford: Oxford UP, 1971), 54, 63, 84, mainly addressing the calamitous years 218–216 BC.

3 Quincy Wright, *A Study of War* (Chicago, IL: University of Chicago, 1942), i. 665, Table 57; these estimates appear to be very tenuous, but can still serve as a rough indicator.

primitive societies. Thus, contrary to perception, only particularly catastrophic spates of state warfare resulted in anything near the twenty-five percent violent mortality rates among adult males that small-scale pre-state societies are recorded to have incurred as a matter of course in their incessant inter- and intra-group violence. As Hobbes rightly saw, only protracted civil war or particularly severe foreign invasions resulted in something akin to a return of pre-state anarchy, pervasive insecurity and great violent mortality.

The substantial *decline* in violent mortality under the leviathan thus runs counter to the view that blames fighting on the state. The state has been aptly likened by Charles Tilly and others to organized crime, in the sense that it monopolized force and compulsorily extracted resources from society for its own profit in return for the promise of protection from both internal and external violence.[4] Indeed, some would further extend the analogy by arguing that the main threat of violence came from the state itself; that it offered a solution to a problem of its own making. However, in view of what we have seen, at least the latter conclusion should be regarded with caution. While violence under the state, as under organized crime, was of greater magnitude and more spectacular, it actually produced *fewer* casualties than pre-state violence. Systematic 'extortion' by the state was economically less disruptive than pre-state violence, and the state offered more protection. Undeniably, though, 'protection money' was channelled upwards, to the state's socio-political-military leadership and elite.

Finally, there is a strong interrelationship forged during modernity between wealth, technology and power, all undergoing tremendous growth with the rise of capitalism and especially with the Industrial Revolution.[5] For the first time in history wealth means power. Poor tribal folk from the barbarian frontiers of wealthy civilizations had regularly taken over by force, but, as Adam Smith discerned, this was no longer possible. Generally, the wealthier a country, the more powerful it is.

But there is also another side to this correlation, and this is that economically developed societies have demonstrated a markedly decreasing tendency to fight each other, and today they scarcely fight each other at all. A so-called 'long peace' has existed among the great powers since 1945, which is often attributed to nuclear deterrence. But the second longest peace among the great powers in fact took place long before the bomb, between 1871 and 1914, and the third longest peace took place between 1815 and 1854 – all occurring in the nineteenth and twentieth centuries, in the wake of industrialization. After 1815, wars among the great powers and other economically advanced countries declined in their frequency to about a *third* of what they had been in the early modern period, when the great powers fought on average in one out of every two or three years. This decline in warfare, relating to

4 Charles Tilly, 'War Making and State Making as Organized Crime', in P. Evans, D. Rueschemeyer, and T. Skocpol (eds), *Bringing the State Back In* (Cambridge: Cambridge University Press, 1985), 169–91.

5 Azar Gat, *Victorious and Vulnerable: Why Democracy Won in the Twentieth Century and How It is still Imperiled* (Lanham MD: Rowman & Littlefield Pub. Inc., 2009).

the advent of industrial society, was already noted in the first half of the nineteenth century by thinkers such as Saint-Simon, Auguste Comte and John Stuart Mill.

People tend to believe that the reason for the decline in warfare between developed countries is that wars have grown more lethal, expensive and destructive, but in reality they haven't – at least not when measured relative to population and wealth. The Second Punic War, or the Thirty Years War in Germany, were no less lethal and destructive than the World Wars, and these are only examples. The *real* change since the takeoff of the industrial revolution is that *peace has become more profitable*. Several interrelated developments are responsible for this. First, in the past, wealth was finite, and the only question was how it was to be divided. But now the Malthusian trap has been broken, and production and wealth have rocketed. Wealth per capita in the developed countries has grown about 30-fold since the beginning of the industrial era, with the acquisition of wealth ceasing to be a zero-sum game. Indeed, more widely recognized is the second development: commercial interdependence for the first time in history makes the enemy's loss also our own, because it depresses the entire system. This reality, already noted by Mill, materialized after World War I, as Keynes predicted it would in his *Economic Consequences of the Peace*. Third, greater economic openness has decreased the likelihood of war by disassociating economic access from the confines of political borders and sovereignty. It is no longer necessary to politically possess a territory in order to benefit from it. Thus, the greater the yields of competitive economic cooperation, the more counterproductive and less attractive conflict becomes.

As previously mentioned the role of sexual competition in sparking pre-state fighting played a major role. It is interesting to look at the effect of the sexual revolution on warfare over the past generation, an effect which comes not from the state level but, no less powerfully, from the grass roots. One of the reasons men enlisted in the past was the allure of sexual adventure away from the restricting norms of their home communities. Much of the adventure took the form of mass rape, something we could still witness horrifically in the wars in the former Yugoslavia, Rwanda and Sudan. Like looting, rape was one of the perks of war. However, with the sexual revolution, the balance changed completely. Young men are that much more reluctant to leave the pleasures of civilian life for the rigour and chastity of the front. It is no coincidence that the slogan of the anti-war campaign of the 1960s, when the sexual revolution took off, was 'make love not war'. There is no need to fully embrace the theories of Freud, Reich and Foucault to appreciate the significance of this factor.

The virtual disappearance of war among developed countries is hugely encouraging. Throughout history, the greatest and most destructive wars have always been those fought among the great powers. By contrast, the only wars that we have today are against relatively weak states, either taking place among non-developed countries, or between them and developed powers. The affluent parts of the world constitute a 'zone of peace', which is a great novelty. And the trend is even stronger among liberal democracies.

For, indeed, liberal and democratic societies have proven most attuned to modernity's pacifying aspects. Although this proposition was long regarded with scepticism, the supporting evidence has become quite overwhelming. Relying on arbitrary coercive force at home, non-liberal and non-democratic countries have found it more natural to use force abroad. By contrast, liberal democratic societies are socialized to peaceful, law-mediated relations at home, and their citizens have grown to expect that the same norms be applied internationally. Living in increasingly tolerant societies, they have grown more receptive to the Other's point of view. Promoting freedom, legal equality and political participation domestically, liberal democratic powers – though initially in possession of vast empires – have found it increasingly difficult to justify rule over foreign peoples without their consent. And sanctifying life, liberty and human rights, they have proven to be failures in forceful repression. Furthermore, with the individual's life and pursuit of happiness elevated above group values, sacrifice of life in war has increasingly lost legitimacy in liberal democratic societies. War retains legitimacy only under narrow and narrowing formal and practical conditions, and is generally viewed as extremely abhorrent and undesirable.

The fruits of these deepening trends and sensibilities have been nothing short of miraculous. Their most striking and widely-noted manifestation is the inter-democratic peace. Although modern democracies have been extensively involved in wars with non-democracies, they hardly ever fight among themselves, and this trend is deepening with growing liberalization, democracy and economic development. Domestically too, on account of their stronger consensual nature, plurality, tolerance and, indeed, a greater legitimacy for peaceful secession, advanced liberal democracies have become practically free of civil wars, the most lethal and destructive type of war.

The inter-democratic peace is increasingly found to be merely the most conspicuous element of a larger whole. The liberal democracies' particularly poor record in counter-insurgency wars is a case in point.

Historically, the crushing of an insurgency necessitated ruthless pressure on the civilian population, which liberal democracies have found increasingly unacceptable. Pre-modern powers, as well as modern authoritarian and totalitarian ones, rarely had a problem with such measures, and overall have proved quite successful in suppression. Suppression is the *sine qua non* of imperial rule. The British and French empires could sustain themselves at a relatively low cost only so long as the imperial powers felt no scruples about applying ruthless measures, as the British, for example, did most memorably in Ireland, the Scottish Highlands and India as late as the 1857 Mutiny. However, as liberalization deepened from the late nineteenth century, the days of formal democratic empires became numbered even while outwardly they were reaching their greatest extent. At the turn of the twentieth century, the British setbacks and eventual compromise settlement in South Africa and withdrawal from Ireland were signs of things to come for other liberal democratic empires as well.

Sceptics might dispute democracy's uniqueness, citing the successful guerrilla war waged against Nazi Germany in Yugoslavia and the Soviet Union. However, there can be little doubt that had Germany won World War II and been able to apply more troops to these troublesome spots, its genocidal methods would have prevailed there too. Russia's failure in Afghanistan is another obvious counterexample, but Afghanistan was the exception, the outlier, rather than the rule in the Soviet imperial system. Chechnya may be more enlightening in this respect. Indeed, the sequence is unmistakable: Soviet methods under Stalin – including mass deportation – were the most brutal and most effective in curbing resistance, while liberal Russia of the 1990s proved to be the least brutal and least effective, with Putin's authoritarian Russia constituting an intermediate case. The Soviet empire only fell apart with the collapse of the Soviet totalitarian system. Indeed, Germany and Japan lost their empires because of their defeats to other great powers in the world wars, and not because of indigenous uprisings. The same logic applies to China, whose continued successful suppression of Tibetan and Moslem nationalism is likely to persist so long as China retains its non-democratic regime. The record of counter-insurgency wars is sharply skewed towards the democracies' post-colonial conflicts, but as Sherlock Holmes pointed out, it was the 'dog that did not bark' under the totalitarian iron hand that was in greater need of an explanation.

This is not to say that the democracies' conduct has been saintly. Atrocities, tacitly sanctioned by political and military authorities or carried out unauthorized by the troops, have regularly been committed against both combatants and non-combatants. All the same, strict restrictions on the use of violence against civilians constitute the legal and normative standard for liberal democracies, radically limiting their powers of suppression, judged by historical and comparative standards. Indeed, legal and normative constraints on the conduct of war – 'lawfare', as they have been called – are increasingly gaining in potency in liberal societies and their international institutional offshoots, influencing the outcome of wars no less than warfare itself.

There is a catch that lies at the root of affluent liberal democracies' torment in conflict situations. Since wars are abhorred in liberal societies as antithetical to both their interests and values, they are sanctioned only as a last resort. Yet it is difficult to determine if all alternative policies have indeed been exhausted, that war has really become unavoidable. A feeling that there may be another way, that there *must* be another way, always lingers on. Errors of omission or commission are ever suspected as being the cause of undesired belligerency. Moreover, it never becomes clear that the democracies come to a conflict with entirely clean hands – morally – because of past or more immediate alleged wrongs, given the inevitable gap that separates ideals from reality.

The democracies' reaction to the Axis' challenge during the 1930s amply epitomizes this predilection. Everything was done to avoid military action, even if this meant allowing Germany to regain its power, grow in confidence owing to the democracies' inaction, and cross the point of no return on its road to expansion. The democracies' feelings of guilt for not having treated Germany in good faith after World War I contributed to their paralysis. To be sure, as critics caution, Hitlers are rare and not every crisis is the 1930s. All the same, the 1930s are a standing reminder that the democracies' strong aversion to war, while immensely beneficial overall, can become a grave problem in serious conflict situations.

In conclusion, an affluent and liberal democratic world promises to be pacific, yet most countries are still far behind on the road to modernization, and some develop in directions which are not liberal and democratic. Huge and fast-industrializing non-democratic China is the most obvious case, with Russia too retreating from democracy and liberalism. Another ominous development is the trickling down of weapons of mass destruction to below the state level. We now face the fearful prospect of unconventional terror, especially biological and nuclear. This cannot be deterred, as between states, because terror organizations are often composed of a small number of zealots who have no clear address against which to retaliate, and who in any case are willing to sacrifice themselves. This is a mighty challenge we are facing today and shall increasingly face tomorrow. War has been receding, but a world without war has yet to materialize.

Gender and the Causes, Tactics and Consequences of War

Debra L. DeLaet

Introduction

Most theories of international relations treat war as a gender neutral phenomenon. Mainstream international relations scholarship, while diverse in many respects, shares a basic assumption that gender is irrelevant to understanding the causes of war, the ways in which wars are fought, and the consequences of war. Despite this assumption of gender neutrality, war is, in fact, a highly gendered phenomenon. Socially constructed notions of masculinity and femininity play a critical role in shaping both *why* and *how* wars are fought by state and non-state actors alike. Similarly, gender norms lead to differential treatment of men and women in war and, thus, dramatically influence the ways in which wars are experienced by people in conflict zones. This chapter explores the gendered nature of war. To this end, it provides an overview of the complicated ways that gender shapes the causes, tactics and consequences of war.

Feminist Perspectives on War and Gender as a Tool of Analysis

To assert that gender fundamentally shapes the nature of war is not, necessarily, to put forth an explicitly feminist analysis of war. Jean Bethke Elshtain puts it simply: "There is no separate feminist tradition on war and peace."[1] Rather, various feminist analyses of war represent different feminist perspectives that have grown out of existing theoretical traditions in the study of international relations, and "… each

1 Jean Bethke Elshtain, "Is There a Feminist Tradition on War and Peace?" in Terry Nardin (ed.), *The Ethics of War and Peace* (Princeton: Princeton University Press, 1996), 214.

articulated feminist position represents either an evolution within, or a breakout from, a previous historic discourse ... What feminist position a particular thinker or advocate endorses will depend upon the tradition in which her feminism is lodged or out of which it emerges."[2]

Thus, different feminist perspectives offer competing views on the ways in which gender influences war. Feminist pacifists might claim that war would be less common if women ruled the world. Similarly, difference feminists put forth claims that women, due to their maternal roles and nurturing natures, are less likely than men to condone or participate in violent conflict within or between states. In general, liberal feminists argue that women leaders do not govern differently than men. Instead of focusing on the ways in which women's political or military service might change the nature of war, liberal feminists identify equality—in governance, in military institutions, and in combat roles and obligations—as a goal in and of itself. In short, no single feminist perspective on war exists.

Although gender analyses have emerged out of feminist traditions of scholarship, feminist and gender analyses are not one and the same. At its most fundamental level, feminism is about the effort to achieve legal, political and social equality for women. Feminist international relations scholarship, in its many variants, is grounded in two basic goals: 1) at an empirical level, feminist scholarship in international relations strives to bring women to the center of the analysis; and 2) at a normative level, feminist international relations scholarship involves advocating perspectives and policies that value women's perspectives and that advance women's interests across the globe. As noted previously, no single feminist perspective exists, and feminists do not necessarily agree on how to interpret women's perspectives or how to advance women's interests. That said, a common emphasis on paying attention to women's voices, issues and interests cuts across competing feminist perspectives.

Gender analysis in international relations scholarship has different foundations. Although most international relations scholarship focusing on gender has been feminist in nature, gender analysis does not require staking out feminist positions on global issues.[3] Indeed, gender, defined as socially constructed notions of masculinity and femininity, is a critical variable for understanding the nature of war both historically and in the contemporary global system whether or not one strives to take an explicitly feminist position on war-related issues in international relations scholarship.

To put it simply, a scholarly focus on gender as a concept has broader implications than a feminist focus on women as a category of analysis. The study of gender involves examining the ways in which socially constructed ideas about masculinity and femininity shape the roles performed by women and men in particular societies, the prevailing identities embraced by women and men, and the ways in which political and social actors mobilize these gender norms toward

2 Ibid.

3 R. Charli Carpenter, "Gender Theory in World Politics: Contributions of a Non-Feminist Standpoint?" *International Studies Review* 4: 3 (Fall 2002), 153–165.

specific political and social objectives. Conceived in this manner, gender analysis may or may not be feminist in nature. Feminist gender analysis seeks to bring women into focus and to advocate on behalf of women's perspectives and interests. Non-feminist gender analysis, although it also typically pays attention to women's perspectives and interests, highlights the ways in which prevailing gender norms may harm the interests of men as well as women and may diminish the importance of critical male perspectives on important global issues.

This chapter's approach is guided by feminist concerns but also acknowledges the utility of non-feminist gender analysis. To this end, this chapter is grounded in a few core assumptions.

A gendered understanding of war must recognize women as well as men as both agents and objects of action in the arena of international relations. Like men, women are agents in regards to war. They can engage in anti-war advocacy. They sometimes serve as combatants. Even when they don't actively participate in combat, women frequently support war and condone wartime violence. Women also are objects in war. They are often innocent civilians who become victims of wartime violence. States manipulate images of women to rationalize decisions to go to war, to increase public support for war and to demonize "the enemy" in gendered propaganda campaigns. The exact same things can be said about men, who are both agents and objects in war. In conflict zones, men are anti-war advocates, combatants and bystanders who support or condone war to greater or lesser degrees. Men suffer as innocent civilians, and prevailing notions of masculinity are mobilized by states to recruit and train soldiers and to generate public support for specific wars and for militarism in general.

Acknowledging the complicated reality of both women's and men's gender roles is critical to understanding the complex relationship between gender and war. In reality, war does not engage rigid dichotomies involving women as innocent civilians and men as combatants. Rather, men and women each serve as combatants/civilians, perpetrators/victims, supporters of war/anti-war advocates, and women and men may find themselves in multiple roles simultaneously. For example, a male in a society with forced conscription could be a combatant who perpetrates violence against an innocent civilian (male or female) in an enemy country while personally opposing the war in which he is participating. A woman might be an "innocent civilian" who is not directly participating in wartime violence but who favors the use of indiscriminate violence against enemy civilians. Perhaps this woman lives safely outside of the immediate zone of conflict and is not in imminent danger of being victimized by wartime violence. In such contexts, the application of labels such as perpetrators and victims or combatants versus innocent civilians seems misleading and overly simplistic.

Gender analysis helps us to make sense of these categories in both practice and theory. In addition to gender, numerous other variables—including ethnicity, class, culture and ideology—shape the orientation of particular individuals to war. Patterns of gendered behavior in war exist, even though the relationship between gender and war does not involve rigid dichotomies. While men may be anti-war advocates and women may serve in combat, the fact that men are far more likely to serve directly in militarized roles and that women are more likely to experience war from the vantage point of support/service roles or civilian status is a pattern that prevails across cultures and countries throughout history. Notwithstanding the great diversity among cultures and societies in both history and in the contemporary era, "... the selection of men as potential combatants (and of women for feminine war support roles) has helped shape the war system." Despite the changing nature of war, this pattern has prevailed across time and space.[4] The concept of gender—and not biological sex—helps us to understand the near-universality of this pattern.

The remainder of this chapter seeks to elaborate upon the ways in which gender analysis is critical to understanding the nature of war both historically and at present. To this end, subsequent sections highlight the ways that gender shapes the war system in three fundamental areas: the causes of war, the tactics of war and the consequences of war.

Gender and the Causes of War

International relations scholarship on the causes of war tends to distinguish between the proximate and the underlying causes of war. Proximate causes deal with the immediate and direct factors that precipitate specific wars. In contrast, underlying causes focus on fundamental forces and conditions that make war, in general, likely. As a general rule, feminist scholarship in international relations tends to discuss gender as an underlying cause of war.

Scholars have hypothesized that gender variables may serve as an underlying cause of war in numerous ways. At the most fundamental level, some scholars contend that gender norms have been critical to legitimizing war. Jean Bethke Elshtain has argued that the idea of a just war depends upon a dichotomous construction of ideal types of masculinity and femininity. She identifies three gender archetypes that have dominated historical discourses on war. The male "just warrior" is a good, Christian man who is willing to fight and die to defend a righteous cause, the honor and security of his country and, notably, the women and children who represent its innocence and virtue. The just warrior's female counterpart is the feminine "beautiful soul" who is vulnerable, self-sacrificing and needs protection from enemy forces. To this end, the just warrior fights to protect the beautiful soul.[5] "What we see when we look at history is that women, cast as society's beautiful souls, have not only *not* succeeded in stopping the wounding and slaughtering of sons, brothers, husbands and fathers, but have more often than not exhorted men to the task and honored them for their deeds."[6] To this gender dichotomy,

---

4 Joshua Goldstein, *War and Gender* (Cambridge: Cambridge University Press, 2001), 9–11.

5 Jean Bethke Elshtain, "On Beautiful Souls, Just Warriors, and Feminist Consciousness," *Women's Studies International Forum* 5: 3–4, 341–348.

6 Jean Bethke Elshtain, "Thinking about Women and International Violence," in Peter R.

Elshtain adds a third gendered archetype: the Spartan Mother. The Spartan Mother is a militant woman who proudly bears and raises children, particularly sons, to be willing to fight and die for the polity. "Her identity is entangled with war's honor and valor of husbands and sons."[7] For Elshtain, these gender archetypes have fundamentally shaped prevailing discourses on war throughout history and have played a critical role in legitimizing war as an institution.

Gender norms that involve "heroic" men going to war to defend a society's women (or their own masculine honor) have been manifest in prevailing discourses on war in other ways. For example, a polity's desire to avenge or defend the honor of its women has been present in historical discourses on the causes of war. A prominent example comes from the story of the Trojan War in Greek mythology. According to the myth, the Achaeans went to war against the city of Troy after Helen left her husband, Menelaus, the King of Sparta, for Paris of Troy. A similar gendered system of values is at stake when male warriors seek to capture women as a prize of war. An obscure yet interesting example involves the Yanomamo people in South America, who contend that they go to war to capture women from other Yanomamo villages; motivations for these raids include a desire to avenge prior wrongs and to increase reproductive possibilities for the raiding village.[8]

The gendered nature of security language also has been identified as a factor that legitimizes war as an institution. Carol Cohn has argued that gender is a ubiquitous force in the world of strategic analysis, both in terms of the overwhelmingly male composition of this world and in the gendered nature of the prevailing security discourse. Cohn's central claim is that the gendered language of defense intellectuals working in the area of nuclear weapons—a language that includes references to "penetration aids," "thrust ratios" and "vertical erector launchers"—reveals that national security elites exploit sexual images to promise domination (sexual, political and otherwise).[9] Such language also distances those who speak it, as well as those who hear it, from the horrible reality of war's destruction. In this way, gendered security discourse serves as a tool for the legitimization of war.

Other scholars emphasize gender inequity as an underlying cause of war. J. Ann Tickner argues that unequal gender hierarchies contribute to violent conflict and make war more likely.[10] The crux of the argument is that inequitable societies make it more likely that force will be used as a mechanism for dispute settlement. Indeed, evidence indicates that violent conflict is more likely in societies that discriminate against women in the economic, political and social spheres. For instance, evidence suggests that countries with lower percentages of women in parliament have a

Beckman and Francine D'Amico (eds), *Women, Gender, and World Politics: Perspectives, Policies, and Prospects* (Westport, CT: Greenwood Publishing Group, 1994), 115.

7 Elshtain, "Is There a Feminist Tradition on War and Peace?", 216.

8 Goldstein, *War and Gender*, 7.

9 Carol Cohn, "Sex and Death in the Rational World of Defense Intellectuals," *Signs: Journal of Women in Culture and Society* 12: 4 (1987), 692–696.

10 J. Ann Tickner, *Gender in International Relations* (New York: Columbia University Press, 1992), 27–66.

higher propensity to go to war. Additionally, a correlation exists between female suffrage in a state and that state's propensity to be involved in interstate conflict. A similar correlation exists between the level of women's participation in the labor force and the likelihood of a state going to war. In general, states with high fertility rates and low levels of gender equity are more likely to be involved in armed conflict.[11]

Another form of gender inequality that contributes to war is distorted sex ratios. Significant gaps between the proportion of men and women in particular societies are rooted in discriminatory gender norms that lead people to value male children more highly than female children. Sex-selective abortion, infanticide and the killing of girls have emerged as widespread practices in societies that devalue women and girls. In such societies, these practices have led to the emergence of a "surplus of men." According to Valerie M. Hudson and Andrea Den Boer, the distorted sex ratios that result from these sex-selective practices and other forms of gender-motivated killing of girls and women serve as a cause of violent conflict within and between states.[12] It is a universal pattern that young men between the ages of 15 and 35 commit most of the violence in a given society. A surplus of men in a particular society exacerbates this tendency in that the men who are most likely to precipitate and participate in violent conflict do not have available partners for marriage; this trend is particularly strong among low-status men who are not able to attract potential marriage partners from among a limited pool of women. The presumption, supported by extensive empirical evidence, is that marriage as an institution makes it less likely that young men will engage in violent behavior.[13] Thus, a surplus of young men is expected to increase rates of violent conflict.

This section has identified a number of gender variables that serve as underlying causes of war: archetypical gender norms and gendered security discourses that legitimate war as an institution, gender inequalities that make it more likely that actors will resort to violence as a mechanism of dispute settlement, and distorted sex ratios that generate a surplus of males with a higher propensity to engage in violence. It is more difficult to specify with confidence gender variables that have served as proximate causes of particular wars. In this regard, one of the major challenges is disentangling the relationship between rhetoric and reality in assessing the extent to which gender considerations contribute to specific wars. States, and other actors, frequently invoke gender as a motivation for war. However, such invocations of gender as a rationale for war often represent mere cynicism on the part of states and non-state actors. Parties to violent conflict frequently deploy gender norms in an effort to mobilize support for their involvement in war. When actors use gender in this manner, gender can be seen as a *strategy* or *tactic* and not a cause of war. The following section takes up the strategic and tactical use of gender in war.

11 Mary Caprioli, "Gendered Conflict," *Journal of Peace Research* 37: 1 (2000), 51–68.

12 Valerie M. Hudson and Andrea Den Boer, "A Surplus of Men, a Deficit of Peace: Security and Sex Ratios in Asia's Largest States," *International Security* 26: 4 (Spring 2002), 5–38.

13 Ibid., 11–16.

The Strategic and Tactical Use of Gender in War

Gender not only helps to explain the underlying causes of war but is also critical to understanding *how* states and non-state actors fight wars. The strategic and tactical use of gender in war manifests in many ways. Militaries, paramilitaries and other groups engaged in organized violence use gender norms as recruitment and training tools. Similarly, gender norms are often at the center of propaganda campaigns intended to increase support for war among relevant publics and/or to divert attention away from criticisms of ongoing war efforts. Gender also can be used in an effort to gain tactical advantage on the battlefield. This section considers these varying strategic and tactical uses of gender by states and non-state actors in wartime.

Gender and the recruitment and training of soldiers

One of the most basic strategic uses of gender in war involves efforts to build effective armed forces. States mobilize widely embraced assumptions about appropriate gender roles for men and women in order to raise armies, and non-state actors deploy gender norms to similar effect in their efforts to developed organized armed forces. The use of gender as a tool for building effective armed forces takes place at two levels: 1) in recruiting soldiers for the armed forces; and 2) in training soldiers to function effectively on the battlefield.

In the area of recruitment, both states and non-state actors invoke gender norms to entice potential soldiers, most frequently young men, to join the armed forces. US military recruitment efforts provide a case in point. The military recruitment advertisements running on American television during the current college football season in 2010 are a striking example. Recent advertisements juxtapose images of strong, tough young men in football gear being tackled on the field with images of similarly strong and tough soldiers in military uniform. The US military's effort to connect these two images of masculinity is neither coincidental nor subtle. American football is the most stereotypically masculine of American sports, and its television audience is largely (though not exclusively) male. Such advertisements are clearly intended to appeal to the masculine aspirations of the young men watching in order to encourage them to join the military. In the United States, the masculinist underpinnings of military recruitment (in print, radio, televised and online media) is consistent. Images of electronic battles in complicated, violent and "realistic" video games, images of unsmiling young men in military uniform representing visions of masculine order, discipline and strength, and images of athleticism and grittiness all signal the importance of idealized masculinity as a tool for recruitment into the armed forces.

The appeal to idealized masculinity is not unique to the United States in particular or to large, modern, industrialized armies in general. The ways in which a different masculinist vision is deployed by Al-Qaeda also demonstrates the ways in which gender norms can be used as a tool for recruitment. In the case of

Al-Qaeda, potential young recruits are reminded that their faith promises them the reward of seventy-two virgins in paradise, a reward that is given to all male Muslim believers but which an Al-Qaeda recruit will receive sooner if he becomes a martyr for Islam. Al-Qaeda also appeals to masculine ideals of heroism and disdain of cowardice as a recruiting tool.[14] Interestingly, Al-Qaeda has not only appealed to idealized masculine norms but also to the subversion of these norms in their recruitment efforts. For example, Al-Qaeda operatives have reportedly raped potential recruits—both men and women—and then turned them over to other Al-Qaeda officials who convince them that they can cleanse themselves of their shame and the affront to their masculinity and femininity by becoming suicide bombers.[15]

Notably, the contrasting cases of US military and Al-Qaeda recruitment illustrate that the particular variants of gender norms that will be used in recruitment efforts vary across cultures. Yet it is notable the gender plays a central role in recruitment efforts despite significant differences in the type of armed force (state or non-state) and in the cultures in which these actors are operating.

States and non-state actors also use gender to train soldiers. Gender norms have been critical in transforming civilians into soldiers willing to kill and die in service to a country, cause or, often, simply to defend their comrades in the armed forces. Contradicting widely held assumptions that men are "natural born killers," it can be incredibly difficult to get human beings to kill one another. Historical studies of war, including the Napoleonic War and the American Civil War, have shown that soldiers often fire over the enemy's head during battle. Such "misfiring" of weapons enables a soldier to posture as a warrior and to "engage" in battle without actually having to kill.[16] Along these lines, a study of the battle behavior of average soldiers during World War II, conducted by US Army historian Army Brigadier General S.I.A. Marshall, found that only approximately 15 to 20 percent of soldiers actually used their weapons in battle.[17] These examples demonstrate that men, as well as women, have a strong innate resistance to killing other human beings; they must be trained to kill. Of course, some men seem to enthusiastically engage in killing in war. Even more men may claim to love war.[18] (Note, however, that "loving" the intensity of the experience of war is not the same thing as claiming that it is easy or desirable to kill in war.) Nevertheless, most men do not kill readily on the battlefield.

Therefore, states and non-state actors must go to great lengths to train soldiers to kill. To this end, a militarized masculinity has been one of the most powerful

14 Anat Berko, *The Path to Paradise: The Inner World of Suicide Bombers and Their Dispatchers* (Westport, CT: Praeger, 2007), 44–49.

15 Deborah Haynes, "Al-Qaeda Damaged by Arrest of 'Rape and Suicide Bomb' Woman," *The Times,* February 5, 2009. Available at: http://www.timesonline.co.uk/tol/news/world/iraq/article5661466.ece. Accessed on September 5, 2010.

16 Lieutenant Colonel Dave Grossman, *On Killing: The Psychological Cost of Learning to Kill in War and Society* (Boston: Little, Brown and Company, 1995), 9–11.

17 Ibid., 3.

18 William Broyles, Jr., "Why Men Love War," *Esquire* (November 1984).

tools used by military organizations to recruit and train soldiers. Militaries seek to condition soldiers to kill through repetitive drilling and discipline. These exercises typically rely on shows of masculine dominance and threats of physical violence for failure to comply.[19] Military forces also encourage male bonding as a mechanism for training soldiers who will be willing to kill and die on the battlefield.[20] Oral histories of combat experiences indicate that soldiers are often motivated to fight in the heat of battle more out of loyalty to their "brothers in arms" than out of commitment to a state, an organization or an ideological cause. Thus, militaries use gender norms to create and maintain unit cohesion. The exclusion of women and gay men from combat roles in most militaries throughout history has been justified, in part, by the importance of male bonding to the creation of effective fighting units. Soldiers also become desensitized to killing through exposure to broader social forces that glorify masculine violence and heroism and denigrate "feminine" submission and non-violence.[21] The specific form that militarized masculinity takes varies across cultures. That said, the militarization of masculinity as a mechanism for creating soldiers willing to kill and fight appears to be a universal phenomenon.

Gender and war propaganda

In addition to using gender to recruit and train armed forces, states and non-state actors use gender as a propaganda tool to increase public support for war. As in the case of the recruitment and training of soldiers, political actors draw on a militarized masculinity—involving strong, dominant, heroic men—to generate public support for war. A demilitarized femininity serves as a necessary counterpoint to militarized masculinity in mobilizing populations for war. States and non-state actors alike have relied on depictions of women as supportive mothers and faithful wives who are especially vulnerable to the effects of war. Although such gendered depictions of women also can be used in anti-war campaigns, the trope of vulnerable women has served as a powerful propaganda tool for mobilizing support for war when the lives of "innocent women" are at stake.

Witness the case of the lead up to the US invasion of Afghanistan in 2001. The Bush Administration highlighted the suffering of Afghani women at the hands of the Taliban as a rationale for going to war. The oppression of women under Taliban rule served as a compelling reason for intervention that appealed to political factions that might otherwise have been inclined to oppose an interventionist US foreign policy. However, it is noteworthy that the Bush Administration only offered this rationale for war when it had a compelling strategic reason—namely, the September 11, 2001 attacks by Al-Qaeda against American targets—to pursue war in Afghanistan. In short, the status of women in Afghanistan was not the real motivation for war in this case. Rather, the Bush Administration made the

19 Grossman, 18.

20 Goldstein, *War and Gender*, 194–199.

21 Mark Baker, *Nam* (New York: Berkley Books, 1981).

decision to strategically engage gender norms as a means of mobilizing popular support for war.

The case of the rescue of Jessica Lynch during the 2003 US war in Iraq also illustrates the ways in which states deploy gender norms for strategic ends. In March of 2003, Jessica Lynch was injured and captured by Iraqi forces following an ambush of a convoy in which she was riding. Initially holding her as a prisoner of war, the Iraqi army took her to a hospital. Subsequently, the US Marines staged a diversionary attack near the hospital and then "rescued" Jessica Lynch. The military's version of this story is that Jessica Lynch had been tortured and raped by the Iraqi forces that captured her, and had subsequently been abused by hospital personnel. The Bush Administration turned this event into a major propaganda campaign, portraying Jessica Lynch as both a victim who needed to be saved from enemy forces and a hero who fought bravely in the face of danger.

Unfortunately, the facts of the case do not support the version put forth by the US military. Iraqi hospital personnel claimed that Lynch had been protected from the military and treated humanely. Jessica Lynch's later public testimony concurs with the claims of the hospital personnel. Moreover, the Iraqi military had fled the hospital prior to Lynch's "rescue," suggesting to some critics that the entire militarized operation was staged for propaganda purposes. The fact that the rescue mission was filmed lends support to arguments that the US military intentionally staged the event.[22] In the end, "[t]he story of the 'dramatic' rescue of a young, vulnerable woman, at a time when the war was not going well for the US, acted as the means by which a controversial war could be talked about in emotional rather than rational terms. Furthermore, constructed as hero, Lynch became a symbol of the West's 'enlightened' attitude towards women, justifying the argument that the US was 'liberating' the people of Iraq."[23]

These two brief examples illustrate a broader pattern. Whether or not gender considerations are truly at stake, gendered images invoking underlying fears about "enemy populations" are a powerful tool of war propaganda. Al-Qaeda uses gender in this way when it decries the "sexual immorality" of Western women and relies on gendered depictions of Western depravity both as a rationale for war and as a tool for mobilizing support for its cause. Gendered propaganda often has explicit racial or ethnic overtones as well, as in the case of US propaganda during World War I and World War II when war propaganda frequently depicted frightened white women as captives of (ethnically different) male enemies.[24] In sum, states and non-state actors routinely invoke gender norms in their inevitable wartime propaganda campaigns.

22 Deepa Kumar, "War Propaganda and the (Ab)uses of Women: Media Constructions of the Jessica Lynch Story," *Feminist Media Studies* 4: 3 (2004), 304–306.

23 Ibid., 297.

24 Ibid., 298.

The tactical use of gender on the battlefield

In addition to drawing on gender norms in military recruitment and training and in war propaganda campaigns, states and non-state actors also deploy gender as a weapon on the battlefield. Wartime sexual violence is probably the most prominent example of the tactical use of gender in battle. Most commonly, wartime sexual violence involves rape perpetrated with strategic intent. Tactical sexual violence in war might also involve sexual mutilation of civilians, prisoners of war or the dead bodies of enemy soldiers or civilians. At times, sexual violence perpetrated during war may be perpetrated by "isolated" individuals acting without sanction from military authorities. However, wartime sexual violence commonly is perpetrated with sanction from above and with the intent of gaining tactical advantage on the battlefield.

The concept of gender is critical to understanding how and why wartime rape can be used as a tactical weapon. In both peacetime and wartime, rape is a crime of power and not of sex.

The centrality of power relations in rape (whenever and wherever it occurs) depends upon a socially constructed notion of a dominant and controlling masculinity in contrast to a vulnerable and submissive femininity. In wartime, this gendered power dynamic becomes especially critical. The rapist and victim serve as symbolic proxies for their respective groups, nations, or states. The rapist represents the dominant and conquering group, nation or state, while the victim (male or female) represents the conquered or submissive party in a conflict. Thus, wartime rape serves as a means of attacking and humiliating the enemy.[25] Because women historically have been viewed as the property of men, the wartime rape of women becomes a way of attacking the men of a group, and wartime rape against women is particularly prevalent. However, sexual violence is also perpetrated against men during war, and the tactical logic is the same. The rape and sexual mutilation of men during war is intended to emasculate, humiliate and subjugate the enemy.

The tactical nature of wartime rape is especially apparent when it is perpetrated in ethnic conflicts. In such conflicts, the gendered power dynamics involved in wartime rape are interconnected with the critical nationalist dimensions of these conflicts. In these cases, wartime rape can be deployed as a weapon designed to diminish the nationalist strength of an "enemy" ethnic group.

The Bosnian War is a case in point. Wartime rape and sexual violence were widespread during this war and were perpetrated most frequently by Serbian forces against Bosnian Muslims. The widespread use of rape not only signified an effort to subjugate and humiliate the enemy but also represented a military policy with a specific tactical objective. In a culture where ethnicity is determined by paternity, the widespread and systematic rape of Bosnian Muslim women involved

25 Rhonda Copelon, "Surfacing Gender: Reconceptualizing Crimes against Women in Time of War," in Lois Ann Lorentzen and Jennifer Turpin (eds), *The Women and War Reader* (New York: New York University Press, 1998), 71.

an underlying objective of impregnating these women with "little Serb babies." In this regard, the policy of tactical rape, which involved not only individual men raping women but systematic rape in the equivalent of rape concentration camps, sought to serve the Serbian military objective of "ethnic cleansing." It is important to note the critical role that gender norms played in allowing for the tactical use of rape in the case of the Bosnian War. The fact that ethnicity is determined by a socially constructed notion of masculinity and paternity in this society enabled Serbian forces to deploy rape as a tactical weapon.

Another example of the ways in which gender can be deployed as a tactical weapon involves unconventional warfare. A variety of non-state actors involved in violent conflicts with state actors or other non-state actors have resorted to relying on female militants. The case of female suicide bombers provides a striking example. Militarized organizations that have relied on female suicide bombers include groups involved in the Iraq insurgency, the Tamil Tigers, Chechen rebels and Hamas. An explicitly gendered logic underlies the use of female suicide bombers. Militants involved in unconventional warfare are not part of a regular army and, thus, cannot be identified by military uniforms; insurgents are indistinguishable from civilians. In this context, state security forces seeking to prevent insurgent violence use profiles of likely insurgents—profiles based on the reality that the preponderance of such violence is perpetrated by young adult men—in efforts to prevent suicide bombings and other forms of unconventional violence. Accordingly, militant groups may recruit female militants to elude security screening. In doing so, they may use highly visible symbols of femininity, for example, pregnancy or the Islamic *burqa*, to evade detection or capture.

Both of these examples—wartime sexual violence and the use of female militants in unconventional warfare—demonstrate the ways in which gender can be used as a tactical weapon in war. Notably, gender is not a marginal influence on tactics in these cases. Rather, gender can play an absolutely central role in the tactics adopted by states and non-state actors in battle.

Gender and the Consequences of War

In addition to contributing to the reasons wars are fought and to serving as a tool for creating functional militaries, mobilizing popular support and gaining tactical advantage on the battlefield, gender fundamentally shapes the consequences of war. War has divergent effects on men and women, who experience war differently at every stage, from a militarized peacetime to the post-war period. The concept of gender is critical to understanding the divergent consequences of war for men and women.

Men are much more likely than women to serve as combatants in war, and the consequences of war for male combatants are fundamentally shaped by this status. In countries with forced conscription, men are more likely than women to face mandatory military service. Male soldiers are more likely to die and suffer from grave injuries in war. The psychological costs of participating in combat can be incredibly high, and men may suffer high rates of post-traumatic stress disorder.

Although the costs of war for men can be severe, men are also more likely to benefit from their participation in war. High-ranking military officers may find that military service can be a path to political power, a path that has been historically denied to women in most countries. Veterans may be treated as "heroes" for their military service, especially in the case of popular wars. War often leads to concrete economic and social benefits for veterans in the aftermath of war. For example, in the United States, the G.I. Bill, initiated at the end of World War II, continues to provide educational and other economic benefits for returning American soldiers.

War's cost–benefit calculus is quite different for most women across the globe. Whereas men are more likely to suffer as combatants, women and children make up a higher proportion of civilian casualties and war refugees. In the current era of global politics, women's suffering as civilian casualties and refugees is especially felt in less-developed countries, where most contemporary wars are waged. Thus, the gendered consequences of war for women also have political, economic and cultural dimensions. For example, unless she has a family member in military service, a civilian woman in the United States can generally expect to be isolated from the direct consequences of ongoing US military operations in Iraq or Afghanistan. An Iraqi or Afghani civilian woman will not generally have that same luxury and may face the real possibility of being a victim of intentional violence or "collateral damage" in the wars in these countries. Thus, it is critical to remember that other variables—class, geographical location, ethnicity, citizenship, among other factors—shape the consequences of war for women and men alike.

Because men are more likely to suffer grave harm as combatants and women are more likely to suffer violence as civilians, there is disagreement as to whether the effects of militarized violence disproportionately harm men or women. Debates over the utility of civilian protection norms in international law illustrate the nature of this debate and the critical ways that the concept of gender can shape the consequences of war for both women and men. International advocacy organizations working on civilian protection issues have relied primarily on the rhetorical frame of "innocent women and children" to denote civilian status. R. Charli Carpenter has argued that this discursive frame harms men by suggesting that they are not typically innocent civilians. Thus, the rhetoric that equates civilian status with "innocent women and children" implies a parallel and implicit assumption that men of "fighting age" are presumed combatants. As a result, Carpenter contends that the framing of civilians as "innocent women and children" makes it more likely that men will be targets of violence during war and will not gain access to international civilian protection regimes.[26] Some feminist scholars of international relations have reached a different conclusion. In their view, the presumed noncombatant status of women, grounded in basic gender stereotypes, actually harms women by promising but failing to

26 R. Charli Carpenter, "Women, Children, and Other Vulnerable Groups: Gender, Strategic Frames, and the Protection of Civilians as a Transnational Issue," *International Studies Quarterly* 49: 2 (2005), 295–335.

provide immunity for civilians at the same time that the idea of protected civilian status legitimates war.[27] In short, critics charge that gendered civilian immunity norms protect neither women as "innocent civilians" nor men as combatants. In either case, it is important to note that the concept of gender plays a critical role in these interpretations of the consequences of war for both men and women.

The divergent consequences of war for men and women result not only from the extent to which they are likely to be victims of military violence but also from the customary wartime roles that are available to women in comparison with men. Although women serve in the military in many countries and may also participate in militant violence on behalf of non-state actors, women have been much more likely than men to serve in supporting or service roles in wartime throughout history. The list of possible service or supporting roles for women in war is long: women are "camp followers" (sometimes as wives or companions of soldiers and sometimes in an individual capacity) who travel with military forces to serve as cooks and laundresses and to meet the other basic needs of male soldiers; women serve as nurses in the war; women are prostitutes and sex workers who serve military men; and women are the mothers and wives of soldiers.[28] Unlike male combatant status, these supporting roles typically do not serve as a path to political or economic power.

Indeed, women who have served in supporting roles tend to have fundamentally different experiences in the aftermath of war than male combatants. As mentioned previously, most women do not participate in combat roles and, as such, do not tend to receive the political or economic rewards associated with military service. Instead, women's status often declines in the aftermath of war. Take the example of "Rosie the Riveter" in the United States during World War II. Rosie the Riveter involved a US propaganda campaign to encourage women to work in military industries during the war to fill the void left by male workers who were serving in the armed forces. Many women fulfilled these roles and, in the process, often realized new-found economic independence and income. After the war, women were typically forced out of these traditionally masculine occupations to open up jobs for returning soldiers and were encouraged either to work in traditionally feminine jobs, such as teaching or secretarial work, or to retreat to the home and to focus their time and energy on domestic work, including childcare, cooking and keeping house. The "domestic goddess" image that featured prominently in American popular culture in the 1950s was a product of the complicated interactions of gender and the war system during this historical period.

The concept of gender is at the crux of explanations for the vulnerability of women in the aftermath of war. The economic, social and political roles available to women in peacetime are connected to the proscribed gender roles generated by the war system. A consequence of this dynamic is that women's lives are often militarized

27 Laura Sjoberg, "Gendered Realities of the Immunity Principle: Why Gender Analysis Needs Feminism," *International Studies Quarterly* 50: 4 (2006), 900–904.

28 Cynthia Enloe, *Maneuvers: The International Politics of Militarizing Women's Lives* (Berkeley: University of California Press, 2000), 40–45.

even during peacetime.[29] Gendered assumptions about men as the appropriate heads of households are a primary reason that women heads of household face significant difficulties in the aftermath of war.[30] Such assumptions underlie policies, including those giving property to demobilized male combatants, job preference programs for veterans, and a lack of economic opportunities in general, that create and reinforce sexual inequality before, during and after conflicts.[31] The wives and daughters of veterans may benefit indirectly from these policies. At the same time, even female family members of returning soldiers will be constrained by legal, political and economic structures that privilege men over women. Furthermore, unmarried women face extraordinary obstacles to their political and economic independence under these circumstances. These gender inequities and challenges are especially pronounced in countries where women do not have access to basic economic, social or political rights.

As this discussion indicates, the divergent effects of war on men and women can be attributed in large part to socially constructed gender identities and behaviors that define men's and women's wartime roles in fundamentally different ways. Men who participate in combat face momentous and grave consequences, but the potential benefits of war for men, especially men with privileged economic status, are also quite high. Fewer women are likely to benefit directly from participation in war. The costs of war for women vary dramatically based on their class, political status and culture. Some women may benefit from war when their interests are connected with privileged male elites and when they are not likely to face the dangers of combat. Other women face the danger of being among the countless civilian casualties of war while at the same time being at risk of exploitation during war and facing the economic hardships confronted by so many women in the aftermath of war.

Conclusions

In contrast to the assumption of gender neutrality that dominates much of international relations scholarship on war, this chapter has shown that war is a fundamentally gendered phenomenon. Every dimension of war is significantly shaped by the concept of gender. For some scholars, the very legitimacy of war as an institution depends on gendered depictions of heroic men and the vulnerable women they must "save" from enemy threats. The discourse of policy elites who make critical decisions about war and peace is infused with gender. Gender

29 Ibid., 1–34.

30 Codou Bop, "Women in Conflicts, Their Gains and Their Losses," in Sheila Meintjes, Anu Pillay and Meredeth Turshen (eds), *The Aftermath: Women in Post-Conflict Transformation* (London: Zed Books, 2001), 27.

31 Ibid., 28–29; Sheila Meintjes, "War and Post-War Shifts in Gender Relations," in Sheila Meintjes, Anu Pillay and Meredeth Turshen (eds), *The Aftermath*, 69.

inequities within societies are correlated with both intrastate and interstate war. The extensive use of gender in military recruitment and training, wartime propaganda and on the battlefield begs the question of whether the building of effective armies and the waging of war would be possible in the absence of reliance on powerful gender identities and norms. Gender norms and gendered political and social structures lead to divergent consequences for men and women during war. Notably, the gendered consequences of war shape the lives of men and women in peacetime as well. Ultimately, gender norms legitimize and reinforce legal, political and cultural forces that contribute to gender inequities in peace as well as war.

Age of Progress or "Age of Extremes?": The Escalation of Warfare in Modern Times and the Nature of its Mutation

Oleg Kobtzeff

> *... And war began, that is to say, an event contrary to human reason and to human nature took place. Millions of people committed against each other crimes so innumerable, deceptions, betrayals, theft, forgery and issuing of forged banknotes, looting, arson and murder, that for centuries, no courts in all the world could suffice to record, and which, at that time, the people committing them would not regard as a crime.*[1]

Tolstoy's sentence announcing the war of 1812 in the novel *War and Peace* was more than a dramatic device in a work of fiction. Only one generation after the events, it echoed the cry of anxiety of millions of Europeans traumatized by the Napoleonic wars that ravaged Europe with an intensity that had rarely been experienced in the history of humanity. Characteristically, one of the central events of *War and Peace* was the Battle of Borodino, called Battle of the "Moskowa" by the French. The scenes describing unheard-of numbers of men slaughtering each other are among the most violent in the history of literature, together with another rendition of the same battle, Prosper Mérimée's gory short story *L'enlèvement de la redoute* (*The Taking of the Redoubt*). This intensity marks a turning point in the history of war, and the fact that Tolstoy's novel has become one of the most universally recognized classics expresses a fear of war in all humanity, a fear made indubitably real by the escalation of violent language used by ideologues, by the massive character of modern armies, by the extent of the areas affected by combat, and by the spectacular and constant improvement of technologies of destruction during and after the industrial revolution.

1 Leo Tolstoy, *War and Peace*, Volume III, Part I, Chapter I (my translation). The expression "the Age of Extremes" is from the well-known title of the book by Eric J. Hobsbawm.

Statistics clearly explain why Borodino was accurately perceived by Tolstoy and his readers as a turning point in the history of warfare. By the end of the War of Independence, the Americans, civilians and soldiers, suffered a total of 25,550 deaths after eight years.[2] After nearly two centuries of debate over the exact numbers, historians concur on the death toll following the Battle of Borodino: no less than 70,000 perished from direct injury or indirect causes (infected wounds, disease, exhaustion, etc.) in a few hours.[3] Borodino is a symbol because it was the bloodiest episode of a period of quasi-permanent warfare that began in the early 1790s and would continue for more than a quarter of a century, ravaging Europe and sending to their graves millions of soldiers and civilians—numbers that, in human memory, had occurred only during the great invasions of Antiquity, at the end of the Middle Ages, and during the religious wars and the Thirty Years War. It impressed the imaginations of the generations that experienced those terrible years because warfare was entering a new phase of escalation: an age of deliberate industrialized mass destruction often justified by ideologies.

We men and women of the twenty-first century perceive ourselves as more humane and tolerant than our ancestors. The perception is not false inasmuch as resistance against the escalation of violence became organized. Azar Gat in the present volume (Chapter 2) correctly conveys what others, like Pierre Chaunu in his numerous books or chapters on collective violence, have been saying about the pre-state societies, i.e. that violent death may have been the cause of up to a quarter of all deaths. Since 1815, a social and international consensus has been building around a "culture of peace"—a new political culture based on international cooperation ruled by a humanitarian regime with new international laws and institutions such as the Red Cross, the League of Nations or the UN. The consensus does not contradict the interests of Realists whose priority was to maintain equilibrium over Napoleonic or Nazi "anarchy" in international relations as in Vienna in 1815, Berlin in 1878, or while sitting at the UN Security Council. These or the more liberal institutions (UNESCO, the EU, humanitarian intergovernmental or private organizations) were a reaction to unparalleled bloodbaths.

The optimistic Liberal interpretation of this evolution is that finally Tolstoy's plea was finally heard: "since all evil people are bound together and form a force, honest people should just do the same. It is as simple as that."[4] But growing pacifism

2 Howard H. Peckham, *The Toll of Independence: Engagements and Battle Casualties of the American Revolution* (Chicago: University of Chicago Press, 1974), quoted as being the most authoritative study on the subject by John W. Gordon, *South Carolina and the American Revolution: A Battlefield History* (Columbia: University of Carolina Press, 2003) (Foreword by John Keegan), 185.

3 The latest estimates of casualties of the war of 1812 are in S.V. Lvov, "The loss of the Russian army at the Battle of Borodino, 24–26 August, 1812," in Gerasimova, Galina and Sergey Lvov (eds), *The Era of Napoleonic Wars: People, Events and Ideas,* Proceedings of the VIth All-Russian Scientific Conference, April 24, 2003, Moscow; and Richard K. Riehn, *Napoleon's Russian Campaign* (New York, Chichester, Brisbane, Toronto, Singapore: John Wiley and Sons, 1991).

4 *War and Peace,* Epilogue, Part I, Chapter XVI (my translation).

and humanism worldwide must not, conceal the reality of the bloodbaths. Pacifism and humanism have achieved such an unprecedented status in world politics because the violence itself has been so unprecedented.[5] Battles almost as terrible as Borodino—Solferino, Gettysburg, Shiloh and many others—were already exerting a terrible levy in human lives while Tolstoy was still writing *War and Peace,* and record-breaking man-made catastrophes were yet to come. In 1871, the men who lay on the battlefield of Sedan outnumbered by three times those killed at Borodino. When counting fatalities after the battles of the twentieth century, six-digit numbers became common; seven-digit figures appeared after Verdun, the Somme, Stalingrad, or the siege of Leningrad. And this does not include the Holocaust and other atrocities perpetrated against unarmed civilians on an unimaginable scale. In the light of these horrors, and although we may agree with Azar Gat, when he claims in his contribution to this volume that "only protracted civil war or particularly severe foreign invasions resulted in something akin to a return of pre-state anarchy," yet remind the reader that 1914–1918 and 1939–1945 have been compared, precisely, to large-scale civil wars. Also, why systematically separate civil war and inter-state war? Our main point here, is that the distinction is becoming increasingly blurry as we advance into modernity.

Like Tolstoy, the following pages will express a strong refusal of the dogma according to which war is an inevitable fatality. Peace *is* a responsibility that we must not avoid through convenient pessimism on human nature. However, this chapter on the escalation of war in the past two centuries and a half will demonstrate that Tolstoy was wrong in thinking that it is "as simple as that." Rather than identifying with certainty the forces that led to the man-made political cataclysms of the industrial and post-industrial ages, we will only try to recapitulate what are the clues obtainable so far. Without offering any theory or model, they could at least explain, in a limited fashion, the weakening of barriers that normally contain acts of massive and extreme violence, and how they failed to prevent "the people committing them" to "not regard" them "as a crime."[6]

5 Coverage by newly accessible means of mass communications (first, periodicals illustrated by drawings, then by photos, then newsreels in theaters, then television) enhanced the perceptions of the violence by the public.

6 We will limit this study essentially to the *roots* of violence from 1789 to the present, a study of its *evolving characteristics* rather than its actual causes or the events themselves, since many questions regarding these subjects will be addressed by other chapters in this volume such as those on alienation and World War I by Hall Gardner (Chapters 1 and 13), and my own chapters on the sixteenth and seventeenth centuries (Chapter 10), World War II (Chapter 15), or on Yugoslavia (Chapter 19).

Inconvenient Numbers: The Exponential Escalation of Violence

> *This is the last time I am bothering you with my terrible figures. But compared with the time taken by the Allied troops to march down the Champs Elysées during the victory parade—about three hours I think—I calculated that under the same conditions, at the same pace in marching formation, the parade of the poor dead of this unforgivable folly would not have lasted less than eleven days and eleven nights.*[7]

The words are from *La vie et rien d'autre* (*Life and Nothing But*), a film on the aftermath of World War I in the devastated war zones of Eastern France. They are the last sentences pronounced by the main character of this sad, yet inspiring, fiction, Commandant Dellaplane (played by Philippe Noiret), a military doctor. In charge of finding soldiers still missing in action in 1920, this forensics expert has made a nuisance of himself by insisting that families still desperately searching for their missing loved ones should find closure by accepting the horrible reality of their deaths, and move on with their lives. He has gained enemies with his distaste for the growing industry of profiteers—bogus forensics investigators selling false hope, petty local politicians or high government officials making themselves the high priests of a necrophiliac cult of the dead war heroes, sculptors collecting contracts to build monuments to the dead. Above all, he has irritated his superiors by counting, and recounting with maniacal statistical precision, the horror discovered in the never-ending number of victims emerging from the mud of Verdun, the Marne or the Somme,[8] and demonstrating that this massacre has been exceptional in human history. This film was meant to put an end to all war memorials. So it would be a paradox to list it, with *La Grande Illusion*, as one of the best monuments to the memory of the suffering endured during the "Great War". It is far more realistic than Leo Tolstoy's optimism, and not a bit less pacifist. Such was the attitude of Gaston Bouthoul. This scholar has made himself a nuisance just as the fictional Doctor Dellaplane, which maybe explains why his name remains unknown to academia outside of France, where he is already fading from memory.

Despite the shocking disappointment caused by the inventory of suffering during the World Wars, liberalism and progressivism remain strongly influential even to this day. This influence probably explains why the work of researchers such as Russian-American pioneer sociologist Pitirim Sorokin (who accused the Bolshevik revolution and Lenin, not only Stalin, of unspeakable atrocities—another violation of liberal political correctness) or Gaston Bouthoul, with his *Traité de*

7 Scenario of *La vie et rien d'autre* (*Life and Nothing But*), 1990 film directed by Bertrand Tavernier, written by Jean Cosmos and Bertrand Tavernier.

8 As well shown in the film, the death toll continues to rise, although slightly, years after the armistice because of accidents caused by unexploded weapons. To this day, civilian victims are added every year to the list of World War I casualties by unexploded shells and grenades. The author of these lines discovered such artifacts several times a week during his hikes across the countryside in the area of Mourmelon in the 1970s.

polémologie (*Treatise on Polemology*) or *Introduction à la polémologie* (*Introduction to Polemology*), can be as annoying as Doctor Dellaplane's endless computations. Like him, they force us to face the evidence that wars are becoming more and more violent, and thus not only do they ruin the optimism of Liberals, but they also threaten the arrogant "quiet strength" of Realists[9] who may think that balance of power and classic military strategies are sufficient to contain violence.

Bouthoul warned about an *escalation* of warfare, an escalation of the technologies of warfare, an escalation of ideological discourse and especially an escalation of what he called the "volume of warfare." Decades of statistical analysis, never fully completed, are barely summarized in such impressive volumes as his *Traité de polémologie* (published most recently by Payot, in Paris in 1992). Much of it was inspired by his discovery, at a young age, of the work conducted by Sorokin.[10] These estimates indicate several important trends for the centuries where quantitative data could be deduced from archival and archeological sources without resorting to speculation. That data retrieved by Sorokin and verified by Bouthoul in the following decades, which appears in Table 4.1 below, represents the number of casualties on a scale of zero (representing zero casualties) to ten:

Table 4.1 Historical casualty rates

Ancient Greece		
5th century BCE	2.9	
4th century BCE	4.5 to 6	
2nd century BCE	0.3 to 0.4	
Ancient Rome	**in Italy**	**in the Empire**
4th century BCE	1.4	
3rd century BCE	6.3	
1st century BCE	2.8	0.3
1st century CE	0.6	0.08
3rd century CE	1.3	0.13

9 The expression "quiet strength" ("*la force tranquille*"), now a household term in France, was the slogan of the 1981 campaign of presidential candidate François Mittérand, definitely a Realist in foreign policy.

10 Pitirim Sorokin, *Contemporary Sociological Theories* (New York: Harper & Bros., 1928), and Pitirim Sorokin, *Crisis of Our Age* (New York: E.P. Dutton, 1941). The casualty figures are reproduced from Gaston Bouthoul, *Traité de polémologie* (Paris: Payot, 1991), 261. See the concurring estimates based on other sources presented by Azar Gat, "The Roots and Evolution of Conflict: From Cain to the Present" (Chapter 2 of the present monograph, footnotes 2 and 3). There is a pressing need to verify these results and the methods used by both authors. Present efforts to establish systematic casualty estimates are still too dispersed.

Europe	
12th century CE	0.2
13th century CE	0.2
14th century CE	0.68
15th century CE	1.04
16th century CE	1.86
17th century CE	2.1
18th century CE	1.8
19th century CE	1.1
1900 to 1925 CE	8.12

Noticeable are the surges corresponding to the sixteenth and seventeenth centuries, the period of the wars of religion and the Thirty Years War. Also noticeable is the relative decrease of the curve corresponding to the nineteenth century which can only be interpreted as a representation of the stability between the Congress of Vienna and World War I, especially the years between 1878 (the treaty of Berlin) and 1914: the period of the highest number of international commercial exchanges (never to be repeated until the 1970s), the period where one could travel "around the world in 80 days" without a passport, the period of great hopes for peace, the period of the formation of the first international organizations, of great technological progress and international scientific cooperation and communication, of Verdi's and Tolstoy's triumphs as global pacifist heroes, of the Impressionists and of the first great international sports events, the period the French called "La Belle Epoque"—the Beautiful Age. Alas, the *most* obvious surge follows immediately. It is the one corresponding to the first decades of the twentieth century, to World War I.

Also evident is the almost regular rise of the curve, culminating in the dramatic surge corresponding to the twentieth century, that only started when Sorokin began his computation. One can imagine how high the curve would rise if we added all the conflicts of the twentieth century. And there is more bad news: according to research recognized as among the most reliable, the numbers of conflicts on the planet have been steadily increasing between 1945 and the beginning of the twenty-first century, the magnitude of the violence of those conflicts rivaling the destruction of the previous two World Wars.[11]

11 We are referring to the Armed Conflict and Intervention (ACI) Project, a joint project of the Center for Systemic Peace and the Center for Global Policy (with the Integrated Network for Societal Conflict Research—INSCR) at George Mason University, and the Polity IV project at the University of Maryland. The main statistical results of the research of this team is made available at: http://www.systemicpeace.org/conflict.htm. Also to be consulted: the book of one of the project's main artisans: Monty G. Marshall, *Third World War* (Lanham, MD: Rowman & Littlefield, 1999).

Attempts to provide detailed reports of all events and estimates of victims of mass murders, crimes against humanity and genocide, although still incomplete, already occupy almost all the pages of book-length publications.[12] This is the historical background against which has erupted a series of acts of genocide or democide and other mass killings.[13] Most of them have occurred after 1492, the overwhelming majority from the end of the nineteenth century to the end of the twentieth:

- Planned eradications of Indian populations in the Americas (intermittently from the early sixteenth to the early twentieth century and beyond): murders difficult to distinguish from the abnormal population decline (roughly 90 percent of pre-contact figures) due to imported diseases (occurring in a population weakened by anomie, famine, loss of hunting grounds, alcoholism and other calamities often resulting from government policies);
- Planned eradication of religious minorities in the sixteenth century (such as the Saint Bartholomew massacre, massive killings of Jews, etc.);
- massacres during the conquest of Congo by King Leopold of Belgium: 3 million victims according to the 1904 Casement Report,[14] 10 million according to more recent scholarly research;[15]

12 Dinah Shelton (ed.), *Encyclopedia of Genocide and Crimes Against Humanity* (3 vols) (Detroit: Macmillan Reference USA, 2005); the *Online Encyclopedia of Mass Violence,* http://www.massviolence.org, by the French graduate school of political sciences "Sciences Po" and the government's Center for the Study of International relations (CERI) of the CNRS; the special pages on R.J. Rummel's website (http://www.hawaii.edu/powerkills/NOTE5.HTM); and journals such as *Genocide Studies and Prevention: An International Journal* (the official journal of the International Association of Genocide Scholars, University of Toronto Press), *Holocaust and Genocide Studies* (United States Holocaust Memorial Museum, Oxford University Press), or the *Journal of Genocide Research* (Routledge) of the International Network of Genocide Scholars. These are only a few examples of projects, organizations and publications among many others.

13 According to article 2 of the 1948 UN Convention on the Prevention and Punishment of the Crime of Genocide, based in essence on Lemkin's definition (which is now the quasi-universally accepted definition), "genocide means any of the following acts committed with intent to destroy, in whole or in part, a national, ethnical, racial or religious group, as such: (a) Killing members of the group; (b) Causing serious bodily or mental harm to members of the group; (c) Deliberately inflicting on the group conditions of life calculated to bring about its physical destruction in whole or in part; (d) Imposing measures intended to prevent births within the group; (e) Forcibly transferring children of the group to another group." The word democide was coined by University of Hawaii specialist in massive casualties of collective violence R.J. Rummel, and has proven to be of convenient use when massive killings (such as those committed by Soviet firing squads or in the Cambodian "killing fields" indiscriminately against all categories of citizens in the territory ruled by the perpetrator) do not target any specific "national, ethnical, racial or religious group."

14 British Parliamentary Papers, 1904, cd. LXII, 1933.

15 Summarized in Adam Hochschild, *King Leopold's Ghost: A Story of Greed, Terror and*

- massacres by German colonial troops in Namibia;
- Armenian genocide: 600,000 to 1.5 million (figures generally accepted by UN, EU and most academic institutions; completely denied by the Turkish government);
- 3 to 10 million civilian victims (mostly Asian) of Japanese war crimes;[16]
- 1933–1945 victims of death camps: 6 to 7 million Jewish victims of the Holocaust (universally admitted rough estimate); 500,000 Gypsies; up to 6 million unarmed civilians (hostages, sexual minorities, political dissidents, victims of eugenic measures, etc.);
- Serbian victims of the Nazis and Ustaše: 300,000 (most conservative estimate admitted by Croat historians) to 800,000 (figure claimed as accurate by Serbian historians); tens of thousands of victims of other nationalities (such as Croats and Bosnians, but excluding Jews and Gypsies already counted above as death camp victims);
- Unarmed civilian victims of Soviet repressions (1918–1987): roughly 20 million (generally admitted conservative estimate, i.e. over 1/6 of the population of the Russian Empire in 1914); Solzhenitsyn's rough estimate, causing scandal after the publication of his *Gulag Archipelago*, was 60 million;
- 400,000 to 1 million victims of the Cultural Revolution in China;
- Democide perpetrated by the Pol Pot regime in Cambodia: 1.7 to 2.2 million (i.e. 20 to 25 percent of the Cambodian population in 1975);[17]
- Rwandan genocide: 1 million.[18]

These are only the best-known episodes. Other lesser-known episodes of massacres are so numerous (the Harmidian massacres of 1894–1896 against Armenians, the unarmed victims of Nazi hostages or victims of terrorist raids against unarmed civilians in communities such as Oradour-sur-Glane, Tulle, Sant'Anna di Stazzema, Marzabotto, Ochota, Frankolovo, the Soviet villages devastated by Operation Cottbus, and numerous other communities in occupied Europe, the German POWs or civilian victims of mass rapes or forced displacement or killings in territories occupied by the Red Army, the indigenous troops loyal to the French tortured and murdered in Algeria in the early 1960s, etc.) or are so difficult to categorize (for example, the difficulty in distinguishing armed insurgents and terrorists from innocent civilians among the nearly one million victims killed by French troops in Algeria between 1945 and 1962) that they merit a special study impossible to fit into the present article.

Heroism in Colonial Africa (Boston and New York: Mariner Books, 1999).

16 Yuki Tanaka, *Hidden Horrors: Japanese War Crimes in World War II* (Boulder, CO: Westview Press, 1996), 72–73.

17 Craig Etcheson, "Khmer Rouge Victim Numbers, Estimating" in Dinah Shelton (ed.), *Encyclopedia of Genocide and Crimes Against Humanity* Vol. 2, 615–617.

18 Milton Leitenberg, *Deaths in Wars and Conflicts in the 20th Century*, Occasional Paper #29, Cornell University Peace Studies Program: Ithaca, 2003, 2005, 2006.

This chronological coincidence between these murders and international conflicts, plus the terrible number of victims of both manifestations of collective violence raises the question: are the acts of genocide and the episodes of modern warfare part of the same societal phenomenon? Rafael Lemkin made the connection from the start, when he coined the term "genocide" in 1943. Commenting on Lemkin, University of Sussex professor Martin Shaw implies the kinship between civil war and interstate conflict fought by regular armies:

> *Beginning with Nazism, it is evident how fundamentally this genocidal movement was defined by the experience of war and militaristic ideology. From its earliest street-fighting days the Nazi party defined political and social groups – Communists, Jews, homosexuals – as enemies linked in gigantic geopolitical conspiracies, to be 'destroyed' in a quasi-military sense. And although genocidal policies began as the Nazi regime consolidated its control over German society, it was only as it moved into aggressive war that the most generally murderous phases began.*[19]

Jean-Paul Charnay makes this correlation in this very volume, from a *geopolitical* perspective. His analysis mentions a *sociological* mutation in the twentieth century that modifies the nature of war itself: "technology and bureaucratic racial *statolatry*" (when the state becomes the object of idolatry—one of the common denominators between Nazism and Stalinism) determined "a qualitative change in the elimination of the Other." The mutation of civilizations are also consistent with a reshaping of the conduct of war according to new geopolitical and geostrategic configurations, which explains why wars have become *total* wars: "*increased firepower reaching civilian populations*, geopolitical remodeling of the vanquished to be reset on the chessboard of balance of power, *psycho-ideological excitement by propaganda leading to the negation of the opponent* which renders necessary measures ranging from demanding unconditional surrender to *ethnic genocide*. Fierce economic competition was *legitimized by the nobility of political philosophy*."[20]

"Hail death!": The World as a Prey

> *'Viva la muerte!'* (rallying cry of the Falangists during the Spanish Civil War)

19 Martin Shaw, "War and Genocide: A Sociological Approach," *Online Encyclopedia of Mass Violence* [online], published on November 4, 2007, accessed October 7, 2010, URL: http://www.massviolence.org/War-and-Genocide-A-Sociological-Approach, ISSN 1961 -9898.

20 Jean-Paul Charnay, "Metastrategy: Sociostrategic Systems in the West" (Chapter 25, p. 540 of the present volume).

> *Polemology should follow epidemiology; peace science should model itself on health science ... The promise of an epidemiology of peace and war rests on the analogy between war and disease, peace and health.*[21]

In history textbooks and documentaries narrating the first days of World War I, there has been extensive use of images of joyful volunteers leaving for the front. "*Nach Paris!*" ("To Paris!"), "*A Berlin avant Noël!*" ("In Berlin before Christmas!") are the slogans written with chalk on the sides of train carriages leaving for the front that can be seen in many photographs taken in the first days of the "Great War." The enthused send-off of conscripts and volunteer troops "with a flower on the rifle" ("*la fleur au fusil*") has almost become a cliché in French historiography. In 1917, George M. Cohan's song "Over There" and images of smiling "doughboys" waving farewell, seemed to have also become symbols of an enthusiastic embrace of warfare in American popular culture. Or was this only propaganda?

This perceived enthusiasm has recently been challenged as a cliché by French military historians and other scholars.[22] The impression that humanity is progressively getting rid of its worst tendencies is not a fantasy. As mentioned above, rapid and widespread adoption of the "culture of peace," no matter how superficial, would have been unthinkable only a few generations ago. The love of peace and fear of war already existed long before. As long as there existed happy, peaceful communities, there existed people who, although not "pacifist" in the modern ideological sense, were concerned above all by the question of how to feed their children and not by the "question of the Orient" or by the borders of Alsace and Lorraine. So the terrible hours when all the church bells of Europe sounded off the tocsin during the harvest of August 1914, forcing farmers and their farmhands to drop everything, bid farewell to their families and rush off to join their regiment, will be remembered forever as a somber event, not a joyful departure for a great adventure.

Nevertheless, it would be careless to dismiss the real traces of enthusiastic behavior among other portions of the population.

The grand illusion: The power of denial

Pacifism was a very powerful movement. But it failed in 1914, even in its attempt to persuade the masses of Socialist leaders and scores of militants to stop the preparation for war. Jean Jaurès, the great French Socialist and pacifist legislator was marginalized even in his own party and eventually succumbed to the bullets of an assassin. Raoul Villain, who shot him in public and assumed his act was

---

21 Francis A. Beer, "The Epidemiology of Peace and War," *International Studies Quarterly* 23: 1 (1979), 45, 46.

22 The debate between historians is analyzed by Jean-Jacques Becker, "'La fleur au fusil': retour sur un mythe" in Christophe Prochasson and Anne Rasmussen (eds), *Vrai et faux dans la Grande Guerre* (Paris: La Découverte, 2004), 152–165.

acquitted while the widow of Jaurès who was suing for damages was dismissed and condemned to pay trial expenses. Others exemplify the exaltation of warfare: the idealistic poet-paladin Charles Péguy (killed on September 5, 1914) who had called for the execution of Jaurès, or the thousands of anonymous extremist nationalists in numerous organizations like the Black Centurions of Russia or the terrorist groups like the one joined by Gavrilo Prinzip (still considered a hero by entire portions of all ex-Yugoslav populations, judging by existing street names), or simply the hundreds of thousands of enthusiastic recruits—volunteers, not enlisted men—like the insignificant failed Austrian painter, Hitler, who finally saw his chance to make something out of his life. Nor should we neglect the utopian representations of war in the twentieth century and a bellicose spirit in popular culture before, during and after the conflicts. After all, if propaganda became an essential feature of modern warfare (from the civic monuments and paintings or popular engravings of the late 1700s and early 1800s to Jamie Shea's daily NATO press conferences during the war in Kosovo, during which the expression "collateral damage" entered our vocabulary), it is because there exists a fertile soil. Newspapers, that were numerous at the time, poems, literature or painters' battle scenes, as well as popular engravings and songs, have left us with a vivid picture of the exalted state of a significant portion of the masses.

After many great wars or periods of conflict, euphoria and optimism flourish in at least a portion of the population struck by warfare. *"C'est la der des ders"* ("it's the last of the last wars") was the slogan that motivated French troops and populations in the rear in the last stages of World War I (the armistice was followed immediately by wars and atrocities against civilians causing millions of victims in the collapsing Russian Empire, the Balkans, and Turkey, while in China, local conflicts that began earlier in the century slipped into the second global conflict almost without transition). "Never again" is what we still hear at commemorations of the victims of the Holocaust.

But as Renoir's film already warned in 1939 in its very title, this confidence is a *grand illusion*. Within two generations following World War II—what could have been considered as a lesson for humanity—the wars of de-colonization, two genocides in Central Africa plus bloody civil wars, and the massive political repressions by Leninist regimes, or successor regime's such as the Yugoslav state under Milošević, continued to claim millions of victims. Denial of such violence reached record-highs as hundreds of Western intellectuals and thousands of youths hailed Mao's Cultural Revolution as a model. Just at the end of the Cold War and just before the Rwandan Genocide, Francis Fukuyama made himself a name by announcing "the end of history." This wishful thinking of those who remained faithful to Kant or to other progressivist ideologies can be explained by a doctrinal representation of these and other abominations as *an abnormality*, an absurdity or an accident. Typical of this idea that conflict is an accidental parenthesis in history is the constant referral to Hitler as a madman and the passionate attacks against any other interpretation. A good example of this form of denial were the attacks against Hannah Arendt's concept of the "banality of evil" which she developed in her articles and book about Eichmann.[23] More recently, Oliver Hirschbiegel's film *Der Untergang* (*Downfall* in English) with actor Bruno Ganz scandalized viewers for depicting Hitler as extremely cruel but not suffering from any medical psychiatric condition. Twenty-six percent of Germans reacted with hostility towards such a portrayal that made the Führer appear "too human."[24]

Denial of horrific events and of our capacity to commit them has not only been a trait of Nazi or Turkish revisionists. Anyone old enough to remember the sensation caused by the release of the Gulag Archipelago will also remember the barrage of extremely hostile language from not only Marxists, but moderately left-wing intellectuals as well, like the otherwise talented Max-Pol Fouchet, going as far as to accuse Solzhenitsyn of being a pro-Nazi propagandist.[25] The same generations will also remember how long it took to convince Western public opinion (right-wingers as well as left-wingers) of the numbers of the dead under the Khmer Rouge regime. In those years, following Vietnam, disgust with Western, especially American, policies, when the new codes of "political correctness" were born, any accusation against a Communist regime appeared as tainted with CIA propaganda and shifted the focus of all "concerned citizens" towards the victims of Chilean general Augusto Pinochet, Argentine's general Jorge Rafael Videla or of the Shah of Iran. Mainstream liberals, even those leaning to the right, could not come to terms with the magnitude of the death tolls and their implications about the Enlightenment's fundamental paradigm of human goodness. Realists, although less optimistic about human nature, could also have been tempted to lower their usual guard by thinking, for example, that the lessons of World War II had been learned, and that the balance of terror created by the nuclear arsenal and a battery of policies for "global security" would be sufficient to avoid further abominations.

---

23 Hannah Arendt, *Eichmann in Jerusalem: A Report on the Banality of Evil* (New York, 1963).

24 Anonymous, "Début de polémique sur 'La chute'," *Le Nouvel Observateur*, January 5, 2005, reproduced on the *Nouvel Observateur* website, http://tempsreel.nouvelobs.com/actualite/culture/20050104.OBS5331/debut-de-polemiquesur-la-chute.html, retrieved October 6, 2010; Anonymous, "Frankreich debattiert über den 'Untergang'," *Stern*, January 6, 2005, reproduced on the website of *Stern* magazine, http://www.stern.de/kultur/film/filmstart-frankreich-debattiert-ueber-den-untergang-534738.html, retrieved October 6, 2010; Roger Ebert, "Downfall," *Chicago Sun-Times*, March 11, 2005, reproduced and edited on the author's web page: http://rogerebert.suntimes.com/apps/pbcs.dll/article?AID=/20050310/REVIEWS/50222002/1023, retrieved October 4, 2010. In many of his publications since the times when Hitler was still alive, Erich Fromm has analyzed the Führer's psychology. Often in agreement with Hannah Arendt's own views on the mental state of Nazi criminals, Fromm's assessment of Third Reich leaders was that their insanity was not so much a *medical* condition as it was a manifestation of a *mentality* forged by an education and social standards, a *culture* that will be explained further in this chapter.

25 During the sensational debate with Jean Daniel, Nikita Struve and other intellectuals in the studio of the famous French channel 3 literary talk show "Apostrophes," immediately after the expulsion of Solzhenitsyn from the USSR.

The orgy of violence in Yugoslavia and Rwanda in the 1990s, erupting just as the world thought itself safe, came as a shock.[26]

The cause of the reflexes of denial lie also in the fact that to question both Liberal and Realist views of the modern history of warfare is often to be assimilated to extremist anti-humanistic ideologues. Indeed, reactionary right-wing ideologies, or, on the contrary, a certain Marxist-Leninist (usually Trotskyite) discourse, are fueled by resentment against a world seen as decadent and dangerous. Whether it is fascism, Stalinism, McCarthyism or vociferous dictators wearing military fatigues or elegant business suits, or "war against terrorism" demagogues, the main rallying themes are always the same: the world is becoming more and more violent and dangerous, all foreigners hate us, we are surrounded by enemies and they are helped by disloyal elements in our society who are the cause of all our interior problems. Such ideologies need paranoia and an apocalyptic vision of history as a driving force in their opposition to the optimism of a liberal bourgeois world order (or Soviet Socialist realism), since the legitimacy of that order is founded upon the representation of a world made safer and wealthier through technological and cultural progress. Mainstream historiography has therefore had many reasons to fear Cassandras like Robert Conquest, Alexander Solzhenitsyn or Gaston Bouthoul. But by doing so it is depriving society of the analytical tools that could resist the apocalypse when some are making it into a self-fulfilling prophecy.

However, to be prudent is to at least listen to the Cassandras. It is never prudent to dismiss warnings about possible dangers when the degree of atrocity has been as high as with the Holocaust or other genocides of the twentieth century, and with the wars that occurred in the same relatively short period.

The reason behind academia's desire to minimize the "myth" of the outburst of joy at the beginning of World War I is plain: to face one's own propensity to violent crime is painful. Yet we are forced to consider the following. The peculiarity of the forms of violence of the nineteenth and twentieth centuries is that they are *massive*, not only in number of victims but also in the fact that entire populations participated in the perpetration of the horrors, if not directly, at least indirectly by lending their moral support. In some cases these popular masses indulged in the murders of other masses while frequently living in denial and developing self-deceiving thought processes. But in other cases, the violence was assumed in full awareness and with conscientious commitment to *necrophiliac urges* as expressed in a whole *subculture*, with its known symbols and codes as military insignia (black uniforms, skulls and bones ...) or slogans like "*Viva la Muerte*!" ("Hail Death!").

It is difficult to measure the degree of commitment of entire populations to these calls for violence. One fact remains: the Nazi program of genocide was explicit enough in the bestselling *Mein Kampf* along with numerous other documents

26 When that happened we were unprepared, and, having refused to face the reality of tens of millions of victims of decades of communism, we were disarmed when the first "ready-to-think" simplistic explanations, such as Huntington's "clash of civilizations," seized the media's attention. See my case study on Kosovo in Chapter 19 of the present volume.

widely available to millions of Germans and Austrians through the new mass media. Moreover, many fascist regimes gained power through a democratic parliamentary process. Another chilling example is the popular response to calls for genocide broadcast by radio stations in Central Africa without anyone trying hard enough to countermand the slogans either locally or from abroad. Ever since Nuremberg, the argument "we did not know" cannot be used as a legal argument. Assuming that the majority of the populations of countries at war in the modern age "did not know" what was happening should not, therefore, be used as a working hypothesis.

That democracy inevitably leads to peace, as defends the otherwise lucid R.J. Rummel, should therefore be seriously reconsidered. We will argue in our conclusion that the missing element that could have prevented democracies from self-destruction and violence would have been a better commitment to a more efficient human rights regime. Our point, therefore, is not to deny the possibility of an interdependence between democracy and peace, in which we believe; it is to understand why democratic governments and *entire societies* under a democratic regime, have willingly accepted or supported unspeakable violence or lied to themselves about its existence.

Societal Changes and the Approval of Collective Violence

The first sociologist to raise this essential question also happens to be one of the founders of sociology: Alexis de Tocqueville. The question is clearly formulated in the title of Chapter XXII of Volume II, Part Three, of his classic *Democracy in America*: "Why democratic people naturally desire peace and democratic armies naturally desire war."

The democratization of violence, or the "Tocqueville paradox"

Alexis de Tocqueville's apology for democracy, equality and freedom in his analysis of American society has made it a model of liberal thought for generations, to this day. It was in fact among the first publications ever to defend the idea that democracy could very well be a good condition for assuring peace, and that a society based on equality generated peaceful manners with social and family stability and cohesion (Volume II, Part Three, Chapter XI). The young Frenchman was one of the first political thinkers to observe rule of law, separation of powers, balance of power, decentralized institutions and independent regulatory institutions like the jury, plus freedom of the press, actually put into practice in a society, and defended these institutions, that we have now taken for granted as the evident foundations of our modern democracies, as counterweights against the negative effects of what he called the "tyranny of the majority" (Volume I, Part Two, Chapter VIII). He thus became recognized by posterity as one of the pillars of democratic thought.

At the same time, his lucidity and the sense of growing danger that tormented him so much made him also a model of realist thought for someone like Raymond Aron, who rehabilitated Tocqueville in France in the 1950s and 1960s where he had become all but forgotten. Tocqueville's awareness of the "tyranny of the majority" did not require much imagination. Not only an excellent observer of slavery and a commentator on racism,[27] he was an eyewitness, during his travels across America, of the misery of the Choctaws during an episode of the genocidal "Trail of Tears." His writings clearly describe a form of violence which is not simply a form of persecution but a new form of warfare. "These savages have not only retreated, they are being destroyed. As the indigenous people retreat and die, in their place comes and ceaselessly grows an immense people. Never among nations has been seen such a prodigious growth, nor such a rapid destruction."[28] German geographers would later theorize about this new form of warfare (and glamorize it) as *Lebensraum*.

Of the numerous contradictions found in democratic societies that tormented Tocqueville (and may have turned him into a disillusioned cynic towards the end of his life), grave contradictions that came to be called "Tocqueville paradoxes" by his commentators, the most obvious, even to a child, was how a democracy founded upon the principle of equality allowed two "unfortunate races ... both suffering the effects of tyranny" to be treated, respectively, like animals or, for the Indians, "to be destroyed if they do not bend to the will" of the European.[29] A century later Orwell put the paradox into other words: all are equal but some "are more equal than others." How can this absurdity be explained? According to Arthur Kaledin, Tocqueville "saw democracy leading to much greater independence for the individual as old social bonds, traditions, and agencies of order faded; yet at the same time, it harbored powerful tendencies that would create a new and more extensive system of constraints."[30] Michel Foucault called these "micropowers." Kaledin's analysis of Tocqueville summarizes with remarkable acuteness the

27 "Racial prejudice seems even stronger in the States that abolished slavery than in those where slavery still exists, and it appears nowhere as strong as in the States where servitude has remained unknown," Alexis de Tocqueville, *Démocratie en Amerique* (Paris: Vrin, 1990) (critical edition edited and commented by Eduardo Nolla), 264 (this and all other excerpts from Tocquveille are translated by myself).

28 Ibid., 250.

29 Ibid., 246, 247.

30 Arthur Kaledin, "Tocqueville's Apocalypse: Culture, Politics and Freedom in *Democracy in America*," in Laurence Guellec (ed.), *Tocqueville et l'esprit de la démocratie* (Paris: The Tocqueville Review/La Revue Tocqueville & Presses de Sciences Po, 2005), 60. This excellent summary of Tocqueville's understanding of the difficulties of society in adapting to change is to be compared with Erich Fromm's analysis of the collective neurosis affecting populations experiencing societal change in the transition from medieval to pre-modern civilization (see my "A 'Time of Troubles': War in an Age of Planetary Upheaval, From the End of the Middle Ages to 1648," Chapter 10 of this volume, footnotes 33 and 49–53).

pioneer sociologist's vision which makes him a precursor to Durkheim, Marshall McLuhan, Hayek, Orwell, Hannah Arendt, Erich Fromm, and arguably Adorno:

> *Below the busy surface of democratic life, with all its swelling affirmations and energy, he had penetrated a restless psychic underworld of fear and anxiety. If incessant change was the chief feature of the democratic social condition ('social state'), uncertainty was the ruling principle of the democratic psyche ('frame of mind'). And it was Tocqueville's growing sense that the loss of social order would be followed by psychological disorder …, and ultimately, of the psychological disorders that seemed endemic to democracy, three stood out: an increasing psychic isolation of the individual, resulting in an impoverishment of the social imagination and a pervasive social indifference …; a phobia about anarchy and disorder resulting in a compulsive grasping for order; and, as a consequence of the anxiety and uncertainty about identity experienced in the fluid egalitarian 'social state', an increasing intolerance of difference, especially manifest in fear and envy of 'others' and in repressiveness toward 'the deviant'. All of these, he thought, would add up to a politics of assent to state power.*[31]

Before analyzing the "phobia about anarchy and disorder resulting in a compulsive grasping for order," and its consequent exaltation and pursuit of warfare, let us understand two aspects of the democratization of violence: the transfer to the people of the regalian right to the use of force plus, the development of large popular armed forces.

The transfer of regal powers to the masses and the right to bear arms

What is meant by the expression "democratization of violence" proposed here? The question may shock many.

The notion must be understood in the light of the classical definition of *legitimate violence* and *regal powers*. It is important to remind the reader that the medieval theorists of state authority led a *campaign against private warfare*. They focused many of their efforts to construct a theory of government in the extremely unstable centuries following the collapse of the Roman Empire and the multiplication of constantly shifting feudal lordships, against the numerous private wars that ravaged villages and cities. Their arguments against private warfare led to the concept of the monopoly of violence, in other words the principle according to which *only a legitimate state*—a king, or any government comparable to the Roman emperor as opposed to a vassal, i.e. a feudal lord—*should have the right to maintain troops and wage war or exercise the use of force* against those who put in danger the population. Such violence exercised by the state against those who disrupt law and order is thus a legitimate form of violence, the only acceptable form of the use

31 Ibid., 62.

of arms. Centuries later, Hobbes and the Enlightenment philosophers continued to explore this philosophy of public force as they developed the idea of the social contract. The members of society, the governed, accept the authority of those who govern in the name of the public common interest (*Res Publica*).

In exchange for the recognition of the government's authority and the right to bear arms, the governed receive the guarantee that public force will be used precisely for the common good, to protect society from enemies exterior and interior, and not for the benefit of any private warlord, local tyrant, foreign invader or criminal band. But with the development of democracy—the concept according to which the people, or the *nation*, are not only governed but govern, and that they constitute the state and simply delegate their authority to representatives—*the right to bear arms becomes the right of all*. Hence such an importance given to hunting rights in French Canada (forbidden by the monopoly of hunting granted to the aristocratic caste until August 1789 in France), the importance, to this day, of hunting and fishing rights and fish and game management in French rural politics or the debate around "the right to bear arms" in the American Constitution and political culture.

We have so much difficulty understanding what a social advance the right to bear arms was in the Americas or in Europe because today we see only the reactionary demagogues running for office. We barely understand how important, how revolutionary and *democratic*, was the notion of the right to carry a weapon. It is important here to be reminded that any form of hunting was usually the last feudal privilege granted to the nobility which was exercised as a monopoly, making any other form of hunting, by anyone who was not a nobleman, a form of poaching. Once the right to own firearms was admitted, it changed the lives of millions of peasants who now had the ability to feed themselves with game and defend their gardens against rodents.

Also, it must be remembered that any objects such as pots and pans, other kitchen utensils, furniture, blankets and other household goods were considered as extremely precious because they were rare, expensive but also remarkably solid and of a very high quality (contrary to the mass-produced disposable objects in our consumer society). Handed down from generation to generation, they were not only respected for the services that they provided, they were valued for the history that they represented and the skills of the artisans that went into their manufacturing. Moreover, in the history of manufactured objects, firearms made between the end of the eighteenth century and World War I remain, to this day, among the most extraordinary examples of craftsmanship and engineering. It is sufficient to look at the catalogues of dealers in second-hand or antique guns to assess the quality and efficiency of some of these weapons, many of which are still in perfect working condition decades or even centuries after their careful and loving manufacture by the gunsmith or blade maker. Thus, the hunting gear sold in the nineteenth century in the catalogues of the "Manufacturer of Arms and Cycles of St. Etienne" in Central France, or of "Sears & Roebuck" in America, or the hunting knives of Tübingen, Germany, Toledo, Spain, or Finland, or the double barrel shotguns crafted in obscure shops in Scotland, Switzerland, or Central Europe, gave their owners an extraordinary sense of respect and power. We should add also a sense of *empowerment*. The first shotgun or rifle or hunting knife appearing in a household thus had its own mystique.

Representations of an exalted wilderness and ownership of a Kentucky rifle forged early American identity. That gun became legendary by terrorizing British troops: for the first time in history, the quality of the rifle allowed one to aim at a target from a very remote distance and hit it efficiently.[32] Thus, the sense of power obtained by the possession of an exceptional object slips progressively towards a sense of power that can be exercised in real action or by simply thinking of what one can do with a weapon. How many people would enjoy a sense of power generated by the democratization of the right to bear arms is very difficult to measure statistically. The evidence in our possession is the mystique of firearms that is so obvious in popular culture in the United States. But in many other countries as well, we can tell from the popularity of specialized publications and documentaries, how there is an enduring interest in individual weapons worldwide.

It is to be noted that one of the great exceptions to this pattern is Russia, where the right to bear arms has been systematically resisted by all past and present governments. The other greatly different situation is to be found in the Muslim World and Asia, in the Middle East and parts of Central Asia and India, particularly in certain tribal societies as we can see in Afghanistan throughout its history or on the famous "Northwestern frontier," where group identity and manhood were intimately linked to the possession of a knife, a sword or a rifle generations before there was any question of the right to bear arms for all in Western societies. Among the Sikh, a man's responsibility for the blade that must be carried at all times except when bathing and sleeping has been elevated to a level of religious duty.

Duty was also the inspiration of the Second Amendment in the United States Constitution. In his ground-breaking analysis of the history of the "right to bear arms" surrounding the present debate around this issue,[33] Saul Cornell explains that not only was possession of firearms a right, in the years immediately following the Declaration of Independence it was a civic *obligation* in many communities: the main concern was not the right to bear arms but maintaining the "well regulated militia." Saul Cornell sees a difference between this historical "civic model" of the individual and the collective-rights models respectively defended today by opponents and defenders of gun control. The aim of the Second Amendment was "protecting the militia against the danger of being disarmed by the government"

32 Until then, and well into the nineteenth century, in the words of the Russian general Suvorov, "the bayonet is smart, the bullet is an idiot," because individual guns were only used at short range, to pierce body armor, or from a distance, only to spray a terrain with random projectiles to discourage the advancement of enemy troops in that terrain, never as a direct means to neutralize a specific target such as a soldier, especially if the target was moving rapidly.

33 Saul Cornell, *A Well-Regulated Militia: The Founding Fathers and the Origins of Gun Control in America* (New York: Oxford University Press, 2006).

rather than "protecting individual citizens' right of personal self-defense."[34] But as he tries to defend this historic fact, demonstrating that the opposition to gun control became a struggle for individual rights, the author simply describes how, for part of the American population, the individual psyche progressively assimilated the possession of firearms first as a necessity, then a civic obligation, then a right and a form of empowerment, then, finally, as part of one's *identity*.[35]

But once the weapon as a sign of identity and empowerment has entered the foundations of a political culture, how will it be possible to resist the development of an arms industry without appearing as an enemy of the fundamental values of society? Especially when the armament industry, as it grows exponentially from the middle of the nineteenth century to the twenty-first century, will offer more and more jobs. The pacifism that should have resulted from liberal democracy, and that had in effect become one of the social norms of a democratic society like America, will become almost impossible—yet another paradox noted by Tocqueville almost two centuries before Michael Moore's *Bowling for Columbine*.

"Band of brothers": The military as a new community or the democratization of the armed force

The great social and economic changes following the American and French Revolutions were preceded by a quiet social revolution, in the eighteenth century, that would change military history forever.

This revolution was the change in the way military personnel were recruited, trained and equipped. The great changes are usually associated by historians with the reforms in the Prussian army under Frederic the Great, to which we should add the spectacular reforms of Peter the Great which represented a cultural revolution in Russian civilization. Very rapidly, in only two or three generations during the eighteenth century, the Prussian model was adopted by most of the states of Europe, and was successful for the following reasons.

First of all, it must be remembered that these were times of great upheaval in the social geography of Europe. Agricultural lands were becoming overpopulated and a greater number of young men suffered from unemployment in rural areas where the division of property between the inheritors of farmers/owners was no longer sustainable due to the reduction of parcels. Specific problems such as enclosure in the British Isles added to the problem. The rural exodus that was going to transform

34 Ibid., 4.

35 What Saul Cornell then describes is how the process of the transfer of regal rights to use legitimate violence to the people evolved, in the nineteenth century, into a transformation of those regal prerogatives into *individual* rights once individualism had become an essential condition of social relations (as predicted by Tocqueville), triggered by legislative initiatives tending towards gun control, which created a fear of losing the empowerment procured by possessing arms.

the majority of Europeans or North Americans into city dwellers over the next two, three or four generations (more in Southern and Eastern Europe) had started.

This was also the last century of the miniature glacial age which continued to cause catastrophes until the very end, such as the famine of 1788 (still seen as one of the great factors of the French Revolution during the following summer). These environmental upheavals wreaked havoc just when technological progress and the new economic concepts of the Enlightenment encouraged populations to believe that their lot could improve. As noted, once again by Tocqueville, populations are ready for a revolution not when their situation is deteriorating but, on the contrary, when it is improving. The interruption of progress or sudden disasters creates unbearable frustration. In this European society which was not yet industrial but which had definitely turned its back on the old rural civilization, the technological progress of the Enlightenment sparked the first urban economies based on manufacturing. This prelude to the industrial revolution was more the affair of craftsmanship and large shops rather than factories, yet these activities were sufficient to absorb some of the unemployed masses. Or at least in part.

Whether they were lost in a totally new cultural environment and starved for not finding work, or whether they did find employment but in the harsh conditions of early industrial capitalism, these men and women who were the first in centuries to leave a rural environment regulated by the heavy rules of village conformity, ritual, religion, moral and economic traditions, were experiencing a cultural shock that can only be compared to traditional tribal societies breaking up in today's Africa or rural Asia to be thrown into our great modern cities. They also gained a new freedom, but one which they could only enjoy if they were particularly creative and confident in themselves. If they were not, disorientation could become extreme and unbearable. The cultural shock of no longer being guided by rules of tradition, and the difficult living conditions of eighteenth-century cities, were described by many novelists at the time like Daniel Defoe, the numerous authors of picaresque novels, or Voltaire in his *l'Ingénu*.

In this changing and unstable world the army, or even the navy, with their order and predictability, became, despite the severely regulated life, or possibly *because* of it, a solution to many problems such as joblessness or loneliness and disorientation. Numerous novels and, later, movies of all sorts, portray life in eighteenth century barracks as marked by harsh discipline, cruel snobbish officers and stick-wielding loud-mouthed sergeants. While this was indeed part of reality, what past and especially present fictions fail to note is that regiment life immediately provided, to homeless youths facing starvation and death, food, clothing and lodgings, sometimes in excellent buildings if one cares to take notice of some of the classic or baroque monuments still standing all over Europe and that used to quarter the military in those times. The simple regulated life, run by the book and by the clock, with its lack of responsibility, guidance at every moment by officers, and its rituals, replaced the codified life of the rural community. The military rituals of the eighteenth century are still with us and continue to impress men and women of all ages and cultures worldwide as evidenced by the crowds gathered to watch all sorts of parades, in front of Buckingham Palace, on the Champs-Elysées, the Red Square, or in New Delhi, as well as in almost all countries of the world. The uniform dress code, often a prestigious outfit that an eighteenth century or nineteenth-century farmer would never have dreamt of even on his wedding day, and the recognition of the authorities symbolized by medals, insignia and eventually promotions—including to the highest ranks as illustrated by James Cook, William Bligh, von Scharnhorst, or Clausewitz—could replace entirely the way of life once experienced in the village community.

Last, but not least, as symbolized in the song *Ich hatt' einen Kamaraden*, the fellowship of men, which is only made stronger during combat as well known by all sociologists and military historians, completely resolves the problem of loneliness which was the challenge of the alien urban environments. Thus, a sense of *identity* could be found in the armies of the generation of Frederic the Great and in centuries to come. In addition, before new regimes would allow the ownership of firearms, a soldier would be responsible for personal gear which included, more and more frequently, a rifle. There is no reason to believe that an eighteenth century soldier did not experience the feeling of empowerment described previously in our discussion of the right to bear arms.

Once again a precursor, this time in the field of diversionary theory, Tocqueville warned (in Chapters XXII and XXIII of Volume II, Part Three of *Democracy in America*) of two problems with an army in a liberal democracy: first of all, that it is constantly threatened with marginalization and obsolescence in a society that values peace as a necessity; secondly, to avoid the danger of military coups (Tocqueville's generation was marked by Napoleon's seizing power in 1799, and witnessed military coups in the new Latin American countries) that it would need exterior wars as an outlet to prevent it from turning against the civilian government.

Psychological Factors

Too many occasions existed to start these wars. The social tensions and consequent deteriorating mental health of large portions of the population in the late eighteenth century, and throughout the industrial age, would provide numerous occasions to start these wars.

From movements of panic to organized military action

Two events can be used as particularly symptomatic and rich case studies, being focal points of several of the social trends that we have evoked in this chapter: collective fear, the right to bear arms and the sense of power that comes from exercising armed force in a democratic society.

The first was a strange event, made famous by Georges Lefebvre, and recognized by such enthusiastic reviews as the one signed in the *Annales* by Marc Bloch, one of its leading figures. This event was the Great Fear—*la Grande Peur*.[36] Although Lefebvre never designated the phenomenon as a factor in igniting the French Revolution, his study is valuable because it demonstrated that this bizarre episode of collective paranoia was, in the words of Bloch's review (maybe borrowed from Tocqueville who often used the terms), "a symptom of a social state" intimately linked to the revolutionary process.

From July 20 to August 5, 1789, an epidemic of mass hysteria spread across French rural communities in different regions of the kingdom. Peasants armed themselves and prepared to fight to defend their lives and their crops as rumors reached them announcing that armed gangs of bandits were terrorizing the countryside. In fact, these bandits proved to be completely imaginary. It appears that the false rumors were created by sparse news of the revolutionary turmoil in the cities, then news about aristocrats allegedly preparing to defend themselves (against those same revolutionary movements). Eventually, the Great Fear became politicized once the rumors evolved into a conspiracy theory accusing the aristocrats of preparing a massive attack against the peasantry, for example by hiring bandits, which rationalized the initial rumors of gangs of bandits; and if they were nowhere in sight, it only proved how well-hidden and prepared they were and how the conspiracy was well planned and malicious. As Marc Bloch commented, anyone that was skeptical about the delusions became a suspect.[37]

Eventually, many peasants launched preemptive attacks against their lords' manors, news of which spread more rumors of bandit attacks among the people who did not realize that their own actions caused the noise. Some of these attacks against manors simply blended in with the general civil conflict that was erupting at the same time and that became the French Revolution. The root causes of this outburst are considered to be hunger (as mentioned earlier, the last manifestations of the miniature ice age were devastating crops), extreme demographic pressure (such expressions of extreme duress as "the numbers of our children are causing our despair" left archival traces),[38] and the tensions between the world of urban civilization and rural communities which were exacerbated when city dwellers traveled to the countryside to buy rarified food staples (which were perceived as raids to deprive the countryside of its supplies and may have caused some of the rumors about bandit predations).

What is most intriguing is that the peasants armed themselves during the most labor-intensive months of the year. Revolts usually occurred only during the Spring months when the peasants had spare time. That peasants actually abandoned working on the harvest in late June and July is probably an indication

36 Georges Lefebvre, *La Grande Peur de 1789* (Armand Colin: Paris, 1988) (the original edition, reprinted several times, dates back to 1932); see Marc Bloch, "L'erreur collective de la 'Grande Peur' comme symptôme d'un état social," in Marc Bloch, *L'histoire, la guerre, la Résistance* (Paris: Gallimard, 2006), 435–441 (originally published in 1933 in the *Annales*).

37 Bloch, "L'erreur collective ..." 437.

38 Ibid., 439.

of their violent determination. The paranoid hysteria disappeared, but its massive character should be more systematically reflected upon and compared to other social events with similar symptoms. As a recent commentator of Lefebvre pointed out, research in this domain is still insufficient.[39] Therefore phenomena of mass hysteria coincidental with a period of preparation for armed conflict deserve deeper research than the usual journalistic pieces.

The second event was the Brunswick Manifesto of 1792, and would set in motion far-reaching historical consequences. In 1792, after two years of transition from an absolute monarchy to a constitutional regime, during which tensions grew between an executive still in the hands of the king and an elected legislative body, France was losing a war against foreign invaders. An army of Austrian and German allies was occupying the eastern provinces of France and, in the middle of the summer, was almost at the gates of Paris. The crown and the parliament appeared as weak in dealing with the military situation as they were in resolving the economic crisis. Public opinion believed the king to sympathize with the invaders. On July 25, Brunswick, Commander-in-Chief of the Austro-German armies, published a manifesto demanding the surrender of Paris. To this ultimatum he added a clear threat: should the French National guard or any armed civilians continue to combat the allied Austro-Prussian armies, they would be treated as irregular rebels and punished accordingly. This basically meant that all prisoners would be executed. The ultimatum also implied violent reprisals against civilians if Paris did not surrender.

Whether it had authentically been signed by Brunswick himself or whether it was a text forged by French counter-revolutionary aristocrats in exile, matters little. Public rumor (again), as a result of the text being circulated, created fear among Parisians and throughout all of France. The entire city and an important portion of the French population had heard of the Manifesto by August 3. At this point the ultimatum triggered the radical revolutionary movement of August 10, and accelerated the convergence towards the capital of volunteers from the West and from the South of France, forming armed militia. What followed was the first great popular movement of men and women in arms since the wars of religion, and what was at the same time a mass insurrection and a powerful military counter-attack resulting in the victory at Valmy, 150 kilometers east of Paris, on September 20. The many documents from the time allow us to measure the degree of fear sparked by Brunswick's ultimatum, and the progressive transformation of this fear into a massive popular reaction ending with an explosion of joy after the first military victories over the enemy.

Newspapers that were already numerous at the time, poems, literature, painters' battle scenes, as well as popular engravings and songs (culminating, later, with the *Marseillaise* which would become the national anthem), all leave us with a vivid picture of the exalted state of the revolutionary masses. It is true, however,

39 Timothy Tackett, "La Grande Peur et le complot aristocratique sous la Révolution française," *Annales historiques de la Révolution française*, 335 (January–March 2004), http://ahrf.revues.org/1298, accessed October 10, 2010.

that the volunteers of 1792 did not stay long, returning home within a year. The revolutionary government had to instate a draft system in 1793. In that way it constituted an army of nearly one million men by Christmas. Many resisted the conscription, often joining the counter-revolutionary insurgent troops, which were entirely composed of volunteers. But those who stayed in the Republican army would be the core of the armed force that would soon (within only slightly more than a decade) be able to conquer almost all of Europe.

At the early stage of the revolutionary wars the power of the troops, an as yet undisciplined muster of armed civilians, rapidly became unstoppable because it was, in very large portion, an army of volunteers who saw themselves as having nothing to lose. The Brunswick ultimatum and the panic that it created caused them to believe that they were fighting for their very lives. Then, the massive character of that assembly of unprofessional but highly motivated soldiers, incomparably higher in numbers than the American minutemen, felt justified as they were greeted as heroes and liberators by entire foreign populations in Belgium, Italy, Switzerland and several German states. This gave them a sense of empowerment, an antidote to the sense of loss and alienation caused by the social and economic upheavals on the eve of the industrial revolution.

If we look at the Brunswick ultimatum not as a spectacular anecdote, but as a significant moment revealing the symptoms of a "social state" (to repeat the words of Tocqueville and Marc Bloch), we may recognize certain essential patterns in the evolution of modern societies. Let us remember that it was literally the moment where all the present symbols of French identity were suddenly created—the national hymn, the blue, white and red flag, and the very concept of "nation". What was a divided assembly of regions on the verge of a civil war suddenly found a sense of unity and identity, across the barriers of religion and regional languages, as it was challenged by an outside enemy.

The fear that had spread among Parisians and then to a portion of the French population, and that produced so many volunteers, was followed by the creation of new military institutions, a form of modern military organization with massive recruitment and generalized conscription, reminding us of the institutionalized exploitation of fear observed by Jean Delumeau when he studied earlier periods of history.[40] Here is what links the process of democratization and the legitimization of armed force, noted for the first time (as we have seen) by Tocqueville. He is also the one who warned about the risk of legitimate armed force becoming a mass phenomenon and then degenerating into mass mob justice, an argument well summarized, once again, by Kaledin as throughout his entire article:

> *The disappearance of a fixed social order would lead to a sense that there were no longer fixed limits to human effort or aspiration and thus to a psychic buoyancy that would be a great source of vitality. It would release power. But increasingly, it was the strains of the endlessly deracinated life that drew Tocqueville's attention. He saw that while democracy would release men's*

40 See Oleg Kobtzeff, "Time of Troubles," Chapter 10 of the present volume.

dreams, the uncertainty and perpetual striving that accompanied the fading of traditional limits and social bonds would exact a high price.[41]

Warfare as a social upheaval

We have seen how spontaneous commotions due to panic may be channeled into popular movements leading to historic political and military changes. One of the main causes of the violence was the extreme fear, studied by Jean Delumeau, caused by unique environmental disasters, and no less disastrous social and economic upheavals, in the difficult transition from the Middle Ages to the pre-modern era. A highly dysfunctional society was born of this pathological transition. The new civilization had put an end to networks of solidarity and plain social bonds without replacing them. At the same time, villages had been wiped out by epidemics or ruined by predators (warlords, or the new capitalist speculators, or the tax collectors of more powerful governments), and the guilds and professional corporations which had defined the social order of urban environments had been rendered obsolete by the new financial and mercantile powers that created a world of competition ruled by market forces. In that new world, many suffered from loneliness, and from fear that the loneliness would become even worse.

The family unit remained, but was constantly threatened by the high mortality rate during epidemics which returned frequently, and also by warfare which was becoming more intense. In such a world where society itself seems hostile, what barriers could prevent a lonely, broken and fearful individual, with nothing to lose, becoming hostile toward anyone symbolizing the new society or threatening new-found solutions such as new religious beliefs? However, one can never admit one's own hostility toward society. It only generates feelings of guilt. But as every psychiatrist knows, guilt generates aggressiveness. It then becomes very easy for any political power to exploit that aggressiveness. Jean Delumeau and Denis Crouzet describe the mechanisms with acute detail. René Girard would add his own theories on the scapegoat mechanisms.[42] Ideologies can conceal the real nature of one's aggressiveness against society, or against one's fellow human beings, while justifying violence.

As we have seen with the phenomenon of the "Great Fear" of 1789 and how that led to armed conflict, extreme fear and collective hysteria were still present in the Enlightenment. The difference with the new transitional period of the end of the eighteenth century and the first half of the nineteenth century observed

41 Kaledin, "Tocqueville's Apocalypse," 71.

42 To analyze René Girard's theories of scapegoating would require not only several more pages, in order to be accurate, but several more articles. Although the author of the present article almost fully subscribes to Girard's philosophical views, he also believes that to serve as evidence of the cause of war they need to be validated by more extensive historical research (as conducted by Jean Delumeau, Denis Crouzet or Alfred Soman) or clinical research such as Hannah Arendt's or Fromm's.

by Tocqueville is that the religious dimension of the fear affecting populations a century or two earlier, or the religious character of the discourse justifying holy wars, were no longer present in the Modern Age. However, what Denis Crouzet calls the prophetic dimension of the discourse of warfare during the wars of religion[43] is replaced with a secular version of a messianic call to arms, in the name of democracy, liberty or abstract historical forces. As mentioned earlier in this chapter, it motivated the French revolutionary troops; it also motivated the Spanish and Russian guerillas in 1812, the volunteers in 1914, the revolutionaries in 1917 or the 15-year-old adolescents voluntarily joining the White army.

All of Russian literature from Dostoyevsky's "Devils" (less accurately translated as "The Possessed") to Alexander Blok's "The Twelve" throbs with apocalyptic themes. Blok's messianic vision is very present in his famous militant poem about twelve Bolshevik soldiers patrolling the streets of a ghostlike devastated Saint Petersbourg scorched by "a jolly and wrathful and happy" winter wind. "In front, marched Jesus Christ" is the last verse. Submitting to the abstract forces of history—whether one gives them a Hegelian metaphysical definition, or calls it dialectic materialism, "manifest destiny" or the unstoppable march of progress and democracy—seems like the same "escape from freedom," the same retreat into the illusion of safety provided by submission to the authority of the angry vengeful God of Savonarola, of the Puritans, or to the power of a fascist leader; it is a sado-masochistic mechanism.

One could argue that a major difference between the pre-modern era and the last 240 years is that fear and violence in the Enlightenment and throughout the industrial age lacked the eschatological dimension of the end of the 1400s to the 1600s. But that would change in 1945 after Hiroshima and Nagasaki and apparently unstoppable developments in arms technology, as well as man-made environmental cataclysms, demonstrated that the destruction of all human life on the planet is feasible. Otherwise, there are very few differences between the transitions that followed the collapse of the European civilization of the Middle Ages and the transition to first the industrial civilization, and now to the post-industrial civilization. Erich Fromm, analyzing the late Middle Ages and Renaissance, and Tocqueville, observing democracy in America, wrote about the same newly acquired freedom, the same anxiety in learning how to manage it, and the same dangers of tyranny when that maturation fails.

Constantly neglected in almost all recent historic research is how the rural exodus became an aggravating factor in the difficulty of adapting to the upheaval in times of historical civilizational transitions. It has been rapid and often brutal. It started in the second half of the eighteenth century, and at the beginning of the twenty-first century the majority of the globe's population has become urban. In some parts of the planet the transition has been extremely sudden. At the end of the nineteenth century, 90 percent of the Russian population lived in the countryside; four generations later 90 percent of the Russian population lives in a city. And for

43 Denis Crouzet, *Les Guerriers de Dieu. La violence au temps des troubles de religion (v. 1525–v. 1610)* (Seyssel: Editions Champ Vallon, 1990), 81.

many this exodus has been brutal: forced collectivization and forced displacement in the USSR (at least one million dead), warfare, economic strife, and now man-made environmental disasters in Africa and other areas labeled (in pure Orwellian manner) "developing" countries. In times when civilizations mutate, experiences such as forced exile, immigration, and travel, or simply living in a rapidly changing world, can make individuals or groups stronger and more tolerant by confronting them with difference, by making them more aware of the "Other."

Being forced to find the richness of "strange new worlds," the one who accepts the difficulty finds his or her reward as an explorer, a pioneer, a discoverer. He or she is in control. Thus, estrangement can be strength, freedom and growth. But because it demands courage and a persistent mental effort of adaptation to change and difference, it is also a discipline. Unfortunately, such discipline cannot always be easily acquired in the deteriorating social conditions described by Karl Marx when he analyzed the phenomenon of alienation. Alienation is the subject of another chapter of this volume (Chapter 1 by Hall Gardner) so we shall look at some other aspects preventing a sane enjoyment of the challenges of a changing world and of one's own freedom.

Social disintegration and anomie

Emile Durkheim, in his landmark study of suicide, was the first to correlate self-destructive behavior (violence against the self) with violence against others. *Integrated* societies, the traditional pre-industrial communities, the village, no longer existed to help individuals organize their social interaction and resolve their conflicts. Moreover, twentieth century societies, poor or rich, experienced rapid economic and social transition while accumulating contradictory cultural values and corresponding codes of social behavior. This forced upon each individual incoherent demands to conform socially.

The examples of these conflicting forms of social pressure leading to *anomie* are numerous: religious conservatism conflicting with secular values, Puritanism conflicting with the exaltation of sex, the values of a consumerist society of leisure conflicting with the Spartan values of competition in education and in the workplace, the daily spectacle of mediatized death and violence (illustrated magazines, pulp fiction and popular theaters in Durkheim's time, then radio and cinema, and now television) conflicting with society's exacerbation of youth and its denial of death, peer pressure conflicting with educational or corporate norms, and, as we advance into the twenty-first century, the harrowing dilemma imposed upon children torn by contradictory loyalties to divorced parents. Material wealth is becoming a more and more demanding tyranny in the rich societies where status means social survival, and in poor societies where a minimum income means plain physical survival. In both cases, material possession is vital, but at the same time it is volatile. This makes employment vital.

Unemployment then, as during the depression or toward the end of the twentieth century, becomes a psychological catastrophe added to the financial catastrophe, since one's profession has become one of the defining traits of identity. The situation also became dangerous after the 1980s, in rich areas, since the high cost of labor perniciously rendered the workforce a liability. In poor areas the extremely low salaries devalued work, and bred alienation as in nineteenth-century European industrial societies. Pressure, uncertainty and confusion became unbearable for increasingly large and fragile segments of society deprived, by these very aggravations, of the favorable mental conditions to creatively respond to the complex and ever-changing forms of socialization. Numerous anthropologists, throughout the twentieth century, particularly those studying populations in ghettos, Indian reservations or other groups living in conditions where dysfunctional families, self-destructive behavior and violence are typical social traits, have insisted that the destruction of social bonds (through colonialism, economic disruption, forceful removal) or the severing of bonds after immigration to a foreign land is a growing problem in industrial and post-industrial societies.

But only a minority of individuals initiate violent actions. Is it then legitimate to correlate the rise of violent crime and suicide with the escalation of violence in warfare? What we must then realize is that for one individual that suffers and does express his or her frustrations through violent behavior, hundreds of others suffering from the same frustrations and fears are still restrained by the last remaining controls of social pressure and self-discipline against what is condemned as a crime, i.e. the taking of life. But should any discourse—nationalistic, populist, Communist, anti-Communist, or other—legitimize the taking of life as self-defense, as the only alternative to resolve a conflict, then the dormant tendencies towards violent behavior may find an outlet. In addition, the dormant violent tendencies are glorified through military ritual, popular festivities and the orgiastic communion described by Fromm which provides the illusion of a restored unified society.[44]

In addition to all the previously mentioned aggravations in the second half of the twentieth century, other frustration mechanisms complicated already confusing standards of social behavior. These frustration mechanisms operated differently under socialism or under capitalism, but resulted in the same feeling of being redundant and powerless which exacerbates narcissism. In one part of the world, societies under Marxist-Leninist regimes failed to deliver the constantly postponed promise of the triumph of a Communist Utopia (a fully equalitarian global society of safety, abundance and peace). Elemér Hankiss explains why anomie was added to the frustration: the exceptional complexity of small bureaucratic hierarchies all competing with one another and therefore making contradictory demands and threats against the citizens who needed them for their daily necessities, or who had to reason in terms of market forces when operating their small private vegetable garden. Thus, a "normal Hungarian ... has to switch discourse, style of speech, modes of behavior, several times a day."[45]

44 See above, footnotes 27 and 37 to 40.

45 Elemér Hankiss, quoted in Jacques Rupnik, *The Other Europe* (London: Weidenfeld & Nicholson, 1988), 169.

This complex type of anomie, according to Elemér Hankiss, creates the proper conditions for collective schizophrenia, which itself is the breeding ground for paranoia.[46] In another part of the world, societies based on consumerism encourage the futile pursuit of an ever-evading hedonistic happiness. Their abundance (sustained mainly with debt, which is now beginning to destroy the system) has made them a model for impoverished populations—the local poor or the Third World—for whom the need to possess is exacerbated inasmuch as the spoils of abundance and consumerism become less and less attainable. The poor are frustrated by never being able to possess what the rich are in constant fear of losing and what, anyway, is never sufficient to provide a feeling of accomplishment, of comfort, since society is ever mutating. Furthermore, in the rich societies, material happiness is no longer even an acceptable norm. The pursuit of happiness and never attaining it is the new norm. *Disenchantment with the world* has become a necessity for the endless pursuit of spiraling consumerism artificially boosted by the addiction to debt. This disenchantment with the world has become a marketing strategy. In the words of the autobiographical satirical novel written by one of France's former top advertising executives (immediately fired after he published this):

> *When, after sweating to make savings, you will be able to buy the car of your dreams, the one I made to look so stunning in my last campaign, I will have already made it look out of fashion. I am always three fads ahead of you and I always arrange it so that you are always frustrated. Glamour is the place where you never get to. I get you hooked on novelty, and the advantage with novelty is that it never stays new. ... in my profession, no one wants your happiness, because happy people do not consume. ... in our jargon we have called this 'post-purchase disappointment'. ... to create needs, we have to foment jealousy, pain, non-fulfillment: that is my ammunition. And my target is you.*[47]

The accumulation of so many frustrations, and the fact that a mature representation of self, of society, of history, of culture and of the Other is therefore such a titanic effort, leaves very few options for the weak and frustrated individual. There is a deficit of alternatives to the self-destructive behavior described by Durkheim and observed in the following decades by scores of sociologists, social psychologists and anthropologists who studied dysfunctional families or groups. For the one who resists the destruction of self, only one option remains to give him the impression of regaining control of his life: the destruction of the Other. One of the easiest solutions, providing immediate gratification—the illusion of control over one's own destiny, over society, or even over history—is a paranoid attitude toward the

46 Elemér Hankiss, personal communication, American University of Paris, November 1999, question and answer session after a lecture during a seminar on *Central and South-Eastern Europe in transition* organized by Elinore Schaffer and Oleg Kobtzeff.

47 Frédéric Beigbedder, *99 francs* (Paris: Bernard Grasset, 2000), 17–18.

changing world. Paranoia feeds on reactionary reflexes and pathological defense mechanisms: conspiracy theories, scapegoat mechanisms or a morbid, obsessive recollection of past historical injustices inflicted by strangers on one's recent or remote ancestors—the Crusaders, Muslims, the rich capitalists, the dangerous ignorant poor, colonialists, the Versailles Treaty winners and the interior traitors who "stabbed Germany in the back," Croats, Serbs, mainland treacherous politicians (De Gaulle, the intelligentsia and the Communists during French decolonization) against European settlers in colonies—or allegedly committed by imaginary interior enemies (Jews, immigrants, Free Masons, "religious fanatics," all sorts of foreigners, in a long list of other imaginary scapegoats). In a leisure society or in societies where unemployment is extremely high, inactivity *offers the time* to imagine, discuss and turn over in one's mind complex mythologies explaining why "life was once so simple" and whose fault it is that it is now so complicated. Therefore, terrorism, racism, primitive nationalism, religious fundamentalism and many forms of collective violence could be the result not of "ancestral hatreds" or poverty alone, but of the failure to provide every individual with the socially acceptable means to risk losing one's identity and thus overcome one's inevitable separateness from the Other. Accepting this separateness is the courageous process of accepting one's terrifying loneliness everywhere, with everyone. For unscrupulous politicians and tabloid publishers, in Berlin in 1933 or Belgrade in 1990, or for educators, political analysts and social actors today, Manichean and simplistic interpretations of conflict are easier to market than the long, complex and subtle "art of loving" the Other defended by Erich Fromm—whose works are characteristically forgotten and neglected in our century.

So why do the weak, the frustrated, the immature, the fearful and the suffering people kill? Because the peaceful and reasonable answer is too complicated to listen to. This is how we should approach the problem.

* * *

Our purpose here was to raise doubts about the way some scholars and political thinkers almost take it for granted that democracy necessarily brings peace. Democracy indeed creates an environment more favorable to peace than dictatorships. But it is far from being infallible: the evolution of the Weimar Republic into the Third Reich, or the massive character of the fascist movements in democratic countries like Austria, Belgium and France should serve as some of the many reminders. Also, at a time when religious jihads waged against democracies become a pretext to launch new colonial adventures under the guise of a *secular jihad* for democracy, and when it has become impossible to oppose a useless and costly war in the democratic way that the war in Vietnam was once opposed, it seems important to raise the question of the complicity or at least the passivity of the masses when war is waged in their name and funded by their taxes. Thus, it is indispensable to add the essential defense against collective violence which majority rule cannot always guarantee: *a commitment to the principles of human rights,* and to the regime that, fortunately, many constitutions and international laws and organizations provide us with.

At the beginning of this chapter, we showed that warfare was on the rise between 1945 and the beginning of the twenty-first century.[48] Let us try to finish on a happier note. The same research indicates very clearly that the number of conflicts is rapidly declining. We could become arrogant, complacent and lazy and proclaim "mission accomplished" or the "end of history." Instead, as after 1648, or as after 1945 when the UN was created and when the political culture of many societies partly embraced the "culture of peace," we should pause and reflect upon the mistakes committed that led to the horrors of the twentieth century, the most promising and at the same time the most violent in human history. We have never enjoyed this chance to pause after the end of the Cold War. The pause would give us the opportunity to measure the effects of nearly seven decades of a human rights regime and diplomatic measures favoring conflict prevention. Their accomplishments are relegated to the last pages of our newspapers: only great failures and violence gets attention. Violence that *does not erupt* cannot be noticed. It is a great time, however to notice.

If democracy is indeed a factor of peace, it will very certainly be demonstrated that a commitment to human rights, not only of governments but of entire societies, is the key to success. "Since all evil people are bound together and form a force, honest people should just do the same." It is *not* that simple. But it is the only option left if we do not want the twenty-first century to become as damned as the twentieth century, and we now have the tools to avoid the worst. Most difficult will be to ask ourselves if we are truly among "the honest people." Facing our own violent tendencies or our laziness in addressing mounting humanitarian disasters (as, recently, during the horrors committed in Central Africa), is the first step to neutralizing it.

48 See footnote 11.

The State as a Cause of War: Anarchist and Autonomist Critiques of War

Andrew Robinson

States are major agents in most wars, and yet in the popular imagination and mainstream theory, anarchy – the absence of the state – is associated indissolubly with war. The state is generally viewed as a source of social peace and, therefore, its destruction is taken to be extremely threatening. The irony is that, while anarchists are stereotyped as archetypally violent bomb-throwers, anarchists are among the most consistent critics of dominant forms of war. The stereotype of the anarchist bomb-thrower contrasts with the reality of state violence and anarchist involvement in anti-militarism.

While the popular image of anarchists dates back to earlier moral panics, the theoretical frame is specifically Hobbesian, with statists imagining the state guaranteeing social peace and mapping their own fears of social war onto their adversaries. For liberals and other statists, anarchists must either want a war of all against all, or else be hopelessly naïve. Nothing could be further from the truth. The view that anarchists oppose the state because they naïvely assume the benevolence of human nature is a straw-man argument constructed mainly by their adversaries (anarchists actually differ drastically on this question, taking positions ranging from social-constructivism, to the insistence that everyone is different, to negative or 'realistic' views of people unable to be trusted with centralised power). On the other hand, anarchists are far less violence-prone than statists. Some anarchists are entirely nonviolent, while others justify the use of force in what might be called 'just war' terms, as a means to defend against the oppressive violence of the state. In general, anarchists identify the state as *inherently* violent and as a *source* of violence, and seek to end (or, failing this, to moderate) its violence through exercises of counter-power. Such counter-power is necessarily exercised by social movements, social networks or ethically-oriented individuals, not by states or other hierarchies, and can either take the form of nonviolent counter-power or of 'popular defence'.

Paradoxically, anarchist accounts of state behaviour often echo the more intransigent wing of statist theory associated with Realism in International

Relations. Classical anarchists often adopt a broadly Realist view on state 'behaviour', accepting the view that states tend to pursue ever greater control of territories and people, and that this inherent logic of the state is a major cause of interstate wars as well as intrastate repression and violence. Where they differ from Realists is in rejecting the view that the state has created social peace *inside* societies. For anarchists, the Hobbesian account is nonsensical: if people cannot be trusted to coexist peacefully in a situation of relative equality, it is absurd to think that a few of them could be trusted with excessive concentrated power. (This is in fact a persistent problem with Hobbesian arguments, and is implicitly resolved in dominant discourse only by its supplementation with theories of more and less 'civilised' groups of people.) Anarchists also reject the Hobbesian view of the state as the origin or guarantor of society and of social order. This rejection of the Hobbesian account is often grounded in knowledge of the history of state-formation and the political anthropology of stateless societies, which problematize the Hobbesian view of life without the state. If a statist responds that Hobbesian theory abstracts from the type of subjectivity dominant today, not from actual historical conditions, an anarchist response would be that the state in fact *produces* conditions of atomisation and mutual rivalry, which require constant activity to decompose social movements and force people into dyadic relationships based on hierarchy. Hence, one finds the state *counterposed* to society in anarchist theory.

The result is that anarchist theorists refuse the division between internal and external, or the framing of war as a phenomenon occurring between already-formed states. Rather, warlike relations often pertain between states and social forces inside societies. Indeed, in a state-controlled society, there is something like a situation of permanent social war. Therefore, the tendencies Realists recognize in interstate behaviour, such as the aggressive pursuit of maximum power and the tendency to exploit others' weakness to grab resources and control, are also taken by classical anarchists to govern state behaviour towards 'its' subjects and towards social movements. Anarchist strategy is thus often directed to forming other loci of power which can balance against or draw energy or forces out of state power.

Anarchism is sometimes criticized for 'essentializing' states or assuming in advance that all states are bad. It is true that anarchists are opposed in principle to the state, and not only (say) to authoritarian states. It is untrue, however, that anarchists cannot deal with diversity among empirical states. This is because the state *as a social logic*, not a particular empirical state, is the main target; empirical states are opposed because they express, to one degree or another, this logic, and since this logic is essentialist, to criticize its rejection as essentialist misses the point. When anarchists talk about the 'state', they are thinking partly of a core group of assemblages and partly of a social logic, a hierarchical way of relating which is not located at any particular site. In Kropotkin's terms for instance, the state is not simply a power over society, but the usurpation of social functions (a 'whole mechanism of legislation and of policing') and the creation of new social relations.[1] Hence, anarchist views of the state are both broader and narrower than

1 Piotr Kropotkin (1897), 'The State: Its Historic Role', Panarchy website, http://www.

conventional views. They are narrower in that the state properly speaking refers only to the 'deep state' of coercive and administrative agencies committed to imposing hierarchy and sovereignty in society, such as the police, army and bureaucracy. Anarchist attitudes to later add-ons in terms of the 'soft state' are at best ambiguous: anarchists criticize the bureaucratic distortions and disempowerment arising from the hierarchical arrangement of the welfare state, but are broadly sympathetic with its social objectives.[2] It is broader in that statist ways of being have always been conceived in anarchism in a sense which prefigures Foucault, as forms of life not limited to the state apparatus but embodied wherever social relations are atomised, alienated and hierarchical. The state is taken to exist in constant conflict with other forces which it decomposes, and to draw on 'reactive' force in a Nietzchean sense. For instance, Colin Ward argues that the state is a constant source of social conflict, but blames others for this conflict.[3]

Classical Anarchism: Bakunin, Kropotkin, Stirner

Bakunin's approach in many ways initiates anarchist approaches to war, linking war to the aggressive and dominatory tendencies of states. Only conquering nations create states, and then only for the purpose of subjugating the conquered, though Bakunin also views some states as more exclusively 'military' than others. The modern state is thoroughly anti-popular, seeking intensified exploitation on behalf of the ruling class. 'The modern state, in its essence and objectives, is necessarily a military state, and a military state necessarily becomes an aggressive state'.[4] Military discipline is the last word of capitalist civilization, and synonymous with its decay, serving as a force for corruption. Strong, centralized states tend to produce moral and intellectual decay, with invention drying up and corruption mushrooming.

Bakunin's grounds for viewing states as incapable of avoiding warlike measures are recognizable from Realism: states must compete with other states or die, political power produces structural effects and strives for its objectives even against the wishes of those in power. Only great powers are truly sovereign. The pressure of international self-preservation prevents democratic reforms from affecting the oppressive nature of the state. What is more, no dictatorship can have a function other than to perpetuate itself. The difference between Bakunin and Realism is that Bakunin puts forward a second pole offering a way out of the statist race to the

panarchy.org/kropotkin/1897.state.html.

2 See, for instance, Colin Ward *Anarchy in Action* (London: Freedom Press, 1973); and Colin Ward, *Social Policy: An Anarchist Response* (London: Freedom Press, 1996).

3 Colin Ward, *Anarchy in Action*, 137.

4 Michael [Mikhail] Bakunin, *Statism and Anarchy*, trans. and ed. Marshall Shatz (Cambridge: Cambridge University Press, 1990 [1873]), 13. Unless otherwise noted, all discussions of Bakunin pertain to this text.

bottom. A revolutionary upsurge of the excluded can put an end to state coercion. The 'people' (referring to the excluded and exploited strata) appear periodically as a separate force in international affairs, even as a distinct military power. They sometimes defeat great powers, and they can take a different degree of interest in wars; in general they only actively embrace wars beyond national borders if these are in a revolutionary or religious cause. A war with active popular support becomes a 'national' war, but the people have a political interest against the strengthening of states. While in previous struggles military force had been held back against social movements, the massacre of the Commune showed the military to be something monstrous and merciless, against which only an equally unrestrained insurrection is sufficient. Bakunin views social hatreds as 'intense and deep', in contrast to political hatreds.[5] Since force not right determines political outcomes, the excluded must organize forces outside the state and against it in order to conquer freedom. Popular defence and insurrection are bywords for this process, which also requires the creation of transversal links among local communities.

Controversially, Bakunin differentiates between states in their degree of propensity for imperialism and war. Germany is taken to particularly embody the statist pole due to the particular class composition of the German state, and to seek to spread this state-form across Europe. Bakunin sometimes seems to use a rather crude conception of national character to explain such differences, but a more subtle reading might imply a nascent theory of social forces in state-formation, similar to later neo-Gramscian approaches, and a theory of identity-formation linked to social fantasies. Bakunin is particularly determined to critique anti-colonial nationalism as imitation of the colonizer, arguing that state-formation renders immobile the very vital forces which achieve liberation. Nationalism is one attribute among others and, like others, should not ground a special or essential status.

The idea of countervailing social logics of statism and autonomy is taken further in Kropotkin's work on the state as 'political principle'. Kropotkin portrays the state as an outgrowth of the 'political principle', a specific function for which the state was created historically. The political principle is associated with decomposition and dyadic relationships, in which each person is only related to a hierarchically empowered superior or disempowered subordinate. This is counterposed to the 'social principle' of horizontal, voluntary connections. In Kropotkin's theory, social networks are portrayed as bubbling with life, whereas states bring death and decay through social fragmentation and the ever-present structural violence of subordination. The moment states emerge, their first act is to 'pillage' surrounding communities. They then seek to decompose horizontal relationships so as to create the scarcity and alienation which then form the pretext for its existence.

In this account, the state is not necessary for defence. Kropotkin views stateless societies as producing diffuse self-help associations to meet needs such as mutual defence or peacebuilding, which, while quasi-obligatory, do not approximate the

5 Mikhail Bakunin (1870), 'The Class War', Marxists.org website, http://www.marxists.org/reference/archive/bakunin/works/writings/ch06.htm.

state form. Dispute-resolution is horizontal and relies on interconnectedness for its power. Outside the state, conflict can be an enriching force which generates a 'new lease of life' from each struggle. The reason that this enrichment does not extend to state wars is the difference in motive. States fight wars to destroy liberties, whereas social networks struggle in order to defend liberties. Even as peace bringer, the state is a negative force both because it imbalances the conflict towards one side and because its imposed peace is a generalized disempowerment in a 'colorless, lifeless whole', to which frank, open conflict is preferable. Though war can exist without states, the rise of states vastly increases and transforms war. Indeed, for Kropotkin, the 'State is synonymous with war'.[6] The state is viewed as growing from privileges arising from military conquest in cases of intergroup war among stateless societies, and from the desire to punish. It is periodically pushed back by rebellions, and the establishment of *fait accompli* it is unable to reverse. Once established, the state wages a constant struggle against tendencies to recompose social life; war is one of the means of this struggle, particularly in terms of 'ruining' countries (i.e. through ecological and economic destruction).

Since the state brings only destruction, it has to rely on plundering societies to survive, by means such as expropriation, land grabs and enclosure. Though located in a cyclical historiography, Kropotkin's frame is rather Manichean, implying a perpetual social war between the two principles. Because the state demands sovereignty – and it 'cannot admit of a "State within a State"', it recognizes only subjects and not unions – it is necessarily in a situation of war with other loci of attachment. Hence, the political and social principles cannot coexist; the political principle must seek to crush the social principle. According to Kropotkin, the first act of the state wherever it has arisen is to crush autonomous social life. 'Thus as soon as the State began to be constituted in the sixteenth century, it sought to destroy all the links which existed among the citizens both in the towns and in the villages'. Kropotkin reads historical events through this frame, with state victory bringing ruin and social victory bringing life.[7]

Max Stirner also refers in passing to the question of war. Stirner views people as unique beings expressed of a singularity which cannot be represented. His work expresses an extensive effort to attack attempts to subordinate people as unique beings to representational categories, termed 'spooks', including those which appear inclusive (such as humanity). In this context, he mounts a critique of both liberal humanism and nationalism. For Stirner, the rights pertaining to humanity or the nation are simply rights of the category which are not applied to concrete people. Stirner is concerned that the state in fact acts as the bearer of the category and hence arrogates the rights it is meant to guarantee. The 'nation' in particular is

6 All quotes in this section are from Piotr Kropotkin, 'The State: Its Historic Role'.

7 For an application of Kropotkin's theory to the 'war on terror', see Andrew Robinson, 'The Oppressive Discourse of Global Exclusion', in Maurice Mullard and Bankole A. Cole (eds), *Globalisation, Citizenship and the War on Terror* (Cheltenham: Edward Elgar, 2007), 253–74.

simply the identification of people with the state.[8] Stirner does not analyse the social form of the state as Kropotkin and Bakunin do, but his account echoes their view in that he portrays 'spooks' as inherently misanthropic and as inherently expansive in their demands and claims. Spooks are based on particular attributes (such as nationality and humanity) which are wrongly taken to be ethically overwhelming in relation to other attributes. Each 'spook' is a way for the state to insist on its self-interest at the expense of others, demanding sacrifices from others for its own benefit. For Stirner, decades before Trotsky, states are inherently substitutionist, taking the place of attributes they claim to represent. By subordinating themselves to spooks, people become alienated from each other, relating only through the abstraction and hence the state. In reality, attributes such as nationality are simply instances among many individual attributes each 'possessed' by the person, which should not in turn be allowed to 'possess' her/him.

Since claims made through abstractions (national interest, common humanity and so on) are necessarily self-interested claims by the state, it follows that the state's demands that people fight at its command for such general goods are really arrogations by the state which carry no authentic ethical obligation. 'The patriots fall in bloody battle or in the fight with hunger and want; what does the nation care for that? By the manure of their corpses the nation comes to "its bloom"! The individuals have died "for the great cause of the nation," and the nation sends some words of thanks after them and—has the profit of it. I call that a paying kind of egoism.'[9] The state is thus engaged in constant self-interested violence at the expense of its subjects, of which war is one variety. On the other hand, the state practice of violence is echoed in its prohibition for others. For Stirner, 'law' and 'crime' are nothing more than the state's names for violence depending whether it is the state or someone else who commits it, and the individual violence of 'crime' is a necessary means to overcome the state's violence by denying it the exclusive privilege of the monopoly on violence.[10] The relationship of each person to the state is thus entirely antagonistic, a conflict of a rival will with the will of the state.

Continuing the Anarchist Critique: Tolstoy, Goldman, Bourne, Rocker

Leo Tolstoy's Christian anarcho-pacifism takes a different slant on the relationship between war and individual acts of resistance, but follows the same broad lines

8 Max Stirner, *The Ego and His Own*, trans. Steven T. Byington (New York: Benjamin R. Tucker, 1907 [1845]), available at the Nonserviam website: http://www.nonserviam.com/egoistarchive/stirner/bookhtml/The_Ego.html. All discussions of Stirner are based on this text.

9 Stirner, *The Ego*, 5, 9.

10 Stirner, *The Ego*, 257–8.

as the other authors discussed here. Tolstoy counterposes the 'law of violence' prevalent in state societies with a 'law of love' derived from Christianity.[11] The law of violence arises from generalized mistrust connected to exploitation, and creates a vicious circle of fear and hate, and social misery. Tolstoy takes violence as such to be reactive, corroding the spirit of those who commit or support it, with pretexts always covering its real basis in 'low passions' such as hatred. In contrast, a Christian ethic of love provides a basis for intense inner peace and happiness. Tolstoy also blames wars on patriotism (nationalism), which he treats as a type of prejudice allowing slaughter of foreigners, encouraged because of its usefulness to certain groups. The prevalence of nationalism leads to a situation where states can enslave people as soldiers and stand like predators waiting for other states to become weak. Tolstoy also alleges that states deliberately antagonize one another to create an enemy against which to pose as defenders.[12] Aside from his Christian ethic of humility and self-denial, this account is similar to the other authors discussed here: active forces are aligned with liberation, but held down beneath a dead weight of reactive forces (evil, hate, fear). Tolstoy calls for love to be treated as the ultimate moral law that tolerates no exception, and for withdrawal of cooperation in violence by means of conscientious objection.

As America prepared to enter the First World War in 1917, anarchists Emma Goldman and Alexander Berkman played a central role in initiating anti-conscription activism. While not pacifists, they held to anarchist principles that the state has no entitlement to wage war. They particularly denounced American involvement as capitalistic, promoting instead socialist internationalism. Patriotism, the ideological basis for war, is viewed by Goldman as a superstition built on conceit, arrogance and egoism of national groups. With each national group displaying the same sense of superiority, it creates a basis for self-interested power-holders to manipulate people into conflict. Instead of fighting each other, elites force their subjects to fight on their behalf. Militarism creates wasteful expenditures at the expense of the people, as well as demanding loss of life. War is deemed hypocritical when violence by the poor is banned, and is also viewed as a spectacle to dazzle the masses. Goldman also raises issues well-known to contemporary feminists, such as the link between armies and sex work and the problem of social dislocation of ex-combatants.[13] Goldman and Berkman were jailed and deported for opposing the draft. Speaking at her trial, Goldman argued that it was hypocritical for America to be claiming to promote democracy abroad while curtailing it at home, a criticism echoing today's crisis of democratic rights.

11 Leo Tolstoy, *The Law of Love and the Law of Violence*, trans. Mary Koutouzow Tolstoy (New York: Rudolph Field, 1948).

12 Leo Tolstoy (1900), 'Patriotism and Government', Anarchy Archives website, http://dwardmac.pitzer.edu/anarchist_archives/bright/Tolstoy/patriotismandgovt.html.

13 Emma Goldman (1917), 'Patriotism: A Menace to Liberty', Women's History website: http://womenshistory.about.com/library/etext/bl_eg_an5_patriotism_menace_to_liberty.htm.

Similarly responding to the irrational outpouring of nationalism in First World War America, Randolph Bourne coined the famous designation that 'war is the health of the state'. This is because it brings to the fore the ideal or logic of the state which is normally hidden or flagging in a liberal society. The state seeks loyalty or 'mystic devotion', and war serves as an opportunity to demand it. 'The State ideal is primarily a sort of blind animal push toward military unity', including 'the terrorization of opinion and the regimentation of life'.[14] In war, the State ideal, what I've above termed the 'deep state' comes into the fore at the expense of 'government' and partisan politics, substituting for the government and the nation in the popular mind. The outcome of the war is almost incidental, since its functionality derives from its emotional effects. Bourne emphasizes the reinvigorating effect of war on reactive attachments, engendering great commitment and sacrifice by mobilizing the 'herd instinct' and 'filial adoration', appealing for a desire to surrender the burden of freedom.

For Bourne, the nation or country is a peaceful, inclusive, life-affirmative concept, whereas the state concept is aggressive, life-destructive and competitive; the society–state split is again fundamental. War is an artificial phenomenon produced by states, indeed, 'the chief function of States', reliant on the military establishment. But the military elite does not generate wars itself. War is specially connected to a reactionary, nationalistic layer of the 'significant classes' who feel most intensely the State ideal. The warmongering elite desire the repressive and homogenizing effects of war as 'just, proper and beautiful'. The state 'represents all the autocratic, arbitrary, coercive, belligerent forces within a social group' against those with creative spirits. Foreign policy is the last stronghold of state power. Legislatures fail to constrain executives in foreign policy which, as a result, is conducted in 'utter privacy and irresponsibility', as the actions of states outside democratic control.

Rudolf Rocker's account, like Bourne's and Goldman's influenced mainly by the First World War, extends the arguments made by earlier authors. Arguing against reductionist forms of Marxism, Rocker argues that motives of political power must be included along with economic motives in explaining war. The foundations of the evil of war reside not in individuals 'but in power politics itself, irrespective of who practices it or what immediate aims it pursues'.[15] This is because power politics is based on an imposed instrumentalism as an outlook. Rocker bemoans that people allowed themselves to be led herd-like to the slaughter, and seeks to explain it in nationalism, an ideology or mythology based on a cult of the state, which raises the good of the state above ethical concerns. Rocker portrays nationalism as a distortion of legitimate particularities into loci of irrational chauvinism, connected to the 'will to power' conceived as a negative, destructive force. He echoes Kropotkin's argument that centralization has destroyed creativity and thereby culture.

14 All quotes are from Randolph Bourne (1918), 'War is the Health of the State', Big Eye website: http://www.bigeye.com/warstate.htm.

15 Rudolf Rocker (1937), *Nationalism and Culture,* Chapter 1, Anarchy Archives website: http://flag.blackened.net/rocker/insuf.htm.

Libertarian Marxist Critiques of War: Reich, Benjamin, Marcuse

With fascism overrunning Europe and Stalinism corroding the plausibility of orthodox Marxism, European Marxists moved in increasingly libertarian directions which intersected with anarchist concerns. Wilhelm Reich's combination of psychoanalysis, sexual liberationism and Marxism is especially conducive to anarchism. Reich argued that fascism and other authoritarian ideologies are enabled by repressive character-structures (psychological and bodily dispositions) conditioned by social control, authoritarian parenting and sexual repression. He put forward a psychoanalytical interpretation of the origins of war, which he blamed on negative emotions connected to the distortion of biopsychological energies by repression. Warmongering leaders such as Hitler could only take power because of their resonances with widespread biopsychological dispositions.

In this context, the substitution of representations for the self is crucial, with Nazi supporters characterized by 'a complete identification with the state power'.[16] Nationalism serves as a cover for identification with state power and with the leader, providing ersatz satisfactions for everyday subordination through vicarious reactive pleasures. This is in turn enabled by a pattern of identifying with the oppressor which is carried over from authoritarian families: 'In the figure of the father the authoritarian state has its representative in every family, so that the family becomes its most important instrument of power.'[17] The authoritarian family serves to bind desire into a limiting frame, blocking wider horizontal connections. This restriction produces a model for the channelling of attachments into nationalism. Repressed sexual energies are channelled through sexualized militarism such as parades and uniforms. Fascist desire is further strengthened by the desire of the oppressed-but-included to differentiate themselves socio-culturally from the mass of people worse-off than themselves, and is strongest in people who are socially isolated within family units. The implication of Reich's theory is that a healthy, non-repressive sexual and bodily disposition would produce a permanently peaceful society.

Reich's approach is taken further by Klaus Theweleit. Theweleit's account, based on the diaries kept by members of the proto-fascist *Freikorps*, analyses fascist desire in broadly Reichean terms, showing how the writers were terrified of being engulfed by amorphous, excessive forces associated with flows, floods, bodily decomposition, the nameless 'mass', and femininity. Faced with such threats, they sought existential security in categories of racial and class purity and in military masculinity.[18] This stance had to be constantly reinforced, however, through

16 Wilhelm Reich, *The Mass Psychology of Fascism* (New York: Farrar, Straus and Giroux, 1970 [1933]), 46.

17 Reich, *Mass Psychology*, 53.

18 Klaus Theweleit, *Male Fantasies* (2 volumes) (Minneapolis: University of Minnesota Press, 1987 [1977]).

violence against others deemed to represent the feared engulfing forces, performed as a reassuring and expressive form of agency. Hence, in this analysis, the structure of fascist desire directly produces war.

Around the same time, Walter Benjamin was formulating a critique of state violence in terms of a typology of three kinds of violence. Benjamin's critique of violence extends the critique of the state in relation to war, suggesting 'violence' (taken to mean threat-advantage) to be central to the state's existence and practice.[19] Law always has violence beneath it, as threat and origin. The basic form of violence, for 'natural' or empirical ends, is termed 'law-making' by Benjamin. Such violence produces outcomes by altering the relation of groups, thereby producing a new law. This is the kind of violence seen in states' goals towards other states in war. States also engage in 'law-preserving' violence to prevent changes in the status quo; this is violence for 'legal' rather than 'natural' ends. Both are ultimately founded in mythical violence which exercises power over life for its own sake, and are hence founded on reactive desire. They are also caught in a vicious circle, since law-preserving violence relies on but disempowers law-making violence. Modern states fear diffuse violence even for 'natural' (law-making) ends, and seek increasingly to strip individuals and groups of any entitlement to use violence, instead substituting 'legal' means connected to law-preserving violence. This is because the state feels threatened at the risk of losing its monopoly on violence. Hence, even criminal violence can seem subversive in the way it corrodes state control and bears the potential to found a new law. The state increasingly lacks confidence in its own violence, and hence becomes increasingly authoritarian, seeking to prohibit such counter-powers from emerging.

A third type of violence, such as Sorel's proletarian general strike, 'sets itself the sole task of destroying state power'.[20] It does not aim for goals, and hence is not a means, but, rather, actually consummates the conditions it aims to bring about. It is thus similar to outbursts of anger which are expressive rather than instrumental. This is forceful action directed against the capability to use violence to make or preserve laws, exercising pure power over all life for the sake of the living. It is sometimes termed purely nonviolent by Benjamin, primarily because it does not reduce others to means as it is not itself a means, and at other times as pure violence, separated from law, or divine violence, beyond the mythology of law-making violence.

Authors in the Frankfurt School tradition connect the increasingly destructive nature of modern war and genocide to the actualization of industrial instrumentalism. This tradition, drawing heavily on the horrors of the Holocaust and Hiroshima, threw into question the benevolence of modernity, suggesting its basis in domination. This included discussions of war which related it to the dynamics of instrumental and 'Establishment' forms of reason similar to the anarchist theory of the state. For instance, Adorno suggests a close link between inter-human

19 All discussions of Benjamin relate to Walter Benjamin, 'Critique of Violence', in *Selected Writings* (Cambridge, MA: Harvard University Press, 1999 [1921]), volume 1, 277–300.

20 Benjamin, 'Critique of Violence', 291.

aggression and the desire to dominate nature, with enemies metamorphosed as animals.[21] Erich Fromm argues that humanity's survival is jeopardized by a peculiarly human kind of malignant aggression which grows cumulatively along with 'civilization'. It emerges as a way of expressing otherwise benign urges to agency which are denied their natural expression due to the alienation of work.[22]

The Frankfurt School theory of war was taken furthest by Herbert Marcuse, reacting to the atrocities of the Vietnam War and the accompanying closure of American public discourse. For Marcuse, the closure of discourse in affluent or overdeveloped societies creates an almost totalitarian effect of self-reproduction. In this context, the aggressive instincts take on an increasingly central role in society, mainly through a frustration-aggression complex.[23] Frustration arising from alienated work and massification is channelled and made socially useful by being connected to military and nationalistic aggression. For Marcuse, wars serve to ward off, not an external threat, but the threat posed to the system by its own irrationality. Loaded and normalizing language similar to Orwellian doublethink is used to integrate war into social normality: 'in the mouth of the enemy, peace means war, and defense is attack, while on the righteous side, escalation is restraint, and saturation bombing prepares for peace'.[24] The established vocabulary not only represents but creates the Enemy 'as he must be in order to perform his function for the Establishment'. The resultant rhetoric reserves the 'bad' words such as 'violence' for the Enemy, using a different vocabulary for pro-systemic actions, and treating anything the enemy does as a crime and anything the approved agents of violence do as above reproach.[25]

War does not satisfy, however, because the replacement of human with technological means of aggression ensures that the pent-up aggressivity is not dissipated in action. Hence, 'the more powerful and "technological" aggression becomes, the less it is apt to satisfy and pacify the primary impulse [to aggression], and the more it tends toward repetition and escalation'.[26] This switch to technological means is a qualitative break which changes the nature of war. War is thus itself frustrating, and therefore tends to escalation. This tendency is reinforced by the corrosion of guilt and the reinforcement of Puritanism arising from the 'clean' distance involved in technological war. This repressive channelling is enforced by means of the prohibition of more effective outlets. Hence, technological violence is deemed acceptable, but affectively engaged violence is increasingly criminalized.

21 See Karsten Fischer, 'In the Beginning was the Murder: Destruction of Nature and Interhuman Violence in Adorno's Critique of Culture', *Journal for Cultural and Religious Theory* 6: 2 (2005), available at: http://www.jcrt.org/archives/06.2/fischer.pdf.

22 Erich Fromm, *The Anatomy of Human Destructiveness* (New York: Holt, Rinehart and Winston, 1973).

23 The discussion in this paragraph is taken from Herbert Marcuse, 'Aggressiveness in Advanced Industrial Society', in *Negations* (London: Free Association Books, 1988 [1968]), 248–68.

24 Marcuse, 'Aggressiveness', 261.

25 Herbert Marcuse, *Essay on Liberation* (Harmondsworth: Penguin, 1969), 75, 78.

26 Marcuse, 'Aggressiveness', 264.

Autonomy and War: Negri, Post-*Autonomia*, Clastres

Also corresponding with the Cold War era of nuclear deterrence and the 1960s/70s protest wave, *autonomia* emerged in Italy as a radical challenge to capitalism and the state. The theory of the state as force of decomposition by means of social war was extended by Antonio Negri in his *autonomia*-era works. In this kind of autonomist theory, the state is not accorded a logic of its own, but is attached to a presumed capitalist or systemic logic echoing earlier anarchist claims about the nature of state violence. The state is associated with the function of the capitalist system to decompose the agency of proletarian or autonomous forces. While such decomposition is central to capitalism in all its phases, it has become particularly crucial in the current period. With no possibility of values being derived from external criteria due to the density of capitalist penetration of social space, the system becomes tautological, and premised on an empty gesture of sovereignty of the system as a whole. This process renders the system increasingly violent, as its ability to legitimate itself is reduced.[27]

As a result, in the neoliberal phase, the state becomes more central in society via the mechanism of 'command', corresponding to the transmutation of capitalism into an increasingly irrational, violent and dominatory form.[28] Command is connected with Schmittian sovereignty and hence is explicitly violent; the implicit violence of economic coercion is increasingly replaced by the explicit violence of state dominance as a means to compel compliance. 'Command becomes ever more fascistic in form, ever more anchored in the simple reproduction of itself, ever more emptied of any rationale other than the reproduction of its own effectiveness'.[29] The state's role is expanded, with obligations diffused as coercion intensifies. Negri terms this the real subsumption of society under the state. Furthermore, a new state-form, the 'crisis-state' based on a permanent state of exception, has emerged in which state power is conditioned on the diffuse ability to produce crises and the ever-present possibility of state terror as a response to the sudden breakdown of command. The crisis-state is connected to the centrality of nuclear power, nuclear weapons and vulnerable energy systems, in which the state is connected to self-reproducing crises and to extreme risks (such as nuclear terror) which it simultaneously engenders and wards off. For Negri, this nuclear-state is the actualization of command at its most terroristic and wasteful.[30]

As a result of the loss of mediation and the rise of command, social antagonism becomes permanent, irreducible and radical. Without a basis in value, state/capitalist command can only be imposed by means of repression and state terror, throwing down a dragnet against the various forms of everyday resistance and

27 Antonio Negri, 'The Constitution of Time', in *Time for Revolution* (London: Continuum, 2003 [1981]), 48–9.

28 Antonio Negri, *Books for Burning* (London: Verso, 2005 [1971–7]), 32–3, 78–80.

29 Antonio Negri (1980), 'Crisis of the Crisis-State', Libcom website: http://libcom.org/library/crisis-state-antonio-negri.

30 Negri, *Time for Revolution*, 70; *Books for Burning*, 263–5.

sabotage. Hence, the divisions between war and policing become blurred. An internal warfare state is constituted as a means of regulating diffuse labour and resistance.[31] Movements of autonomy, or 'proletarian' movements, emerge from self-valorization and 'constitutive power.' These terms refer to the reclamation of subjectivity to create the world. Between capitalism and the state on the one hand, and movements of autonomy on the other, there is an irreducible, insoluble conflict. The relationship between autonomous movements and the state becomes a relation of war, not because of the movements' objections but as a direct result of the requirements of exploitation.[32] Forcible defence of spaces of autonomy emerges as a necessary response to repression.[33] They ultimately form an armed society counterposed to the state, organizing a counter-violence to prevent the suppression of spaces of autonomy by the state.

The thesis of radical antagonism fades in Negri's more recent works, including his collaborations with Michael Hardt. Yet the thesis of *Empire*, that war is obsolete in a globalized world, required rethinking as the 'war on terror' emerged. Hardt and Negri analyze the 'war on terror' as a global civil war arising from a generalized state of exception, connected to the latest phase of the global economy. They argue that a 'war against terror', because unwinnable and unlimited in spatial and temporal terms, is a 'continuous and uninterrupted exercise of power' indistinguishable from policing.[34] War is now fought biopolitically, with wars waged against social phenomena (drugs, terrorism, etc.) and claims of justice and democracy abandoned. On the other hand, state violence now seems less legitimate than before, due to discourses such as human rights. The appearance of legitimacy, now dependent largely on systemic effects, is extremely unstable. Increasingly networked state warriors collide with asymmetrical adversaries in a field of netwar.

Other post-*autonomia* theorists have offered more clearly autonomist accounts of contemporary wars. For instance, the Midnight Notes collective analyzes the Iraq war as connected to a discourse which seeks to maintain dependence by denying technological equality under the pretext of preventing WMD proliferation.[35] George Caffentzis has published a range of pieces analyzing post-Cold War conflicts in terms of resource wars.[36] Negri's discussion of states of exception echoes strongly with the work of Giorgio Agamben, who over a series of works advances a theory of contemporary state power as the terminal stage of a long degeneration towards permanent exception. While Agamben does not discuss war directly, the logic of exception clearly applies to the figure of the enemy in situations where adversaries

31 Antonio Negri (1982), 'Archaeology and Project: The Mass Worker and the Social Worker', Libcom website: http://libcom.org/library/archaeology-project-negri.

32 Negri, *Time for Revolution*, 99, 124.

33 Negri, *Books for Burning*, 200.

34 Michael Hardt and Antonio Negri, *Multitude* (Harmondsworth: Penguin, 2004), 1.

35 Midnight Notes Collective (2004), *Respect Your Enemies: The First Rule of Peace*, Midnight Notes website: http://www.midnightnotes.org/pamphlet_sept11enemies.html.

36 George Caffentzis (2005), *No Blood for Oil! Energy, Class Struggle and War, 1998–2004*, self-published: http://www.radicalpolytics.org/caffentzis/no_blood_for_oil-entire_book.pdf.

are dehumanized, and Agamben's work refers explicitly to situations such as Guantanamo Bay.

A different angle on autonomy can be advanced by discussing states and war in relation to indigenous societies. From its earliest days, political anthropology has had affinities with anarchism, because the fact of stateless or acephalous societies is a perennial challenge to the Hobbesian account. One can trace anarchist anthropology from the early reflections of Kropotkin, through scholars such as Marshall Sahlins and Richard Lee, to contemporary authors such as Howard Barclay and David Graeber. Many anarchist anthropologists observe how some (but by no means all) indigenous societies are extremely peaceful and less prone to war-making than state societies (the Bushmen and Mbuti are most often cited). Drawing on such cases, and on an alleged lack of archaeological evidence for Palaeolithic warfare and the ritualized and non-lethal nature of much indigenous warfare, John Zerzan argues that war is an effect of domestication and hence of hierarchical, statist societies.[37] Others observe how the low-level fights occurring in stateless settings have far less serious effects than wars between states. For instance, Richard Lee has observed that states may reduce some kinds of minor violence such as fist-fights, but then cause other forms of violence such as war.[38]

In this rich field of scholarship, Pierre Clastres stands out for having theorized the mechanisms of warfare in indigenous societies in a way which contributes to anarchist theory. Clastres links the function of war in indigenous societies to the prevention of coalescences of power which could create domination. Indigenous societies are constructed around an ideal of economic and political independence in which alienation is actively refused. War is thus an obstacle placed in the way of state-formation. As an autonomous group, each group is in a permanent situation of potential war with others, since it depends on the ability to resist any particular attempt at domination by others. 'What is the function of primitive war?' asks Clastres, replying, '[t]o assure the permanence of the dispersion, the parceling, the atomization of the groups. Primitive war is the work of a *centrifugal logic*, a logic of separation'.[39] War prevents unification, and the indigenous war-machine operates to maintain dispersion. On the other hand, generalized war is precluded since it would eliminate the basic equality of societies and people. Indigenous war is also fought without introducing power-divisions and military command.

Other factors are needed to explain the extraordinary brutality and genocidal impulses of modern war. Clastres looks for such explanations in the state. While indigenous and metropolitan cultures can both be ethnocentric, only metropolitan cultures are ethnocidal. Clastres argues that ethnocide arises from the characteristic of being a society with a state. Ethnocide is 'inscribed in the nature and functioning

37 John Zerzan (2006), 'On the Origins of War', Green Anarchy website: http://www.greenanarchy.org/index.php?action=viewwritingdetail&returnto=viewjournal&printIssueId=18&writingId=543.

38 Richard B. Lee, *The !Kung San: Men, Women and Work in a Foraging Society* (Cambridge: Cambridge University Press, 1979), 398–9.

39 Pierre Clastres, *Archaeology of Violence* (New York: Semiotext(e), 1994 [1980]), 164.

of the state machine'.[40] Ethnocide and genocide arise from the desire to destroy difference, based on a hierarchy of superior and inferior cultures. While all states express this desire, modern states are particularly dangerous because of their intersection with the borderlessness of capitalism.

Contemporary Critiques: Chomsky, Bonanno, *La Ruta Pacifica*

Perhaps the best-known anarchist writer on war is also the least likely to be known as an anarchist: Noam Chomsky. Chomsky has written on most of the major wars of his time, from the 1954 American invasion of Guatemala, through Vietnam, Nicaragua and ex-Yugoslavia, to Iraq and Afghanistan. In addition, he exposes violent suppression due to American complicity in brutal regimes and involvement in proxy wars. Chomsky is an anarchist, although his primarily anti-imperialist critiques of war are motivated by humanism and have a wide resonance with critics of US foreign policy regardless of their stance on the state. His approach is empiricist, focusing on the contrast between the facts of American foreign policy and the false claims perpetuated by the media, political leaders and mainstream academics.

In explaining wars, Chomsky focuses on economic causes. He views the economic self-interest of elites as the primary reason for American warmongering. Sometimes this is a matter of directly defending investments or seizing resources, and Chomsky often draws attention to the most immediate of economic motives such as control of oil, counterposing these to high-minded rhetoric. He often portrays the military–industrial complex as akin to a racket, with public money diverted into subsidies and state spending. He also has a theory to explain cases (such as Laos, Grenada and Nicaragua) where such direct interests are limited. America wishes to ensure the continued subordination of other countries into a global system operating for the benefit of elites. Wars are thus motivated by the 'threat of a good example', the fear that a country, however small, will succeed in demonstrating an alternative development trajectory which could set in motion a domino effect.[41] War serves to prevent such 'good examples' by economically devastating societies which would otherwise serve as examples, and through the spread of terror and torture to such a degree as to demoralize resistance. Such wars are necessarily extremely brutal. According to Chomsky, the only way to defeat

40 Ibid., 49.

41 Noam Chomsky, *What Uncle Sam Really Wants* (Tucson, AZ: Odonian Press, 1993), chapter available at: http://www.thirdworldtraveler.com/Chomsky/ChomOdon_Example.html.

a people's war with broad popular support is to destroy the people by means of genocide and forced displacement.[42]

Although the underlying causes of war are economic, they require a system of ideological justification Chomsky terms the 'propaganda model'. The media functions to promote war through methods of soft thought control such as anathematizing enemies. Signifiers such as communism and terrorism are mobilized to obtain popular acquiescence through systematic misrepresentation and strategic omission, playing on desires to avoid painful knowledge about complicity in atrocities. The war on terror, for instance, targets the 'retail' or small-scale terrorism of small groups but ignores the 'wholesale terrorism' of states.[43] Such ideologies can be self-perpetuating, beyond the interests they mostly serve. Ideologies can thus escape their economic motives, forcing continued war even when it ceases to serve economic interests and becomes disproportionate, with officials caught in their own fantasies, illusions and justifications. This is particularly clear in the mentality of the 'backroom boys', the apparatchiks of warmongering whose day-to-day operational calculations create sustained pressure for war beyond its economic causes.[44] Such planners operate with an entrenched ideological frame which often takes false claims as axiomatic and defines ethical critique as irrational. Hence, Chomsky's overall account includes at least three levels of causality: direct economic interests, maintenance of systemic dominance, and ideological overreach.

Among contemporary anarchists, the theme of war reappears as the idea of social war in the insurrectionist tendency. Leading insurrectionist theorist Alfredo Bonanno uses the concept of social war to refer to an irreducible antagonism in which affirmation goes hand-in-hand with assault on structures of power. His autonomist-inflected anarchism is based on the rejection of alienation, and in particular of the reduction of life to production. Resistance must thus be something playful, restoring excitement to life while freeing oneself from the death-machine of the dominant system. 'Unfortunately civil war is an obligatory road which must be passed in any historical moment of profound, radical transformation'.[45] It is the point of explosion of accumulated discontent. The nature of social war should not be blurred by the lack of lethality. The reactive forces do not use massacres frequently today, to avoid disrupting the Spectacle. Instead they rely on subtler but more effective forms of therapeutic control. Bonanno calls for a stance of permanent conflictuality towards the dominant system.

42 Noam Chomsky, *At War with Asia: Essays on Indochina* (Edinburgh: AK Press, 2004 [1970]), 196.

43 Noam Chomsky (1991), 'International Terrorism: Image and Reality', Chomsky Info website: http://www.chomsky.info/articles/199112--02.htm.

44 Noam Chomsky, 'The Backroom Boys', in *For Reasons of State* (New York: New Press, 2003 [1973]), 3–171.

45 Alfredo Bonanno, 'For an Anti-Authoritarian Insurrectionist International', The Anarchist Library website: http://theanarchistlibrary.org/HTML/Alfredo_M._Bonanno__For_An_Anti-authoritarian_Insurrectionalist_International.html.

Insurrection emerges as a response to capitalist–statist dominance. Power realizes itself in control over the physical space of social reality, and hence can be attacked in its presence in physical space. One should also aim to create networks 'capable of holding together and creating consequences in the struggle against power'.[46] One must be careful with armed struggle, because it often imposes the 'dominion of death'.[47] It is important, in other words, that the desire-structure of militarism not be reproduced. One part of avoiding this is to treat means as simply tools, avoiding enslavement to them. Another is to avoid roles and fixed identities. It is also necessary to refuse reactive forces such as the desire for revenge.

While this accounts for social struggles and systemic violence, it leaves the question of ethno-religious wars in the current period. Bonanno analyzes nationalism and fundamentalism as reactive formations arising when categories of oppressor and oppressed are mistakenly mapped onto ingroup–outgroup dichotomies. They are a way of defusing the potential for revolt through the freezing of life into fixed categories. They do this by drawing on widespread reactive emotions. Wars such as those in the former Yugoslavia are channelled by elites to perpetuate domination, while others, as in the Muslim world, become complicated as reactive categories are mapped onto imperialist relations. Wars channelled along nationalist lines are taken to be a way of defusing the 'powder-keg' of social unrest to reduce the risk of rebellion.[48]

There are also many autonomous social movement networks focused on nonviolent responses to war, based on theorizing war as social decomposition and refusing it through re-composition. Groups such as the Colombian network *La Ruta Pacifica de las Mujeres* seek forms of effective social action working through direct composition which disempower violent agents in general. *La Ruta* is a network of feminist peace groups across Colombia, which seeks to counterpose positive energies of hope and social weaving to the pervasiveness of war and violence. Weaving is a powerful metaphor, signifying the re-composition of social relations and forms of connectedness disrupted by the effects of war, and the creative cycle of life and death. Many of their activities focus on emotional repair and morale-boosting, such as collective mourning and working through fear. The group uses an elaborate symbolism comprising rituals, colours and performances so as to create combinations of energies powerful enough to interrupt militarism. Participants often report the recovery of inner strength as an effect of such activities. In addition, they carry out large-scale reoccupations of public space to overcome fear and challenge military territorialities.[49] The implicit assumption is that social power

46 Alfredo Bonanno, 'The Anarchist Tension', The Anarchist Library website: http://theanarchistlibrary.org/HTML/Alfredo_M._Bonanno__The_Anarchist_Tension.html.

47 Alfredo Bonanno, 'Armed Joy', The Anarchist Library website: http://theanarchistlibrary.org/HTML/Alfredo_M._Bonanno__Armed_Joy.html.

48 Alfredo Bonanno, 'The Insurrectional Project', The Anarchist Library website: http://theanarchistlibrary.org/HTML/Alfredo_M._Bonanno__The_Insurrectional_Project.html.

49 Martha Colorado, 'In Colombia, in the midst of war, women weave love knots',

does not flow from violence; rather, violence is socially decompository, breaking down relations. Hence, power can be exercised by re-composing relations.

Poststructuralist Autonomism? Deleuze/Guattari, Virilio, Baudrillard

Views of a distinct logic of state power as crucial in the causality of contemporary wars also arise in the works of a number of poststructuralist authors, including Deleuze and Guattari, Virilio and Baudrillard. Deleuze and Guattari modify Clastres' theory of the war-machine to theorize how states continue to produce war in spite of their antagonism with war-machines. In their theory, the state is identified with regimes of overcoding and despotic signification, and hence with transcendentalism, reactive desire and 'antiproduction' (decomposition). In particular, it reduces the density of horizontal connections and replaces them with vertical connections. This places the state in fundamental opposition to the Clastrean war-machine as a machine of metamorphosis and deterritorialisation. The state does not transcend war, though it rejects the transformative functions of war-machines. It incarnates a kind of perpetual war through which society is dominated. Hence, state violence is 'a violence that posits itself as pre-accomplished, even though it is reactivated every day'.[50] It is constantly violent because of its preference for total destruction over the risk of anything slipping outside its total integration.[51] While the state has its origins in classical despotism, it persists in modernity in alliance with capitalism, which reawakens the state to ward off the flows it has released.

This situation produces a new phenomenon: the state war-machine. One way the state handles potentially disruptive flows is by bringing war-machines inside itself.[52] Unlike the Clastrean war-machine, the statist war-machine adopts war as its object. It reissues from states as a global war-machine which turns states into its parts, constructing an unspecified enemy against which to fight, and spreading reactive structures throughout societies. This process has reached its epitome

Educational Insights 8: 1 (2003), http://www.ccfi.educ.ubc.ca/publication/insights/v08n01/praxis/colorado/english.html_; Martha Colorado, 'Colombian Women: Survival Amidst War', *Women's Health Collection* 7: 99 (2003), http://www.peacewomen.org/resources/Colombia/WomensSurvival2003.html; Cynthia Cockburn, 'Violence Came Here Yesterday: The Women's Movement Against War in Colombia', research paper (2005), http://www.cynthiacockburn.org/ColombiaEngblog.pdf.

50 Gilles Deleuze and Félix Guattari, *A Thousand Plateaus*, trans. Brian Massumi (New York: Continuum, 1987), 447. Previous discussions refer to this work, pp. 428 and 433, and to Gilles Deleuze and Félix Guattari, *Anti-Oedipus*, trans. Helen R. Lane and Robert Hurley (London: Athlone, 1983), 34.

51 Deleuze and Guattari, *Anti-Oedipus*, 213.

52 Deleuze and Guattari, *A Thousand Plateaus*, 450.

today, with the statist war-machine seeking a total 'peace of terror and survival' based on generalized deterrence, and constituting the world as a single, globalized space. 'Total war is surpassed, toward a form of peace more terrifying still. The war machine has taken control of the aim, worldwide order, and the States are no more than objects or means adapted to that machine'.[53] This produces the potential for its own negation: as the state clenches its fist, grabbing more resources, minoritarian forces re-form to resist it. The state produces war in its inability to recognize difference as well as its reproduction of anti-production as daily violence.

Virilio similarly theorizes the state or military logic as central to current global changes, including the spread of globalized 'permanent war'. Virilio views the military class as an entity expressing a social logic or essence above and beyond its instrumental goals. This logic consists in eliminating chance and contingency, and hence also freedom and dialogue, from social relations, to force others to be passive, to interrupt their movement.[54] Hence, societies are recomposed as 'logistical' entities, with value determined by instrumentality.[55] The method of the military class is not simply to defeat the enemy but to control and rearrange space in order to disempower enemies in advance, or to sap their energy in advance of any conflict. It thus involves the creation of ecologically hostile spaces in which it is difficult for non-military actors to operate, 'an earth made *uninhabitable* by the military predator', in place of the dense ecosystems propitious to popular defence.[56]

In Virilio's theory, military projects initially directed outwards escape the control of civilian powers by reacting back on the arrangement of power and space. Colonial projects react back upon the colonizer as 'endocolonialism', in which the military logic colonizes the colonizer; this process reaches its zenith in global nuclear deterrence covering the entire world, and is also expressed in totalitarian 'national security' regimes and forms of technological control which Virilio terms a 'deliberately terroristic manipulation of the need for security'.[57] Virilio does not see states as necessary to defend civilians, who could historically defend themselves through popular defence. Today, popular defence (or assault) is recomposed as insurrection, as in Vietnam and Palestine, and has as its goal resistance to the reduction of social situations to military control. Hence, 'the Palestinian tragedy ... is the way of the future'.[58] Deceleration, countering the speed of modern war, is also advocated as a form of defence.[59]

---

53 Deleuze and Guattari, *A Thousand Plateaus*, 421.

54 Paul Virilio, *Popular Defense and Ecological Struggles* (New York: Semiotext(e), 1990), 13–14, 18–19.

55 For a detailed application of the idea of 'logistical society' to the 'war on terror', see Julian Reid, *The Biopolitics of the War on Terror* (Manchester: Manchester University Press, 2006). Similar issues are raised in Michel Foucault, *Society Must Be Defended* (Harmondsworth: Penguin, 2003).

56 Virilio, *Popular Defense*, 50.

57 Paul Virilio, *Speed and Politics* (New York: Semiotext(e), 1977), 122.

58 Virilio, *Popular Defense*, 57.

59 Virilio, *Speed and Politics*, 137.

Baudrillard advances a similar argument, claiming that states cause not 'war' (which implies an adversarial and symbolic element), but 'non-war', a phenomenon similar to war in its destructiveness but not its discursive construction. Non-war is aspirationally a policing activity based on a 'spiral of unconditional repression' which depoliticizes conflicts and which has as its goal the liquidation or domestication of challenges to the dominant system. It forecloses the enemy, reconceived as a kind of informational noise, in a 'suffocating and mechanic performance' of impersonal disempowerment.[60] The strategy of Northern non-war is to deter, not simply defeat, adversaries, and its ultimate goal is to preserve an emotionally empty, repressive consensus by means of generalized deterrence of active forces, a process in which the flow of media images is crucial. This consensus is a kind of foreclosure of history in which everything is already decided and overwhelming force is the guarantor of meaning. The situation does not become a war because the adversaries remain separated by radical discursive difference. Unable to recognize the Other, 'The Americans can only imagine and combat an enemy in their own image'.[61] Non-wars often fail because there is not enough communication for the enemies to interact on the same field. Enemies tend to resist being 'deterred' and slip outside control because they are able to continue war by other, asymmetrical means. In particular, terrorism epitomizes an unrepresentable act which annuls the regime of control in a moment of excess, counterposing absolute uncertainty to the certainty of deterrence.[62]

Conclusion: The Anarchist/Autonomist Critique of War

We have thus surveyed a sizeable literature providing a number of different angles of critique of the role of the state in war. Despite their differences, certain leitmotifs run through these accounts. Reconstructing an anarchist-autonomist critique of the state from these various texts, one can repeat a number of claims which arise in various forms throughout the literature discussed. A series of claims can be advanced.

States express a social logic which drives towards domination, control and authoritarian social relations, and which thus produces violence as a necessary effect (Virilio's logistical or military logic, Kropotkin's political principle, Bourne, Benjamin, Reich, Negri, Clastres, Deleuze/Guattari).

States are formed in order to conquer others, and/or are formed through conquest (Bakunin, Benjamin, Clastres).

60 Jean Baudrillard, *The Gulf War Did Not Take Place* (Sydney: Power Publications, 1995), 53–4, 64, 85–6.

61 Baudrillard, *Gulf War*, 36.

62 Jean Baudrillard, *In the Shadow of the Silent Majorities* (New York: Semiotext(e), 1983), 51–3.

States find themselves in permanent conflict with societies because of their status as self-contained or separate social entities seeking dominance over society (Kropotkin, Clastres, Deleuze/Guattari, Bakunin, Bourne, Chomsky's 'backroom' boys).

States contain certain logics or capabilities which, while often adopted as means, tend to take over their holders as ends in themselves. States are taken to have a logic of competition, expansion and power-politics which, as in Realism, predisposes them to war (Bakunin, Rocker, Virilio).

States do not fight in the interests of the nation or society, but in the state elite's own interests or those of its elite supporters and allies (Chomsky, Bourne, Stirner).

States, bound up with dominant elites, rely on 'plunder', extraction and dispossession for the continuation of their power, leading to repeated wars in order to grab resources or to maintain the systemic subjection and dependency of other societies (Chomsky, Caffentzis, Goldman, Kropotkin, Baudrillard, Deleuze/ Guattari).

Wars are a form of crisis management through which the state manages the instability of the dominant system (Negri, Baudrillard, Bonanno, Marcuse, Deleuze/ Guattari).

States also need enemies so that their subjects continue to desire state protection and tolerate state machinations (Tolstoy, Bourne, Chomsky).

States are based on, or else reproduce, negative emotional dispositions dominated by fear, hatred, vengeance, aggression and 'herd instincts' (Reich, Marcuse, Deleuze/Guattari, *La Ruta*, Benjamin, Tolstoy, Bourne, Bonanno).

Strong states tend to induce identity-formations based on nationalism, which reproduce warlike attitudes (Bakunin, Tolstoy, Chomsky, Bonanno).

Nationalism is an essentialist (mis)representation which takes a certain partial attribute to be of overwhelming importance, when in fact it is one attribute among many (Stirner, Rocker, Goldman).

The state is an 'antiproduction machine', seeking to disempower and sap energies from other social forces, and to prevent social composition (Deleuze/ Guattari, Virilio, Baudrillard's deterrence, Tolstoy, Bonanno, Reich, *La Ruta*).

States tend to produce stultifying, disempowered societies lacking in vital energies (Bakunin, Kropotkin, Reich, Baudrillard).

States are not necessary for defence, which can instead occur as popular defence; rather, state militaries exist for purposes of attack and destruction (Virilio, Clastres, Kropotkin, Deleuze/Guattari).

States are hypocritical in condemning others' violence as criminal while celebrating their own (Stirner, Benjamin).

A counter-power exercised by social forces outside the state offers a potential external limit to and line of flight beyond state power. This counter-power, while radically different from states, exists in something akin to a situation of war with the state (Bonanno, Kropotkin, Negri, Deleuze/Guattari, Bakunin, Benjamin, Clastres).

Counter-power can also be exercised as a withdrawal of energy from the state and its reactive underpinnings, in favour of alternative, active attachments (Tolstoy, Stirner, Negri, *La Ruta*).

What about wars between stateless societies, such as the periodic intergroup conflicts affecting the Papuans, Amazonian peoples such as the Guarani, and the clans of Somalia? In general, anarchists are quick to relativize and downplay non-state warfare. It is claimed to be far less devastating than war among states, to orient to relative position rather than extermination of adversaries, to be balanced by conflict-resolution mechanisms, to be highly ritualized in ways which minimize its frequency or lethality, and to have goals and structures which render it fundamentally distinct from wars among states. Clastres and Kropotkin discuss conflict between stateless societies, while Bakunin, Deleuze, Negri, Bonanno and Virilio discuss the capability of non-state social forces for defence or conflict. One possible argument is that stateless wars arise from herd-moralities similar to those of state societies, as a kind of latent statism within stateless societies; this is the implication of Bourne's and Tolstoy's approaches. Another is that wars among stateless societies occur for entirely different reasons to those found among states, in particular as forms of lively self-expression or unrepressed emotions: people who aren't enslaved will sometimes come to blows. Non-state war is either a valuable force or an unavoidable evil, but is turned into something far more malignant through the rise of the state. This is Kropotkin's view, and is echoed by Benjamin and Fromm. A third possibility is that stateless war as a means to prevent, or 'ward off', concentrations of power which could coalesce in states or other apparatuses of concentrated power. This is the position taken by Clastres, and echoed by Deleuze.

Notwithstanding its extensive appeal through time, there are certain respects in which this approach is particularly apposite today. One of these is that the situation of permanent emergency entailed by the 'war on terror' is blurring boundaries in ways which break down categories of International Relations theory. In an age of transnational armed opposition networks for example, it becomes increasingly indefensible to assume that states are the sole agents of war, or that internally composed societies are united in social peace behind states oriented for war only externally. With wars fought through internal policing measures and social repression, it becomes necessary to theorize war between states and networks. Most wars today are intrastate and asymmetrical, operating in a field defined clearly in anarchist-autonomist theories but traditionally neglected in statist thought.

In addition, neoliberalism has broken down many of the established forms of social insertion of states, and social movements are globally tending to adopt autonomous, networked and horizontal forms of agency which place them broadly within an anarchist-autonomist frame. What is more, states increasingly act as transmission belts for neoliberalism. The liberal view of the state as representative of society is increasingly untenable as states act on behalf of international forces, and social movements are pitted against states.[63] Anarchist and autonomist approaches also find echoes in recent work on the political economy of war, in peacebuilding

63 See for instance M. Anne Pitcher, *Transforming Mozambique: The Politics of Privatization* (Cambridge: Cambridge University Press, 2002); John Crabtree, *Patterns of Protest: Politics and Social Movements in Bolivia* (London: Latin America Bureau, 2005); Marina Sitrin, *Horizontalism: Voices of Popular Power in Argentina* (Edinburgh: AK Press, 2006).

approaches emphasizing transformation in everyday life, and in feminist critiques of the social construction of military masculinities.

Nevertheless, the approach needs to address certain issues in order to understand the entire range of phenomena it faces today. The challenge posed by comparativists such as Joel Migdal is that actually-existing states do not always conform to the modernist model implied in anarchist critiques. The challenge for anarchist and autonomist theory is thus how to maintain the thesis of society against the state, while recognizing the fact that states very often operate within society. This apparently insuperable contradiction can be overcome by distinguishing between the state as social logic – what authors discussed above have termed the 'political principle', the 'State ideal', or 'command' – and particular states as assemblages, which combine the state social logic with other logics to produce syncretic social forms.[64] In practice, one has to view the (deep-)state social logic as one logic among others, which is constantly hedged-in, warded-off, restricted, contaminated and syncretized by other social logics. It is this field of contestation between principles which produces hybrid forms such as the welfare state, the African rhizo-state and so on. This does not, however, mean that the state logic does not exist. In certain conjunctures and sites, it gains an unusual virulence which brings its entire potentiality to the fore. Situations of war and 'states of emergency', particularly when combined in a field of endless warfare against an unspecifiable enemy, create one such situation where the state logic expresses itself in an unusually unmediated form.

The current state of critical theory also allows a clearer definition both of the popular agency counterposed to the state, and of the structures of desire and affect connected to statist and oppositional logics. The 'social principle' can today be re-theorized as the logic of affinity-networks, as a separate social logic irreducible to and antagonistic with the state logic. Similarly, theories of active and reactive desire can be deployed to understand the construction of the emotional underpinnings of state power. These more subtle concepts allow us to move forward in theorizing an 'outside' to state power without falling back on an image of society or of classes as pre-formed entities.

To conclude, therefore, anarchist and autonomist critiques of the state coalesce around certain key claims which are counterposed to those of dominant tendencies in International Relations and war studies, and which provide a powerful alternative to statist ways of seeing. In showing the complicities of state power in the production of war, and in theorizing the co-extensiveness of interstate war with other forms of social conflict, they challenge the Westphalian biases built into dominant perspectives, pointing towards an alternative theorization which enables both a more effective understanding and an ethical alternative to the reproduction of wars by states. In a world where the 'health of the state' puts the future of freedom, of human life, and even of the ecosystem in jeopardy, this critique is more vital than ever.

64 This approach is set out in Chapter 1 of Athina Karatzogianni and Andrew Robinson, *Power, Conflict and Resistance in the Contemporary World* (London: Routledge, 2010).

Majority Rule – A Cause of War?

Peter Emerson

In a plural society the approach to politics as a zero-sum game is immoral and impracticable.

Words like "winning" and "losing" have to be banished from the political vocabulary of a plural society.[1]

Introduction

Democracy, as practised in most countries nowadays, is adversarial. The fact that presidents and/or policies are changed as the result of a vote rather than as a consequence of a bloody revolution or war is to be welcomed. Sometimes, however, the distinction between peaceful and violent change is muddied. Indeed, the use of divisive voting procedures has often exacerbated tensions in societies that have then tumbled into violence. Win-or-lose voting procedures – i.e. yes-or-no majority votes in decision-making and single preference systems in elections – along with a practice which is based on these voting methodologies, namely, that of majority rule, have often been part of the problem.

An Historical Background

In earlier times and in many cultures, the democratic process was rather more inclusive. In yesterday's America, for example, 'The tribes ... gathered and discussed the issue at hand, listening and speaking until common understanding had been reached. ... [a] practice of government by consensus ...'.[2] Similarly, in parts of

---

1 Sir W. Arthur Lewis, *Politics in West Africa* (London: George Allen and Unwin, 1965), 66–7.

2 Keny Nerburn (ed.), *The Wisdom of the Native Americans* (Novato, CA: New World Library, 1999), 138.

Africa, 'Majority rule was a foreign notion',[3] and instead, to quote Tanzania's first President, Julius Nyerere, 'The elders ... talk until they agree. This "talking until you agree" is the essential of the traditional African concept of democracy'.[4] The sub-Saharan attitude to conflict resolution was also rather different: to quote an example from Ethiopia, 'If someone ... is quarrelling with someone else, then the court ... will set itself the sole task of ending the conflict ... while granting to each that he is in the right'.[5]

A more exclusive form of democracy emerged in Europe and America where philosophers like Jean-Jacques Rousseau and Jeremy Bentham spoke of '*le volonté général*, the general will' and 'the greatest good of the greatest number'. Having seen the errors and horrors caused by various forms of minority rule, especially those of the absolute monarchs, it seemed obvious to them and many others that rule by a majority would be a huge improvement. Hence, today, we have statements such as 'Democracy rests upon the principles of majority rule'.[6] Few would disagree.

There then came a huge mistake: it was assumed that a majority opinion could be identified by a majority vote. Despite huge advances in so many fields of science, including that of social choice, this assumption still holds. Thus many believe that 'Democracy works on the basis of a decision by the majority',[7] and that 'Democracy is based on majority decision'.[8] The world – or at least the Western world – appears to be committed to the majority vote.[9]

In many countries, then, current democratic structures are based on a majority vote form of decision-making. As a direct consequence of this, most parliaments divide into two – government and opposition – either under single-party rule or in majority/grand coalitions. While some international organizations try to operate in consensus – and by that is meant a *verbal* consensus – many forums base their decisions on a (simple, weighted, qualified or consociational) majority vote. Such

3 Nelson Mandela, *The Long Walk to Freedom* (Boston, MA: Little Brown and Company, 1994), 25.

4 Paul E. Sigmund, *The Ideologies of Developing Nations* (New York: Frederick A. Praeger, 1966), 197.

5 Ryszard Kapuściński, *The Shadow of the Sun* (London: Penguin, 2001), 315.

6 US Dept of State: http://usinfo.state.gov/products/pubs/principles/what.htm.

7 Report of the Constitution Review Group, Government of Ireland, 1996, 398.

8 International UNESCO Education Server for Civic, Peace and Human Rights Education: http://www.dadalos.org/int/Demokratie/Demokratie/Grundkurs5/mehrheitsprinzip.htm.

9 By 'the world' I mean not only the political sphere of media, academia and practitioner, but also the business community. 'The statutory contract succinctly lays down the basics of the legal relationship between the company, its members, and the members inter se. In consequence, a member agrees to be bound by the decisions of the majority taken at a general meeting of the company. ... just like in a parliamentary democracy when the cabinet has taken a decision ... a dissenting member will nevertheless be bound by it.' Clement Chigbo, *An Examination Of Majority Rule Principle And The Remedies Available To Shareholders,* May 12, 2006: http://www.jonesbahamas.com/?c=135&a=8793.

an adversarial democratic structure is perhaps adequate in some jurisdictions; elsewhere, however, it has been a recipe for division, violence and, at worst, war.

This chapter will examine some of the dreadful consequences which have resulted from this Western, if not now universal, practice of majoritarianism,[10] first in decision-making, secondly, in elections. In addition, it will question the logic of such a polity with particular reference to conflict resolution work.

Majoritarianism in Decision-Making

There are a number of instances where simplistic voting procedures and/or inadequate democratic structures have at the very least exacerbated tensions in society. They include i) majority vote plebiscites on secession; and ii) majority votes taken in parliaments and international organizations.

i) Majority vote plebiscites on secession

'All peoples have the right to self-determination. By virtue of that right they freely determine their political status ...'[11] But people cannot 'freely' determine anything if someone has already reduced that which should have been a free (i.e. multi-option) choice to a binary dilemma of just two options. Unfortunately, none of the international declarations on self-determination stipulate the particular methodology by which 'a people' should actually determine their status. In the absence of such, most politicians have resorted to the binary majority vote, not least because this voting procedure allows them to control the agenda.[12]

A simple instance comes from Canada, where M. Parizeau decided that the people of Quebec should be presented with the choice of either the *status quo* or independence. In effect, it was a contest between the Anglophones and the Francophones, and the question all but disenfranchised the indigenous Cree Indians. In the 1980 referendum, Parizeau lost by 59.6 percent to 40.4 percent. So he

10 The original Russian word for 'majoritarianism', by the way, is 'bolshevism'. The word was coined in London in 1903 when the Social Democrats split into two: the bolsheviks (bolsheviki), the members of the majority (bolshinstvo) of that party, all 19 of them, and the 17 mensheviki of the minority (menshinstvo). There were three abstentions, while others, the Jewish Bund, had already walked out.

11 Article 1.1, The International Covenant on Civil and Political Rights, adopted by the UN in 1996. This 'right' was first devised by President Woodrow Wilson who, in his later years, reflected, 'I never knew there were a million Germans in Bohemia', quoted in Abba Eban, *Diplomacy for the Next Century* (New Haven: Yale University Press, 1998), 38.

12 In like manner, many dictators, including Napoleon, Lenin, Mussolini, Hitler, Duvalier, Khomeini, Mugabe and Saddam Hussein, have all used majority votes – Peter Emerson, *Defining Democracy* (Belfast: The de Borda Institute, 2002), 104–10.

waited for a few years before having another go: in 1992, there was a second defeat, this time by 56.7 percent to 43.3 percent. Accordingly, he chose to hang on for a little longer, and then he held yet another ballot in 1995 – the referendum process was rapidly becoming a *'never-end-'em'*[13] – and he now lost by a mere whisker, 50.6 percent to 49.4 percent. He blamed the hapless Cree Indians – who voted for the devil they knew – so he had to resign.

As this and other examples show, the right of self-determination, at least as currently defined, is a provocation. It might have been an adequate principle for those who wished to resolve the *external* problem of colonialism. As a means of settling *internal* disputes, however, it was and still is a recipe for mayhem. Consider the following: if a minority does not want the *status quo,* and if a majority of that minority chooses to opt out, then of course it may. If, however, a minority of that minority disagrees, and if a majority of that minority of the minority so chooses, then it may choose another path. The logic of this is as nonsensical as the consequences have often been bloody: there will be peace and tranquillity on this planet when every pair of individuals is an independent nation-state of just two persons.

The British Isles

This form of balkanization started in the UK in 1920 when Ireland opted out. So Northern Ireland opted out of opting out and opted back in again. The eventual consequence was violence, the thirty years of 'the Troubles'.

The Caucasus

Lessons were unlearnt. When Mikhail Gorbachev introduced his policy of *perestroika,* the USSR started to break up. In 1991, Georgia opted out.[14] Whereupon Abkhazia and South Ossetia, and very nearly Ajaria as well, opted out of Georgia: on October 3, 1999 in Abkhazia, in an 88 percent turnout, 97 percent supported independence; and on February 18, 2007, South Ossetia followed with a 99 percent margin. Hence more violence: in 1992, it was between the region and the nation; in 2008, it was outright war between Georgia and Russia.[15] Meanwhile, in Nagorno-Karabakh on December 10, 1991, i.e. *after* its war, a majority of 99.89 percent voted for independence, and only 24 individuals bravely said 'no'.[16]

In today's Russian Federation, the prospect of a similar chain of events has come to be called 'matrioshka nationalism',[17] named after the famous Russian

13 The phrase was first coined by some wit on BBC Radio 4.

14 In 1990, the author gave a press conference in Tbilisi on the need for a non-majoritarian, more inclusive form of governance. The speech was well received.

15 The author was a member of the EU monitoring mission for South Ossetia from September 2008 to January 2009.

16 When the first post-Soviet ethnic dispute broke out in 1988, in Nagorno-Karabakh, the headline in Pravda was Наш Ольстер, 'This is our Northern Ireland'.

17 Anna Reid, *The Shaman's Coat* (London: Phoenix, 2002), 136.

dolls: lurking within every majority is another minority; within that, a smaller one; in every smaller one, a tiny one; and inside that again, a miniscule one. There is the fear, certainly among some Russian politicians, that if one part of the Federation were to opt out – like Chechnya – others would surely follow; which partly explains why the two recent wars in that Republic were so bitter.

The Balkans

In Yugoslavia in the 1990s, a major political question was the dichotomy, 'Are you Serb or Croat?' For many people in what is now Croatia, this closed question was of course unanswerable: the partners in or children of a mixed marriage; those who were neither Orthodox nor Catholic; those like the Vlahs or the Roma who were of another ethnic minority; and most tragically of all perhaps, those Slavs and others who were trying to move beyond any sort of antagonistic nationalism. Alas, the binary format was chosen for the Croatian referendum question in May 1991: independence, 'yes or no?' The Serb minority in the *Krajina* – three areas in Croatia which had a predominantly Serb population – organized their own ballot one week earlier. In the latter poll, a turn-out of 95 percent voted by a 90 percent margin to stay in Yugoslavia. The international community did not recognize it. In the other poll, 93 percent of an 84 percent turnout voted for Croatian independence. This result was recognized. The other consequence of both votes was war.

In November 1991, the EU set up the Badinter Commission, a team of international lawyers to consider the problems of Yugoslavia.[18] It endorsed the referendum and, as a result, there was a spate of such ballots: votes in Slovenia, Croatia, Macedonia (polls which preceded the work of the commission) and Bosnia were recognised; votes which did not produce the required result, as in Montenegro, were repeated; and votes which were not wanted were just not recognized – Republika Srpska, Herzeg-Bosna, Sandžak, and Kosovo – unless or until, of course, the West changed its mind.[19] The Commission did not, however,

18 On his first visit to Yugoslavia in 1990–1, the author had written on the possible benefits of a consensual polity and of the inherent weaknesses of any form of majoritarianism. At home, in October 1991, and one month before the commission under Robert Badinter was inaugurated, he invited a native of Sarajevo to a cross-community conference in Belfast, not least to warn of the dangers of using a majority vote referendum in Bosnia (Peter Emerson, *From Belfast to the Balkans* (Belfast: The de Borda Institute, 1999), 47). The Commission recommended use of the referendum for any people seeking self-determination, although it did suggest the outcome of the Bosnian referendum 'would be valid only if respectable numbers from all three communities of the republic approved' (Susan Woodward, *Balkan Tragedy* (Washington: Brookings Institute, 1995), 280). There again, it did not define the word 'respectable'. On the day of the vote, March 1, 1992, the Bosnian Serbs boycotted and the barricades went up in Sarajevo; it was soon full scale war (Misha Glenny, *The Fall of Yugoslavia* (London: Penguin, 1992), 163). Nine months later, the author returned as a freelance war correspondent.

19 The 1991 poll in Kosovo – 99 percent in favour – did not qualify. Eight years later, at Rambouillet, the international community decided that it would now recognize a

question the methodology of the majority vote. The final comment comes from Sarajevo's now legendary newspaper, *Oslobodjenje*: 'all the wars in the former Yugoslavia started with a referendum'.[20]

Rwanda

In the 1930s, the Belgian authorities issued everyone in Rwanda with an ID card. The question was closed: 'Are you Hutu or Tutsi?' Now Rwandan society was quite unlike Kenya's, for example, where different tribal groups speak different languages. In Rwanda, in contrast, they all live cheek by jowl and everyone speaks the one language, Kinyarwanda. The Belgians nevertheless decided to split everyone into two, and thus they converted a social distinction into a tribal one. Those who were tall were called Tutsi, the small were the Hutu (the Twa were ignored), and anyone of average build was asked if they had ten or more cows; in effect, another closed question. Those who said 'yes' were classified as Tutsi; the 'no's were Hutu.[21]

Having supported a form of *minority* rule – in which the tiniest minority, the colonialists, were on top – the authorities post-WWII changed their minds and argued for its opposite, *majority* rule. So the losers of yesterday could become the winners of tomorrow. Little wonder, then, that when the *Interahamwe* launched their gruesome genocide in 1994, the slogan they used was '*Rubanda Nyamwinshi*', 'the majority people'.[22]

Sudan

In all the words that have been written on the Rwandan genocide, few have questioned the practice of interpreting the *principle* of majority rule to mean the *practice* of majority voting. Instead, as in the Balkans, so too in Africa, the international community blunders on. In July 2002, British diplomats were present in Kenya when the two sides to the civil war in Sudan, the Khartoum government

referendum in Kosovo. Slobodan Milošević refused to sign (just as Vojislav Koštunica would have done). As a result, NATO launched its 2000 offensive. Still he refused. So the Russian Foreign Minister, Viktor Chernomyrdin, went to Belgrade to negotiate a settlement, and he removed the referendum clause. Whereupon Milošević did sign. So the war came to an end. In this regard, then, the war achieved nothing.

20 '… su svi ratovi u bivšoj Jugoslaviji počeli nekim referendumom' (author's translation), *Oslobodjenje*, February 7, 1999, 11.

21 John Reader, *Africa* (London: Penguin, 1998), 616.

22 In a brave attempt to overcome the legacy of that genocide, the Rwandan government has initiated a series of 'mini Truth and Reconciliation Commissions' called *gacacas*. A European company undertook a social survey on this policy and asked a series of binary questions. At a subsequent press conference in Kigali which the author attended, it presented its findings. In the questions which followed, one participant observed, 'Asking yes-or-no questions is very unAfrican'. Gérard Prunier, *The Rwanda Crisis: History of a Genocide* (London: Hurst, 1995), 183.

and the Sudanese People's Liberation Movement, SPLM,[23] signed the Machakos Protocol to end that war. *Inter alia*, this agreement promised a binary referendum in South Sudan on whether or not the latter could have independence. Here was the answer the South wanted. But if one part of the country could gain its objectives by waging a war, and then negotiating a settlement to hold a referendum, then why not another? In a word, balkanization. '[I]f the South were to secede, this would open a Pandora's box in the whole of Sudan.'[24]

The first consequence took place in what was already a very turbulent region: Darfur. Literally within months, in early 2003, another 'movement', the Sudanese Liberation Movement, SLM,[25] again with an army to match, resorted to violence. The government responded with its now notorious *Janjaweed*.[26]

The South Sudan referendum was held in January 2011 and, on the whole, it passed off peacefully; the actual handover of power is scheduled for July 2011. A second consequence of Machakos, however, is that further referendums are already in the pipeline. Within Sudan, the oil-rich region of Abyei on the border between North and South is due to hold its own referendum shortly; there has already been some violence there. Meanwhile, South Kurdufan and Blue Nile are due to hold consultations on the initial basis of an either/or question.

Consequences further afield may be even more traumatic. Somaliland now wants a referendum to secede from the rest of Somalia. They are the first. And elsewhere? Will the Moslem North try to secede from the Christian South in Nigeria, a land that has often seen violence in Kaduna, which straddles the two, including post-election violence in April 2011? Will something similar happen in Ivory Coast, a land which saw two rivals from a previous civil war participate in a win-or-lose election and which, partly as a result, descended into more violence? (see below). Consequences in the DRC could be even worse.

In effect, the international community has now exported the right of self-determination and the practice of binary plebiscites to a continent of almost unlimited ethnic or tribal diversity. Little could be more unwise. South Sudan may be just the calm before the storm. I hope not; suffice here to say, however, that when Slovenia held its referendum in 1990, many people did not realize that there would very soon be ghastly consequences elsewhere in Yugoslavia.

Multi-option referendums

The referendum has been suggested and/or used in a number of other conflict zones, often with serious consequences. One was proposed for Kashmir in 1947, but never implemented; Northern Ireland held its border poll in 1972, to which the

23 Its military wing is the SPLA, the SPL Army.

24 Timothy Othieno, 'Democracy and Security in East Africa', in *Challenges of Conflict, Democracy and Development in Africa* (Johannesburg: EISA, 2007), 280–1.

25 This also has its military wing, the SLA, and has often had direct links with the SPLM.

26 The *janjaweed* first appeared in 1988, but only now did these 'evil horsemen' come to prominence.

mainly Catholic Social Democratic and Labour Party, SDLP, organized a boycott;[27] and in 1999, East Timor's referendum led to massive post-ballot violence.

In contrast, some jurisdictions have used multi-option ballots: Newfoundland held a three-option poll in 1948, Singapore did the same in 1962 and so did Puerto Rico five years later; while in Guam in 1982, there were six options on the ballot paper, and just in case that was not enough, a further seventh option was left blank, for any other suggestion.[28] All of these polls were held under a form of two-round voting, TRS, and all passed off peacefully. Preferential polls could be even more inclusive.

ii) Majority votes taken in parliaments and international organizations

Majoritarianism was, without doubt, a major part of the problem in (Northern) Ireland: the Unionists claimed the right of majority rule in a six-county jurisdiction, while the Republicans wanted a 32-county structure. It was also 'a major cause of ethnic conflict' in Sri Lanka,[29] in Cyprus, in the Middle East, and elsewhere. It is, indeed, a catalyst of division if not a cause of war, not only within the nation state, but also in international organizations.

Take, for example, the UN Security Council vote on Iraq. In October 2002, the 15-member Council voted on Resolution 1441; it was another closed question, for-or-against.[30] France, for one, did not like the draft on offer; in particular, she objected to the inclusion of the phrase 'serious consequences'.[31] And yet she voted in favour. Now why would anyone vote for something they did not like? Well, in majority voting, this happens quite frequently. People and countries vote for 'this', either i) because they definitely support 'this'; or ii) because they think, on balance, 'this' is better than the alternative 'that'; or maybe iii), as in the case in point, because they are giving priority to some other consideration like the need for international solidarity. Sadly, as is now widely acknowledged, the passing of that Resolution was one more step on the road to war in Iraq.

There are, in fact, many reasons why, both in parliamentary votes and in regional/national referendums, the two-option majority vote is an inaccurate measure of the

27 In 1988, the same SDLP supported the use of a two-option referendum to endorse the Belfast Agreement, a document which allows for NI to hold a referendum every seven years or so on whether or not to join a united Ireland. If the poll is lost, NI stays in the UK for another seven years; if it is won, it joins a united Ireland for ever! It is another instance of a 'never-end-'em'.

28 Emerson, *Defining Democracy*, 119–20.

29 Rudhika Coomaraswamy, 'The Politics of Institutional Design' in Sunil Bastian and Robin Luckham, *Can Democracy be Designed?* (London: Zed Books, 2003), 146.

30 And this in the wake of another closed question, George W. Bush's now infamous, 'Are you with me or against me?'

31 In Article 13, the Security Council 'Recalls ... that the Council has repeatedly warned Iraq that it will face serious consequences as a result of its continued violations of its obligations'.

collective will.[32] Furthermore, there are other voting procedures by which decisions can be taken: apart from simple, weighted, twin and consociational majority votes, there are some multi-option forms: these include a) plurality voting (which is like first-past-the-post, FPP); b) the alternative vote, AV, otherwise known as the single transferable vote, STV, or instant run-off vote, IRV; c) the two-round system, TRS; d) a Borda count, BC; e) approval voting; and f) a Condorcet count. If one of these methodologies is more accurate,[33] and if therefore there is little justification for using majority voting, then the justification for the current practice of majority rule based on majority voting falls. Unfortunately, the 'public is deeply imbued with the mystique of the majority',[34] and '... there is a surprisingly strong and persistent tendency in political science to equate democracy solely with majoritarian democracy and to fail to recognize consensual democracy as an alternative and equally legitimate type'.[35]

Thus, for the moment, the international community tends to promote a majority vote interpretation of majority rule, and only when such a policy goes wrong does the argument swing in favour of that which in some ways is a complete opposite – power-sharing. The latter is usually considered to be an extraordinary measure and a temporary expedient whereas, if it were more widely practised, it could help to prevent that which it is only used to cure. The list of countries where such a polity is practised, or has at least been considered, grows longer by the day: Afghanistan, Belgium, Bosnia, Honduras, Iraq, Kenya, Lebanon, Northern Ireland and Zimbabwe all come to mind. Of stable countries, however, only Switzerland uses such an inclusive all-party polity.

Majoritarianism in Electoral Systems

Electoral systems, too, can be a cause of war, especially if they are inaccurate. Such outcomes, and the divisive campaigns which precede the votes under the more adversarial systems, often promote competing antagonisms within a stable society, that or they exacerbate divisions in an unstable one. Indeed, many single-preference voting procedures often act as instruments of ethnic cleansing.

32 Peter Emerson, Proportionality without transference: the merits of the Quota Borda System (QBS), *Representation* 46: 2, 197–209.

33 The most accurate measure of a majority opinion is a Condorcet count, while a similarly precise measure of the collective will is a modified Borda count, MBC. In many circumstances, a Condorcet winner is also the MBC winner.

34 Michael Dummett, *The Principles of Electoral Reform* (Oxford: OUP, 1997), 81.

35 Arend Lijphart, *Patterns of Democracy* (New Haven: Yale 1999), 6. Lijphart is talking here more of consociationalism, which though a huge improvement on straight majoritarianism, is still adversarial. Both in decision-making and in elections, voting procedures in a 'consensual polity' would be preferential.

Ukraine

Ukraine consists of a predominantly Russian-speaking and Orthodox East, and a mainly Ukrainian-speaking Catholic or Uniate West. But there is no hard and fast line between the two. Both are Slav and both are Christian. To use a divisive electoral system to elect just one individual, with the possibility of a small margin of victory, would obviously be unwise. Yet such was the scenario in the final of a TRS election in 2010: Viktor Yanukovich won with 48.7 percent to Julia Timoshenko's 45.7 percent. Thankfully, Ukraine has not spilt over into violence, and nor did it do so at the time of the Orange Revolution in 2004.[36]

Kenya

Other countries have not fared so well. As a result of the democratic process in many emerging democracies, 'Ethnic parties developed, majorities took power, and minorities took shelter'.[37] A recent example of this sort of chaos was seen in the violence that followed the December 2007 FPP elections in Kenya. In such a land, where the people are divided not only into various tribes but also into two ethnic groups – the Bantu and the Nilotic – a win-or-lose electoral system that pits one person against another would obviously be unwise. Needless to say, of the two main candidates contesting the presidency, one was a Kikuyu, a member of the dominant Bantu tribe, and the other was a Luo, the major Nilotic tribe.

Eventually, a power-sharing settlement was negotiated. It could all have been so easily obviated if the electoral system itself had been of a win-win variety. This could still be done before the next contest, either by a system of PR or even by the very simple expedient that was used initially in the USA: the winner becomes the president and the runner-up becomes the vice-president.

Côte d'Ivoire

Not every divisive election leads to war. Indeed, Sierra Leone had a civil war in 2002 but a reasonably peaceful election under FPP in 2007. Côte d'Ivoire also had a civil war in 2002, but its story is now one of renewed violence. The 2010 TRS elections brought together two of the previous combatants: Laurent Gbagbo from the South and his old rival, Alassane Ouattara, from the North. As in Kenya, the results were disputed, and violence erupted. Eventually, after hundreds of deaths, Ouattara's forces marched into Abidjan, albeit with outside assistance. For the long-term stability of the country, some form of power-sharing would again be advisable.

36 The author was an OSCE election observer on both occasions.

37 Donald L. Horowitz, *Ethnic Groups in Conflict* (Berkeley, CA: University of California Press, 2000), 629.

Other electoral systems

Some systems are more suitable for plural societies: the more sophisticated forms of PR-list, for example. But any system that allows the voter to cast only one preference – i.e. which thus restricts the voter's ability to express his/her full opinion – is not much better than the majority vote used in decision-making. Elections under such a system in places like Bosnia and Kosova are often little more than sectarian headcounts.[38] The same was true in elections for the Council of Representatives in Iraq and for the House of People (*Wolesi Jirga*) in Afghanistan, the former under PR-list, the latter under the single non-transferable vote, SNTV; again, both are single preference forms of voting. Therefore they are often inaccurate. Furthermore they are divisive if not indeed dangerous.

Some countries have tried to devise a more inclusive, or at least a non-sectarian, electoral system. Lebanon uses a multiple form of FPP such that, in any constituency where there are say, 25 percent Maronite, 50 percent Shia and 25 percent Sunni, any party wishing to stand must nominate four candidates: 1 Maronite, 2 Shia and 1 Sunni. Furthermore, the voter must vote for 1 + 2 + 1 candidates, and the easiest way for them to do so is to just vote for the party ticket.[39] Dagestan has devised another non-sectarian system: in one constituency, all the candidates of every party must be of one religion; in a second constituency, they are all of another faith; and so on.[40] Of those other electoral systems which allow the voter to cast his/her preferences across the sectarian divide, the most inclusive are probably PR-STV and the quota Borda system, QBS, not least because they are not based on party labels or 'designations'.[41]

Conflict Resolution

If the political process is to be a means by which disputes can be resolved peacefully, it should be one of mediation, just as it was originally in Africa. Now in seeking to arbitrate any dispute, domestic, industrial or political, professional mediators seldom ask questions which are *closed*. Rather, they talk to both or all parties, firstly

38 On its own initiative, the Organization for Security and Co-operation in Europe, OSCE, used a single-preference form of PR-list in both Bosnia in 1996 and Kosovo in 2001; admittedly, the latter system involved set-aside seats for the major ethnic minorities. The author was an observer for both of these contests. Věra Stojarová and Peter Emerson, *Party Politics in the Western Balkans* (Abingdon: Routledge, 2010), 11–3.

39 Farid el Khazen, *Prospects for Lebanon* (Oxford: Centre for Lebanese Studies, 1998), 15 et seq.

40 Anna Matveeva, *The North Caucasus* (London: Royal Institute of International Affairs, 1999), 18.

41 The form of consociationalism prescribed in the Belfast Agreement requires all the elected representatives to 'designate' themselves as 'unionist', 'nationalist' or 'other'; thus the very agreement entrenches sectarianism.

to find out just what options exist. Next, in a process that is sometimes called shuttle diplomacy, they try to improve on some of these options, to make them more acceptable to the other parties. And finally, they aim to identify that option which enjoys the widest level of support from all concerned. In any instances of violence, internal or external, the eventual resolution will often consist of a compromise. In other words, as in Africa, the mediators will treat both or all sides (at least initially) as if they were both or all (at least to some extent) right.

In the Middle East, the best option on offer is probably the two-state solution, not least because neither Palestine nor Israel would contemplate the ideal – the one-state solution – for as long as either believes in majority rule.

In other scenarios, too, the outcome must often involve a form of power-sharing. This should consist of the following:

- decision-making which caters for open questions, i.e. multi-option voting, so that, as an absolute minimum, compromise options can be on both the agenda and the ballot paper;
- elections that are both preferential and proportional;
- power-sharing in governance which allows not only for post-sharing but also for compromise in decision-making.

An inclusive polity

In short, voting procedures should be 'peace-ful', that is, the democratic process should be a vital part of the peace process, and the relevant voting procedures should enable the voters to use the democratic process as an act of reconciliation, if of course such is their wish.

In decision-making, the very structure of the ballot should allow the voter to cast a preference for those options that have been proposed by the erstwhile foe. If a negotiation relies on a majoritarian process, the various participants might well keep their cards fairly close to their chests. After all, 'Once your fall-back positions are published, you have already fallen back to them'.[42] If the final vote were to be preferential, however – and this author would suggest the MBC – then those concerned could reveal and even adjust their own preferences at any stage of the debate.[43]

In elections, the voter in NI should be able to vote for both a Catholic and a Protestant.[44] Likewise those in Bosnia should be able to cast their preferences

42 Abba Eban, *Diplomacy for the Next Century*, 81.

43 Peter Emerson, *Designing an All-Inclusive Democracy* (New York: Springer, 2007), 15 et seq.

44 The present system, PR-STV, does allow the voter to vote for both the Catholic and the Protestant; unfortunately, in the count, the vote may be transferred, and thus it may go to either the Catholic or the Protestant. A more inclusive voting procedure is called the Quota Borda System, QBS (Peter Emerson, *Designing an All-Inclusive Democracy*,

across the gender, the party and the ethno-religious divides. As noted above, the Lebanese have an electoral system that could, in theory, take religion out of politics. Unfortunately, their system of governance actually entrenches sectarianism, just like the consociational decision-making in the Belfast Agreement.[45]

Finally, in governance, just as the people elect the parliament (and, if the electoral system is a good one, that parliament will represent *all* the people), so too parliament should elect the government, again by a system of PR, such that the government represents the *entire* parliament. The only methodology for a parliament to elect a cabinet in which each member undertakes a different responsibility and yet which, overall, represents the given parliament proportionally, is the matrix vote.[46] In many conflict zones – Kenya and Zimbabwe, for example – negotiations on power-sharing have been both problematic and protracted. In Belgium, too, forming a government can involve seemingly endless negotiations, and on February 17, 2011, Brussels inherited the mantle previously held by Baghdad and before that Amsterdam, for the longest running talks on forming a government: 250 days.[47]

Conclusions

A voting procedure is inadequate if the choice of options/candidates and/or the ability of the voter to express a full opinion has been excessively restricted. In some conflict zones, the outcome of a vote has been different to that of the opinion poll. In Northern Ireland, for example, throughout the troubles, the public consistently expressed support for integrated education, mixed housing and power-sharing; alas, in elections, in large part, those same individuals chose politicians who at the time opposed such measures. Furthermore, while the British government realized that the choice of electoral system was very important – and hence, they re-introduced PR-STV in 1972 – they continued to use FPP elections for Westminster. In February 1974, the Unionists got 53 percent of the vote and 92 percent of the seats. So the FPP voting system was part of the discrimination that caused and then fed 'the Troubles'.

Likewise in Bosnia, 'public opinion polls in May and June 1990, and again in November 1991, also showed overwhelming majorities (in the range of 70 to 90 percent) against separation from Yugoslavia and against an ethnically divided republic'.[48] Yet the single-preference electoral system used in 1990, TRS, gave the voters little choice, and the result was an overwhelming victory for the three ethnic parties.

39–60), and the two electoral systems are compared in Peter Emerson, 'Proportionality without transference'.

45 Footnote 41

46 Peter Emerson, *Designing an All-Inclusive Democracy*, 61 et seq.

47 Peter Emerson, *Defining Democracy*, chronology.

48 Susan Woodward, *Balkan Tragedy*, 228.

Sometimes, then, the use of majority voting in parliaments and referendums, the use of simplistic single-preference electoral systems, and the absence of any voting mechanism by which a parliament can elect a government, have given results which, as a minimum, do not reflect the general will; and at worst, they have been a cause of war. There are, however, better more inclusive voting procedures.

Appendix: Abbreviations

AV	alternative vote (= IRV = STV)
BC	Borda count
DRC	Democratic Republic of Congo
FPP	first past the post
IRV	instant run-off voting (= AV = STV)
MBC	modified Borda count
NI	Northern Ireland
OSCE	Organisation for Security and Co-operation in Europe
PR	proportional representation
QBS	quota Borda system
SDLP	Social Democratic and Labour Party
SLM	Sudanese Liberation Movement
SNTV	single non-transferable vote
SPLM	Sudanese People's Liberation Movement
STV	single transferable vote
TRS	two-round system
UNESCO	UN Educational, Scientific and Cultural Organisation
WWII	World War II

The Legitimacy of War – Toward a New Principle for Intervention, with its Application to the Iraq War of 2003

Troy Davis

Introduction

One of the major problems of war is how to decide whether it is legitimate. This problem is one of the manifold issues addressed in Just War Theory (JWT). JWT finds its grounding in the general philosophical concepts of natural law used by the ancient Greeks and then Cicero, the great Roman writer, philosopher and statesman, who strived to determine the perfect form of government and to revive the Roman Republic. Early and medieval Christian theologians like St Augustine of Hippo and St Thomas Aquinas further refined these concepts, while the Dutch theologian and jurist Hugo Grotius published in Paris in 1625 the first extensive and detailed discussion of the subject,[1] laying down as his introduction a general theory of international law as emanating from a natural law common to all humanity:

> *After examining the sources of right, the first and most general question that occurs, is whether any war is just, or if it is ever lawful to make war. But this question like many others that follow, must in the first place be compared with the rights of nature. Cicero in the third book of his Bounds of Good and Evil ... proves with great erudition from the writings of the Stoics, that there are certain first principles of nature, called by the Greeks the first natural impressions, which are succeeded by other principles of obligation superior even to the first impressions themselves.*[2]

1 Hugo Grotius, *On the Law of War and Peace*, translated from "De Jure Belli ac Pacis" by A.C. Campbell, AM (Kitchener ON: Batoche Books, 2001).

2 Ibid., 18.

While JWT is therefore a Western tradition and is grounded in Greco-Roman roots and Christianity, other civilizations and religious traditions also have their own versions of the morality of war.[3] Philosophically, JWT is generally opposed to Realism, which says that states only have interests and should not encumber themselves with morality, and to Pacifism which says that wars are intrinsically immoral and can never be justified. It seeks to determine how to reconcile morality and the need for justice (without which it is impossible to have a society worth living in), and the rough reality of a chaotic world. This dilemma was most acutely felt by Christians who are bound to follow the commandment "Thou shalt not kill" (catholic version), hence the predominance of theologians in its definition and development. This is important to keep in mind for the later discussion since ultimately most JWT is based on assumptions of divine will and argued from biblical sources.

JWT is traditionally divided into *jus ad bellum*, the justice and morality of going to war in the first place, *jus in bello*, the norms of justice during war, and the recent *jus post bello*, which determines the justice and criteria for post-war situations, the re-establishment of peaceful government, etc. Thus JWT seeks to make explicit the strict and explicit criteria or conditions under which war might be moral. These criteria have evolved over time but in general are considered to include the following (for *jus ad bellum* only):[4]

1. Having just cause;
2. Being declared by a proper authority;
3. Possessing right intention;
4. Having a reasonable chance of success;
5. Being a last resort;
6. The end being proportional to the means used.

The problem is that JWT is not that easily applicable and its criteria are subject to endless disputes and interpretations, based on one's underlying assumptions and philosophical theories, and even its practitioners can come to diametrical conclusions on some of its most central elements. A good example is given in the critique[5] Carrie-Ann Biondi makes of the book *Rethinking the Just War Tradition*.[6] She shows persuasively how different authors (Walzer and Kaufman), seemingly starting from the same premises, come to opposite conclusions concerning the so-

3 For a discussion of this topic from the viewpoints of Sunni and Shi'i Islam, as well as Hindu, Chinese and Japanese perspectives, see Howard M. Hensel (ed.), *The Prism of Just War, Asian and Western Perspectives on the Legitimate Use of Military Force* Justice, International Law and Global Security (Farnham: Ashgate, 2009).

4 Just War Theory, Internet Encyclopedia of Philosophy: http://www.iep.utm.edu/justwar/, accessed on September 26, 2010.

5 Carrie-Ann Biondi, "'Dirty Hands' and Just War Theory," *Democratiya* 22 (Summer 2008).

6 Michael W. Brough, John W. Lango, Harry van der Linden (eds), *Rethinking the Just War Tradition* (New York: State University of New York Press, 2007).

called Doctrine of Double Effect which seeks to differentiate between people whom it is legitimate to target and others. She does the same with other pairs of JWT theorists: Spatt and Hoag coming to opposite conclusions about the UN role in authorizing intervention,[7] or Lango and Brough coming to opposite conclusions about the moral equivalence of combatants.[8] She thus justifiably concludes that

> *JWT ultimately is unsuccessful on account of its protean nature. It is not really clear that JWT can be re-thought, given that there is not a JWT but rather a number of JWT's in 'the tradition'. The one attempt in this volume to rethink what seems to be central to all JWT variants leads one to the conclusion that JWT is perhaps best jettisoned. It does not follow that war should not be conducted justly, but that a new ethic of war is required that can meet the demands of justification, consistency, and justice.*[9]

A New Principle: "No intervention without legitimacy"

Biondi's well argued conclusion is a major challenge. How do we address it? I would like to offer one element of response by digging deep into one of the criteria, and suggesting that if we can maximize it, then we can solve several problems at once, without changing the overall superstructure. That criterion is the criterion of *legitimate authority*. I argue that in an increasingly interconnected, democratic and transparent world where individuals expand the boundaries of their "moral bubble" to other nations, peoples or even to the entire world and to all its citizens (thus emotionally or philosophically connecting as "world citizens"), where they base their decisions of endorsement, allegiance or action on their perception of whether a particular decision is moral or legitimate, that *the criterion of legitimate authority should be the major criterion*. But this is only possible if it becomes much stricter and more specific. In adopting a much stronger legitimacy principle, one truly based on the corollaries of the natural law of the fundamental equality of the human dignity of all,[10] we will solve at one stroke several of the problems associated with the variability of interpretation of several other criteria. By doing so, we will see that existing criteria of legitimacy, in particular following the UN charter, are not good enough, and even harmful as they waste precious time and by

7 Biondi, "Dirty Hands," 121.
8 Biondi, "Dirty Hands," 122.
9 Biondi, "Dirty Hands," 125.
10 The statement that all humans possess equal dignity is one of the two axioms of democracy engineering theory, and comes from a variety of historical sources, including the American Declaration of Independence of 1776, the Déclaration des Droits de l'Homme et du Citoyen of 1789 and Article 1 of the Universal Declaration of Human Rights of 1948. See note 23 for more details.

preventing principled and creative thinking, actually *encourage* war (this is exactly what happened with the Iraq war and occupation, which could have been avoided).

Focusing on this criterion, we are led to formulate a new principle about the criteria for intervention. The new principle, logically and conceptually equivalent to the well-known principle of "No taxation without representation," is "No intervention without legitimacy." It should therefore appeal to a wide range of people who accept the first sentence, people from different political persuasions. It is equivalent to "No taxation without representation" because taxation is a definite "intervention" (by the state) in one's pocketbook, and representation by definition creates legitimacy, so the statement "No intervention without legitimacy" is a more abstract and general way of saying "No taxation without representation."

First we will see why it is so important that we focus on legitimacy, then explain some problems with the present situation, then finally apply this new principle to the case of the Iraq war in 2003, or rather, to the pre-war situation with Saddam Hussein still in power, using the technique of democracy engineering to create a maximally legitimate authority.[11]

Why is Greater Legitimacy Necessary for Human Survival?

It is claimed that by having a true, maximal legitimate authority, we obviate many of the problems associated with other criteria of JWT, and are even able to eliminate one altogether: the principle of having a reasonable chance of success. This is because this principle is the most subjective one and the one whose real-world variability is the highest; and it also happens to be, in the increasingly transparent world we live in, the one most directly correlated with the recognition of the intrinsic legitimacy of the authority declaring war.[12] In other words, a maximally legitimate authority will automatically create the highest chance of success because people will adjust their behavior and will follow the line of action which is perceived as having maximal legitimacy, even in the face of large odds. Why would people do that? A reasonable hypothesis would be that people wish to live their lives in a

11 Note that at times this principle would not work fast enough to justify intervention in extreme cases like the Rwanda genocide, therefore quick intervention in that case must be grounded on other principles. In the long run, a maximally legitimate authority must be built globally to deal with potential genocides.

12 The criterion of success in JWT is even less useful than here stated because even if success is not attained quickly, action justified on the other criteria very possibly could create a dynamic that undermines the murderous regime, and in any case, it creates at least some positive political capital because politicians followed principle instead of convenience. Just as activists such as Gandhi, Martin Luther King, Nelson Mandela may not win immediately, the fact that they are widely perceived as being on the side of political morality means that their cause usually wins in the long run. States could benefit from the same phenomenon of the ordinary man adhering to politically moral actions if they followed the same logic of moral behavior.

coherent fashion, and to see the same values applied in the political sphere which they have been told apply in the private sphere. This reduces cognitive dissonance and helps to provide meaning to the world, to society, to life in general. With the decline of organized religion in most industrialized countries, with the advance of technology and the dominance of economic thinking, the lack of meaning in people's lives is one of the greatest challenges of society today, and politics should strive to provide meaning, rather than wantonly destroy it.

If I can see with my own eyes that politics is amoral or immoral, while I have been taught since I was a child that I should not hit my schoolmates, I should not lie, I should not cheat, I should not steal etc., why should I care about the good of society, the good of Mankind? If politicians lie or cheat on the most important issues, why should I behave well? Why should I vote? Just as people care in their private dealings about the legitimacy of decisions, it is logical that they would care even more about the legitimacy of decisions when it comes to going to war. One can probably find the origin of the widespread reduction of electoral participation (especially for distant elections like elections to the European Parliament), and the increasing cynicism about politics and politicians, in the "disenchantment" of people for politics. This is all the more worrisome because solving global problems requires global cooperation and global governance. But if people do not trust politics (and since the level of trust is inversely proportional to political distance), by not behaving in a way that maximizes the trust of the public we are endangering the needed global cooperation to solve Humanity's problems.

Therefore the future of Humanity is in even greater danger than it would otherwise be because of short-term thinking by politicians who do not realize that they need to build a "surplus" of political capital to build a trusted structure of global governance. This political capital surplus can only be built by systematically choosing the path of greatest legitimacy for every important political decision (like creating a constitution for Europe, solving the Israeli-Palestinian conflict, creating global governance, and of course, going to war or intervening militarily), which in practice will often mean to deliberately design/create a novel process.[13] Following this line of thought, if we underwent catastrophic environmental collapse this century, which with hindsight we trace back to the inaction in the 2000s, historians may ascribe one more cost to the Iraq war and occupation of 2003–2010 (in addition to the trillion dollars so far): the incalculable cost of billions of lives disrupted because the USA was too busy with a minor war and obsessed with Islamist terror that it did not cooperate with the rest of the world to deal with truly important global problems. Hence another complication today compared to the past when one applies JWT in the age of globalization: we need to think of the opportunity cost in an interconnected and interdependent world. That opportunity cost may in some cases be the greatest factor, and obliges us to think more creatively to apply our principles, rather than follow the easy path of war.

13 All the specific political problems mentioned are real case studies that can be solved by democracy engineering theory.

If the opportunity cost becomes so high so as to affect the future of Humanity as a whole, which is a distinct possibility today, then we have the moral imperative to find a way to reconcile people and politics. Since necessity compels us to create *some sort* of global governance, if we fail to create a democratic version of it early enough, the "perfect storm" of dire global problems, disenchantment and mistrust of citizens toward politicians and the severity of measures needed to deal with global problems means that we could end up with global tyranny[14] instead of global democracy. And since global problems are likely to last longer the later we start addressing them, global tyranny (instead of being a short-lived phenomenon like it was in ancient Greece and Rome when tyrants were elected for a short period of time to deal with emergencies like war) could last for generations, leading to the permanent enslavement of Mankind. The final aggravating factor is that since large numbers of people would be likely to revolt, the global ruling elite would be forced to resort to violence, hence deepening the mistrust of citizens, and creating a vicious circle of control and repression.

Hence we connect JWT to globalization and show that the problem of legitimacy is a crucial one to solve if we want to avoid nightmare scenarios of Mankind living in the sort of dystopia so popular in SF novels.

The Present Situation: Conflicting Paradigms

A first general observation is that most scholars embrace, no matter their origin or ideology, the general idea that law shall provide the guide by which wars are to be decided. While it seems to make sense, it also causes a major problem, because it treats law nearly ahistorically, as if the present state of law reflected some sort of ideal state, disregarding the bloody history of the establishment of laws, both domestically and internationally.[15]

This emphasis on law, meaning *whatever law happens to be at the present moment*, has obvious weaknesses and has the major flaw of being too static (despite

14 This nightmare scenario in the case of global environmental collapse is called "global green fascism." The alternative is a return to a world of warring nations, as well as the collapse of global trade, and thus of consumption and of environmental impact, thus correcting the growing environmental impact of present world society though at the probable cost of hundreds of millions of lives.

15 I will use here for simplicity's sake the traditional term of international law, though the use of the noun "law" to describe the set of treaties that rule international behavior is misleading, since law requires a legislature, which exists neither internationally nor globally. The terms used in other languages represent reality more accurately: in French "droit international," in German "internationales Recht," in Italian "diritto internazionale," etc. It is unfortunate that the English language, which possesses the world's largest vocabulary, uses the same term for two totally different concepts, when an accurate understanding of those differences is of vital import to the pursuit of world peace.

lip service to the dynamic nature of the world system, which evolves at glacial speed compared to the evolution of technology and threats). It is as if present law represented the ultimate accomplishment of what is possible, notwithstanding the fact that it may enshrine a previously "unlawful" situation which has only become lawful by the chance of war and history.

This is most obvious in the quasi-universal deference towards the UN Security Council (UN SC), whose composition has nothing to do with the natural law principles which found JWT, but everything to do with the vagaries of force and of victory in war. There is therefore a clear contradiction between the paradigm that undergirds JWT and the paradigm that gave birth to the UN Security Council.

David Kennedy underlines this: "Legality is almost always a matter of more or less, and legal legitimacy is in the eye of the beholder. Indeed, as law has become an ever more important yardstick for legitimacy, legal categories have become far too spongy to permit clear resolution of the most important questions."[16] That is why we cannot hide systematically behind existing international law, but must follow clearly stated principles. Yet many scholars still reflexively hide behind international law, and in practice the UN Security Council, when it comes to determining what that law is, even though they also recognize the inherent shortcomings of the international system.

An extreme view of what I call the absolutist legalistic position is expressed by Jianming Shen, who concludes an article about the principle of non-intervention and the issue of humanitarian intervention under international law by stating that: "both within and outside the context of the United Nations, no state may engage in so-called humanitarian intervention by resorting to the threat or use of force without the Security Council's authorization" and "The principles of non-intervention and respect for State sovereignty are so fundamental to the maintenance of peace and justice and so inseparable from one another that they constitute *jus cogens* principles both as a matter of treaty and customary international law."[17]

An obvious confusion here is made about the fact that States have no intrinsic purposes but to protect real human beings, and hence their sovereignty is only valid insofar as they protect the sovereignty of their own citizens. In addition, peace and justice are concepts which are irrelevant to "States" as abstract entities but are only valid as to their application to real flesh-and-blood humans. A State is a legal fiction which is composed of and lives for and by the people, for which peace and justice exist or are sought after.

Here we have what is called the New Paradigm, based on human dignity and the sovereignty of the people, in contradiction with the Old Paradigm based on "state dignity" and state sovereignty.[18]

16 David Kennedy, *Modern War and Modern Law* (*International Legal Theory*, Harvard Law School, vol. 12, Fall 2006).

17 Jianming Shen, "The non-intervention principle and humanitarian interventions under international law," *International Legal Theory* 7: 1 (Spring 2001).

18 On the face of it, "state dignity" sounds absurd since a state is not a person. But state defenders use that term to justify non-intervention. It is true that some people use the

JWT is obviously based at its roots on the New Paradigm (NP), even though its operational criteria were created using Old Paradigm (OP) concepts, since in practice States are the ones who wage wars (or who have mostly up to the end of the twentieth century). But it is important to always keep in mind the origins of our differing paradigms because the contradiction between differing paradigms in implementing JWT is where many of the problems of interpretation and implementation lie. The conflicting interpretation about UN roles are grounded in such conflicts of paradigms between the stated aim of peace (a NP feature) and the political structure based on States instead of humans (an OP feature). If we systematically tried to apply the NP, we would progress much faster, and much contemporary work in comparative justice, *jus post bello*, etc. is in essence just that, i.e. trying to systematically implement NP principles of human dignity and legitimacy.

So when we speak of law, we should always mean that the only truly legitimate law is not necessarily whatever law happens to exist today (i.e. the UN Charter which enshrines a particular historical outcome), but whatever is compatible with what is commonly considered natural law and the New Paradigm. All descriptions of natural law (whether religious or secular) emphasize the fundamental idea that humans are equal, that they possess equal human dignity etc., so they are a New Paradigm way which is by definition the human dignity-centered paradigm.

A philosophically equivalent way of stating the idea of the New Paradigm is the concept of equal moral regard, used by Jean Bethke Elshtain in her work.[19]

Fundamental Axioms

For clarity's sake and to avoid the sort of problems that plague contemporary debate about Just War Theory (which is sometimes described as a "moral goulash"[20] or as

concept at times but this is rooted in old mental habits dating from the time that States were all physically incarnated in a King/Emperor, so the dignity of the King, a person, was identified as and "carried over" to the "dignity" of the State. To see the incongruity of the concept, one only has to apply it to other legal fictions, such as a corporation. No one speaks of the "dignity" of a corporation, even in the worst case when an activist might accuse a corporation of corruption or complicity in murder. Why not? Because we all rightly understand that a company as a legal invention created by humans is not a person and cannot possess dignity, which is a purely human attribute. Dignity is reserved for people, or at most by extension to a specific political position which is occupied by a person (e.g. dignity of the office of the President).

19 Jean Bethke Elshtain, "International Justice as Equal Regard and the Use of Force," *Ethics & International Affairs* 17: 2 (Fall 2003).

20 Biondi, "Dirty Hands," 121: "JWT emerges – as the title of the volume indicates – from a specific moral tradition. However, 'the just war tradition' encompasses many complex principles, has undergone change over the centuries, and has manifested both religious and secular variants blending deontological and consequentialist approaches

"the chaos of just war theory"[21]) and the often contradictory conclusions authors arrive at while using JWT theory, we will state explicitly the assumptions and the theory which will be used: the theory of democracy engineering with its axioms and laws. Here, allow us to remind the reader that an axiom by definition is a statement which one cannot prove, as this will spare us much trouble later.

The often unstated problem in *clearly choosing* and therefore *clearly stating* fundamental axioms is that at the very source, one has to either accept a metaphysical foundation, hence which has to be taken on faith, or a physical (i.e. materialistic) foundation, which many people reject instinctively because it leads to the slippery slope of moral relativism and of the justification of the ends by the means.

Part of the problem in JWT is the attempt to ground the theory in some incontrovertible foundation that can somehow be proven from first principles and by a logical chain of reasoning. While principles and logic are necessary, it is impossible to find principles which can be proven absolutely without an act of faith at the very beginning. At root, the most fundamental statements are merely assumptions. No one can "prove" that humans possess dignity (whatever that means), much less that they do so equally. But we assume so because not to do so would leave the door open to the justification of systems based on domination and oppression of one group over another, as has happened throughout history, and of stratified closed societies without freedoms, with different categories of humans, some superior, some inferior.

Traditionally the problem of ultimate authority was solved by reference to religion and scriptures, and recourse to the concept of natural law, without feeling the need to elaborate further. Indeed, since natural law itself is subject to the same impossibility of proof, the use of natural law or of religious references is equivalent to using an arbitrary axiom.

Yet secular philosophers who try to ground JWT in a non-theological context, if they are not extremely clear from the start about this, end up creating circumvoluted and circular references that on the one hand obfuscate and on the other frustrate the very intent of having a JWT theory to start with, i.e. to be able to determine the conditions under which war might be moral.[22]

We will solve that problem by adopting secular philosophical axioms to guide us, albeit tempered in their potentially absolutist interpretations by a set of pragmatic laws inspired by observations of human nature.

Here are the two fundamental axioms[23] which we shall assume:

---

to morality. Indeed, this is so evident that Brough, one of JWT's supporters, refers to 'the body of just war thought' as 'a moral goulash'" (p. 162).

21 Irfan Khawaja, "Contingent Pacifism: A Critique," *Democratiya* 11 (Winter 2007), 92.

22 Khawaja, "Contingent Pacifism," 107.

23 Readers will recognize that the first axiom finds its source in ancient religious beliefs, in particular Christian beliefs, but that it has been adopted nearly universally as a fundamental political principle, including in the Declaration of Independence of 1776 ("We hold these truths to be self-evident that all men are created equal ..."), in the Déclaration des Droits de l'Homme et du Citoyen of the French Revolution of 1789 ("Les hommes naissent et demeurent libres et égaux en droits."), in the Universal Declaration

- All humans possess equal human dignity;
- Sovereignty belongs to the people.

The Example of Iraq

How do we proceed to the original problem of creating maximal legitimacy in decisions about war or intervention?

Let us take the concrete example of Iraq in 2002–2003 to illustrate our discussion. Following the embargo and several UN SC resolutions about Iraq, in which the UN took various measures to determine whether Iraq had weapons of mass destruction, the US started to publicly argue for military intervention because diplomacy had "failed," and tried for more than a year to convince its allies and the world at large that, first, this was the right thing to do (because of the supposed danger of allowing Saddam Hussein to remain in power), and secondly, the legal thing to do (by getting the imprimatur of the UN SC). President Bush even sent his Secretary of State Colin Powell on an (ultimately unsuccessful) whirlwind visit of Arab states to convince them that it was necessary to attack Iraq, and to rally once again a large coalition as had been the case in the first Gulf War in 1991.

The world remained unconvinced, and apart from a few national leaders (most notably from the UK, Spain and Portugal), few agreed to war. Yet independent from what may have been ulterior motives, to have a free democratic government in Iraq is a worthy, even moral aim, since it fulfills both basic axioms: humans who happen to be Iraqi possess the same dignity as any other humans and the Iraqi people are as sovereign as any other people.

Finally, France, a permanent member of the Security Council, objected with its veto,[24] so the US retracted the resolution by which it had sought to have UN authorization to attack Iraq. Therefore, according to the UN and standing international law, the Iraq war of 2003 was illegal. But did we really need the UN to tell us that? Was there no better way?

Kofi Annan, then UN Secretary General, stated the obvious according to international law, when he said that the war did not fulfill the requirements of international legality.[25] But as we stated before, present international law is not

of Human Rights of 1948 where it is the first article, and to give just one example in modern constitutions, in the German Grundgesetz of 1949, the Fundamental Law – chosen because of its solemn beauty – where there too it is the first article : "Article 1 (1) Die Würde des Menschen ist unantastbar" ("Human dignity shall be inviolable," or literally, "untouchable").

24 On March 19, 2003, following a memorable speech by Dominique de Villepin, then Foreign Minister of France: http://www.ambafrance-uk.org/Speech-by-M-Dominique-de-Villepin,4917.html.

25 "I have indicated it is not in conformity with the UN Charter, from our point of view and from the Charter point of view it was illegal," *BBC News Interview* (September 16, 2004): http://news.bbc.co.uk/2/hi/middle_east/3661640.stm.

based on fundamental axioms derived from natural law, or on the centrality of human dignity. It is a state-centric system, as is evident from the charter which claims the sovereign equality of States.[26] Note that the legitimacy of UN member States is independent from the consent of their citizens. Therefore the UN is *structurally* amoral, a superstructure of states which themselves derive their legitimacy not from democratic principles but from the recognition of other states, whether democratic or not. The system itself is based on the historical vagaries of who managed to gain power, since there is no independent standard by which a country is judged to have a "legitimate" government. A state is legitimate if it is recognized by other states. Whether the state being recognized is thuggish, or whether the state doing the recognition is thuggish, has no bearing. Therefore if we seek to find legitimacy in a process which is grounded in our fundamental axioms, we have to look elsewhere than the UN when we think about intervention. Focusing on basic principles possesses a hidden silver lining. It allows us to avoid getting caught up in the power politics of the moment, in pressure politics, in potential corruption, spying, in diplomatic wrangles, etc. In the real case of Iraq, applying this new approach earlier could have reduced by several years the duration of the cruel world embargo which made millions of Iraqis suffer. A truly principled approach does not necessarily waste time, it can make us gain time, as it would have had we followed such an approach in the case of Iraq. A truly principled approach that would have successfully created democracy in Iraq earlier would have helped to resolve other conflicts in the region, and might even have prevented 9/11 and its catastrophic consequences. Thus, acting with principles is not as Realists would have us believe, utopian, but on the contrary, in the real world, often the most pragmatic way to act.

Wavering Principles and a Slippery Slope

Yet the US wavered between a truly principled position of helping an oppressed people (based on some sort of process which we will try to determine), and the position of playing by the amoral rules of the international law system, and therefore the UN. It used the moral rhetoric of getting rid of a tyrant when it spoke to its own public opinion, and more largely to the world, but it got bogged down in the nitty-gritty of UN procedures and lost there despite exerting tremendous diplomatic pressures. The US did not ask the fundamental question: how could intervention in the affairs of Iraq be legitimized in a way that follows the axioms? Or at least, it stopped thinking about the question once it lost the battle inside the UN system.

It then acted unilaterally, because it could, thus following the principle that Might makes Right, which of course must be rejected as a normative principle and

26 Article 2, section 1 of the UN Charter: "The Organization is based on the principle of the sovereign equality of all its Members."

is incompatible with our axioms (since if all men possess equal dignity, that is contradictory with my imposing my will by force on a fellow human being). Yet, had it understood that the UN as such does not provide any intrinsic legitimacy since it itself is not based on our axioms, it would have gained time and could have implemented a truly efficient strategy. The US did make efforts to justify war using at least one of the traditional arguments of JWT, which is self-defense. To that end, it stretched and sometimes fabricated evidence, made claims that proved to be wrong (the presence of WMDs, Blair's assertion that Saddam Hussein could launch a WMD attack within 20–45 minutes of ordering it, the assertion of cooperation with Al-Qaeda, the emotional use of images of a mushroom cloud over an American city, etc.).

The US administration argued throughout that the old rules did not apply anymore because of the changed circumstances, and that therefore, preemptive measures, including preemptive war, were necessary. This disregarded the fact that an official doctrine of preemptive war is the ultimate slippery slope. Anyone wishing to convince himself of the necessity of going to war could argue that if he does not do it right away, his neighbor will attack before he does and that then it will be too late. If the US was allowed to successfully defend such a theory of preemptive war in the case of Iraq, which is thousands of kilometers away and had no means to attack the US, then what would prevent India or Pakistan, who share an unstable border, who have already had several wars, and who both officially possess nuclear weapons, from using the same justification? Either country would be on much stronger grounds to justify an all-out preemptive attack on its neighbor than the US was in attacking Iraq which shares no borders with the US, has never attacked the US before and has no nuclear weapons.

Yet comments from JWT experts and from the Catholic Church made clear at the time that this justification of the decision to go to war could not be used in this case, as there was no proven clear danger of imminent attack from Iraq on the US. Yet the argument of the US that Iraq may enable a proxy attack by terrorists is a strong one, so what to do?

Diplomatic "Oxygen" and Who Legitimately Represents Iraq?

But all this is ultimately irrelevant if you adopt a principled approach based on our axioms, since the legitimacy for action does not rest anymore on uncertainties, but on the sound assumption of equal human dignity which does not change from day to day. Let us assume we are in the summer of 2002. Ideally, how do we justify intervention in Iraq? According to our principles, the Iraqi people are sovereign, and have the right to decide their political destiny, yet the President of Iraq is a murderous dictator. No one can decide for the people of Iraq but the people of Iraq. But what to do if they cannot express themselves?

The problem is the following: who legitimately represents Iraq? In the present world order based on States rather than people, of which the UN is the epitome,

Saddam is legitimate as long as he is recognized by other States as being the legitimate ruler. Whether he kills his own people is irrelevant. In theory, it would be easy to get rid of him, since his only legitimacy derives from the recognition from other States: simply stop recognizing him as the legitimate authority of Iraq and his legitimacy falls to zero. This is equivalent to withdrawing the diplomatic "oxygen" without which no government can survive for long. But that is also the atomic bomb of diplomacy, since once you have done that, you cannot deal with the government anymore, since doing so would legitimize them again. You might talk to them as the police talk to gangsters or to terrorists who have taken people hostage, but that is all. That is also the reason why this solution is only ever used by diplomats just prior to a war, but even that is unusual. And this only works if everyone does it, if all States withdraw the "oxygen," but since a majority of States worldwide is not considered democratic, they don't have any incentive to stop recognizing one of their own.

Competing Legitimacies

But the dynamics of the problem would be completely changed, and much simpler for diplomats, if there were a choice, i.e. if Saddam's government was not the sole government claiming legitimacy. If somehow there could be another government, a more democratic one that would compete for the claim of legitimacy, then countries of the world would be forced to choose, since it is impossible to recognize more than one government at a time. Hence the problem becomes one of creating a more legitimate competing government than Saddam's, then convincing the world's countries to transfer their recognition from Saddam's government to the new government. This is a tall order, but it is possible. At least, it is something that is better than war, and since war is the last resort, we have a moral duty to try this line of attack. Remember we saw previously that the cost of war is not just the immense direct suffering and trillions spent, but also the still incalculable opportunity cost of not cooperating to solve truly dire global problems.

We also have to remember a fact that makes that idea quite plausible: global public opinion was strongly against the war, and even those in favor of the war were not for war because it was their first choice, but because they believed that there was no other way. Unfortunately no surveys were made at the time, but one can easily imagine that had people been asked in 2002 the following question (assuming the existence of two Iraqi governments, one provisional democratic one and the one in power in Baghdad): "Do you believe that our country should stop recognizing the regime of Saddam Hussein and recognize the provisional democratic government of Iraq?", the overwhelming majority, well aware of preparations for war, would have agreed. Thus even countries which might be reluctant to recognize a novice government out of power because of Realpolitik considerations might be compelled by their war-adverse public opinions to do so.

Since Saddam was in power and did not allow democratic elections, that government could not, of course, sit in Baghdad. It must be located elsewhere. Where? In theory it could be anywhere, and examples of governments in exile are well known, like the Polish government in London during WWII. In practice, it would be better if it were on some portion of Iraqi territory, and this was possible as at the time Saddam's air force was forbidden to fly north of the 36th parallel by the imposition of the Northern No-Fly Zone enforced by the US and Great Britain. Thus the Kurdish region of Iraq was a possible location.

Let us first assume this democratic government existed; how might it solve the problem of the legitimacy of war in Iraq? Let us also assume that this government was trusted by global public opinion as being as democratic as possible under the circumstances. We will see later the practical question of how to create it.

A Plausible Political Scenario: How to Avoid a Unilateral Illegal War

Call the alternative government the Transitional Democratic Iraqi Government (TDIG), and the one in power Saddam's government (SG). Here a plausible political scenario: the TDIG publicly denounces the SG, declares to the world, via TV, internet, etc. that it is the legitimate government of Iraq, and calls upon Saddam to step down, promising him a fair trial. If he accepts, game over. But he would probably refuse, so the TDIG calls publicly for all officials in the Iraqi government to arrest Saddam, arguing that he is no longer the rightful head of Iraq, is an impostor and is usurping power illegally. The provisional leader of the TDIG can even call the individuals around Saddam by name, ordering them publicly via the media to arrest Saddam. Knowing the psychology of Saddam, who imprisoned or executed anyone far and wide whom he thought *might become* a threat to him, this would destabilize him, because he would become even more paranoid and trust even fewer people. The TDIG leader would publicly promise great rewards to the man or men arresting Saddam, in effect putting a bounty on his head. Would this threat to Saddam matter at all? Yes, because of the real situation on the ground, with American forces at the ready. Remember this is an example in the real-world situation of 2002–2003, not an ideal model applicable in all circumstances. But even if US forces were not in the Gulf, the threat could be credible. Then if Saddam still did not step down voluntarily and no one arrested or killed him, then the TDIG would issue an ultimatum, backed by the threat of the intervention of its allies. This threat would be made publicly, so as to constantly remind people of the openness of the process. But the key point is that *intervention would not occur unless there were a specific request from the TDIG*, which, we must remember, at that point is the most representative government of Iraq.

Therefore, in the very worst case, war would still occur, but we would not have a unilateral and illegal war, but a war between two governments competing for the same State. If the US then invaded Iraq at the request of one of the governments, legally it would be a government calling another government for help, not a unilateral decision. In that way, one prevents the dangerous precedent of preemptive war and its disastrous consequences for the world. Coming to the help of someone else is allowed under natural law or international law. War would have proceeded as it did, for about a month, but the most important outcome would have been that instead of a foreign occupation, Iraq would immediately have had an indigenous government. That more than anything would have been the key historical deal-changer. Before going on, note that the very existence of an alternative government would probably have shortened the already short war, as it would have encouraged officials, commanders, soldiers, etc. to switch allegiances as progress was made. This is a practical application of the hypothesis we stated at the beginning that increasing legitimacy raises the chances of success, to the point that it becomes irrelevant to even abstractly consider chances of success if maximal legitimacy is achieved (except in very rare cases).

Allow us to recapitulate where we are before moving on. Either Saddam stepped down, was arrested, was killed by someone close, or lost a war. In all cases:

- unilateral war and what some call a war of aggression is avoided;
- the disastrous precedent of preemptive war is avoided;
- most importantly, occupation by foreign powers is avoided, which means that we can actually take advantage of the short honeymoon created by the fall of Saddam;
- the TDIG can push through unprecedented political reforms and compromises because it is composed of representatives of all segments of societies, and personal bonds were forged in the adversity of the process of replacing Saddam, thus creating the precious personal and political capital that is needed to govern a fractious country;
- we reduce the risk of foreign elements, especially ideological ones like Al-Qaeda, from streaming into Iraq and getting support from the local population;
- security is better, so we can focus on reconstruction faster and get it done cheaper, get results faster too, increase the wellbeing of the general public, therefore improving security etc., creating a virtuous circle;
- we have lesser overall levels of violence, allowing a more serene political process on the one hand, and on the other a more serene judicial process;
- and we can take advantage of the short window of political flux to have an unprecedented democratic constitution-making process.

Finally, the benefit side for the helping (rather than invading) power, the USA, is also tremendous:

- it gains precious time and avoids political squabbles with allies and non-allies since it does not have to spend time convincing them go to war unilaterally or to vote in favor of it at the UN, while at the same time avoiding "making the

UN irrelevant" as was often claimed at the time (since the US went ahead in spite of the lack of explicit UN SC authorization);

- it avoids ridiculing itself by using false information (like the information Colin Powell presented to the UN SC to try to justify the invasion of Iraq);
- it avoids the sterile bureaucratic debates about which excuse is best to support going to war;
- it saves the lives of many of its soldiers;
- it avoids the polarization of politics back home;
- it avoids the tremendous expenditures caused by the war, thereby reducing public debt and the deficit, and reducing the risks of later financial crises;
- it reduces the corruption caused by such huge military expenditures and no-bid or "cost-plus" contracts;
- it reduces the trend toward the privatization of defense in the USA, the use of private contractors and mercenaries, and reduces the global growth of the private army sector;
- and maybe most importantly for the US public, it allows the US to focus on their number one enemy, Osama Bin-Laden and Al-Qaeda, instead of being distracted in the sands of Iraq.

And since the USA would have been *asked* to help, it would not have needed any coalition partners (since they were only needed to provide a fig leaf of legitimacy), it would not have had to resort to pressure, bribery, etc. to get such allies, and it would have gained time and money here too. No other country would have lost soldiers in Iraq.

All in all, the chances of creating a virtuous circle where an indigenous transitional government takes charge and then creates Phase Two of the process are increased. Phase Two repeats the process which we will now describe: democratic constitution-making, meaning the design and implementation of a democratic and participative process to create a constitution[27] and a government.

An Iterative Democratic Constitutional Process

The overall process of creating a transitional democratic government, demanding Saddam Hussein step down, if necessary calling in allied governments to oust him by force, assuming power and finally creating a democratic government with the entire nation of Iraq can be called "preemptive democracy"[28] as opposed

27 Professor Vivien Hart of the University of Sussex has independently described precedents for and examples of democratic constitution-making in Special Report 107, July 2003, of the US Institute of Peace: http://www.usip.org/resources/democratic-constitution-making.

28 Not to be confused with another (negative) connotation of the expression used to describe oligarchic tendencies in deliberative democratic processes.

to preemptive war. Of course, the actual details of this process determine the probability of success. The process described below is specific to Iraq and the very real historical circumstances of that time, and could not necessarily be used as such in other situations. Situations are unique even if some general principles apply.

The key factor of success is the actual and perceived legitimacy of the original competing government: the transitional government must be trusted by global public opinion as being the most legitimate possible government under the circumstances. How would that happen? It would have to be seen as the most democratic government possible. How can we create that government?

The general idea is to use an iterative democratic process, an open and transparent constitutional process, that involves as many people as possible under the circumstances of the time, and that culminates in a final democratic constitution-making process and democratic elections in Iraq under the authority of the previous transitional government.

Creating a democratic government when a dictator is in power is not easy. Once again, specific historical circumstances of Iraq made it possible. Saddam Hussein had over the years alienated nearly every group in Iraq, had driven out people from all ethnic and religious backgrounds, including Sunni generals who sometimes barely escaped with their lives because Saddam believed that they were a threat to his absolute power. Millions of Iraqis lived abroad, representing all segments of society, all united by their fear of Saddam, and by at least a nominal and public commitment to working towards a democratic and often federal Iraq. We saw how we needed to create a Transitional Democratic Iraqi Government to actually take on Saddam Hussein with the most potent of weapons: the weapon of legitimacy.

To create that legitimacy, we suggest that the Iraqis outside of Saddam's direct control, i.e. the worldwide diaspora, possibly with the Iraqis in the No Fly Zones, should have implemented a transparent constitutional process to decide upon a provisional constitution and a transitional government. That process would have lasted several months, under the watchful eyes of the world media, and therefore global public opinion. It would thus have a pedagogic impact as well on the region, doing more for democrats and democracy in other Arab countries than any military adventure could. Iraq had a complicated constitutional history in the twentieth century and we could spend much time discussing it. But this is beyond the scope of this chapter.

The most difficult part of the process would have been the initial selection of the members of the constitutional assembly. The main criterion is that the Assembly possess maximal credibility, meaning that all groups should be represented. The suggested number of representatives is between 300 and 400 people, enough to have wide representation, while still a manageable number.

The new constitution would have been negotiated openly, under the glare of the media. One can imagine the huge media attention given to a group of Iraqi men and women of all political persuasions, ethnic and religious groups, meeting and discussing fundamental questions of basic rights, women's rights, the place of Islam, the issue of federalism, the distribution of power, etc. It is likely that accredited journalists would have outnumbered the constitutional representatives.

Such a public display of political discussion and courage would have been the most riveting political reality show, possibly drawing hundreds of millions of viewers across the world, mostly in Arab countries via Al-Jazeera, Al-Arabiya, and national satellite channels. That in itself would have deflated anti-democracy propaganda from Islamist extremists who brand democracy an evil Western import.

While Saddam would have surely tried to prevent this from happening, including by force, it would have been justified for other countries to provide security. The cost of such security, and of the conference itself, pales in comparison to the huge gain in legitimacy. Hence such expenditure on a transparent political process provides the highest return on investment in terms of later savings in reduced warfare, etc. But the presence of journalists from all over the world and the daily broadcast of proceedings would have been the best protection, as Saddam would have been totally discredited even among his last allies if he acted violently against a peaceful assembly speaking about democracy under the eyes of the world.

It is not better not to go into much detail, since pragmatism dictates the humility of deciding as one goes along and being flexible in managing the process. We have come full circle, by using our axioms centered on human dignity to design maximally legitimate institutions, and therefore reduce as much as possible the risk of war, but if war happens, to ensure that it is legitimate. Thus, we have described another sort of moral war apart from the well known moral wars of self-defense, a war that might be called a war of self-liberation, called as last resort by the people against the tyrant that oppresses it. The validity of this approach is dependent on the existence of mass media hungry enough for novel approaches that by broadcasting far and wide the proceedings, they are least likely to be hijacked by hidden interests. This was absolutely the case for Iraq in 2002.

Midwives of Democracy: A New Profession?

This brings us to one last consideration: how do we maximize legitimacy? How can such processes start? How can they keep their integrity? How do we prevent the participants from falling back into bad old habits? How to minimize the risk of manipulation, or simply errors? To whom can people turn for help in starting such processes? The usual sources, governments and NGOs, do not necessarily have the requisite expertise, but even if they did, there would always be the suspicion that they may possess some ulterior or hidden motives, political, economic, strategic, diplomatic, etc. Since NGOs are mostly funded by governments, and since they have been largely instrumentalized by governments, especially since the Iraq war, they have unfortunately collectively lost much credibility, and the number of murders of humanitarian personnel has rapidly risen in the last few years.

Maybe the solution is to use the power of competition and of free-market principles, and invent a new profession of "midwives of democracy," independent consultants, like high-level coaches for athletes, who would provide democracy services to anyone wishing to hire them. Such "democracy engineers" would specialize in designing and implementing with relevant stakeholders maximally legitimate processes. Independent democracy engineers, beholden only to their professional pride in designing and implementing democratic solutions to political problems, would help people find their own solutions.

Just as war spawned an entire class of professional "war-makers" thousands of years ago, maybe it is time to spawn a new profession of democratic peacemakers. The founders of JWT, nearly all theologians, would probably have appreciated the concept.

ASHGATE
RESEARCH
COMPANION

PART II
Major Wars in History

From Innovative Democracy to Warfare State: Ancient Athens as a Model of Hegemonic Decline

Athina Karatzogianni

> *The attempt to write another book on a subject so old and so often treated requires an explanation, perhaps even a defense ... Each generation needs to write its history for itself. Our questions are likely to differ from our fathers' and grandfathers.*[1]

This chapter focuses on less popularized aspects of Athenian hegemony and decline, starting from the capture of hegemony after the Persian wars, exploring specific strengths and weaknesses of the Athenian system, and debating the causes and the effects of that violent architect of hegemonic decline, the Peloponnesian war. The chapter sheds light on the disastrous effects of the hunt for regional hegemony and power for Ancient Greek city states, the role of political innovation through the establishment of knowledge networks in Ancient Athens, both as an enabling force to capture hegemony, but also as a factor for inciting fear and suspicion in Athens' own allies, in their fluctuating relationship with Sparta and elsewhere, especially with the halt of that innovation by war, resulting in Athenian hegemonic decline.

Hegemonic Transition after the Persian Wars

By 525 BC Spartan hegemony was established, with Sparta eventually leading the other Greek states during the Persian wars. The original Spartan alliance (Peloponnesian League) extended to the Isthmus of Corinth and included all the Peloponnesians, except Argos and Achaea. The allied states promised to have the same friends and enemies and to follow Spartan leadership. The alliance was primarily based on the fear of Argos, and the fear of having oligarchies replaced

1 Donald Kagan, *The Outbreak of the Peloponnesian War* (Ithaca and London: Cornell University Press, 1969), vii.

by tyrannies.[2] There were weaknesses to the alliance both internally and externally. Internally, conflicts between Spartan kings, the rotation of ephors and conflicts between kings and ephors caused instability, which, paradoxically for a strong constitution, produced a very unstable foreign policy. The always present threat of helot rebellion, or Argive attack, meant that the Spartans could never leave the Peloponnese for long, which meant losing allies that depended on them to defend them.[3] It is worth noting that the Peloponnesian League did not meet throughout the fifteen years of the First Peloponnesian War, until 432 when the Spartans had to call a meeting before launching the second war. By the fourth century, with Athens defeated, Sparta did not consult her allies and found herself at war with stronger ones, such as Corinth and Thebes:

> *As an Athenian spokesman complained to the Spartans in 371, "You declare enemies for yourselves without consulting your allies whom you lead against them. The result is that often people who are said to be autonomous are forced to fight against their own friends."* (Xen. Hell. 7.3.8)[4]

During the Persian wars, the Hellenic Alliance united the Greek states, but it was different to the Peloponnesian League, as it was not based on separate treaties between states and a hegemonic power, but a general covenant freely accepted, while not permitting secession. Sparta was the hegemonic power, as a Spartan was the commander in chief, but the consent of the allied states' generals was nevertheless necessary.[5] Meanwhile, the alliance between Sparta and Athens was never an alliance of states, but of factions. As Kagan notes, it was the faction of Cimon in Athens and the faction of King Archidamos in Sparta that were willing to accept limits to the hegemonic claims of their states, but in each state there were factions that were not: 'The Spartans simply were not yet prepared to share hegemony with Athens, nor were the Athenians prepared to accept Spartan checks on their ambitions.'[6]

After the Greek victories against the Persians at Plataea and Mycale in 479, Spartan unwillingness or inability to extend its borders and take responsibility of the Aegean, combined with the behaviour of the Spartan Pausanias towards the Ionians and the islanders meant that the allies were seeking a new leader. Although the initiative for the Athenian-dominated Delian league was not taken by Athenians themselves, 'Herodotus spoke the simple truth when he said that the Athenians "offered the hubris of Pausanias as a pretext" when they took away the Spartan hegemony'.[7] In the early period of the Delian League, hegemony was not domination, as Athenians exercised 'what Thucydides called a "hegemony

2 Ibid., 1.
3 Ibid., 30.
4 Ibid., 20.
5 Ibid., 34.
6 Ibid., 78.
7 Ibid., 31.

over autonomous allies who participated in common synods" ... In this synod, all members, including Athens, the hegemon, had only one vote'.[8]

The transition of hegemony from Sparta to Athens after the Persian wars was a significant factor contributing to the Athenian overall pre-eminence in the early fifth century, while the defeat of the Persians was seen as a triumph for their democratic system of government. Athenians remembered Marathon and Salamis as their finest hours, with Marathon won by the hoplites, from the wealthier classes, and Salamis won by the rowers of the triremes, from the poorer classes.[9] In Demosthenes' time, Athenians considered their 'golden age' to be different eras, depending on the issue. The Persian wars and the Delian League were the golden age in foreign policy terms, while they went further back when discussing their constitution to Cleisthenes, Solon and even the mythical king Theseus.[10]

Lastly, suffice to say, both Sparta and Athens, 'despite their rival protestations that they stood for the autonomy of the Hellenes or liberty and democracy, in fact used their leagues to secure their own political supremacy'.[11]

Political Strengths of the Athenian System

The political strengths of the Athenian system of government were many, but here the focus is on specific innovative aspects, which contributed to the supremacy of their governance structure and their hegemonic status among Greek states. These innovations are important to explore here, as they both inspired admiration in other ancient Greek states and led to Athenian hegemony, but also inspired the fear and suspicion which alienated Athens' allies and enemies alike. These aspects can be summarized as follows: political networks and *philotimia*,[12] the public–private distinction,[13] hostility to political professionalism,[14] and more importantly the innovation-promoting and learning-based context of democratic institutions and culture.[15]

Political networks were operating on the basis of what all Athenians were expected to do: help their friends and harm their enemies. At the same time, honour and power were the benefits of leadership, beside the obvious material rewards.[16]

8 Ibid., 42.

9 John Thorley, *Athenian Democracy*, 2nd edn (London and New York: Routledge, 2004), 4.

10 Hans Mogens Hansen, *The Athenian Democracy in the Age of Demosthenes: Structure, Principles and Ideology*, trans. J.A. Crook (London: Bristol Classical Press, 1999), 297.

11 Arnold Hugh Martin Jones, *Athenian Democracy* (Oxford: Basil Blackwell, 1960), 60.

12 Robert K. Sinclair, *Democracy and Participation in Athens* (Cambridge: Cambridge University Press, 1988).

13 Hansen, *The Athenian Democracy*.

14 Ibid.

15 Josiah Ober, *Democracy and Knowledge: Innovation and Learning in Classical Athens* (Princeton and Oxford: Princeton University Press, 2008).

16 Sinclair, *Democracy and Participation*, 186–7.

Personal relationships were central to these networks and there was an evolution of significance, from family and marriage ties in the mid-fifth century to wider ties of friendship in the fourth, while as 'the individual extended the basis of his support, vertical alignments were likely to become more important than horizontal dimensions'.[17]

The love of honour or *philotimia* was another central element in Athenian politics. Although *philotimia* was an ambiguous term, used to mean honour in the polis derived from good services done to the polis, but also, as part of the ethos of aristocratic competition conflicting with the ethos of the polis or the family. In its worst incarnation, it meant excessive, selfish or misdirected ambition, and it was used this way in Thucydides' analysis of the policies of Pericles' successors.[18] At the time of decline, in the middle of the fourth century, it was used in honorific degrees to encourage the Athenians to emulate the example of *philotimia* and the readiness of Demos to return favours, especially in periods when 'serious doubts had arisen among the well-to-do whether love of honor expressed in benefits to the polis still brought fitting rewards'.[19] 'We often meet in the sources the idea that reward and punishment are the two main driving forces of democracy: willingness to take the initiative must be encouraged with rewards, but promoting private interest to the detriment of the common weal must be punished by the law's harshest penalties'. [20]

Another political strength of the Athenian polis was a strict distinction between the private and the public. As Hansen explains:

> *... according to the Greek conception, most clearly formulated by Aristotle, a polis was, "a community [koinonia] of citizens [politai] with regard to the constitution [politeia]"* (Arist. Pol. 1276b1), *and politeia is further defined as the "organization of political institutions, in particular the highest political institution".* (Arist. Pol. 1278b8-10)[21]

In this conception of the polis, territory was not left out by chance, because it was always the people who were stressed and not the territory – it was not Athens and Sparta that went to war, but always 'the Athenians and the Lacedaemonians' (Thuc. 5.25.1).[22]

More importantly, the Athenians were firmly against the polis controlling every aspect of the life of its citizens. Athenians could do as they pleased in their private lives as long as they obeyed the laws and did not do any harm to their fellow citizens.[23] In fact, the polis, which was the political community, was separate from the private sphere in ancient Athens. The private sphere was not focused on the

17 Ibid., 187.
18 Ibid., 188.
19 Ibid., 190.
20 Hansen, *The Athenian Democracy*, 310.
21 Ibid., 58.
22 Ibid.
23 Ibid., 62.

individual. It encompassed not only family life and personal relationships, but the overall society, business, industry and religion. The public sphere was essentially the polis, the public administration, governance and political community of the city-state. Therefore, the Athenians distinguished not between the individual and the state, but the individual as a private person and the individual as a citizen;[24] in contrast to contemporary societies where the state prevails overall, despite privatization.[25] One only has to look to the UK as a contemporary example with its hyper-surveillance and over-regulation of the life of its citizens and failure to effectively manage finance and trade.

Innovation and the Knowledge-Based Organizational Model in Athens

Ober, in his recent work *Democracy and Knowledge: Innovation and Learning in Classical Athens* (2010), puts forward the hypothesis that democratic Athens competed successfully over time against hierarchical rivals:

> *... because the costs of participatory political practices were overbalanced by superior returns to social cooperation, resulting from useful knowledge, as it was organized and deployed in the simultaneously innovation-promoting and learning-based context of democratic institutions and culture.*[26]

Various intriguing arguments stemming from social network and organizational theory are utilised by Ober to understand the superiority of the Athenian system. For instance, process innovation, highly valued by knowledge-based organizations, is thought to be impeded by hierarchies, as they favour learning-as-routinizing at the expense of learning-as-innovation, which is paramount in highly competitive environments.[27] The Athenian system was operating in flexible small teams in horizontal governmental structures, where the mentality of peer production prevailed over rigid hierarchy. Moreover, in the participatory Athenian context, social knowledge served as a sorting device: 'Experienced citizens learned habits of discrimination, of recognizing whom to attend to and whose opinion to trust in what context'.[28]

Consequently, Ober identifies two features of Athenian decision-making institutions which conjoin the innovation-promoting and routinizing aspects of organizational learning: social/knowledge networks and task-specific work

24 Ibid., 80.
25 Ibid., 64.
26 Ober, *Democracy and Knowledge*, 37.
27 Ibid., 106.
28 Ibid., 120.

teams.[29] By serving in rotation and being educated in the democratic machine, the Athenian system made experts out of life-long learning amateurs: 'Through its day-to-day operations, the Athenian system sought to identify and make effective use of experts in many different knowledge domains'.[30] Moreover:

> *By participating in "working the machine" of democracy, the individual Athenian was both encouraged to share his own useful knowledge, and given the chance to develop and deepen various shorts of politically relevant expertise ... learning as socialization helped to sustain democracy by granting it ideological legitimacy in the eyes of the citizenry. Predictability, standardization, and legitimacy all lowered transaction costs and thereby reduced the friction inherent within every complex system.*[31]

Cleisthenes is credited with making the most important of the democratic reforms in Athens, significantly enabling the golden era of participatory democracy by creating institutions built for knowledge sharing, lowering communication costs, and context-sensitive information sorting. He did that first of all by creating the weak ties that were crucial to uniting local strong ties across Attica, bridges that were essential to knowledge aggregation. The blatantly artificial tribes in his system drew from three areas, communities located in coastal, inland and urbanized regions of Athenian territory, whereby a stable local identity was linked to a desired national identity, that of the participatory citizen of Athens.[32] As Ober puts it:

> *The system very literally "intermixed" Athenians from different geographic/ economic zones in a variety of psychologically powerful activities. The experience of marching, fighting, sacrificing, eating, and dancing, together in this newly "intermixed" grouping, would, according to Cleisthenes' plan, lead to a strengthened collective identity at the level of the polis.*[33]

Athenian citizens acted as individuals bridging holes in the network, linking sub-networks and gaining social capital in the process, similarly to the process described in Ron Burt's theory on structural holes.[34] The ability of the Athenian citizen to learn and be part of a knowledge network runs contrary to the 'ignorant mob' assumption often put forward by historians.[35]

29 Ibid., 123.
30 Ibid.
31 Ibid., 273.
32 Ibid., 138.
33 Ibid., 142.
34 Ibid., 146. Cites Ronald S. Burt, *Structural Holes: The Social Structure of Competition* (Cambridge, MA: Harvard University Press, 1992).
35 Ibid., 162.

In terms of leadership, the Athenian polis did not depend on authoritarian leaders or hierarchy, and rejected Spartan-style hyper-socialization. Athens relied instead on choices freely made by free citizens to gain its public ends: 'intermixing the four mechanisms ... for facilitating complex coordination: first choice, informed leader, procedural rules, and credible commitments'.[36] For example leadership would shift readily depending on which individuals or groups happened to know something useful, with consensus following from plurality and alignment.[37] Balancing elite and non-elite preferences meant, for example, that someone like Themistocles in the debates leading to the Athenian decision to fight at Salamis in 480 were able to assume leadership roles by advocating and carrying through innovative policies.[38] It was indeed his genius in Salamis of using the change of wind direction at a specific time of the day and the agile small Greek ships to devastate the large inflexible Persian ships. Lastly, Athenians valued innovation as a good in itself, because it was a manifestation of their communal identity, of what they supposed was special and excellent about themselves as a people – as well as valuing innovation as an instrument.[39]

Weaknesses in the Athenian Democracy

The crisis of the late fifth century saw a short-term collapse of democratic institutions, disruption of the Athenian social equilibrium and a corresponding decline in productive capacity.[40]

It is well known that slaves, women and foreigners were excluded from citizenship, and this is seen as a major weakness of the Athenian state. There were also age limits, such as the rule that jurors, legislators and magistrates had to be over thirty years of age. This meant that, because one third of all citizens were between the ages of eighteen and thirty, every third Athenian had only limited citizen rights. They could attend the Assembly and vote, but could not be a juror or magistrate.[41] Another frequently mentioned flaw in the Athenian democracy is the growth of professionalism in politics, especially toward the end of Athenian hegemony. In the golden age of Athens all citizens were encouraged to take part in running the state, but all were to be amateurs, and professionalism and democracy were regarded as, at bottom, contradictory.[42] As Hansen explains, Athenians were hostile to professionalism, because they believed political participation as compatible with one's everyday job, as something citizens did in their spare time, and participation

36 Ibid., 179.
37 Ibid., 174.
38 Ibid., 182.
39 Ibid., 275.
40 Ibid.
41 Hansen, *The Athenian Democracy*, 89.
42 Ibid., 308.

was, at the height of democracy, astonishingly high.[43] In the fourth century, the ordinary citizen was much less able to lead in the political field, as the recruitment of leaders was from the wealthier classes, as well as due to the increasing numbers of professional politicians.[44]

Often mentioned as the reason for success and high political activity is slavery. Although having a slave would have made political activity easier, Athenians did work for a living, and often slaves were wealthy and had their own slaves. It is not easy to explain away the rise of the Athenian democracy through slavery as a necessary precondition.[45] Neither was the Athenian democracy dependant on the proceeds of a naval empire. Some historians have coincided hegemonic decline with the collapse of democratic institutions. Hansen explains that such an assertion is not valid, because payments to attend the Assembly, and to the *theorika,* were introduced in the fourth century, when Athens had already lost her empire. Furthermore, democracy in Demosthenes' time was far costlier than in that of Pericles.[46] A more likely explanation for the bankruptcy was the endless wars Athens was engaged in rather than the political payments and the running of the democratic institutions.[47]

To conclude the discussion on Athenian strengths and weaknesses, the presence of unique and mostly organically evolved traditions enhanced the political environment in Athens, and enabled its citizens to advance new ways to govern themselves as people, and to attract the initial admiration of other city-states and Athenian leadership in war, politics, culture and trade. The same traditions were responsible for the major advances in terms of innovation in the creation of knowledge networks to support the Athenian ideal of humans as political beings seeking honour and power, as *philotimia* for the common good; acting within established knowledge networks of participation, power and civil engagement, resulting in innovative democratic mechanisms of government as peoples and not as territories. In conjunction, the decadence of the traditions, which came during war, when politics became more professionalized, and more about individual advancement rather than the common good of the community and the polis, contributed to the breakdown of innovation and the eventual decline of Athens. It is this time one needs to turn to, a time when Athens was at its greatest hegemonic peak, in order to understand how this great innovative power was suddenly overwhelmed with the weaknesses identified above, and how these failures in the democratic system rendered the innovative aspects not only useless, but also a symbol of fear of Athens by allies and enemies alike.

43 Ibid., 309.
44 Ibid., 276.
45 Ibid., 318.
46 Ibid., 318.
47 Ibid., 317.

Empire Decline: The Transition to Warfare State

Athens' loss of her empire to Sparta in 404 BC came after a catastrophic expedition to Sicily, which started in 415 with Athens attempting to expand her hegemony with loyal support from her allies. By the autumn of 413, though, their force in Sicily was annihilated, while oligarchic factions in allied states, encouraged by Sparta, revolted. The Spartans, with Persian funding, built triremes with which they challenged the Athenian naval supremacy and eventually won the Ionian War (412–405) at Aigospotamoi in the Hellespont. The Athenians were starved to death having lost their navy, without which they could not protect the grain from the Black Sea.[48]

One of the popular explanations for the two Peloponnesian wars, with a thirty-year peace in between, was the fear of Sparta for Athens. De Ste. Croix mentions the now famous Thucydides quote: 'It was the growth of the power of Athens, and the alarm which this inspired in Lacedaemon, which made war inevitable' Thuc. i.23.6.[49] After the defeat of the Persians, Athens and her allies, the Delian League, pursued aggressive wars against the Persians and succeeded in driving them out of the Aegean, with a *Pentecontaetia* or fifty years of Athenian Empire established, transforming allies into paying subjects of the empire and using their money to fund public works glorifying the Athenian state. This caused fear in Sparta and resentment in Athens' allies, especially when Cleon, who replaced Pericles, persuaded Athenians to let their lazy subjects eastwards and westwards pay for the privilege of Athenian rule. Athens broke the Charter of her Empire – the contract drawn up two generations before by Aristeidis the Just between Athens and her Allies – and doubled the rates of tribute.[50] In 413 the *phoros* was replaced by a 5 percent duty on seaborne goods.

In the massive bibliography on the Peloponnesian wars, the wars that contributed to, if not outright caused, Athenian hegemonic decline, there is plenty of controversy over what were the underlying causes, whether war could have been avoided and where lay the responsibility for war. It has been argued that, setting the fear and resentment of Athens aside, the two powers were dragged into war due to alliance commitments and rivalries between allied states, such as Corcyra and Corinth, that could not be left unchallenged lest the balance of power be disturbed. In fact, the role of proxies/allies in the break-up of Athenian hegemony cannot be overstated. In the first war in 459, during a rivalry between Corinth and Megara, both Spartan allies, Athens took the Megarian side, and a 15-year war ensued which only ended when Attica was invaded and both sides agreed to respect each others' territories and alliances. A 30-year peace was signed in 446, which broke when Corinth was defeated by their colony Corcyra. Athens supported Corcyra and then

48 Anton Powell, *Athens and Sparta: Constructing Greek Political Social History from 478 BC* (London: Routledge, 1988), 74.

49 Zimmern 1915, quoted in Geoffrey Ernest Maurice de Ste. Croix, *The Origins of the Peloponnesian War* (London: Gerald Duckworth, 1972).

50 De Ste. Croix, *Origins*, 437.

in turn faced Corinth's wrath and meddling in the Athenian ally Potidaea, which was encouraged to revolt against Athens. The Corinthians ferociously pressed the Spartans at a meeting of the Peloponnesian League to declare war on Athens, while Athens was asking Sparta to resolve the conflict through arbitration as the Peace treaty allowed for.

The role of the Corinthians was paramount, as they used propaganda and kept urging individual states to convince them to vote for war. They argued that war was absolutely necessary, as Athens was so powerful that she could defeat all the Greek states one by one, so the only chance was to unite against her. They argued that submission to Athens was an automatic establishment of a tyranny.[51] The Spartans did decide that the Athenians broke the peace and by default went to war, the Archidamian War (431–421). A plague in Athens, during which Pericles died, was a contributing factor to the initial defeat which was overturned when Athens made a comeback with Cleon.

Eventually a truce was signed, the Peace of Nicias, which was broken this time when Argos, an independent state in the Peloponnese, decided with the help of the Athenians to create a coalition of democratic states in the region. This culminated in the battle of Mantinea, the biggest land battle of the war, which Sparta won, breaking up the new democratic alliance in the Peloponnese and restoring her regional hegemony. The Sicilian expedition, which followed, was linked again to an Athenian ally in Sicily being under attack from Syracuse. The biggest protagonist, the Athenian Alcibiades, who encouraged the expedition, was accused of religious crimes in Athens, but was allowed to leave for Sicily without being tried, and was subsequently immediately recalled upon arrival. He defected to Sparta and encouraged them to aid the enemies of Athens, with Athens eventually defeated by the Syracusans and their allies. As Kagan explains:

> *The personal rivalries, factional disputes, and general distrust that swirled around this unique figure in Athenian life did cause his city great harm and had much to do with Athens' loss of the war. The most serious consequence of Alcibiades disgrace was that it removed his friends and associates from influence and command when their military and political skills were most needed.*[52]

Ober makes the point that the costs to an organization before making major policy choices can be extraordinary high, especially as knowledge collection becomes more complicated.[53] The Alcibiades story points to such failures, even after Athens managed to recover parts of the Empire between 410 and 406, when Alcibiades returned, despite having been condemned as a traitor. Finally, Athens lost their navy and their empire in a freak and unexpected Spartan naval victory

51 Kagan, *Outbreak*, 314.

52 Donald Kagan, *The Fall of the Athenian Empire* (Ithaca and London: Cornell University Press, 1987), 420.

53 Ober, *Democracy and Knowledge*, 120.

in Aigospotamoi, never to recover. An oligarchic regime was set up by Sparta, the Thirty Tyrants, but democracy was restored by Thrasyboulos in 403. Thebes, and later on the Macedonians, sealed the fall of Athens from a hegemonic power.

Aside from the role of the interlocking alliances intensifying the struggle for hegemony between Athens and Sparta, the role of factions inside allied states was also critical. From one moment to the next, factions in allied cities would take over and pledge allegiance to the Spartans, if oligarchic, and to Athenians, if democratic. Within the Athenian empire, patterns often corresponded to the fluctuating successes of military fortunes, with J. de Romilly even suggesting that there was an exact correspondence between Athens' military position and the power of the democratic and oligarchic factions in the allied states of the Empire.[54] There were some not wholly unselfish incidents, for example in Samos, when a coup was mounted in favour of *democratia*, but later Samos itself became oligarchic. Even worse, in Athens in 411 the plotters were willing to hand the keys to the city and the empire to Sparta, provided their lives were spared. There is therefore a connection to be made about the internal politics of allied states' contributions to the break-up of hegemony, and the constant pressure for the hegemon to intervene. 'The link between internal politics between factions and external military balances also has parallels in the positions of pro-US and communist forces in various countries in the Cold War.'[55]

Furthermore, the effects of civil strife across the Ancient Greek world at the time are reflected in chilling accounts by Thucydides, Xenophon and Diodorus, with anger, frustration and violent outbreaks of vengeance accelerating as the war dragged on:

> *... they produced a progression of atrocities that reached the point of maiming and killing captured opponents, throwing them into pits to die of thirst, starvation, and exposure, hurling them into the sea to drown, enslaving and killing women and children, and destroying cities with their entire populations. This war, more than most, was "a violent teacher".*[56]

When the war started, the Athenians, having had success at little cost to human life, were reluctant to do what was necessary for victory.[57] Kagan argues that the innovative, swift, aggressive Athenians were less able to adjust to the new situation than the slow, traditional, unimaginative Spartans the Corinthians described. More importantly, with warfare in democracies where everything must be openly debated, it is harder to adjust to the necessities of war than in less open societies.[58] This type of argument was famously made also in the case of the US during the Vietnam War.

54 Powell, *Athens and Sparta*, 81.
55 Andrew Robinson, email communication with the author, July 2010.
56 Kagan, *Fall of the Athenian Empire*, 415.
57 Ibid., 426.
58 Ibid.

In the final analysis a lot can be said about the transition of Ancient Athens from an innovative democracy to a warfare state. Perhaps the elements that made Athens great, as described above, political networks and *philotimia*, were breaking down due to factional disputes and a spell into oligarchy on the way to decline. The public–private distinction was blurred due to the wars, while political professionalism became the order of the day during hegemonic decline. The innovation-promoting and learning-based context of democratic institutions and culture was also disrupted severely due to the wars and their devastating effects on society and culture.

The Problem with Thucydides

One of the most notorious moments for the Athenians was the Melian incident, also described as a war crime and a massacre. Thucydides has penned a dialogue between the Athenians and the Melians which raises issues about Athenian hegemony but, moreover, about Thucydides' motivations and understanding of the event and the Athenian warfare state. The Melians in Thucydides' account are pleading with the Athenians not to destroy their city:

> *It is expedient that you should not destroy what is our common protection, the privilege of being allowed in danger to invoke what is fair and right. Surely you are as much concerned in this as any, since your fall would be a signal for the heaviest vengeance, and an example to all the world.*[59]

The Athenians supposedly reply to this:

> *"We feel no unease about the end our Empire, even if end it should," came back the proud answer, as if to challenge the high gods; "a fellow Empire, like Lacedaemon – though it is not she who is our real enemy – is not so terrible to the vanquished as subjects who by themselves attack and overpower their rulers. This, however, is a risk that we are content to take."*[60]

It is worth pointing out that Thucydides cannot have had any information on the Melian debate, because it was held behind closed doors between the Athenian commissioners and the Melian government, who were subsequently executed, so it must be regarded as a free composition. It is essentially what Thucydides imagined to have been said between the two parties before the massacre took place.[61] In fact, Arnold Jones rightly challenges these speeches. He finds it remarkable that the Athenians of the fifth century would openly admit in this manner that their

59 De Ste. Croix, *Origins*, 439.
60 Ibid.
61 Jones, *Athenian Democracy*, 66.

policy was guided purely by selfish considerations and that they had no regard for political morality, but also that they underwent a complete transformation in the fourth century, from which we possess genuine speeches.[62] Jones actually has a point when he concludes that Thucydides put into the mouths of Athenian spokesmen what he considered to be their real sentiments, stripped of rhetorical claptrap, in effect revealing his own opinion of the empire:

> *His view was that Athens was universally hated by her allies or subjects, who were held down by fear or force only, and were eager to revolt at every possible opportunity – this thesis he twice states in his own person apart from the speeches – and that Athens was wrong in "enslaving" them, by her refusal to allow them to secede from the league and by her interference in their internal government. Furthermore, that the Athenians, to enforce their tyranny (as with Mitylene) or to enlarge it (as with Melos) committed or very nearly committed acts of the grossest brutality.*[63]

Melos was not an unoffending neutral, as it was an ally of Sparta at the beginning of the war, subscribing to her war fund and sheltering her fleet in 427. Athens had been at war with Melos since 426.[64] Athenians were accused of barbaric acts against the Mitylenaeans and Scionaeans, who in their eyes were traitors, while Melians assisted their enemies. By contrast, in the Spartan massacre against the Plataeans, the Plataeans were defending their own city from being attacked by Thebes in peacetime. Jones puts forward the thesis that Thucydides lived in exile in oligarchic circles, and he appeared to have really believed the Athenians were hated by their allies, whereas the Peloponnesian League was a free association of cities. For a historian of Thucydides' calibre, though, this is quite hard to believe. It is easier to believe that he greatly desired to find a moral justification for the fall of Athens: 'It was not enough to say that it was due to the folly of the democratic politicians whom he so much disliked. It must have been deserved. Athens had suffered grievously; this could not have been so if she had not sinned greatly.'[65] Powell explains that Thucydides had a low opinion of the conduct of the war by both great powers, but he believed that Athens was brought down not by Spartan genius, but by Athenian mistakes.[66]

Moreover, Thucydides' explanation that war was inevitable because of the growth of Athens and its insatiable demand for expansion, has not persuaded everyone. Kagan, for example, argues that:

> *Athenian power did not grow between 445 and 435, that the imperial appetite of Athens was not insatiable and gave good evidence of being satisfied, that*

62 Ibid.
63 Ibid., 67.
64 Ibid, 71.
65 Ibid., 72.
66 Powell, *Athens and Sparta*, 198–9.

the Spartans as a state seem not to have been unduly afraid of Athenians, at least until the crisis had developed very far, that there was good reason to think that the two great powers and their allies could live side by side in peace indefinitely, and that it was not the underlying causes but the immediate crisis that produced the war.[67]

Similarly, Kagan counters the belief that the causes of war can be found between Athens and Sparta, as the Greek world between the years of the Persian and the Peloponnesian wars was not bipolar. Even if Athens dominated her allies, Sparta did not with Thebes and Corinth (who had a special role to play throughout the Peloponnesian War) being free agents.[68] The Spartan alliance permitted a power of the second magnitude to drag the regional hegemon into war for its own interests.[69] As explained above, the interlocking alliances affected the Athenians, who could not permit Corinth to attack Corcyra and add that fleet to the Spartan alliance therefore questioning naval supremacy, the basis of Athenian security.

The question remains, though, whether these alliances were more of a factor for the war, rather than the fear of Athens by allies and enemies alike. Despite certain weaknesses in Thucydides' account identified above, his take remains the most plausible explanation for the cause of the war, and supports this author's thesis that both allies and enemies of Athens were suspicious of and intimidated by Athenian innovations and overpowering influence during her hegemony. In turn, innovation can make it difficult for leadership to assess capabilities and this is exacerbated when factions alternate in favour of different political systems, both internally and in the allied states. As Hall Gardner argues elsewhere in this volume:

Alienation among states is further extended (in a psychological sense) by situations and conditions in which socio-political, political-economic and/or military-technological innovations of differing forms take place, often making it even more difficult to properly assess force capabilities ... Changing socio-political circumstances make it more difficult to determine precisely which political faction might be in command and thus what is the true intent of the leadership. Alienation is likewise exacerbated if political factions (which may or may not be able to obtain strong domestic or external supports) use violence, strikes or acts of "terrorism" in order to assert their interests or gain recognition for their cause. If deterrence or dissuasion fails, then conflict can accordingly break out. (Gardner, Chapter 1)

67 Kagan, *Outbreak*, 345.

68 Ibid., 350.

69 Ibid., 353.

Concluding Remarks

After the Peloponnesian Wars Athenian political culture transformed, with loyalty to friends being valued before loyalty to country and ambition before morality; it was natural to seek scapegoats to save face, while vengeance was not seen as merely acceptable, but a moral obligation. Between 403 and 386, over twenty generals, magistrates, ambassadors and orators stood trial in Athens, with more than half convicted and five executed – the Athenian leaders were more likely to fall in domestic politics than in battle during the Corinthian war.[70] As Strauss argues, post-war politics were shaped by the loss of empire and confiscation of property, and the demographic consequences of battle: casualties, epidemic, hunger, poverty and perhaps birth control.

The democratic institutions that promoted knowledge networks and the innovative organizational model, which enabled Athenian hegemony, recurred as the ideal of the 'ancestral constitution' that the Athenians dreamt of re-establishing every time Athens lost a war.[71] Hansen points to the constitutional debate being almost a function in foreign policy:

> *In 404 the Athenians suffered the most serious defeat in their history, and in the years that followed first the Thirty (in pretense) and then the returned democrats themselves attempted to restore the "ancestral constitution"; in 355 Athens lost her war against the Allies who had revolted and again the dream of the "ancestral constitution" turns up in the sources; in 338 Athens was finally defeated by Philip, and at once we meet reforms designed to restore "the ancestral constitution"; in 322 Athens was actually captured by the Macedonians, who set about re-creating the "ancestral constitution" – only this time since it was dependent on the military power of Macedon, it appeared in an oligarchic guise, whereas changes up to then had never been more than modifications of the democracy.*[72]

Besides the failure to re-establish the democratic institutions and the innovative model of organization accompanied by the devastation of war as an explanation of Athenian hegemonic decline, there might be more similarities with contemporary global politics than initially meet the eye. As the bipolarity thesis is not credible in the Peloponnesian War, an analysis would need to go beyond the obvious Cold War comparisons. In that fashion, as Ober argues, the multi-state Greek world in certain ways seems a miniature version of the post-1989 contemporary world:

> *... in both cases we find several hundred states of various sizes engaged in interstate economic and military competition. Borders are, on the whole, fairly*

70 Barry S. Strauss, *Athens After the Peloponnesian War: Class, Faction and Policy 402–386 BC* (London and Sydney: Croom Helm, 1986), 172.

71 Hansen, *The Athenian Democracy*, 300.

72 Ibid.

> *stable, and systematic border violations by aggressor-states trigger vigorous military response. Many of the most successful states feature republican/democratic institutions and relatively open-access economies.*[73]

Ancient Athens owed its hegemony to innovation in knowledge networks permeated by democratic traditions organically involved in a specific space and time. When this innovation and the concentration of wealth and power resulted in hybris and threatened the other city states, the decline inevitably came through the halt of innovation and the overpowering of the Athenian system by its weaknesses.

The contemporary move toward sustainable ecology, ethical economy, fair trade and alternative capitalism through horizontal peer production, knowledge and social networks promoting innovation with successful examples in the technology industry, might be pointing to an organizational environment that is 'characterized by diversity rather than homogenization, distributed knowledge rather than centralized expertise, democracy and choice rather than command and control'.[74] This appears to be an emergent environment similar to the one responsible for the admirable Athenian golden age. It remains to be seen if it will prove, in our times, one capable of overpowering the weaknesses of democracy and capitalism to escape global catastrophic decline.

73 Ober, *Democracy and Knowledge*, 278.

74 Ibid., 279.

Origins of Arab-Israeli Conflicts in the "Greater Middle-East"

Marco Rimanelli

Since Antiquity the geo-strategic role of the "Greater Middle-East" region (from Morocco in the West to Arabia and Iran in the East) has grown in world trade with its vital land-, sea- and air-bridges to Europe, North Africa, Central Asia and India via the Straits of Gibraltar to the Suez Canal and Turkish Straits. Also, the geo-economic importance of the Middle-East as the energy "life-line" of the global economy has skyrocketed since the 1960s, as it is the main oil/gas source with two-thirds of all oil reserves and 50 percent of oil imports by the United States (US) and West (Europe, Japan, Canada, Australia). Thus, the old vague geo-political term of "Middle-East" has slowly expanded from its traditionally narrow historic "Green Crescent/Middle-East" (Israel, Egypt, Jordan, Lebanon, Syria and Iraq) to the geo-strategic "Middle-East/Gulf" of the Cold War (1946–90) and the current "Greater Middle-East" (2004 G-8 Summit) now encompassing the entire "Arab World" from North Africa's Maghreb to oil-rich Arabia/Persian Gulf (or "Gulf" since 1980) and regional non-Arab Muslim states (Iran, Turkey, Afghanistan), plus Israel.

Throughout history, control of the "Greater Middle-East" has been contested by each era's Great Powers (Ancient Egypt, Persian Empire, Hellenistic Empire, Carthage, Ancient Rome, Venice, Ottoman-Turkish Empire, Spain–Austria, Great Britain, France, Russia, Germany), up to the US and USSR superpowers during the Cold War, but rarely has any one Power succeeded in unifying the entire geo-strategic region under a single long-term empire (Ancient Rome, the Muslim Omayyad-Abbasid and Ottoman-Turkish Empires). Only rarely, is this geo-strategic region also expanded to all ex-Soviet Muslim states (Kazakhstan, Uzbekistan, Kyrgyzstan, Turkmenistan, Tajikistan and Azerbaijan) to reflect this parallel area of operation of the US Central Command, but such broad geo-strategic additions are irrelevant to Middle-East politics.[1]

---

1 Arthur Goldschmidt Jr. and Lawrence Davidson, *A Concise History of the Middle East* (Boulder, CO: Westview, 2006), 1–85; Tom Clancy and Gen. Tony Zinni, *Battle Ready* (New York: Putnam, 2004).

Finally, the region's volatile politics are impacted by its historico-spiritual significance as the birthplace of the world's three rival Monotheist religions—Judaism, Christianity and Islam—each claiming exclusive spiritual interpretation of God's will and a hold on the "Holy Lands." After Christianity, in the twenty-first century Islam is the second world religion with 1.3 billion Muslims split among Sunni (84 percent) and Shi'a (15 percent) in the Middle-East/Gulf, North Africa, West Africa, parts of East Africa and Asia (ex-Soviet Central Asia, Pakistan, Indonesia, Bangladesh, Malaysia), with minorities in India, China, Russia and the Balkans, plus immigrant communities in Europe and the Americas. Yet, by 2010 only 20 percent of Muslims live in the Arab "Greater Middle-East", while 30 percent are in the Indian subcontinent, and 18 percent in Indonesia and Malaysia.

The Arab "Greater Middle-East" totals 350–420 million people (with large Christian minorities under Greek Orthodox, Greek Catholic and Maronite churches, plus Druze factions) and Pan-Arabism as a common identity is represented by the Arab League. Arab Muslims are Sunni (85 percent) with Shi'a minorities (13 percent in non-Arab Iran/Persia and Azerbaijan, plus minorities in Arab Southern Iraq, Bahrain, Saudi Arabia, Lebanon, Kuwait, Syria, Oman, Yemen, and minorities also in non-Arab Turkey). The largest Arab Sunni countries are Egypt, Iraq, Algeria, Syria and Morocco, followed by other Arab states of the Arabian Peninsula, Jordan, Lebanon, North Africa's Maghreb and Sudan. Outside the "Greater Middle-East," 12 million Arabs live in Brazil, 2 million in Iran, 1,600,000 in the US, 1,414,000 (Palestinians) in Israel, 1 million in Mexico, and 26 other nations have minorities of 100,000 Arabs each.[2]

The Ancient Middle-East had often been dominated by early-Semitic pagan raiders who overran the first non-Semitic civilizations of Mesopotamia and intermixed with them; later, all were subjugated by non-regional non-Semitic Empires (Persia, Alexander's Hellenistic Empire and the Roman Empire), although the entire region long used the Semitic Aramaic as its *lingua franca* alongside Greek and Roman. Today, most people of the "Greater Middle-East" derive from Semitic-Arab populations which emigrated out of *Jazirat al-Arab* ("island of Arabs" or the Arabian peninsula) in successive waves from 2,000 BC to 700–1000 AD and were absorbed religiously/culturally by the expanding Arab-Islamic Faith created by Muhammad to unite warrior Arab tribes against Persia and the Christian Byzantine Empire (the successor to Ancient Rome).

In the 600s AD Muhammad in Arabia proclaimed himself Prophet, fighting against local Paganism by promoting Islam as his new-founded Monotheistic religion based on his *Qur'an* (holy book of the word of a single God/*Allāh*). Muhammad united Mecca, Medina and the entire Arabian peninsula in a unitarian Islamic socio-political community (*Umma*). At his death, the Caliphs' elected political rule and Arab traditional quest for new lands and booty under Islam's religious fervor led them to conquer and convert the "Greater Middle-East," exploiting the weakness

2 John L. Esposito and Dalia Mogahed, *Who Speaks for Islam? What a Billion Muslims Really Think* (New York: Gallup Press, 2007); A. Goldschmidt Jr. and L. Davidson, *A Concise History of the Middle East*, 5–100.

of both the rival Byzantine and Sassanid-Persian Empires, whose long wars (603–628) and Byzantine victory had exhausted both. These first and second expansions of Islam in Middle-Eastern lands rapidly overran the Byzantine Christian Middle-East (Palestine, Syria, Armenia, eastern Anatolia and Egypt) in 634–645 AD, plus the Sassanid-Persian Empire (Persia and Mesopotamia/Iraq) in 633–651, as well as Byzantine and Vandal North Africa in 647–709, in 711–718 the Iberian Peninsula and Mediterranean (Cyprus, 647 AD; two Sieges of Constantinople, 674–678, 717–718; Crete, 820; Sicily and Southern Italy, 827–1091, 847–982), creating the ever-expanding Islamic Arab Omayyad and Abbasid Empires in 640s–1000s. Over many centuries Muslim pirate raids in the Mediterranean and Islamic conquests spread terror among "non-Believers," conquering Central Asia (662–709 AD), Caucasus (711–750), India (664–712, 1,000 AD), Nubia/Sudan (700–1609) and Indonesia/Malaysia, with cyclical invasions of Europe's periphery by sea and land (Franks' Kingdom in 721–732 and Anatolia in 1060–1360), while expanding the Arab world though migrations within the "Greater Middle-East". Islam's expansion and two great Arab Empires (Omayyad and Abbasid) led to the consolidation of these disparate lands and peoples until the 1200s.[3]

Yet Islam's expansion was always beset by internal religious factionalism and strife, especially the rebellion by the growing Shi'a minority (dominating Persia/Iran). Shi'a Islam also follows the *Qur'an*, but with a completely independent system of religious interpretation and political authority it holds that the prophet Muhammad's family, and some of his individual descendants, were divinely sanctioned as "Imam" successors of Muhammad to rule over the *Umma*, starting with Muhammad's cousin and son-in-law 'Alī, but were deprived of power by the dominant Sunnis' "elected" Caliphs who denied that Muhammad had appointed 'Alī as his successor. Only 35 years later, in 656 AD, did 'Alī become the fourth Muslim Caliph, but his rule remained contested with many unsuccessful wars until he was murdered in 661 AD, while in 680 AD at the Battle of Karbala his son Hussein and 71 loyalists were killed by the Caliph.

Islam's long-term expansion was based on several factors: a) the Arabs' upbringing as nomadic desert warriors who could rapidly hit the entire Fertile Crescent (the Middle-East from Persia to Egypt) from their desert strongholds and were united for the first time by Muhammad in the great politico-religious *Umma* of all Islamic "true-Believers," who readily turned their weapons and faith against all "infidels" (pagans, Jews, Christians and even dissident Muslims sects such as the Shi'a, Druze and Bahai'i); b) Muhammad's successors, the "Caliphs," conquered first the Middle-East in 50 years, then in the following century the entire southern Mediterranean region up to the Maghreb (Libya, Tunisia, Algeria, Morocco, Mauritania) and Iberian peninsula in the West, plus reaching Central Asia and India by 1000 AD, with Arab traders and fleets converting Indonesia, Malaysia and coastal East Africa; c) Arab conquests created the multi-ethnic Omayyad and Abbasid Empires, where the Arabs monopolized the government, military and religious functions, using Christians and Persians to forge vast

3 Efraim Karsh, *Islamic Imperialism: A History* (New Haven: Yale University Press, 2001).

bureaucracies as ruling structures during the Islamic "Golden Age;" d) as Muslim expansion consolidated, subjects converted to Islam (Muslims paid no taxes), as did invading Turks.

Finally, from 945 to 1260 internal politico-religious factionalism and strife fatally weakened the Omayyad and Abbasid Empires, which collapsed in stages once they were overrun by internal and external invasions: a) in 945 the Persian Buyids captured Baghdad and destroyed the Abbasid Empire, while reviving Persian influence with its distinctive cultural-religious Shi'a influence within the Islamic world; b) later Islamized Sunni Turkic nomads coalesced around the Seljuk-Turks who, from the Central Asian borders, attacked the weak Persian Buyids, conquered Baghdad and the Middle-East in 1055, reestablishing the Abbasids in power under the Seljuk-Turkish Empire, which also wiped out the Byzantine forces at Manzikert in 1071, slaughtering the local population, conquering and Islamizing most of Christian Anatolia (today's Turkey); c) as the Seljuk-Turk Empire also collapsed by 1092, it retrenched in newly Islamized Anatolia as Rum-Seljuks until 1248 when it too was overrun in 1218–27 by the Mongol invasions of Genghis Khan who conquered and depopulated all of Muslim Central Asia, killing almost a million Islamized Turks and Arabs; d) exploiting the Arab crisis, Byzantium and the Christian Kings of Europe cooperated in launching the Crusades to deflect further Muslim attacks on the Byzantine Empire, while liberating the Holy Lands in 1097–99 until the Muslim re-conquest of Jerusalem in 1187–92 by Sultan Salah al-Din (Saladin); e) the death of Genghis Khan only slowed the later Mongol invasion of the decaying Abbasid Empire in 1256–60 and the conquest of Persia, Iraq, Syria and Middle-East by Hulagu Khan, who destroyed the Abbasid Caliphate in 1258 and threatened Egypt in 1260, before an inter-Mongol succession crisis forced the Mongols back to Central Asia leaving behind another million Muslims dead.[4]

The Arab-Islamic world of the Middle Ages never recovered from the collapse of the Arab-Turkish neo-Abbasid Empire under the Mongols in the 1250s–60s. This left Arabs and Persians in fragmented regional states for 200 hundred years, while regional chaos and infighting allowed the liberation of Spain and Portugal from Arab rule (*Reconquista*), as well as the commercial penetration of competing Italian Maritime states (Venice, Genoa, Amalfi and Pisa) breaking open the Muslim monopoly of trade to Europe of expensive and exotic spices from the Far-East (India, Indonesia and China). However, within two centuries the Islamic world was reunited in a formidable empire (but a non-Arab one), threatening Europe again with conquest and oppression under the Islamized Sunni Ottoman-Turks from 380s to 1918.

Since the late 1300s the surviving remnants of the Central Asian Islamized Seljuk-Turks, plus other nomadic Turkish tribes, were reorganized under Osman into the Ottoman-Turkish Empire, which kept alive both nomadic traditions and constant raids against the Byzantines and other Europeans. In a century, Osman and his able ten Ottoman Sultan successors expanded their empire to all of Anatolia and the Balkans (Greece, Bulgaria, Albania, Macedonia, Serbia and

4 A. Goldschmidt Jr. and L. Davidson, *A Concise History of the Middle East*, 85–130.

Bosnia) in Europe, destroying the Byzantine Empire (Constantinople fell last to the Turks by 1453). This completed the first historical "Islamic duty" of destroying the regional Christian successors of Rome, a task in which neither the Arab Empires nor the Seljuk-Turks had succeeded. The Ottoman-Turkish Empire made Istanbul/ Constantinople its new capital, imposed a Turkish Blockade to stop trade with Europe and fostered pirate sea-raids.

Then, in the late 1400s/1500s, the Ottoman-Turkish Empire fulfilled its second historical "Islamic duty" by extending its unifying Islamic rule and nomadic kingdom over the "Greater Middle-East", annexing Persia and the Arab world (Syria, Palestine, Iraq, Arabia, Egypt, Libya and Tunisia), plus the Maghreb's Barbary pirate states of Algeria, as Turkish vassals (only Morocco, Yemen and Oman remained independent Arab states). The Ottoman-Turks' Muslim conquests and pirate raids, which terrorized Christian Europe in 1500s–1689 and subjugated also the higher Balkans (Romania, Croatia, Transylvania and Hungary) and Ukraine, were finally stopped by the Spanish-Austrian Empire in the Mediterranean with the naval Battle of Lepanto in 1571, followed by the Two Sieges of Vienna in 1566/80 (when terrible Winter conditions doomed the Turks) and 1688 (when an Austrian-led Christian European coalition defeated the Turks).[5]

The Ottoman-Turkish Empire's successes were due to a number of factors: a) forcible recruitment of conquered men to expand Turkish invading armies, centered on the feared élite Janissaries (traditionally recruited from Christian slave child-soldiers) as fanatical Islamic soldiers; b) Turkish speed, mass and valor, while also being the first to employ gunpowder and cannons since the Siege of Constantinople; c) the Islamized Turks' conquests and international panic made them into the new standard-bearer of Islam, replacing the Omayyad-Abbasid Empires in reunifying the Arab and Muslim world, although Turkish nomadic traditions kept them secluded from settled life for centuries; d) the Ottoman-Turkish Empire remained only a huge bureaucratic-warrior state, never emulating the Arab Empire's old "Golden Era" of advancements in the arts, science and exploration, while centuries of Turkish decay and neglect weakened domestic affairs. European technological and military advances in the 1600s surpassed the Ottoman-Turks and militarily defeated their last invasion of Europe and the Second Siege of Vienna in 1688, followed by Austrian counter-offensives liberating the Northern Balkans in 1690s–1740s.

The long decline of the Ottoman-Turkish Empire from 1700 to World War I (1914–18) aggravated its traditional loose system of government and sparked rebellions by oppressed ethnic groups (mostly Christians in Armenia and the Balkans). At the same time, the Turkish Empire's decline ("Sick Man of Europe") influenced Europe's Balance of Power by attracting the Great Powers' interventions to carve out peripheral lands following Austria's 1680s–1740s successes: a) along

5 Lord Kinross, *The Ottoman Centuries: The Rise and Fall of the Turkish Empire* (New York: Morrow Quill, 1977); Ludwig Dehio, *The Precarious Balance: Four Centuries of the European Power Struggle* (New York: Vintage, 1962); A. Goldschmidt Jr. and L. Davidson, *A Concise History of the Middle East*, 130–200.

the Black Sea, Czarist Russia liberated the Ukraine in the 1600s/1700s, then in the 1700s/1800s expanded over the Caucasus and Persian borders, followed in the late 1800s by Turkic Central Asia and Afghanistan; b) France and Great Britain fought for dominance over geo-strategic Egypt as a gateway to the Middle-East and India, with Napoleon's 1799–1801 Egyptian Campaign against the Anglo-Turks until their re-conquest in 1802, then again in the 1830s/60s French economic penetration of autonomous Egypt culminating with the 1869 Suez Canal, followed in 1882 by a failed Anglo-French naval blockade of Egypt which allowed Great Britain's colonial conquest of Egypt and Sudan in the 1890s, rolling aside French rival claims to the Nile at the risk of a European war; c) Greece's Independence War (1820s–30) against the Ottoman-Turks inspired Christian secessionism and Great Powers' rivalries to control the Turkish Balkans ("Near-East") and geo-strategic Turkish Straits with the Crimean War (1854–56) and 1877–78 Russo-Turkish War (only Anglo-Franco-Austrian intervention saved the Turks from defeat and annexation to Russia).

The Ottoman-Turkish Empire's politico-economic and military decline sparked cyclical revolts of its Balkan Christian provinces from the 1840s until their independence in 1913, while the Turks' failed Westernization left them vulnerable to the European Great Powers' rival colonial claims over the entire "Greater Middle East." Only the Great Powers' inability to jointly partition the Ottoman-Turkish Empire let it survive until World War I (1914–18), while playing the Powers against each other: a) using Great Britain against Russia in 1830 to thwart Russia's penetration of newly independent Greece and the weakened Turkish Empire; b) then Turkey precipitated the Crimean War against Russia in 1854–56 and forced Great Britain, France, Piedmont and Austria into combat to protect her; c) later securing again the diplomatic protection of Great Britain, Austria–Hungary and Germany against Russia at the 1878 Congress of Berlin after losing the 1877–78 Russo-Turkish War; d) European revulsion at Turkish atrocities on Christian rebels left Constantinople isolated in the 1911–12 Italo-Turkish War with Italy's conquest of Libya and Rhodes, followed by the 1912–13 Balkan Wars when Russia's secret Balkan League (Serbia, Greece, Romania, Bulgaria and Montenegro) annexed the last Turkish European lands; e) finally, during World War I in 1914–18 the Turkish-Ottoman Empire joined Germany, Austria–Hungary and Bulgaria as Central Powers, but was defeated by the Allies (Great Britain, France, Russia, Serbia, Japan, Portugal, Greece, Italy, Romania, USA), losing its Arab lands (to Lawrence of Arabia's guerrilla attacks and the Anglo-French Mandates of the League of Nations) and disintegrating at the 1919–20 Versailles/Lausanne Peace Treaties.[6]

With this historical background, the 90-year-old Arab-Israeli clash has emerged as the most entrenched international and regional politico-religious problem in both the "Greater Middle-East/Gulf" and in Western security, and six Arab-Israeli wars in 25 years have redefined desert armored combat and hastened the superpowers' intervention. Also, global access to regional Middle-East oil and gas resources exerts

6 Lord Kinross, *The Ottoman Centuries*, 100–200; A.J.P. Taylor, *The Struggle for Mastery in Europe, 1848–1914* (London, 1950); A. Goldschmidt Jr. and L. Davidson, *A Concise History of the Middle East*, 130–200.

powerful pressures on international politics and Western security. The Arab-Israeli clash has multiple political roots: a) the rival historico-religious claims of Jews, Christians and Muslims to the same "Holy Lands"/"Palestine;" b) the 1890s–1950s Jewish Zionist Movement (which resettled Jews in Turkish Palestine and promoted the creation of Israel as a Western state); c) pan-Arab nationalism since the Ottoman-Turkish Empire's collapse; d) World War I's 1917 Balfour Declaration for a Jewish Homeland in Palestine; e) the 1920–47 British Mandate on Palestine, beset by local Arab anti-Jewish hostility and British anti-Jewish immigration restrictions; f) the Nazis' Holocaust of six million Jews in World War II; g) the failed 1947 United Nations (UN) Partition Plan for mixed Israeli-Palestinian states; h) and the 1948 Israeli Independence which sparked several Arab-Israeli Wars. The Arab-Israeli conflicts embody also a complex "clash of civilizations" between Western-inspired Modernity among Christians and Jews vs Third-World economic stagnation and Islamic traditionalism with civil wars between rival sects of Islam (Lebanon, Algeria, Palestine, Iraq, Afghanistan, Pakistan, Somalia) and terrorism against both Muslims and the non-Muslim West.

Tragically, both Arabs and Jews use Semitic languages, and both have historical claims to Palestine, while the long Arab-Israeli conflict was fueled by Arab and Palestinian nationalism against the collapsing Ottoman-Turkish Empire and the Jewish Zionist policy of "return to Israel." On one hand, the historical Jewish national-religious identity remained rooted in the Old Testament/Torah, with its Monotheistic belief in one God who promised to the Jews as his "chosen people" the lands of Israel and Judea as their own states, centered on the city of Jerusalem. Since Moses' return of the early "Diaspora" Jews from their captivity in Ancient Egypt, the original 12 nomadic Jewish tribes formed several mini-states, coalescing into the Kingdoms of Israel under the unifying rule of Kings Saul, David and Salomon. However, centuries of political strife, wars with neighbors and general decline ended with the Second "Diaspora" to Babylon after the Assyrian conquest, followed by a later return to Israel under a more benevolent Persian rule. Further internecine strife weakened the Kingdoms of Israel and Judea until they fell under Roman rule under Pompey the Great. In the Second Century AD, the Bar Kokhba Revolt in Israel/Palestine against the Roman Empire (Emperor Titus) led to its complete destruction and the exile of most Jews from their ancestral land ("Diaspora"). With only a minority of Jews left in Palestine under Roman, Arab and later Turkish Islamic rule, the Jews emigrated during 2000 years of "Diaspora" to the Middle-East, North Africa, Europe, Russia, Australia, North and South Americas, keeping their ethno-religious identity and spiritual ties to ancestral Israel. Despite cycles of centuries of discrimination and persecution vs conversions to Christianity and Western assimilation in Europe or to Islam in Muslim lands, most Ashkenazi European Jews and Sephardic Middle-Eastern Jews retained their faith, while persecuted Jews of East Europe, Czarist Russia and the Ottoman-Turkish Empire fled to the US and Europe.[7]

7 Israel Pocket Library, *History until 1880* (Jerusalem: Keter, 1973); Israel Pocket Library, *Anti-Semitism* (Jerusalem: Keter, 1974); Howard M. Sachar, *A History of Israel From the*

On the other hand, in the late 1800s the Hungarian Jewish journalist Theodor Herzl (1860–1904) founded the Zionist movement, which has reshaped the Jewish national-religious identity from its traditional roots (as God's "chosen people" in Israel and Mount Zion's fortress near Jerusalem, which since King David symbolized both the city and Israel) into a modern politico-secular mission for all Jews to become a single people after 2000 years of "Diaspora" away from Israel. As one Nation, all Jews must cooperate to survive in a hostile world against anti-Semitism, pogroms, or full "assimilation" (conversion to Christianity), so that all Jews can return to Israel as their own new Westernized and democratic Jewish state. Zionism attracted supporters among Jews and Christians to rebuild a new Jewish state of Israel (*Eretz Yisra'el*) in Turkish Palestine in the late 1800s through agrarian communities (communist-style Kibbutz) by buying lands from the Ottoman-Turkish Empire and feudal Arab landlords, which provoked landlessness and Arab unrest (often led by the same landlords who had sold the lands). Later, the Ottoman-Turkish government blocked further Jewish settlements for fear of adding another nationality to its volatile break-away Balkan and Arab lands. Early Jewish European settlers revived Hebrew as their national language, but Palestine remained a marginal symbolic destination compared to the popular US for persecuted Jews, while Zionism became the main Jewish political movement only with both World Wars. Yet not all Jews are Zionists, with millions preferring to live as Jews in their birth-countries rather than emigrate to Israel: by 2000, 40 percent of the world Jewish population was in Israel, 40 percent in the US, and 20 percent in Russia, Europe and Latin America.

In World War I most US/European Jews supported Germany, given the collective hatred of Czarist Russia's tyranny, until Russian-Jewish Chaim Weizmann, who led the Zionist Movement until 1948, influenced the British government to endorse with the 1917 Balfour Declaration the creation of a Jewish Homeland in Palestine and a pro-British Zionist military corps. However, the Balfour Declaration never specified any specific borders or timetable, while London also promoted regional Arab guerrilla insurgency under Lawrence of Arabia and King Faysal of Arabia against the Ottoman-Turks. Thus, after the war, the Anglo-Egyptian Force and Faysal's Arab army both occupied Palestine, while the British government became embroiled in the growing rivalry between local Jewish settlers and Palestinian-Arab resentment. The League of Nations assigned the Palestinian Mandate to Great Britain with the express purpose of applying the Balfour Declaration to create a Jewish national home with civil-religious rights for all the inhabitants of Palestine, irrespective of race and religion. However, British policy toward both Jews and Arabs remained contradictory, with London internationally supporting the Jews, while in Palestine British officials soon favored the Arabs (e.g. 1922 Colonial Secretary Winston Churchill's White Paper restricting Jewish immigration to Palestine), although most Jews supported the British in local Jewish and Arab institutions. Then in 1922 London split the Palestinian Mandate to create a Palestinian-majority Arab Emirate of Trans-Jordan under King Abdullah as a concession to the Arabs,

---

Rise of Zionism to Our Time (New York: Knopf, 1976).

but British rule became unpopular with both communities: some Jews turned to violent resistance, while Arab fears of losing their local ethnic majority led to the Arab riots in 1920, 1921 and 1929, with massacres of local non-Zionist Orthodox Jews. In reaction to Arab violence, London stopped Jewish immigration and Arab land sales in 1929, angering the aggrieved Jews and Zionists.[8]

Since 1933, the rise of Nazism provoked massive anti-Semitic persecutions in Germany and among her Axis satellites in Europe, leading to further Jewish emigration to Palestine. Nazi and Italian Fascist propaganda during the 1930s and World War II condemned relentlessly Jewish emigration to "Arab" Palestine, seeking to stoke Arab resentment of both the Jews and Anglo-French control of the Middle-East. At the same time, London's nomination of Arab nationalist Hajj Amin as Chief Mufti of Jerusalem backfired once he led the Palestinian-Arab nationalists (Committee for Arab Palestine) during the 1936–39 Arab revolts in Palestine. Great Britain clamped down on both Arabs and Jews, while its 1936 Peel Commission called for a "Two-States solution:" north and central Palestine to the Jews, and the rest to the Arabs, with transfer of populations from mixed areas to stop local Arab violence. Both Jews and Arabs condemned the British Peel partition plan: Jews saw their lands as too small for them and for resettled Jews from Nazi Europe, while Arabs condemned losing Palestine to alien Westerners building a State of Israel in "the heart of the Arab world." British expulsion of Hajj Amin in 1937 worsened matters as he went to Nazi Germany and advocated Hitler's anti-British and anti-Jewish propaganda in the Arab world. By 1939 a British White Paper recanted the Peel partition plan, but fixed an end to the Palestinian Mandate in 1947 with a nebulous independence.

During World War II (1939–45) British courting of regional Arab support against Germany and Italy was unsuccessful, as most Arab countries remained neutral and others openly sided with Hitler to free themselves of both the Anglo-French and Jews: the pro-German Arab revolts of Iraq and Syria in World War II were both repressed by British forces and occupation, while Egyptian forces refused to fight the Axis although the 1936 Anglo-Egyptian Treaty allowed the British free use of Egypt and the geo-strategic Suez Canal in World War II for common security. The Jews, instead, fought again in World War II under British forces, but most looked for US political support since the 1942 Biltmore Program by American Zionists demanded that Great Britain rescind its 1939 White Paper and make Palestine entirely a Jewish state. The US government supported the Biltmore Program, while the World Zionist Organization was sidelined by the Jewish Agency for Palestine under David Ben-Gurion, which was backed by the US Zionist funds of the America Palestine Committee. As the Holocaust exterminated six million European Jews in World War II, Holocaust survivors emigrated to Palestine in defiance of the British blockade, imprisonment or deportation back to the British-controlled Allied Occupation Zone in Germany (1947 *Exodus* ship). Under Zionist influence the US

8 Goldschmidt Jr. and L. Davidson, *A Concise History of Middle East*, 291–328; Israel Pocket Library, *Zionism* (Jerusalem: Keter, 1972); Ian Westwell, The Ultimate Illustrated History of *World War I* (London: Hermes House, 2008).

Congress refused economic aid to Great Britain, while Zionist paramilitary groups attacked British forces in Palestine and rejected a British plan for a joint Arab-Jewish Palestine, which neither side wanted.[9]

Nearing bankruptcy after World War II and with its empire crumbling, London relinquished its old League of Nations' Palestinian Mandate to the UN in 1947, due to escalating rival Jewish and Arab opposition and terrorism undermining British control of Palestine (for example the 91 dead in the 1946 Jewish bombing of the King David Hotel as a site of the British Mandatory government and Headquarters of British Forces in Palestine and Trans-Jordan). To stop this spiral of violence, bombings and fighting, the UN promoted the 1947 Palestine Partition Plan with cantonization and intermixing of small Jewish and Arab communities, which displeased both ethnic groups (in 1947 there were 608,000 Jews and 1,237,000 Arabs), especially the provision of holding back independence for 10 years under a UN trusteeship. As inter-ethnic fighting intensified, the Jewish Agency accepted the UN Partition Plan on May 14, 1948, declaring all Jewish-controlled areas of Palestine as the new state of Israel, freed from old British restrictions on Jewish immigration and land sales. The newly independent Israel was immediately recognized by both the US and Soviet Union (USSR), each vying for influence, followed by most countries at the UN, but the entire Muslim world rejected any Jewish state over any portion of Palestine. The next day, Egypt, Trans-Jordan, Syria, Lebanon, Saudi Arabia and Iraq invaded Israel with the backing of other Arab states in the First Arab-Israeli War. Israel's independence sparked several wars during 90+ years of unrelenting regional hostility against Jewish settlers and Israel, condemned as a Western democratic state "in the heart of the Muslim world" by most Arab states and Palestinians in the "Greater Middle-East/Gulf." Ultimately, Israel's independence and victory over vast Arab armies was seen by the Jews as a revolutionary vindication of Zionism and the recreation of the ancestral Jewish Homeland over Nazi, British and pan-Arab hostility.

Instead, Arabs and Arab-Palestinians, with the support of pan-Islamic political cooperation against Zionism and Israel, advocated their rival territorial/religious rights since the Muslims' First Caliph conquest in the 600s AD of the "Holy Lands" and rejected both the return of the Jews to their long-lost ancestral Palestine to form a Jewish state over dispossessed Arab-Palestinian lands and losing East Jerusalem. For Pan-Arabism, the Palestinian cause and existence of Israel as a "Western" non-Muslim democratic state in the heart of the Muslim Middle-East are the politico-religious sources of most regional political instabilities and the 90-year Arab-Israeli conflict. Rival land claims after World War I were exacerbated by the six Arab-Israeli Wars and two "Catastrophes" (the 1948 and 1967 Palestinian flights of hundreds of thousands Palestinians as stateless refugees in the Middle-East), plus Arab terrorist attacks.

9 Charles D. Smith, *Palestine and the Arab-Israeli Conflict* (Boston: Bedford, 2007); Netanel Lorch, *One Long War* (Jerusalem: Keter, 1976); Gerhard Weinberg, *A World at Arms: A Global History of World War II* (Cambridge: Cambridge University Press, 1994).

Pan-Arabism as a politico-cultural nationalist movement (created in Turkish Lebanon in 1911 against the Italo-Turkish War) supports the reunification of the entire Arab World in the "Greater Middle-East/Gulf" and is closely connected to Arab nationalist advocacy of all Arabs as a single nation, despite their post-1919 division into multiple independent states. Pan-Arabism opposed first Italo-Anglo-French colonialism after World War I and later US-Western political influence in the Arab world during the Cold War (1946–90), while also seeking the destruction of Israel on behalf of Arab-Palestinians. In contrast to religiously-based pan-Islamism, pan-Arabism is secular, nationalist and regionally exclusive to the "Greater Middle-East" (minus its non-Arab Muslim states of Iran and Turkey), promoting security among its Arab states through regional alliances and the 1949 Arab League symbolic politico-security organization. Despite de-emphasizing the religious role of Islam, pan-Arabism as an ideology still fosters prejudice in racially mixed Arab countries against non-Arab Muslims (North African Berbers, Turks, Persians, Kurds and Black Africans) and Arab-Christians (multi-ethnic and Christian Lebanon in turn rejects its past pan-Arab identity, due to rivalries with its oppressive neighbor Syria).[10]

As all Arab states opposed the 1947 UN Partition Plan creating Israel, and all Arab League members vowed to fight, pan-Arab victory seemed inevitable, but political divisiveness, lack of committed troops from other Arab states and lack of military coordination among the six Arab coalition forces undermined their offensive against Israel in the First Arab-Israeli War (1948–49). Still, the numerically superior Arab armies of Egypt, Trans-Jordan, Syria, Lebanon, Saudi Arabia and Iraq successfully invaded Israel, backed by the anti-Jewish attacks of local Palestinian-Arab villages. Yet the Israelis were prepared for such a coalition attack and their security plan sought to repulse all invading forces and extend Israeli borders beyond the UN Partition lines to more defensible positions, while destroying hostile Palestinian-Arab strongholds among the Arab villages bombing the vital Jerusalem road stranglehold. Trans-Jordan fielded the strongest and most organized Arab army, which quickly seized the entire West Bank and East Jerusalem, while the numerically stronger Egyptian army only conquered the Gaza Strip and northernmost coastline, plus parts of the Negev Desert. Syria and Lebanon also penetrated deep in Northern Israel/Galilee, but were soon stopped. Indeed, the pan-Arab forces neither coordinated their invasion nor cooperated during the war, and when the 100,000-strong Israel Defense Force (IDF) finally pushed back the Arabs, their morale collapsed.

The Arabs hoped for some support from Great Britain, due to her interests in Middle-East oil, but London's reliance on US military/economic support dashed Arab plans, while many in the US and Europe supported Israel to atone for the Holocaust. Both superpowers also favored Israel over the Arabs, with diplomatic

10 Ian Bickerton and Carla Klausner, *A History of the Arab-Israeli Conflict* (Upper Saddle River, NJ: Prentice Hall, 2007), 65–111; C.D. Smith, *Palestine and the Arab-Israeli Conflict*, 170–225; N. Lorch, *One Long War*, 48–81; Kenneth Pollack, *Arabs at War: Military Effectiveness, 1948–1991* (Lincoln, NE: University of Nebraska Press, 2004).

support, funds and arms. US President Harry Truman supported a Jewish Homeland (despite pro-Arab geo-strategic arguments at the Pentagon), while seeking the support of the Jewish-American Lobby to redress his impending tight re-election in 1948. Instead, everybody was shocked by the USSR's backing of Israel, but Stalin believed it would weaken British influence in the Middle-East and that a Jewish state would be Communist (given its Kibbutz traditions), discrediting "feudal and bourgeois" Arab regimes.[11]

The UN tried to mediate and stop the fighting with multiple cease-fires, routinely violated on the ground by the combatants, while Arab refusal to negotiate directly with Israel led to Ralph Bunche's 1949 "proximity talks" on the far-away Rhodes Island, finally securing the 1949 Armistice between Israel and its Arab enemies, formalizing Israeli control over both the UN area assigned to the Jewish state plus over 50 percent of the area assigned to the Arab state. As the only democracy in the region, Israelis saw their War for Independence as a struggle by oppressed people for freedom and against extinction from the horrors of the Nazi Holocaust and pan-Arab hostility, while rebuilding their war-torn country as a modern nation founded on their old ancestral homeland and absorbing thousands of Jewish refugees from Europe and the Arab world. During the war two-thirds of Palestinian-Arabs (630,000–711,000) fled or were expelled from areas under Israeli control, never to return. They became permanent stateless refugees in Egyptian-ruled Gaza and in refugee camps throughout Lebanon, Trans-Jordan and Syria, never allowed to leave and mix with local Arabs; the remaining 150,000 Palestinian-Arabs became Israeli-Arab citizens. Also, smaller numbers of Jews in areas captured by the Arabs were expelled. Through the decades Israel always rejected the Palestinian-Arabs' claim of their right to return to Israel, claiming them to be hostile enemies who would also fundamentally alter Israel's Jewishness, while the Arab states' expulsion of their ancestral Jewish populations (900,000) brought two-thirds of them to Israel and the rest to France, the US, Western or Latin American countries. Israel argued that the parallel Palestinian and Jewish exoduses represented a population exchange between Arab nations and Israel. Yet Zionist Israel never truly accepted the moral responsibility that the long Arab-Israeli clashes had created another group of oppressed people—Palestinian refugees—with exclusivist anti-Jewish nationalist claims backed by terrorist attacks (internally, on the border and internationally), which constantly marred any quest for regional peace, while Israel's victory in the 1967 Six-Day War added more Palestinians as new internal civilians (West Bank and Gaza) and even more refugees abroad.

The creation of a Zionist Israel as homeland for all Jews led to immediate reprisals against both Jews and Christian non-Muslim minorities in the Islamic world, where Jews lived centuries before the "Diaspora" (especially in Iraq and Yemen) as "Arabs of Jewish Faith, rather than a separate race."[12] Instead, after

11 Samuel Katz, *Battleground: Fact and Fantasy in Palestine* (New York: Bantam, 1973); I. Bickerton and C. Klausner, *A History of Arab-Israeli Conflict*, 65–111; C.D. Smith, *Palestine and the Arab-Israeli Conflict*, 170–225.

12 Philip Mendes, "The Forgotten Refugees: the Causes of the Post-1948 Jewish Exodus from

repeated anti-Jewish Arab riots in British Palestine in the 1920s–30s and Iraq in the 1930s and 1940s, most Jews were expelled from Arab states between 1948 and the 1960s, emigrating to Israel, France and the US, with small Jewish communities left in Morocco, Tunisia and Iran. Christians represent 5.5 percent of the population of the Arab world, mostly in Lebanon (39 percent, although Lebanese Christians reject identification as Arab-Christians since the 1975–88 Civil War), Palestine (24 percent), Syria (6 percent), Iraq and Egypt. Yet local Middle-Eastern politico-religious turmoil since Israel's creation has led to steady declines until 2010, due to persecution and emigration abroad. In Israel/Palestine, the proportion of Arab-Christians shrank to 3.8 percent (with 1.7 percent in Israel as 9 percent of the Israeli-Arab population, 8 percent in the West Bank and 0.8 percent in Gaza), while 6 percent emigrated to Jordan since 1948 and most Arab-Christians emigrated to the US, South America and Australia. Many more Christians have more recently fled religious persecution in Lebanon, Syria, Iran and Iraq since the 2003 Second Gulf War and sectarian strife under the Coalition's Occupation (2003–11).[13]

From the Arab viewpoint, Israel's survival and victory was condemned as a humiliation of the entire Arab and Muslim worlds, with their belief that all Palestine should be exclusively Arab. No Muslim country recognized Israel, which was ostracized as an "agent of Western imperialism," while the defeated Arab states accepted only an armistice with no post-war land adjustments solution or Arab help for the Arab-Palestinian population to create any Palestinian state. Instead, during the war the majority of Arab-Palestinians (700,000) fled to the Gaza Strip, which was annexed by Egypt, 400,000 fled to the West Bank and East Jerusalem, both annexed by Trans-Jordan, with tens of thousands more forming refugee camps in Lebanon and Syria, while in Israel-proper 150,000 Palestinians became Israeli citizens against their will. Thus, Israel's intelligence and public view before the 1956 Suez Canal War and 1967 Six-Day War was that it was surrounded by Arab states seeking to destroy it, while Nasser's provocative war mobilizations in both 1956 and 1967 drove the Israeli politico-military leadership into preemptive strikes to avert any imminent demise by the Arab coalition (despite the US temporizing in both crises).

During the Cold War the US had inherited Great Britain's regional role in containing Soviet expansion and protecting the international oil trade vital to the survival of both Japan and the North Atlantic Treaty Organization (NATO). US President Dwight Eisenhower cooled ties with Israel and pursued a pro-Arab US

Arab Countries" (lecture March 2002) in http://www.palestineremembered.com/Articles/General/Story2127.html#Iraqi%20Jews%20in%20the%20pre-1948%20period (retrieved 4-4-2010). See also: Khazaria Info Center, "Jewish Genetics: Abstracts and Summaries" (2011), at: http://www.khazaria.com/genetics/abstracts.html; Shaker Khalid, "Yemeni Jews ... from Creation to The Magic Carpet Operation" in *Yemen Post* (February 9, 2009), at: http://www.yemenpost.net/62/Reports/20082.htm; and Martin Gilbert, *The Arab-Israeli Conflict: Its History in Maps* (London: Weidenfeld & Nicolson, 1979), 49–51.

13 Israel Pocket Library, *Immigration and Settlement* (Jerusalem: Keter Books, 1973); Michael Prior, *Zionism and the State of Israel: A Moral Inquiry* (London: Routledge, 1999); John Quigley, *The Case for Palestine: An International Law Perspective* (Durham, NC: Duke University Press, 2005).

policy towards the Middle-East aimed at containing Soviet penetration through Arab nationalism and weak US-led pro-Western regional alliances, supporting the 1954–55 Middle-East Treaty Organization (METO: Great Britain, Iran, Iraq, Pakistan, Turkey and the US) and its 1955–79 successor the Central Treaty Organization (CENTO) (after its only Arab member, Iraq, left). CENTO provided a strategic link to NATO against a Soviet invasion from the Caucasus and Central Asia towards Turkey, Iran and the oil-rich Gulf. Moreover, the US, Turkey and Iran had close bilateral military/economic ties with Israel; Iran contained the USSR and pro-Soviet Iraq; Turkey contained pro-Soviet Arab client-states Syria and Iraq; and Pakistan contained the USSR inroads into Afghanistan and India.[14]

However, Eisenhower's pro-Arab Middle-East policy and links with "Nasserite" Egypt jeopardized the regional security interests of America's main Allies, Great Britain and France, who jointly controlled the Suez Canal, which was vital in both World Wars and for future use to transport troops from Europe to/from Asia/ Australia in case of any World War III, while since 1955 two-thirds of Western Europe's global oil shipments passed through Suez. Instead, Arab domestic unrest to cut all ties with London, plus the October 1954 assassination attempt, convinced Nasser to forge pan-Arab ties against Anglo-French colonialism using secret Soviet weapons sales. However, the US soon discovered Nasser's secret Soviet arms and stopped Western and World Bank aid for building, on the Nile River, the world's largest dam (Aswan Dam). To save his pride and political life, Nasser had the USSR build Aswan, while he nationalized the Anglo-French Suez Canal on July 26, 1956 and secretly armed anti-French rebels during the 1954–62 Algerian War.

As President Eisenhower was focused domestically on his tight re-election campaign and on condemning at the UN the bloody Soviet repression of the 1956 Hungarian Revolution, London and Paris felt abandoned by the failed US diplomacy of appeasement with pro-Soviet Nasser. Thus, the Anglo-French alliance decided to topple Nasser's dictatorship and retake the Suez Canal by force, and sought to deflect US and Arab anger through the secret Sèvres Pact between Israel, Great Britain and France: Israel would invade the Sinai and defeat Egyptian forces, while the Anglo-French would "intervene" to separate Israeli-Egyptian forces and "reseize" the Suez Canal, while toppling Nasser (the pact also allowed Israel to secretly receive French nuclear technology), presenting the US with a *fait accompli*.[15]

Israel's October 29, 1956 invasion of the Sinai under legendary Chief-of-Staff Major-General Moshe Dayan focused on seven military goals: 1) a diversionary attack on the Gaza Strip to destroy Palestinian terrorist camps and Egyptian forces poised to attack advancing Israeli troops fighting Egyptian forces in the northern

14 John Spanier, *American Foreign Policy* (Washington, DC: C.Q. Press, 1987); Marco Rimanelli, *NATO and Other International Security Organizations: Historical Dictionary* (Lanham, MD: Scarecrow/Rowman, Littlefield, 2009).

15 I. Bickerton and C. Klausner, *A History of Arab-Israeli Conflict*, 226–60; Erskine B. Childers, *The Road to Suez: A Study of Western–Arab Relations* (London: MacGibbon & Kee, 1962); Peter Hercombe and Arnaud Hamelin, *L'Affaire de Suez: le Pacte Secret* (Paris: France 5/Transparence, 2006).

Sinai; 2) a primary Israeli armor strike bypassing in the north the heavily defended Gaza and Rafah, staging a deep turning maneuver to flank and destroy the Egyptian right flank on the central Sinai route, then moving north to double-envelop the Egyptian forces at Umm-Gatef and Abu-Agheila who were under attack by another Israeli border offensive from Auja/Nitzana; 3) after routing the Egyptian Army in Northern Sinai and destroying its retreating units before they could reach the Mitla Pass and Suez Canal, Israel reached the Canal by November 2, just as Anglo-French forces landed at Port Said; 4) on October 29 a fourth parallel and larger Israeli turning thrust under Colonel Ariel Sharon struck first central Sinai, pushing west towards the Mitla Passes and then rapidly conquering the rest of the Sinai; 5) by November 1–5 a fifth Israeli offensive in southern Sinai seized Sharm el-Sheikh to stop Egypt's economic blockade of Israeli shipping in the Red Sea and Tiran Strait; 6) Israeli forces on the Jordanian border deterred it from joining the war; 7) Israel gained air supremacy through surprise attacks, followed by Anglo-French naval air attacks.[16]

However, the Suez Canal War turned into a political disaster for the winners once condemnation from the US and USSR at the UN forced them to withdraw, symbolizing the Anglo-French complete failure to politically and strategically regain control of the Suez Canal and survive as independent Empires. Eisenhower was angered that this crisis risked scuttling his impending Presidential re-election and global campaign against the USSR over its repression of the 1956 Hungarian Revolution. Washington engineered an international financial crisis that together with Saudi Arabia's oil embargo forced the Anglo-French evacuation (the US and NATO members refused to fill the energy gap until the Anglo-French withdrawal). Anglo-French resentment at the US left a deep crisis in NATO, with a very slow recovery of the Anglo-American "Special Relationship," while Franco-American tensions increased until by the 1960s President General Charles de Gaulle decided that France could not rely on either of its Anglo-American Allies and in 1966 withdrew from NATO's integrated military command.

On November 7, 1956, UN Secretary-General Dag Hammarskjøld formed the first UN Emergency Force (UNEF) to demilitarize the Sinai and force both Israelis and Egyptians out until 1967. America's pro-Arab policy failed to win the gratitude of either Egypt or the Muslim world to side with the West against Communism. Instead, in the 1950s–70s Egypt, Syria and South Yemen became Soviet client-states with advisors, weapons and aid, while influencing also Iraq, Algeria and Libya. Parallel to this, the 1964 Arab League Summit created the Palestinian Liberation Organization (PLO) to attack Israel and create a Palestinian homeland through terrorist attacks along its borders and the hijacking of international commercial

16 Robert A. Doughty and Ira D. Gruber (eds), *Warfare in the Western World: Military Operations Since 1871, Volume II* (Lexington, MA: Heath, 1996), 674–86, 756–63, 935–48; Jonathan M. House, *Combined Arms Warfare in the Twentieth Century* (Lawrence, KS: University Press of Kansas, 2001), 122–7, 226–38; Kennett Love, Suez: *The Twice-Fought War; A History* (New York: McGraw Hill, 1969); David Tal (ed.), *The 1956 War: Collusion and Rivalry in the Middle East* (London: F. Cass, 2001).

planes. Israel, instead, enhanced its military air-land operations against superior enemies through a tank-heavy strike force, with massive air power substituting for artillery cover in combat.[17]

Renewed Arab-Israeli tensions escalated a decade later when Nasser reimposed in May 1967 the sea blockade of Israeli trade through the Suez Canal and Tiran Straits, then forced a depleted UNEF to evacuate the Sinai by May 30. As Nasser remilitarized the Sinai against Israel with 100,000 Egyptian forces, he constantly called for a new pan-Arab anti-Jewish war by an Egypt-Syria-Jordan-Iraqi alliance. These events sparked Israel's preemptive June 1967 Six-Day War: 240,000 Israeli forces attacked first the larger Egyptian forces in the Sinai, quickly wiping out most Arab air forces to secure full air superiority, although the Arabs interpreted Israeli air-raids as "proof" that Israel was helped by US/British aircraft instead. Egyptian forces in the Sinai deployed the Soviet doctrine of mobile defensive armor at strategic depth to protect infantry on the heavily defended Abu-Ageila/Umm-Gatef/Kusseima central route, the Kuntilla/Thamad/Nakhl southern route and the Jebel Libni/Bir-Hasana strategic reserve area. Israel surprised the Egyptian forces by timing its air strikes on Egyptian airfields with combined-force flanking armor attacks through northern and central Sinai, contrary to Egyptian expectations of a repeat of the 1956 Suez Canal War.

Israel slowly conquered the Gaza Strip, Rafah and El-Arish by June 5–7. Combined Israeli armor, paratroopers, infantry and artillery attacked the Egyptians from the front and flanks, while Israeli paratroopers dropped to the rear to cut off the enemy. In just four days of combat Israel defeated the largest Arab army, forcing an Egyptian panic withdrawal to the fortified Suez Canal, reached by the Israelis on June 8, plus Sharm el-Sheikh and Tiran Islands on June 7. On June 5 Syria and Jordan shelled West Jerusalem, Natanya and Tel Aviv, while Israel's Air Force destroyed a Jordanian Brigade en route from Jericho to East Jerusalem, as other Israeli armored brigades conquered East Jerusalem and the West Bank from the Jordanians. Lastly, by June 9 Israel crushed the demoralized Syrians on the fortified Golan Heights.[18]

On May 23, 1967, President Johnson secretly authorized air supplies to Israel of weapons systems, while an arms embargo was placed on the Middle-East. Also, the USSR supported its Arab allies with an unprecedented deployment of their Black Sea Fifth Eskadra naval forces into the Eastern Mediterranean, while US Secretary of Defense Robert McNamara revealed in 1983 to the *Boston Globe* that "We damn near had war" when the US Sixth Fleet sent a carrier battle-group from Gibraltar to the Eastern Mediterranean to defend Israel if the USSR supported Syria's attack on Israel, while the Soviets threatened an East–West war. But after six days of war,

17 Isaac Alteras, *Eisenhower and Israel: U.S.-Israeli Relations, 1953–60* (Gainesville, FL: University Press of Florida, 1993).

18 Henry Kissinger, *Diplomacy* (New York: Touchstone, 1994); George W. Gawrych, *The Albatross of Decisive Victory: War and Policy Between Egypt and Israel in the 1967 and 1973 Arab-Israeli Wars* (Hartford, CN: Greenwood, 2000); Michael Oren, *Six Days of War: June 1967 and the Making of the Modern Middle East* (Oxford: Oxford University Press, 2002).

Israel emerged as the most powerful military country in the Middle-East, defeating all combined Arab enemy coalitions and trebling its size by capturing the Gaza Strip, Sinai, West Bank, East Jerusalem, Golan Heights, the Strait of Tiran and half of the Suez Canal, with one million Arabs under Israeli control (out of one million Palestinians in the West Bank, 300,000 fled to Jordan, increasing local unrest).[19]

Egypt's humiliated Nasser rejected defeat by launching the undeclared 1967–70 War of Attrition (small-scale deep incursions into the Sinai, limited artillery duels along the Suez Canal against the Israeli Bar-Lev Line and aerial dog-fights), seeking to inflict major losses and undermine Israel's morale and economy until a war-weary Jewish government would return the Sinai to Egypt. Fearing an East–West escalation of the Attrition War, US President Richard Nixon sent Secretary of State William Rogers to negotiate a 90-day cease-fire and a framework peace based on UN Resolution 242. Despite Egyptian public acceptance, the Rogers Plan failed once the Egyptian Army launched massive artillery shelling along the Suez Canal on March 8, 1969, deep commando raids in the Sinai and extensive air battles, relying on Soviet assistance (weapons, jets, 20,000 Soviet trainers and artillery positions, plus Soviet pilots flying Egyptian planes). Yet Israel repeatedly defeated the Egyptian air force, pushing it deep inside Egypt to fight above its capital Cairo and strategic facilities along the Nile River, while airborne raids and air strikes pounded Egyptian positions on the Suez Canal. US and Soviet aero-naval forces increasingly confronted each other in the Eastern Mediterranean in support of their respective allies, until the risk of a superpower show-down led both to force Nasser to stop the Attrition War on August 1970 with the Rogers cease-fire, which left all frontiers unchanged.

During the 1967–70 Attrition War, Nasser failed to force the USSR to finally supply Egypt with advanced weaponry compared to the sub-standard Soviet arms supplied after Egypt's 1956 and 1967 losses. Since the Attrition War, the USSR instead sought to avoid a new Arab-Israeli war and a confrontation with the US with the East–West relaxation of tensions during Détente (1969–79). Following Nasser's sudden death, his successor General Anwar Sadat started preparing in absolute secrecy, from 1971 to 1973, a new Arab coalition war against Israel without Soviet support, while Syria too undertook a massive Soviet-supported military build-up to reconquer the Golan. A military victory would allow Sadat to overcome the national shame of the Arab-Israeli defeats and implement unpopular reforms. Then, as winners, Syria and Egypt could impose peace talks to force Israel out of the West Bank and Gaza.[20]

---

19 Tom Segev, *1967: Israel, the War, and the Year That Transformed the Middle East* (Paris: Denoël, 2007); "McNamara: U.S. Near War in '67," *Boston Globe* (September 16, 1983).

20 Jeremy Bowen, *Six Days: How the 1967 War Shaped the Middle East* (London: Simon & Schuster, 2003); Moshe Shemesh, *Arab Politics, Palestinian Nationalism and the Six Day War* (London: Sussex Academic Press, 2008); R.A. Doughty and I.D. Gruber (eds), *Warfare in the Western World*, 935–48; G.W. Gawrych, *The Albatross of Decisive Victory*, 123–257.

By Fall 1973, Sadat's diplomatic offensive secured the political support of 100 states from the Arab League, Organization of African Unity, Non-Aligned Movement and Soviet Bloc, plus a new Arab military coalition (Egypt and Syria, with support from a symbolic Arab Expeditionary Force of troops from Iraq, Jordan, Saudi Arabia, Libya, Sudan, Morocco, Algeria, Kuwait, Cuba and North Korea). The Yom Kippur War of October 6–26, 1973, started on the main Jewish holiday with a surprise joint two-front attack by Egypt and Syria against Israeli-held Sinai and the Golan Heights. Sadat planned a massive Egyptian amphibious attack across the heavily fortified Israeli Bar-Lev Line on the Suez Canal, then a move inland by 50 km to dig in an Egyptian defensive barrier with 100,000 men and Soviet missiles against Israeli armor and air power, while Syria's armor would also seize the Golan.

Instead, Israel's Military Intelligence discounted the likelihood of a new Arab war, despite all proofs to the contrary. Only after 13 war warnings, and just hours before the Arab attack, did Israeli Chief-of-Staff David Elazar finally realize that war was imminent and issued a partial call-up of reserves on October 5. But his requests for a general call-up of reserves and preemptive air strikes against Syrian airfields, missiles and ground forces the next day were rejected by Premier Golda Meir, Defense Minister Moshe Dayan and Chief of Military Intelligence Major-General Eli Zeir. Although Golda Meir agreed on Elazar's later mobilization of the entire Air Force and 120,000 armored troops, she saw Israel as totally dependent on the US for military resupply and did not want to be blamed for starting the war as in 1956 and 1967, while Europe and other industrialized countries dependent on OPEC oil had already stopped supplying Israel ammunitions, given the threats of an Arab oil embargo (First Oil Shock) and trade boycott. The Arab attack caught the US by surprise, while US Secretary of State Henry Kissinger futilely sought the USSR's and Arab countries' help to prevent war.[21]

In the first days of combat, Chief-of-Staff Elazar was one of very few Israeli commanders who kept his cool, while Israel's political leadership was crushed by fears that the country would be wiped out altogether (with Defense Minister Dayan warning on October 8–9 of a last-ditch "Samson Solution" with Israel's secret 13 20-kiloton tactical nuclear weapons). For several days, Israel's Air Force shifted focus from the Sinai to the more threatening Syrian invasion of the Golan which, under cover of their SAM batteries and Soviet anti-tank weapons was making major inroads against the outnumbered Israelis. Mobile Israeli armored defenses contained a night-time Syrian breakthrough, while by October 8–10 Israeli forces repulsed the Syrians to the pre-war Purple Line border and the Air Force destroyed an Egyptian attack south of Suez.[22]

21 J.M. House, *Combined Arms Warfare*, 226–38; C.D. Smith, *Palestine and the Arab-Israeli Conflict*, 306–53.

22 Martin van Creveld, *Military Lessons of the Yom Kippur War: Historical Perspectives (The Washington Papers)* (London: Sage-CSIS, 1975); J.M. House, *Combined Arms Warfare*, 226–38; N. Lorch, *One Long War*, 138–252.

Israel redeployed armored forces to the Sinai, while Premier Golda Meir politically decided to continue combat deep into Syria for four more days (October 11–14) to retake Mount Hermon and restore Israel's image as the regional military power in the Middle-East. On the diplomatic level, Syria's defeat and the Egyptian reversals increased Arab pressures on Jordan to join the war against Israel, while OPEC launched a successful oil embargo on NATO countries (First Oil Shock) to keep them from helping the US military resupply of Israel. A Jordanian expeditionary force was sent into Syria, while the US was secretly asked to inform Israel that Jordan would not attack along the Jordan River border. Then as Egypt's massive armored offensive on October 14 to relieve the defeated Syrians was checkmated by the Israelis, General Sharon launched on October 15 a massive surprise counter-offensive across the Suez Canal at the northern edge of Great Bitter Lake, rapidly crossing into Egypt proper and attacking supply convoys, SAM sites and logistics centers. After three days of bitter fighting, the Israelis defeated the numerically superior and entrenched Egyptians overlooking the Suez Canal. Then Israeli divisions operated a wide turning offensive southwards along the east bank of the Great Bitter Lake and Suez Canal (1,600 square km), double-encircling the entire Egyptian Third Army at Suez (30,000 men) and Egypt's narrow Sinai invasion strip (1,200 square km with 70,000 Egyptian troops and 720 tanks) up to 101 km from Egypt's capital.

The Egyptian General Staff slowly realized the extent of its defeat only when confronted by Soviet satellite imagery of Israeli forces operating inside Egypt proper, but Sadat refused to stop fighting. During the Yom Kippur War, the USSR undertook its most ambitious international air- and sea-lift of military supplies to Egypt and Syria (78,000 tons in total), replacing the heavy Arab combat losses with Soviet equipment taken directly from Soviet/Warsaw Pact dépôts, while other Arab states also provided weapons and financing. But all of this was surpassed by the massive US air- and sea-lift to Israel (112,395 tons total), which stopped secret talks of a "Samson Solution"[23] using tactical nuclear weaponry.

Just as a US-USSR mediated UN cease-fire came into effect on October 22, Israel completed its encirclement of the Egyptian Third Army in Suez City with tacit US support: Kissinger saw Israel's last-minute military gains as an incredible opportunity for the US to draw Egypt away from the USSR if Washington prevented Israel from destroying the trapped Egyptian army and then mediated between the warring forces. Tremendous US diplomatic pressures on Israel stopped further attacks. But by October 24–26 the Yom Kippur War risked escalating into a superpower World War III: Kissinger's diplomacy with the Egyptians, Israelis and Soviets stalled once Sadat and Soviet Premier Leonid Brezhnev jointly appealed for US-Soviet peacekeepers to oversee the UN cease-fire against Israeli violations. US rejections of the proposal met Soviet threats of unilateral intervention on Egypt's

---

23 Jacques Derogy and Jean-Noël Gurgand, *La Morte in Faccia: Il Giorno Piu Lungo D'Israele* (Milan: Rizzoli, 1975); Juan-Carlos Ferrari, *La Guerra del Petrolio: Il Quarto Conflitto Arabo-Israeliano* (Roma: Coines, 1974); Ahron Bregman, Israel's Wars*: A History Since* 1947 (London: Routledge, 2002).

side to enforce the UN cease-fire. As the USSR prepared to air-lift seven airborne divisions to Syria and Egypt, while deploying in the Mediterranean the Fifth Eskadra to escort seven amphibious vessels with 40,000 naval infantry, Kissinger sent a conciliatory message to Brezhnev, while also raising the national Defense Condition (DEFCON) to three (an increased readiness level reached only during the 1962 Cuban Missile Crisis, the 1973 Yom Kippur War and in 2001 during the "9/11" attacks by Al-Qaeda). Finally, another US message to Sadat threatened US intervention on behalf of Israel if the Soviets were to intervene. DEFCON-3 and the US Sixth Fleet readiness in the Eastern Mediterranean to block Soviet flight routes and warships left the Kremlin afraid that the US would fight World War III with the USSR over the marginal Middle-East (rather than in Europe). As the USSR and Egypt stepped down from the crisis, the US also forced Israel to stop its offensive.

On the international level, Israel scored an incredible upset victory by defeating all her Arab enemies after the surprise two-front 1973 Yom Kippur War and staggering casualties, but left the country deeply vulnerable psychologically and ended its post-1967 sense of military supremacy in the Middle-East. The Arab World, humiliated by the 1967 Six-Day War rout, felt psychologically vindicated by the initial 1973 successes, while both Jordan and Egypt abandoned their claims to the West Bank and Gaza. With the US as key peace-maker in the Middle-East, Kissinger's "shuttle diplomacy" eventually led Israel to leave Egypt by March 5, 1974, giving Egypt complete control of the Suez Canal, while also retreating to the Purple Line on the Golan by May 31, 1974 with a local UN peace-keeping buffer zone vs Syria. Kissinger's diplomacy drew into the US' orbit both Jordan and Egypt (marginalizing the USSR) with the promise of a "land-for-peace" settlement, which later under President Jimmy Carter led to the 1978 Camp David Accords, with the return of the entire demilitarized Sinai to Egypt (patrolled ever since by the US-led non-UN Multinational Force Organization/MFO) and the first diplomatic recognition of Israel by an Arab country (Egypt). Yet Arab anger at Camp David led to Egypt's ostracization for years by all Arab League states, and Sadat's assassination by Islamic Fundamentalists on October 6, 1981. Also, anger in Israel against the government led to the collapse of Labour's traditional hold on power on behalf of the right-wing Likud Party under Premier Menachem Begin with Sharon as Defense Minister, whose uncompromising nationalism led to the annexation of East Jerusalem, Golan and Jewish settlements in the West Bank. Likud triumphed in the Lebanese War of 1982–85, finally destroying also the PLO, but only replaced one threat with others given the rise of nuclear proliferant Islamic Iran supporting Syria, Palestinian Islamic Hamas and Lebanon's Hizb'allah (as proven in the brief 2008 "Hizb'allah Missile War").[24]

Since 1948 peace in the Middle-East has been traditionally undermined by a bloody history of Arab-Israeli rival nationalisms and opposite exclusivist religious claims, followed by Arab and Palestinian attempts to destroy Israel during the Arab-Israeli Wars (1948 Independence of Israel; 1956 Suez Canal War; 1967 Six-Day

24 H. Kissinger, *Diplomacy*, 423–72; C.D. Smith, *Palestine and the Arab-Israeli Conflict*, 354–450.

War, 1967–70 Attrition War; 1973 Yom Kippur War) and Palestinian international terrorism, as well as pan-Arab diplomatic ostracism until the 1978 Camp David Accord. Also, Israel's drive to attain absolute national security against all hostile Arab neighbors (since 1973) and Islamic Iran (since 1979) has fallen short and been reluctantly rebalanced with half-hearted regional peaceful co-existence strategies under US–European and internal pressures based on "land-for-peace" with Arab neighbors and a future "Two-States Solution" for a Palestinian mini-state carved out of some Israeli "Occupied Lands" (Gaza Strip and 90 percent of the West Bank, excluding East Jerusalem and Golan).

Yet US diplomatic plans to solve the Arab-Israeli conflict have routinely reached stalemate over mutual Israeli-Palestinian intransigence on the 1967 borders, Palestinian despair and cyclical Arab-Palestinian suicide terrorism (PLO, Hamas, Hezbollah). Indeed, as Middle-Eastern tensions keep widening with regional wars (Iran–Iraq War, Two Gulf Wars and Two Afghan Wars), civil wars (Lebanon, Algeria, Iraq, Somalia, Afghanistan), Islamic Fundamentalism (Iran, Afghanistan, Lebanon, Sudan, Algeria, Somalia, Pakistan), nuclear proliferation (Iraq, Libya, Iran), Islamist terrorists with suicide bombers (Muslim Brotherhood, Hamas, Al-Qaeda), and finally also the 2010–11 Arab grass-roots democratic revolts against dictatorial regimes (Iran, Tunisia, Egypt, Bahrain, Yemen, Libya and Syria), any Arab-Israeli Peace Process remains hostage to Likud's intransigence on security vs inter-Palestinian civil war between the PLO/Palestinian Authority in the West Bank and Islamist Hamas in Gaza.[25]

25 A. Bregman, *Israel's Wars*; Lawrence Meyer, *Israel Now: Portrait of a Troubled Land* (New York: Delacorte, 1982); Camille Mansour, *Beyond Alliance: Israel and U.S. Foreign Policy* (New York: Columbia University Press, 1994).

A "Time of Troubles": War in an Age of Planetary Upheaval, from the End of the Middle Ages to 1648

Oleg Kobtzeff

A fame, a peste, a bello, libera nos, Domine. (Late medieval prayer)

Pierre Chaunu, the recently deceased founder of "serial history," devoted a lifetime to the study of world demographics, particularly between 1492 and 1789. His prolific bibliography abounds with materials about death and the cataclysmic decline of the world's population. Speaking of this transition between the Middle Ages and the next centuries, Chaunu summoned up the main factors of one of history's greatest upheavals: climate change, devastating epidemics and war. Doomsday could appear as a realistic scenario for the most reasonable inhabitants of the Old World. For the indigenous people of the Americas who were experiencing first contact at the time, the end of the world was a reality: 90 percent of any given Indian population would be wiped out by disease, massacres and mistreatment, economic disaster and cultural ruin in less than a generation. Some Europeans were sympathetic to the Indians after reading Bartolomeo de Las Casas. His report on the genocide substantiated the belief in the rise of evil forces. The conquistadors and their supporters, on the other hand, rationalized their violence and constructed philosophical justifications which would consequently sanction the carnage of Christians by Christians. Chaunu qualified those decades as a time of "flight, aggressiveness, delusions, hallucinations"[1]—a state of panic that caused yet more conflicts. The Russians refer to roughly the same period as the "Time of Troubles," one of the darkest eras of their history just between the French Wars of Religion and the Thirty Years War, a term that Denis Crouzet found appropriate as a title for his monumental thesis on the "warriors of God" in the West.[2]

1 Pierre Chaunu, *Trois million d'années* (Paris: Robert Laffont, 1990), 228.

2 Denis Crouzet, *Les Guerriers de Dieu. La violence au temps des troubles de religion*

Given the rapid succession of conflicts throughout this era, the intensity of their violence (witnesses unanimously described it as spectacularly extreme), and their frequently ideological dimension, should we consider them as one continuous experience instead of separate events?

A New Type of War

The Hundred Years war coincided with the great epidemic of the fourteenth century. A major characteristic of that war was its dimension as a popular political and mystical movement. The conflict was more than the usual military actions of troops on a theater of operations. An uprising had surged and it was massive. It was driven by prophetic inspiration and an unusual energy that mobilized crowds of simple folk for crusades, such as the one fought by Joan of Arc—that fanatic adolescent whose exaltation was so unusual and whose revolutionary quality (a girl dressed as a man, prophesying and leading troops) was so subversive, that she was burnt as a witch. But this threatening quality may have been her main weapon in frightening the enemy and in channeling the revolutionary vigor of poor peasants who took arms because she presented them with a sublime cause: to take control of history, with the blessing of Heavenly forces. The war waged by Catholic fighters only three or four generations later, in the sixteenth century, and the energy which they invested in it, was a *prophetic* movement.[3] It is not a coincidence that the sixteenth century Catholic League—the generation of the great-great-grandchildren of Joan's companions—chose the "Maiden of Orleans" as their symbol. She was the icon of a pivotal moment in history.

The year 1492 marks the end of the Muslim state of Grenada and the conclusion of the *Reconquista*. What followed immediately were the forced conversions and massive reprisals against Jews and Muslims. It also marks the beginning of the conquest of the Americas and its own atrocities, a novelty for the times: a phenomenon that we today call genocide, the planned eradication of a targeted population. In the next century, during the lifetime of the same generation, began a string of wars which will remain among the most traumatic in human memory.

In 1494 began the Italian wars that would last until 1559, with only a few interruptions (38 years of war over a period of 65 years), involving France, the Holy Roman Empire, the states of Italy, England, Scotland, Spain, the Ottoman Empire, Switzerland, and Saxony. During the same period, Germany experienced

(v. 1525–v. 1610) (Seyssel: Editions Champ Vallon, 1990).

3 Ibid. Guizot affirmed, as early as 1828, i.e. long before Engels, that "the crisis of the 16th century was not simply reformism, it was essentially revolutionary" (ibid., 18). Based on hundreds of primary sources, Crouzet's research confirms that the violence of the French Catholics in the wars of religion was "a prophetic collective conscience that tries to represent the union between the collective violence and the violence of God and proceeds in a symbolic way" (ibid., 81).

the Peasants' War (1524–1525), the "Pilgrimage of Grace" of 1536, the Schmalkaldic Wars (1546–1547, 1552–1555), and the Cologne War (1582–1583). In Switzerland, a conflict was fought over the Anabaptist state of Münster in 1534–1535. In the same century, in France, in addition to the 1545 massacre of Mérindol and French intervention in the German wars, between 1562 and 1598 religious civil wars ravaged the kingdom with only months of peace between the fighting. At the same time, the war over the Netherlands began in 1567. It would last until 1609, only nine years before the beginning of the Thirty Years War (1618–1648), a war that overlapped chronologically with the English Civil War (1642–1651).

At the same time, wars on the borders of Russia were on the rise.[4] The chronological coincidence has been overlooked for too long. Even fewer have noticed the chronology (1598–1613) of the Time of Troubles, a period of civil war, social turmoil, and inter-state conflict.[5] Only those fifteen years are traditionally designated as the Time of Troubles. However, the chaos was the result of profound instability provoked by many inner conflicts surrounding the tormented rise of the new Muscovite state at the expense of the republic of Novgorod (almost all the Northern half of Russia) and the painful birth of a new society unsettled by growing serfdom, centralized autocratic tyranny and religious tensions (the "heresy of the Judaizers" and the conflict between partisans and opponents of monasteries holding large properties). Social and religious tensions continued throughout the seventeenth century. Their culmination was the unrest following the Church reforms imposed by Patriarch Nikon and Tsar Alexis (equivalent in scope to the Tridentine Reformation), an unrest marked by episodes of extreme punishment of traditionalist communities and collective suicide by fire of large groups of the "Old Believers." Before that, Tsar Alexis had faced a peasant war led by the legendary Sten'ka Razin. Meanwhile, countless other conflicts had ravaged the borderlands of the Ukraine, Poland and the Baltic region.

4 Using John Fennell's demonstration in his *Crisis of Medieval Russia, 1200–1304,* I have shown how the escalation of warfare between the Republic of Novgorod and its neighbors, and the representations of earlier medieval "battles" and conflicts, transformed what were possibly no more than repeated skirmishes and border incidents into grandiose wars. From the exaggerations of early chroniclers to the films of Eisenstein, the inflation of the spectacular dimension of these representations is proportional to the escalation of violence and the progression of warfare into modernity—see Oleg Kobtzeff, "Espaces et cultures du Bassin de la Neva: représentations mythiques et réalités géopolitiques" in Walter Zidaric (ed.), *Saint-Petersbourg: 1703–2003* (Nantes: CRINI, 2004), 19–50; and John Fennell, *The Crisis of Medieval Russia* (London and New York: Longman, 1983), chapter 5.

5 The instability begins with the crowning of Boris Godunov after the death of Tsar Feodor, son of Ivan the Terrible, and ends with the election by a nationwide assembly of the first Romanov tsar, Michael. This is as far East as this chapter will journey. Asia could not have been unaffected by the cataclysmic upheavals that ravaged the entire planet in the fourteenth century and beyond. Historians should begin investigating the history of warfare in Asia from a comparative perspective to explore the differences and the coincidences of historic trends.

In this all too brief evocation of the conflicts of this period, we will not indulge in the somewhat outdated positivist attempts to reconstruct cause-and-effect chains of events, focusing on border disputes, on commercial agreements gone sour, or on legal battles between monarchs or pretenders over issues of succession to one throne or another. Although the importance of the Marxist economic paradigm and class struggle is fully acknowledged here, it is sufficiently known through Marxist literature since Engels' history of the Peasants' War, and need not be discussed in detail. The Marxist interpretation will be endorsed here, in the search for the causes of the wars of the sixteenth and seventeenth centuries, with one reservation: the economic causes of conflicts must be presented in association with, not in opposition to, the psychological dimension of the conflicts. A historiographic tradition that began almost a century ago with Lucien Febvre's *Luther*, and continues today with Jean Delumeau, has shown us that mentalities were too deeply affected by extreme environmental and social changes to be simply confined to the "suprastructure" and not be taken seriously as a driving force of the history of pre-modern collective violence. We must consider how economic, geopolitical, psychological and religious factors accumulated to create an extreme crisis of civilization, the symptoms of which were war, civil unrest and other forms of collective violence.

The witch hunts are a revealing phenomenon of those times. Michel Foucault was the first to point out in his *Discipline and Punish* that a phenomenon much larger than a simple episode of mass hysteria and superstition was at play. Likewise, violence from the fifteenth to the eighteenth centuries was one of the subjects of Pierre Chaunu's ongoing research, particularly owing to the exceptional efforts of an American scholar, Alfred Soman, who invested decades of his life to the study of witch hunts.[6] His work is too rich to be summarized easily.[7] It is better to focus on some of its aspects. The study on the witch hunt began with the meticulous study of the prison records of the Parliament of Paris.[8] What becomes evident after

---

6 Alfred Soman, "La sorcellerie devant le Parlement de Paris (1565–1640)" *Annales ESC*, 32: 4 (1977), 790–814, was the article published at the beginning of his investigation and remains a landmark. His major monograph is Alfred Soman, *Sorcellerie et justice criminelle: le Parlement de Paris, 16e–18e siècles* (Aldershot: Ashgate-Variorum, 1992).

7 … and it would always be a betrayal because of the innumerable nuances of his rich and highly sophisticated arguments. Fortunately, a very informative autobiographical and scholarly essay helps non-specialists understand his thesis: Alfred Soman, "Sorcellerie, justice criminelle et société dans la France moderne (l'ego-histoire d'un Américain à Paris)," *Histoire, économie et société* 12: 2 (1993), 177–217. Acknowledgement of my master's thesis on appellants convicted for arson came as a surprise given its errors due to the impatient excitement of youth, and lack of experience in paleography. May Dr. Soman find here the expression of my gratitude. If I have ever published anything worth reading, I owe much of it to him and especially to the dearly regretted Pierre Chaunu.

8 See Soman, "La sorcellerie devant le Parlement de Paris …": prisoners were booked upon arrival in Paris where they were to stand trial after an appeal. Between 1564 and 1638 (before the method of recording entries was modified), a register (*registre d'écrou*) recorded every arriving appellant. The document systematically also registered

quantifying the cases of appellants convicted of witchcraft are the two abnormal surges in the numbers of such prisoners brought between 1574 and the late 1580s and again between 1594 and 1629. The sudden variations corroborate the sources witnessing the phenomenon which we know as "the witch hunt." But then, the same records reveal similar surges, in the same years, for other prisoners, those convicted of arson, infanticide, bigamy and spouse murder, and possibly sodomy. Appellants of sentences for such crimes were brought to justice in unusually high numbers almost at the same time. What this may suggest is that the witch hunt could be a phenomenon related to *lynching* and that it was not isolated. A much larger and complex phenomenon of violence was transpiring.

The gap representing the differences between Michel Foucault and Alfred Soman or Pierre Chaunu in their respective historical analyses of violence is much wider than the Rue Saint Jacques that once separated their offices; it is a wide canyon. Nevertheless, not unlike Foucault, and at a time when René Girard was forming his own thoughts on violence and scapegoats, Alfred Soman's investigations insinuated that the line between collective murder, lynching and local justice was a very fine one. Contrary to Foucault, his work on sixteenth and seventeenth century justice showed how it understood this fine line, and resisted the tendency to legitimize lynching through judiciary decisions. The judiciary institution tried to contain collective violence rather than indulge it.[9] But what was the nature of this violence that needed to be contained? Could it be the same violence that would then be released in the outbursts of warfare occurring in the sixteenth and seventeenth centuries?

One of the common denominators of all the conflicts mentioned above is their *massive character*, i.e. how popular masses could initiate them, and how determined they were in carrying them out in a way contemporary sources unanimously describe as atrocious. Engels' *Peasant War in Germany* has long lifted any doubts about the revolutionary dimension of the wars of religion of the sixteenth century. Yves-Marie Bercé counted 450 to 500 armed popular uprisings between 1590 and 1715 in Aquitaine alone.[10] Later, between 1715 and 1787, the tensions receded: only 100 such uprisings occurred in all of France.[11] The years

the date of arrival, the identity of the prisoner, the original charges brought against him or her, the sentence pronounced by the local jurisdiction, and the location of the jurisdiction. After the appeal was judged by the court in Paris, the new sentence was recorded. This makes the *registre d'écrou* a very rich source for research in serial history.

9 Decriminalizing sorcery was a way to mark the power of the great monarchies over the social pressures which, in the Holy Empire, resulted in widespread popular repressions against witches. That Foucault never noticed this problem is another of the fundamental differences between his *Surveiller et Punir* and the work of the authors in Chaunu's team.

10 Yves-Marie Bercé, *Histoire des Croquants: étude des soulèvements populaires au XVIIe siècle dans le sud-ouest de la France* (Geneva, Paris: Droz, 1974), Vol. II, 674–81, quoted in Jean Delumeau, *La Civilization de la Renaissance* (Paris: Arthaud, 1984), 188.

11 Daniel Mornet, *Les Origines intellectuelles de la Révolution française* (Paris: Armand Colin, 1934), 443–6; there were only 275 in the entire English countryside between 1735 and

1590 to 1715 coincide with the years between the beginning of the wars of religion and the end of the Thirty Year War.

A New Type of Violence

Another common denominator in the "Time of Troubles" is the rise in casualties, which was probably proportional to the progress of military technology and tactics: field artillery, muskets and combined arms tactics.

A reliable death toll has not yet been established for every war fought between 1493 and 1648. Enough is known to consider that casualties were extreme. Robert J. Knecht, one of America's foremost specialists of French history and a specialist of the Wars of Religion in France, considers that the rough estimate is two to four million dead.[12] This is approximately 11 percent to 25 percent of the French population of the fifteenth century (16 to 18 million inhabitants). The Thirty Years War has long been considered, and is still considered by many, as the deadliest in human history before World War I. The 40 percent to 50 percent casualty rate often proposed until recently is believed today to be a perception distorted by contemporary sources, first of all by the unprecedented degree of atrocity, often gratuitous, described by witnesses, and secondly, by the fact that the plague returned with a vengeance during the same years. The astounding statistics in areas having suffered the most between 1618 and 1648 have understandably created the apocalyptic vision: the Altmark province lost 40 percent of its rural inhabitants and 50 percent of its urban dwellers; Mecklemburg-Pomerania lost 65 percent of its population, and the Rhenish Palatinate 70 percent;[13] in Wissembourg (Alsace), the pre-war population of 1,500 was reduced to 150; in Spandau, the numbers were 3,600 to 1,500; in Frankfurt (Oder), 13,000 to 2,400; and in Cologne, 12,000 to 6,200.[14]

Today, the death toll has been reduced to 15 percent to 20 percent according to most historians.[15] Yet others like Henry Bogdan continue to claim that Europe has lost up to "60% of its population during the period of the Thirty Year War."[16] Also, famines and pestilence may have been less devastating had they not been in part caused by warfare. Famines were caused directly by the war when crops were burnt, or indirectly when masses of refugees hid in forests or flooded areas where food was scarce, or because the armies and the refugees carried germs and spread

1800 according to George Rudé, *The Crowd in History: A Study of Popular Disturbances in France and England, 1730–1848* (New York, London: Wiley, 1964) and George Rudé, *Violence and Civil Disorder in Italian Cities, 1200–1500* (Berkeley: University of California Press, 1972).

12 Robert J. Knecht, *The French Religious Wars* (Oxford: Osprey, 2002), 91.

13 Henry Bogdan, *La Guerre de Trente Ans. 1618–1648* (Paris: Librairie Académique Perrin, 2006), 276–7.

14 Ibid.

15 As reviewed in Yves Krumenacker, *La Guerre de Trente Ans* (Paris: Ellipses, 2008), 167.

16 Bogdan, *La Guerre de Trente Ans*, 279.

the plague, dysentery or typhoid fever, which in turn caused more crop failure by depriving the harvests of the indispensable workforce.[17]

It is not a coincidence that this is also the period of Machiavelli, a period of new philosophical reflection over government and the legitimate use of force which revolutionized Western culture by liberating it from religious moral standards, or any moral standards for that matter, other than the pursuit of power in the name of security and prosperity. Mentioned by many specialists of the period, these new wars mark the end of the dream of a united Christendom and the end of an ideal of Christian governance. Machiavelli perfectly exemplifies the change. But for many, that change becomes itself a cause of the chaos against which the *Prince* is expected to protect the modern state.. The ideological change with which the modern territorial state is associated crosses a new threshold in legitimizing violence and rejecting an entire worldview, the Roman ideal, the hope that lasted for an entire millennium, a restored *Pax Romana*—the ideal that shaped an entire Medieval European civilization and had even maintained the last bonds between the West and the Byzantine world. The Roman ideal went up in smoke not only in 1453, when Constantinople fell, but once again in 1532, when *The Prince* was printed. It is the end of an ideal of catholicity (i.e. the representation of universality, called *sobornost'* in the Slavic world) and the beginning of a fragmented Europe personified by Machiavelli, Louis XI of France or Ivan the Terrible.[18] We shall see further how the ideological changes contributed to the confusion that had such negative effects on the mentalities of the times.

It would be unfair, however, to forget that the thoughts of a Machiavelli, and later of Realism, emerged as a reaction to what Hobbes called the "passions" of their times. Their call for state violence was a reaction to the chaos and extreme violence of mercenaries, bandits or fanatical popular movements such as the peasant movements mentioned above or the unrest started by Savonarola in late medieval Italy. As already mentioned, hundreds of contemporary sources were unanimous in describing a violence that reached a degree of atrocity hardly imaginable, when cities were sacked or farms were attacked. Scenes of particularly cruel executions and tortures became a routine part of the new literary genre—the picaresque novel—that spanned the entire pre-modern age from Grimmelshausen's *Simplicius Simplicissimus* to Voltaire's *Candide*. Even worse was the justification of the violence. While one category of accounts of the massacres on Saint Bartholomew's night lament the horror of innocent men, women and children being butchered as they were dragged out of their sleep, other documents exalt the same acts as heroic feats. Frescoes still visible today in the Vatican belong to this category. Bartolomeo de Las Casas' frightening testimony about the enslavement and eradication of indigenous American populations met opposition, particularly during the Valladolid controversy, not because such detractors as the humanist Juan Ginés de Sepúlveda denied the reality of the violence, but because they *defended it* as a

17 Ibid., 277.

18 The tsar's kinship with Machiavellian thought is a well documented key theme of Jean-François Colossimo's brilliant *L'Apocalypse Russe* (Paris: Flammarion, 2009), 125–33.

necessity. Eventually, despite much ambiguity and internal quarrels, the position that prevailed within the Catholic Church after the papal bull of 1537 (*Sublimis Dei*) and the 1542 *Leyes Nuevas* was a reaffirmation of the "catholicity" of the Church, the roots of which are in the Hellenic-Roman representation of universality (which also inspired Byzantine Christianity and Islam) and the unity of human nature. The perverse effect of affirming that all humans are equal in the eyes of God is that it forced those who profited from slavery to develop racist theories to make it "impossible for us to suppose that these creatures are men," as was once said by Montesquieu, because, "should we suppose that they were men, one would begin to believe that we ourselves are not Christians."[19] The ideological construction necessary to dehumanize "races" of which morality would then allow enslavement or extermination finds its origins here. This process of dehumanization can then be applied to any enemy. It is the perfect ideological weapon, applicable, over the centuries, to any scapegoat or competitor, starting with Jews and culminating, in the great age of the nation-state, with the demonizing of any foreigner in the name of "race" and "civilization."

However, ideology alone cannot explain the fury expressed by entire European crowds when they invested so much energy in killing each other. The conditions favorable to such a man-made disaster must have been extreme to render society so dysfunctional that it generated such unspeakable violence as we have alluded to above.

Extreme Conditions of Life as Factors of Aberrant Tension

In 1381, there were only 60 million inhabitants left in China, in a land that had counted more than 123 million less than two centuries earlier.[20] Famines had taken their toll, but another earth-shattering cataclysm, one of unheard-of proportions, had reaped a frightening number of victims: the Great Plague or "the Great Mortality"—one of the worst, if not the worst, epidemics known in human history, to this day.

Reported in the vicinity of Lake Balkach in 1338, the disease spread in nine years across Central Asia, to the Western shores of the Caspian Sea, to Crimea, the Northern shores of the Black Sea and along the Volga River.[21] In 1348 it progressed into the Balkans, throughout Asia Minor, along all the shores of the Northern Mediterranean, along the Tunisian coast and into Italy, south-western France, and most of Spain, also along the Seine and the Loire, across the basin of the Jordan

19 Charles de Secondat de Montesquieu, *De l'esprit des lois* (London: Garnier, 1777), Book XV, Chapter V, 68–69 (my translation). Montesquieu specialists argue that this is a satirical segment, not to be interpreted literally as reflecting Montesquieu's own views.

20 In 1193–95; see Pierre Chaunu, *Un futur sans avenir: histoire et population* (Paris: Calmann-Levy, 1979), 108.

21 Ibid., Georges Duby, *L'Europe au Moyen-Age* (Paris: Flammarion, 1984), 194–5.

River, and the Nile's delta—and along all of the southern shores of England.[22] In that terrible year, a large territory around Mecca was contaminated.[23] Within the next four years the plague had spread across all of Europe and the greater part of the North African Coast, the entire Nile basin and most of Mesopotamia. The Indus valley was infected in 1350, and the south-eastern tip of the Arabian peninsula in 1351.[24] Except for only a few preserved areas,[25] almost all of the European continent including Russia, half of the coast of Northern Africa, plus many of the regions around the great cities of the Islamic world, had lost a number of inhabitants difficult to fathom.

This epidemic of earth-shattering proportions, according to the most conservative estimates, exterminated at least one third of the population of the contaminated territory, maybe half, possibly even more.[26] The average numbers for the totality of the population should not eclipse local situations that were even far worse: entire communities were totally eradicated. While some pockets of human habitat were preserved, it is not difficult to imagine the fear that these survivors had experienced hearing about less lucky villages or towns driven into extinction. Moreover, the bubonic plague remained present in the same zones until the eighteenth century, returning at each new generation to take its horrendous harvest.

Another cataclysm was yet to ravage the entire planet without discrimination. For the second time since Antiquity, a period of global cooling progressively changed the geography of most habitats on earth, causing confusion and forcing large populations to rapidly adapt to these changes of cosmic proportions in order to survive. Probably the most extreme example of the geographic mutation was Greenland: its name bears witness to its appearance during the centuries of global warming that lasted roughly during the second half of the Middle Ages. The Scandinavian settlers had raised cattle and even grown a limited number of crops in its green pastures that had resembled today's Northern Ireland, Northern Scotland, or Iceland, and not a wasteland covered by snow and ice. Climatic change that turned the entire island into an extreme polar environment put an end to its medieval history. Throughout the world, agricultural activities and entire ways of life were also coming to an end—vineyards and wineries in England and in other

22 Ibid.

23 Ibid.

24 Ibid.

25 A spot in south-eastern Ireland, almost half of Wales and Scotland, the Central mountains of France, an important region around the Rhine delta, roughly half of the Danube basin (roughly, present day Romania), a large portion of Ukraine and Belarus between the Dnepr and the Dnestr basins, and the scarcely inhabited Lapland, Finland and Carelia. Ibid.

26 See Chaunu, *Un futur sans avenir*, 108–11; Chaunu, *Trois million d'années*, 221–33; the comparative chart published by the US Census Bureau listing a variety of sources, particularly Jean Biraben's research frequently referred to by Chaunu, on the rise and decline of the global population (http://www.census.gov/ipc/www/worldhis.html, accessed October 11, 2010); Jean-Noël Biraben, *Les hommes et la peste* (The Hague: Mouton, 1975).

parts of Northern Europe (a commercial resource but also an essential source of calories for the consumers), contact between villages in the coldest areas, maritime transportation in Nordic regions and numerous other aspects of daily life were affected.[27] Russia, already surrounded by hostile political forces, became even more isolated as its Nordic harbors were trapped by ice for longer and longer periods.

Historians like Jean-Noël Biraben and Pierre Chaunu noted the rapid demographic recovery. By two or three generations, Europe's population grew back to the number of Europeans living on the eve of the plague. But the psychological recovery from the trauma of such horror could not have been so rapid.

Fear and Disorientation

> *Cowardice, alone of all the vices, is purely painful—horrible to anticipate, horrible to feel, horrible to remember; hatred has its pleasures. It is therefore often the compensation by which a frightened man reimburses himself for the miseries of fear.*[28]

After the cataclysms experienced in those times, terror had become a normal part of late medieval and Renaissance mentalities. A constant perception of the presence of death, as pointed out by numerous artistic expressions of the times, graphic or literary, was studied by historians, more systematically in the 1970s and 1980s by the French "New Historians" (identified as the *Annales* school by Anglo-Saxons) like those already mentioned and also by Philippe Ariès, Emmanuel Leroy-Ladurie and particularly Jean Delumeau.

A culmination of twentieth-century research launched by Lucien Febvre's *Luther*, Jean Delumeau's work first analyzed the causes of the social tensions resulting in the birth of Protestantism and the mutation of Catholicism.[29] He then produced the world-famous monograph on fear which will remain for posterity as one of the most important monuments to the history of mentalities. Which of those—the social tensions or the phenomenon of collective fear—came first and generated the other is a chicken-and-egg question too complicated to answer now, and seems to matter little as long as it is clear that the two are inseparable. More importantly, how did the general climate of extreme anxiety then evolve into a general climate of extreme warfare?

The brutal and widespread decline of the population caused by the Black Death and climate change created a unique situation of *anomie*. It dislocated the

27 Emmanuel Leroi-Ladurie, *Histoire humaine et comparée du climat. Tome 1: Canicules et glaciers XIIIe–XVIIIe siècles* (Paris: Fayard, 2004).

28 C.S. Lewis, *The Screwtape Letters* (West Chicago: Lord and King Associates, 1976), 134.

29 Jean Delumeau, *Naissance et affirmation de la Réforme* (Paris: Presses Universitaires de France, 2003) and Jean Delumeau, *Le catholicisme entre Luther et Voltaire* (Paris: Presses Universitaires de France, 2010) are the most recent editions of these two classics.

fundamental structures of sacramental life and the representation of life on Earth and in the Hereafter. The upsurge of mortality meant that an important number of priests, proportional to the general population and its mortality rate, had also disappeared. Moreover, during an epidemic of such proportions, when it becomes clear that a healthy person's chances of living until next year have been reduced to two chances out of three or even one chance out of two, the clergy was needed more than ever. Corruption, absenteeism, simony, or plain brutal bed manners, or any other form of incompetence that had been very common so far, was now unbearable. The sick and the dying who were most in need of sacraments, particularly the last rites, were deprived of those in a particularly unusual way: not only were the habitually incompetent or the mediocre even less available than ever, but those few priests who had had the courage to visit contaminated homes became the most exposed to the disease and died, thus depriving society of the best of their caste. The other priests either feared contagion or were banned from approaching the sick by local decrees imposing quarantines. One of the most painful memories of the survivors, according to numerous sources from those terrifying times, was that three-fourths of the dying passed away without the assistance of the last rites. To make matters worse, in quarantined households where the entire family had been wiped out, dead bodies were left to decompose without a Christian burial, a treatment reserved normally for heretics or excommunicated sinners. Not only would this cause great distress but also extreme anomic dilemmas since the Catholic authorities had become very legalistic in their insistence on certain rituals as an indispensable validation of one's passage from this life to the afterlife.[30]

The disasters accelerated a process that already started: the profound reshaping of social and economic structures. This aggravated an already complicated situation.

In medieval society, individuals were not free but they were not aware of it. Medieval society, wrote Erich Fromm, "did not deprive the individual of his freedom, because the 'individual' did not yet exist; he was still related to the world by primary ties."[31] By primary ties, Fromm meant the social pressures that shape an individual in childhood (the discipline that comes through parenting) and, later, the adult in traditional tribal or rural societies. Such bonds defined the relations between persons in the urban societies of the middle ages, within guilds and corporations acting as clans maintaining group solidarity and identity.[32]

These primary ties let him or her recognize himself or herself and others only through the medium of his or her or their participation in a clan, a social or religious community, and not as human beings; in other words, they block his or her

30 Unfortunately, very little attention to these problems has been paid by historians studying the Orthodox populations. From the crisis of the schism of Old Believers in seventeenth-century Russia, and how the faithful themselves were ready to die for the conservation of the formal ritualistic aspect of salvation, we can speculate that this conflict between the necessity to perform certain rites and the absence of clergy was just as acute as in the West, if not even more dramatic.

31 Erich Fromm, *Escape from Freedom* (New York: Avon Books, 1969), 59.

32 Ibid., 39–51.

development as a free, self-determining, productive individual. But although this is one aspect, there is another one. This identity with nature, clan, religion, gives the individual security. He or she belongs to, he or she is rooted in, a structuralized whole in which he or she has an unquestionable place. He or she may suffer from hunger or suppression, but does not suffer from the worst of all pains—complete aloneness and doubt.[33]

The environmental and demographic upheavals and the rise of capitalism and market forces were to destroy this "structuralized whole," and with it the sense of security and identity provided by this *highly integrated* medieval society. These changes were the rapid decline of the ancient caste system which had defined group identity and personal identity for centuries, a decline caused by a faster accumulation of capital, changes in the organization of urban societies (the liberalization of markets destabilizing the ancient system of guilds and corporations) and, as mentioned before, the obsolescence of the church as a driving force of social interaction.

The new society had also become a society of "winners" and "losers" (to use terms very popular in the early 1990s after the transition in the former Soviet bloc). All winners and losers now had to adapt to a society based on competition and where the measure of success was individual gain of wealth and power, and no longer the performance of one's duties as set by tradition. "The individual was left alone; everything depended on his own effort, and not on the security of his traditional status."[34] The reduction of the population caused by the calamities of the time played a role in constituting new middle classes or, on the contrary, new pockets of poverty. Craftsmen and peasants living in areas where the demand for their skills, rarefied by the deaths of their colleagues, exceeded supply, could negotiate for better contracts and salaries. This workforce thrived on supplying the new rich—essentially the rising merchant class and a few noblemen—with the skills needed for housing and for new consumer goods such as fashionable clothing, spirits and long conservation foods.[35] Elsewhere, the opposite situation may have occurred. Supply of labor exceeded the demand either because entire markets had been wiped out by the death of customers or, on the contrary, because the area had been preserved from disease or wars and had an excess of labor. In both cases, unemployment reduced entire crowds to poverty. Poverty was also the result of growingly intense competition between capital and market forces. For the losers, it created uncertainty, adding to the already existing anxiety. But for the winners as well, financial gain only created a new dilemma since the Church had always condemned excessive material wealth.

Jean Delumeau, in his now classic *Fear in the West*, uncovered an escalation of a *discourse* constructing collective representations of the Other as an extreme

33 Ibid., 51.

34 Ibid., 77.

35 Spices, tea, jams and jellies, candy, tobacco, furniture and artistic objects, and many other goods, all of which had either been completely unknown, or present in small quantities, outside the richest households of monarchs and powerful feudal rulers.

danger; then, taking his investigation one step further in *Sin and Fear: the Inception of Guilt in the West, 13th–18th Centuries*, he showed how this mechanism created conditions favorable to aggressive behavior. His thesis is that these representations and the tensions that they created escalated violence in the centuries from the second half of the Middle Ages to the beginning of the Enlightenment.[36] The populations of Christendom were trapped in one of the most extreme situations of anomie in history.

The reasons for millions of Europeans living in terror were quite evident given the extent of the environmental and social-economic upheavals they were trying to adapt to. We also just mentioned the problem of the insufficient presence of an educated and devoted clergy. But at the same time, two more factors of extreme stress made the situation worse in a time when the Final Judgment seemed to be at hand, and when the believer needed to be reassured about salvation for self and for loved ones.[37] On the eve of the Reformation, Christian communities and families facing climatic change and the plague suffered from two additional forms of stress. First of all, Christianity kept placing growingly unrealistic demands upon individuals to atone for their sins and to vouchsafe a favorable passage into the Kingdom to come. Secondly, Christianity let itself be torn apart on the question of how to attain that goal.

For the last two centuries of the Middle Ages, Christian consciences were distressed by uncertainty about the future of the Church as it appeared to be torn apart by many internal divisions. Above all, century-old memories of inter-Christian wars had accumulated and damaged their assurance in the benevolence of the Church. Every time a war was fought between any Christian monarch or feudal lord, since local Church authorities almost always bestowed their blessings upon the armies as Voltaire would point out in his eighteenth century satirical novel *Candide*, an anomic situation was created: since one of the armies was carrying out a just war and the other was not, one of the local Churches was necessarily wrong. But how could that be compatible with the indivisible unity of the Church, its sanctity and its infallibility? Also, numerous dilemmas arose for those whose families or communities were divided by new schisms. The first major schism since Antiquity was the break between East and West in 1054.[38] Barely noticed at the

36 Jean Delumeau, *Le péché et la peur. La culpabilisation en Occident, XIIIe-XVIIIe siècles* (Paris: Fayard, 1983); English translation by Eric Nicholson, *Sin and Fear: The Emergence of a Western Guilt Culture, 13th–18th Centuries* (New York: St Martin Press, 1990).

37 This tragic fear of eternal damnation and the search for a solution or a consolation resulted in the historic process revealed in yet another classic of the French *Annales* school as the invention or the birth of Purgatory: Jacques Legoff, *La naissance du purgatoire* (Paris: Gallimard, 1981), translated by Arthur Goldhammer as *The Birth of Purgatory* (Chicago: University of Chicago Press, 1986).

38 The differences remain, essentially, to this day, monarchical papal authority versus a decentralized governance of a confederation of dioceses, the cultural differences between Western and Middle-Eastern spiritualities (influencing Balkanic and Slavic worldviews, e.g. the view of salvation as an *ontological* process as opposed to the Catholic or Protestant doctrines of "justification" rooted in Western societies' evolving

time by anyone outside the circles of royal courts or the highest ranking clergy, awareness of this schism grew with the increase of wars, after the thirteenth century and beyond, opposing Western armies and Eastern ("Orthodox") Christians under Byzantine or Arab rule. From that time on, where communities of both Latin and Byzantine-Middle Eastern traditions coexisted, Europeans became torn between loyalties.[39]

The tensions would increase with the conversion, in the late Middle Ages and in the sixteenth century, of those that are still called today "Uniates"—former Orthodox Christians practicing Eastern rituals but under Roman Catholic papal authority.[40] The other great division among Christians began in the tenth century in the Balkans when a new Manichean movement, the "Bogomils," began to spread throughout Europe, creating tensions culminating in the twelfth century in France with the violent war against the Cathares. In Central Europe in the fourteenth century, the movement of Jan Hus and the turmoil that followed became the first stage of the Reformation. At the same time, throughout Catholic lands, the faithful were disturbed by the political intrigues that moved the headquarters of the papacy to Avignon in France. The consequence was the most anomic situation ever experienced by Catholics: the social and political divisions caused by a schism between a succession of popes, and a succession of anti-popes, divided in turn into splinter groups. How could one possibly know which was the right side? Yet, all had been taught that outside the Church there was no salvation.

After 1492, the new uncertainty about the foundations of a triumphant and safe Christian world was also due to the discovery of not only New Worlds, but of alien populations so large that Christians should now be seen as a small minority on the globe. The information filtering from what was beginning to be known about the splendors of China signaled that these new non-Christian lands could even be far more advanced in civilization than Europe. This was a genuine cultural shock. Why not see this above all as an opportunity to enlarge the Christian world like the missionaries or the Conquistadors? That many perceived this as a negative shock

culture of civil and criminal law), compulsory clerical celibacy versus married clergy, and small but numerous customs that shaped the rituals. The use of Latin versus the use of the colloquial was resolved after Vatican II.

39 Latin and Byzantine traditions had coexisted peacefully long after 1054 in such hubs of cultural and commercial exchange like most of the great cities of the Eastern half of the Danube basin, or the ports of the Baltic Sea, or the Russian merchant city of Novgorod, or in pilgrimage destinations like Jerusalem. The religious border between the Roman Churches and the Eastern Patriarchates is, roughly, a line cutting across what is now Eastern Finland, Eastern Poland and Belarus, Eastern Slovakia and Bosnia. The Latin and Eastern religious traditions had also coexisted in Southern Italy and on the Island of Corsica where the Byzantine liturgy still survives to this day in remote parishes.

40 See Raymond Janin, *Les Eglises orientales et les rites orientaux* (Paris: Letouzay & Ané, 1997). Originally published in 1922, it is the reference providing one of the most detailed typologies and cultural geographies of all non-Latin Christian denominations, including the groups that converted to Roman Catholicism while retaining their Byzantine or Middle Eastern customs including married clergy.

is understandable because it coincided with the expansion of the Islamic world. The advance of the Turks deeper into Eastern and Central Europe was soon to reach the gates of Vienna. This is why 1453 became a particularly symbolic date in Christians' representations.

The Byzantine Empire of the fifteenth century, a small territory limited to a few regions in Greece and to the Bosporus, was only a shadow of its past glory when it fell. The impact of that event may have been entirely symbolic (the territory of the Byzantine Empire was no more than a few small dots on the map by the time the Ottomans seized it), but at the time, the worldview of most Europeans, insists Denis Crouzet, was fundamentally based on symbols.[41] The fall of Constantinople was such a symbol for both Western and Eastern Churches that many early twentieth-century history books still chose 1453 as the date of the end of medieval civilization rather than 1492. The event of 1453 was felt as apocalyptic because of the geostrategic march of Islam, but also, as will never be repeated enough, because it was the end of the last official remnant of the Roman state, the end of almost two millennia of history. If the West represented 1453 as a cosmic tragedy, it is then easy to imagine how this event was experienced by the populations of the Orthodox lands, where the "second Rome," no matter how decadent, was not only a political model but a real presence and a cultural reality in daily life. Western historians have neglected this last aspect, as if the millions of Slavs and Greeks now under Ottoman rule suddenly went to sleep, and as if Moscow grabbed this opportunity only to invent a new expansionist ideology. Hence the profound misconceptions and tenacious clichés about the ideology of the "Third Rome."[42]

The term, probably formulated immediately after 1453, was a rallying call for a last battle before Judgment Day. It had nothing to do with nineteenth-century ideologies behind modern Russian expansionism. When, in 1510, one of the ideologues of the "Third Rome," the monk Philotheus of Pskov, proclaimed that "two Romes have fallen, the third stands, and there will be no fourth," no one in the Kremlin or in wooden cabins in villages imagined that Russia's future was to become a great world power. "The Third stands" meant that this was the last stand of the last Orthodox Christian state at Armageddon. After that, there could be no future governments on Earth. Since this is a battle to the end and there is nothing to lose, we may understand why Russians could fight with such energy. It is the "energy of the desperate," a sense that all is already lost, and that without any hope for a future life, clinging to life, hoping to survive, is futile. The only positive thing that one can do is fight evil with one's last energies. All military historians, psychologists or experienced soldiers will confirm that the best fighter on the battlefield is the one who has given up any idea of survival. Then fear of

41 Crouzet, *Les Guerriers de Dieu*, 81.

42 These clichés, spread like a virus by Arnold Toynbee, resulted from the fact that years of Russian historical research on the subject never reached the West. Most recently, Colossimo (*L'Apocalypse Russe*, op. cit. 115–124) sets the record straight in the best possible concise clarification, and provides a good bibliography composed almost entirely of titles in English (ibid., 349–350).

death disappears, once death is seen as inevitable, while a feeling of injustice can spark an extreme compulsion to take revenge against the enemy. This can explain the energy of the volunteers who fought in the ranks of Minin's militia during the Polish invasion of Russia in the "Time of Troubles." It could also explain later events such as the collective suicides by fire of the Old Believers losing their last stands against the troops of the tsars trying to impose religious reform, or how usually passive and humble peasants set fire to their houses and crops in 1812 before joining the guerillas to fight the invading "Antichrist" or "infidels" (*basurmany*, as the poet Lermontov quoted the veterans of the Borodino battle). Historians should reconsider the history of conflict in Western Europe from this eschatological perspective and ask themselves (as Denis Crouzet?) if the Russian peasant is not the model of the combatant in many conflicts throughout Christendom before the industrial age.

The second cause of instability pointed out by Professor Delumeau, aggravating all other factors of anomie and anxiety, was a *collective neurosis* resulting from the clergy's increasing surveillance and control over individuals and communities. From the 13th to the 18th centuries official missionary campaigns or spontaneous popular mystical movements were launched to infiltrate society, to connect each aspect of daily life to the Church, to make Christianity the center of popular culture. This was a solution to the insufficient number of clerics, but also to the geopolitical or spiritual threats posed by influential alternative religious movements. For ecclesiastical authorities, there was even more at stake. In many areas of Europe, the population had never fully assimilated the precepts of Christianity. Moreover, as the Church was becoming less relevant in a world being shaped by market forces, it needed to reestablish a new authority. For the strong and imaginative characters, change was a challenge. But in the general atmosphere of panic, such characters were the rare visionaries and saints like Las Casas, Saint François-Xavier, Michel de l'Hopital or Nil Sorskiy. For many others in the clergy, as well as among the simple folk, change was yet one more source of anxiety. It was easier to exploit the ambient fear rather than be more creative and sophisticated in attempting to reduce it. Also, the missionaries were themselves a product of the times where the environment was favorable to sado-masochistic tendencies. Therefore, the missionary campaigns relied on violent discourses about widespread sin (especially sexual), death and fear of eternal fire. This fire-and-brimstone style produced terror in the remotest rural communities. Local priests, certain groups of monks, official or self-appointed missionaries, traveling preachers or mystics, groups of penitents, flagellants, artists, or other groups or individuals trying to reach out to the populations, imposed upon already anxious crowds representations of the end of times, with images of damnation by a vengeful God, rather than speak of salvation and of a loving and forgiving God. The categorization of human weaknesses into seven capital sins dates from that movement of spiritual *Reconquista* of Latin Europe. It spread to the Orthodox lands through the Ukraine and through the religious academic institutions in Kiev.[43] This codification of sin was to simplify a believer's efforts in

43 See George Florovsky, *Ways of Russian Theology* (Belmond: Nordland Publishing, 1979),

self-discipline. Instead, it only overwhelmed and frustrated him or her (especially *her*) with a lengthy and complicated catalog of situations and occasions in life leading to sin. An entire literary genre mixing casuistics and psychology influenced many writings, like Dante's poetry. The spiritual *Reconquista* spread through sermons, personal injunctions to families and individuals, rituals, morality plays, and also through two novel features of religious life: confession (made compulsory once a year from 1215) and, soon, the spreading of the printed word.

Thus, an accumulation of factors of fear and anomie created an extremely dangerous situation which became the fertile soil in which the seeds of violence could grow. But what force actually *made* them grow?

From Disorientation to Violence

One answer to this question is that there may be no demonstrable cause-to-effect chain between the social tensions or the ideologies encouraging violence and the acts of violence themselves. Our only certainty is that somewhere, somehow, there occurred a breakdown of the balance of power between individuals and groups and the social pressures that normally contain violence. In more simple words: those who killed did it because they could. Witnesses who had observed the violence at the time, and historians who analyzed it centuries later, agree that once the population felt completely hopeless facing inevitable death, once it was clear that there was nothing to lose, normal individuals could either offer their lives and energies in exceptional self-sacrifice, while others, on the contrary—and this is the interesting part for a better understanding of the phenomena of warfare and gratuitous violence—could live out their darkest fantasies.[44]

Another simple explanation is that solutions to the crisis were found and immediately forbidden, creating unbearable frustration leading to violence. Reformers like the Protestants, the Catholic Tridentine renovators or pro-Nikon partisans and Old Believers in Russia,[45] found what they and millions of their followers discovered as the ideal ideological construction alleviating the fear discussed above. But these rival ideologies annulled each other. For the loyal Catholics or the Orthodox, supporting the old traditions, the rituals, the sacraments, the authority of the clergy, the artistic heritage and the folklore, the

Chapter II.

44 Chaunu, *Un futur sans avenir,* 110; see also Elizabeth Carpentier, *Une ville devant la peste. Orvieto et la peste noire* (Paris: SEVPEN, 1962). Among many other authors, Boccaccio or Defoe noted this; they had been preceded, centuries earlier, by Thucydides.

45 The great paradox was that the extreme conservatism of the Old Believers forced one of their factions to become radical reformists. Viewing all clergy as dangerous reformers, or simply deprived of clergy in remote provinces where they were hiding, the *bezpopovtsy* (literally *anapresbyterians*—"those without priests"), developed a doctrine abolishing the priesthood as radically as Calvinists or Baptists.

"good works"—acts of charity, pilgrimages, tithing, or other ethical actions—were what had always been reassuring. It embellished daily life, it meant continuity in history and a link between the generations, and it created group identity. If anyone tried to change this, be it Roman Catholic renovators or patriarch Nikon attempting to bring popular Russian religious practice closer to its original Greek sources, or Protestants who put all rituals and hierarchy in question, they were perceived as a threat. Where the breakdown of traditional cultural structures had rendered meaningless the pursuit of salvation through "good works" and rituals, rules and regulations (because the clergy was absent to sanction these "good works"), the solution was to rely on the gratuitous mercy of God and nothing else. Those who wanted to bring back tradition and "good works" were then the threat. When negotiation is impossible the use of force becomes the only solution to eliminate the threat. This still does not explain the passion with which violence was exercised, especially since killing other Christians—including harmless women, elderly people and children—denied all of the beliefs the fanatics held so dearly.

The merit of the French "New Historians," and particularly Jean Delumeau, was to have dug profoundly into the history of mentalities to find an alternative to Max Weber's or the Marxist interpretations of the history of the Reformation era which had almost become clichés. Denis Crouzet's work can be seen as a capstone of Delumeau's work. *Warriors of God* provides an explanation of the violence. He distinguishes two different forms of violence: Protestant and Catholic. They were different in nature. Reviewing the monograph at the time of its release, Gabriel Audisio provided a very accurate summary of Crouzet's thesis. Protestant violence

> *is expressed especially in iconoclastic movements and actions. It is about cleansing divine service. The violence of the Huguenots is antithetic to Catholic violence. The former is turned towards a waiting for God; the latter looks back and wants to restore a lost primitive purity. ... The Reform's principal effect was to disintegrate the anxiety of man regarding his salvation, as the author repeats it unceasingly. But this Huguenot violence had a subversive, utopian, millenarianist component; violence could have been the sign of a ritually controlled revolutionary crisis, a discharge of energy. ... D. Crouzet is opposed here to P. Deyon: it is not 'social protest' in the Reform which causes religious conversion, it is the contrary.*[46]

Catholic violence was different inasmuch as it

> *was the result of a long process, an impulse of crusading, of holy violence against the heretics, and then against the king seen as having become complacent. However, the massacres were unsuccessful in eradicating heresy, in purifying the kingdom, to satisfy God. This acknowledgement of*

46 Gabriel Audisio, "D. Crouzet. Les guerriers de Dieu. La violence au temps des troubles de religion, vers 1525 – vers 1610," *Revue de l'histoire des religions*, 211: 1 (1994), 109. My translation.

> *failure possibly induced a feeling of guilt and at the same time a diversion of the violence. ... Predatory warfare replaced ritual massacre. The wave of the white processions of 1583–1584, the assassination of Henri III, are to be understood as a reemergence of the collective anxiety, millenarianism, the eschatological spirit of the crusades and holy war.*[47]

The classic Marxist analysis of the infrastructure and other interpretations stressing the power of culture and ideology over mentalities, is somewhat reconciled in the works of psychoanalyst Erich Fromm. Written during World War II, his *Escape From Freedom* is, in great part, a historical analysis of the socio-psychological roots of Nazism. A great portion of this classic is devoted to the sixteenth century and draws on a great variety of late nineteenth-century and early twentieth-century innovators in German academia. Of course there was Max Weber, but there was a great variety of other social scientists whose common denominator was their interdisciplinary approach to history, particularly those who had researched that period prefiguring the *Annales* school, such as Karl Lamprecht who evidently influenced Lucien Febvre's famous biography of Luther.[48] Writing about the new capitalists of the fifteenth–sixteenth centuries, Fromm remarked that

> *their economic activity and their wealth gave them a feeling of freedom and a sense of individuality. But at the same time these same people had lost something: the security and feeling of belonging which the medieval social structure had offered. They were more free, but they were also more alone.*[49]

Violence was also the only means of conflict resolution because the institutions of society had not had the time to adapt to the new infrastructure: market mechanisms, credit, and especially the demise of the old professional guilds. The power and the wealth plus violence over the weak provided an illusion of control when solitude had become unbearable. At the same time, not only was guilt unavoidable for the violence that they inflicted, but in addition, the new capitalists suffered yet another anomic situation: for centuries the Church had instilled the idea that material wealth was sinful if not profoundly evil if it exceeded the amount strictly necessary for survival. But the "losers" also felt guilt: resentment or jealousy towards the newly rich were difficult to resist and those sentiments were reprehensible. All this guilt was aggravated by the fire-and-brimstone missionary campaigns analyzed

47 Ibid.

48 Compare Karl Lamprecht, "Zum Verständniss der wirtschaftlichen und sozialen Wandlungen in Deutschland vom 14. zum 16. Jahrhundert" in *Zeitschrift für Social- und Wirthschaftsgeschichte*, Bd. 1 (Freiburg, Leipzig: 1893) and Lucien Febvre, *Un Destin. Martin Luther* (Paris: P.U.F., 1928). Among these scholars, the most influential were theologian Ernst Troeltsch, sociologist and economist Wereer Sombart, and historian Willy Andreas; Fromm also acknowledges the English historian Richard Henry Tawney.

49 Fromm, *Escape from Freedom*, 63–64.

by Delumeau. But then, generations of social psychologists have established that repressed guilt leads to aggression. To repress guilt, one resorts to mechanisms of denial. One of these is to project one's own repressed violent tendencies upon the Other. It is what political science calls mirror imaging and what psychiatrists view as a symptom of paranoid tendencies in an individual. The next step is to take violent action as if it were a preventive measure against the perceived violence of the Other. There are tensions, but I am innocent; it is them and not me harboring violent feelings. But instead of attacking the capitalist, since this would be acknowledging one's sinful jealousy of the capitalist's wealth, those who felt oppressed or disoriented, peasants or members of the middle classes, would project their aggression on a scapegoat—Jews, heretics or simply foreigners. Also, the angry mobs were more easily manipulated by those who wanted to create diversions, redirecting revolutionary energies.[50] We have mentioned the importance of charismatic exalted leaders like Joan of Arc at the end of the Middle Ages. This type of leadership was characteristic of the sixteenth and seventeenth centuries as well, as shown by the legends surrounding figures like "Good Queen Bess" facing the threat of the Spanish Armada, "Good King Henry" the Fourth of France (originally a Protestant leader), the Conquistadores' leaders, Cromwell, Russia's Minin and Pojarsky, or any of the founders of religious movements. Submission to a leader, a mechanism of sado-masochism, resolves the problem of insecurity created by individual freedom in a too-rapidly changing world. The leader can then launch crowds against rivals, competitors or those who, in his own paranoid delusions, he sees as threatening—potential enemies.

Thus, Machiavelli's *Prince* may have been a rationalization of the predatory nature of the new social-economic relations. Although he wanted to be part of a solution to the violence of his times, he may have been part of the problem.

* * *

In 1648, the envoys of the states that had been at war with one another, all at once or separately since 1618, gathered in the city of Westphalia to sign a series of peace treaties which put an end to a conflict that had maintained Europe in a state of war for more than a generation. Reason seemed to finally prevail as the diplomats agreed upon the documents that established the legal notion of sovereignty and the foundations of international law. Christendom was to enjoy a period of relative peace and economic progress that partly laid the ground for the development of the Enlightenment and the future industrial age. Rapidly filling the gaps left open by the large number of deaths, an unprecedented demographic growth was to increase the number of Europeans dramatically.

It seems as if exhaustion had been the main cause of the end of the Thirty Years War. The war could not continue endlessly. The world had profoundly changed in the twelve generations between the great plague and 1648. A new civilization had emerged and needed to settle down, look at itself and understand some of

50 Ibid.

its own fundamental features: the rise of new world powers such as England and France at the expense of Spain, Portugal and the Holy Roman Empire, the new power of mercantile and financial enterprises, the opportunity that represented the expansion of their networks of trade and transportation (hampered by war), yet another redistribution of wealth (the violence had aggravated the landlords' dominance over farmers who had often traded their independence for security), the emergence of a new spirituality with its profusion of new religious monuments (classic or baroque depending on the region or the denomination), festivities, music, theological debates and literature, while at the same time secularism was rising. To digest all of this, the society of seventeenth-century Europe needed to rest, study and reorganize itself to be able to face the challenges of this new world. In Russia, the adaptation would be more dramatic than anywhere else after Peter the Great. Such adaptations needed the equilibrium that the realists had argued for. This is what was achieved by the peace signed in Westphalia. But something else was going to emerge on this date: accepting a system of international law opened the door to novel concepts of supra-national standards in international relations. The possibility of prolonging the peace that resulted from that agreement was an idea the seeds of which were now planted in European minds. The Westphalia treaty was signed only 46 years before the birth of Voltaire and 70 years before the publication of *Projet pour rendre la paix perpétuelle en Europe* (*Project to restore perpetual peace to Europe*) by Charles-Irénée Castel de Saint-Pierre, which appeared 82 years before Kant's *Perpetual Peace*.

Jean Delumeau concludes his long investigations with the Enlightenment. Equalitarianism or the notion of happiness (as in "the pursuit of happiness" in the American Bill of Rights) seemed to have become more appealing ideologies and may have drawn popular attention away from the morbid fears and the temptation of violence which had provided only an illusory resolution to their perception of the dangers surrounding them. The end of the world that had seemed so near never happened, and the fears that had tormented generations of Europeans were no longer justified. The prophecies of doom had been essentially self-fulfilling prophecies. It was maybe time to salvage whatever could be found in the wreckage, adapt to the new civilization, which was said to be a world of progress, and, as the characters concluded in the last pages of Voltaire's novel *Candide*, "to cultivate one's garden."

But would the stability last? For six generations after Westphalia yet other great mutations would begin reshaping the history of civilizations. Had Western mentalities gained the maturity that they had lacked in the 15th and 16th centuries to adapt to the upheavals in a peaceful, humane and cooperative way? Professor Delumeau (as well as René Girard) recognized that the poisonous seeds sown in the ugliest moments of what should have been only the joy and tolerance of the Renaissance had never been completely removed from European consciousness. That Adolf Hitler published millions of copies of the abominable diatribes of Martin Luther against the Jews should serve as a warning.[51]

51 Jean Delumeau, "L'Édit de Nantes dans son contexte historique," in *Comptes-rendus des séances de l'année ... – Académie des inscriptions et belles-lettres* 142: 4 (1998), 1070.

Napoleonic Wars: Art of War, Diplomacy and Imperialism

Marco Rimanelli

[O]nly by destroying an opponent's field forces could he be induced to give up the struggle. (Napoleon)[1]

France's General and Emperor Napoleon Bonaparte has been widely admired as a "Genius of War" and the best commander since Friedrich the Great, Julius Caesar, Hannibal and Alexander the Great, whose victories he had carefully studied, masterfully applying their lessons to his own battlefields with his famous instinctive "eagle's glance" of lightning decision-making. From the 1789 French Revolution to his ultimate demise at Waterloo in 1815, Napoleon led the armies of France to victory and glory, conquering all of Europe and unifying it through an imperial network of land annexations, pro-French satellite states (Italy, Germany, Poland, Denmark, Spain, Portugal), unequal alliances subjugating ex-enemies (Austria, Prussia, Sweden, Russia) and an economic Continental Bloc against Great Britain's own anti-French blockade.

Although Napoleon stated he had "not invented anything new," his mastery of the "art of war" blended in innovative ways all known forms of strategy, tactics and logistics from Antiquity to Friedrich the Great's impact on the eighteenth-century *Ancien Régime*, while relying on personal leadership, aggressive strategic and tactical offensives with "real-time" battle-plans uniquely devised for each combat. Napoleon's masterful all-encompassing war-planning always envisaged coherent, sequential, strategic maneuvers and crushing victories in fast-paced wars of annihilation of the enemy, which then allowed his "diplomacy of force" to politically and territorially reshape Europe under his imperial mastery. Politically styling himself as the "heir" of the Revolution, it was Napoleon's battlefield victories that completed diplomatically the French Revolution's popular politico-economic overthrow of the decadent absolutist *Ancien Régime* and its fragmented Balance of Power system (1500s–1940s), replacing it with a vast French Empire and an all-encompassing Napoleonic European System (1805–1813).[2]

1 David G. Chandler, *The Campaigns of Napoleon* (London: Macmillan, 1966), 162.
2 Will and Ariel Durant, *The Age of Napoleon* (New York: MJF Books, 1975); "Napoleone

From France's Hegemonic Wars for "Natural Frontiers" to the French Revolution's War

Napoleonic France's hegemony over Europe was the culmination of 300 years of international conflicts and the Thirty Years' War (1618–48) to secure three traditional geo-strategic goals: a) consolidate France's emerging hegemony in Europe over the collapsing Spanish and Austrian Empires; b) expand France's lands up to her "natural frontiers" (the Pyrenees in the southwest; the Alps in the southeast; German Alsace, Rhineland and Luxemburg in the east; rich Netherlands and Belgium in the northeast); and c) dominate Europe politically, culturally and economically as its new hegemon. Louis XIV ("Sun King") came closest to achieving France's hegemony over Europe 150 years before Napoleon by exploiting the 1659 politico-military and economic collapse of Habsburg Spain and its long succession crisis. The "Sun King's" international conquests and power started as he personally led his army to conquer the rich Spanish Flanders (Belgium) and Rhine River estuary controlling Europe's inland trade (1667–68 War of Devolution and Peace of Aix-la-Chapelle).

However, France's hegemonic threat to the Balance of Power system spurred the Netherlands and Great Britain to contain her through counter-alliances with other rival European Great Powers, given the threat the "Sun King" posed now for their own national interests and even survival. Louis XIV reacted by seizing Spain's Lorraine (1670) and unleashing the 1672 Holland War to annex the Netherlands and control Northern European trade through the Rhine. The defeated Dutch stopped Louis XIV only by breaking their dykes with massive flooding, while expanding their anti-French coalition to Spain, Austria, Prussia and other German principalities. Stymied in the northeast, Louis XIV expanded France's areas of operations by conquering from Spain the border areas of Burgundy/Franche-Comté and Alsace, then sent a fleet to Spanish Sicily to support a local revolt, defeating the Mediterranean Dutch and Spanish fleets, while tying down Prussia and Austria by supporting against them political instability in Poland and Hungary, plus a brief war by pro-French Sweden. Although France was ultimately contained by Spain and domestic revolts, the "Sun King" kept his conquests (1678 Peace of Nijmegen) and in the 1680s annexed Saar, Luxemburg and Alsace, while pursuing territorial inroads in Northern Italy, the Mediterranean and North America.[3]

At home, Louis XIV had also strengthened his religious-political "Divine Rights" Absolutist rule by both limiting Papal authority and expelling French Protestants (Huguenots), while on the politico-propaganda level the "Sun King" built, with a great équipe of top-notch artists, the most luxurious, largest European Court and

Bonaparte," *Storia Illustrata* 115 (Milan: Mondadori, 1967); David G. Chandler, *The Campaigns of Napoleon*.

3 Winston S. Churchill, *Marlborough: His Life and Times* (Milan: Mondadori, 1973); Larry H. Addington, *The Patterns of War Since the Eighteenth Century* (Bloomington: Indiana University Press, 1990); Rosario Villari, *Storia Moderna* (Roma: Laterza, 1974), 157–316.

gardens of Versailles (outside Paris) as the new glorious "capital of Europe" in both arts and architecture and a global symbol of France's Absolutist state. Abroad, Louis XIV also renewed the old Franco-Turkish cooperation (since the 1500s against the Austro-Spanish hegemony of Charles V), despite European condemnations that French pro-Turkish policies threatened the survival of all Christianity against the Muslim threat during the last Turkish-Ottoman invasions of Europe in 1663–99 and the Siege of Vienna (1683). Only *in extremis* were the Turks defeated at Kahlenberg by a Polish-led international army (the 1684 Holy Alliance of Austria, Poland, Venice, Russia and the Papacy), followed by Austria's 1684–1718 liberation wars to free from Muslim oppression Hungary, Transylvania, Slovakia, Croatia, Serbia and Romania (the 1699 Carlowitz and 1718 Passarowitz Peace Treaties). Louis XIV's efforts to support the Turkish invasion by attacking Austria's German Palatine (Pfaltz) precipitated instead the 1688–97 War of the League of Augsburg/Palatine against a new anti-French European Coalition (Austria, Spain, Netherlands and Sweden), renamed the Grand Alliance in 1689 once Piedmont/Savoy and Great Britain joined after its 1688 Glorious Revolution replaced the pro-French Stuart rulers with the joint Anglo-Dutch crowns of Mary Stuart and William of Orange/Netherlands. This diplomatic reversal geostrategically cemented the Anglo-Dutch-Austrian axis against France, and after a nine-year war fought simultaneously in the Netherlands, Palatine, Piedmont, North America and at sea, France's economic collapse forced Louis XIV to the 1697 Peace of Rijswijk, losing most of the lands gained with the 1678 Peace of Nijmegen.

Despite these defeats, in 1700 Louis XIV was again poised to dominate Europe with the master-stroke of positioning his grand-nephew Philippe d'Anjou as ruler after the death of Spain's last Habsburg King, although the Spanish testament denied any future merger of both thrones. As all European Powers and France had previously schemed to partition Spain, the "Sun King's" stunning political control of Spain exacerbated widespread fears that France would eventually annex dynastically the entire Spanish Empire and so dominate all of Europe and most of the global colonial areas. This precipitated the devastating Spanish Succession War (1700–14) against the Franco-Spanish imperial merger, challenged by the Grand Alliance of the Hague (Great Britain, Netherlands, Austria, Prussia, Hannover and German principalities, plus Portugal and Piedmont/Savoy who switched sides abandoning France and Spain).[4]

Under Anglo-Dutch leadership with the united command of John Churchill, Duke of Marlborough, and his Deputy Eugène of Savoy for Austria, the Grand Alliance fought globally to a standstill the Franco-Spanish in the Netherlands, Piedmont/Italy, Germany, colonies and at sea in the equivalent of a "first world war." The Duke of Marlborough's persistent campaign of sieges and victories (Gibraltar 1704, Blenheim 1704, Höchstaedt 1705, Turin 1705, Ramilliers 1706, Oudenaarde 1708, Malplaquet 1709) steadily pushed back France to its earlier

4 Henry A. Kissinger, *Diplomacy* (New York: Simon & Schuster, 1994), 5–77; Geoffrey Parker, *The Thirty Years' War* (New York: Military Heritage Press, 1987); Will and Ariel Durant, *The Age of Louis XIV* (New York: MJF Books, 1973).

frontiers, cracking the great system of fortified cities that Marshal Vauban had created to anchor French forces. Despite some French victories, Louis XIV was cornered by the threatened allied invasion of France, while France reeled again with impending economic collapse.

In the end, paradoxically, it was the inability of the Balance of Power's traditional geo-political structure of multiple rival Great Powers to dominate Europe that saved Louis XIV from total defeat, despite his earlier efforts to destroy that system with France's own hegemony. As France and Spain faced collapse, the unexpected Austrian rival claim to the Spanish throne forced all anti-French Powers to stop the war: none could countenance either a new Franco-Spanish Bourbon world hegemony, or a return to the old sixteenth-century Austro-Spanish Habsburg world hegemony. Thus, the 1713/14 Peace Agreements of Utrecht and Rastadt allowed both defeated France and Spain to continue as parallel but independent Bourbon Kingdoms (although Paris unofficially dominated Madrid throughout the 1700s), while Great Britain became the world's new maritime hegemon (dominating colonial trade, Netherlands and Portugal), and pro-British Austria reemerged as a multi-ethnic Empire over the Balkans against Turkey, Germany against France, and Italy against Spain. Despite this crushing set-back for France's hegemony in Europe, followed by the 1715 death of Louis XIV, the Balance of Power continued to be rocked by the struggle between France and European coalitions until the French Revolution and the Napoleonic Empire's second era of French hegemony in 1789–1815.[5]

France's collapse in 1788–89 was due to the failure of the eighteenth-century Enlightenment reforms and financial/economic bankruptcy after a century of disastrous wars that first shattered Louis XIV's 400,000-strong army and hegemonic plans to dominate Europe (Spanish Succession War, 1700–14; Austrian Succession War, 1740–48; French and Indians/Seven Years War, 1754–63), and ended up also losing the first French Colonial Empire by 1763. Instead of rebuilding the country, the lure of quick revenge to destroy the arch-enemy, the British colonial Empire, propelled Bourbon France to commit its last economic/military resources in the American Revolution (1774–83) to save a grateful secessionist British-America, despite the long-term ideological threat posed to the *Ancien Régime*'s legitimacy by any democratic republican revolution. Thus, just few years later in 1788/9, a bankrupt, corrupt Bourbon France collapsed under its own domestic version of the American republican sedition: the French Revolution's powerful domestic trinity of democratic "liberty" from the *Ancien Régime*'s absolutist political oppression, "equality" between socio-economic classes against aristocratic dominance, and revolutionary democratic "fraternity" among all oppressed peoples of the world. The French Revolution's violent political turmoil demoted King Louis XVI from

5 Pierre Goubert, *Louis XIV and Twenty Million Frenchmen* (New York: Vintage, 1966); Michel De Jaeghere (ed.), *Versailles. Le Château* (Paris: Le Figaro Collection n. 1, December 2005), 1–146; E.N. Williams, *The Ancien Régime in Europe: Government and Society in the Major States, 1648–1789* (New York: Harper & Row, 1970).

absolute rule to constitutional monarchy and ultimately execution, as a popular democratic National Assembly erased old nobility and clergy privileges.

But if Bourbon France and Spain could countenance and support in 1775–83 a revolutionary democratic America from the safety of an ocean's breadth, by 1790 the Great Powers of Europe could not tolerate in their midst the contagion of France's Revolution threatening both the absolutist legitimacy of their *Ancien Régime* and the Balance of Power system of limited wars. Thus, British financial aid cemented all the Great Powers in seven Coalitions seeking the common diplomatic/military goals of isolating and destroying the French Revolution and Napoleon to restore the "legitimate" absolutist Bourbon monarchy (under either Louis XVI or finally Louis XVIII). In 1789–91, threatened at home from a Royalist civil war (Vendée) and abroad by French aristocratic *émigré* armies backed by the Coalitions of all the European Great Powers seeking to restore Louis XVI (1791 Declaration of Pillnitz), Revolutionary France fought for its survival:

- with its reforms challenged by political turmoil and famine, France's ex-Royalist army shrunk to 130,000 men by 1790, losing 97 percent of its aristocratic officers (despite the influx of 100,000 volunteers in 1791 and 220,000 more in 1792);
- the execution as traitor of Louis XVI failed to quell Royalist rebellions or the 1794 foreign invasion by the Ist Coalition (Austria, Prussia, Great Britain, Netherlands, Spain and Piedmont/Sardinia);
- only dictator Maximilien Robespierre's 1793–94 Great Terror succeeded in stamping out all internal enemies (37,000 killed and 500,000 prisoners), while his call for the "Nation at arms" with revolutionary *levées en masse* of 750,000 soldiers by 1794 still faced see-saws of defeats and victories (Valmy and Jemappes in 1792, Tourcoing and Fleurus in 1794) against larger, slower Coalition armies;
- the French Revolution then strove to defeat the Coalition by liberating neighboring states from their absolutist rulers, but again brief victories and frequent defeats on the Dutch-Rhine-Italian borders (1793–1800) led to limited territorial annexations, until the corrupt Directoire's seizure of power by executing Robespierre left the Revolutionary army neglected (shrinking it to 400,000 by 1796);
- only Napoleon's genius for war and military-imperial drive saved the Revolution by deposing its collapsing Directoire and giving glory to French armies through years of victories, while crafting a Napoleonic Empire over Europe (1805–13);
- in the end, Napoleon's military overreach with the disastrous Peninsular (1808–13) and Russian Campaigns (1812) allowed anti-French Reactionary Coalitions to defeat and depose him twice in 1813–15, while the 1815 Congress of Vienna's Restoration diplomatically erased both the Revolutionary and Napoleonic legacies.[6]

6 Robert Doughty and Ira Gruber (eds), *Warfare in the Western World: Military Operations*

Building the Empire: Napoleonic Wars in Italy, Egypt, Germany, Prussia and Austria (1796–1809)

Napoleon Bonaparte was born on August 15, 1769, at Ajaccio in Corsica, after its sale to France by the sub-Italian state of Genova, and spent his youth in French military schools, graduating as an artillery officer with an uncommon knowledge of ancient and modern warfare. As a young officer he joined French Revolutionary politics, siding with the radical Jacobins against the federalist Girondins in the midst of Robespierre's Great Terror against Royalist revolts and the Ist Coalition's invasion of France. As Captain in 1793 at the Siege of Toulon, he secured the reconquest of the main French naval base against 15,000 Royalist and Coalition forces; rewarded in 1794 with promotion to Brigadier-General he was later briefly imprisoned for two weeks due to his Jacobin political ties when the July 1794 Thermidorian coup by the Directoire executed Robespierre. Although not fully trusted, Napoleon was made Commander-in-Chief of the Artillery in the French Army of Italy (1794/95); then in October 1795 he was the only General present in Paris and quickly saved the Republic using artillery fire against a Royalist counter-revolution by 30,000 rebels. This event, and his new political connections (by marrying Josephine de Beauharnais, the ex-lover of Paul-François Barras, Head of the Directoire) led to Napoleon's promotion in spring 1796 at only 26 to General-in-Chief of France's 50,000-strong Army of Italy.

Italy was a diversionary theater for French forces focused on conquering Belgium, Netherlands and Germany against the Reactionary Coalition Powers (Great Britain, the Austro-German Empire, Prussia, Spain, Piedmont/Sardinia), but Napoleon immediately led his famished forces through the Western Alps into Northern Italy in a war of movement that split and quickly defeated the Kingdom of Piedmont/Sardinia (Ceva, Mondoví) and the Austro-German Empire (Voltri, Dego, Lodi). While France's main war effort in Germany was stalled, Napoleon conquered Central and Southern Italy, then defeated four Austrian counter-offensives on both sides of Lake Garda, at Arcole and Rivoli, to finally take Mantova. By 1797 he conquered the weak Republic of Venice and from its borders invaded Austria up to within 100 miles of Vienna. The ensuing Campo Formio Peace Treaty (October 17, 1797), unilaterally imposed by Napoleon on both the governments of Vienna and Paris, reorganized the fragmented Italy into a few republican satellite-states under France, with France also annexing Savoy, Netherlands, Belgium and the Austro-German Rhineland (securing her "natural borders" on the Rhine River and Alps), while all of Venice was annexed by Austria.[7]

from 1600 to 1871 (Lexington, MA: Heath, 1996), 63–295; Jackson Spielvogel, *Western Civilization: Since 1500*, Volume 2 (Belmont CA: Thomson, 2005), 500–79; Albert Mathiez and Georges Lefebvre, *La Rivoluzione Francese*, 2 volumes (Turin: Einaudi, 1950–52); Albert Soboul, *La Rivoluzione Francese* (Roma: Newton Compton, 1974); Georges Lefebvre, *Napoleone* (Bari: Laterza, 1971).

7 Nicolas Boussard, "Napoleon, Héros Hegelien" in *Souvenir Napoléonien* 400 (Paris:

As the "Savior of the French Revolution," Napoleon soon surpassed all other Revolutionary generals and adversaries with his magnetism, combat genius and ability to fully exploit his politico-military successes. The young Corsican General fighting for France's new struggling republican democracy was well aware that the explosive impact of the French Revolution (1789–1800s) with its whirlwind of new political symbols had cracked open the old aristocratic world of the *Ancien Régime*. Thus, one of the key historical differences between Napoleon and other French Revolutionary Generals was his awareness and ability to build his political success upon his own iconography of war and a privileged relationship with the greatest artists of his time who crafted just such images. Bonaparte was the only one to be fully aware of the impact of the arts as new mass-media, together with national newspapers, on politics, propaganda and personal image. Since 1796 he had already built around himself a real artistic propaganda machine using the greatest artists of his time to craft a personal propaganda iconography in impressive war-paintings, etchings, lithographs and newspapers, where everything was keenly debated, studied, corrected and perfected to produce just the right political message aimed at enhancing Napoleon's own national and international political image as an unrivaled leader blessed by Glory and Fortune.

Napoleon's conquest of Europe and reforms at home brought forth a politico-propaganda figure based on middle-class and popular nationalist views and a merit-based democratic equality, rather than the decrepit aristocratic *Ancien Régime* of the "Sun King" dominating Europe by "Divine Right" from domestic politics to religion, from complex set-piece battles to artistic monuments extolling political propaganda (like the creation at Versailles of the largest European Court palace complex, as world symbol of France's Absolutist state and international hegemony). But here as well, Napoleon was able to outdistance the pictorial art-of-war propaganda of Louis XIV, through the modernity of his own artistic politico-propaganda message and the capillarity of its reach throughout Europe.[8]

In military terms, Napoleonic warfare differed sharply from the *Ancien Régime*'s doctrine of well-disciplined, steady moving forces as reorganized by Great Britain's Duke of Marlborough and the widely copied masterful Prussian King Friedrich the Great, who introduced long-service national professional militaries led by an aristocratic officer corps, with a strict disciplinary system and ésprit-de-corps over lower-class enlisted men. Friedrichian armies massed 50,000 professional Prussian

March 1995), 8–21; W. and A. Durant, *The Age of Napoleon*, 3–95, 237–321; J.C. Quennevat, *Atlante della 'Grande Armée'* (Roma: Giannini, 1966), 13–22, 227–95; Archer Jones, *The Art of War in the Western World* (Oxford: Oxford University Press, 1987), 274–377; D.G. Chandler, *The Campaigns of Napoleon*, 13–190; R. Doughty and I. Gruber (eds), *Warfare in Western World*, 173–207.

8 P.-M. Laurent de l'Ardèche, *Histoire de l'Empereur Napoléon* (Paris: Dubochet, 1840); *Napoléon en Italie; A Travers Les Aquarelles de Giuseppe Pietro Bagetti et les Chroniques de Stendhal et d'Adolphe Thiers* (Milan: Ricci, 2001); Yveline Cantarel-Besson, Claire Constans and Bruno Foucart, *Napoléon, Images et Histoire: Peintures du Chateau de Versailles, 1789–1815* (Paris: Musées Nationaux, 2000); Max Gallo, *L'album de l'Empereur* (Paris: Laffont, 1997).

soldiers (the best in Europe, augmented in wartime by recalling all ex-soldiers from civilian life as reserves), with logistics based on a network of dépôts supplemented by field-foraging. Instead, Napoleonic armies often reached 100,000 soldiers with logistics based mostly on field-foraging and pre-arranged paid foraging from villages, plus meager dépôts and transports. Friedrich the Great had refined centralized military organizations based on linear infantry tactics with rigidly disciplined fast deployments under enemy fire from marching columns into firing lines (*ordre mince*/"line order"), anchored on pre-deployed fixed heavy artillery. He also relied on a military strategy of exhaustion of his enemies with sieges of vital fortresses/cities, while avoiding major battles except when ready to crush the enemy decisively, followed by limited politico-diplomatic gains (except his annexation of Austrian Silesia).

The 1789–1800s French Revolution lost its aristocratic officers, who joined foreign *émigré* armies and the European Great Powers against France. But this devastating politico-military loss was replaced through untried mass conscription of citizen-soldiers for long enlistments, led by a new officer corps drawn from all classes on the basis of merit and political loyalty in a new patriotic disciplinary system. As a master theorist Napoleon blended all combat tactics from Classical history of war to the post-1760s reforms adopted by the French Revolution for a modern, decentralized, mobile national army of citizen-soldiers organized in independent combat/administrative Divisions of several regiments and Revolutionary *démi-brigades* of three battalions (one-third ex-Royalist troops and two-thirds Revolutionary volunteers, with added divisional cavalry and artillery units), later regrouped into independent Army Corps. Napoleon's army also used the Revolutionary *ordre mixte*/"mixed order" infantry tactics (combining Friedrich the Great's Prussian *ordre mince* with the French *ordre profond*/"deep order" of rapidly attacking massed columns), supported by mobile artillery or large batteries.

Thus, Napoleon blended combined-arms maneuverability, high speed of deployment and annihilation of the enemies through targeted, flexible "decisive battles," based on persistent attacks on interior lines and turning maneuvers to seize the "central position" against enemies, following Friedrich the Great's emphasis on this tactic and belief that winning or losing any "decisive battle" of annihilation would influence the entire military campaign. The blending of these warfare reform theories with the supreme articulation of French troops and unity of command over a close-knit General Staff led to Napoleon's trademark relentless offensive strategy of annihilation by pursuing and utterly destroying France's enemies through either a crescendo of clashes or a few decisive battles, followed by total politico-diplomatic domination of the defeated enemies through "diplomacy of force" and punitive peace treaties to create a system of satellite alliances and a Napoleonic Empire over Europe (1805–13).[9]

9 J. Christopher Herold, *L'Etá di Napoleone* (Milan: Saggiatore, 1967); R. Doughty and I. Gruber (eds), *Warfare in the Western World*, 63–101; D.G. Chandler, *The Campaigns of Napoleon*, 13–100, 195–265.

Yet Napoleon's military strategy and tactics are also difficult to define: in the biographical *Mémorial de Sainte Hélène* (1821) by Emmanuel de Las Cases, the Emperor confirmed how he always quickly improvised battle-plans on the field after careful study of the terrain and enemy forces, rapidly assessing the immediate risks to secure victory, then fighting only if the chance of winning was 70 percent, while relying on a vague mix of historical Principles of War. These were refined by both Prussian General Carl von Clausewitz (1780–1831) who fought Napoleon in both the Prussian and Russian armies and wrote the masterpiece *On War* (1830s), and by Antoine-Henri de Jomini (1779–1869) who fought for Napoleon until the 1812 Russian Campaign and wrote *The Art of War* (1838). By 1830–70, theorist Dennis H. Mahan introduced Napoleonic-Jominian strategy to the West Point US Military Academy, teaching most US Civil War generals and Confederate Robert Lee, as these concepts were adopted by the US military.

As general, Napoleon relied both on his military genius and the natural political support of the French Revolution's achievements, whose three principles of liberty, equality and fraternity appealed to most of Europe's middle classes. As a European statesman he also politically blended Revolutionary principles with religious tolerance, legal justice, administrative order and replacing the *Ancien Régime*'s aristocracy with a new national merit-based, loyal, egalitarian élite open to all classes and conquered nationalities (legally equated as citizens). During the *Ancien Régime* (1500s–1789), armies mixed professional soldiers and mercenaries, and their combat effectiveness was often hampered by inefficient supply lines, lack of funds and low discipline. Wars consisted mostly of sieges of enemy fortresses, garrisoning city-dépôts and a few major battles by whole armies in rigid formations to maximize cohesion and firepower. Instead, Napoleon organized France's Revolutionary and Imperial armies with mixed volunteers/mass conscripts, strict discipline, a military strategy of relentless attacks, quick articulation of combined-arms forces (coordinating heavy infantry, light infantry, heavy cavalry, light cavalry and mobile field artillery), fast maneuvers of troops by columns of battalions with immediate combat redeployment from march to line formations, well-executed turning maneuvers and promotions based on bravery.[10]

While using his General Staff for organization, decision-making and intelligence, the *Grande Armée* was always under the Emperor's firm touch, relying mostly on his lightning personal style of command, exhaustive planning alone in the secret of his mind over maps and available intelligence to tailor strategy and tactics with last-minute alterations and restless front-line surveys. Napoleon used his Army Corps and smaller formations in tightly coordinated parallel forced-march maneuvers (within supporting range of each other in case

10 Quoted in Emmanuel de Las Cases, *Le Mémorial de Sainte Hélène*, 2 volumes (1821, reprinted Paris: Flammarion, 1983); John W. Chambers II, "Principles of War" in *The Oxford Companion to American Military History* (Oxford: Oxford University Press, 2000); R. Doughty and I. Gruber (eds), *Warfare in Western World*, 173–98, 291–328; Carl von Clausewitz, "Principles of War" (1812, reprinted 1942); Carl von Clausewitz, *On War* (1840, reprinted 1873).

of either Corps being attacked by surprise), screened by reconnaissance cavalry, until he could swiftly concentrate all formations in a single day against the main enemy forces through simultaneous attacks and envelopments, with battlefield autonomy given to his impetuous brave Marshals, generals and combat units. Yet Napoleon was also famous for his furious tirades excoriating his Marshals if they failed to arrive on time at prearranged *rendezvous* points, or did not display resourcefulness and initiative in combat (strict orders were to always march to the sound of the cannon as a supporting force), or did not vigorously pursue a retreating enemy, or did not keep in constant touch with the Emperor, who continually altered marching plans based on the enemy's changes without throwing the *Grande Armée* into any hopeless confusion. All of this left his slow-moving enemies consistently surprised, divided and defeated until 1813.

Napoleon's persistent offensive strategy of relentless attacks until the enemy collapsed ensured military victory and his politico-diplomatic supremacy over Europe until 1813; as a strategist and theoretician, he determined when to accept battle and how to fight, while as a statesman his strategy of fast annihilation of all enemy forces reinforced his overarching political goals and "diplomacy of force." Napoleon's great victories also reveal two fundamental combat scenarios on both strategic and battlefield levels, although ultimate success still depended always on his personal genius to instantly understand the battlefield: 1) either fight with superior or equal French forces vs the enemy; or 2) fight with French forces well inferior to the enemy ones.

Napoleon's wars of movement constantly widened the theater of operations from areas as large as Northern Italy or France to the whole of Germany/Austria and even Russia. Until 1805, when on campaign Napoleon often did not keep an official Reserve, but relied on the vastness of his theater of operations to carefully husband his unity of command and open lines of communication via his aide-de-camp couriers to recall immediately any of his individual sub-formations (two Italian Campaigns of 1796–97 and 1800), or, since 1805, entire armies (like the strategic turning maneuver of Ulm) to aid the main French offensive anywhere within a half-day's march. By 1805 Napoleon also relied psychologically as official Reserve on both large cavalry formations for the final mass charge to break the enemy and on his formidable Imperial Guard of veterans, although in time he became paradoxically reluctant to commit to combat (and then dramatically lost it at Waterloo when surprise British firepower coupled with enveloping Prussian enemies finally broke the Imperial Guard's attack and morale).[11]

Napoleon's strategy for quick battles and fast campaigns also presented chronic logistical problems. Though French soldiers carried much of their supplies, they relied on existing transportation networks of roads, rivers and canals to transport additional food, supplies and ammunition. By 1805, Napoleon believed in the

11 Albert S. Britt III, *The Wars of Napoleon: West Point Military History Series* (Wayne, NJ: Avery/West Point USMA, 1985); J.C. Quennevat, *Atlante della 'Grande Armée'*, 265–72; Thomas E. Greiss, *West Point Atlas for the Wars of Napoleon* (Wayne, NJ: Avery/West Point USMA, 1985).

concept of constant re-supply of his armies, while ammunition came via wagons and boats (supplemented on the battlefield by captured enemy artillery, munitions and stores). On one hand, French engineers worked miracles building multiple bridges of boats or pontoons to support fast army advances or as lines-of-retreat, with the most impressive being against Austria (both the 1805 French crossing of the Rhine for the turning maneuver of Ulm, and the bridge from Vienna's Lobau Island with multiple smaller parallel ones launched for the 1809 Battle of Wagram).

On the other, the faster the military campaign, the less likely his soldiers would be to run out of provisions, but his logistical strategy combined heavy dependence on foraging in enemy lands (Italy, Austria, Prussia and Russia), with the occasional carrying of supplies with logistics caravans (Russia and Waterloo). Napoleon's forces since the French Revolution needed fewer wagons and could move more quickly than their enemies because they did not carry the aristocratic officers' excessive baggage as did the Reactionary Coalition armies. Napoleon consistently opposed looting, preferring to ask local villages and towns for contributions, but the reality of combat often left his forces with 3–4 days' worth of supplies and in daily need of organized raids to requisition food along the road, while the cavalry had to roam 10 km around their bases for foraging. Napoleon would accept logistical disorder and shortfalls as the price for speedy operations inside enemy territories, but in the end such chronic French logistical disorder also hastened the *Grande Armée*'s undoing during the military defeats of 1812–14.[12]

After Italy, Napoleon's international heroic reputation rose, but the French Republic's corrupt Directoire feared Napoleon's popularity and, in 1798, readily agreed to Napoleon's quest to emulate Alexander the Great by sailing a French army across the Mediterranean Sea to conquer the decaying Ottoman-Turkish Empire's Egypt and the Middle East, dig a navigable channel to the Red Sea and threaten British India. After eluding the British Mediterranean Fleet under Admiral Nelson, Napoleon swiftly conquered strategic Malta and then Egypt by seizing Alexandria in a day and then Cairo after the epic Battle of the Pyramids (21 July). Yet, just as suddenly, this lightning victory was turned into a strategic stalemate once British Admiral Nelson conquered Malta and sank the entire French invasion fleet at Alexandria (Battle of Aboukir, 2 August), isolating Napoleon. He countered this by swiftly defeating Nelson's amphibious landing at the Battle of the Nile, and through local Arab guerrilla warfare in the desert. Then a French military offensive into Palestine and Syria countered the impending Anglo-Turkish invasion army (Battles of El-Arish, Gaza, Jaffa, Acre and Mount Tabor). Yet with bubonic plague ravaging the French army at Jaffa in 1799, Napoleon retreated to Egypt, then secretly returned to France to overthrow the collapsing Directoire with the 18 Brumaire Coup of 1800, becoming military dictator as First Consul and portraying himself in politico-artistic propaganda as the military savior of the Nation. The Directoire's loss in 1799 of all earlier Napoleonic conquests forced

12 A. Jones, *The Art of War in the Western World,* 274–377; D.G. Chandler, *The Campaigns of Napoleon,* 13–100, 195–265; Hans Pohle, *Geschichte in Bildern: Die napoleonischen Kriege* (Leipzig: Pohle, 2003).

Napoleon to lead a French army in the IInd Italian Campaign, secretly crossing five snow-covered Alpine passes in May 1800 to strike from the rear the 100,000-strong Austrian army, which he destroyed finally at the Battle of Marengo in July. Vienna was now diplomatically forced to return to France, Italy, Belgium and the Austro-German Rhineland (Lunéville Treaty, February 9, 1801), followed by the Peace of Amiens with Great Britain (March 27, 1802).[13]

With these successes, Napoleon consolidated his power, turning the Consulate into an Empire by 1804 (crowning himself and Josephine), but this precipitated another war in 1805 by the Reactionary IIIrd Coalition (Great Britain, Austro-German Empire, Russia, Sweden). With Great Britain facing the threat of a French invasion across the Channel, Austria expected the main war effort to focus a third time on Austria's forces in Northern Italy/Veneto, thus it launched an invasion of pro-French Bavaria and Southern Germany to block all Alpine passes, without waiting for the slow-moving 40,000-strong Russian army under Czar Alexander and General Mikhail Kutusov. Napoleon instead concentrated his military and diplomatic efforts on destroying in detail his divided enemies: in just a month and a half he secretly force-marched his 165,000-strong Imperial *Grande Armée* (in 7–8 flexible Army Corps) from the Channel all the way to the Rhine and then into Southern Germany to the Danube in two spectacular turning maneuvers along a twin strategic arc of 200 miles. The Austrians were trapped and defeated at Ulm, then lost Germany, the capital of Vienna and were finally wiped out with the Russian armies at the Battle of Austerlitz (December 1805). After this 100-day 1,000-mile trek from the Atlantic to Moravia and Vienna, Napoleon's *Grande Armée* controlled most of Europe (Treaty of Pressburg, December 26, 1805), annexing Veneto, Dalmatia and South Germany, and dissolving the Austro-German Holy Roman Empire.[14]

Napoleon's attempt to diplomatically win over Russia by showing clemency precipitated instead the unexpected bloody 1806–07 winter Campaigns against Prussia, Russia and Great Britain: Napoleon swiftly occupied key cities of Northern Germany (Prussia, Hannover, Saxony) and then conquered Prussia, Berlin and Russian-Poland, recreating an independent pro-French Poland in response to the continuing winter fighting with the Prusso-Russian armies. Finally, Napoleon's bloody defeats of Prussia and Russia (Jena, Auerstaedt, Eylau, Friedland) led to a Franco-Russian strategic alliance (Treaty of Tilsit, July 7, 1807), leaving the French Empire in control of both Germany and Italy, while a halved, humiliated Prussia and a sharply reduced Austria seethed powerlessly.[15]

---

13 Paul George Guitry, *L'Armée de Bonaparte en Égypte, 1798–1799* (1801, reprinted Paris: Flammarion, 1897); D.G. Chandler, *The Campaigns of Napoleon*, 275–410; Filippo Bonfant, *Da Parigi a Marengo, 1800* (Aosta: Valdostana, 1993); A. Jones, *The Art of War*, 338–41; Max Gallo, *Napoleon* (Milan: Mondadori, 2000).

14 David G. Chandler, *Austerlitz, 1805* (London: Osprey, 1990); J.C. Quennevat, *Atlante della 'Grande Armée'*, 13–47; D.G. Chandler, *The Campaigns of Napoleon*, 410–544.

15 J.C. Quennevat, *Atlante della 'Grande Armée'*, 48–68; D.G. Chandler, *The Campaigns of Napoleon*, 545–620; David G. Chandler, *On the Napoleonic Wars* (Barnsley: Greenhill Books, 1999), 181–213; W. and A. Durant, *The Age of Napoleon*, 193–220, 554–604, 659–92.

Napoleon now ruled most of Europe as a twin system based on a vast French Empire (annexing Belgium, Netherlands, Austro-German Rhineland and the Northern coast, half of Switzerland and a third of Italy including ex-Papal Rome) and a confederation of pro-French satellite states (Kingdom of Northern Italy, Neapolitan Italy, Germany's Confederation of the Rhine, a reduced Switzerland, Denmark, Poland, Sweden and Spain), but his increasingly authoritarian hold on power was constantly challenged at home and abroad. Austria and Prussia, as defeated Reactionary conservative Powers, still opposed French hegemony over Europe with timid appeals to anti-French pan-German nationalism, although leadership of such a common future pan-German national destiny remained politically controversial, especially for multi-ethnic Austria where its several sullen major nationalities (Hungarians, Czechs, Slovaks, Croats and Rumanians) could respond by seceding and disintegrate the Habsburg Empire. Finally, in February 1809, the Austrian war party at Court decided to seize political leadership of Germany through open war against France by conquering pro-French Bavaria and Southern Germany, while appealing to Prussia and Great Britain to join her (instead, both remained neutral). The surprise Austrian attack in April left the French defenses of Germany disorganized with contradictory orders from Napoleon who arrived in mid-April with 175,000 men and dispersed French forces in forced marches, hampered by bad intelligence and miserable weather. But Austria's chance to defeat France was suddenly swept away by massive deluges and logistical breakdowns in the Austrian offensive. Napoleon quickly regrouped, defeated the Austrians in Southern Germany and reconquered Vienna, but was forced to keep fighting, just across the Danube, the bloody Battles of Aspern-Essling and Wagram (May–July 1809). The *Grande Armée*'s victory was clouded by heavy losses and 30 dead generals, aggravating its gradual decline, while Austria was forced into a new punitive Peace losing a sixth of its lands. By 1810, Austrian Ambassador Prince Klemens von Metternich brokered a Franco-Austrian alliance through Napoleon's second marriage (to have a male heir) to the Austrian Emperor's daughter Marie-Louise, while exploiting Russia's refusal to have him marry instead the Czar's sister.[16]

Breaking the Empire: The Peninsula, Russia, Germany, France and Waterloo (1808–1815)

In 1812 Napoleon was at the height of his power with all of Continental Europe under his control, also annexing to the French Empire Croatia-Illyria and one-tenth of Spain, plus adding Portugal as a French satellite state, and keeping ex-enemy

16 François-Guy Hourtoulle, *Wagram: Apogée de l'Empire* (Paris: Histoire & Collections, 2002); D.G. Chandler, *The Campaigns of Napoleon*, 710–885; J.C. Quennevat, *Atlante della 'Grande Armée'*, 91–125; D.G. Chandler, *On the Napoleonic Wars*, 115–29.

Powers (Austria, Prussia, Sweden) subjugated in unequal alliance treaties with France, while Great Britain remained an isolated foe.

But as Napoleon's thirst for power and glory over Europe became boundless he met the road-block of the Peninsular War (October 1807–April 1814) following France's conquest of Portugal and Spain in 1807–08 and simmering local guerrilla warfare until 1813. The inability of large French armies to pacify the vast, rugged and rebellious Iberian Peninsula was compounded by poor roads, combat losses of veterans during the parallel 1809 and 1812 Campaigns against Austria and Russia, French occupation duties throughout Europe, large-scale Spanish popular guerrilla struggles against French occupation forces and vulnerable supply lines, and Anglo-Portuguese diversionary strikes against the French. Most importantly, Napoleon and his Marshals all consistently underestimated the deadly impact of popular hostility and the local guerrilla bands that replaced the routinely defeated Spanish armies: 300,000 French troops spread out over the entire peninsula could never overcome fanatical guerrillas who controlled rural areas and the night, while French forces secured only the territory they occupied in daylight and had to disperse half their troops just to patrol long lines of communication against guerrilla attacks, while searching for scarce food/forage. The undisciplined guerrillas and their Spanish civilian supporters used "hit-and-run" terrorist tactics and unspeakable brutality to kill and maim many French and their Spanish sympathizers, terrorizing any town that opposed such "patriotic" banditry, while France retaliated with increasing violence but was unable to concentrate persistently a sufficiently overwhelming power to crush the Anglo-Spanish armies. In the end, the guerrillas did not win, but they consistently weakened French troops and logistics until the British forces under General Duke Wellington could finally defeat and push out the *Grande Armée* from both Portugal and Spain.[17]

As the Peninsular War still raged on, Napoleon had to respond to Czar Alexander's withdrawal from the Continental Bloc, which precipitated the 1812 French invasion of Russia for European control. Napoleon had meticulously prepared the provisions for the *Grande Armée*'s invasion, but the traditionally inadequate French logistics and "field-foraging" could not sustain long operations in thinly populated and agriculturally sparse hostile regions, while French forced marches often made the troops move well beyond their logistics trains, with consequent thirst and hunger slowly weakening both men and horses over a network of dusty broken dirt roads, which became unmanageable mud-pits under autumn rains and finally frozen ruts in sub-zero winters. In a 1,000 km chase of the constantly retreating Russian armies who sought to avoid being destroyed in battle by France's 285,000 strike force (out of a total invasion force of 690,000–800,000 men), Napoleon was forced to reach deep into Russia as far as Moscow. But the Russian use of "scorched earth" tactics

17 David G. Chandler, *Napoleon's Marshals* (London: Weidenfeld & Nicolson, 1998), 247–51; Jean Tranié, Jean-Claude Carmignani and Henry Lachouque, Napoleon's *War in Spain: The French Peninsular Campaigns, 1807–1814* (London: Arms & Armour, 1993); R. Doughty and I. Gruber (eds), *Warfare in the Western World*, 244–53; D.G. Chandler, *The Campaigns of Napoleon*, 750–95.

cost Napoleon the loss of 50 percent of his troops in the first eight weeks of the invasion due to fatigue, starvation, desertion, typhus and clashes with local peasantry and guerrillas, plus the loss of most cavalry to mud pits in summer/autumn 1812, killing more French soldiers than all the combined battles won over the Russians (Vitebsk, Smolensk, Borodino, Moskowa). Finally, the Battle of Borodino (7 September) was the largest and bloodiest combat of the Napoleonic Wars, with 250,000 troops and 70,000 casualties, but Napoleon still failed to fully destroy the Russian army and force Czar Alexander and Marshall Kutusov into peace.

By winter 1812, Napoleon's invasion of Russia had become one of the worst military disasters in history: had he taken the painful choice of withdrawing his half-starved forces earlier (a strategic option still viable and debated shortly after the fall of Moscow when no Russian surrender came) then despite such a diplomatic disaster his severely weakened *Grande Armée* would have remained still formidable enough to retreat into Poland ahead of the Russian winter, and there, supplemented by fresh French recruits, it could have successfully defeated any counter-invasion by a weakened Russian army supported only by treasonous Prussia and Austria. Instead, Russia's winter "victory" wiped out the "invincible" *Grande Armée* as a military force: starved, snow-bound and frozen, weak French forces were defeated by the Russians (Vyazma, Krasnoi and Polotsk), but succeeded in crossing the Berezina River despite horrendous casualties, thus avoiding a fatal encirclement and capture by the Russians. In the end, France lost 370,000 men to fighting, starvation, typhus and freezing weather, plus 80,000 Germans, 72,000 Poles, 50,000 Italians, 61,000 from other nations, 200,000 horses and 1,000 artillery pieces, while 200,000 men were captured. Russia lost 210,000 men and 40,000 militia (with 40,000 wounded militia redrafted), while civilian losses along the devastated war corridor reached a million. Only by mid-December had the *Grande Armée* remnants exited Russia, leaving in tatters Napoleon's own European hegemony.[18]

After the disastrous 1812 winter retreat from Russia and irreplaceable French losses, nobody in Europe expected Napoleon to recover, but he stunned all by swiftly regrouping green troops for a mobile defense of Poland and Germany before the inevitable spring 1813 renewed two-front distant war against Russia and Great Britain (threatening France from Spain's Pyrenees border). Czar Alexander had also lost 400,000 men and relied on mostly depleted green recruits, but Prussian nationalists and Russian forces rapidly fomented anti-French revolts in early 1813 in Prussia (seeking her lost territories of Westphalia, Silesia and Poznania) and Germany's Confederation of the Rhine, with many of its satellite states abandoning France to join the "German War of Liberation" and the new VIth Coalition (Great Britain, Russia, Portugal, Spain, Prussia, minor German states and Sweden who annexed Danish Norway), while the key politician who had secretly been building

18 Paul B. Austin, *1812: The March on Moscow* (London: Leventhal, 1993); R. Doughty and I. Gruber (eds), *Warfare in the Western World*, 268–76; D.G. Chandler, *The Campaigns of Napoleon*, 923–48, 970–1020; Stephan Talty, *The Illustrious Dead: The Terrifying Story of how Typhus Killed Napoleon's Greatest Army* (New York: Crowne, 2009); J.C. Quennevat, *Atlante della 'Grande Armée'*, 132–5, 139–149.

such an anti-French Coalition among rival Powers since March 1813 was Austrian Foreign Minister von Metternich.

Napoleon planned to defeat the VIth Coalition as in the past, by quickly attacking each enemy army separately before they had a chance to join forces. But his ability to locate enemy forces and defeat them was badly curtailed by three key problems: 1) the combat inexperience of his new French army; 2) the lack of French cavalry, most of which had perished in Russia, depriving him of intelligence and screening; 3) the fact that his Coalition enemies had finally learned to band together to prevent Napoleon from defeating them in isolation. Thus, on May 2, 1813, Napoleon defeated a Prussian force at Lützen, outside Leipzig, but the lack of cavalry meant that he was blind-sided by an enemy force on his flank. He beat them as well, and also occupied Leipzig, but could not rout the Prussians. The French then captured Dresden and defeated the Russians at Bautzen (May 20–21, 1813), again without destroying them fully. Yet soon Napoleon learned of new large Coalition armies seeking to surround him, so on the political strength of his earlier victories he forced the Coalition into a two-month armistice to earn breathing space to gather more reinforcements (4 June–13 August 1813). But his enemies also raised more forces and openly enlisted Austria in the war, tipping the scales against Napoleon, whose own German troops were defecting to the Coalition. Napoleon could really count only on 450,000 reliable troops in Germany, where he was outnumbered two-to-one by the Coalition, while neither Italian troops nor French forces on the Pyrenees could disengage and repulse new threats by Austria in Italy and the Anglo-Spanish on France's southern border.

To break this diplomatic/military impasse, on 26 June von Metternich offered Napoleon a lasting Coalition peace if France retreated to her "natural borders" (Pyrenees, Alps and the western bank of the Rhine), evacuating most of Germany, Netherlands, Italy, Croatia-Illyria and Poland. But Napoleon refused as he could never compromise his glory and imperial hegemony, while believing that his personal Fate would always hold bigger victories for him. Napoleon used a French Corps to defend the vital port of Hamburg and threaten Prussia, while another French Corps from Dresden threatened Austrian Bohemia. At Dresden Napoleon defeated Prince von Schwarzenberg's superior Coalition army, inflicting major casualties compared with fewer French losses, but again Napoleon's lack of cavalry made it impossible to rout the defeated enemies. Finally, the Coalition's Trachenburg Plan to fight only Napoleon's Marshals and avoid any battle with the Emperor led to the defeat of French forces in Germany by late 1813.[19]

Forced back over the Rhine, Napoleon fought the ensuing 1814 Campaign of France with only 70,000 green troops against the 500,000-strong Coalition armies' multiple-pronged invasion. Yet he still inflicted heavy losses on the Coalition in a whirlwind of battles, while slowly retreating towards Paris. Weakened by

19 F. Loraine Petre, *Napoleon's Last Campaign in Germany, 1813* (London: Leventhal, 1992); R. Doughty and I. Gruber (eds), *Warfare in the Western World*, 277–80; J.C. Quennevat, *Atlante della 'Grande Armée'*, 150–8, 163; D.G. Chandler, *The Campaigns of Napoleon*, 1020–75; W. and A. Durant, *The Age of Napoleon*, 712–27.

Napoleon's constant combat, the hesitant Coalition Powers split and openly discussed retreating (especially Austria, Sweden and Prussia), while all were alarmed by Russia's ambiguous political attempts at hegemony. Only British funds and the diplomatic efforts of Viscount Robert Stewart Castlereagh kept the Powers together (Treaty of Chaumont, March 1–9, 1814) until France's total defeat. Castlereagh then offered Napoleon a lasting peace with France contained to her old 1789 borders (Châtillon Conference, mid-March), but Napoleon again refused any humiliating peace tarnishing his legendary glory.

Soon after, his fortunes collapsed: the Battle of Laon turned into a French defeat and the Coalition finally captured Paris on March 30. Imitating the Austrians' and Russians' past refusal to surrender once their capitals had fallen in 1805, 1809 and 1812, Napoleon continued fighting, while occasional French actions also in Italy, Spain and Holland throughout spring 1814 tied down Coalition reserves. Napoleon boldly planned to retake Paris with the support of an army eager for revenge, but now his own Marshals and senior officers mutinied on April 4, forcing him to abdicate (Fontainebleau, April 6, 1814) on behalf of his infant child (François) Napoleon II, the "King of Rome." He, in turn, was immediately deposed and exiled to Vienna ("Duke of Reichstadt") under von Metternich's control: the Coalition feared that in France he would become a symbol of Bonapartist resistance, as happened with the 1830 French Revolution, but he died of tuberculosis, still hostage at Schönbrunn, in July 1832. Thereafter, the victorious Coalition restored after 25 years of exile the unpopular King Louis XVIII of Bourbon to rule a monarchist France within the old 1790 borders, while they exiled a prematurely aged Napoleon to the Italian Island of Elba.[20]

The Coalition Powers' Peace Celebrations in London (June) led to the Congress of Vienna (November 1814–June 1815) that reversed all European political changes since the 1789 French Revolution, redrew the map of Europe and imposed a Reactionary aristocratic peace against any future Napoleonic return, revolution or nationalist revolt anywhere in Europe. Yet the Congress of Vienna soon degenerated almost into war between rival Coalition Powers over their conflicting exorbitant land claims: Russia and Prussia vs Great Britain, Austria, Bourbon France, Spain and Portugal. On one hand, Russia wanted to annex much of pro-French Poland as a vital buffer against further invasions from Europe and also extend its political influence over the Reaction to replace Napoleonic France as the Continent's new hegemon, while pro-Russian Prussia secured both her lost lands and sought to annex pro-French Saxony to consolidate her politico-military domination over Northern Germany against Austria's wishes. On the other, British Viscount Castlereagh and Austria's von Metternich diplomatically isolated and opposed all Russo-Prussian land claims for fear that pro-British Austria and Netherlands would soon be squeezed as well. Thus, disregarding the British Parliament's non-involvement directives, Castlereagh

20 Jacques-Olivier Boudon, "La campagne de France (15 février–13 mars 1814): De Montereau à Reims," *Napoleon 1er* 20 (May/June 2003), 18–29; J.C. Quennevat, *Atlante della 'Grande Armée'*, 166–206; D.G. Chandler, *The Campaigns of Napoleon*, 1115–90; D.G. Chandler, *On the Napoleonic Wars*, 115–29.

secured British annexation of South Africa, continued economic domination of Spain's Latin American colonies and political control over weaker pro-British states (a united Netherlands/Belgium, Portugal, Spain, Bourbon Naples, Ottoman Turkey and a weak Bourbon France) to jointly contain Prussia and Russia (secret treaty of 3 January 1815), while supporting Austria as Great Britain's key European ally (Vienna regained all her lost imperial lands and became again the leader of the German Confederation, plus the dominant Power in Italy as an all-encompassing politico-military bulwark containing France in the West, Prussia in Germany, and Russia from Poland to the Balkans). This secret diplomacy almost precipitated war among Russia, Great Britain and Austria when the King of Prussia revealed he had been secretly offered Anglo-Austrian support to annex Saxony if Prussia backed an independent smaller Poland as buffer against Russian expansionism. After the British Parliament openly censured Castlereagh's controversial diplomacy, the Coalition's politico-diplomatic disarray among ex-allies seemed complete.

Napoleon's humiliating ten-month exile on Elba Island (1814–15) was bolstered by his beliefs that soon he could escape back to France and revamp from the ashes a modicum of his old Empire's power. On one hand, France seethed under the mistreatment by the unpopular Bourbon Royalist nobility (*Ultras*) of the people and respected French veterans repatriated from Coalition prisons (Russia, Great Britain, Austria, Prussia, Spain), who Napoleon saw also as a future patriotic, combat-ready, new army, larger than the one of the 1814 Campaign of France. On the other, Europe was exhausted after decades of constant wars and Napoleon sought to exploit the raging political clashes among victorious Coalition Powers at the Congress of Vienna, plus British entrapment in the far-away 1812–14 Anglo-American War, to secure a general peace accepting the Emperor's "inevitable" return from exile to power (as in 1800 after the Egyptian Campaign). This very fear of a Napoleonic return was shared by French Royalists, Castlereagh and other Powers, who debated at Vienna how to deport him from Elba to Portugal's distant Atlantic Azores Islands or the British Island of Saint Helena in the South Atlantic, while others hinted at assassination.[21]

Thus, Napoleon chose not to wait for the Congress of Vienna to dissolve and suddenly escaped from Elba on February 26, 1815, landing in France near Antibes on March 1 promising constitutional reforms and elections. During the "Hundred Days" the French population enthusiastically welcomed him back as their Emperor, while Bourbon troops (ex-Napoleonic veterans) joined him *en masse* without a shot fired. Napoleon entered Paris in triumph on March 20 after King Louis XVIII and the Bourbon Royalists fled back into exile (only Vendée revolted). The "Hundred Days" shocked the divided Coalition Powers into cooperating again against their "common enemy" and rejecting Napoleon's peace offers. Now Napoleon knew his fate, like that of Murat, was sealed by the Congress of Vienna: after Napoleon's 1813 defeat at Leipzig, Murat had abandoned him in an accord with Austria to save his throne of Naples, only to be betrayed by the Congress of Vienna's decision to restore Naples' old Bourbon king. Murat reacted by declaring

21 Henry A. Kissinger, *A World Restored: Metternich, Castlereagh and the Problem of Peace, 1812–1822* (Boston: Houghton Mifflin, 1955); H.A. Kissinger, *Diplomacy*, 77–103.

war on Austria (March 15, 1815, during Napoleon's return to Paris), and with the Rimini Proclamation advocated Italy's national unity under his rule, but Austria's reinforced armies defeated Murat at Tolentino (May 2–3, 1815) and restored the Neapolitan Bourbons (Murat was executed later). Napoleon decried this unilateral war, which undermined his own tenuous power and deprived him of any ally, while only in the long run did Austria's occupation of Italy spark cyclical nationalist revolts and Unification Wars (1820–70).[22]

Thereafter, Napoleon feverishly mobilized against the Reactionary VIIth Coalition (Great Britain, Russia, Austria, Prussia, Netherlands, Denmark, Spain, Switzerland, Naples and Portugal) formed on March 17–25, 1815 on the pledge of raising 500,000 men with British funds to depose Napoleon in another multi-pronged invasion of France by July 1. Despite British forces being scattered in bases and colonies world-wide from the United States and Canada to India, the only immediate threat to Napoleonic France came from the Duke of Wellington's forward-deployed 93,000 Anglo-Dutch forces around Brussels in Southern Netherlands/Belgium, and the Prussian armies of Field Marshal von Blücher (190,000 men) and General von Kleist (Prusso-German Federal Corps), while a Russian army of 250,000 men under Barclay de Tolly, plus two more Austro-German joint armies under Archduke Charles and Prince Karl von Schwartzenberg very slowly marched towards the Rhine. Other Coalition forces were mobilized for national border defense against France: two more joint Austrian-Italian armies (in Piedmont and Naples), supported by the Swiss Army; an Anglo-Sicilian Army shipped by the Royal Navy to land in Southern France; a Reserve Russian Army and a Reserve Prussian Army to defend the Western German borders; a Danish Auxiliary Corps; an Anglo-Hanseatic contingent; two ineffective Spanish armies under British command along France's Pyrenees border. Napoleon had to decide between a defensive war (to reenact the 1814 French Campaign with larger French forces and flanks anchored on the fortified cities of Paris and Lyons, plus French guerrillas), or an offensive "Hundred Days" Campaign (to preemptively strike the disorganized Coalition armies before they could coordinate their invasion) seeking victory in Belgium and Germany to force the Coalition into a permanent peace with Napoleon as recognized Emperor of a "smaller" national French Empire along its 1790 borders, possibly augmented by political secession of French-speaking Brussels and South Belgium/Wallonia against the Netherlands. If such preemptive French military victories and peace offers were still rejected by the Coalition, Napoleon would have to fight a longer war defeating in detail the rest of the Coalition armies (like the 1805 and 1806–07 Campaigns against the Austrians, Prussians and Russians).

However, the June 1815 French invasion of Belgium was hampered by bad weather and command confusion: French forces defeated both the Prussians and British (Ligny, Quatre Bras) without crushing the retreating foes, allowing

22 W. and A. Durant, *The Age of Napoleon*, 727–57; J. Spielvogel, *Western Civilization*, 609–705; H.A. Kissinger, *A World Restored*, 7–203; Jonathan North (ed.), *The Napoleon Options: Alternate Decisions of the Napoleonic Wars* (London: Greenhill, 2000).

Wellington's Anglo-Dutch troops to hunker down on the heights of Waterloo, secretly plotting for Prussian forces to double back and rejoin him at the rear of Napoleon's army, while terrible unseasonable thunderstorms prevented Napoleon from attacking at Waterloo (June 18, 1815) from dawn until 1pm when the mud resolidified. Napoleon sought to separate Wellington's army from the Prussians by attacking his centre, with secondary attacks on *Hougoumont* farm to draw in British reserves against the French threat to their line of retreat to the sea. However, Anglo-French cavalry mêlées stalled the 20,000-strong French infantry attack with heavy casualties, while Napoleon's bout of severe hemorrhoids fatally debilitated his command-control, leaving Marshal Ney to launch, without infantry or artillery support, 12 bloody charges by the French heavy cavalry against Wellington's army entrenched in squares. Napoleon also did not succeed in recalling his absent 33,000-man Reserve Army under Marshal Grouchy, which had been sent after the retreating Prussians: despite hearing distant artillery fire from Waterloo, Grouchy kept moving away towards Wavre where he defeated the Prussian rearguard (June 19) at the fatal cost of holding up fresh French troops from winning at Waterloo against von Blücher's second Prussian Army, whose surprise attack on the French rear saved the hard-pressed Anglo-Dutch from defeat. Napoleon's crushing defeat at Waterloo doomed his strategy of keeping the VIIth Coalition armies divided: now all Coalition armies coordinated their invasion of France spearheaded by Wellington's and von Blücher's joint forces, while both Napoleon and Grouchy withdrew. Napoleon still hoped to organize a national resistance to the Coalition, but his remaining French forces were not strong enough for prolonged combat, although they had crushed the Vendée Royalist revolt. In this second chaotic downfall of his Empire and Paris, Napoleon abdicated again and fled to Bordeaux to seek political asylum in the democratic United States, only to be thwarted by a British naval blockade which took him to his final exile at Saint Helena.[23]

Conclusion: The Enduring Napoleonic Myth

Europe since the 1789 French Revolution had been consumed by 25 years of near-constant wars and international tensions between Revolutionary and Napoleonic France vs all the Reactionary Great Powers in seven loose Coalitions financed by Great Britain. Napoleon's military campaigns of annihilation repeatedly defeated on the battlefield most enemy Powers and extended politically his Empire to the entire European Continent in 1805–13 through his "diplomacy of force," achieving all that King Louis XIV had failed to attain in his 1690s–1713 wars for European

23 Alan Schom, *One Hundred Days: Napoleon's Road to Waterloo* (New York: Atheneum, 1992); David G. Chandler, *Waterloo* (Milan: Rizzoli, 1982), 12–13; Geoffrey Wootten, *Waterloo 1815* (London: Osprey, 1992); Henry Lachouque, *Waterloo* (Paris, Rome: Stock/Ciarrapico, 1974/75); J.C. Quennevat, *Atlante della 'Grande Armée'*, 207–21; D.G. Chandler, *The Campaigns of Napoleon*, 1195–285.

hegemony. Only Napoleon's 1812 catastrophic retreat from Russia with the loss of the once magnificent *Grande Armée* of over 600,000 men to sickness, hunger, combat and winter blizzards finally led to his narrow strategic defeats in both the 1813 German and 1814 French Campaigns against larger Reactionary Coalition forces, culminating with his twin abdications and exiles (1814 to Elba; 1815 to Saint Helena), framed by the glorious 1815 "Hundred Days" and Waterloo's doom.

After Waterloo, Prussia and a few other Powers sought excessive fines and major land cessions, plus permanent Coalition tutelage of twice-restored Bourbon France to completely crush Paris as a Power, but Castlereagh's diplomacy softened the Peace of Paris (November 20, 1815) which formally ended the Napoleonic Wars: Bourbon France reverted to her 1790 borders with heavy indemnities and five years' occupation by 150,000 Coalition troops; the Congress of Vienna made the Reaction permanent through a Quadruple Alliance (Great Britain, Russia, Austria and Prussia) to jointly attack any renewed Napoleonic or Revolutionary or nationalist threats (Italy, Poland, Germany, Hungary) to the 1815–59 Metternichian Balance of Power over Europe. But the Reaction could never make European peoples forget that the Coalition's final victory had come only because the Napoleonic Empire had self-destructed in 1812–15—what alternative fate would have befallen all Coalition Powers if Napoleon had won in 1815 or risen once more thereafter? Thus, the 1820s' new cycle of pro-French and democratic revolutions in Europe (Naples, Piedmont/Sardinia, Spain, Spanish Latin American colonies and Greece) saw the Reaction's adoption of the 1821 Troppau Doctrine of automatic unilateral military "interventions" in the internal affairs of other states to crush the 1820s' Revolutions, despite British opposition by Castlereagh and Wellington at the 1822 Conference of Verona. The August 1822 suicide of a deranged Castlereagh accelerated London's drift away from the Holy Alliance, leaving von Metternich as the lone despised Reactionary "balancer" to contain Russia in the Balkans, Prussia in Germany and France, until the nationalist 1830s and 1848–49 European Revolutions finally also doomed von Metternich's diplomacy of Reaction.

These European revolutions and wars of independence against the Metternichian Balance of Power's Reactionary oppression were sparked by both the values of the French Revolution and by Napoleon's magnificent politico-artistic propaganda celebrating his victories since 1796 and cementing globally his myth of national glory, military valor and loyalty by alternating his images as general and "God of War" (Arcole, Rivoli, Pyramids, Austerlitz), or successful European Statesman lending a diplomatic "hand of peace" (Schönbrunn, Tilsit), or as concerned national leader symbolically embracing as "father" all wounded soldiers to his bosom in their shared doomed glory (Jaffa, Eylau, Beresina, Waterloo). Indeed, from his final exile in the empty vastness of the Atlantic, Napoleon still managed a last "revisionist" political propaganda coup of recollecting his political thoughts and military reminiscences in Doctor Las Cases' popular biography, *Memorial of St. Helena,* where he also reinvented himself as the friend of all oppressed peoples by reclaiming his symbolic leadership of the French Revolution and recasting the Napoleonic Empire as precursor to Europe's future unification. Finally, Napoleon's own mysterious death on May 5, 1821 (poisoned by the British, or for an *affaire* with an Aide's wife?) immortalized his tragic, glorious myth for all ages culminating with his December 1840 full honors reburial at the Hôtel des Invalides in Paris, where French and international tourists now flock to him. Thus, if France as Europe's imperial hegemon is no more, and Louis XIV holds court only at his sumptuous Versailles, Napoleon still endures as one of History's most impressive conquerors and myths of power.[24]

24 Benjamin Constant, *Mémoires Intimes de Napoléon Ier par Constant son Valet de Chambre* (1830, reprint Paris: Mercure de France, 1967); "Napoléon, ses Femmes, ses Manies, sa Mortre, son Héritage," *Figaro Magazine* (Paris: 27 November 1999), 56–76; René Maury, *L'Assassin de Napoléon, ou le mystère de Sainte Hélène* (Paris: Albin Michel, 1994); Luigi Mascilli-Migliorini, *Le Mythe du Héros: France et Italie après la Chute de Napoléon* (Paris: Noveau Monde, 2002).

War and the Sacred: Russian-Ottoman Conflict, 1876–1878

Ilya Platov

The aim of this article is to examine the connection between war and the sacred in the modern world within a methodological framework informed by both anthropology and history. In the hope to understand how war is lived and perceived by *homo religiosus*, I will put aside its strategic, political, economic or technological dimensions. As an experience of the limits of humanity, war contains a dimension which is at the same time intense, fascinating and terrible, and therefore contains a sacred or numinous element. The French historian René Rémond observed that war can never be fully accounted in rational terms only, since nothing can possibly justify why an individual would accept the loss of his life.[1] In the modern world, it is the sacrifice for the nation that is often invested with a religious significance, a point I will try to illustrate with the example of the unprecedented mass mobilization in favor of the war that occurred during the Russian-Ottoman conflict of 1876–1878.

It is possible to identify at least three domains where war experience intersects with the experience of the sacred: 1) the first is what might be called the process of *sacralization* of war: great myths or narratives whose function is to interpret, explain and justify the sacrifices of combatants and civilians; collective rituals which re-enact and articulate these myths in a dramatic form. It can range from providing a religious sanction to war to the extreme case of the sacralization of war for its own sake (e.g. Ernst Jünger in *War as an Inner Experience*). In this article, the example of the collective mobilization before and during the Russian-Turkish war of 1876–1878 will be studied as an early example of such a process; 2) the way in which religion, orthodox or heterodox, becomes an important resource for ordinary believers, clergymen and combatants faced with the omnipresence of death (religion as a recourse); and 3) commemorative practices, the cult of the fallen soldier and of the glorious dead, the celebration of sacrifice for a higher cause through monuments

1 R. Remond, "Du politique," in René Remond (ed.), *Pour une histoire politique* (Paris: Editions du Seuil, 1996), 384.

and remembrance ceremonies, etc. This article will focus mostly on point 1 (the sacralization of war).

In traditional or pre-modern societies, war, like all other human undertakings, usually acquires significance within a preexisting mythological framework. The data provided by ethnologists suggests that the states of war and peace are not so clearly differentiated, and that war is somehow considered as an inevitable part of life, a feature of a preexisting order established by higher powers in the "higher times" of the origins. The mythical structure supports here a relatively stable frame perception of reality and sustains the differentiation of such crucial categories as friend/enemy, war/peace, defines the goals of warfare and also sets limits to the permissible violence. According to Viviana Pâque, war in traditional societies is always accompanied by a ritual and follows carefully a preexisting scenario known to all belligerents, marking both the beginning and the end of hostilities.[2] From this point of view, "real" wars could be seen as actualizations of a higher mythical structure endowed with a cosmic significance.

Can we then consider that this close connection between war and the sacred remains in a sense valid for modern "secular" societies as well, with their complex differentiation and separation of various spheres of activities, and which are often characterized by a certain recession of the sacred into the background of social life? The evidence provided by historians and anthropologists mentioned in this paper seems to challenge the popular narrative of secularization and disenchantement which predicts the extinction of religion under the pressure of technical–scientific rationality. Instead, secularization can rather be seen as a process which involves a transfer, an explosion or a reinvestment of the sacred in areas hitherto considered as "profane." Pressed with the necessity of explaining outpourings of mass fervor characteristic of modern conflicts of the first half of the twentieth century, historians borrowed from anthropology such notions as myth, ritual or liminality. In the following pages, I will review two major intellectual contributions that attempted to account for this phenomenon within a functionalist anthropological and historical framework: those of Roger Caillois and Alphonse Dupront.

For the French anthropologist Roger Caillois (1913–1978), the sacralization of war is essentially a modern phenomenon. While in traditional societies the frontier between war and peace remained ill-defined, the total mobilization of men and resources required by modern war transforms the state of war into an antithesis of the ordinary peacetime society with its ethics of mutual benefit. The excessive rationalization of war leads paradoxically to the spectacular effervescence of the irrational and of mythical modes of thinking. As an event that marks a paroxysm of social existence, war becomes the functional equivalent of the *festival* in traditional societies studied by ethnologists. War in our societies acquires the dimensions of an extraordinary event leading to the reevaluation of all values. Like the festival, war takes place within "higher times" that are set apart from ordinary or profane times.

2 V. Pâques, "La guerre dans les sociétés traditionnelles," Jean Poirier (ed.), *Histoire des mœurs III: Thèmes et systèmes culturels* (Paris: Gallimard, 1991), volume 1, 347–81.

There is an intimate connection between carnage and the sacred, both being related to the idea of human sacrifice without an immediate or apparent utility:

> *War well represents a paroxysm in the life of modern society. … It is a phase of extreme tension in collective life, a great mobilization of masses and effort. … War offers satisfaction to the instincts that are opposed to civilization and that, under its patronage, take a glorious revenge, a total annihilation and destruction. … A monstrous societal brew and climax of existence, a time of sacrifice but also of violation of every rule, a time of mortal peril but yet sanctifying, a time of abnegation and also of license – war has every right to take the place of the festival in the modern world and to excite the same fascination and fervor. It is inhuman, and it is sufficient to be deemed divine. … A feeling of special reverence imbues the believer, which fortifies his faith against critical inquiry, makes it immune to discussion, and places it outside and beyond reason.*[3]

With his idea of war as a substitute for the festival, Roger Caillois agrees with the argument formulated earlier by Sigmund Freud who interpreted modern conflicts as both a consequence and as a negation of the "civilizing process" analyzed by Norbert Elias. For Roger Caillois, the phenomenon of modern warfare must be apprehended in the framework of an anthropology of the sacred, as it involves a transvaluation of values, an overcoming of egoistical impulses for the sake of an act of collective salvation. It is therefore necessary to examine the connection between the development of modern warfare on one hand, and the Durkheimian affirmation of the idea of the nation as a novel form of social transcendence on the other. The battle of Valmy (September 20, 1792), an insignificant cannonade from a certain standpoint, had a profound impact on the reevaluation of the significance of war from another point of view (of French revolutionaries).

The work of Alphonse Dupront (1905–1990) on the historical transformation of the meaning of warfare in modern times and on the connection between war and the sacred was an important, if unacknowledged, contribution. Alphonse Dupront, a major representative of the French historical "Annales school," focused mainly on the study of collective psychology. His monumental thesis *The Myth of the Crusade* set out to explain the religious underpinnings of the crusades and the survival through centuries of this "socio-mythical complex." For this purpose he developed a "convergent" methodological approach which studies simultaneously the crusade as an event embedded in its social and historical context, and as a transhistorical essence investigated from a phenomenological standpoint. The crusade he studies is a grass-root phenomenon: he establishes the analogy between crusade and pilgrimage, two expressions of a "religion of the extraordinary" according to him. The crusade is thus likened to a paroxystic pilgrimage, an idea similar to that of Roger Caillois with his analogy between war and the festival. For

---

3 R. Caillois, *Man and the Sacred* (Champaign, IL: University of Illinois Press, 2001), 177 et seq.

Alphonse Dupront, the crusade reflects a certain "condition" ("Etat") of society, characterized by a kind of dramatic intensity. His concept of the crusade implies a sacred space and time: the striving to reach the Holy Land, the thirst for salvation and deliverance, the invocation of a transcendent cause, the participation of God in battle, apparitions, omen, etc. Through the analysis of various "transfers" and "resurgences" of the myth of the crusade beyond its original historic context, he concludes at the ubiquity of this myth in the history of the West, with periodic resurgences in events such as revolutions and wars, to name some:

> *Nouns, verbs, images, referential choices, all of these carry a trace of actual historic experiences; through the play of attributive adjectives, the psycho-affective is laid bare, and crystallizes slowly into a system of correlative associations – the language of the myth is made of images both immediate and adhesive to the real – of elementary oppositions and consecrating fictions … Every collective image with the power to compel, whether as an inner call or an awakening – isn't this the very essence of myth?*[4]

The crusade has the power to create an alternative form of socialization which he calls the *society of the crusade* (La société de croisade):

> *Deep within the impulse of the crusade lays a society of the end of times or of a conquest of eternity, of an eternity where the differences, inequalities and enclosures no longer exist, and where there is instead a common plenitude of the Reign.*[5]

The crusade is thus considered as a paroxystic pilgrimage that takes place within a mythical space, both qualitative and discontinuous: it is dependent on the awareness of the existence of a "holy place," Jerusalem, the center of both Christendom and the universe. Through their quest to recover Jerusalem, the crusaders were guided by essentially religious motives: "the pursuit of the goal necessitates here a sacred action. Twice sacred, in truth: the places themselves and the sacrificial fervor to reach them, and, to the men who bear the cross, an act of sacralization."[6] Like Roger Caillois, Alphonse Dupront stresses the importance of this "extraordinary" dimension of the crusade. The crusade as an ideal-type represents a paroxysm of social existence, hence the link with pilgrimages and other practices characteristic of a religion of the "extraordinary" as opposed to a religion that consecrates ordinary times:

> *Changes of place, a tense expectation of deliverance, the separation with the ordinary through the actualization of the unheard of – through prodigies and*

4 A. Dupront, *Du Sacré. Croisades et pèlerinages. Images et langages* (Paris: Gallimard, 1987), 16; *Le mythe de croisade* (Paris, Gallimard, 1997), volume 1, 18.

5 Ibid., 18.

6 Ibid., 23.

> *miracles, the performance of gestures and rituals psychologically associated with the discovery of another reality – a whole gamut of actions distinct from the everyday and which attest for the reality of the extraordinary.*[7]

Excesses of imagination, prophecies, pilgrimages and collective rituals are the landmarks of this religion which derives its appeal from periodic paroxystic returns to the origins, and reenacts the great collective myths of a given society. To the *cognitive* and *explanatory* functions of myths and symbols he adds a third: the *mobilizing* function, measured by its capacity to reunite the society around a collective project. In times marked by a confusion of values, modern myths of the nation express a quest for authenticity and call for a social and mental restructuring to give meaning to individual and collective existences.

Historians have already used these anthropological concepts in order to study the connection between war and the sacred in modern times. The historiography of the Great War, considered by some to be the crucible of the twentieth century, has proved to be a particularly fertile testing ground for such approaches. The works of Modris Eksteins, Eric J. Leed, George L. Mosse, Annette Becker, Stéphane Audoin-Rouzeau and Jay Winter are characterized by a common concern to understand the way in which the experience of modern war possesses affinities with the experience of the sacred, as well as the religious dimension of collective mobilization of European societies during World War I. They tend to demonstrate that the sacralization of war from the point of view of the ordinary believer not only did sustain a determination to fight and the necessity of sacrifice, but also allowed the combatants to endure the hardships of the front through such practices as prayers, devotions and intercessions, together with less orthodox practices such as the use of amulets and talismans.[8] The renewed role of religion and even superstitions was studied by Annette Becker, Eric J. Leed and Paul Fussell.[9] Annette Becker points thus to important continuities with traditional religious practices even within such supposedly "modern" conflicts as World War I. Jay Winter shows that this conflict was marked by a revival of an interest in the occult and spiritualism as an attempt by the survivors to cope with loss and trauma.[10] The sacralization of war continues after the conflict as well. These and other authors also studied various commemorative practices, war memorials and pilgrimages to the battlefields through which the sacred memory of the event was carried in the post-war years.

It is worthwhile in this context to mention the erstwhile investigation of George L. Mosse (1918–1999) into the religious dimension of the twentieth-century

7 Ibid., 423.

8 See, in particular, A. Becker, *La guerre et la foi. De la mort à la mémoire, 1914–1930* (Paris, Armand Colin).

9 P. Fussell, *The Great War and Modern Memory* (London: Oxford University Press, 1977); E.J. Leed, *No Man's Land: Combat and Identity in World War I* (Cambridge: Cambridge University Press, 1979).

10 J. Winter, *Sites of Memory, Sites of Mourning. The Great War in European Cultural History* (Cambridge: Cambridge University Press, 2000).

conflicts. For him, the Great War itself represents the culmination of a process the origin of which can be traced back to the French Revolution, which was the first to consecrate the alliance between war and national identity. George L. Mosse identifies what he calls a "Myth of the War Experience," which looked back upon the war as a meaningful and sacred event, and which emerged progressively in the course of the nineteenth century and engendered a true war religion in its wake.[11] Alphonse Dupront would have interpreted it as a transfer of the crusading complex onto the struggle for nation and freedom. The "political cults" which emerged in the nineteenth century were meant to actualize these myths in an expressive and tangible form and to mobilize the active participation of the masses through ritual ceremonies, myths and symbols deployed within a space specially designed for the circumstance.[12] For George L. Mosse, war became since the French Revolution an attractive alternative form of life to those who sought liberation from the stiff norms of the bourgeois society, as well as a new understanding of death on the battlefield. In modern societies, the time of war is characterized by a radical discontinuity with the norms and values that regulate social existence in peacetime, a conception similar to that of Roger Caillois. He also points to the crucial importance of the process of sacralization for the reshaping of the memory of the event after the conflict, with its subsequent and ubiquitous cult of the fallen soldier.

The American historian and anthropologist Eric J. Leed studied "the ways in which experience of combat altered the status, self-conceptions, attitudes, and fantasy lives of participants" during World War I.[13] Starting with a postulate of an essential discontinuity between war and peace in modern wars, he proceeds to make an inventory of myths and symbols which mark the separation of the combatants with their peacetime condition. For Eric J. Leed, the war mystique represents less an explosion of irrational energies, but represents rather an inversion of preexisting social norms and values characteristic of the "normal" society. The features of the *communitas* of 1914 (a concept borrowed from the anthropologist Victor Turner) is very similar to those of the "society of the crusade" of Alphonse Dupront. The time of war is characterized by a "flow state," an experience that sets the individual apart from the normal social structure (in Victor Turner's acceptance of the term), and is reminiscent of a *rite de passage* which actualizes a preexisting scenario where all ambiguities disappear: "… a 'flow state' is not a release from rules, procedures, or order. It presumes the existence of a structure, of a 'script' that both challenges the potential actor and restricts the range of possibilities facing him, rendering his objectives coherent, noncontradictory, and obtainable."[14] It has a psychologically comforting consequence, since the uncertainty and autonomy characteristic of

11 G.L. Mosse, *Fallen Soldiers. Reshaping the Memory of the World Wars* (New York, Oxford University Press, 1991).

12 G.L. Mosse, *The Nationalization of the Masses. Political Symbolism and Mass Movements in Germany from the Napoleonic Wars to the Third Reich* (Ithaca and London: Cornell University Press, 1996).

13 E.J. Leed, *No Man's Land*, 36.

14 Ibid., 55.

peacetime is superseded by a state where the individual is uplifted on a superior level of existence at the price of a surrender of his freedom. Through this logic of inversion, the subordination and the experience of the loss of self in the collective mobilization is transfigured into an act of liberation and actualization of the self.[15]

As we have seen, this understanding of mass mobilization in religious terms, with concepts derived from anthropology, was so far most successfully used to study the great world conflicts of the first half of the twentieth century. The perception of the conflict in religious terms and the crusading zeal on the behalf of one's nation (Stéphane Audoin-Rouzeau and Annette Becker used the title "The Crusade" for one of the chapters of a book dedicated to the wartime culture of WWI) with or without the support of traditional religious authorities is not an atavism or a flashback of a bygone age, but a manifestation of perennial human religiosity in a society marked by a progress of secularization. Of course, not all modern and contemporary wars necessarily unleash such passions. It is clearly less relevant to the study of low-intensity conflicts, colonial wars, military interventions, etc. However, it remains pertinent for the understanding of the contemporary recurrence of national and religious conflicts around the world, and of the link between religion, nation and war.

The Russian-Ottoman Conflict of 1876–1878 and the Birth of a War Religion in Russia

The process of sacralization can be illustrated with the example of the Russian-Ottoman conflict of 1876–1878.[16] There is a certain singularity about this war that sets it apart from other similar conflicts which opposed the Russian and the Ottoman empires from the late seventeenth to the nineteenth centuries. While it was frequently called a "crusade" by the contemporaries themselves, it was a crusade in the modern sense as identified in the works of the above-mentioned authors. In comparison with these early conflicts, its historic context was very different: since the Great Reforms of Alexander II, Russia was drawn into a whirlwind of rapid modernization and social transformations. The war of 1876–1878 took place between the twilight of the reform era, the populist "march to the people," and the assassination of Alexander II in 1881. It is a period marked by the utopian expectations of the populists and by the crisis of reformist policies of the government. The nascent civil society (*obschestvennost'*) is also then progressively emerging as an autonomous social and political actor. The political mobilization

15 Ibid.

16 It is important to stress in this study that the participation of Russian volunteers to the Serbian-Turkish war of 1876 and the subsequent Russian-Turkish War of 1877–1878 form a single historic period. Therefore, we will always refer only to the Russian-Turkish War of 1876–1878.

of the masses in this war sets it apart from earlier conflicts. The war is imagined, anticipated and accepted by the reading public which becomes an authentic political actor. The conflict thus represents an essential milestone in the emergence of a modern national consciousness in Russia. It is important to stress that this "nationalization of the masses" in 1876–1878 occurred outside any attempt at government-sponsored propaganda.[17]

During this period, the new idea of the political nation structures the political imaginaries of the Russians, and conducts to the sacralization of war through common beliefs, ritual conducts and public festivals. At the very core of this mobilization, there is a set of representations forming a "myth of the Slavic crusade:" the liberation of the Orthodox Slavs of the Balkans from the Turkish yoke, leading also to moral and spiritual regeneration of the Russian people. This myth is a complex and heterogeneous coexistence of images and ideas, characteristic of a society caught between tradition and modernity. Through its main themes, this myth is situated, according to the incisive observation of Alphonse Dupront, in an ideological *chiaroscuro*:

> *The Christian liberation is the traditional language, valid for all; the liberation of nationalities is the modern language which takes its origin in the universalism of the revolutionaries; liberal, this language celebrates diversity while at the same time scrupulously maintains the fiction of a common 'Europeanization'.*[18]

For many contemporaries, this war represented a conflict between civilization and barbarism, the most dramatic phase of an antagonism which promised to create a world purged of evil. It assumes the identification of Christianity with a certain superior civilizational order, a sense of superiority over "Asiatic" Islam. The enemy is not only godless, but also a barbarian, an inferior. This sense of a link between a civilizational and moral order and divine law is characteristic of the "Age of Mobilization" of the nineteenth–twentieth centuries.[19]

On the one hand, the crusade "against:" the representation of the enemy as a barbarian and an infidel is at the core of this wartime imaginary. In order to account for the intensity of public outrage at the outset of the Balkan crisis in 1875, it is necessary to pay close attention to the tales of cruelty describing bodily mutilations, a major theme circulating in the press and in the other media. On the other hand, the war prompted great expectations about a coming spiritual regeneration in contrast with the crisis of traditional social and religious values before the war. The period between 1876 and 1878 is marked by the widespread conviction of the coming end

17 The term "nationalization of the masses" is the title of a book by G.L. Mosse in which he describes the entry of the masses on the political stage via myths and symbols, as a well as through their participation in collective rituals. See G.L. Mosse, *The Nationalization of the Masses*.

18 A. Dupront, *Le mythe de croisade*, volume 1, 533.

19 C. Taylor, *A Secular Age* (Cambridge: Belknap Press, 2007), 423–72.

of an era, of a beginning of a period of wars and tribulations, as well as of renewed expectations about an advent of a new era of human happiness on Earth. Various individual utopias reflect these expectations with a millennial connotation (V. Solovyev, F. Dostoyevsky, N. Fyodorov). The eschatological time of the crusade is concentrated and closed on itself, marked by the expectation of the imminence of an end of history. It is a kairotic time where the ordinary course of things is interrupted, superseded by a religion of the extraordinary, with a prescience of an imminent catastrophe of cosmic proportions. The mobilizing power of the myth of the Slavic crusade derives to a great extent from this belief that the war conducts to a total upheaval of normal existence. The poet Yakov P. Polonsky observed in his diary in 1876 that "We are living in a remarkable time [znamenatel'noe vremja] in the history of our country, and no one shall remain silent in times like these ... Such events will have great consequences, and they enclose the promise of a Russian future."[20] The rejection of the morally unacceptable was coupled with an appeal to recover a renewed social and national identity, through a vision of a regenerated society emerging from a universal conflagration. The proliferation of millennial hopes, of prophecies announcing an imminent, collective, terrestrial, total and miraculous transformation were then very common, and not only among fervent Christian believers. The war is not then conceived in political, but first and foremost in religious terms, as a redemptive sacrifice. Taken together, these representations are at the core of the myth of the Slavic crusade.

The myth of the Slavic crusade informs also new patriotic rituals, an entirely unheard of form of political mobilization in Russia at the time. The awakening of the political consciousness of the masses occurs through the appropriation of the traditional national and religious myths. While the phenomenon of the sacralization of war goes well beyond the frame of the institutional religion, it would be a mistake to neglect altogether the role played by the Orthodox clergy. The patriotic mobilization is amplified through its legitimization by traditional religious authorities, which played also a role of a cultural intermediary between pan-Slav intellectuals and the masses with a discourse which sought to endow the faithful with a theological interpretation of the war. Through their sermons and appeals, but also through the liturgical framing of patriotic gatherings, a great number of Orthodox clergymen made a conscious effort to mobilize the masses for the support of the Slavic cause. In certain cases, members of various "Slavic committees" in Russia's regions addressed the parishioners gathered in the churches, with the explicit approval of the local clergy. The church thus became a space where religious faith and patriotic fervor became closely intertwined.

In a country where the traditional religion was still the religion of the majority, the priest was an indispensable mediator who informed, harangued and legitimized the struggle for the liberation of the Slavs, under the watchful and suspicious eye of the Holy Synod. For the great majority of the Russians, being a Russian was synonymous with being an Orthodox Christian. The patriotic mobilization on the

20 Ya. P. Polonsky, "Dnevnik. Rossiya v 1876 g.," *Na chuzhoy storone* 4 (1924), 88.

behalf of the "brothers of blood and faith" (*edinovernye i edinokrovnye*, a consecrated formula at the times) was amplified by the clergy.

The homiletic texts of the period are commonly composed of three main parts: in the first part, an overall view of the situation is presented, dominated by a sense of a crisis—the suffering of the Slavs, their oppression by a barbarian and godless enemy, as well as their struggle to shake off the yoke. What is stressed here is the extraordinary nature of the ongoing crisis: Theodosius, the bishop of Taganrog and Ekaterinoslav, speaks thus of "terrifying and unheard of" news from the Balkans in his sermon of September 1876.[21] After the celebration of a liturgy at the cathedral of Ekaterinoslav, the bishop pronounced a sermon on the public square in front of the cathedral; a prayer was then said for the victory of the Southern Slavs in the presence of a great crowd. In his sermon, the events in the Balkans are presented as extraordinary, as a struggle between life and death. The insurrection of the Slavs is interpreted as the final episode in the struggle between two secular enemies. In order to strike the imagination of the faithful, the preacher reverted to theriomorphic images and compared the enemies to lions, tigers and hyenas. The horrors of Turkish massacres were described in great detail. In the second part of the sermon, the preacher usually called for a holy war and backed it with arguments designed to explain why the fate of these remote populations were of such pressing concern to the Russians, chiefly through historical references that were designed to demonstrate the vital link between *us* and *them*. In the third part, the bishop enumerated the necessary actions that good Christians must perform in order to fulfill their religious and patriotic duties. Thus, the predication of the crusade deploys a language of images and forms which aims at generating conviction and the performance of certain actions. Through the mediation of the clergy, the formerly abstract notions won the acceptance of the literate and the illiterate alike, for whom the war became nothing less than a war for the defense of the cross.

The movement of the volunteers represented the high point of this mobilization of the masses. The year 1876 in Russia was marked by the sudden apparition of a new model of an active patriot: the volunteer. Between July and November 1876, several thousand Russian volunteers from diverse social backgrounds answered the call of clergymen, of pan-Slav intellectuals and of the press, and embarked for the Balkans to fight alongside the Serbs. Their contemporaries often displayed a sense of both incredulity and fervor in front of such an unprecedented phenomenon. The appearance of volunteers was a symbol of a profound cultural change that affected the notion of war during this period. For George L. Mosse, this change was generated by the myth of the war experience, which infused the sacrifice for the country with a sacred value, and transformed the combatant into a citizen-soldier on the behalf of the nation.

Russian society was greatly impressed by the movement of the volunteers, seen as "soldiers of Christ" and exemplary heroes of a national war. The scope of this

21 Feodosy [Theodosius], "Slovo preosvyashchennogo Feodosiya episkopa Ekaterinoslavskogo i Taganrogskogo," *Pravoslavnoe obozreniye*, September 1876, 212.

phenomenon reveals a profound revaluation which affected the social status of the combatant. The testimonies of the volunteers provide an invaluable source of information about their motivation, and are impregnated by the overall conviction of the religious significance of their cause. The ceremonies of their departures to the front quickly turned into great exhibitions of patriotism, true "patriotic liturgies" in the big cities and in the regions. The cult of the volunteers anticipates to a certain extent the cult of the fallen soldiers which will emerge after the war.

The old Russian word *dobrovolets* (volunteer) received at that time an additional meaning which drew its force from the link established between the ideas of freedom and the moral good. It endowed the war of national liberation of the Slavs with a religious meaning of a struggle of good versus evil, of light with darkness. This semantic transformation was noticed by some keen observers of the time. In a letter intercepted by the tsarist police, the political émigré V. Pecherin considers the word *dobrovolets* to be a neologism which reflects a new state of mind of the Russian society, and finds there a politically subversive meaning:

> *A new word recently gained circulation in your country:* dobrovolets *(between us, I would say that Nikolay Pavlovich [Nicolas I] would have never allowed such a word in a Russian dictionary). I would like here to make some observations of an etymological nature:* dobrovolets *designates somebody who goes, following his own good will, to fight for the independence of his neighbors and brothers ... However, the love of independence might prove to be extremely contagious; the man who shed his blood for the independence of his neighbor will inevitably consider about establishing it in his own country.*[22]

For the volunteer, traditional social hierarchy pales in comparison with the holiness of the struggle which makes all the volunteers both free and equal. The noun possesses strong egalitarian connotations, since it is applied to individuals of different social origin whose acts are prompted by their moral convictions. It is a form of commitment in the name of the society, of the Slavic idea, of nation and faith, without any coercion from above.[23] The meaning of the volunteer's response to the call lies not in a decree of the sovereign, but in the moral indignation in the face of abuses and sufferings of the Slav peoples.[24]

22 *Osvobozhdenie Bolgarii ot turetskogo iga. Sbornik dokumentov*, t. 1, M., An Sssr, 1964, 470.

23 The volunteer Kliment Voronich explains the significance of the term "volunteer" by stressing the absence of any form of coercion from state authorities, in the context of a discussion about the departure of the general Chernyaev for Serbia: "they [the passers-by] said that the government did nothing and that his departure was purely voluntary [*dobrovolnaya*]," in K.I. Voronich, *Iznanka Serbskoy voyny*, Spb., 1879, t. 3, 509.

24 This was well understood by the most acute observers of the period, such as the writer Vsevolod Garshin, who wrote a poem in September 1876 in which he stated very clearly the motivations of the volunteers (fearing censorship, the newspaper *Novoe Vremya* refused to publish it): "We go, though not to please a Sovereign's caprice, / To suffer and to die" (translation in P. Henry, *Vsevolod Garshin. The Man, his Work, and*

Seen from this angle, volunteering can be interpreted as an effort directed at a psychological and social regeneration. The act of self-sacrifice is viewed as a spiritual elevation, as an access to immortality through identification with the nation. In a short story written by the populist writer Gleb Uspensky dedicated to Serbian volunteers, revealingly entitled *He Is Not Resurrected* (*Ne voskres*), the main hero, Dolbezhnikov, a member of the Russian *intelligentsia* with "radical opinions," tormented by a sense of guilt for the oppressed people, undergoes a similar transformation symbolized by the endorsement of the uniform:

> *I was extremely surprised at the sight of this sickly and tormented man ... dressed in a volunteer uniform with a long saber. What impressed me most was his face, now blooming with a new vigor, and bearing no trace of the sufferings that tormented him so much before ... He now appeared to me as an authentic 'new man'.*[25]

A patriotic ritual designed to celebrate the fallen heroes soon emerged, and the collective rallies at the occasion of the departure of volunteers to the front attracted large crowds. A ceremony of departure could typically consist of a church service, followed by a procession of crowds to the train stations where patriotic and religious hymns were sung, and an inflammatory harangue pronounced by representatives of Slavic committees or by other prominent local figures. These ceremonies were particularly impressive in big cities such as Moscow or Saint-Petersburg. Despite their ephemeral nature and an amateurish organization, the genuine enthusiasm displayed by the participants greatly impressed the contemporary witnesses, enough at least to alarm the state authorities suspicious of any uncontrolled outburst of popular emotions. Many participants described an intense feeling of communion, a welding of consciousness, an "ekstasis" of the identification with the people and the national cause. While at the beginning these rituals were mainly confined to the big cities, it soon spread to provincial towns and even to the countryside.

It is through such ritual that the "religion of the extraordinary" as defined by Alphonse Dupront manifests itself most explicitly: the time of war against the time

his Milieu (Oxford: Willem A. Meeuws, 1983), 37). V.M. Garshin himself hoped to join the volunteers, but was not able to do so in 1876. He then volunteered in the regular army after the official declaration of war in April 1877. "On April 12th, 1877, the day when Russia declared war, he and Vasily Afanasyev were revising their examination. When someone brought in a copy of the Manifesto they left their books open on their desks and went off to enlist there and then": V.M. Garshin, *Sochineniya*, M.-L., 1960, 343. His letters written during this period demonstrate that he was sensitive to the crusading call: indignation in the face of atrocities committed in the Balkans, the quest for self-liberation through war. His writings, often considered as pacifist, are in reality a testimony of his quest to unveil the moral and metaphysical roots of war. War is sacred: loathed or accepted, it remains this extraordinary event that claims the commitment of the whole being of man.

25 G.I. Uspensky, *Sobraniye Sochineniy*, M., 1956, t.3, 204.

of peace, the time of world-defining events, properly "historic" as against a time of neutral, private or insignificant quotidian events. Participation in these rituals engenders in participants a sense of communion of great emotional intensity, also different from the everyday religious ceremony. Their function is to dramatize the *passage* from the ordinary social structure to the *communitas* or "the society of the crusade," and, for the volunteers, access to a new condition and an identification with the will of the nation. The overall significance of these rituals is to stage the ritual sacrifice. Their strength derives from their capacity to reinforce and to recreate collective imaginaries.

The phenomenon of volunteering confirms the way the "modern" sacred functions: it appeals to the moral consciousness of individuals, ready to mobilize for a cause that uplifts them and takes them beyond their existence as private individuals (the essence of the crusade according to Alphonse Dupront). This is also the reason why the volunteers and the patriotic rallies were considered subversive by the state authorities, even when the participants sang the hymn "God Save The Tsar."

The sacralization of war continued also after the war through diverse commemorative practices designed to celebrate the sacrifice for the Slavic cause. A great number of churches, military cemeteries, and monuments were built in Russia and in the Balkans between 1878 and 1914. The political religion centered on commemoration of the fallen soldiers gained a new life in Bulgaria at the turn of the twentieth century. The Bulgarian state played a leading role in the design of these "holy places" of Slavdom on former battlefields such as Shipka or Plevna. The eschatological hopes which sustained the Slavic crusade between 1876 and 1878 were revived and transferred onto a mythical Bulgaria, with its promise of a future great Slavic unity.

There is an obvious continuity between the ideas and values of 1876–1878 and the ones that supported the initial enthusiasm of public opinion at the outbreak of war in 1914. The collective representations which structured the wartime culture of 1876–1878—the image of a godless and barbarian enemy, the oppressed and suffering Slavs, the mirage of Constantinople—were still very much alive in 1914. In 1912, the outpouring of patriotism at the occasion of the departure of Russian volunteers for the Balkan wars to fight for the "brothers of faith and blood" leaves the impression of *déjà vu*. Recognition of the power of public opinion was a significant factor in the Russian government's resolution to stay firm during the July Crisis of 1914. The popular press already accused the Russian government of being "antinational" in its passive and conciliatory attitude in 1912. Public opinion was ready for the war and considered unacceptable any concessions made in order to preserve peace at all costs. During his visit to Moscow on August 5, 1914, the tsar Nicolas II imitated the words and gestures of Alexander II in 1877, and pledged to the crowd gathered in front of him to act as a defender of Slavdom. The patriotic communion between the tsar and the people would be, however, short-lived. As in 1876–1878, many prominent representatives of the Russian *intelligentsia* expected a spiritual renewal from the war. The philosopher V. Rozanov thus saw the war as a "youthful surge," a *podvig* (heroic feat) on behalf of the "brotherly peoples."[26]

Today, the continuing commemorations of the Great Patriotic War, with giant monuments erected in all the major cities, with parades and public meetings, points to the persistence and even strengthening of the cult of the fallen soldier, a source of patriotic dignity and pride. The war of 1876–1877 is also not forgotten at a time when Russians explore with a renewed interest their historic and religious traditions. The Slavic crusade is the object of a minor cult within the ranks of the nationalists, with its embedded nostalgia for a period when men sacrificed themselves for a great cause. The surge of widespread pro-Serb sympathy in 1999 during the Kosovo crisis and its vehemence has surprised many western observers. The Russian Orthodox Church performs today the role of guardian of the memory of this war. Numerous monuments of this war, neglected during the Soviet times, are being repaired or rebuilt. Commemorative religious services are now regularly conducted in the restored chapel of the monuments to the heroes of Plevna in Moscow. These initiatives are actively encouraged by the state or municipal authorities.

The example of the process of sacralization of war in Russia between 1876 and 1878 allows us to return to our original question about the connection between war and the sacred. The appeal to traditional historic and religious myths and symbols during this mobilization may cause one to overlook the fact that this period witnessed a profound alteration of the relationship between the sacred and war in Russia. The most spectacular phenomenon is the emergence alongside a traditional sacred, in which war was seen as an act of a transcendent and mysterious deity, sowing rewards and punishments on men, and in which church and state were closely interwoven in a hierarchical cosmic order, of a modern sacred centered on the nation. The modernity of this war is not so much in the appearance of new *themes*, but rather in the emergence of a new interpretive *frame*. The sovereign was an essential *locus* of the sacred, a mediator between the divine and human order, and it is against this unchanging background that he declared war and established peace. A mobilization in this context was solely an initiative of the divinely appointed authorities. In the course of the mobilization in 1876–1878, the role and place of religion in the society was challenged by a mobilization of a new type. The advent of modernity (the claim for autonomy by various social actors) is realized through the appeal to pre-existing myths and practices, now being appropriated and reinterpreted by a nascent civil society. This process (sometimes called the "transfer of the sacred") allowed the "nationalization" of war in Russia during this period. War is no longer passively accepted by the population as an expression of the transcendent will of the sovereign, but is instead appropriated by the nascent civil society (*obshestvennost'*) and invested with the higher meaning of a struggle for a just cause (justice, nation, freedom) which requires the active participation of the entire nation. The eschatological goal of the Slavic crusade—the capture of Constantinople, the utopian vision

26 V.V. Rozanov, *Posledniye listya*, M., "Respublika," 2000, 256–7.

of a regenerated Orthodox empire, or the liberation of the Slavs and their reunion in a common body—takes place within an "immanent frame," to use the expression of the philosopher Charles Taylor.[27] It is a fully worldly oriented, or rather an intraworldly, form of transcendence. The resistance of traditional symbols associated with a religion of extraworldly transcendence during this war intermeshes with a very "modern" feature, the essence of which is the claim for autonomy: the war is no longer imposed from above but expresses rather a new collective project which concerns the nation as a whole. The sacralization of war is thus closely dependent on the power of appeal of the national idea, and the mobilization clearly reinforced such a sense of a common identity.

All these evidences of the emergence of a "modern sacred" must nevertheless be tempered in the sense that this was not an act of sudden conversion of a whole society, but rather a process of unequal intensity distributed among different social groups. The myth of the Slavic crusade is ambivalent, and the study of its reception among the peasant masses rather suggests a high capacity of resistance of the traditional sacred. The majority of the Russian peasants of the time still viewed the war as a divine punishment; for them, the departure of General Chernyaev to Serbia was not an act of his will, but an order from the tsar; they lacked a conception of the unity of the Slavs and remained largely unresponsive to purely pan-Slav ideals and slogans dissociated from a religious interpretation. Certain observers from the *intelligentsia* noticed that the peasants did not respond with an outpouring of enthusiasm to the proclamation of the Manifesto of April 12th, but listened to it with a serious and grave attitude. This *gravitas* which characterized the popular response was often noticed by external observers. Some journalists considered it a specific popular reaction different from the noisy patriotic jingoism of the city-dwellers: the peasants thus considered the war as a religious duty and not as a festival or "promenade" to Constantinople. War according to them was far too serious a thing to be taken lightly. For some other journalists, this attitude was simply proof of a profound indifference. These misunderstandings reveal two profoundly different conceptions of order. Besides, the malleability of the constitutive themes of the myth of the Slavic crusade facilitated its appropriation, after some hesitations, by state authorities after the official declaration of war (a phenomenon akin to the "institutionalization of the crusade" described by Alphonse Dupront). The official interpretation during and after the war stressed the decisive role played by the tsar and the members of the imperial family.

The redefinition of the place of religion in society as revealed by the war of 1876–1878 seems to corroborate the idea that war was invested with a new meaning in modern times due to the shift in the social significance of religion itself. To use the term of Charles Taylor, the Russian masses had a foretaste of the "age of mobilization," a process through which "people are persuaded or dragooned into new forms of society, not only to adopt new structures but to alter their social imaginaries and sense of legitimacy, about what is crucially important in their lives

27 C. Taylor, *A Secular Age*, 539–93.

in society."[28] It requires a voluntarist action and a sense of religious belonging that is central to political identity. If anything, the war revealed the gap between the nation and the empire, an endemic tension in the nineteenth century in Russia. The passivity or incompetence of state authorities could now be easily presented as "anti-national." The wave of revolutionary terrorism between 1878 and 1881 suggests a transfer of the crusading zeal to the inside against the "bashibuzuks of Saint-Petersburg," thus fulfilling the prophecy of Pecherin. This seems to confirm the close connection between the national and revolutionary wars observed by some scholars: the demand for mobilization and patriotic fervor, the necessity to maintain the link between the army, the nation and the state, may easily lead to passivity being considered as a form of treason.[29] This is the birth of the idea of the "enemy within," a marked shift in the value system and the social imaginaries in Russia of this period.

Also, as the example of the Russian mobilization in 1876–1878 seems to suggest, there are numerous cases of overlapping and coexistence between what we may call a "traditional" and a "modern" sacred. We cannot say that nationalism, however, totally absorbed religion or even became an *ersatz* religion, as is the case in totalitarian regimes. The example studied in this article suggests a more complex picture of the interaction between the traditional and modern sacred in this and other conflicts. In our case, we can argue that the Orthodox Church and a more widely cultural background shaped by Christianity certainly provided the main interpretative framework that justified the patriotic mobilization and individual sacrifice; it also provided restraint on the more radical pan-Slav elements through a continuous stress on the necessary loyalty to the traditional institutions, as well as through a necessity to submit to the divine will and the relativity of earthly gains. From this standpoint, the sacrifice of the volunteers, the messianic idea of the capture of Constantinople, and even the need to save the Slavs, could be interpreted within a traditional religious framework. While nationalist pan-Slav intellectuals reached back in an effort to appropriate older myths and symbols, Orthodox clergymen and believers also seized on and reinterpreted such new ideas as civilization, progress and the nation.

This particular form of the "modern" sacred, a feature of the "age of mobilization," is characterized by a relative decline of extraworldly authority and an investment in the immanent value of human action, both terrestrial and anthropocentric, with a heavy emphasis on the value of the autonomous human will. Of course, the issue of a "decline" or "transfer" closely depends on our definition of the "religious" and the "sacred," between a substantive ("a quest for an individual and collective salvation in a cosmic and supraempirical cosmos that guides and controls our world") and a functional approach ("a system of beliefs and practices that distinguishes the sacred and the profane and unites its adherents in a single moral community of

28 C. Taylor, *A Secular Age*, p. 445

29 J. Freund, "La guerre dans les sociétés moderns," *Histoire des mœurs III*, volume 1, 382–458.

the faithful").[30] It seems that the majority of theoretical contributions considered in this article were rather functional or neo-Durkheimian in inspiration (this I believe is even valid for Alphonse Dupront, for whom the quest for transcendence is a defining feature of the "myth of the crusade," but which leaves open the question of the *type* of transcendence involved—in-worldly or extra-worldly). It also appears that, in modern times, national identity becomes a vital link between religion and war. The concept of "sacred foundations" of a nation developed by Anthony D. Smith might prove to be useful at this point: the basic elements of the belief-system of nationalism (myth, symbols, values, etc.) regarded as sacred by the community, which furnish "deep cultural resources" on which members of the nation can draw in the maintenance of their national identities.[31] Modern national identity involves not so much creation, most often sterile, as re-creation. This certainly explains the close association between religion and war in many contemporary conflicts. The link between war and the sacred seems strongest in the contexts where religion remains a vital source of national identity.

As the example of the mobilization for the war of 1876–1878 suggests, the religious investment in the war took place during a period of dramatic social restructuring, a period characterized by an ongoing erosion of traditional social and religious references. It is all those who felt "uprooted" within different social groups who responded most readily to the call for the crusade. The myth of the Slavic crusade embodied the dream of a better and happier life, and provided a vision of a future seen through the lenses of a restored mythical past. Its impact was nonetheless revolutionary, since history is often set into motion by such calls for a "return" and "regeneration." The example of Russia is certainly a particularly dramatic case of a transformation that occurred or is currently occurring in other places of the world, and the ongoing pertinence of the link between war and the sacred.

30 A.D. Smith, *Chosen Peoples. Sacred Sources of National Identity* (Oxford: Oxford University Press, 2003), 5.

31 Ibid., 25–36.

The Failure to Prevent World War I

Hall Gardner

Two things, in my opinion two extreme things, would produce conflict. One is the attempt by us to isolate Germany. No nation of her standing and her position would stand a policy of isolation assumed by neighbouring powers ... Another thing which would certainly produce conflict would be the isolation of England, the isolation of England attempted by any great Continental Power so as to dominate and dictate the policy of the Continent. That has always been so in history. The reasons which have caused it in history would cause it again.[1] (Sir Edward Grey, March 1909)

Introduction: Negotiations and Mutual Geostrategic Interests[2]

An examination of the multiple countervailing forces involved in global conflict indicates that the origins of major power warfare can be found in the failure to resolve outstanding disputes that directly (or indirectly) possess "vital" geostrategic, military-technological, political-economic and socio-cultural ramifications for the elites and populations of the major powers involved. This chapter argues against the views that German expansionism, rising nationalism, a growing economy, its burgeoning conventional force and rising naval capabilities necessarily caused World War I. Imperial Germany did represent a *potential* military threat to France and Europe after the Franco-Prussian war, but the formation of the Franco-Russian Dual Alliance tended to exacerbate that potential threat. Contrary to more traditional views, the war was caused by the failure of Britain and France to find irenic ways to accommodate Germany's rise to major power status, and to peacefully forge a "United States of Europe," much as was proposed by Chancellor Caprivi in the 1890s and then by Kaiser Wilhelm in 1912.

1 Quoted in E.L. Woodward, *Great Britain and the German Navy* (Handen, CT: Archon Press, 1964), 232.

2 This chapter updates points made in my PhD dissertation, Hall Gardner, *Alternatives to Global War* (Washington, DC: Johns Hopkins SAIS, 1987).

World War I can be said to have resulted from the failure of Britain and Prussia/Germany to build a close alliance (as advocated by Friedrich List, for example, in the early nineteenth century), which would have, in effect, revised the Congress of Vienna. Prussia's humiliating defeat to Napoleon at the Battle of Jena (1806) led the Prussian monarchy to accept "democratic" reforms adopted from the French Revolution in the effort to reform its laws and educational system, and to rebuild its military organization (introducing competition and abandoning privileges of the nobility in the officer corps) so as to propel French troops out of Germany.[3] In 1813, Prussia joined Austria and Russia to defeat Napoleon at the Battle of Leipzig. Later, after its subsequent humiliation at Olmütz in 1850—in which Prussia was compelled by Britain, France and Russia to submit to Austrian interests—the Prussian elite were once again impelled to enhance or modernize Prussia's political-economic and military-technological capabilities and seek German unification by "blood and iron" while concurrently attempting to first repress, then co-opt, burgeoning social democratic movements with advanced welfare systems and parliamentary participation. Rather than attempting to forge an Anglo-French-German entente versus Russia (that would, in effect, have brought together the more "progressive" democratic states of the time against Tsarist Russia), Britain attempted to engage in "balance of power" politics. In effect, while Bismarck sought to play the interests of Britain, France, Russia and Austria-Hungary against one another in such a way as to not permit any alliance system to gain a geostrategic advantage over Germany, Britain sought to divide these same powers from a position of insular superiority, but was less interested than Germany in breaking the Franco-Russian alliance, and was, ultimately, impelled to align with the Dual Alliance despite Britain's historical rivalry with both France and Russia.

In many ways, the seeds of World War I itself were first inseminated prior to the 1870–71 Franco-Prussian war in large part due to the inability of the French, the Germans and the British (and the Russians) to come to terms over the strategically important and mineral-rich region of Alsace-Lorraine, as well as the Balkans and the Ottoman Straits, not to overlook the question of Belgian neutrality. Repeated efforts to bring the three major powers, France, Germany and Great Britain, into a closer *entente* or alliance relationship after the Franco-Prussian war failed. Instead of seeking to mediate between France and Germany, Britain continued to play "holder of the balance" by playing Germany, France and Russia against each other. Yet with the formation of the Franco-Russian Dual Alliance in 1894, it became increasingly impossible to play the Triple Alliance against the Dual Alliance: Britain felt compelled to side with the latter. Franco-German-Russian tensions would subsequently wax and wane over the next 43 years. No "honorable peace" would be forged between France and Germany (mediated by the British).

By 1907, as Imperial Germany continued to engage in its quest to obtain parity on all levels with Great Britain, Berlin found itself in a largely self-fulfilled prophecy,

3 On Prussian reforms and Napoleon, see Library of Congress, *Country Studies: Germany*, available at: http://memory.loc.gov/frd/cs/detoc.html#de0022; http://memory.loc.gov/cgi-bin/query/r?frd/cstdy:@field(DOCID+de0022).

confronted precisely with what Bismarck had warned against: the "nightmare of coalitions." Imperial Germany was not only confronted with the "encircling" Franco-Russian Dual Alliance, which it had been unable to splinter, but also with the fact that Great Britain had unexpectedly joined that alliance, forging the Triple Entente of Great Britain, France and Russia. A single act of terrorism (the assassination of the Austrian Archduke Ferdinand in 1914) then set the two contending alliances, the Triple Alliance and Triple Entente, against each another, sparking a major power conflict that the Germans expected to last only six months.

Largely due to space limitations, this chapter will focus primarily on the geostrategic failure to forge a British-German-French entente and the factors that led to German and British alienation from one another in the period 1870 to 1894. It will not focus greatly on other dimensions of alienation resulting from the naval race or from political-economic rivalry. It is consequently the purpose of this chapter to reveal some of the diplomatic options that were either ignored, or not taken fully into consideration, in the pre-World War I period in which Berlin moved from *preventing* the "nightmare of coalitions" (under Bismarck) to a post-Caprivi policy of seeking to forcibly *break apart* the 1894 Franco-Russian Dual Alliance. Berlin then sought to break apart the 1904 Anglo-French entente and the tightening 1907 Anglo-Russian entente. Concurrently, while Britain did make efforts to reach out for an Anglo-German entente, Britain failed to prevent the formation of the 1894 Franco-Russian alliance, and then, largely by default, continued to back that Alliance with the formation of the Triple Entente by 1907.

The efforts of French and German industrialists, plus French political leaders, such as Prime Minister Joseph Caillaux and Socialist Jean Jaurès (among others seeking reconciliation with Germany), to forge a compromise years before the war proved futile due to opposition by *revanchards* such as Raymond Poincaré and Théophile Delcassé (who helped forge the Anglo-French entente in 1904 and who sought British naval support against the ascendant German navy), among others. But, more fundamentally, the failure or refusal of Great Britain to go beyond its policy of "holding the balance of power" in Europe and thus move away from a position of aloof superiority and "splendid isolation" can be regarded as a major factor in permitting the war to erupt. By contrast with Lord Grey's qualified optimism as expressed in 1909 and cited at the beginning of this chapter, the "isolation" (or alienation) of Britain and Germany had already begun to weigh upon the foreign policy and military strategy of both states during the period 1894 to 1914. The forceful efforts of Imperial Germany to break out of this isolation, and the British failure to recast alliance relationships—by working to forge an Anglo-French-German entente or alliance in a new global equilibrium that would take into account both the rising United States and Tsarist Russia through the formation of a United States of Europe—set the geohistorical groundwork for disaster in the Balkans in August 1914.

The Franco-Prussian War

Following the revolutions of 1848, which in essence sought to unify Germany by peaceful means, the November 1850 "humiliation of Olmütz"[4] provided the irredentist rationalization that Prussia needed to augment its military power in the effort to unify the German "nation" by force, if need be, and free Germans from Austrian hegemony. In effect, the 1850 "humiliation of Olmütz," in which Prussia accepted the revival of the German Confederation under Austrian leadership, was linked to the support of Great Britain, France and Russia for Denmark's claims to rule over Holstein, against the claims of Prussia. Prussia's humiliation and its subsequent efforts to revitalize its military capabilities led to the "deliberate rivalry in armaments of all the great powers in Europe"[5] and to nationwide conscription by most of the continental powers.[6] Yet contrary to the expectations of the Imperial German elite, which did hope to establish an entente with Britain, by the threat to use force, if necessary, it would once again be essentially the same combination of powers, Britain, France and Russia, that would forge an alliance to check Prussian/German ambition in 1914.

As Prussia turned to force to unify Germany and break Austrian hegemony, Bismarck rapidly defeated Austria in 1866, and then pressed the Austrian empire into the Prussian-led confederation. Yet despite Bismarck's rapid assimilation of "Germanic" peoples, the German empire did not entirely live up to its image of a united "nation-state" forged with "blood and iron." Rather, by 1900, Imperial Germany (after the forced assimilation of Alsace-Lorraine and other regions in 1871) could be characterized as an unstable pluri-national conglomerate of roughly 52 million Germans, 3 million Poles, 140,000 Danes, 100,000 Lithuanians,[7] with 1.5 million inhabitants of Alsace-Lorraine, in addition to alienated southern Catholics of Bavaria and other southern states, not to overlook the generally repressed social-democratic working classes. Austria-Hungary represented an even more shaky structure of Magyars, Italians, Rumanians, Czechs, Slovaks, Poles, Serbs, Croats, Slovenes, among other essentially alienated ethnic and religious collectives, including Jews and Moslems.

Despite Prussia's efforts to unify Germany by force, Great Britain's primary concern, as an insular-core power with overseas interests, was Russian continental expansion into Central Asia and pressure on India and on Turkey and the Black

4 For an interpretation of the humiliation, see Halford MacKinder, *Democratic Ideals and Reality* (Westport, CT: Greenword Press, 1981), 123.

5 J.P.T. Bury (ed.), *New Cambridge Modern History Volume 10: The Zenith of European Power 1830–70* (Cambridge: Cambridge University Press, 1960), 504. Anti-monarchist liberals and radicals, such as Carl Schurz, Karl Marx and Julius Froebel, the so-called "filth of the year of shame," were sent into exile. As Julius Froebel put it in 1859: "Germany was sick of principles and doctrines, literary existence and theoretical greatness. What it want[ed] [was] Power, Power, Power!" (cited in ibid.).

6 Great Britain only began to debate the issue of conscription at the time of the war's outbreak.

7 http://www.verwaltungsgeschichte.de/fremdsprachen.html.

Sea as Eurasian rimland states, as well as the Far East. However, in the period 1869–70 (just prior to the Franco-Prussian War), Britain did begin confidential negotiations to persuade Bismarck to reduce "those monster standing armies." Yet Berlin refused to budge: London's proposals for troop reductions on the European continent were interpreted by Berlin as serving primarily French interests. The announcement that the Spanish government would offer the crown of Spain to Prince Leopold of Hohenzollern unexpectedly (from the British viewpoint) set off a crisis in which France believed it would find itself "encircled" by Prussian influence. Bismarck's infamous manipulation of the secret Ems telegram then drew Napoleon III into his trap, sparking a war with France for which Bismarck was well-prepared. (Interestingly, once he became British Foreign Secretary in 1870, Lord Granville had been informed by the Permanent Undersecretary that there were no major issues that Britain needed to contend with. Three weeks later the major powers, France and Prussia, engaged in warfare!)[8]

Much as Karl Marx had argued in 1870 in one of his few commentaries on geopolitics, the roots of Bismarck's "nightmare of coalitions" can be traced to the Franco-Prussian war. Marx had argued that war between Germany and Russia could precede the possibility of a social revolution in Russia, but only if it were not possible to forge an "honorable peace" between France and Germany over Alsace-Lorraine. If, however, Alsace-Lorraine was to be placed exclusively under Prussian control, Marx argued "then France will later make war on Germany in conjunction with Russia. It is unnecessary to go into the unholy consequences." Marx thus accurately predicted the future formation of the Franco-Russian Dual Alliance, and the possible "unholy consequences" of that alliance.[9]

On the other hand, if a compromise over Alsace-Lorraine could have been formulated, as Marx argued, then the Franco-Prussian war would play a role in "emancipating" Europe from the Russian dictatorship, and thus permit Prussia to "merge into Germany" and likewise permit western Europe to develop more peacefully. A Franco-German accord would additionally help to spur a social revolution in Russia. In Marx's view, Europe's future was consequently in the hands of the German victors, who would determine whether the Franco-Prussian war would be "useful or damaging."[10]

The issue not raised by Marx, however, was that it appeared dubious that France and Germany could have formulated an "honorable peace" over Alsace *by means of their own effort and volition*. In essence, they needed the backing of the predominant capitalist state, Great Britain. French and German politicians and press at the time did call for British assistance in helping to resolve the conflict, and there was some consideration of joint Anglo-Russian mediation. Gladstone's proposals for mediation that would seek either the neutralization of Alsace-Lorraine or else

8 See Hans Morgenthau, *Politics Among Nations: The Struggle for Power and Peace* (New York: McGraw Hill, 1993), 23.

9 Karl Marx in David Fernbach (ed.), *The First International and After* (Middlesex: Penguin Books, 1974), volume 3, 178–9.

10 Karl Marx, ibid.

its demilitarization (which would sustain French sovereignty over the territory but destroy French fortresses) were not accepted by the British Foreign Minister Lord Granville. The latter argued that the decision should be left to the Kaiser and Bismarck, in the fear that British mediation might exacerbate the conflict. The British cabinet even refused to protest the German annexation, and to intervene in any significant way in the conflict between Louis Napoleon and Bismarck. Britain's primary concern was to make sure that the Franco-Prussian war would remain "localized" (yesteryear's euphemism for "limited").

At that time, Britain feared the possibility that France might attack Prussia through the critical buffer state of Belgium, and potentially become a threat to England across the straits. Britain was, in fact, more concerned with the immediate French effort to take over the Belgian railways in 1869 than with the potential rise of a German territorial state forcibly unified by Bismarck. Hence, Britain concluded bilateral treaties with both France and the North German confederation pledging to enter the war against the side that violated Belgian neutrality. In this perspective, the seeds of World War I were already planted in the 1870–71 Franco-Prussian war. By 1914, in its confrontation with Imperial Germany, the Franco-Russian Dual Alliance would obtain British military assistance (which was not part of Karl Marx's scenario) once Imperial Germany had violated Belgian neutrality (but ignoring the violation of neutral Luxembourg.)

The Anglo-Russian refusal to mediate between France and Russia in 1870–71 can also be explained by Bismarck's manipulations, and not merely by internal British bureaucratic politics. Bismarck skillfully attempted to skirt outside mediation by accepting British demands to lower the war indemnity, and by pressing his generals not to take all possible territory. At the same time, the Russians (encouraged by Bismarck) refused to play a role as a neutral mediator along with Britain. Saint Petersburg accordingly chose to take advantage of the Franco-Prussian war by pressing Britain for a revision of the Black Sea treaty that governed the Ottoman Straits.

The key issue raised here is that had Gladstone persisted in seeking to mediate between France and Germany (overcoming opposition from Lord Granville), and had Britain dropped its policy of "splendid isolation" at an earlier date and asserted its own interests in attempting to mediate, it may have been possible to forge an Anglo-French-German concert that would have worked to forge a "United States of Europe," and which likewise could have attempted to check Russian ambitions in the Ottoman Straits. One could furthermore speculate that such an Anglo-French-German alliance could have helped foster a very different form of social revolution in Russia than the one that took place during World War I under Lenin's dictatorship, more like that Marx had actually predicted. Or, on the other hand, if war did break out, it would most likely have been waged against Tsarist Russia, and not Imperial Germany.

After the Franco-Prussian War

Following the 1870–71 Franco-Prussian war, the government of British Foreign Minister Benjamin Disraeli essentially argued that the *Dreikaiserbund* (an alliance between Prussia, Austria and Russia) served as a strategic bulwark to keep Russia away from a possible alliance with France. Germany was seen as a mediator (or "honest broker") between Britain's two major enemies since it was generally in Germany's interests to limit Tsarist expansion, and prevent Russia from conflicting with Austria-Hungary. By 1875, however, Britain began to push Austria-Hungary and Russia into potential conflict, largely in an effort to snap Russia from its embrace with the *Dreikaiserbund* and to "dam the Russian current"—the nineteenth century British equivalent to the American policy of containment in the effort to check the Russian quest for amphibious status as a "blue water" naval power.

Between 1875 and 1877 Britain and Germany began serious alliance discussions; both Britain and Russia intervened during the 1875 war scare to prevent Bismarck from overrunning France once again. Neither Britain nor Russia wanted France so weakened that it could not serve as a counterbalance to German power. On the other hand, Britain went so far as to seek a possible entente with Germany as a means to prevent the possible resurgence of France. Bismarck, however, was slow to reciprocate; he hoped Britain would remain neutral in case an anti-German coalition formed. Bismarck's primary concern was the "nightmare of coalitions." Bismarck believed that an alliance with Britain at that time would alienate Russia and could actually precipitate just such a possibility.

Though not allied, Britain and Germany did consult over the Balkan and Turkish questions. Bismarck sought to form an Anglo-German condominium over the Turkish rimland and had already foreseen the possibility of a break-up of the *Dreikaiserbund*. This appeared plausible as Russia and Austria-Hungary both refused to give up their Balkan rivalry, particularly following British and German efforts to reverse the Treaty of San Stefano, which had originally given Russia a clear advantage in the Balkan region.

At the same time, since Bismarck did not rule out the possibility of an Anglo-Russian understanding (which had been considered by Disraeli), he thus maintained links with Russia through the *Dreikaiserbund*. Consequently, a stronger Anglo-German relationship was ironically prevented by Germany's continued ties with Tsarist Russia. The predominant factor governing the foreign policy of both powers was the fear of being pushed into conflict with the Russian "bear." Berlin believed that London wanted Imperial Germany to do the dirty work; Bismarck in particular hated Gladstone and the Liberals for refusing to commit British troops to check Russian advances and pan-Slav ambitions.

At the Congress of Berlin in 1878, Bismarck found himself pressed to side with the Austrian position versus Russian interests. Accordingly, in 1879, an Austro-German alliance was formed, in part to restrain Austria and prevent it from siding with Britain or France, but in many ways, contrary to Bismarck's global "balance of power" strategy, the alliance with Austria would also draw Germany into defense of Austrian and pan-German demands against Russia. British membership was considered, but ruled out, for the new Austrian-German alliance for fear of further antagonizing St Petersburg. Thus, Bismarck sought to revive the *Dreikaiserbund* in 1881 but, by 1883, it was clear that Austro-Russian rivalry—*which ultimately fractured the continental entente between Germany and Russia*—had not subsided in the Balkans.[11]

The dilemma was that Bismarck found himself in a constant game of what has been called a "zig-zag policy" or "tacking" between Britain and Russia.[12] Thus, when Russia pressed closer to the Afghan buffer (occupying Merv in 1884) in the "Great Game," and as it pushed toward the Penjdeh in 1885, while likewise threatening Sofia in Bulgaria, Imperial Germany then moved closer to London. But then Germany also moved away from its courtship of France out of fear that an Anglo-Russian rapprochement might be in the offing. Yet, contrary to Bismarck's fears, as Russia pressured India, Britain succumbed to a fit of "Mervousness" and proceeded to bolster security in India and Persia. Britain likewise seized Port Hamilton in 1884–85 as a means to counter Vladivostok on Russia's vulnerable far-eastern flank. By taking preliminary steps to encircle Russia in 1884, Britain also looked toward an alliance with the Qing dynasty in China. At the same time, Britain believed that Imperial Germany represented its best hope to discourage further Russian expansion toward the Eurasian rimland.

As Britain looked for ways to check Tsarist Russia, it consequently looked to Imperial Germany. At the same time, so as not to alienate Russia, Bismarck sought to block British entry into the Black Sea once Russia threatened Afghanistan and India. Although Bismarck sought close contact with Great Britain, he also sought to sustain the *Dreikaiserbund* (Germany, Austria and Russia) intact. Bismarck feared that Anglo-Russian conflict could either degenerate into a major power war, or else result in a far-reaching Anglo-Russian entente (which, in effect, it did after the Russo-Japanese war of 1904–05). In May 1885, Bismarck warned against what he saw as Gladstone's program, as declared in the House of Commons:

> *Should [an Anglo-Russian alliance] be realized with pretended Christian and anti-Turkish, but in reality pan-Slavic and radical tendencies, the possibilities would be open that this alliance could if necessary be strengthened at any time by the addition of France, should an Anglo-Russia policy meet with German resistance. It would constitute the basis of a coalition against us, of which nothing could be more dangerous to Germany.*[13]

11 Paul Kennedy, *The Rise of the Anglo-German Antagonism* (London, Boston: George Allen and Unwin, 1979), 29–35.

12 On a form of strategic leveraging, which Brandenburg calls calculated "tacking," see Erich Brandenburg, *From Bismarck to the World War* (London: Oxford University Press, 1938), 206–7. But Britain likewise sought "a zig-zag course" between the Dual and Triple Alliances!

13 Kennedy, *The Rise of the Anglo-German Antagonism*, 186.

Bismarck's statement indicated the deep interconnection between external alliances and domestic socio-political movements: Russian pan-Slavism would undermine the Austro-Hungarian empire through the support of ethnic Slavs, while France would seek to undermine the German occupation of Alsace. Bismarck's strategy thus sought a truly flexible "balance of power" (unlike Great Britain's position of superiority as the "holder of the balance"). In this sense, Bismarck sought to play the interests of Britain, France, Russia and Austria-Hungary against one another in such a way as to prevent any alliance system gaining a geostrategic advantage over Germany. By contrast, Britain sought to divide these same powers from a position of insular superiority, but was less interested in breaking apart the Franco-Russian alliance than was Imperial Germany.

Despite the interest in a closer relationship with Britain, Germany would not sell itself cheaply: Berlin hoped to impel Britain into alliance on the best possible terms. Thus, after Great Britain opted for essentially unilateral military intervention in Egypt in 1882, Berlin began to play its strategic-economic lever, the *bâton égyptien,* as a diplomatic mediating stance that used political and economic pressures in order to impel Britain to yield concessions to both France and Germany in Egypt (following the costly unilateral British intervention). Germany also attempted to check British efforts to obtain hegemony over the strategically and economically vital Suez Canal that would then permit London to obtain paramountcy from the Cape to Cairo.

Ironically, the French had opted not to intervene in Egypt at the last moment due to Deputy George Clemenceau's unwillingness to help Britain "pull its own chestnuts out of the fire" and to engage in a major occupation of the country. The French were likewise in fear of expending too many resources overseas, thus weakening their defense of the continent against Germany. Here, it can be argued that the 1882 Egyptian crisis (which came after the 1881 formation of the Triple Alliance of Germany, Austria and Hungary) was one of the major reasons that led France to look toward Tsarist Russia for an alliance in the late nineteenth century. The British intervention overseas raised French fears that Great Britain would be bogged down in Egypt and unwilling to defend French interests on the continent or elsewhere. France thus needed to counterbalance Imperial German pressures by means of an alliance with Russia, as it could no longer trust British supports. (The unilateral British intervention in Egypt consequently led to the rupture of the post-1871 Anglo-French *Entente Cordiale,* at least from 1882 to 1904. The intervention also harmed British relations with Germany, Austria, Russia and Italy.)

As Britain expanded its colonial empire, Germany, which was considered the lesser rival in comparison to France and Russia, attempted to establish a mini-colonial empire by 1884, in part by hiding behind the French "civilizing mission." (Berlin would, however, soon begin to challenge Britain in South Africa by supporting the Boers against English rule, in addition to playing the Egyptian card.) Moreover, Bismarck's opening to London at the time did not prevent him from supporting Russia in its conflict with Turkey: Bismarck was able to play upon the antagonism of the Anglo-Russian "Great Game" for Germany's benefit.[14] In many ways, Britain's (mis)adventure into overseas imperialism from the Suez to the Cape would tend to draw London's attention away from the conflict brewing in southeastern Europe and between France and Germany, and between Austria and Russia.

French Revanche

In the period 1886–89, Bismarck feared the rise of the populist and revanchist movement led by French General Georges Boulanger ("General Revanche"), who sought to revitalize French colonialism, but who also sought to regain Alsace-Lorraine as well. Until this point, Bismarck had encouraged French colonial expansion as a means to deflect France away from its claims to Alsace and to press it into rivalry with Great Britain. As Bismarck likewise feared steps toward a Franco-Russian alliance, he sought to forge the secret "Reinsurance Treaty" in 1887 with Russia, designed to prevent France and Russia from aligning, while he concurrently sought an entente with Britain. Bismarck's steps toward Russia had the effect of alienating Austria-Hungary, which increasingly looked to Britain for diplomatic, naval and financial support. The problem was that Germany did not believe that, once the Liberal Party came to power, it would necessarily continue to sustain an entente with Germany. Secondly, Germany concluded that close association with Britain would upset Russia. (Germany's eastern border was considered "long, exposed, and vulnerable.")[15]

In the period 1887–90, Britain and Germany engaged in entente or alliance negotiations. While Britain refused a direct alliance offer by Bismarck, it did agree to implement partial measures of the Mediterranean Agreements in order to secure that region from Russian expansion, thus representing a "quasi-alliance." The first secret agreement of February–March 1887 between Britain, Italy and Austria sought to preserve the status quo versus France in the Mediterranean, Aegean and the Black Sea. In the Second Mediterranean Agreement, Britain secretly promised to work with Vienna and Rome for the defense of Turkey and the Balkans against Russian encroachment, and this depended upon the non-intervention of France to work.[16]

14 Ibid., 176. On German support of France to change British policy toward Germany, Holstein affirmed: "We will only gain England's good will – albeit a good will accompanied by a gnashing of teeth – by way of alliance with France."

15 Rose Louise Greaves, *Persia and the Defense of India* (London: Athlone, 1959), 93–4.

16 A.J.P. Taylor suggested that the Mediterranean Accords represented "more nearly an alliance with a group of Great Powers than any Great Britain had ever made in time of peace and more formal than any agreement made with France or Russia twenty years later." A.J.P. Taylor, *The Struggle for Mastery in Europe* (London: Oxford University Press, 1969), 321, 352–9 and passim.

Thus, despite the fact that no full Anglo-German entente or alliance was formed in 1887, Germany did find a means to complement its interests with those of Britain. By tolerating the Mediterranean Agreements between Britain, Italy and Austria-Hungary, Berlin tacitly supported British insular efforts to "dam" (or "contain" in twentieth century language) the Russian "current" in the Black Sea. Britain also hoped to use the Mediterranean Agreements as a means to encircle France. In particular, London sought the support of Italy, but Bismarck counseled Italy against joining any British efforts against France unless Britain formally joined the Triple Alliance. At the same time, however, Germany did not actually join the Mediterranean Agreements (and thus forge a full-scale alliance with Britain) out of fear of antagonizing Russia. Britain, for its part, turned down Bismarck's famous series of alliance offers in the period 1887–90, largely due to its adamant belief in "splendid isolation." Britain likewise believed the German alliance offer was a ruse to crush France once again. Moreover, the pact would not have provided adequate German security guarantees for the Indian sub-continental rimland against Russian encroachment.[17] How to deal with Russia in particular continued to plague both parties.

The year 1890 represented the height of Anglo-German détente once Germany was able to obtain from Britain the strategically placed island of Heligoland in the North Sea in exchange for concessions in East Africa—the surrender of German claims to Uganda and Zanzibar. Although he arranged the negotiations, it was not Bismarck who negotiated this treaty, but Leo von Caprivi, the new chancellor. Very few in Britain predicted the long-term danger that such a policy of retraction, or the nineteenth century form of "appeasement," would entail at that time. Most regarded the deal as a means to counterbalance the Triple Entente versus both France and Russia.[18] Yet British concessions would ultimately permit the amphibious naval capabilities of the German fleet to expand significantly in the North Sea with the acquisition of Heligoland, *but only because the two powers were ultimately unable to forge an entente or alliance relationship.*

Following Bismarck's dismissal, the new German leadership was more willing to risk conflict with the Russians than Bismarck had been. The year 1890—also the time of the heights of Anglo-German détente—marked the breakdown of Bismarck's reinsurance treaty with Russia and the first steps toward Franco-Russian alignment. British intervention in Egypt had already alienated the French, who felt they could not trust the British to help defend them against potential German threats. In addition, it was believed that London was seeking a separate accord with Berlin that did not incorporate French interests.

---

17 Fritz Fischer, *War of Illusions* (New York: W.W. Norton, 1974), 44–45. At this time, Germany stressed the threat of France, but Britain was not moved. On the other hand, Britain moved toward Germany due to conflict with the USA over Bering Sea fisheries and colonial conflict with France.

18 Eric Brandenburg, *From Bismarck to World War*, 34; Arthur J. Marder, *The Anatomy of British Sea Power* (Hamden, CT: Archon Books, 1964), 288.

The Caprivi era (1890–94) represented an era of attempted reforms in German domestic and foreign policy. In essence, Caprivi sought a "little German" policy which would retract imperial claims abroad and seek social reconciliation at home (ending anti-Socialist legislation). With Baron von Holstein as the architect of the "New Course," the Caprivi administration opposed the traditional Bismarckian policy of "blackmail" and "balance of power." The Caprivi government argued that Germany should develop an army sufficient for defensive purposes on the continent, but argued against establishing a blue-water navy that could protect Germany's overseas interests. In arguing that Europe might one day be dominated by American hegemony, Caprivi argued for a liberal European trading bloc and a reversal of German protectionism. This meant seeking an alliance with Great Britain. (One could argue that Caprivi's foreign policy with regard to Great Britain paralleled that of reformist German social democrats who began to argue against orthodox Marxism in support of an alliance with capitalist Britain.)

Unlike Bismarck's policy, which sought to impel Britain into an entente by the threat of a German-Russian alliance, among other measures, Caprivi's policy sought to draw Britain into an entente relationship through positive concessions and negotiated trade-offs, and move toward an essentially defensive strategy. These "concessions" included breaking off relations with Tsarist Russia; they also included a decision not to confront Britain by seeking to expand German overseas interests or attempt the development of a blue-water navy. In November 1892, Caprivi thus claimed that Heligoland[19] represented the last piece of territory that Germany claimed. Once Britain returned the island, then the security of the German empire, in his view, would be best served by positive relations with Britain (as well as with France).[20]

The new German government was accordingly more willing to risk conflict with Russia than Bismarck was (the secret Reinsurance treaty was not renewed), and more willing to reconcile with France.[21] The fact of the matter is that Bismarck himself had already begun to alienate Russia by cutting Russia from the German Bourse in 1887, thus pressing a needy Russia to seek eager French investment. Moreover, Bismarck's "balance of power" policy was largely doomed as long as Germany continued its amphibious expansion, in effect bringing French and Russian (and

19 The 1890 Heligoland–Zanzibar Treaty gave Germany the island of Heligoland in the North Sea, as well as access to the Zambezi River and thus to German colonies in East Africa. In exchange, Britain took control of the island of Zanzibar in East Africa.

20 Admiral von Muller argued that Germany had a historic choice to make: "... our motto must be all or nothing. Either we harness the total strength of the nation quite ruthlessly, even if this means accepting the risk of a major war, or we confine ourselves to continental power alone." The latter course would "admittedly bring the present nation comfortable days without serious conflicts and excitement, but as soon as our exports start declining, the artificial economic edifice would start to crumble and life therein would become very unpleasant indeed." Admiral von Muller in a letter to Prince Heinrich, the Kaiser's brother, from Gorlitz, *Der Kaiser*, 36–41, quoted in John C. Rohl, *Germany without Bismarck* (London: Batsford, 1967), 162.

21 Bismarck continued to denounce the government after his dismissal in 1890.

British) interests closer together, and as long as there was no negotiated settlement over the Alsace question. By contrast with Bismarck, Caprivi believed that a tighter German-Russian rapprochement would alienate both Italy and Austria, in addition to damaging relations with Britain. Caprivi further downplayed the possibility of an Anglo-Russian alliance, arguing that it would undermine Russian relations with France and that Bismarck's fear of the "nightmare of coalitions" was not possible due to the fact that Britain and France would clash over their differing interests in the Mediterranean (Egypt, Tunisia, Morocco), and that the British fleet could safeguard German interests against those of France.

The non-renewal of Bismarck's 1887 Reinsurance Treaty with Russia in 1890 has been mentioned as a major strategic error that ultimately led to the Franco-Russian alliance; yet an alternative perspective points to the failure or refusal of the British and Germans to incorporate France into an entente vis-à-vis Russia. On the one hand, the Caprivi government downplayed the possibility of a Franco-Russian alliance (and its potential links with Great Britain). On the other hand, Caprivi overplayed the belief that Germany and Britain possessed significant enough common interests in the Mediterranean to counter those of France and Russia. The Caprivi government failed to foresee that the tacit Anglo-German alliance (based upon the 1887 Mediterranean Accords) would not be sustained in the aftermath of the formation of the Dual Alliance in the period 1891–94, and that it was possible—counter to most expectations at the time—for Britain to come to terms with its historical rivals.

Once the 1887 Reinsurance treaty with Russia was dropped in 1890, the Russian Foreign Minister Giers believed that Russia would be completely isolated, particularly if Britain did pursue a closer relationship with the German-led Triple Alliance. Thus Russia looked to France. Following French allocation of financial credits to Russia in 1890 (the year in which the French navy visited Kronstadt), France and Russia signed a military convention in 1892; by 1894, they were formally aligned. The Dual Alliance thus linked French support for Russian concerns in the Dardanelles and Russian support for French revanchist claims to the Saar and Alsace-Lorraine. Russia and France additionally had common interests in checking British (and German) influence in China/Asia. Likewise, France began to finance the Trans-Siberian railway, with both significant strategic and economic implications (raising Halford MacKinder's concerns with respect to the power of railroads in assisting the force capabilities of land powers). It was in the period 1891–94 that Germany formulated the Schlieffen plan involving a two-front war versus the Dual Alliance, and began to consider a battle fleet by 1892, against French and Russian naval capabilities (not British).

It was, furthermore, in December 1892 that German pressure on the Ottoman empire and the attempt to use the *bâton égyptien* against British railroad interests in the Ottoman empire that began to alienate both Lord Rosebery and Sir Edward Grey. Whereas Caprivi had initially begun to work with Britain in Egypt and with regard to the Ottoman Empire, the German government began to develop a harder stance with regard to railway concessions once English interests attempted to check German plans to build what was later called the Berlin-Baghdad railway (a concession won from the Ottomans in 1893). After this, Germany no longer tended to support British and Austrian plans in the Near East, but began to take a look again toward Russia. From this point of view, having failed to establish an alliance with Great Britain, Caprivi began to look back toward Russia, but too late. Ironically, the Caprivi government would ultimately alienate both Britain and Russia.

Despite German pressures in the Black Sea in support of the Ottoman Empire, Britain was still willing to consider the alliance option with Germany once it was confronted with the danger of Anglo-French conflict in Indo-China in 1893, as well as the danger of the combined Franco-Russian fleet in the Mediterranean. The Franco-Russian naval demonstrations at Toulon in 1893, for example, were ostensibly directed primarily against Britain, not Germany. Lord Rosebery spoke of a "quadruple alliance" (Britain and the Triple Alliance), but Germany refused to pressure France for fear of also going to war with Russia. Furthermore, both Britain and Germany also sought colonial concessions from France in Africa, at the same time that Germany sought a secret agreement with Britain against Russian influence in Austria and the Ottoman Straits.

Yet rather than seek out a firm entente with Germany, Britain opted to build its own naval capabilities. The 1892 Franco-Russian military convention led to the 1893 Naval Scare in Britain and the Spencer Program, which led to the conversion of the Liberal "Little Navy" cabinet to the policy of the "Big Navy." Not only did the British increase the size of their fleet in order to establish a three power pause (enough force to deter three powers from challenging British interests), but London also augmented its Persian defenses, and continued to tilt toward the Triple Alliance without actually joining it.

As the Spencer program would not be completed for a number of years, Britain still needed a temporary ally to defend against the Dual Alliance in the Mediterranean. Austria Hungary was by then considered too weak and unstable (as was Italy), and Germany did not want to commit itself militarily against the Franco-Russian alliance *without strong British security guarantees*. Here, the Kaiser argued that even if Germany did join in an alliance with England, the two powers together would still be weaker than the Franco-Russian combination. This represented the original impetus for the development of the German fleet—*to assure the alliance potential of Germany with England*. The combination of the Franco-Russian alliance, followed by the Spencer naval program, however, consequently set off an "insecurity–security dialectic" of spiraling domestic tensions combined with geopolitical and arms rivalries.[22]

Thus, despite Anglo-German efforts to forge an entente, the period of détente thus began to break down between 1892–94, when Britain, and then Germany,

22 See Hall Gardner, Chapter 1. In my view, the Franco-Russian dyad represented the most important, if not the determining dyad, in helping to provoke World War I; yet William Thompson inadvertently overlooks it in his list of dyadic interactions prior to the Great War. See Thompson, Chapter 27. Thompson's overall argument is nevertheless very well taken and points to the fact that dyadic relationships can disrupt the presumed flexibility of the "balance of power" in realist conceptions.

intensified their rivalry to attract Russian allegiance. Both sides tried to draw Russia to their side: Germany threatened a Russo-German combination so as to block both an Anglo-Russian and French-Russian alliance. In 1894, Lord Rosebery sought an Anglo-Russian agreement over the Pamirs, in the same year of the Franco-Russian alliance commitment. The year of 1894 thus represented a year of "serious estrangement" between Britain and Germany. *That relatively quiet year of 1894 represented a major turning point in nineteenth century history as Britain, in effect, abandoned its traditional policy of imperial projection vis-à-vis Tsarist Russia and followed instead a path of retraction and mediation with Russia.*[23]

It was consequently in 1894 that Britain began to tilt toward the Dual Alliance of France and Russia, *prior to the major Imperial German naval build-up of 1897*. In effect, Britain reversed its previous course following the 1877 Russo-Turk war (see Platov, Chapter 12) and Treaty of Berlin that had attempted "to dam the Russian current" in tacit cooperation with Imperial Germany. Britain thus changed tack: rather than attempting to "contain" Tsarist Russia, Britain intensified diplomatic, economic and (ultimately financial) efforts to draw Russia into an entente.

By contrast, during the Sino-Japanese war of 1894–95, Berlin tried to pull both France and Russia into a continental alliance against Japan, Britain's soon-to-become insular ally in the Pacific. At the same time, the German elite hoped that its efforts to form a continental alliance would ultimately impel Britain to look toward Germany for support. This duplicitous form of strategic leveraging—or "tacking" or a "zig-zag course" between Britain and the rival Dual Alliance—would in many ways prove to be the roots of the great disaster of 1914.[24] In addition to threatening a continental alliance, Germany began to pressure British overseas interests. At this point, Britain accused the Germans of pursuing a "French policy" in Africa, particularly in the Transvaal region where Germany was seen as providing support for the Dutch Boers.

German support for the Boers conflicted with British interests in the South African perimeter: Lord Kimberley asserted in 1894 that South Africa was more important than Malta or Gibraltar (for the defense of world trade), and expressed fear that Germany could establish a base in South Africa to challenge its sea routes

23 As A.J.P. Taylor put it, "9 July 1894 was a historic date: It marked the end of Anglo-Austrian cooperation against Russia: a policy which the British had begun at the Congress of Vienna (or perhaps even in 1792), failed to achieve during the Crimean War, and on which they had staked much in 1878 and everything in 1887 (Year of the Mediterranean Accords). In fact, it was the day on which British 'isolation' began." See A.J.P. Taylor, *The Struggle for Mastery in Europe*, 321, 352–9 and passim. Due to the formation of the Franco-Russian Dual Alliance, Britain could not obtain assistance from France, as Germany feared becoming involved with Russia at the same time. Hence the Mediterranean Accords became meaningless. Germany could no longer pressure France into neutrality as it could in the days of Bismarck's last alliance bid of 1889. Hence Germany strove to create continental alliance, while Britain looked to forge relations with the Dual Alliance. See ibid., 348. On the significance of 1894 for Germany, see Erich Brandenburg, *From Bismarck to the World War*, 38–43.

24 See footnote 11.

to India. In addition, Germany continued to support the French colonial effort as a lever to pressure Britain. Yet by showing Germany that a British rapprochement with the flanking powers of Russia (and hence France) was a real possibility, Lord Rosebery sought to impel the Germans to back off their support for the Boers, but in exchange for concessions elsewhere. At the same time, Lord Rosebery still did not want to become entangled in the affairs of the Franco-Russian Dual Alliance.

With the Liberal government of William Gladstone and Lord Rosebery leaving office in 1895, Rosebery's decision to tilt toward the Franco-Russian Dual Alliance was extended by Lord Robert Salisbury, despite his disclaimers. Lord Salisbury began to question the necessity for the Mediterranean Accords (as had Lord Rosebery), largely due to the unreliability of Italy and Austria-Hungary as checks against Russian expansion. In addition, continued German efforts to play Britain and the Dual Alliance against each other angered London. Thus, by 1895, Britain sent out signals to both Paris and St Petersburg and began to work in concert with these states over the ongoing crisis in Armenia and Macedonia, as well as over the Central Asian Pamirs. In effect, Britain began to retract its support for the Sick Man of Europe, the Ottoman empire. By July 1895, Lord Salisbury stated that he would "let the Russians have Constantinople *avec tout ce qui s'ensuit*."

It was thus in the period 1892–95 that Britain considered a reversal of alliances, with 1894 being the key date of mutual alienation. Britain thus began to tilt away from the tacit alliance with Imperial Germany and the Triple Alliance that it had established in 1887 and then moved closer toward the Franco-Russian Dual Alliance. Britain consequently took steps toward the "encirclement" of Imperial Germany *prior to the development of the Imperial German "risk fleet" and the "Made in Germany" protectionist campaign*, just as Admiral Tirpitz had argued in his *Memoirs*. London had also taken steps toward the Franco-Russian Dual Alliance not in response to the relative military and economic strength of the Triple Alliance, *but due to its relative weakness* (as was manifest in the mid-1890s). At the same time, however, the perceived threat that Imperial Germany could somehow resolve differences with both France and Russia in the effort to forge a continental alliance against Britain would have greatly augmented Imperial German capabilities if such a threat could have been carried out.

On the German side, it was also the realization of the Triple Alliance's relative weakness vis-à-vis France and Russia that would spur Admiral Tirpitz's naval program "to enhance the alliance potential of the country" as well as boost Germany's technological and economic capabilities. Both military and economic efforts were thus expended by the elite to stave off Germany's decline from major power status (if not the collapse of its empire) and to break the "encirclement" by France and Russia, who could possibly be allied to Britain.

Whereas the effort to *prevent* the "nightmare of coalitions" had dominated Bismarck's foreign policy, the attempt to *break* the vise of encirclement dominated German policy from 1894 until the outbreak of the Great War in 1914. In effect, the formation of the Dual Alliance with tacit links to Britain in the period 1891–94 represented the actualization of the contingency that the Prussian elite had feared, at least since the 1850 humiliation of Olmütz, the very contingency that Caprivi had denounced as impossible.[25] The irony of the post-Caprivi period is that the Germans hoped that their risk fleet and "mailed fist" would pressure the British into an alliance, while simultaneously countering France and Russia.

Moreover, the completion of the Kiel Canal in 1895 enhanced German amphibious naval capabilities. Matters were made worse (from the British perspective) by the Kaiser's insistence that the French navy participate in the proceedings. German fears that Britain had provided tacit support for insular Japan in the 1894–95 war versus continental China (which had been striving for amphibious status) led Germany to support Russia as a strategic bulwark versus Japanese expansion—the "yellow peril" in Asia. With conflict over the Transvaal intensifying, the January 1896 Kruger Telegram plus Germany's unloading of weapons at Walvis Bay angered Britain. Germany once again appeared to be threatening to form a continental league linking the Triple Alliance of Imperial Germany, Austria-Hungary and Italy with the Dual Alliance of France and Russia against Britain—unless the British formed a secret alliance "binding England to go to war under certain conditions."[26]

As Anglo-German antagonism intensified, in 1897 Britain withdrew its support for the Mediterranean Accords of 1887 and 1888 with both Italy and Austria-Hungary, and retrenched to secure Egypt. Britain also terminated its trade agreement with Germany (dating from 1865), but renewed Germany's status on a yearly basis: Economic relations were no longer automatic. Berlin then threatened to abandon Constantinople to Russia if Britain did not uphold the Mediterranean Agreements, as a form of tacit alliance with Germany.[27]

Likewise, in 1897, Berlin found it could not obtain French support against Britain and its control of Egypt, nor could it forge a continental league with France and Russia versus Britain. It was then that Berlin opted for an increase in its naval expenditure, as a "political power factor" to influence British behavior. Berlin followed the military-technological policy of Tirpitz, who sought to *impel* Britain into entente upon the threat of force. Not all Germans were in agreement: Both Metternich and von Bülow wanted to reduce naval estimates and thus make some concessions to Britain in order to gain geopolitical and economic concessions elsewhere, yet these proposals were repulsed by the Kaiser and Tirpitz.[28]

With Anglo-French tensions reaching a near-boiling point at Fashoda, from 1898–1901, but not breaking into a war between two democratic states,[29] Great

25 In 1851, the same three powers, Britain, France and Russia, had checked Prussian efforts to achieve hegemony over the German confederation in its conflict with Austria.

26 Eric Brandenburg, *From Bismarck to World War*, 65–72.

27 The year 1897 also saw the renunciation of British trade agreements with Germany, placing trade agreements on a year-to-year basis: The Kaiser called the act "a knife against our throat." See Paul Kennedy, *The Rise of the Anglo-German Antagonism*, 231–2; Laurence David Lafore, *The Long Fuse* (London: Weidenfeld and Nicolson, 1966), 84–97.

28 Jonathan Steinberg, "The Copenhagen Complex," *Journal of Contemporary History* 1: 3–4 (July 1966), passim. See also, Kennedy, *The Rise of the Anglo-German Antagonism*, 224.

29 See Christopher Layne, "Kant or Cant: The Myth of Democratic Peace" in *Debating the Democratic Peace*, eds, Michael E. Brown, Sean M. Lynn-Jones and Steven E. Miller (Cambridge: MIT Press, 1999) for the argument that Britain was prepared to carry

Britain under Joseph Chamberlain's guidance sought a possible alliance with Germany, hoping to break the possibility of a continental alliance pitted against Britain while simultaneously drawing the ascendant US into an entente. Hence, Chamberlain began to take steps to formulate what he stated would represent an Anglo-German-American entente. To accomplish this, he first began to appease both German and American claims in Africa and Latin America respectively; yet, in fact, his arrangements resulted in a tilt toward the rising US. (Unable to check the rise of the US after the Civil War, Britain, by and large, sought to "appease" American interests by retracting British interests in North and South America.)

On August 30, 1898, Joseph Chamberlain signed a secret agreement with Germany in which Germany promised not to support the Boers in exchange for Portuguese territories which were to be obtained (but never were) once Portugal collapsed. But then, on the contrary, with the advent of the Boer War, on October 14, 1899, Chamberlain signed a secret treaty with Portugal guaranteeing Portuguese territory and which allowed the Portuguese to permit exportation of arms to the Transvaal. When von Bülow found out about the treaty, he denounced it. Following the resolution of the Samoa crisis—in favor of the US over Germany—Chamberlain signed the Hay-Paunceforte Treaty of 1901 which in effect represented a *retractionist* policy that surrendered British claims to the Panama Canal.

Germany regarded the tilt toward the US as a means to counter German influence in both Central and Latin America, as well as in Asia, as the US was rapidly becoming a new countervailing force in both regions. Other evidence of "perfidious" behavior by Britain surfaced: von Bülow soon learned that just three months prior to Chamberlain's overtures to Germany, Chamberlain had made proposals for a general Asiatic settlement covering both China and Turkey with Russia. Although negotiations were cut off following the Russian fleet's move toward Port Arthur in an effort to counter Germany's seizure of bay of Kiao Chow (in Shandong Province[30]) in 1897 in the scramble to control the China market, Germany began to fear the possibility of an encircling Anglo-Russian alliance against her.[31]

out threats to use force against democratic France in the Sudan, if the latter did not capitulate. For the view that common democratic values prevented war, see Bruce Russet, "The Fact of Democratic Peace," ibid.

30 German-contolled Shandong would be handed over to Japan at the Versailles Treaty, setting off the revolutionary May 4th 1919 movement in China. See Hall Gardner, Chapter 29.

31 Paul Kennedy, *The Rise of the Anglo-German Antagonism*, 231 and passim. On Portugal, see D.K. Fieldhouse, *Economics and Empire* (London: Weidenfeld and Nicolson, 1973), 360; G.P. Gooch, *Before the War: Studies in Diplomacy* (New York: Russell and Russell, 1938/1967), volume 1, 203–11. G.P. Gooch and Harold Temperley (eds), *British Documents on the Origins of the War 1898–1914* (London: HMSO, 1926), volume 1, 331. In 1898, Chamberlain's fear of Russian expansion in the Far East led him to seek a German alliance, and "compensate Germany for the abandonment of her interest in the Transvaal." Cooperation in Africa was to lead to compensation in other regions of the world in order to block Franco-German cooperation. See C.J. Lowe, *The Reluctant*

Britain had hoped to secure a quasi-condominium with Germany versus Russian encroachment, particularly in Asia. (Britain did not, however, support the German seizure of Kiao Chow until the Russians seized Port Arthur.) Germany, however, believed that Britain was on the defensive; it opted for neutrality in order not to be drawn into Asia as Britain's surrogate versus Russia, as the Germans saw Russia as a strategic bulwark versus the development of an Anglo-Japanese alliance in Asia. The Germans likewise hoped that the Fashoda crisis of 1898 would re-instigate Anglo-French conflict, and once again open up the possibility of a continental alliance. Britain feared isolation, but Germany's insistence that France recognize the status quo in Alsace-Lorraine (there had been some, though limited, French support for Franco-German reconciliation), largely prevented such a possibility.[32]

In addition to these concerns, the "atmospherics" were not right for such an Anglo-German accord. From the perspective of the German foreign office, Paul von Hatzfeldt, the German ambassador to London, believed these talks had failed, as public opinion in both countries was vehemently opposed to such a deal, particularly as the Germans criticized the British role in the Boer War, and vice-versa. (The German attitude during the Boer War helped to press Britain toward France.) Accordingly, there seemed no way for Britain to obtain parliamentary ratification in such a climate. (Joseph Chamberlain had promised that it was still possible, however.) A public announcement of such a treaty, and then its possible denunciation by Parliament, would also have had severe repercussions in France and Russia, perhaps leading to war. Here, domestic democratic politics appeared to check the possibilities of Anglo-German diplomatic compromise. Moreover, it was argued that Germany's position of relative neutrality between Britain and the Franco-Russian Alliance helped to deter France—there was no need for an alliance at that time. (If Germany did not oppose Russia in Asia, there would be no reason for the Russians to support the French to regain Alsace-Lorraine. In fact, Germany was secretly committed to support Russia in the Far East following the defeat of China by Japan.)

Chamberlain, on the other hand, argued against the proposal that Germany and England should be content to settle only minor bones of contention. Minor accords were worthless without a general agreement, according to Chamberlain: If an entente was not soon forthcoming, Britain would align with France and Russia. Chancellor Bernhard von Bülow saw this as an empty threat: "as a hideous specter invented to terrify us."[33] He believed that British setbacks around the world would ultimately impel Britain to return to the bargaining table; at that point, Germany would possess even greater negotiating leverage.

The German Ambassador to London, Count Wolff-Metternich, was, however, more in favor of concluding an alliance with Britain than von Bülow or Holstein.

Imperialists (London: Routledge and Kegan Paul, 1967), volume 1, 219.

32 Gooch, *Before the War*, 230–31. In 1901, von Bülow had stated: "The indispensible condition of an Anglo-German alliance in the present European situation is its extension to Europe, in plain language the guarantee of our territories." Ibid., 201.

33 Brandenburg, *From Bismarck to the World War*, 157. Gooch, *Before the War*, 231.

First, Russia would never be completely won over as it was to its advantage to play the game of *tertius gaudens*, and to play Germany and Britain against each other. Second, if Austria were to break apart, Germany would have to fight Russia over the remnants. Third, Germany could no longer pressure Britain through Turkey, since the British withdrawal to the Suez. Fourth, without an alliance, Germany would continue to be confronted by both France and Russia. Fifth, Britain could help German amphibious-core growth versus the USA. Sixth, there was really no immediate danger to British colonial holdings that would involve Germany in a war. And lastly, an Anglo-German alliance in China would forestall the Russian advance, counter the burgeoning US march to industrial supremacy, and help develop German industry. This fundamental split between Holstein and Count Metternich appeared to affect German policy up to the war, with the dogmatic Holstein gaining favor.[34]

In 1899, in part in response to fears of a closer Anglo-German relationship, French Foreign Minister Théophile Delcassé sought to tighten alliance relations with Tsarist Russia. The goal was to assure French, Russian and Italian hegemony over the forecasted collapse of the Austro-Hungarian empire and to prevent Imperial Germany from seizing key territories and obtaining undue influence in eastern Europe, especially Trieste, with its outlet to the Adriatic and Mediterranean. It is possible that France may also have hoped to grant Germany some Austrian territories in case of the latter's collapse, in the hope that Germany would give up Alsace.[35] What appears clear was that the major powers were beginning to consider the destabilizing impact of Austrian (and Ottoman) decline. While Germany sought to sustain both Austria and the Ottoman Empire for as long as possible (and work with Russia where feasible), it was clear that France and Russia were beginning to combine against Austria (Germany's key ally), as well as against Imperial Germany itself, exacerbating the latter's fears of encirclement and alienation.

In 1900, with the passing of the Second Naval Law of Germany, an additional factor in Anglo-German relations came to the forefront: The British director of naval intelligence argued that the German fleet would become more powerful than that of the Russian fleet after 1906.[36] The fear that Germany was intent on superseding

34 On Germany's view of duplicitous British behavior, see Brandenburg, *From Bismarck to the World War*, 154–6, 160–181. The hardliner, Holstein, believed that British policy intended "to get others to pull the chestnuts out of the fire for her" and thus wanted Great Britain to support the Triple Alliance. Moreover, in June 1901, Holstein argued that "Neither Yunnan nor Morocco are important enough for Germany to risk a war, or to be compelled to seek support." The Kaiser agreed. See also D.K. Fieldhouse, *Economics and Empire*, 428–30, on Salisbury's efforts to back Tsarist expansion into China by establishing multipolar "partitions of preponderance."

35 Paul W. Schroeder, "Life and Death of a Short Fuse," in Raimo Väyrynen (ed.), *The Waning of Major War* (London: Routledge, 2006), 53–4.

36 In 1900, Lord Selbourne had argued that a formal alliance with Germany was "possibly the only alternative to an ever-increasing navy and ever-increasing naval estimates," but this dilemma was seemingly resolved by "burden sharing" and a pre-WWI version of the "multilateralization" of containment or "encirclement" of Imperial Germany.

the British in commercial and naval capabilities led the British Foreign Office to argue for rapprochement with Russia.[37] Thus, in January 1902, Britain made the incredible announcement that it recognized the superior position of Russia in northern Persia, and that it would not object to a purely commercial Russian outlet on the Persian Gulf.[38] (The Russians themselves believed that such a port would be indefensible, and thus were more concerned with gaining access to the Ottoman Straits—an aim opposed by Britain until 1915.)

Following the January 1902 pronouncement, which in effect threatened an entente with Russia, Britain halfheartedly sought an accord with Germany, as it likewise began to consider forging an entente with France. Following the Anglo-Japanese Alliance of 1902, Britain suggested a new pact with Germany and sought a joint effort to settle the Moroccan question. But as no understanding arose here, Britain turned to France, despite conflict over issues ranging from the Newfoundland fisheries to Siamese frontiers. Depending upon the perspective of the British elite, the primary intent of the move toward an Anglo-French entente would be to resolve colonial disputes, yet its secondary aspect would be to "counterbalance" Germany in Europe, at the same time that an entente with France was not regarded as a step toward committing Britain to assist France's ally, Russia.

In 1902, even the joint Anglo-German intervention in Venezuela did more to antagonize than to ameliorate tensions, particularly as the Germans perceived that the British were unwilling to back them against the ascendant US, which could threaten British interests throughout the hemisphere. Moreover, just as Anglo-German tensions began to rise, Russian hostility toward Germany likewise increased. The continued German insistence that, if an alliance was to be formed, then Britain must also defend the interests of Austria-Hungary and Italy, proved an anathema to Britain, as did the German demand that Britain remain neutral in the case of a Franco-German war.[39] Moreover, the 1902 Anglo-Japanese alliance, coupled with steps toward an Anglo-French rapprochement, prevented France from assisting Tsarist Russia in the 1904–05 war, as this could have brought France and Britain into conflict.

37 By December 1901 at the latest, Chamberlain believed that the effort to deal with Germany was a mistake, and that he would not try again: Gooch, *Before the War*, 244–5. Following the Yangtze Treaty in October 1900, Germany had no real interest in checking Russian expansion in Asia as Britain hoped. See Paul Kennedy, *The Rise of the Anglo-German Antagonism*, 244–5; John F.V. Keiger, *France and the Origins of the First World War* (New York: St. Martin's Press, 1983), 16; J.L. Garvin and Julian Amery, *The Life of Joseph Chamberlain 1901–03* (London: Macmillan and Co., 1951), volume 4, 170–71). The alliance project had died several months before December 1901, when Lord Landsdowne formally intimated its demise.

38 B.H. Sumner, *Tsardom and Imperialism* (Hamden, CT: Archon Books, 1968), 22.

39 Hamilton argued that cooperation with Germany was "endangering our good relations with the United States" and that the December 31, 1902 joint action in Venezuela, "... conclusively disposes of any ideas of our being able to form or make any alliance with (Germany) for the future." Quoted in George Monger, *The End of Isolation* (Toronto, New York: Thomas Nelson and Sons, 1963), 105–7.

As British and Russian quarrels over the Afghan, Persian and Chinese rimlands prevented the improvement of relations with Russia, Britain sought to improve relations with its ally France in order to reach Russia, at the same time applying pressure on Russia by supporting Japan in the Russo-Japanese War. This "carrot and stick" approach ultimately proved successful, to the dismay of Germany, which had hoped to capitalize on the 1904 Dogger Bank Incident to either bring Russia into a Continental League, or else instigate an Anglo-Russian war. Both the French and British realized that Anglo-Russian conflict would serve the interests only of Imperial Germany. An Anglo-French entente provided an enticing lure to a Tsarist Russia faced with defeat abroad and revolution at home; yet despite Russian defeat, British strategists, unlike the British Left, did not expect the Tsar to topple any time soon; rather, the British elite in power generally expected a defeated Russia to devote its resources toward rebuilding its navy, regaining its lost prestige, and extending its influence on the Eurasian continent.[40]

The first tentative steps toward an Anglo-Russian alliance were taken over the extended shatterbelt conflict in Macedonia in 1903, in which an international concert of forces attempted (albeit largely unsuccessfully) to maintain peace. Likewise in 1903, Britain and Russia began to settle the question of Persian loans and the Berlin-Baghdad Railroad. When negotiations resumed following the Russo-Japanese War, Lord Kitchener agreed in part with the warnings of Lord Curzon and Rudyard Kipling; thus, in 1906, he reiterated the view opposing amphibious Russian naval entrance into the Arabian Sea, but added that a purely commercial port in the north end of the Gulf "would not, in my opinion, be so grave a danger as some people suppose."[41] At the same time, Russia continued to extend its influence over northern Persia and, in the first stages of the Russo-Japanese War, it refused to accept an Anglo-Russian partition over Persia; yet the Anglo-French entente may have altered the Russian perspective, particularly with the promise of loans from both Britain and *rentier* France.

The 1904 Anglo-French Entente Cordiale was, to a large extent, based upon the reconciliation and partition of colonial holdings between France and Britain in Egypt and Morocco, as the two began to consider a stronger entente relationship

40 In 1904, the British Committee on Imperial Defence concluded that India could not be defended successfully from an attack by Russia without either conscription or alliance with Germany, both impossible conditions. See J. McDermott, "The Revolution in British Military Thinking from the Boer War to the Moroccan Crisis" in Paul Kennedy (ed.), *War Plans of the Great Powers* (London, Boston: George Allen and Unwin, 1982). See also Keith M. Wilson, *The Policy of the Entente* (Cambridge: Cambridge University Press, 1985).

41 Lord Curzon was opposed by Sir Thomas Holdich who argued that as long as Britain held command of the sea, a purely commercial port in the Persian Gulf would not be harmful (and such a port would not be particularly easy to defend). Likewise, Kitchener warned against the Russian navy, but a port in the north end of the Gulf "would not, in my opinion, be so grave a danger as some people suppose." See Briton Cooper Busch, *Britain and the Persian Gulf, 1894–1914* (Berkeley and LA: University of California, 1967), 114–20, 237–48, 347. See also Brandenburg, *The Rise of the Anglo-German Antagonism*, 214, on the "Continental League."

to counter Imperial Germany.[42] The Entente Cordiale accordingly gave Britain a free hand in Egypt, isolating Germany, and likewise gave France a free hand in Morocco, also alienating Germany. Germany thus responded by increasing its efforts to separate France and Britain (the 1905 Moroccan crisis) and by attempting to draw Russia toward it and away from France. Yet despite German support for Russia against Japan during the Russo-Japanese War of 1904–05, that war did not impel Russia toward Germany as Chancellor von Bülow had hoped, due to a large extent to the 1905 revolution which impelled the Tsar to seek domestic reforms. Here, at the beginning of the Russo-Japanese war, with social unrest in both Russia and Austria, Germany urged Russia to make a statement in opposition to any territorial aggrandizement at the expense of Austria. Yet Russia refused, without seeing the possible relationship that the feared collapse of the Austrian Habsburg monarchy might also lead to the collapse of the Russian Romanovs.[43] In many ways, the domestic conflict in both countries was intensifying due to opposition to monarchist claims to legitimacy.

Imperial Germany not only began to counter its increasing isolation through militarization of its naval program, and support for Tsarist Russia, but it also began to support burgeoning pan-Islamic movements in Morocco, the Ottoman Empire and elsewhere. The Berlin-Baghdad-Basra railway was designed, in part, to compete with the Suez Canal. The 1905 and then 1911 crises over Morocco represented shows of force intended to split Anglo-French interests, and open the door for those of Germany.

Despite the 1905 Treaty of Bjorko farce in which the Tsar signed an alliance with the "craftier" Kaiser (only to be rejected by the Russian foreign ministers a few days later), Russia did not align with Germany for fear of losing French military and financial support, and for fear of provoking Britain into war. (Von Bülow and Admiral Tirpitz also opposed the German-Russian accord for fear of provoking war with Britain.) Furthermore, Russia was concerned that Germany would attempt to dominate Russian policy as it had during the period of the *Dreikaiserbund* (of Germany, Austria, and Russia).

Russia thus gravitated toward Britain out of a quest to rebuild its military capabilities following the Russo-Japanese War, while it still bolstered its interests in the Ottoman Straits. The Russian foreign ministry decided it was better to work for British concessions on the Straits (a goal that was not obtained in the pre-WWI epoch, although it would be promised by Britain in 1915 as an enticement

---

42 The 1888 Convention of Constantinople, which came into force in 1904, the same year that the Entente Cordiale between Britain and France had been forged, had declared the canal a neutral zone under British protection. In ratifying it, the Ottoman Empire agreed to permit international shipping to pass freely through the canal, in time of war and peace. Yet, despite its entente with France, Britain was able to block the use of Suez Canal for the Russian fleet, France's ally, forcing Russia to sail around the coast of Africa during the 1904–05 Russo-Japanese war. (The 1904 Anglo-French entente parenthetically helped to stabilize Egyptian finances.)

43 Paul W. Schroeder, "Life and Death of a Short Fuse."

to keep Tsarist Russia in the war) and float loans on the London money market. At the same time, however, Russia played the game of *tertius gaudens* by making specific accords with Germany and by duplicitously threatening the possibility of realignment with Germany to gain more British favors.[44]

The year 1907 was one of relative calm between Britain and Germany, despite the inability of the two powers to come to terms over the Berlin-Baghdad Railway, among other disputes. It was also the year of the Second Hague Conference, which held out at least a momentary glimmer of the possibilities of peace. Between 1906 and 1908, Britain had embarked on a largely unilateral reduction in naval building, meant to signal a thaw in tensions, under the Liberal Campbell-Bannerman government. Yet despite the apparent calm in Europe, new schemes were being hatched that would largely doom the possibilities of an amelioration of geopolitical tensions: Britain proceeded to forge an entente with Russia.

Within the British Foreign Office, in a famous series of memoranda in 1907, Crowe and Sanderson debated the relative merits of a tilt toward Russia: Crowe favored a "balance of power" approach versus Germany, while Sanderson decried the anti-German myopia of the Foreign Office. Gaining converts while Sanderson became isolated who had to come out of retirement to oppose Crowe, Crowe won in the intra-bureaucratic rivalry. Anglo-Russian Conventions, signed in 1907, sought to put an end to the "Great Game"—Anglo-Russian disputes over Tibet as well as Afghanistan, Baluchistan and rimland Persia. (The two powers continued their dispute over their partition of the latter country, however, despite, if not because of, the ongoing Persian Constitutional reforms taking place, which, in turn, had been stimulated by Moslem participation in the 1905 Russian revolution.)[45] Yet the fact that the 1907 Anglo-Russian Entente did not incorporate the Ottoman Straits led Saint Petersburg to seek a separate deal with Austria-Hungary, resulting in Austria's provocative annexation of Bosnia-Herzegovina. Following the humiliating Petersburg Dispatch in 1908, the Bosnian crisis resulted in a revision of the 1878 Treaty of Berlin, and led to the alienation of Italy from Austria within the Triple Alliance, but also the permanent alienation of German-backed Austria and Russia.[46]

By 1908 German newspapers began to openly denounce "encirclement." In October 1908, however, Germany sought to obtain naval and political agreements

44 L.V. Bestuzhev, "Russian Foreign Policy, February – June 1914" in Walter Laqueur and George L. Mosse (eds), *1914: Coming of the First World War* (New York: Harper and Row, 1969), 89–90; Gooch, *Before the War*, 256.

45 Antony D'Agostino, "Global Origins of World War I, Part II: A Chain of Revolutionary Events Across the World Island" Historia Actual Online, Núm. 13 (Primavera, 2007), 61–77

46 See Crowe–Sanderson Memoranda, *British Documents on the Origins of the War*, volume 3, 390, 397–431. After 1901, the anti-German tide set in. The Assistant Under-Secretary Bertie opposed the Permanent Under-Secretary Thomas Sanderson. In addition, Bertie, Mallet and Harding all saw the French entente as being directed against Germany. Crowe, Nicolson and Spencer soon joined the anti-German party after Sanderson's retirement in 1906. See also George W. Monger, *The End of Isolation*, 99–102.

in which it was prepared to "relax the tempo" of its naval program without consulting the Reichstag, but this was to be accompanied by a political agreement with Britain that would guarantee its control over Alsace-Lorraine—a deal, if accepted by London, which was hoped would break up the British entente with France. Britain replied that it would accept a frank exchange of naval information between the German and British admiralties, but that it would not agree to a political settlement.[47]

Through the 1909 naval scare, up to the first Balkan War of 1912, negotiations consisted of a "vicious circle," in Admiral Tirpitz's phrase. The Germans insisted upon British neutrality toward France, while Britain insisted that "specific colonial understandings, but not a general political entente, were quite in order yet they must be accompanied by talks upon naval expenditure."[48] The Germans, however, would not discuss disarmament proposals until a general political settlement had been reached. Proposals by Metternich and von Bülow were shelved: Mere colonial trade-offs were not considered sufficient by Tirpitz and the Kaiser, as there was no guarantee Britain would live up to its accords, particularly as its allies and friends, South Africa, the US, Japan, Australia, New Zealand and France were all opposed to an extension of German influence in their respective regions of interest. Yet in 1909 a draft formula was adopted which stated that Britain would not support France if France attacked Germany, and Germany would agree not to support Russia versus England and Japan. Britain was not to join Russia versus Austria-Hungary and Germany.

Following the 1911 Agadir Crisis, new proposals were presented. Sir Edward Grey said that he would officially drop the two-power naval ratio and permit a 16:10 ratio with Germany. He likewise proposed a general agreement to limit or diminish naval expenditure, an annual understanding dealing with naval budgets,

47 On the "Baltic Agreements" see Arthur J. Marder, *From Dreadnought to Scapa Flow* (London: Oxford University Press, 1966), volume 1, 174–5.

48 Gooch, *Before the War*, volume 1, 261–68. Metternich argued to his superiors in 1907, "If we categorically and forever reject any understanding about the fleet, even in the many unofficial private conversations which every Ambassador must have, the ill-feeling in England will wax geometrical progression ... the prospects of a peaceful development of events are more or less bound up with the Liberal Cabinet." Kennedy, *The Rise of the Anglo-German Antagonism*, 447. Sir Edward Grey expressed fears that a condominium arrangement with Germany would leave Britain isolated from France and Russia, as these powers would look to Germany for guidance. "We cannot enter into a political understanding with Germany which would separate us from Russia and France and leave us isolated while the rest of Europe would be obliged to look to Germany."

As Fritz Fischer (*War of Illusions*, 51) pointed out, "the question of the British alliance, later at any rate of British neutrality, runs right through the policy of the German government from 1877 to 1914. Hemmed in by France and Russia ... Germany lived permanently under the threat of war on two fronts. This again made an alliance or neutrality agreement with Britain seem all the more the urgent to the Foreign Ministry." Though this is true, the question is whether or not Britain could have worked to direct German hostility toward Russia alone, and away from France.

and an exchange of information about naval construction with mutual controls. Germany's offer was that of a "temporary retardation" of the naval build up, coupled with an exclusive entente, guaranteeing its territories. Although Tirpitz was not opposed to naval inspections with some limitations imposed, he was overruled by the Kaiser. Germany did not want to be left defenseless believing that Britain's "good intentions" could change as radically as the nature of its Parliament. Once again Grey opposed the concept of a general entente in the belief that any major concessions to Germany would harm both the Anglo-French and Anglo-Russian ententes.[49]

In 1912 the Haldane Mission was doomed from the start as a supplemental naval bill was published by the British parliament the day before Haldane arrived in Berlin, at roughly the same time that Britain was increasing the numbers of its active fleet. The Haldane Mission never possessed official status in the first place, and thus proposals for a reduction in the rate of increase of the fleet and promises of colonial concessions to Germany in Persia and elsewhere did not carry weight. Following the Haldane Mission, the Kaiser wrote to King Edward, explaining the course of the Haldane discussions, and suggested "an offensive and defensive alliance with France as a partner and open to other powers to enter *ad libitum*."[50]

The proposal was apparently not taken seriously by the British, and dismissed as a danger to Austria-Hungary by the German Foreign Office. That same year, 1912, the Kaiser had also suggested the possibility of bringing France into a "United States of Europe" in order to counterbalance the real global rivals in his view—the ascendant US and insular Japan, which were both rapidly moving toward core status, as was Russia. Influenced by Walter Rathenau and probably by General Bernhardi, the Kaiser sought to create a "United States of Europe" which would counterbalance the rise of American power, and which would include five states plus France. The Kaiser also proposed the concept to Austro-Hungarian foreign minister Berchtold in March 1912. In addition, a customs union would be formed that might or might not include France. (By August 1914, however, these Mitteleuropa plans were revised—the Central European Customs Union was to be aimed against the USA, Britain and Russia, yet agrarians were opposed to new competition, and "industrialists sought a world, not merely a European, market.")[51]

In 1912 and 1913 Churchill proposed the possibility of a détente with Germany to oppose Russia. Churchill's "naval holiday" proved unacceptable to the Kaiser unless accompanied by a political entente. From the German perspective it would have frozen British (and the Triple Entente's) naval superiority. From the British

49 Woodward, *Great Britain and the German Navy*, 192, 234–8, 273.

50 Woodward, *Great Britain and the German Navy*, 351; Fritz Fischer, *War of Illusions*, 137–40. As Tirpitz pointed out, Germany opposed a collective approach to the Balkan Crisis that might alienate Austria in regard to its Serbian interests: see Alfred von Tirpitz, *My Memoirs* (New York: Dodd, Mead and Co., 1919), volume 1, 327–8.

51 Woodward, *Great Britain and the German Navy*, 351. Fritz Fischer, *War of Illusions*, 137–40. See also Fritz Fischer, *World Power or Decline* (New York: W.W. Norton and Co., 1974), 10–19.

perspective, it was unacceptable in that it did not also include a proposal to reduce non-naval forces. Although usually dismissed as propaganda, Churchill's proposal did possess a legitimate strategic rationale. In a letter to Sir Edward Grey, Churchill argued that although the Triple Entente was in fact outbuilding the Triple Alliance, "that is no consolation, only a half consolation to us, who see in the revival of the Russian fleet a counterpoise to Germany, wh[ich] might suddenly become a makeweight. It is a profound British interest to procure a halt, and worth running serious risks for."[52] In other words, Churchill began to fear that Russia, as *tertius gaudens* or *key pivot state*, might suddenly decide to shift sides, and align with Germany, much as Halford Mackinder had argued.

Yet, Churchill's arguments were in effect countered by those of Sir Arthur Nicolson, who had already argued to Grey in May 1912 that moves toward an alliance with Germany were strategically and economically infeasible. Nicolson outlined three alternative strategies: 1) an increase in the naval budget; 2) an alliance with Germany; and 3) a naval understanding with France. An increase in the naval budget was considered "too heavy a charge on the Estimates"; a naval understanding with France was deemed the "cheapest, simplest and safest solution." The second choice, an alliance with Germany, was ruled out on three grounds. First, it would place Britain in an inferior naval position to Germany (Nicolson does not provide the rationale for this argument); second, the three Scandinavian countries, Belgium and Holland would be dominated by Germany; and third, France and Russia would be "at least cold and unfriendly." Nicolson believed that the British position in the Middle East and on the Indian frontier "would be seriously shaken and imperiled" if Britain were to break its entente with Russia and move toward alliance with Germany.[53]

"The Vicious Circle" had been Admiral Tirpitz's accurate expression for the pattern of Anglo-German negotiations from late 1909 to 1912, which failed to bring positive results. Yet the years 1912 to 1914 were ironically years of a relative détente between the two powers, at least in appearance. Diplomatic bargaining was still in the works: Germany hoped Anglo-Russian squabbles in Persia would break up the Anglo-Russian entente, and that France would not be assured of British support. On the British side, Sir Edward Grey appeared to tilt toward the Triple Alliance during the Balkan Wars, although pretending to play "honest broker." This appeared to be true even though the English had begun to form joint contingency plans for the defense of France, and were being pushed by France for naval talks with Russia. The British tilt toward the Triple Alliance, resulting in decisions to benefit Austria-Hungary, led Prince Lichnowsky to argue that Germany should reduce its

52 See *British Documents on the Origins of the War,* volume 10, part 2, 721.

53 Nicholson's arguments in effect superseded those of Churchill: Anglo-German engagement would not only seriously impair Franco-German relations, but would also react on British relations with Russia. See *British Documents on the Origins of the War,* volume 10, part 2, number 385, 585. See also Nicolson's letter to Sir E. Goschen, ibid., volume 6, number 575, 747. Arguments of Lord Sanderson were largely ignored. See Keith M. Wilson, *The Policy of the Entente,* 41, 78–9, 93.

naval estimates to gain British favor and that Germany should pressure Austria-Hungary "to make concessions to Russia" in order to prevent war. Both the Kaiser and Admiral Tirpitz were opposed. In the spring of 1912, Germany attempted to obtain a guarantee of British neutrality but this was once again turned down on the grounds that France might be provoked.

As late as June 16, 1914, despite evidence that Russia might be inclined to attack Germany at some point, Chancellor Bethmann-Hollweg argued that a European war was not at all necessary ...

> *Whether it will come to a European conflagration will depend entirely upon the attitude of Germany and England. If we two stand united as guarantors of European peace, from which, provided that from the first we pursue this aim on an agreed plan, we are precluded neither by the Triplice nor by the Entente obligations, then war can be averted. Failing this, any insignificant conflict of interests between Russia and Austria-Hungary may set the torch of war alight. A far-sighted policy must keep a watchful eye on this eventuality before it is too late.*[54]

Unfortunately, no far-sighted policy kept its eyes open.

Rather belatedly, with the threat of a general strike in the background, and with burgeoning tensions in Ireland, the British Left began to demand a statesman who could bring Britain, Germany and France into an entente. These demands were predicated on the assumption that, despite Russian expansion (such as seizing Mongolia in 1912, for example, and continuing to threaten the South Asian rimland), the Tsarist grasp on the throne was too precarious, and would soon collapse, particularly if Russia were continually pressured by a concert of Western forces.[55] Whether or not such a strategy would have worked may well have depended on French willingness to accept joint Anglo-German guarantees once it reneged its alliance with Russia, and upon the coordination of Britain and Germany in the Eurasian rimland versus Tsarist Russian expansion. Here, Japanese

54 Quoted in Luigi Albertini, *The Origins of the War of 1914* (London: Oxford University Press, 1952–57), volume 1, 576–7.

55 The fundamental dilemma had been dictated by the Committee on Imperial Defence in 1904. The British realized that they could not hope to defeat Russia at war, without either conscription or an alliance with Germany (Japan had defeated, but not subjugated, Russia), which were considered "impossible conditions." Hence, as Germany refused to counter Russian advances in Asia, and as Russia would prove to be the more powerful rival versus British interests, Britain would pursue a path of "appeasement" or retraction in order to gain Russian favor, ultimately promising it the reward of Constantinople in 1915 (under the pressure of global war). J. McDermott, "The Revolution in British Military Thinking from the Boer War to the Moroccan Crisis" in Paul Kennedy, *War Plans of the Great Powers*, 108. As Monger pointed out in *End of Isolation*, 109, in 1903, although Curzon opposed the entente with Russia, Hamilton effectively countered, "Time is on Russia's side; the longer we delay coming to an agreement, the worse the settlement for us will be."

assistance could have been assured to counter Tsarist expansion, particularly in the Far East.[56]

Such a policy, if it could have been implemented, might have been able to avert global war, and perhaps might have truly limited Austro-German-Russian conflict to the Balkans. The formation of a Balkan "Yugoslavia" as proposed by Marx, and later Trotsky, would likewise have aided in "damming the Russian current."[57] And, finally, concerted pressure may have been able to bring about the political ascendancy of the Mensheviks as opposed to the Bolsheviks. But for this to occur, Britain would have had to have been assured that a partially weakened Russia would not fall into re-alliance with Germany, contrary to British interests.[58] Perhaps, had Tsarist Russia remained the focus of British ire as it had throughout the nineteenth century up to 1894–1907, the Bolsheviks would likewise never have gained power, and the world would not have been confronted with the Soviet Union.

Speculation aside, during the July Crisis of 1914—what was hoped to represent a third "localized" Balkan war—the conflict became a major power war once Britain realized that it could no longer float above the two rival alliances in its dream of "splendid isolation" by playing the so-called "holder of balance" game. In effect, Britain realized that it had to choose sides—a choice largely *pre-conditioned* by the form of political ententes already undertaken. Those Anglo-French and Anglo-Russian ententes supported Germany's flanking enemies, and consequently blocked Germany's overseas access and ability to sustain its core status—*despite denials by the British elite of its intent to "encircle" Germany.*

Paul Kennedy has argued that Britain's alienation from Germany was primarily due to the "vital" threat posed by Admiral Tirpitz's risk fleet in the North Sea coupled with German threats to seize the trade and the industrial potential along

56 Despite its isolationist tendencies, the US would have had an interest in blocking Tsarist continental expansion into China. And despite US-German rivalry, Britain could have at least attempted to ameliorate tensions between the two.

57 Karl Marx in Paul W. Blackstock (ed.), *The Russian Menace to Europe* (Glencoe, IL: The Free Press, 1952). Leon Trotsky, *The War Correspondence of Leon Trotsky: The Balkan Wars 1912–13* (New York: Pathfinder Press, 1980), 12. On December 8, 1912, the Kaiser stated: "The possibility mentioned by the Chief of the Admiralty Staff ... of a war with Russia alone cannot now, after Haldane's statement, be taken into account." By April 1, 1913, the "great plan of operations for the east" was discontinued. Only the Schlieffen Plan, which was revised considerably since its inception in 1905, remained. See Wilson, *The Policy of the Entente*, 98–9.

58 Unlike the British Left, the British elite (both Conservatives and Liberals) were not optimistic as to the possibilities of a revolution transforming Russia's expansionist outreach toward China, Mongolia, the Far East, Central Asia and the Ottoman Straits, even following Russia's defeat by Japan. Hence the entente with Russia came about *more due to Russian strength than weakness, in addition to the fear that Russia might reach a separate agreement with Germany.* See Wilson, *The Policy of the Entente*, passim. See also John F.V. Keiger, *France and the Origins of the First World War*, 23; Fritz Fischer, *War of Illusions*, 175–6.

the river Meuse. (The Meuse represented the western border of the Holy Roman Empire from its creation in the ninth century until the annexation of most of Alsace and Lorraine by France after the 1648 Treaty of Westphalia). Yet, as argued here, Kennedy's point appears misplaced. First, military technological factors were *not* the issue that initially alienated Britain and Germany in 1894, which was prior to Germany's first naval law in 1897. The year of Anglo-German alienation or estrangement was also *prior to* Germany's rapid economic expansion, which occurred in the years after 1900, and *prior to* the second Boer war from 1899 to 1902. Furthermore, a possible German-British political-economic accommodation within a United States of Europe had been proposed by Chancellor Caprivi, and later the Kaiser, to build European capacities against their real rivals, the US, Japan and Russia. Proposed by Victor Hugo, among others, it was an idea way ahead of its time. The key point raised here is that German naval advances, the rise of German economic capabilities, coupled with Germany's newfound colonial *Weltpolitik*, were not the *primary* factors that led to war. The key factor was the inflexible nature of the "encircling" alliances that prevented Germany and Britain from coming to terms in the period after 1890. The formation of these alliances (the Dual Alliance, followed by the Triple Entente) were preconditioned in the period from 1870 to 1890 largely due to the failure of France and Germany (with the proposed help of British mediation) to come to terms over Alsace-Lorraine.

Postscript

Much contemporary research has focused on the potential role that international civil society movements could play as a means to preventing war; yet the period before World War I does not bring much optimism about the possibilities of transnational cooperation. Late nineteenth-century civil society movements appeared unable to achieve transnational solidarity. Liberal internationalism and socialism was unable to stem the tide of pan-nationalism and irredentism. There was split between a number of conflicting socio-political movements that included the rise of middle class professionals such as lawyers as an independent class at least since the 1870s, the women's movement, as well as the socialist international versus pan-nationalist and irredentist movements.

The Interparliamentary Union was formed in Paris in 1888 to bring together members of the various parliaments in the cause of peace, but the *Alldeutsche Verband* (pan-German League) was also founded three years later in 1891.[59] As British policy vacillated between pressing for an alliance with either Imperial Germany or with the Dual Alliance of France and Russia (finally opting for an alliance with the latter in the period 1904–07), the Anglo-German club was formed in 1905 supported by business interests, who opposed a tariff war. The peace movement, with its proposals for arms control and disarmament, was backed by industrialist Andrew

59 Barbara Tuchman, *The Proud Tower* (New York: Macmillan, 1966), 196.

Carnegie who provided funding for the construction of the Peace Palace at The Hague. Although not at all in agreement with disarmament proposals, Secretary of State John Hay strongly supported the concept of arbitration as a means to resolve international conflict, but no such arbitration took place.[60]

The Socialist International had initially been divided between Karl Marx and those who sought to sustain French-German solidarity against British imperial insular hegemony versus those "revisionists" who later sought an Anglo-German rapprochement that would incorporate the French as well, to the exclusion of the Russians, as the latter represented the state with the least-developed civil society, as compared to Bismarck's Germany. Social Democratic Party leaders Eduard Bernstein and August Bebel consequently moved away from the traditional Marxist line in support of European unity by seeking an alliance with Britain, a position that was opposed by a number of hardline German Socialists, who saw British imperialism as the main factor blocking the rise of Germany, particularly after the 1907 electoral setback for Socialists and the second 1911 Moroccan crisis.

At the same time, the voting of war credits in August 1914 was not entirely aimed at England or France, but Tsarist Russia: German Socialist leader Frederich Ebert led the German socialists to vote almost unanimously in favor of war appropriations, largely in the (false) belief that the war represented a defensive action against Tsarist Russia. The British Labour Party, which was generally radical or pacifist, was generally not anti-German; British Labour opposed the 1907 Anglo-Russian entente, and pressed to put an end to the naval race. Bertrand Russell, for example, advocated an Anglo-French-German entente versus Tsarist Russia just before the war broke out.[61] Viscount Cecil and other internationalists belatedly bemoaned the lack of a League of Nations that would prepare "the principles, technique and machinery of peacekeeping" before the outbreak of potential conflict.[62]

The failure of state leaderships (with or without the majority support of civil society movements) to achieve an Anglo-German-French entente can be seen as the underlying cause of the Great War. Such an entente may or may not have resulted in war with Tsarist Russia, but if so, it would have probably been a very different war that could have transformed Tsarist Russian society in much the way Karl Marx had predicted in 1870 along Menshevik social democratic (as opposed to Bolshevik communist) lines. And such a war would most likely not have resulted

60 Avalon Project, The Hague: http://www.yale.edu/lawweb/avalon/lawofwar/hague99/hag99-03.htm.

61 The dream of the British Left, which largely went unarticulated prior to the war, was expressed by a British Labour leader on April 30, 1914: "As long as we maintain our present close relationship with France, inevitably we are less friendly with Germany. When is a statesman going to arise who is capable of creating a Triple Federation of Germany, Great Britain and France." Quoted in A.J. Anthony Morris, *Radicalism Against War, 1906–14* (London: Longman, 1972), 359. See Bertrand Russell, *The Policy of the Entente* (London: National Labour Press, circa 1916); George Bernard Shaw, *What I Really Wrote About the War* (New York: Brentano's, 1932), passim.

62 Viscount Cecil, "The League as a Road to Peace" in Leonard Woolf (ed.), *The Intelligent Man's Way to Prevent War* (London: Victor Gollancz: 1933), 259i.

in a horrific European fratricide, to be followed by a global conflict with a *revanchist* Nazi Germany whose elites (and populations) sought revenge for its "humiliation" at Versailles—a phenomenon of alienation which can be compared and contrasted to Prussia's "humiliation" at both Olmütz in 1850 and Jena in 1806—but which resulted in a violent outburst that was even more extreme and grotesque.

14 ASHGATE RESEARCH COMPANION

World War I in World History

Anthony D'Agostino

"What A said to B, what C heard from D, that's diplomatic history." This little ditty seems to say it all about writing from diplomatic documents about the foreign policy of the great powers and the origins of great events such as the world wars. It reflects a common attitude of historians of the 1920s who were not among those who engendered the great controversy about the origins of the First World War, the Great War, as it was called. In view of these unfortunate events, a small school of diplomatic historians, mostly American and British, had dominated and almost remolded the history writing of their generation in the pursuit of presumed revelations from diplomatic documents, for the most part correspondence between officials of the various foreign ministries of the great powers. The biographer of Stanley Baldwin, G.M. Young, called it a history based on "what one clerk wrote to another clerk." This withering judgment rang in the ears of the generation who were inventing the field of diplomatic history, so much so that their students of the next generation would insist on renaming the field, calling it "international history," and reorienting its approach to its subject matter. Fernand Braudel, founder of the *Annales* school and inspirer of the World Systems Theory of Immanuel Wallerstein, Andre Gunder Frank, Giovanni Arrighi and others, proposed to found a new history in flight from this history of facts and dates, of documents and events, a "delusive smoke that fills the minds of its contemporaries." He struggled to get free of the events of the difficult years (1940–45) when he was a prisoner of the Germans, to release himself and the minds of all the French from gloomy contemplation of the tide of *histoire evénementielle*.[1]

Despite Braudel, we are still not free from this terrible history, this contemplation of the origins of wars. We sense that one necessary task of history is to allow us to grasp the rationality of great events by study at close range of small facts. And we are driven by the hope that something can be learned to inform present practice, to avoid the horrors of the past and open a path to better things in the future. One can readily understand the motive of the reader of the diplomatic history of the Great War in looking to uncover the secret of the recent cataclysm. Even for the generations who follow this fascination continues. Now, however, after a whole

1 Fernand Braudel, *On History* (Chicago: University of Chicago Press, 1980), 28, 48.

century, they have to wonder not only how it happened but how it was told, not only how the Great War was unleashed, but what myths have been conjured in writing its history.

The history of the origins of the Second World War is of an entirely different order. There, the historian who is not a revisionist is taking to task all those who did not prepare to fight almost immediately. He proceeds toward war as a lover running to a tryst. He knows it is inevitable in view of Hitler's blueprint for conquest. He knows also that a happy ending awaits. It is a good war. The Great War was not at all the same for the historians. At its end, regret was the predominant theme, even for the victor nations, who had lost millions of their young men. They could not point to a great victory. The losers were not defeated and occupied, the territorial gains outside Europe could not be boasted about in view of their great cost. There was no cause that could be cited as an end for which the vast sacrifices had been offered.

Yet there had to be a vast dispute. The Versailles peace committed the Germans, as those guilty for having stared the war, to pay reparations virtually into perpetuity, or, according to the agreement later reached in the Young Plan of 1929, into the 1980s. If the case for the reparations was the war guilt of the Germans, there had to be contestation about this guilt. How had the catastrophe happened? Lloyd George said later that the war was "something into which we glided, or rather stumbled and staggered." Could the whole thing have been an accident? That, in effect, was what the historians concluded after all of their investigations and towering works of literary imagination. After the great controversy of the twenties, the disputes among the historians centered more and more on finer investigation of the events of 1914, always within the pattern of inquiry that sought out the guilty party or, for those who increasingly doubted the German war guilt thesis and accepted the thesis of accidental war, the scale of guilt among the parties.

Is it time to abandon the thesis of war guilt as a way of understanding the origins of the Great War? Can the thesis of an accidental war also be set aside, not so much for its lack of utility in explaining the crisis of 1914, but for its lack of broader perspective? After the dust from the Second World War had settled and the historians of the 1960s were opening up a controversy about the origins of the Cold War, a "social imperialist" school of thought emerged to try to integrate understanding about the domestic situation of the great powers and its relation to their war decisions. At the same time the nuclear arms race and its immediate threat of accidental war called up 1914 as a parable. But all these thoughts soon passed, eclipsed by the drama of the fall of the Soviet bloc and the relatively peaceful end of the Cold War. How should the twenty-first century look back on the Great War and its myths? Should we regret it as its contemporaries did? Should we contemplate counterfactuals of a happier coloring? For example: no war, but a peaceful acceptance of Imperial Germany as an organizer of Europe in the manner of the current European Central Bank?[2] Or can we see it more clearly by stepping back a bit from the analysis of daily events? It will be evident that I am not

2 This is the theme of Niall Ferguson, *The Pity of War* (London: Penguin, 1998).

addressing these questions primarily by offering new material or scholarship. Nor am I undertaking a review of the latest scholarly efforts and trends. I am merely trying to assess the impact of historians' work on pundit opinion, to estimate what history underlies the common assumptions of the intelligent public, to gauge how certain images of an event have serviced ideas about other things, to take note of these other contemporary influences on the selection of perceptions of the historical event. This would mean a broad consideration of the intellectual history of the historians' Great War, not as Braudel would have liked, but at least in the realm of what might be called middle-level generalization. Not according to a *longue durée* but perhaps a *moyen durée*.

Before the war was over, there was already a struggle to define its meaning. It did not end cleanly with one side clearly defeated as in 1945. The fighting broke down as mutinies made it impossible to continue. The Russian front broke in 1917 and was almost joined by a French revolution against the offensive led by General Robert Nivelle. Within a half-year the Bolsheviks had come to power in Russia and they were negotiating to leave the war unilaterally, at the same time that their allies were organizing their domestic opponents into a civil war to restore the eastern front. At the time of the Armistice in November 1918, Germany was in the throes of a revolution that appeared to be an imitation of the Russian revolution. A few months later Hungary would have a six-month Soviet regime in an attempt to defy the presumed victors. Newly independent Poland would send troops to annex lands in White Russia (Belarus) and Lithuania, and Soviet Russia would do battle with the White armies and Entente troops on four fronts. The victor powers had adopted war aims that envisioned a democratic revolution against the Habsburg Empire and a fundamental reshaping of Central Europe. Absolute monarchy seemed to have suddenly become a thing of the past. If Francis Fukuyama had been on the scene he would have called it the end of history. The dialogue on the future of society had been pushed to the left by several degrees.

A rival interpretation of the Entente war aims was advanced by the Bolsheviks who called it an imperialist war for the redivision of the world. They went into the Tsarist archives and published the texts of secret treaties made by the British and French. These included promises that the Russians could have the Straits, the Romanians Transylvania, the Italians Trieste, Dalmatian Croatia, a corner of Asia Minor and part of the German colonies in Africa. Britain had promised Germany's Far Eastern possessions to Japan. Russia, in return for a free hand in Poland, had guaranteed to France not only Alsace-Lorraine and the Saar, but an independent French puppet state in the Rhineland. As historian A.J.P. Taylor was later to put it, "France abandoned Poland to Russia for the Rhine frontier." In the press of the new international Communist movement launched by the Russian Bolsheviks, the democratic utterances of Wilson and Lloyd George were mercilessly attacked as a new face of Entente imperialism.

The Germans got a certain benefit from this view of the war's victors, but they wanted to put it in their own way. They published a White Book presenting their own professions of innocence of any intention to unleash the war in 1914. Karl Kautsky, the dean of the Social Democratic Marxists, called it a "white wash book."

Policy documents were released under his editorship, and the Austrian documents were published. On the basis of these materials, Smith College historian Sidney B. Fay published two articles presenting the case that Austria, not Germany, bore the responsibility for the outbreak of war.[3] In the next few years, William L. Langer and Harry Elmer Barnes, joined by G.P. Gooch in Britain, were to present a fuller case directed against the French. Against the orthodox Versailles view that Germany and Austria began everything by their wanton attack on Serbia, they found Russia's mobilization in defense of Serbia the real cause of the war. The mobilization could not have been met by any other recourse than a German declaration of war on the Franco-Russian alliance and the outbreak of hostilities. "Mobilization equals war" was the case against Russia and particularly against Prime Minister Raymond Poincaré of France for supporting her. Weimar Germany formed a subsection of the Foreign Ministry, the *Kriegsschuldreferat* (War Guilt Section), to direct and finance scholarship and publication, with a periodical *Kriegsschuldfrage* (The War Guilt Question).[4] The effort included a multivolume collection of diplomatic documents, *Die grosse Politik der europäischen Kabinette* (The High Politics of the European Cabinets). Leading scholars of this revisionist school used these materials to recast the crisis of 1914: Britain had set German nerves on edge in 1912 by her naval conventions with France and Russia, inadvertently giving life to the German nightmare of encirclement. Russian military policy had duped the Tsar into a mobilization against Austria that had turned into a mobilization against Germany, and Germany had answered with war against France and Russia. German leaders were not criminals plotting war, but victims of circumstances.

To bolster the point there developed a special appreciation of Bismarck, not the Bismarck who made war all over Europe to unify Germany in the 1860s, but the Bismarck who directed German policy, 1871–1890, the time of the "Bismarckian peace," the good European crisis manager. Bismarck's bid for colonies in 1884–85 represented an aberration, caused by ripples in domestic policy or a temporary desire to humor France.[5] Where Bismarck manipulated a complex alliance system and always cultivated Britain, even to the point where he and Salisbury had a kind of condominium during the Near Eastern Crisis of 1887, the hapless Kaiser who dismissed him immediately set out on a naval program and antagonized the British, tempting them to the encirclement strategy that resulted in Germany's

---

3 Sidney B. Fay, "New Light on the Origins of the World War," *American Historical Review* (July and October 1920). Karl Kautsky, *Wie der Weltkrieg enstand* (Berlin: Paul Cassirer, 1919). Richard Gooss, *Diplomatische Aktenstucke zur Vorgeschichte des Krieges 1914*, 3 vols (Vienna: Staatsdruckerei, 1919).

4 Holger Herwig, "Of Men and Myths: the Use and Abuse of History and the Great War," in Geoffrey Parker and Mary Habeck (eds), *The Great War and the Twentieth Century* (New Haven CT: Yale University Press, 2000), 301–3.

5 Erich Brandenburg, *Von Bismarck zum Weltkriege* (Berlin: Deustche Verlagsgesellschaft fur Politik und Geschichte, 1925). For Bismarck and the colonies, a later study under the *Grosse Politik* influence is A.J.P. Taylor, *Germany's First Bid for Colonies, 1884–1885* (London: Macmillan, 1938).

isolation and lurch into war. It was the wise Bismarck who kept the peace against the foolish Kaiser who put Germany's head into a noose.

This was history that fit the times. The Kaiser's navy was out of the way, so Germany could be reconsidered in the light of the prior history. Sympathy for the French cause in Britain and America was waning rapidly. Almost from the time of the Armistice, there was a falling out between the British and the French. British leaders started to view the German situation with more understanding. The British differed with the French over the division of claims in the Mideast, quarrelling about the oil-producing area of Mosul in northern Mesopotamia. France deserted the British cause in its struggle against Kemal in Turkey. When Lloyd George resolved on war in 1922, French leaders opted out and even helped Kemal. A year later, when the French invaded Germany to collect their payments in kind, agreed by the German authorities but not promptly paid, it was the turn of the British to refuse their support. Poincaré himself was back in power by then, threatening by his occupation the possible breakup of Germany. French historians to this day are convinced that the British forced the French to give up the Ruhr occupation by causing a run on the French currency, selling francs on the Amsterdam exchange.[6] In the era of reconciliation after the Locarno treaty of 1925, the French were still regarded by British and American opinion as excessively belligerent. This continued right up to the time when Hitler came to power, and even then the German threat only eclipsed the presumed threat from France by degrees.

The revisionist historians' case was that Poincaré was the cause of it all in 1914 and again in 1923. As the Versaillists studied the same documents on the crisis that unleashed the war, they began to complain that Fay, Barnes and Gooch relied too much on secondary sources. That set the stage for the last round of the historians' controversy. In 1928, Fay published *The Origins of the World War* in two volumes. He built upon the original argument, based on the Kautsky documents and the *Austrian Red Book*, that the War Guilt Clause of the Treaty of Versailles was without foundation. Bernadotte Schmitt, for the Versailles position, answered in 1930 with *The Coming of the War*, which earned him a Pulitzer Prize. He argued that Germany and Austria tried to take Serbia by surprise with the aim of rearranging the Balkans. Everyone made terrible miscalculations, said Schmitt, but Germany in the end counted them all out by insisting on carrying out the war according to the strategy of the Schlieffen Plan, which called for an attack on France prior to the attack on Russia: A defense of the Versailles treaty. Winston Churchill called Schmitt's work "a masterly book which made the anti-Versaillists sick at heart."[7] The great scholars' duel yielded an ambiguous result, however. Neither Fay nor Schmitt had

6 Jean-Baptiste Duroselle, "Entente and Mésentente," in Douglas Johnson and François Crouzet (eds), *Britain and France: Ten Centuries* (Folkestone, Kent: Wm. Dawson and Son, 1980), 278. Jean-Noel Jeanneney, "De la spéculation financière comme arme diplomatique," *Rélations Internationales* 13 (1978), 5–27.

7 Quoted in Michael Cochran, *Germany: Not Guilty in 1914* (Boston: Stratford, 1931), 4. Sidney B. Fay, *Origins of the World War*, 2 vols (New York: Macmillan, 1929). Bernadotte Schmitt, *The Coming of the War, 1914*, 2 vols (New York: Scribner, 1930).

blown his adversary out of the water. Those who had argued for one side found that they were unmoved by all the new citations of evidence by the other. One charitable interpretation was that scholars, when they present evidence honestly, do not select only those facts that make their case, but in effect make it possible to use their evidence to argue for something other than what they intended. But it was surprising to many that earnest study of the same documents could yield two opposite conclusions. The real lesson may have been that evidence, even primary material from archives, does not always speak for itself.

The controversy over the war guilt of the Germans petered out right after the great scholars' duel. The depression set in. Germany stopped making her reparations payments. President Hoover declared a "moratorium" in 1931 that turned out to be the effective end of the regime of reparations and war debts, even if the United States did not want to acknowledge the fact formally. When reparations ended so did the German innocence propaganda. The journal *Kriegsschuldfrage* closed down and the historians turned to other things. The era of the *Grosse Politik* was over.[8] It had apparently been driven in the main by German opposition to the reparations regime.

The parallel literature on the origins of the American entry into the Great War had a slightly longer run. Walter Millis, Charles Callan Tansill and others produced very influential accounts of the American entry, stressing the enmeshment of American policy by the British blockade. The US sold to Britain; Britain needed credit; the loans were made; British victory became an American interest; the American troops were sent. In the words of John dos Passos in his novel *1919*, it was "a war for the Morgan loans."[9] Millis's book appeared at the time of the Congressional hearings into the same question by the Nye committee, which came to much the same conclusion. It was a powerful influence on Roosevelt and a prompting for neutrality legislation.

However, FDR was to turn against all of this. He rejected with vehemence the isolationist perspectives that seemed to follow. He invented his own kind of non-intervention policy, one based on naval and air rearmament. He sold the public on the idea that defense spending safeguarded the peace and that active support for other nations who fought fascism would keep war from America's door. We are so used to this sort of argumentation from the era of the Cold War and nuclear deterrence that we often fail to appreciate how unique Roosevelt's presentation of it was in the years before Pearl Harbor. Nevertheless, it did represent turning one's back on revisionism, isolationism and appeasement. So much so that after the war foundation support was ample for works sympathetic to Roosevelt's viewpoint and the American policy, to a great extent in the fear of a new school of revisionism

8 For a fuller account see Anthony D'Agostino, "The Revisionist Tradition in European Diplomatic History," *Journal of the Historical Society* 4: 2 (Spring 2004).

9 Walter Millis, *The Road to War: America, 1914–1917* (New York: Houghton Mifflin, 1935) and *Viewed without Alarm* (Boston: Houghton Mifflin, 1937); Charles Callan Tansill, *America Goes to War* (Boston: Little, Brown and Co., 1938); C. Hartley Grattan, *Why We Fought* (New York: The Vanguard Press, 1929).

developing on the causes of the Second World War. Harry Elmer Barnes, who personified the continuity of the revisionist trend on both world wars, complained of an official "blackout."[10] Barnes's complaints seemed to underscore the influence of the revisionism of the twenties. A.J.P. Taylor said that "The *Grosse Politik* was a great political stroke. It was not the least of the factors which made possible Hitler's destruction of the Versailles system."[11]

The time of the great diplomatic historians continued to cast a spell over the next generation. A.J.P. Taylor himself, of all people, ended the "blackout" and created a school of revisionist historians on the Second World War with his *Origins of the Second World War* of 1961.[12] A reconsideration of appeasement, Taylor's book stressed not Hitler's intentions on the war path of 1933–39, but accidents and miscalculations, of which there were many. The idea was to use the same method that had been so brilliantly used for the origins of the Great War. There was a definite element of accident at work in 1938–39, minor in comparison with 1914, on which Taylor played fascinating variations. But one suspects that there may also have been a stronger aesthetic imperative to recapitulate the magnificent successes of the earlier literature.

As Taylor wrote, John F. Kennedy was entering the most dangerous phase of the Cold War, with international crises that brought the world closer to nuclear war than it has ever come, in Berlin in 1961 and Cuba in 1962. Kennedy, who led the world through these crises, was mindful, to say the least, of the trope of accidental war and the presumed cause of 1914. He sent copies of Barbara Tuchman's *The Guns of August* to his staff.[13] Theorists of international relations thought they saw in the nuclear balance of that time the menace of the "security dilemma." War might not need a hostile intention but might come as the result of defensive moves against perceived threats, moves understood by the other side in the opposite spirit, as threats, and responded to by fresh defensive moves (threats). Windows of opportunity for preventive war might be perceived. War might come despite the original intentions of the actors. Or, just as bad, the whole syndrome might be perceived by one or more of the actors who might try to game the situation. Kennedy's deft handling of the crises over Berlin and Cuba suggested a model of happy crisis management. As to the anti-model, there was 1914.[14]

In fact, the nature of the nuclear weapons of the period was such as to reinforce this idea. America was depending on an intermediate-range ballistic missile (IRBM) deployment on encircling bases in Britain, Italy and Turkey, as response to the presumed buildup of Soviet inter-continental ballistic missiles (ICBMs) that

10 Harry Elmer Barnes, *Perpetual War for Perpetual Peace* (Caldwell, ID: Caxton, 1953).

11 A.J.P. Taylor, *The Struggle for Mastery in Europe, 1848–1918* (Oxford: Oxford University Press, 1957), 573.

12 A.J.P. Taylor, *The Origins of the Second World War* (London: Penguin, 1961).

13 Barbara Tuchman, *The Guns of August* (New York: Macmillan, 1962).

14 Morton Kaplan, *System and Process in International Politics* (New York: John Wiley & Sons, 1957); Kenneth Waltz, *Man, the State, and War* (New York: Columbia University Press, 1959); Raymond Aron, *Paix et guerre entre les nations* (Paris: Calmann-Lévy, 1962).

was expected to produce a missile gap. IRBMs were vulnerable themselves, so they would have had to be used, perhaps launched on warning, to avoid being taken out by a surprise attack. Two books, penned at the start of the nuclear era, had described the alternatives for states that owned nuclear weapons. Bernard Brodie, in his 1946 book *The Absolute Weapon*, had argued that nuclear weapons were so terrible that no military mission could be served by them; they were only good for deterrence. Two states having them would be naturally deterred from attacking each other. The other book appearing that same year was William Borden's *There Will Be No Time*, with the opposite message: nuclear weapons were like other weapons which had always been used in counterforce missions: counter-battery fire for artillery, tank battles, duels of battleships in the era of the great navies, duels of fighter aircraft. Nuclear weapons would always be used against other nuclear weapons and would be used to try to surprise each other: "There will be no time." One could argue that the history of nuclear weapons has seen an alternance of the paradigms of Brodie and Borden, periods of seemingly automatic deterrence and periods when weapons firing counterforce missions threatened surprise attack. The time from the Soviet ICBM test in 1957 and the US IRBM strategy to the mid-sixties was a time when Borden's picture was the more vivid.[15]

In the discussion of the security dilemma, 1914 was reduced to a case study of a generalization, however topical, to be expressed in the present tense. This is the inevitable tension between international history and international relations theory. The former is seeking to discern the actual movement of unique moments across a graph and the logic of their path. The latter is not interested in the uniqueness so much as the aspect of recurrence and potential material for analogy. The two can coexist, as Marc Trachtenberg and Robert Jervis have argued persuasively.[16] Yet reduction to the status of a case study causes the Great War to lose its personality. One almost sees it entirely as a parable of the folly of strategic calculation. The history gets rather lost. Did it have anything to do with imperialism? During the Cold War this idea was dismissed as Leninist propaganda. Raymond Aron, in his *Century of Total War*, weighed the argument and then laughed it off. The Portuguese colonies and the Berlin to Baghdad Railway were the central issues between England and Germany and they were all settled in the months prior to the war,[17] 1914 was cleaned up and polished for service to an era of confrontation.

15 Bernard Brodie, *The Absolute Weapon: Atomic Power and the World Order* (New York: Harcourt, Brace and Co., 1946); William Borden, *There Will Be No Time: the Revolution in Strategy* (New York: Macmillan, 1946). On the alternance, see Gregg Herken, "The Not-Quite-Absolute Weapon," in Roman Kolkowicz (ed.), *Dilemmas of Nuclear Strategy* (London: Routledge, 1987), 15–22.

16 Marc Trachtenberg, *The Craft of International History: a Guide to Method* (Princeton: Princeton University Press, 2006); Robert Jervis, "International Politics and Diplomatic History: Fruitful Differences," H-DIPLO, 1 June 2009. On the security dilemma, Robert Jervis, "Cooperation under the Security Dilemma," *World Politics* (January, 1978).

17 Raymond Aron, *The Century of Total War* (Boston: Beacon Press, 1960), Chapter 3; on the other hand, Karl Polanyi, *The Great Transformation* (1944) (Boston: Beacon Press, 1967), Chapter 1.

A few years later it was called to service in another capacity. Confrontations with the Soviets lost the center-stage place they had occupied during the Cold War, 1947–63. The Soviets seemed to retire from international life, "acting like a second rate power," in General de Gaulle's pithy phrase. American struggles for civil rights, the Cuban revolution, the Cultural Revolution in China and the Vietnam War rose up in place of the superpower confrontation. Entirely different questions occupied serious minds. Alienation even in the midst of prosperity was found to exist. The international questions that were really interesting, it was thought, were not the ones involving war and peace among the nations, except insofar as those questions can be found to have roots elsewhere. Barbara Tuchman came back to the origins of the First World War in an entirely different mood. In her luminous volume about the turn of the twentieth century, *The Proud Tower*, she returned to the causes of the cataclysm that introduced the Russian revolution, Communism, fascism. "The *Grosse Politik* approach has been used up," she said, "the diplomatic origins, so-called, are only the fever chart of the patient. They do not tell us what caused the fever."[18] The Great War could only be understood as the culmination of a terrible crisis of European culture, rent by class struggle, violent racial nationalism, and other kinds of spiritual turmoil. It had not popped up suddenly, "like a rut in the road," in the famous phrase of Virginia Woolf. It was part of a continuum of crisis.

Communism and fascism were found to have their roots in the prewar period. True enough, prior to the war there had existed Lenin, Trotsky, Rosa Luxemburg and a left wing of the parliamentary social democracy, revolutionary syndicalism in France, Spain and Italy. If you read novels such as Thomas Mann's *The Magic Mountain* or Hermann Hesse's *Demian* you could see the unrest of western civilization and its dissatisfactions. Part of this unrest might reveal a primal love of war among men who were to find life at the front so intoxicating that they could not accept a return to ordinary life.[19] When you contemplated the movement against the Vietnam War, you thought of the pre-1914 anti-war politics of Jean Jaurès. Politicians then, as in the late sixties, must have realized that their positions were vulnerable before the onslaught of mass movements. One had to take a second look at the decisions for war or imperial adventures in terms of the decision makers' fear of threats to their own power from below. Decisions for war must be considered as a kind of *fuite en avant*. Had Tirpitz not spoken of the German navy as a "prophylactic against the Social Democracy"? Had Cecil Rhodes not spoken of imperialism as an outlet for the poor of London? Did the Tsar not make his crucial decision for mobilization with the Saint Petersburg workers fighting against police across barricades?[20] Even the older generation of historians had already turned to

---

18 Barbara Tuchman, *The Proud Tower* (London: Hamish Hamilton, 1966), xvi.

19 Robert Wohl, *The Generation of 1914* (Cambridge, MA: Harvard University Press, 1979).

20 Arno Mayer, "Domestic Causes of the First World War," in Leonard Krieger and Fritz Stern (eds), *The Responsibility of Power: Historical Essays in Honor of Hajo Holborn* (Garden City, NJ: Doubleday, 1967); Hans Ulrich Wehler, *The German Empire, 1871–1918* (Leamington Spa, Dover, NH: Berg, 1985); Fritz Fischer, *World Power or Decline: The Controversy over German War Aims in the First World War* (New York: W.W. Norton,

the same thought. William L. Langer's Lowell lectures at Harvard in 1940, "The Conflagration of Ideas in Prewar Europe," reflecting on the Great War, focused on new revolutionary ideas, integral nationalism, anti-Semitism and cultural despair, developments arising ultimately, Langer thought, from the conflict between science and religion.[21] Raymond Sontag had ended his *Germany and England: Background of Conflict, 1848–1894* of 1938 with the thought that "Greater Britain, like Weltpolitik, was in part an escape or, as men thought, an answer to domestic discontent."[22] "We must turn," wrote A.J.P. Taylor, "from the foreign offices to the more profound forces which shape the destinies of men."[23]

But these passions soon subsided. The Vietnam War became history. Maoism turned into an ally against Soviet Communism and eventually a medium of capitalist restoration, of a sort, in China itself. The Cold War disappeared when the Soviet bloc collapsed. The Great War of 1914–18 continued to be the subject of study for international historians of old and new schools of thought. But it was passing from European history into world history or, more broadly, being judged as a moment when European history was being subsumed into world history. A reflex of those who questioned the old assumption that European history was a gateway to the world was to assume the opposite, that it was a kind of barrier. College freshmen were no longer required to take their existential Western Civilization course but were offered world history instead. Many, especially those who were not European by descent, saw no harm in this substitution.

How did the Great War look to them as world history? In order to address this it would be necessary to consider some of the non-European reasons for the clash of the powers in 1914, specifically their expansionist policies in Africa, the Middle East, Asia and the South Pacific. The blunders and misjudgments of the path to war seem almost incomprehensible if they are seen to be driven only by alliance commitments and the vagaries of the railroad timetables. Continental quarrels such as Alsace-Lorraine were things that French and Germans had lived with for over forty years. When A.J.P. Taylor chalked them up to a struggle for mastery in Europe this made no sense in a purely continental context.[24] It must have been a question of mastery over more than Europe. The diplomatic historians educated in the *Grosse Politik* tradition had taught their undergraduates that European powers saw their national interests in terms of a minimum security requirement with only vague and secondary commitments outside Europe. On the contrary, it makes better sense to consider the opposite proposition, that they were all more

1974); D.C.B. Lieven, *Russia and the Origins of War in 1914* (New York: St Martin's, 1983); Volker Rolf Berghahn, *Germany and the Approach of War in 1914* (New York: St Martin's, 1973).

21 William L. Langer, *In and Out of the Ivory Tower* (New York: Neale Watson Academic Publications, 1977), 178–9.

22 Raymond Sontag, *Germany and England: Background of Conflict, 1848–1894* (New York: Appleton-Century, 1938), 341–2.

23 A.J.P. Taylor, *Englishmen and Others* (London: Hamish Hamilton, 1956), 84.

24 A.J.P. Taylor, *The Struggle for Mastery in Europe*.

or less deliriously expansionist on a global scale. Even if Britain and France came to consider themselves sated with their huge gains after the war, others who had not done as well, seeing themselves as "have-nots" in the great colonial scramble, maintained the delirium.

The background of the war was European to be sure, as the studies of Bernadotte Schmitt and Charles Seymour had explained in lavish detail.[25] But the global background fills out the picture. The powers intersected when the Far Eastern Triplice (Germany and the Franco-Russian alliance) restrained the Japanese after the latter's victory in the Sino-Japanese war of 1895. As compensation they staked out their claims to territories in China. The British answered with the Anglo-Japanese alliance of 1902. With Britain at their backs, the Japanese defeated Russia in 1904–05. Britain headed off war with France, the ally of Russia, in the Anglo-French entente of 1904. But in doing so she promised Morocco to the French, to which Germany objected. In addition to settling the war in the Far East, the Moroccan question had to be settled as well. This was done at Algeciras in 1906, where a new alignment emerged, with the British, French, Russians and Americans arrayed against Germany and Austria-Hungary. This was the lineup that went to war in 1914. The Germans drew up their Schlieffen Plan for war against the Franco-Russian alliance. If you assume that the alliance system was a cause of trouble in 1914, you have to ask how it originated. In fact it came out of the Far Eastern war and the world crisis of 1904–06.

It only remained for this alignment to find a way to its rendezvous in Europe at Sarajevo. This came out of the Russo-Japanese war and the Russian revolution of 1905 that resulted from the Russian defeat. This revolution inspired the Persian revolution of 1905–06. The Russian and Persian events and fear of the Anglo-Russian entente, designed to curb the enthusiasm of the Persians, inspired the Turks to overpower the Sultan in the Young Turk revolution of 1908. Austria and Russia tried to take advantage of the Turks. Austria seized Bosnia, Russia retaliated by encouraging the Balkan League in the first of the Balkan wars. Thus, the Far Eastern crisis, by this chain of events, transferred the conflict around the globe to Europe. This was not the only source of the contagion. The French seizure of Morocco prompted the Italian seizure of Tripolitania (Libya), which prompted the Greeks and their Balkan allies to action against the Ottoman Empire. This was, in effect, a second chain of annexations leading to the Balkans and Sarajevo.[26]

---

25 Bernadotte Schmitt, *Triple Alliance and Triple Entente* (New York: H. Holt and Company, 1934); Charles Seymour, *The Diplomatic Background of the War, 1870–1914* (New Haven: Yale University Press, 1916). Seymour's book went through ten printings by November 1918.

26 This is advanced in more detail in Anthony D'Agostino, "Global Origins of World War One: the World Crisis over Concessions in China," *Historia Actual* 12 (2007); "Global Origins of World War One: a Chain of Revolutionary Events across the World Island," *Historia Actual* 13 (2007); and in *The Rise of Global Powers: International Politics in the Era of the World Wars* (Cambridge: Cambridge University Press, forthcoming).

The annexations continued during the war. Britain and France took the German colonies in Africa and divided the Ottoman Arab possessions in the Sykes-Picot agreement of 1916. Japan took the German possessions in the Pacific; Australia took German New Guinea; New Zealand took German Samoa. Other promises were made in the secret treaties. The world history texts nevertheless often assumed that the *Grosse Politik* conception of the Great War was still in full force. The renowned world historian William H. McNeill wrote confidently that "World War One broke out by accident," owing to the alliance system and to the railway timetables. You would hardly know from his account that anything outside of Europe was involved.[27] This is also the case with Eric Hobsbawm, *The Age of Extremes*, even with its fine attention to the social and cultural crisis of European civilization.[28] The competition for McNeill was the textbook of Leften Stavrianos, *A Global History*, where one finds, in a listing of factors contributing to the great war, that colonial rivalries repeatedly clashed all over the world and were a motive forging the alliance system.

World systems theory in the work of Immanuel Wallerstein put the Great War into the historical context of a succession of hegemonies, each of them secured by a thirty-year-long world war. World War Alpha was the Thirty Years war, 1618–1648, when the Dutch won their complete independence from the Spanish and for a short time essayed a career as a great naval power. World War Beta was the Napoleonic period, 1792–1815, when the French bid to unseat the British failed and the great globalization of the world under the British pound ensued. As to the last one, "World War Gamma was the long Euroasian wars from 1914–45 when US interests triumphed over German."[29] It is a choice of successors to Britain that figures as the main stake in the world wars. In effect, as the Germans liked to think of it, the Wars of British Succession. Was this the stake in the Great War? Was Britain already finished as a hegemon? Her role as a workshop of the world was on the wane by the time the Great War broke out, but her role as a clearing house of the world economy was far from exhausted. We are now getting the idea that this is not at all good enough. If you are no longer a great producer and exporter, you cannot make it up by being a great *rentier*. Eventually the successor power who assumes the role of the great producer will get you by the scruff of the neck. Thus it was with the fading Britain and the rising Germany and America. World Systems Theory says it is the same with the America who cannot yield the role of the world's workshop to China, and still rule the roost through financialization. In this perspective, the Great War was not so crucial. To be sure, territorially Britain was at a high point at the end of the war. It had its chain of African lands for a Cape-to-Cairo railway, the dream of Cecil Rhodes. It had made protectorates of the key Middle Eastern oil-producing countries. It retained, on account of its alliance with Japan, its possessions in the

27 William H. McNeill, *A World History* (Oxford: Oxford University Press, 1971), 482–3.

28 E.J. Hobsbawm, *The Age of Extremes: A History of the World, 1914–1991* (New York: Vintage, 1994), 24.

29 Immanuel Wallerstein, "The Three Instances of Hegemony," in *The Essential Wallerstein* (New York: New Press, 2000), 258.

Far East. The end of *Pax Britannica* was not in the world war, one might think, but in the great depression. This is the view of Giovanni Arrighi, who passes right by the Great War and goes directly to the Great Depression for the benchmark dates of the end of the "Age of the Rothschilds, 1866–1931."[30]

And let us not forget to note the importance of the Great War for the War on Terror, in which we have been engaged since the events of September 11, 2001. The Great War is the beginning of the current phase of the jihad, which owes, at least in the view of Osama Bin Laden, to the depressed condition of the world's Moslems since the fall of the Ottoman Caliphate. It all presumably comes down to one decision, the decision of the Sultan Mehmed the Fifth to throw the dice and enter the war as an ally of Imperial Germany. It brought about the final collapse of the empire and its transformation into modern Turkey. It created artificial states dividing the Islamic world against itself. It let loose various Muslim tendencies striving to restore the greatness lost in the Great War.[31]

No matter how we interpret its meaning, the Great War continues its rule over our minds. We do not know if it belongs merely to European history or to World History, or to both. Yet we need it as a chapter, not only in our account of the end of the story of the European great powers, colonialism, absolute monarchy, the *Pax Britannica*, but also for the beginning of the story of Communism, fascism, the American century, the second globalization, the rise of the non-European world. The world historian K.M. Pannikkar remarked that the Great War was not so much a world war as "from the Asian point of view a civil war within the European community of nations."[32] The scope and dimensions of the tragedy, or even whether it is a tragedy at all, will continue to depend, as always, on the teller of the tale.

30 Giovanni Arrighi, *The Long Twentieth Century: Money, Power, and the Origins of Our Times* (London: Verso, 1994), 169.

31 Anthony Pagden, *Worlds at War: The 2,500-year Struggle between East and West* (New York: Random House, 2008), 454–5.

32 Leften Stavrianos, *A Global History: From Prehistory to the 21st Century* (Englewood Cliffs, NJ: Prentice Hall, 1983), 443.

15 ASHGATE RESEARCH COMPANION

Totalitarian Times—Total War, Global War: The Roots of World War II and the Nature of the Conflict

Oleg Kobtzeff

Fifty to sixty million dead, maybe more. Almost a decade of warfare: the conflict began in 1937 for the more than half-billion Chinese invaded by the Japanese. World War II thus remains to this day, and hopefully (if humanity is responsible enough not to renew the experience) will forever remain, the war that inflicted devastation on the largest scale in human history. Not one population on the planet (except for some rural communities or tribes living in extremely remote areas) was able to escape direct or indirect involvement in the conflict. The intensity of danger varied between two extremes. The highest degree of violence fell upon the victims of genocide and organized mass murder in Europe (between 6 and 7 million Jews, 500,000 Gypsies, plus 6 to 7 million others, executed as Slavs, members of other ethnic or sexual minorities, political dissidents), in Asia (3 to 10 million Chinese or other Asian and European civilians, victims of massacres, forced marches, forced labor and atrocious conditions in concentration camps)[1] and, finally, soldiers made prisoners of war by the Japanese or Soviet armies, who did not apply the Geneva conventions.

Among those who also suffered the most were the inhabitants of martyred areas like the ghetto of Warsaw or the too often forgotten Leningrad (900 days of siege, one million victims of famine and combat) or of heavily bombed cities like Warsaw, Rotterdam, London, Stalingrad, many German cities, Hiroshima, Nagasaki and hundreds more. On the other hand, far away from the combat zones, others only suffered from fear of an extension of the conflict to their area. But even neutral countries such as Switzerland, Ireland, Portugal, Sweden and

1 Yuki Tanaka, *Hidden Horrors: Japanese War Crimes in World War II* (Boulder, CO: Westview Press, 1996), 72–73.

several Latin American countries had cultural, diplomatic, political and especially economic ties to at least one of the belligerents. Exchanges between the neutrals and the Axis or the Allies were sometimes even accelerated because of the war; no society on the planet remained unaffected by what was far more than the "Great War" of 1914–18: it was a real *world* war. The expression appeared at the time and, interestingly, the word "globalization" also appeared then (it entered the Webster dictionary in its 1943 edition). With its widespread and intense exploitation of means of rapid transcontinental transportation and communication, World War II was therefore not only a new stage in the history of warfare, it was a crucial conjunction of social, geostrategic, economic, technological and cultural forces which created a geopolitical situation defined by total warfare, the confrontation between totalitarianism and the resistance against it, and globalization. It was a continuation of many of the dynamics at play during World War I, and it set in motion the mechanisms of the Cold War. Thus, it defined the very civilizations in which we lived today.

One of the characteristics that World War II shares with the world in which we live is its intensive coverage by modern mass media, which explains the proliferation of professional and amateur historians writing on the subject: the primary sources are inexhaustible. But that is also what makes it so difficult to sort out the essential characteristics of that conflict from a jungle of more and less important data, and to briefly assess what was really specific and historically significant about that conflict. This is what will be attempted here.

The Classical Approach: International Relations and Geopolitics

Despite their limitations, political science and diplomatic history should not be neglected now that broader ranges of social sciences have opened new horizons in World War I and World War II historiography. Works such as those written immediately before or immediately after the war by the likes of E.H. Carr, Hans Morgenthau, George F. Kennan or Pierre Renouvin, are still indispensable because no aspect can be neglected in attempting to explain a horror of the dimensions of World War II. It should not necessarily contradict the long-term geographical, economic and cultural analytical approach of the *Annales* school. Marc Bloch, one of its founders, had a long career behind him when he reenlisted as a 53-year-old captain and witnessed the 1940 debacle of the French army. In contrast to everything that the *Annales* school was known for, he wrote about that war. But his *Strange Defeat* was still typical of the interdisciplinarity that he had defended for decades—it was his interpretation of the French defeat of 1940. Focusing on the dysfunctional command structures of the French military and political communities, he presented them as a model of the entire social organization of French society, with deep incursions into domains as varied as education, culture and mentalities, communication theory, logistics, and management theory.[2] After joining the Resistance, Bloch wrote several articles for the underground press, only months before being arrested, tortured and executed.[3] They are all about war, politics and international relations. Bloch has proven that social scientists focusing on anthropological, sociological, cultural, economic and geographic-environmental facts will not betray their disciplines if they cross the threshold of diplomatic and military history.

From Versailles to Munich: "The war that should have had a great peace"[4] and didn't

E.H. Carr's recollections of the times of demobilization following the "Great War" illustrate how certain elites may have completely failed to assess the magnitude of the catastrophe that had just happened. For Carr, at first, "even the shock of the First World War and the Bolshevik Revolution did little to disturb" the balance of "a sheltered environment," a world "solid and stable," where "prices remained the same, where everybody knew their station in life, in which things changed only slowly and then only for the better."[5] Despite the global upheavals, Carr recalled how, in his social class, few "had thought of it as the end of an epoch."[6] For Realists, it was back to the routine of international intrigue and games; they did not bother to detect anything more in Nazism than classical nationalism. How then could allied intelligence services have functioned properly during the war? asked Marc Bloch in 1940, since among elites the taste for being accurately informed was waning: if they had read *Mein Kampf*, they had been too lazy intellectually to find in it anything unusually dangerous. They covered their "ignorance with the delightful word 'realism.'"[7] Liberals believed that 1914–18 was only an accident, terrible indeed, but so terrible that the naturally good human nature would never let it happen again. The result was insufficient vision and insufficient planning.

This near-sightedness of elites progressively dug a rift between the decision-makers and the rest of the population. While government elites in Europe, the Americas or Japan managed international relations as if they lived in the nineteenth century, populations living in the twentieth had to deal with rapidly accelerating technological and social change already forestalling the 1950s. This rift, plus lack of confidence in leadership created by the economic and social crises (which are

2 Marc Bloch, *L'Étrange Défaite* (Paris: Gallimard, 1990) is one of the latest editions. It includes several articles written by Bloch for the underground press.

3 Published in ibid., 213–68.

4 Isaac Frederick Marcosson, *Turbulent Years* (Manchester, NH: Ayer Co., Facsimile of 1938 edition, 1969), 326 (quoting Clemenceau).

5 Michael Cox, "Introduction," in E.H. Carr, *The Twenty Years' Crisis, 1919–1939* (New York: Palgrave MacMillan, 2001), xiv.

6 Quoted in ibid.

7 Bloch, *L'Étrange Défaite*, 180.

discussed below), would eventually divide nations and create tensions leading to both domestic and international instability. Fear of a new catastrophe was legitimate, but too often this fear either encouraged pacifism or, on the contrary, militaristic nationalism, which led to increasingly fearful or sympathetic attitudes toward either fascism or communism or both. Opinions were often volatile; it is a subject that has not been fully measured by historians since World War II, with the exception of authors like the perceptive Henri Amouroux in his extensive history of daily life in France between 1939 and 1945.[8] But it can easily be illustrated by the fluctuating positions of hundreds of famous intellectuals in the 1930s and 1940s, resolutely converting from one ideology to another or hesitating between extremism and moderation.[9] These unstable public opinions, now at the height of their influence in democracies old or new, put a pressure on democratic governments that prevented them formulating coherent foreign policies (as also noted by Marc Bloch). They were aggressive when the most urgent need was to pacify the world in 1918 and 1919, and submissive in 1938 when the survival of millions depended on firmness.

The hope that the armistice and peace treaties of 1919 and 1920 put an end to the Great War was nothing but a *grand illusion* (the title of one of the most famous movies of the 1930s). Not only did the peace treaties resolve none of the causes of conflict, they only aggravated them. First of all, the allied diplomats did not seem to realize how close they themselves had been to defeat. The reason the conflict had dragged on for so long was that the forces present were almost equal. As one of those allied diplomats, Clemenceau, observed, this was a duel to the last man standing. It was only a slight tactical advantage (the use of tanks) that allowed the French and the Anglo-Saxon armies to push back the German offensive of the summer of 1918. But what the near-sightedness of many diplomats prevented them from realizing (even today's textbooks almost avoid the subject) was that the real imbalance was caused by the situation in the rear, where revolutions were shaking the foundations of society. The fierce repressions of the mutinies in the French army in 1917 gave the French government a slight tactical advantage. The peace treaties of Versailles, Saint Germain, Neuilly (1919), and of the Trianon, and Sèvres (1920), were meant to *punish* those who had asked for an armistice, not to establish good relations with them. The most extremist among the allies, essentially the French, intended to break the back of Germany's and Hungary's economies, to wipe out competition, which was almost openly demanded by

8 Henri Amouroux, *La Grande Histoire des Français sous l'Occupation*, 10 volumes (Paris: Robert Laffont, 1976–1993). Amouroux, who pursued a career as a journalist under the Vichy regime, was himself not above suspicion, but as historian and Nazi hunter Serge Klarsfeld declared about the quality of his monumental work, the honesty of the scholar finally prevailed over his personal ambiguities (quoted in Laurent Joffrin, "Fin d'histoire pour Henri Amouroux," *Libération*, August 7, 2007, accessed May 17, 2011: http://www.liberation.fr/culture/0101108555-fin-d-histoire-pour-henri-amouroux).

9 Including Dos Passos, Céline, Bernanos and many others.

Clemenceau.[10] And Clemenceau was considered too weak by the military and the right wing of his legislature.

This was to be even more fatal than the marginalization of France by Bismarck after the victory of Prussia in 1871, for not only does the marginalization of a country force it to seek to destabilize a peaceful international community in order to polarize it and offer itself as an ally to one or another faction (one of the main causes of World War I), but this time there was not even the Realist ambition of establishing some sort of system that would stabilize international relations comparable to what Bismarck had done in 1878.

To be fair, the League of Nations was not as impotent as was suggested at the time, nor as textbooks present it today. In diplomacy, successes in peacemaking are far less spectacular, and therefore less noticeable, than violent events. Among the accomplishments of the organization were the resolutions of dangerous situations such as the Aland crisis between Finland and Sweden, the prevention of war between Greece and Italy over Corfu, the Turkish-Iraqi dispute over Mosul and the 1925 conflict between Greece and Bulgaria; it brokered numerous multilateral agreements, organized and influenced the great world economic summits of Genoa (1922) and Geneva (1927), helped find solutions for millions of refugees (the Nansen passport and relief measures), advanced humanitarian governance and the work of the Red Cross, advanced international law-making and the work of the Permanent Court of Justice (now the ICJ) in the Hague, advanced social and economic rights and the work of the International Labour Organization, and laid the ground for the future development of the European inter-governmental organizations and the United Nations after the war. The League of Nations could have been a tool for peace.

Few had the will to use it, which explains its great malfunctions, i.e. its failure to stop Japanese, Italian, German and Soviet aggressions. This discredited the League of Nations, as a weak and useless institution, and, by extension, all efforts for peace. Yet peace existed as an option had there been a political will: the Common Market was created only twelve years after the war, with many of the same diplomats who could have done more for peace in the 1920s and 1930s had they been prompted by a greater sense of urgency or had they been supported by populations unanimously committed to reconcile and rebuild.[11]

The evolution from the arrogant triumphalism of the victors to the cowardice shown in Munich can be explained by the inner contradictions of democratic societies. France, Belgium, Luxembourg, the Netherlands, the Scandinavian countries and Great Britain were torn between anti-German or anti-Soviet sentiment expressed

10 See John Maynard Keynes, *The Economic Consequences of the Peace* (New York: Harcourt, Brace and Howe, 1920), Chapter V: Reparation.

11 Not to mention scores of lesser-known government members and diplomats. Jean Monnet was present at the 1919 Paris conference of the "Big Four," before Versailles, and became, at age 31, the first Deputy Secretary General of the League of Nations; de Gasperi was a legislator in 1921; Schuman and Bech as early as 1919, and the latter was Prime Minister of Luxembourg from 1926 to 1937...

by parts of their electorate, and pacifism, expressed by others who were inspired either by fear of war or by sympathy toward either fascist powers or the USSR. The same problem existed in the United States, with additional complications. First of all, the United States hosted many minorities, and especially a population of millions of first-generation European or Japanese immigrants representing the entire political spectrum of their country of origin from Jewish associations to the pro-Nazi German-American *Bund,* from the Knights of Columbus to Hungarian Free Masons and every kind of attitude ranging from hostility to full support of various foreign regimes. Families settled in the United States from much earlier times had great difficulty understanding the complex geopolitics that could raise passions among more recent communities. The perplexity and fear of "old country" politics spilling over into American affairs, added to the terrible experience of World War I (a fraction of the death toll of European belligerents, but a six-digit figure in eighteen months, nevertheless) certainly explains in part the isolationism adopted by the Republican governments of the 1920s (Harding, Coolidge and Hoover). Hence the hesitations of French and British foreign policy on the eve of the Munich agreements, and Swedish and American neutrality and ambiguities even during the war, such as the fact that the United States continued to recognize Pétain's Vichy regime until D-Day.

The American journalist Isaac Marcosson, calling for American society to prepare for what we would call today a globalized economy, had warned that "peace will be as great a shock as War." Hence the need for "preparedness to meet the inevitable conflict for Universal Trade. We—as a nation—are as unready for this emergency as we are to meet the clash of actual physical combat."[12] Unfortunately, the international leadership of 1919, and maybe the societies of which they were a product, were not sufficiently committed to understanding the situation they were in. Just as they had been excessively confident in winning a short war in August 1914, these leaders were also unprepared to weather the "Winds of War." Meanwhile, the criminally irresponsible were once again sure that they would win that short war ... and they almost did, as proven by the rapid fall of China in 1937, Poland in 1939, the Netherlands, Belgium, Luxembourg, Norway and France in 1940, and the Philippines.

The solitary rise of Japan

Japanese expansionism was the logical outcome of a geopolitical situation where alternative options were extremely limited for a nation that had been almost completely isolated from the rest of the world and was now developing rapidly. Whatever the ideology or the nature of the regime responsible for Japanese society, peace had been made the most difficult of all choices.

12 Isaac Frederick Marcosson, *The War After the War* (New York: John Lane Company, 1917), 8.

In the middle of the nineteenth century, the Japanese were transformed by what became known as the Meiji revolution, a radical mutation which was not only political but also technological, economic, cultural and geopolitical. Until then, the archipelago had lived in isolation for a great part of its history, especially after the seventeenth century, when it refused any further contact with the outside world which European powers had tried to impose. The extraordinary economic growth that followed this era of industrialization was accompanied by an exceptional demographic growth: there were 35 million Japanese in 1873 and 70 million in 1935.[13] The problem with these figures was that Japan was extremely limited in natural resources. The spectacular development of Japan and its opening up to international relations and foreign trade allowed, however, a hope to resolve the imbalance. Now, with modern know-how and factories, Japan could export manufactured goods (mostly textiles and canned fish) for urgently needed raw materials and, above all, food and fuel. It had also become possible to build a modern Navy to protect the trade routes and protect the homeland from foreign dangers. Such dangers were clear and present.

Western powers, that had always been perceived as a menace, had not only disseminated their colonies throughout Asia and the Pacific, they were now dangerously approaching the shores of Japan. While the Dutch clung to Indonesia, in the second half of the nineteenth century the British, the Americans, the French, the Russians, and even the Germans (in the Marshall islands) conducted annexations of Pacific islands and coastal regions of Asia at a very rapid pace: India (in its final stage of colonization), Alaska, Malaysia, Burma, Indochina, all of Polynesia and Oceania, and the Philippines (seized by the United States from Spain). China was experiencing unbearable pressure. Between 1842 and 1887, the Qing dynasty was forced to sign a series of treaties—the "unequal treaties"—which, among numerous threats to China's sovereignty, involved the loss of Hong Kong, and the abandonment of any hope of regaining past influence in Siberia when the Amur River basin and the Stanovoi mountains were permanently recognized as Russian. The European powers had started the colonization of China. The Russian Empire, the greatest contender, had obvious territorial ambitions in Manchuria and Korea with settlements dangerously close to Japanese coasts: by 1855, Russian settlers were occupying half of the Kurile Islands and Sakhalin (less than 50 miles off the coast of Hokkaido). But a real *crisis* was set in motion in 1853, when a United States vessel appeared in Uraga Bay, "armed beyond anything that the Japanese could launch, and delivered an ultimatum demanding an opening of the Japanese trade market."[14]

In such a geopolitical environment, for the Japanese to renounce military power would have been suicidal. Isolation or inaction would only have encouraged greater Western expansion. Also, why abandon China to the Westerners? The neighboring

13 Hirakawa Hitoshi and Shimizu Hiroshi, *Japan and Singapore in the world economy: Japan's Economic Advance into Singapore, 1870–1965* (London & New York: Routledge, 1999), 17.

14 Philip Bobbitt, *The Shield of Achilles: War, Peace and the Course of History* (New York: Anchor Books, 1995), 41.

Empire was so weak that its becoming a Western colony was only a question of time. It was logical to ask only two questions: which of the Western powers would conquer China before the others, and wouldn't it be better if Japan seized it before any Westerners? This is the logic behind the Sino-Japanese war of 1894–95 and the Russo-Japanese war of 1904–05. The anti-Russian attitude of Great Britain and the United States, fearing Russian competition in Asia and the Pacific, encouraged Japan in that second war, never taking the new modern nation seriously, and never realizing the consequences.

The constant demographic problem aggravated the tensions between Japan and Western powers. Japan's demographic growth was constantly accelerated by its economic growth. Although statistics were not reliable before the first census of 1920, all estimates agree that the rate of population growth rose from 0.5 percent per annum in 1870 to 1 percent around 1900.[15] An average of 1 million Japanese were being born every year between 1863 and 1920.[16] After World War I, Japan enjoyed an unprecedented favorable position in global commerce: together with the United States, it turned out to be the great winner of World War I and could already be seen as one of the great economic powers of the world. Gains had been spectacular.[17] Japan had sold weapons and equipment of all sorts to the allies while its fleet had profited from the absence of the Europeans in the Pacific during the conflict. Its industrial production progressed by 78 percent, its merchant fleet doubled and numerous other economic indicators literally exploded. Not only was Japan's commercial balance positive but it possessed important reserves of foreign currency. Silk and chemical products were among its best exports and India and the US had become among its two major importers. The abundance of Japanese manpower, its low salaries and excellent management organization made Japan what it still is today.[18] But this could not have lasted.

We will discuss later the consequences of the economic crises of the 1920s and 1930s. The particular Japanese problem was that the policies of protectionism and anti-immigration laws prevented it not only from exporting the kind of products that had made its economic success, but also from feeding the additional mouths being born every year. Without exports the Japanese population was facing unemployment and hunger. One of the solutions was emigration. To avoid overpopulation, the Japanese government considered that it needed to send abroad 100,000 to 150,000 emigrants per year.[19] As many as 700,000 Japanese were

15 Hiroshi Ohbuchi, "Demographic Transition in the Process of Japanese Industrialization," in Hugh T. Patrick and Larry Meissner (eds), *Japanese Industrialization and its Social Consequences* (Berkeley, Los Angeles and London: University of California Press, 1976), 333. See eventually (for its bibliography) Carl Mosk, "Demographic Transition in Japan, 1920–1960," *The Journal of Economic History* 38 (1978), 285–6.

16 Hiroshi Ohbuchi, "Demographic Transition in the Process of Japanese Industrialization," 330.

17 Michel Crouzet (ed.), *Histoire générale des civilizations,* Vol. VII. *L'époque contemporaine. A la recherche d'une nouvelle civilization* (Paris: P.U.F., 1957), 48–9.

18 Ibid.

19 Pierre Renouvin and Jean-Baptiste Duroselle, *Introduction à l'histoire des relations*

living overseas by 1934. But the doors of host countries progressively closed in the first quarter of the twentieth century. New Zealand in 1900 and Australia in 1901 circumvented the empire's relatively liberal laws on immigration by imposing a literacy test in the Latin alphabet, preventing non-Europeans (or Greeks and Slavs) from entering the country.[20] In 1908, Canada restricted Japanese entries to only 150 immigrants per year.[21] After many discriminatory laws, the United States finally closed its borders completely to Japanese immigration in 1924.[22] Brazil, that had accepted between 25,000 and 30,000 Japanese immigrants between 1924 and 1934, also imposed drastic restrictions.[23] By 1934, no less than forty nations had imposed a ban or severe restrictions on Japanese immigration.[24]

It was only legitimate that the Japanese perceive these foreign attitudes as racist, especially after the principle of racial equality demanded by Japan at the 1919 Paris peace conference was vetoed by the great powers. Barriers against trade and migration not only created economic hardship, they humiliated Japanese public opinion. The Japanese press carried this indignation in great campaigns which put serious pressure on the government. In July 1925 and 1927, debates organized by the Institute of Pacific Relations failed. When the Japanese delegates presented the quandary they were in, Western powers replied with Malthusian arguments, accusing Japan of "demographic unpreparedness."[25] Such attitudes then fueled the discourse of nationalist and militaristic agitators.

This no-win situation in which Japan was stuck is a perfect argument in favor of Gaston Bouthoul's Malthusian interpretations. The scholar who coined the term *polemology* profoundly believed that problems of overpopulation and resource scarcity constitute one of the most important factors of war.[26] Robert Kaplan says nothing less today.[27] But let us beware: Malthusianism was part of the problem in the 1920s. By refusing the solution of free trade, Western powers postponed the resolution of Japan's problems until the 1950s or the 1960s, when it became obvious that high development brings fertility rates down—trends already perceptible in the 1930s or even the 1920s.[28] Malthusianism became a self-fulfilling prophecy as it boosted the popularity of German theories of vital space (*Lebensraum*) in Japanese scholarly circles and beyond.

---

internationales (Paris: Armand Colin, 1990), 62.

20 Ibid.

21 Ibid.

22 Ibid., 61.

23 Ibid., 62.

24 Richard Dubreuil, "Japon," in Serge Cordelier (ed.), *Le dictionnaire historique et géopolitique du 20e siècle* (Paris: La Découverte, 2002), 381.

25 Renouvin and Duroselle, *Introduction à l'histoire des relations internationales*, 63.

26 On the legacy of Bouthoul, see the General Introduction to this volume.

27 Robert Kaplan, *The Coming Anarchy* (New York: Random House, 2001), Chapter 1.

28 See Ohbuchi, "Demographic Transition in the Process of Japanese Industrialization," and Mosk, "Demographic Transition in Japan, 1920–1960."

Backing the former Axis into a corner: The recipe for its revival

The Versailles Treaty's harsh punitive measures against Germany are well known (loss of all African colonies and of Alsace and Lorraine, occupation of the Ruhr, reduction of the German armed forces to an army of a fairytale kingdom, unreasonable financial reparations) and it can easily be understood why the populations perceived that they had suffered from a "knife stabbed in their back." Paranoia and aggressiveness was to be fueled by this resentment.

Insult was added to injury in the form of derogatory language and extremely negative representations in official discourse or popular culture in the camp of the victors. Historians and specialists in communication theory note that this was an evolution characteristic of World War I. The Other was not only a rival, he was the incarnation of absolute evil, not a soldier but a "Hun" or a "boche" swarming over the countryside as part of a "barbaric horde." The propaganda machine of every belligerent had launched extreme chauvinistic campaigns using this sort of language and frightening imagery during the conflict; if the winners gave it up after November 11, 1918, they would have had the embarrassing task of explaining how little had been gained at the unprecedented cost of millions of lives.

The demonization of yesterday's enemy continued in Western democracies in novels, movies, newspaper articles, history books, political speeches (numerous political careers were built upon a veteran's status)[29] and especially in the great quasi-religious rituals commemorating the dead (parades, memorial services, and the commissioning of monuments to the dead which became a real industry). That this Manichaeism was perceived as particularly unfair by the populations of the former Axis powers was legitimate since, from an objective point of view, all the actors of the Great War had had a practically equal share of responsibility in starting the conflict. But that was the problem: objective points of view, in this climate of passionate triumphalism versus vengeful resentment, could not be calmly debated. When former Axis diplomats, statesmen or intellectuals spoke of reconciliation or peaceful postwar cooperation, they were treated as suspicious by the noble victorious allies and as traitors by a growing number of their own. Also, the allied self-righteousness could only provoke a symmetrical reaction: chauvinistic propaganda that would soon descend to a level of violence, paranoia and vulgarity never previously heard of. It spread spontaneously in beer parlors all over Germany, Austria and Hungary among disoriented, frustrated and unemployed veterans.

29 The "horizon blue legislature" (*"la chambre bleu-horizon"*) was the 1919 French legislative assembly so nicknamed because of the great number of veterans elected to a parliamentary seat (the French uniform after 1915 was of a grayish-blue color called "horizon blue").

The Capuchins' Crypt: The end of Austro-Hungarian cosmopolitanism

Whatever the flaws and injustices of the Habsburg regime (and they were numerous), the Austro-Hungarian Empire can be credited for having contributed to the modern world a sense of cosmopolitanism. Its great Central European cities had grown and nurtured the most exciting and open-minded musicians, psychiatrists, writers, philosophers and graphic artists. It was not a perfectly pluralistic society, but managing diversity, the notion that a state was built upon a population "wanting to live together" as defined by the philosopher Ernest Renan (*"vouloir vivre ensemble"*), had existed as a standard of government and social relations.

In his gigantic monograph on the end of the Austro-Hungarian Empire, Franco-Hungarian political analyst Francois Fetjö,[30] who can certainly not be accused of any Austro-Germanic sympathies, demonstrated that there was a *planned* destruction of the Habsburg state. Neither could the novelist Joseph Roth be blamed for crying over the defunct empire. He was not a nostalgic of the domination over Czechoslovakia, Hungary, Croatia, Slovenia or Bosnia, or over Poland and Italy. What he regretted was the baby thrown out with the water from the bath, the construction of powerful barriers between peoples that had learned to live together, to trade and to exchange ideas. The tragedy is that the new nations had been built without the slightest plan of guaranteeing peaceful relations with the former Austro-Hungarian Empire, nor even among themselves—this would have to wait until the creation of the Council of Europe or the Common Market. Alas, in 1918, Central European identity was poisoned by the spirit of division and demonization developed by the propaganda machines which praised Czechoslovakia, Greece, Poland, Romania and Yugoslavia as "good" while others like Austria, Bulgaria, Turkey and Austria and Hungary were "evil" and deserving to have some of their territories taken away. Aggravating the situation was the design of new borders. While some tried to follow the representation of Europe as a patchwork of nations deserving independence and sovereignty (the principle of self-determination), the French geographic school and its representation of "natural borders" influenced the designers of the new maps, particularly Emanuel de Martonne at the Trianon conference of 1920. The contradictions between the geopolitical doctrines of the allies are still causing problems in the Balkans today.

30 François Fetjö, *Requiem pour un empire defunt* (Paris: Seuil, 1993). Fetjö Ferenc (1909–2008) as he is known in Hungary (originally Fischel Ferenc was his real name) had been a prominent anti-fascist militant in 1930s Hungary where he had been close to the Communist and Socialist movements. When the Nazis invaded the country to which he had fled after persecutions in Hungary, he joined the French underground Resistance. Although a leading journalist after 1945 and a friend of Camus, Emmanuel Mounier and Malraux, he was not well-known by the general public but was recognized nevertheless by academics like Raymond Aron as one of the best scholars writing on Central European twentieth-century history.

The Soviet Union and its mixed signals

The Soviet state, inheritor of the geopolitical assets and liabilities of one of the greatest world powers, the Russian Empire, became the homeland of Marxist Leninism and a focal point for tens of millions of its supporters throughout the world after the creation, in 1918, of the Third International or *Komintern*. By carrying the hopes for liberation of so many laborers in every country of the planet, the geopolitical weight of the Soviet Union in world affairs became incomparably more important than what the last tsars had tried to achieve by rallying the Orthodox populations in the Balkans. So much responsibility complicated the foreign policy of the USSR and considerably disturbed international relations in general.

What were the true plans of the Soviet Union in regards to its neighbors and other countries? Whether the Soviet regime had become isolationist ("socialism in one country"), or one of the most potentially aggressive powers on the planet (as the champion of a worldwide proletarian revolution), or whether there was a more mundane continuity in the policies of the old Russian Empire, renamed USSR, is still a matter of debate. That historians today are still trying to interpret the mixed signals coming out of the Kremlin, and continue to disagree with each other, illustrates how difficult it was for governments and diplomats to deal with the completely new and unpredictable situation that arose after October 1917. The Komintern was openly committed to destroying the entire social and economic system, and even the culture, that defined most countries in the world. But even if the Soviets intended some day to invade the territory of other states, would they have the necessary technology, firepower and manpower? If yes, when would they be ready? But would the regime last?[31] Was it worth investing in preparing for an armed confrontation with an enemy that could actually soon disappear? This question was further complicated by public opinion in one's own country. Attitudes in all countries regardless of borders varied between exacerbated fear of communism at one extreme to militant support of all Soviet policies by members of a communist party at the other. Therefore, the perceptions by populations and their governments of the foreign policy of the Soviet Union were of crucial importance.

What created such mixed signals? On one hand, the Soviet government was responsible for several decisions of utmost importance in creating the impression that it was now retiring from the chessboard of international intrigue. Above all, pacifism was a foundation of Marxist ideology; world peace was not just a slogan to attract sympathizers, it was the ultimate goal of everything that Marxists in general, and Bolsheviks in particular, stood for. The first important act in Soviet foreign policy was to remove itself from World War I (the Brest-Litovsk peace

31 It must be stressed that until World War II, especially in light of conflicts between factions and events like the great purges of 1937, also given the absolute novelty of the regime, Soviet power did not appear at all as stable as it would appear to later generations who could analyze its strengths, retrospectively, after the victory of 1945, and after its spectacular technological and economic achievements as the second world power.

treaty). To those arguing that world proletarian revolution was still officially on the agenda of the Soviet regime, it was easy to reply with obvious examples proving that the Soviet Union had postponed it. The conflict between the Social Revolutionary party (the SR) and the Bolshevik party that had together created the Soviet regime, and the elimination of the former by the latter, was accelerated by Lenin's insistence that, at least for a while, in order to survive the USSR needed to be at peace with foreign powers.

Although he himself led the delegation that traveled to Brest-Litovsk to sign the separate peace with Germany, Trotsky's dream of a world proletarian revolution in the short term already began digging a gap between him and the other influential Bolsheviks. Stalin eventually prevailed over Trotsky because the Georgian militant was a partisan of "socialism in one country." The doctrine could be interpreted as a form of isolationism. What we know today, based on a retrospective long-term analysis of the history of the Soviet Union, is that this was also the adoption by the Kremlin of the standards of *Realpolitik*. It was perfectly consistent with the influence of Clausewitz over Lenin's thought and it might have been the wisest choice of a regime that was still very new, fragile and surrounded by enemies. And this element of realism in the Soviet doctrine may explain how the USSR survived as long as it did, despite chronic economic and social problems.

On the other hand, there existed important evidence of the extremely violent nature of the Soviet regime. When opportunities for seizing territories or gaining influence occurred, pacifism never stopped the Soviet leaders from acting boldly. The USSR played a great role in the rise of Communism in China, and although it remained behind the scenes during the extremely violent civil war between Mao and Chang Kai Chek, the USSR was obviously an actor. As later events would prove, the USSR never accepted the independence of the provinces that had separated from the Russian Empire without ever joining the USSR: Finland, Poland, Lithuania, Latvia, Estonia, and Moldavia. All of these would either be annexed partially or entirely by the USSR in the late 1930s. A fierce war was fought on Finnish soil in 1939 in an attempt to annex it. In the years preceding World War II, millions of witnesses, refugees labeled as Monarchist "White Russians" (although they could be Armenian or Socialist), could attest to the atrocities that had become a method of government in the Soviet Union. Their accounts of atrocities, instead of creating a democratic debate, created divisions among Socialists, or between Liberals and radical right-wing politicians looking for a good electoral platform.

In any case, these mixed signals complicated the task of democratic governments who needed to assess and exploit or resist the geopolitical power of such a mysterious actor with such vast territories, potential resources and international prestige among the proletariat. But as mentioned by Carr and Bloch, the mediocrity of minds prevented any visionary planning or any clear assessment of the international scene.[32] The overestimation of Soviet aggressiveness lead many

32 A good example of unpreparedness in facing the challenge of understanding the Soviet Union was the fact that most great Slavic studies departments or programs were created in Western European universities only after 1945.

politicians into the arms of fascism, while underestimating it encouraged Hitler in committing the madness of Operation Barbarossa and prevented Roosevelt and Churchill from being fully prepared for what awaited them at Yalta. This was the weakness used by the Soviet Union which, in terms of territorial gains, became the greatest winner of World War II.

The Economic Instability

The reason why the French called the 30 or 40 years preceding World War I the "Belle Époque" (the beautiful age) is the financial stability and the growth of that period. It was those who did not have to toil in a factory who could look back with nostalgia at those times. The roaring 20s knew so much growth and prosperity, thanks essentially to the new methods of production theorized by Taylor and implemented by Ford and, after him, thousands of factory owners, and thanks to the unprecedented scale economy that opened the gates of consumerism even to the lower classes, that we tend to forget how extremely difficult were the last years of World War I and the early 1920s. Some of the massive damage would never be repaired.

Because of the war, then because the Soviet Union was blocking all of the roads between Asia and Europe (until the death of Stalin it would remain a quasi-forbidden territory, open to only a minimal number of foreign visitors and diplomats strictly for conducting exceptional business) and especially because of the new system of enforcing strong national borders, customs and issuing of passports and delivery of visas, international trade collapsed. It would recover only in the 1970s! It was not only large corporations that had invested in countries that became enemy territories that went bankrupt with all their employees, and not only tens of thousands of shareholders who lost their lifetime savings and most of their income when the Bolshevik government refused to recognize government bonds and stock issued before October 1917, even small shopkeepers and peddlers suffered and were often ruined by the new borders that suddenly surrounded every country and divided Europe. Those who are old enough to have heard the testimony of their elders who lived and traveled extensively before World War I without ever needing a passport (except in Russia where obtaining a visa was anyway a very simple formality) need only their personal memories for evidence. Younger generations can grasp what was lost in 1918 with a simple image: the chestnut seller in Joseph Roth's novel The Emperor's Tomb (in German, *Die Kapuzinergruft—The Capuchins' Crypt*). He was a man who was not rich but who could live above poverty by traveling from town to town in the Austro-Hungarian Empire selling chestnuts from a pushcart. After World War I his business had been completely wiped out by the division of his usual route by the borders of several new independent countries and the new borders of Austria. At the end of the story he is found completely indigent.

The financial consequence of World War I: The crises of the late 1910s and early 1920s

There is little mention in history books about the recession of 1920–21 in the United States. There was a price to pay for that country to have emerged as the great economic winner (with Japan) of the Great War. As a result of the end of the wartime effort to boost manufacturing and food production, inventory accumulated and prices declined. Also, the sudden return of the soldiers raised the rate of offer of labor compared to demand as never before in history, which provoked soaring unemployment. In fact, if we consider the numbers, the deflation of those years was in fact worse than during the Great Depression following 1929.[33] It receded faster. But meanwhile the rest of the world had suffered a much longer and catastrophic recession with exactly opposite financial effects.

The catastrophic hyperinflation that wreaked havoc over the world in the last years of the war and in the following years is a better-known story, with its images of thirteen-digit numbers and wheelbarrows of cash needed to buy basic daily necessities. What happened in Germany immediately after the end of the conflict provides examples that have become classics in history books worldwide. Salary increases would often follow the soar of prices, but not fast enough for millions of families to maintain their standard of living or even to avoid bankruptcy. One category of income earners was completely wiped out by hyperinflation. These were the people who either thanks to their fortune or several decades of hard work had invested in trustworthy and stable financial products which allowed them to either stay above the level of poverty (to live as "petty bourgeois") or even not work at all. For many at the time, this was the only form of retirement plan before the great Keynesian social reforms. In France, where they were called *rentiers*, they constituted a significant portion of the entire population; by 1922 they no longer existed as a class.

As mentioned before, the scale economy would compensate for the loss of jobs and income. Investments such as gold or real estate saved the lucky ones or even created new fortunes. The most fortunate were those who had invested in armament, vehicles, textiles and other materials subject to wartime government contracts. But the road to progress of the 1920s had left too many victims of economic disaster on the side. Some would never recover and formed an underclass of homeless or destitute worse off than the proletariat (those who lived in the "zone" in the southern periphery of Paris and other slums and shanty towns across Europe, and those who were called the *Lumpenproletariat* in Germany)—typical figures of German cinema during the Weimar republic or of the terrifying paintings of Otto Dix. The gap between the winners and the losers of the economic changes during World War I was an additional cause of social and political tension. Although the opportunities for economic progress were real they were tainted by what was perceived as dishonest gains. The United States, almost untouched by economic

33 J.R. Vernon, "The 1920–1921 Deflation: the Role of Aggregate Supply," *Economic Inquiry* 29: 3 (July 1991), 572–80.

upheaval and, on the contrary, the nation that, with Japan and other emerging powers, had benefited from the changes, lived with the bad feeling that some of the prosperity of the 1920s was owed to the most unscrupulous businessmen; American literature of the times carried that discomfort in novels like *East of Eden* or other lost generation works of fiction. For those who read less sophisticated literature, the figure of the wartime profiteer (especially when associated with anti-Semitic imagery) became the boogey man of all extremist discourse and conspiracy theories.

Less than a decade before the crisis was over and those who had suffered from it were able to understand what had hit them, the stock market collapsed in October 1929 on Wall Street.

After 1929

The prosperity of the 1920s had created what we call today a "bubble." Too much consumption, and new consumers living above their means encouraged by credit too easily granted combined with speculation and the overrating of stocks, followed by a sudden decline in consumption when consumers reached the limits of their purchasing power, provoked the crash of 1929 and chains of bankruptcies of small businesses and large corporations. This time it was catastrophic deflation that ravaged the finances of the families in all social classes. These constituted a dangerous and fragile population whether they had already suffered from the economic crisis of a decade earlier or had just lost everything during this new crisis.

But what about the others? What about those who were still employed—the vast majority of the population—and who had maintained their standard of living and whose jobs were not yet threatened? Who were the additional voters thanks to whom Hitler obtained a quasi-majority? They were those who may still have been employed and enjoying wealth, or at least an acceptable standard of living, but who feared the immediate future. In a decade that had seen not only political upheaval but two major economic catastrophes, it was not even reasonable to believe that one's job and one's property was safe. Thus, there were those who were dangerous because they had lost everything and, having nothing more to lose, whose violent tendencies were no longer restrained by the responsibilities that go along with a secure position in society, and there were those who could become violent because of the fear of loss (which we will discuss further in the light of the research of Erich Fromm at the time) and the need that they perceived to take action, possibly violent action. The fear of declassification—proletarians fearing homelessness and middle classes fearing becoming proletarians—certainly explains the triumph of fascism in several countries or outbursts of racism and chauvinism with the presence of movements comparable to fascism in societies as diverse as rich or poor European countries, Japanese cities, or North American communities (see the paragraphs on ideology below).

The road to serfdom

Whether to prevent this rise of fascism or, on the contrary, the growing popularity of Communism, or any other threat to the social order, different governments worldwide found a solution in applying the rescue plan proposed by the economist John Maynard Keynes: restore job security by launching massive government construction programs and other socially useful activity (roads, airports, dams, facilities for sports and tourism, schools, hospitals, the arts, etc.); restore confidence in the future through a restoration of consumer confidence and raise standards of living for the lower middle classes and the poor by implementing welfare programs (more construction programs with housing projects, free health and disability insurance, free unemployment insurance and retirement benefits, free education, etc.). This in turn boosted demand through consumption while reducing supply and inventory, which in turn required hiring personnel. Rise in productivity followed the rise of demand. Workers returning to factories while offices and corporations increased their sales and consumers returned to the stores provided income taxes, corporate taxes or sales taxes—a new revenue re-injected into the government budget so that this virtuous circle could be rebooted again and again. Implemented in roughly the same form in a great number of countries, including Japan (the economy of which was under the leadership of Takahashi Korekiyo), with only some of the details changed, it had spectacular effects.

But in the 1930s, fatal mistakes were committed because statesmen did not understand what rules needed to be respected in order for the Keynesian plan to function. Neither did they measure the political costs decried by economists Friedrich Hayek or Ludwig von Mises. What should have accompanied the Keynesian plan to ensure not only domestic stability but international stability as well, and probably even greater prosperity (the kind that the West would experience soon, immediately after the War), was the free circulation of goods and labor. On the contrary, to satisfy popular demand that could not understand that free trade was the condition for its return to prosperity, almost every nation on the globe strengthened immigration laws (in accordance with growing racist and xenophobic currents) and reinforced the already rigid trade barriers implemented after World War I. These trade wars are the real beginning of World War II. They created the bipolar division in the capitalist world. On one side, nation states with vast territories overflowing with raw materials, and on the other side, those in desperate need of those raw materials. It has already been mentioned that this was an aggravating factor in Japan's foreign relations. Italy's problem had suddenly become very comparable after the economic crisis; its quasi-medieval social system in the southern half of the peninsula maintained agriculture at a level which could not fulfill the demand of a growing population. Millions of Italians were forced to emigrate in the years that followed World War I. With one of the highest standards of living in the world (despite the economic hardships) Germany was not desperate for imports as it had several important raw materials itself, and an excellent agricultural system. However, it was not obvious that German farms could recover fast enough from hyperinflation then deflation to be able to respond to the growing demand of German consumers now able to afford their daily bread. Also, Germany had lost all of its tropical colonies and with it a quasi-limitless supply of fruits or goods less essential to health but very strategic nevertheless, such as coffee and chocolate (which had been an essential product of export to countries such as Switzerland).

But what weighed in the balance creating the rift between the future axis and the future allies was oil. Whereas horses were still a part of life before World War I, in the 1920s and the 1930s the age of the automobile had become as real as it is today. Then, as now, a modern country could not properly function without a regular and massive supply of crude oil. The United States, France, Great Britain (in partnership with the Netherlands through Shell) and the USSR owned almost all of the oil wells in the world. Germany and Japan did not; Italy had a few and naturally attracted its two future allies of the axis. Protectionism in the form of anti-immigration laws, constantly increasing tariffs and constant devaluation made it more and more difficult for every country to export anything, but since all were protectionist, imports and exports should have been evened out except precisely for crude oil and a few other raw materials.

The more Germany, Italy and Japan chose to manifest aggressive postures, to defend their position in international trade and on the geopolitical chessboard in general, the more they had to backup their bravado with military programs. Otherwise, these would have been empty bluffs. The more they acquired military equipment, the more they needed fuel to run it. The more they needed that fuel, the more they became aggressive. Somewhere in the process the trade wars had degenerated into a military conflict. Did World War II actually start with the Japanese invasion of Manchuria, or other provinces of China, or Italy's invasion of Abyssinia, or with Germany's march toward the east in Czechoslovakia in 1938, a prelude to the invasion of Poland, rich in agricultural lands, and to the invasion of the USSR, also rich in oil? The fact remains that the campaigns of World War II were fought to grab energy resources and other raw materials: all raw materials in China, coal and steel in Norway, the Netherlands, Belgium, Luxembourg and France, oil fields in northern Africa and the USSR, and the trans-Atlantic or trans-Pacific shipping lanes during naval combat.[34]

The price tag upon the Keynesian economic reforms was the return to power of the industrial barons that had exploited the working class in the nineteenth century and had been in great part responsible for World War I. They understood that the period of hyperinflation had in fact destroyed the economy of the Russian Empire and, except for the United States and Japan or other emerging regions, had placed most capitalist countries, especially in central Europe, on the verge of economic bankruptcy and Communist revolution. 1929 was another emergency and, in the absence of any other recipes, Keynesianism may have appeared to some of the barons as a short term solution for damage control to avoid Communism or radical reformist governments like the Popular Fronts of Spain and France.

34 See Daniel Yergin, *The Prize: The Epic Quest for Oil, Money and Power* (New York: Free Press, 2008), Chapters 16–19.

But the great corporations would not accept being heavily taxed, and neither would they accept unprecedented social reform without anything in return. At best, reform programs such as the New Deal obtained concessions from the great corporations by revoking the laws against the monopolies passed under another Roosevelt presidency. What the great corporations were getting out of their tax contribution was protection of their interests at home and conquest of markets abroad. This was true in the early 1930s in Japan, but this would be true during the war and immediately after the war with the Marshall plan for the United States, Canada and Latin American allies, whose subsidized agricultural and industrial products poured into liberated Europe. To subsidize these products, the tax burden or the consequences of deficit spending was not only on corporations but spread out through entire populations. Therefore, large subsidized farms or large corporations won in the end: their tax burden (already made lighter by placing it on the entire population and small businesses as well) was relieved by a real return on investment after they had conquered vast new markets or military contracts. It was clearly Nazi Germany, Italy and Japan who put in place this relationship between big business and warfare. Democracies had not been tempted until the very last minute, the late 1930s, to include a war effort as part of the great construction programs of the Keynesian restructuring (a last minute rearmament in the United States after decades of neglecting the armed forces; the building of the Maginot Line and of a modern French aviation industry—too late to be efficient—in France; and the new arsenals such as the Spitfires and the Hurricanes in England). De Gaulle had lobbied for it at a very early stage but was never heard, and had to wait until 1958 to have his way. Hitler had been dreaming about it since the 1920s, but this was not an original thought since it was, in fact, one of the foundations of Prussian ideology on the eve of World War I.[35]

This actually united certain forms of socialism and conservatism in Germany on the eve of World War I, and explains why national discipline prevailed in 1914 over socialist internationalism in the same way as it did in Russia (with the exception of the Bolsheviks who rejected the war and broke their last ties with the Russian Social Democratic Labor Party). This also explains how, while Japan was still a parliamentary democracy, consensus was found between the militaristic factions and democratic forces. Winning over the Conservatives in 1933 was not difficult for Hitler: he had only to tone down the anti-Capitalist discourse of his party. The cost was the elimination of the "Socialist" elements in the national-socialist movement, particularly the SA (eliminated physically during the "night of the long knives" in 1934). In Italy, the complicity between business and fascism against labor originates in the use of the first proto-fascist elements as private militia against strikers and revolutionaries during the social unrest in Italy following World War I and hyperinflation.

Finally, under Nazism, Italian Fascism and different Japanese regimes, the greatest of the projects of the reconstruction effort, absorbing large portions

35 Friedrich Hayek, *The Road to Serfdom* (Chicago: University of Chicago Press, 1994), Chapter 12.

of the unemployed population, turned out to be the arms race. Since the entire population was solicited to pay higher taxes to fund the pet projects of military industries, taxpayers (corporations not benefiting from subsidies and individuals) could eventually have rebelled, unless there was a real danger of war, and thus a need for rearmament. Hence the escalation of militaristic and paranoid xenophobic discourse.

The tragedy of the economic reforms of the 1930s was that by saving millions of jobs and creating millions more jobs, it gave legitimacy and popularity to any government that had been successful. If the New Deal remains a happy experience in American memory, as pleasant as the old movies of Frank Capra, and while many Frenchmen, even right-wingers, remain grateful to the government of the Popular Front for the improvement that they experienced in their daily life, unfortunately, the same kind of good people were also grateful for Stalin's new factories, or Hitler's highways or Mussolini's stadiums and "trains arriving on time" that dramatically reduced unemployment. These and other dictators bought their popularity with artificial prosperity, the hidden price of which was long-term debt, economic imbalance, freedom and the lives of 60 million victims to the war machine being put in place.

Collective Mentalities and the Power of Ideology

Few historians will deny that the power of fear over the masses played a crucial role in exacerbating extremist ideologies and thus destabilizing international relations. Not since the decades that preceded the Westphalia treaty of 1648 had trans-national and supra-national ideological affiliations made policy-making so difficult for governments. Again, as feared by all Realists, the structure of the nation-state and its legitimacy could be put in question by ideology. The extremely confusing situation may have weakened the positions of Western democracies whose populations were divided over this impossible choice between aligning with Germany against the Soviet Union or aligning with the Soviet Union against Germany, or trying to accommodate both or fight both at the same time.

The decline of democracies

The great asset of the form of capitalistic society that emerged in the twentieth century was that it needed no ideology to justify its organization and authority. The revolution of Taylor and Ford and the scale economy created the best instrument of social cohesion for great urban societies: consumerism. The reduction of prices and rise of salaries created a strong impression, for the first time in history, that one could climb the social ladder through hard work. This was not completely illusory since standards of living truly improved from the twentieth century, and an entire middle class was to emerge and gather the great majority of members of society. It was a new form of social contract.

But consumerism was capitalism's greatest liability as well. If consumer goods suddenly became inaccessible, and if descending the ladder rather than climbing it appeared the most likely scenario for households, the social contract was doomed. Which is why two major economic crises affecting half of the 1920s and 1930s caused severe damage to the democratic regimes founded on consumerism, and turned millions away from democracy to more exalting home-grown or foreign ideologies promising revenge and greatness. The social democratic movement could have been an alternative. But it had been discredited, in the eyes of many, its own militants as well as many potential sympathizers, in many ways.

First of all, and this was its major handicap throughout the first half of the twentieth century, it had failed to unite workers internationally to stop the tide of World War I. On the contrary, too many Socialist militants and their leaders let themselves be influenced by nationalism, which in turn infuriated the most radical elements of the global Socialist movement who, like Lenin, no longer demanded peace alone but armed revolutionary action. This led to the disintegration of the Second International in 1916.

In Russia, in August 1917, different types of Socialists (except Bolsheviks and left-wing SRs) dominated the Provisional Government when the Commander in Chief of the armed forces, General Kornilov, attempted a military coup. The Provisional Government prevented this by reactivating the Red Guards (the Bolshevik militia) who had gone underground after their own failed coup in July. In October, after the Soviets seized power, moderate Socialists continued to lead the Provisional Government in exile and maintained their authority over the Russian regular army which the Soviets labeled "White Army" and "insurrectional," "counter-revolutionary" forces. Right-wingers worldwide could not forgive the Russian liberals and Socialists for using the Red Guards to save themselves from an attempted extreme-right wing coup. Left-wingers, on the contrary, would associate Russian moderate Socialism with the "White" movement.

The third problem was fear of Communism, which divided the ranks of all Social-Democrats worldwide. Many knew of the extreme degree of violence used by the Bolsheviks immediately upon seizing power. Their best source of information was from inside Communism, from those who had refused the violent methods imposed by Lenin, men such as Pierre Pascal, Maxim Gorky (whose period of exile is all but forgotten by historians), or the thousands of Russian Social Democrats, veterans of the February 1917 Revolution or the SR, who had formed the first Soviets and who had become the targets of Bolshevik repressions and massacres, no less than tsarist extremists. Even the German Communist leader Rosa Luxembourg expressed doubt about the idealization of October 1917 (which lasts to this day among millions in Western countries). The disagreement on the interpretation of the Russian Revolution and its controversial methods caused the schism between traditional social-democratic parties and those who formed the world's "Communist" parties federated by the *Komintern* in Moscow.

The fourth problem was that Social Democrats were discredited in the eyes of millions of their partisans, and also in the eyes of many moderate liberals, when the new Weimar and Viennese republics saved themselves in 1918 and 1919 by unleashing paramilitary forces against the German Spartacists and the Austrian Bolsheviks. The violence of the repressions, during which Walter Liebknecht and Rosa Luxembourg became martyrs, radicalized the Socialists' left wing and radicalized the right wing as well, especially the most reactionary elements inside the army. The civil conflicts that erupted in Austria and Germany, as well as in Hungary, Poland, Finland, Bulgaria and Romania, bi-polarized radical factions in the countries, making Socialism irrelevant. The radicalization and militarization of the extremes laid the foundations for the takeover of government either by Fascist organizations in the immediate future, or by Communist forces in the second half of the 1940s.

Alliances between the Socialists and the Communists, such as the 1936 Popular Front in France, were only episodic and only fueled the hostility against social democracy from both sides who blamed them for either the failure of the alliance with Communism or, on the contrary, for initiating them in the first place. For both the right-wingers (moderate and radical) and the radical left, Socialists were nothing but ambiguous, treacherous and therefore dangerous. Democratic resistance against the rising tide of Fascism could not use any of their authority, which had been discredited.

The totalitarian regimes

Although the ideologies were profoundly different, there existed many common denominators between the totalitarian regimes on the eve of World War II, and to this day.[36] Above all, Stalinism, like Fascist regimes, can be defined by the commitment to build an all-powerful State,[37] with no political opposition, the main task of which is to control every aspect of life in the name of rational and efficient organization, in which nothing can remain private. Hence the importance of propaganda and total control of all means of communication. These are meant for re-engineering society's entire culture—rewriting all history books, demolition of ancient monuments, replacement of entire city quarters by futuristic architectural constructions, reeducation of children and youths through compulsory membership in special organizations, grandiose collective liturgies, and the imposition of new social norms: invasive use of flags, symbols and logos, outlawing religion entirely

36 On the debate between historians over the comparison between Stalinism and Nazism see Alain Besançon, *Le malheur du siècle. Sur le communisme, le nazisme et l'unicité de la Shoah* (Paris: Fayard, 1998), which argues that the phenomena have the same roots, and, more recently, a view opposed to that analysis: Fr.-X. Coquin, "Réflexions sur l'assimilation du stalinisme à l'hitlérisme," *Europe* (janvier–fevrier 2006), 283–307.

37 See Emilio Gentile, *Il culto del littorio. La sacralizzazione della politica nell' Italia fascista* (Roma and Bari: Laterza, 2009).

(as in the Soviet Union) or making it socially incorrect (not under regimes like Spain, but certainly in Nazi Germany which attempted to replace Christianity with some form of neo-paganism), and the introduction of scores of new words and expressions to transform language: "Heil Hitler" instead of "hello" or "goodbye," "Comrade" or "Citizen" instead of "Sir" or "Madam," "all is functioning according to norms" instead of "everything is all right," and countless new words formed by abbreviation like "Gestapo" for "Geheime Staatspolizei," "Stalag" for "Stammlager" or Alexander Solzhenitsyn's favorite: "PomZamKomPoMorDe" for "Pomoshnik Zamestitelia Komissara po Morskim Delam"—"Assistant of the Substitute Commissar for Maritime Affairs." What Orwell called "Newspeak" had an essential function of politicizing and controlling reality as expressed by language in the beehive not just as a tool of power to support foreign policy, but as a model of organization for the entire society.

It has been noted that the model of the new modern authoritarian state began to take shape at the approach of the twentieth century under the Second Reich. It was the model of an urban state generated by, and existing for, industrial progress. Its ultimate goal: a perfectly policed society, functioning as smoothly and efficiently as a beehive; which means establishing specialized control over every aspect of individual and collective activity through strong bureaucracies and military-like institutions, surveillance and repressive organs to guarantee discipline and the participation of all members of society in the accomplishment of the noble goal. It is Clausewitz meets Taylor. This passion for efficiency, organization, and consensus results in a profound fear of individualism and of its subversive nature. There existed, however, fundamental differences between Nazism and Stalinism, although one may argue that both had common historical roots. Despite the Soviet Union becoming, in practice, less and less equalitarian, its official doctrine never betrayed the fundamentals of Marxism. The militaristic organization of State and society was still a strategy, at least officially, not the ultimate goal. The essential purpose of the "dictatorship of the proletariat" was to provide happiness for all.

The Nazi regime was neither interested in "all," nor was it actually interested in happiness. Its purpose was to serve what it understood as "nature," from its interpretation of certain German geographic doctrines, mainly Ratzel and Haushoffer, and its extremist reinterpretation of social Darwinism and Nietzsche. The Nazis believed that their mission was not the happiness of humanity but its improvement by letting the abstract forces of nature do their job. According to Ratzel's "organic" theory of the state, simplified by Hitler in *Mein Kampf*, human societies were like organisms that either expanded if they were strong or were destroyed by predators if they were not. Such values as peace, international cooperation and most humanistic values were obstacles to the "laws of nature." Moreover, in his paranoid vision of history, Hitler viewed the imposition of peaceful and humanistic values as part of a global conspiracy to weaken the naturally strong societies like the German race. This was the basis of his racist doctrine. The extremist social Darwinism and sado-masochist submission to the abstract forces of "nature" defines the core of all Fascist social doctrines: violence for the sake of violence. That is what distinguishes fascism not only from communism but from the realism of Machiavelli and Clausewitz themselves (although *The Prince* and the *Art of War* may be at the top of the genealogical tree of *Mein Kampf*).

There remained in Soviet doctrine, as well as in all of the Communist movements in the world federated by the *Komintern*, sufficient humanitarian values and enough men and women who had not been corrupted by power and a taste for violent action, to convincingly stand up against fascism in the name of those humanitarian values. The Jewish industrialist and conservative right-wing politician Marcel Dassault, as well as many others like him, repeatedly paid their respects to such men and women whose sense of human solidarity and generosity allowed them to survive the experience of the death camps. We should never forget them, nor those who died in the Gulag.

The main culprit: Fascism in its different incarnations

During World War I, modern civilization was shaped by different forms of nationalism served by modern means of communication. The roots of fascism are there.

In the last two years of World War I, the allies had presented their fight as a struggle for democracy. But after their victory, it is not the democratic model that imposed itself throughout Europe and throughout the world, but, on the contrary, different forms of authoritarian and often dictatorial regimes: Italy, Spain (first with Primo della Rivera, then Franco), Salazar's Portugal, Admiral Horthy's Hungary, Turkey under Mustafa Kemal, Greece during the regime of Metaxa, and Albania under self-proclaimed King Zog. Also, almost every country in the world outside the USSR had one or several different groups or parties that could be described as fascist. New forms of government that had been established as parliamentary democracies experienced either radical change after a military coup like Japan, or after an election as in Germany in 1933. Other parliamentary regimes evolved progressively toward dictatorships, in stages, like Japan, Italy, Poland and Bulgaria, or like Yugoslavia where the situation was complicated by ethnic divisions. Whatever the form of authoritarian regime, these were no longer the monarchies of the nineteenth century—with more or less autocracy, more or less constitutional law, a more or less elementary rule of law, and more or less civil liberties—mainly concerned with preserving an ancient order based on a hierarchy according to caste, rural values and religion.

The passion for efficiency, organization and consensus mentioned earlier, and the resulting fear of individualism in its subversive nature, explains the hostility toward civil rights. Civil rights had been the answer to the rise of the merchant society clashing with the remnants of the feudal system—protectionist barriers around small feudal territories and retention of the labor force (serfs) on those same territories. It prevented the free circulation of goods and workers needed for the development of industry. Revolutions of the Enlightenment and of the nineteenth century demolished those obstacles to free trade. But the growth of heavy industry and large corporations (that we are still experiencing today), contrary to the assumptions of many liberals, did not necessarily develop in harmony with the philosophical notion of freedom.

The main contradiction of capitalism (not noticed by Marxists) is the clash between the different priorities of the merchant class and the class of manufacturers. The interests of the merchant class are invested above all in trade, in the export and import of goods and services which necessitates a free and rapid circulation of goods and people, imagination, creativity and initiative. These values are the values of Liberalism. While factory owners can fuel the trade of the merchant class, their corporate political culture and managerial experience does not necessarily obey the same imperatives. The larger the factory, the more its vocation is mass production and rational division of labor (especially after the implementation of Taylor's principles), the more its corporate culture demands routine, efficiency, obedience to the collective effort, and not individualism or ideologies based on freedom. Eventually, after World War II, the Capitalist world resolved this conflict between the Liberalism of the merchant class and the Taylorism of manufacturers through a compromise in the emergence of the consumer society—consensus and discipline through desire for glamorized objects rather than consensus through coercion. Before they realized that, the industrial bourgeoisie, but also many Socialist thinkers, had indulged in militaristic organizational theories, a form of Taylorism without the notion of consumerism, and conceiving scale economies serving the collective rather than the individual consumer.

Fear of individualism also resulted from those who still occupied, or whose ancestors had once occupied and lost, prominent or at least secured positions in what survived of the ancient rural and feudal societies such as the clergy, all representatives of the law from judge to clerk or law enforcer, or employees of agricultural and small manufacturing trade associations, aristocrats or other land owners and different professions that depended upon them such as high-ranking house servants, overseers, game wardens, tradesmen and suppliers of all sorts from saddle-makers to interior decorators. There were also populations of soldiers made redundant by a change of regime, like the *demi-soldes*, the demobilized veterans of Napoleon's army in 1820s–40s France. Centralizing authoritarian regimes, although they made traditional pre-Enlightenment rural societies more and more redundant, appealed to those whose cultures and mentalities still belonged to that ancient world and who despised the liberalism of the merchant society or the industrial world. Centralizing authoritarianism offered them the illusion of returning to the feudal past. But Monarchist parties, like horse-drawn carriages, were disappearing everywhere.

Fascism was able to rally many of these relics of the ancient traditional rural societies by praising their aesthetical values. It explains the contradiction in fascist ideologies and movements between their commitment to the interests of heavy industry (often their financial support) and one of the major themes of their propaganda which was a call to a "return to the earth" (a slogan of the Vichy regime) and even an early form of environmentalism.[38]

---

38 The rural romantic theme is another of the most profound differences between fascism

Caution should be used when defining ideologies or regimes as "fascist:" there were great variations in the elements of the doctrines. The various positions on organized religion are a good example. Although Hitler took precautions not to antagonize sympathetic clerical authorities, he, as well as numerous Nazi leaders, was profoundly anti-Christian and neo-pagan, while Italian fascism was more indifferent and Franco or the Iron Guard in Romania defended the Church as a foundation of their society. Pétain promoted rural life with great insistence, Hitler much less. Anti-Semitism was an indispensible element of French, German and central European extreme rightist thought, whereas Italian fascists only paid lip-service to anti-Semitism as a courtesy to their German allies. The Ku Klux Klan, which experienced an upturn in popularity in the 1920s, was violently anti-Catholic, yet a typical figure of an American form of fascism was the radio star, Reverend Coughlin, a Catholic. More importantly, there were those who formed the Axis and went to war—the Nazis, the Japanese militarists, the Italian fascists, and the Hungarian, Romanian and Bulgarian regimes—and those who managed to stay neutral: the Spanish, Portuguese and Turkish regimes, whose choice allowed them to survive until the 1970s. But despite the differences there existed fundamental common denominators that justify a common label on all these ideologies, identified with one of the most successful of their kind: fascism. These common denominators are a total devotion of all energies of society to transform it into a perfect mechanism, as already mentioned, and a violent rejection of all humanistic values inherited from Judeo-Christianity to the Enlightenment, with a profound need to vent frustrations against scapegoats—either a minority or foreign countries.

The majority of regimes based on such foundations make the question "why did World War II occur?" almost irrelevant. We should rather ask ourselves "how could World War II *not* have occured?" with entire populations under such an influence. However, there remains one important question: how could tens of millions of individuals not have realized the risk of a major conflagration, especially after experiencing the Great War, and blindly follow the demagogues on their way to suicide? All the geopolitical, economic and ideological circumstances put together cannot explain the ferocity of the fighting in World War II and the industrial killing of masses of innocent and harmless individuals. Social psychology should better explain the common denominators of the totalitarian regimes by digging deeper into the dysfunctions of individuals' capacity to socialize as well as the collective aggressive tendencies.

---

and Marxism and Leninism. However, in some of its aspects, certain tendencies in Maoism, particularly the Khmer Rouge, exalted rural values. They were, however, the exception. On the nostalgia for nature in the extreme rightist aesthetics and ideologies and its comparison to Marxist attitudes towards nature, see Luc Ferry, *The New Ecological Order* (Chicago: University of Chicago Press, 1995) and O. Kobtzeff, "Environmental Security and Civil Society," in Hall Gardner (ed.), *Central and South-central Europe in Transition* (Westport, CT: Praeger, 2000), 219–96.

The "escape from freedom"

Hall Gardner in his chapter on alienation (Chapter 1) and myself (in Chapters 4 and 10) have already analyzed the accumulation, in the long-term, of social dysfunctions leading to imbalance in international relations and collective homicidal tendencies. Research conducted from the 1940s to the 1970s by social psychologists like Hannah Arendt, Eric Fromm—author of *Escape from Freedom*—or Takeo Doi—author of the *Anatomy of Dependence* (*Amae no Kōzō*)—help us to understand such dysfunctions in the particular context of the first half of the twentieth century and beyond.

The conditions for these dysfunctions were the exploitation of the permanent state of panic generated by the catastrophes and economic revolution of the late Middle Ages and pre-Modern era, and the anomic situations created by the expulsion, during the industrial revolutions, of 80 percent of Europe's population out of stable and integrated rural communities into urban societies in a constant state of change and social disintegration. We have seen how prophetic movements in the wars of religion or revolutionary movements in the Enlightenment were a compensation for fear and loneliness. Violence provides the illusion of empowerment, of *taking control of history*. Control is the key to understanding the violence of the twentieth century that culminated in the political movements leading to World War II.

But why react by hurting others instead of taking control of history in more creative ways? Reacting violently because one feels frustrated from not earning the right place in society despite one's diligence in work and one's efforts to adapt to the social environment is a sign of immaturity, a childish reaction to one's incapacity to either seek better communication strategies or to free oneself from the need to conform and lead one's own life no matter what the social and economic sacrifice (social stigma, obstacles to the advancement of one's professional career, etc.). But this level of maturity is made very difficult by the evolution of contemporary industrial societies. A "having" mode, according to Erich Fromm, is induced by parents, educators or other figures of authority who, in order to enforce the standards of socialization upon the child, the adolescent or the young adult, will use it as a method to break that individual's will.

In Japan, this seems to have been a very ancient feature of socialization. "Amae," in the words of Japanese psychiatrist Takeo Doi (who almost seems to defend it as a quality), is to "make up to" one's parents, "behave childishly in the assumption that [one's] parents would indulge."[39] Note the revealing word "passive" as Doi explains that the "existence of an everyday word for passive love—amae—was an indicator of the nature of Japanese society and culture."[40] To *amae*, Doi opposes *ki ga sumanai*, which structures personality in a way that the individual "can be said to have grown out of the infantile type of *amae* and to rank among the more autonomous type of individual in a society such as Japan's (which) is permeated

39 Takeo Doi, *The Anatomy of Dependence* (Tokyo, New York and San Francisco: Kodansha International Ltd., 1984), 18.

40 Ibid., 21.

through and through with *amae*."[41] From Doi's studies it is difficult, however, to detect whether this form of passive attitude toward the world which he calls *amae* was something that prevailed as a result of centuries of Japanese education and social conformity, or whether it is a sign of psychological regression caused by the social disruptions and anomie caused by the Meiji revolution.

In the West, Fromm calls the "having mode" what Doi describes as *amae*. It could reflect the new corporate culture of extreme efficiency: a fetishist devotion to technocracy and to what Jan Patočka called "titanism"—the common target and common denominator between Charlie Chaplin's *Modern Times*, the Socialist Fromm's *Escape From Freedom* and the Libertarian Hayek's *Road to Serfdom* (which dared not investigate Taylor or Ford, however, as root causes of the idolization of State and collective organization). This devotion to automated efficiency was imposed not only by Michel Foucault's prisons and hospitals but especially by the regimented educational institutions of the nineteenth century: classrooms, boarding schools, military academies, colleges or other places of education and the new mass media. Such Taylorism promoted to a social philosophy can only bring a passive attitude toward others, toward society and toward life itself. It could only impede a future citizen in developing his or her personality based on a sense of autonomy, initiative and the sense of responsibility which is inseparable from the sense of true freedom.[42] With responsibility would have come maturity, and with maturity the desire for what is truly enriching, in other words what makes one grow, what brings new challenges that make a person stronger even though pain can come with it through trial and error. This outlook on life is prevented from developing in many normal households in times of war and economic uncertainty as the world has witnessed between 1914 and 1929. The "being mode," the acceptance of the challenge of diversity, of alterity and of constructive tensions, only brings terrible anxiety to personalities forged through terror into the "having mode" in which the lessons of socialization and maturation have been interiorized without any understanding of their ultimate goal.

The personality in the "having mode" is profoundly irresponsible, not only because his or her education excludes autonomous decision-making and choice, but because it involves facing one's greatest fear (taking a decision alone) and because responsibility carries a great risk, the risk of making a mistake and bearing the consequences. But because the training of the anal character or the personality in the "having mode" involved so much punishment and pain, the thought of the mistake is unbearable. Let others make decisions for me—parents, priests, community leaders, politicians, bosses of all sorts or simply the common rules of social behavior.

Or better, we delegate the great decisions to a leader, a "duce" (in Italian), a "führer." Surfing on the waves of popular frustration and "escape from freedom"

41 Ibid., 110.

42 The rigidity of nineteenth-century educational systems persisted in France well into the 1960s, which explains the reaction to it in May 1968.

and national geopolitical interests, dictators could put together a political platform. Unfortunately, war, for all the reasons listed above, must be part of the "package."

Total War: The Origins of Our World

We did not review the actual military operations and campaigns of the war—the 1937 invasion of China and Chinese resistance, the "strange war" (*Drôle de guerre*) in the winter of 1939–40 on the Franco-German border or the Blitzkrieg that followed, nor the Battle of the Atlantic, nor the events leading from Pearl Harbor to Iwo Jima. These events are widely known and well-analyzed in thousands of publications starting with the most basic encyclopedias. Our interest was in understanding the profound causes and, accordingly, the real stakes involved in World War II.

There exists a school of thought that believes that World War I, World War II, the Cold War and all the smaller conflicts between 1914 and 1991 were one continuous phenomenon. Philip Bobbitt calls it the "Long War." If this is true, World War II could be seen as a culminating point of that experience, the moment when the typical forms of our civilization, a civilization shaped by this Long War, can be seen in crude light.

The geography of World War II: The origins of globalization

World War II reveals how globalization has redefined all of the rules of geostrategy and therefore global geopolitics. The projection across continents of millions of soldiers rewrote the social and cultural geography of all countries involved in the war. These young men of a wide range of geographic and cultural horizons mixed together, forging a new supra-regional culture. But then they were exposed to new, foreign and sometimes very exotic cultures. This double challenge created new identities, often a new form of national identity and, in many cases, a greater awareness of the geography and human diversity of the globe. Returning from the front, the transformed identities redefined social geography, as millions of demobilized soldiers adopted new life styles, found new jobs after acquiring technical skills in the military, brought with them a foreign war bride, chose not to return home but to grow roots in a new region or in an urban rather than rural environment, or constituted entirely new communities and sub-cultures, like the first African or Asian immigrants to Europe, tens of thousands of settlers in Alaska, the Canadian Northwest or other remote regions of the Pacific (developed because of their strategic value during the war), American ex-patriots in Paris and its Afro-American sub-culture, or the first large gay communities in large cities. As after World War I, millions of refugees, called "DPs" (for displaced persons) since 1945, sought a new home in foreign lands. The best-known migratory movement of this sort, with the longest-lasting geopolitical consequences, was toward Palestine.

Not a single ocean, not a single sea, including the coastal areas of neutral countries, was left out of the war. Every maritime region had been affected by combat and had seen the passage of massive convoys. The war ended but the transcontinental shipping lanes remained. Aviation was now a normal part of the geography of transportation. Thus, remote tropical regions and the Arctic became essential theaters of operation. They would remain even more highly strategic regions on the geopolitical chessboard during the Cold War (South-East Asia, North Africa and the Sahara, the Arctic Ocean).

Short-wave radio now supplemented the telegraph in providing instant information across the globe. The rockets used by the space programs of the US and Russia or the builders of ICBMs are directly derived from the suborbital Nazi V2 rocket. All German projects and their conceivers were recycled by NASA and the Soviet IKI. Outer-space strategy completely rewrote the political geography of the planet. But this rewriting is not so much a migration of war into an extra-terrestrial environment as it is a process that makes Earth a smaller world. Soon after World War II, long-range weapons were to allow satellites to orbit the planet in a few minutes, mapping it entirely and surveying natural and human activity from a bird's-eye view. No spot on the globe could be isolated or safe anymore: seen from outer space, natural borders are meaningless. Like nuclear strategy, this will also have revolutionary effects on political geography as each state on the planet is forced to *think globally* in conceiving its foreign policy. This is a new rule of the geopolitical game that defines globalization.

Nuclear weapons were fought over in Norway, prepared in Los Alamos for years and finally used, for the only time in history, to end the war. Nuclear strategy, plus the leadership of the United States and the United Kingdom as the two most powerful allies in the capitalist world, forced countries to rally behind their leadership in military integrated command structures which eventually lead to NATO and the concept of the "nuclear umbrella." This would then bi-polarize the world between those who chose the umbrella of one superpower or the umbrella of the other, the USSR. Every government on the planet was forced to make this choice between the two. The world of the Cold War may have been bi-polar but the Cold War was a global situation. We should call it World Cold War I. It was again a factor of globalization, and its roots were in the confrontation of two potential nuclear powers in World War II.

A new political culture

Everywhere, as feared by Hayek and von Mises, the war effort had promoted central planning in the economy and in many other aspects of life. In exchange for economic reform plans in the 1930s, then, because of wartime emergencies, governments obtained from their parliaments, or through other means, administrative powers and influence that corresponded to the ultimate goals of the likes of Napoleon, and nineteenth century Czars and Kaisers.

The Keynesian model of economic policy was adopted on a very large global scale after the 1944 Bretton Woods conference, the creation at the time of our present international financial institutions (the IMF, the World Bank and the GATT, out of which would emerge the WTO) and the Marshall Plan. This was the second rise of the providential state which was to bring 30 years of prosperity to industrialized capitalist countries until the oil crises of the 1970s.

Many large corporations regained the monopolistic positions that had been limited by legislation since the turn of the century (part of the New Deal agreements between the federal government and corporations was that the Roosevelt administration of the 1930s abolished many of the anti-trust measures of the prior Roosevelt administration of 1901–09). The corporate giants also gained considerable power through government contracts throughout the war and beyond, thanks to the Marshal plan. The best example of the expansion of corporate power is Japan, where the *zaibatsu*—corporate giants like Mitsubishi, Mitsui, Yasuda or Sumimoto—rose to unprecedented power. It is President Dwight D. Eisenhower, another great witness of World War II, who warned us about the regime in which we are living and which began in World War II: the power of a military-industrial complex. It is still with us today, in many countries in the world, and it is still growing.

Never had any government invested such resources and efforts into espionage and counter-espionage, surveillance of one's own population, censorship and repression. The Gestapo, Abwehr or Kempeitai are fortunately defunct. But they were promptly replaced in the GDR by the STASI, and allied powers had to implement, as counter-measures, interior and exterior security institutions almost as powerful and influential as their Soviet model—the former Che-Ka, KGB and now FSB. Societies are now dependent on it and on the CIA, MI5, MI6 and all similar organs worldwide that grew during the war. They have become more than a tool of government. They have become a foundation of state and society to the point that heads of state now emerge from the higher ranks of such organs like G. H. Bush, Andropov or Putin. Others like J. Edgar Hoover would weigh in on state affairs like Talleyrand had never dreamed of. Secrecy, stealth operations, sabotage and political assassinations—what we call today "black ops"—had been essential tactics to combat the Nazis in occupied Europe. The tactics, the personnel and the political culture of covert operations remained after the war, creating situations dangerous to democracy such as the coming to power of Mao, Ho Chi Minh, Tito and of the Communists in Eastern Europe.

The French resistance in France had heroically helped the allies liberate the territory from occupation. But there was a darker side to its history, ever since the betrayal of resistance chief Jean Moulin and other casualties of internal conflicts between factions inside the resistance. This dark side left a legacy of dubious organizations and semi-covert groups like the SAC (Service d'Action Civique, literally a private Gaullist militia), the "France-Afrique" network of Jacques Foccart (that still controls the neo-colonial economies of former French African colonies), or other networks of militants. All were involved in such events as the reconstruction of the French secret services, bringing De Gaulle back to power in

1958 (a coup d'état was planned and could have been successful if legal channels had not brought the general back to government), or violent operations against Algerian separatists or other dissidents. In the United States, the collusion between intelligence organizations and the mafia (needed to liberate and reorganize Italy) opened the gate that led to the most unavowable aspect of US foreign and domestic policy, punctuated by military coups (against Mossadegh, Bao Dai of Vietnam, Sihanouk in Cambodia, etc.), assassinations and outbursts of massive violence (in Salvador, Chile and many other parts of the world). We need more research in the history of security forces and espionage/counter-espionage structures in a larger variety of countries (and this would be an economic, sociological and cultural analysis, not journalistic sensationalist stories about some double agent or a sabotage operation).

Also giving exceptional power to governments and other political forces were the modern means of communication. These include newspapers and magazines (which had attained the highest numbers of readers and one of the highest levels of content quality for text, layout and photos in history), and television: its viewers were only a few hundred privileged members of the elite but others enjoyed a much greater choice of illustrated magazines and listened to the radio for many hours. Hitler's vociferations (soon parodied by Charlie Chaplin), the suave baritone of Radio Moscow's speaker or sound bites like "never in the field of human conflict was so much owed, by so many, to so few," "a date which will live in infamy," or "Ici Londres" ("London calling") still resonate in our collective memories. All that was missing at the time was the Internet, although it must be added that hundreds of thousands of surviving witnesses lived long enough to express themselves through that technology as well (a technology indirectly produced by the research on computers developed during the conflict).

Finally, dramatic changes have taken place in the colonies. In the two world wars, millions of men indigenous to the colonies served and died in British, French and Dutch uniforms. The economies of colonies became essential to the war effort. Of course, raw materials played a highly strategic part, but there was more. The foundations of India's present economic power were laid during the world wars when British manufacturing was outsourced. In French Africa, what was outsourced were the government itself and its administrative services: most of the colonies recognized the exiled government of overseas resistance, and in 1941 Algiers had become the capital of "Free France"—the future legitimate French government in place today. Most of its civilians in military institutions have been posted at one time or another in French Africa, with the help of local civilian and military authorities, many of whom were indigenous Africans. In Southeast Asia, local resistance formed by indigenous guerrilla warriors had organized themselves against Japanese occupants and had liberated their territory by themselves, particularly in Indochina. In Indonesia, partly empowered by Japanese policies, an indigenous force took control of the Archipelago and would not give it back to the Netherlands.

The colonial powers had come to rely on their colonies for manpower. These soldiers were considered expendable during World War I, as evidenced by their frequent assignment to the most dangerous areas of the front line and by their higher rates of casualties. Yet they have proven loyal and highly efficient. During the Second World War, the armies of "Free France" had been composed essentially of colonial troops—Arabs, Berbers, sub-Saharan Africans, Caribbean Islanders, Guyanais, Pacific Islanders or other indigenous soldiers who were almost all volunteers. From the moral point of view, Western civilization had discredited itself in its own eyes after the trenches of World War I and especially after Auschwitz. After it produced such horrors, which were witnessed by the millions of colonial troopers just mentioned, how could it ever again claim to bring "civilization" to "enlighten" any other human population?

Many indigenous people in the colonies felt that the liberation of the world from Nazism and Japanese militarism was owed in great part to them, and that the European colonizers no longer had the legitimacy to impose their authority. Concessions or promises had already been made by colonial powers, to be immediately put in question after the victory. On May 8, 1945, upon proclamation of VE Day, the Algerians of Sétif added to their celebrations of victory against the Axis an independentist theme. The situation degenerated and according to their own sources, the European authorities shot dead over a thousand demonstrators. Other sources claim that ten to forty times more victims were killed during what remains in memory as the "massacres of Sétif and Guelma." This date should mark the beginning of the end of colonial rule in Africa and Asia.

"Never again"?

Fortunately the extreme violence and threats to fundamental liberties and civilization have caused a positive reaction. This reaction has also reshaped world geopolitics.

The United Nations, although increasingly decried in our times, was a dramatic improvement compared to the League Nations. As mentioned earlier, successful peacemaking goes unnoticed. A superficial analysis of the Korean War yields evidence of the organization's usefulness: as a permanent forum, it slowed down the escalation of tensions between the two potential belligerents who were able to assess that they had no interest in continuing conflict.

The Red Cross and humanitarian organizations (they are multiplying exponentially) have reached a level of maturity that made them true actors in the international arena. What UNESCO once labeled as a "culture of peace" prevails in public opinion in most countries in a world where, despite the simplifications or propaganda poured on them by the media of consumerist or authoritarian societies, populations value a critical approach to politics and authority. Basically, war has become unacceptable as a solution to political problems, in most social circles in the world. The new values of peace and cooperation have even been able to take root behind the "Iron Curtain" where (except in Yugoslavia and Romania) a bloodless transition took place from Marxist-Leninist regimes to democracy or semi-democracy. In India, where the process of independence, only a few dozen months after the end of the world war, was extremely violent (700,000 dead), there is still a miracle to be seen: the interesting question is not why India has been so violent in its recent past, the interesting question to ask is about how it managed to maintain a relatively functioning democracy for six decades under extremely difficult conditions of poverty and cultural diversity, and become at the same time one of the future great world powers.

The power of the lesson of World War II—its extreme cost—relayed by our educational systems and the modern means of mass communication, may be the explanation for this acceptance of a "culture of peace." Which can give hope that not all people on this planet can be hypnotized and led into perdition as those who we have described at length in this and three other articles of the present volume. This should give us some hope. But must we always wait for hope to act responsibly and resist evil?

* * *

Classic historians have sufficient material to blame the war on nearsighted nationalism. Economic theories can blame it on unemployment. Marxists have gone further by accusing capitalism and its shortcomings (greedy seizure of markets, incapacity to limit overproduction, etc.) and pointing out diversionary tactics (scapegoating Jews/minorities). Realists blame the ideologues for perverting their principles into an exaltation of violence for the sake of violence and for selling out to the private interests of arms manufacturers and other corporations. Classical liberals have extensively blamed the centralized planning of the economy by governments, including the realists' fear of internationalism and obstacles to free trade. Keynesian liberals defend themselves legitimately by claiming that they were misunderstood when Keynes' "virtuous cycle" was launched without reducing protectionism. Wilsonian liberals lament the lack of commitment toward durable peace-politics and toward the League of Nations. The opponents of Wilson's principles blame their idealism for weakening democracies' positions towards authoritarianism and creating havoc in Central Europe with the principle of self-determination. Other critics such as E.H. Carr saw the international organization for peace or humanitarian causes as a Trojan Horse for establishing American influence at the expense of other nations' sovereignty. To this we have added the interpretation of philosophers, psychologists and geographers. These and many other schools of thought remain, to this day, compartmentalized and dogmatic in their positions. But the reflections over only a few selected facts in these few pages are enough to show that each of these interpretations is founded upon elements of the whole truth.

Given the gravity of what was World War II, no stone should be left unturned in attempting to understand this abomination. Therefore, it is great time to synthesize all the approaches.

PART III
Cold War and Beyond …

NATO as a War-Preventive Organization: Cold War vs World War III

Marco Rimanelli

At 60+ years (April 4, 2009) the North Atlantic Treaty Organization (NATO) remains the longest-lived alliance in history compared to all other international security organizations and pacts. What are the reasons for its longevity and military success in protecting its members until today?

Historically, most military alliances were temporary, focused on wartime emergencies, but had little to no inter-alliance military coordination during war, while often serving the political interests of their hegemonic leader. This applied also to the Allies in both World War I (1914–18) and World War II (1939–45) as ideologico-politico-military forerunners of NATO, but eventually the Allies were forced to develop strong inter-alliance military coordination either at the end (1918 for World War I) or by the mid-course (1943 for World War II).[1]

During the Cold War (1946–90), the bipolar ideologico-political-military division of Europe and the world between the Soviet Union's (USSR) Communist Bloc and a United States-led (US) West/NATO threatened the transatlantic Allies with an apocalyptic conventional and/or nuclear World War III. This forced the creation of permanent US peacetime military alliances, but only NATO (1949) and its enemy the Warsaw Pact (1955–90) developed peacetime integrated military structures and inter-ally governmental political coordination at the level of international organizations, because both blocs saw in each other the "common enemy" led by a rival nuclear hegemonic superpower. Only by 1990 did East–West arms control and Soviet imperial collapse usher in the post-Cold War era with the dissolution of both the Warsaw Pact and USSR, while reconfirming NATO's long-term success in

1 A.J.P. Taylor, *The Struggle for Mastery in Europe, 1848–1918* (Oxford: Oxford University Press, 1954); Gordon A. Craig and Alexander L. George, *Force and Statecraft: Diplomatic Challenges of Our Time* (New York: Oxford University Press, 1990).

defending its allies and new partners in a wider transatlantic region with limited worldwide "Out-of-Area" peacekeeping operations.[2]

Cold War Transatlantic Security vs World War III Conventional/Nuclear Scenarios, 1949–1990

Since the end of Ancient Rome's Empire (600s AD), Europe has been torn by centuries of wars and unsuccessful cyclical drives for regional supremacy. Especially in the 1500–1945 period, European security depended both on the strength of national armies to preserve the independence of all Great Powers, and on the diplomatic impact of a semi-automatic Balance of Power system to lessen warfare's destructiveness and preserve Europe's national fragmentation by pitting occasional countervailing coalitions of all fearful Great Powers against whichever one became the Continent's new imperial hegemon. Yet the Balance of Power automatism only preserved the Great Powers' national independence and imperialism within a context of frequent wars and limited geopolitical gains. Smaller states instead remained just chattels for the winners.[3] Finally, the global destructions of two World Wars (1914–18, 1939–45) fought by opposing alliances until the total collapse of the enemy at a global cost of almost 100 million dead dissolved both the false hopes of the Balance of Power's automated security system and any exclusivist national defense: World War I tallied 16.5 million deaths (9.7 million military and 6.8 million civilians, including 1.6 million in the Armenian Genocide, but excluding millions of deaths in the Russian Civil War and Great Influenza), while World War II deaths based on varying data reached 52-to-54 million deaths, with 48,232,000 wartime deaths (8,268,000 Axis vs. 39,964,000 Allies, of which over 20 million only in the USSR, plus 6,582,000 Axis soldiers and 1,686,000 civilians vs. 14,277,000 Allied soldiers and 25,687,000 civilians), and 6 million Jews exterminated in the Holocaust. Even higher figures reaching 66+ million dead tally Chinese deaths under Japan's invasion in 1937-39 prior to the official beginning of World War II.[4]

World War I had erupted mostly as result of the Entente/Allies' (France, Russia, Great Britain and Serbia) challenge to Germany's old continental hegemony under Chancellor Otto von Bismarck's Balance of Power system (1871–1914) centered mostly on the Triple Alliance (Germany, Austria-Hungary and Italy). Wartime, however, widened the opposing alliances: the Triple Alliance became the Central

2 Gustav Schmidt (ed.), *A History of NATO: The First Fifty Years*, 3 volumes (New York: Palgrave/St Martin's, 2001).

3 Ludwig Dehio, *The Precarious Balance: Four Centuries of the European Power Struggle* (New York: Alfred A. Knopf, 1962); Richard Rosecrance, *Action and Reaction in World Politics* (Westport, CN: Greenwood, 1977).

4 http://warchronicle.com/numbers/WWII/deaths.htm;http://www.historyplace.com/worldwar2/timeline/statistics.htm; http://necrometrics.com/20c5m.htm#Second.

Powers (once Italy left and was replaced by Ottoman-Turkey and Bulgaria) and the Entente became the Allies (with Belgium, Japan, Portugal and Greece; in 1915 Italy, but with Serbia defeated; in 1916 Romania, but also defeated; in 1917 the USA, but losing Russia to the Communist Revolution). Although the Central Powers won in the open Eastern Front (1917) and almost defeated the Allies in the trench-fortified Italian and Western Fronts (1917–18), ultimately US economic and belated military interventions shattered Ottoman-Turkey and Austria-Hungary, followed by Germany under the humiliating punitive Versailles Peace.[5]

By World War II, despite politico-ideological rivalries, the Allies (Great Britain, France, USSR, USA and Nationalist China) were drawn together against Adolf Hitler's Axis (Nazi Germany, Fascist Italy, Imperial Japan, Hungary, Romania, Bulgaria, Slovakia, Finland and Croatia with external support from Francoist Spain). The Axis conquered Europe, North Africa and invaded Soviet Russia up to the Leningrad-Moscow-Volga River-Caucasus line, while in Asia Japan conquered China's coasts, South-East Asia and Pacific islands. The Allies' ultimate victory relied on the immense economic and human power of the US sustaining the British Empire, and the USSR's larger military. Still, the Allied victory left Europe shattered between devastation, reconstruction, millions of refugees and the Continent's split between two hostile superpowers (USSR vs USA) during 45 years of Cold War (1946–90). Again, as after the first war, Europe could only grasp the same false promise of a US-led idealist system, with peaceful global Collective Security for all states in the League of Nations (1919–39) and United Nations (UN: 1945–now).[6]

The 45-year-long East–West Cold War started immediately after World War II, as the USSR expanded its Communist military satellization of Eastern Europe (1946–49), threatened Turkey and backed Communist insurgency in Greece. The collapse of the wartime Grand Alliance and Soviet expansionism in the power vacuum left by the destruction of the Nazi New Order in Europe directly threatened both Western Europe and a victorious Isolationist US with the risk of renewed hegemonic wars soon after the end of the previous global conflict. Anglo-American protests at the UN in 1946–47 were blocked by Soviet vetoes unmasking the fragility of Collective Security, while only the US's final rejection of Isolationism and its renewed leadership of the West (Great Britain, France, Canada, West Europe, Latin America) with the 1947 Truman Doctrine, Marshall Plan and 1948 Organization of American States (OAS) assured the politico-economic containment of the Soviet Bloc, forcing Stalin to evacuate Iran and stop threats against Turkey and Greece.[7]

5 Ian Westwell, *The Complete Illustrated History of World War I* (London: Anness, 2008); H.P. Willmott, *World War I* (New York: Covent Garden/Penguin, 2003); Henry Kissinger, *Diplomacy* (New York: Touchstone, 1994), 17–245.

6 Gerhard Weinberg, *A World at Arms: A Global History of World War II* (Cambridge: Cambridge University Press, 1994); John Keegan, *The Second World War* (New York: Viking, 1990); Robert Osgood, *Ideals and Self-Interests in America's Foreign Relations* (Princeton, NJ: Princeton University Press, 1950); H. Kissinger, *Diplomacy*, 218–422; Peter Baehr and Leon Gordenker, *The United Nations in the 1990s* (New York: St. Martin's, 1998).

7 Stephen Ambrose and Donald Brinkley, *Rise to Globalism: American Foreign Policy since*

Within this bipolar geostrategic context, Western states sought to replace their old exclusive equation of national security against any external threat with the urgency of new never-tried-before regional collective defense institutions to permanently fasten all members' national survival in a Cold War international security architecture and political "community" of shared regional democratic identities and peaceful pan-European unity. Such regional integration grew out of a twin parallel process: on one hand, since 1949, politico-military security and regional integration in both Europe and the wider transatlantic (Euro-American) region as guaranteed by the US-led NATO Alliance; on the other, since 1950/57, cyclical quasi-confederal politico-economic regional integration by the European Community/Union (EC/EU), whose rapid economic growth overcame the long Franco-German animosity at the core of both World Wars. The Cold War fear of a Soviet-led World War III was the best "glue" for US-Western solidarity and German democratic reintegration in Europe, but it was a close call: alternative history studies speculate that had diehard Nazis succeeded in sustaining their 1945 stop-go guerrilla action against Allied Occupation, and had the neo-Isolationist Republicans won the 1948 Presidential elections, then the US would have been forced out of Germany and Europe regardless of Soviet threats (a not too far-fetched concept in 1946–48 when compared to the 2010s US military withdrawals from post-Saddam Iraq and post-Taliban Afghanistan due to local insurgencies and political instability).

At the same time, a central innovation in the history of alliances was NATO's growing sense of "community" centered on its identity as the core of the West or "Free World" against the global Communist threat, and, since 1975, on its common democratic values. Thus, during the Cold War the US-European partnership relied equally on NATO's regional security alliance and war prevention, guaranteed by bipartisan interventionist US governments, forward-deployed US combat forces and their global nuclear umbrella, while Europe's parallel regional economic integration also took hold and intertwined with NATO in a joint new identity as a "community" of nations. Only close cooperation within an interventionist US-led democratic industrial West and NATO allowed these twin transatlantic and European regional integrations to both keep at bay past nationalist hatreds among Allies and contain the USSR's ideologico-military threat, while a confederal EC/EU was seen as the last hope to restore Europe's old international politico-economic role without risks of renewed German imperialist hegemony. NATO's role as a vital regional organization for war prevention and conflict resolution over old territorial disputes among Allies continues unabated from the Cold War (Germany vs France and Great Britain, Italy vs France, or Greece vs Turkey) to post-Cold War enlargements (Germany vs Poland

1938 (New York: Penguin, 1999); H. Kissinger, *Diplomacy*, 394–456; John Gaddis, *The United States and the End of Cold War* (Oxford: Oxford University Press, 1992); John Spanier, *American Foreign Policy Since World War II* (Washington, DC: C.Q. Press, 1987); Alfred Grosser, *Western Alliance* (New York: Vintage, 1980), 1–153; Jerald A. Combs, *Nationalist, Realist and Radical: Three Views of American Diplomacy* (New York: Harper & Row, 1988); John Stoessinger, *Crusaders and Pragmatists* (New York: Norton, 1978).

and the Czech Republic over ex-German territories, or Eastern European new Allies vs Hungary's minorities, or Poland vs Lithuania).[8]

At the military level, the 45-year-long East–West Cold War sparked a series of war plans on both sides to fight any catastrophic conventional and/or nuclear World War III, which contradicts the spurious old debate about whether the USSR was really ever a threat. From 1947 into the 1950s, the US Pentagon military command forecasted war scenarios all based on quick Soviet conventional conquests of West Germany, Italy, Scandinavia, Western Europe and Greece, squeezing NATO defenses to the Pyrenees Line (Spain and Portugal), North Atlantic (Canada, Great Britain, Ireland and Iceland) and the geostrategic Mediterranean islands (Sardinia, Corsica, Sicily, Malta, Crete and Cyprus). Actually, the first 1946–48 Western war plans were even more pessimistic, stressing that if the Pyrenees Line was lost, then the entire Iberian peninsula and Gibraltar would fall, with Soviet forces conquering also Turkey, the Middle East and North Africa to create a vaster Communist version of the German-Axis *Festung Europa* of World War II. All early scenarios of World War III planned years of combat, and only after a massive US rearmament with forward bases in the British Isles could a new Allied invasion of Europe destroy the Soviet "Empire" at the cost of Europe's own total devastation (a French politician told US leaders in 1948: "We know you will come again to liberate Europe [if conquered by the USSR] as you did in both World Wars, but this time you will liberate only a corpse").

This untenable strategic situation forced the US to reluctantly assume the permanent, active leadership of the "Free World" and protect Europe through the April 4, 1949 NATO Treaty with 11 other Allies (Belgium, Canada, Denmark, France, Great Britain, Iceland, Italy, Luxemburg, Netherlands, Norway and Portugal). But the West's success in finally militarily containing the aggressive Soviet Bloc was short-lived: in October 1949 the USSR developed nuclear weapons, neutralizing America's atomic monopoly held since Hiroshima in 1945, and this was followed by the November 1949 Chinese Communist Revolution and the June 1950–53 Korean War. Facing this "ever-expanding" Communist Bloc, a newborn NATO was still too weak to assure Europe's defense if the Korean War expanded into a wider World War III, while Communist China's involvement meant that all of East Asia would also be dragged into the conflict just as in World War II when Japan was the aggressor. Thus, America's post-1950 global military containment against the USSR and a potential new global conflict pushed President Harry Truman to adopt the NSC-68 plan for a US peacetime military rearmament to the levels of World War II to allow it the resources to fight a two-front World War III against the USSR in Europe and Red China in Asia. The US also twice pushed unsuccessfully in early

8 Andrew DePorte, *Europe Between the Superpowers: The Enduring Balance* (New Haven, CT: Yale University Press, 1986); Peter Stirk and David Willis (eds), *Shaping Postwar Europe: European Unity and Disunity 1945–1957* (New York: St Martin's, 1991); A. Grosser, *Western Alliance*, 1–113; S. Ambrose and D. Brinkley, *Rise to Globalism*, Chapters 3–4; David Lewis, *The Road to Europe* (New York: Lang, 1993); Harry Truman, *Memoirs*, 2 volumes (New York: Signet, 1955); Harry Turtledove, *The Man with the Iron Heart* (New York: Del Rey, 2009).

1949 (NATO) and September 1950 ("Bomb on the Plaza") to strengthen European defenses by seeking Allied acceptance for rearming West Germany and adding it with Fascist Spain as equal new NATO Allies, but France held the strongest veto.[9]

In the early 1950s NATO was transformed under aggressive US leadership (with General Dwight "Ike" Eisenhower as its first Supreme Alliance Commander-Europe/SACEUR, then as US President replacing Truman), becoming the only truly integrated Euro-Atlantic military pact able to defend Europe against the USSR (compared to the still-born 1948 Brussels Pact or 1950 European Defense Community), both at the conventional and tactical nuclear levels, while preserving each ally's national control and dampening ethno-nationalist tensions within Europe. The US also finally got NATO enlarged in 1952 to include Greece and Turkey, and in 1955 also West Germany (fully rearmed under NATO command), with vital support from Francoist Spain in case of World War III (US-Spanish Treaty, 1952).

The price for regional security was enhanced European politico-psychological dependency on the US "nuclear umbrella" and global alliance network, plus US insistence that Allied defense plans for Europe be enhanced with the 1952 NATO Lisbon Summit's force goal of 90 NATO divisions. Yet these ambitious NATO force goals were soon scaled down by 60 percent (conversely enhancing the vital role of West German rearmament, which would provide 30 percent of total NATO peacetime forces), and ultimately shelved once fears of World War III waned with the mysterious 1953 death of Soviet dictator Josef Stalin and the end of the Korean War thereafter. Moreover, NATO's traditional conventional forces' weakness was also the direct result of most European Allied governments short-changing conventional defense budgets, due to both domestic pressures for social services over defense and, in part, a desire to retain a heavy dependence on US extended nuclear deterrence as a more effective deterrent to any Soviet World War III attack.

Thus, NATO's response to threats was mediated by its diplomatic consensus-based integration of five missions during the Cold War: 1) conventional/nuclear self-defense against the Soviet-Warsaw Pact threat of a future World War III; 2) regional alliance-building and forces standardization; 3) NATO's evolving innovative vital role in the history of alliances as a permanent peacetime regional organization and closely-knit "community" of democratic Allies assuring also war prevention in Europe among its member states, despite long histories of animosity; 4) integration of new members capable of supporting NATO against the USSR (1952, Greece and Turkey; 1955, Germany; 1981, Spain); 5) nominal (case-by-case acceptance of unpopular "Out-of-Area" missions if unavoidable vs. regular informal support of non-NATO coalitions in 1986–2010s).[10]

9 Gian Gentile, "Planning for Preventive War, 1945–50," *Joint Force Quarterly* (Spring 2000), 68–74; Marco Rimanelli, *Italy between Europe and Mediterranean: Diplomacy and Naval Strategy from Unification to NATO, 1800s–2000* (New York: Lang, 1997), 1, 100; A. Grosser, *Western Alliance*, 59–178; H. Kissinger, *Diplomacy*, 456–521; H. Truman, *Memoirs*, Volume 2.

10 Francis H. Heller and John R. Gilligham (eds), *NATO: The Founding of the Atlantic*

The Cold War's East–West military confrontation and fear of a World War III invasion of Western Europe by the Soviet Bloc was enhanced by the degree of complete politico-ideologico-military and economic control that the USSR wielded over its Communist satellite states of Eastern Europe and Mongolia with the 1945–46 bilateral pacts. Having already consolidated since 1946 total Soviet hegemony over Eastern Europe as a forward military "buffer" zone against the West, Premier Nikita Khrushchëv later rationalized this patchwork of bilateral alliances in the Warsaw Pact Treaty (May 1955), but it was mostly an international Communist propaganda stunt against NATO's 1955 enlargement to include West Germany, condemned as "neo-Nazi". The Warsaw Pact did not significantly increase the existing Soviet Bloc military threat against NATO, because since 1948 the USSR held total military control over all satellized Eastern European national militaries, each individually integrated in the Soviet Red Army. Thus, in 1955–65 the Warsaw Pact existed in a vacuum of common military purposes with a shadow organization plugged inside Red Army command, control, communications and intelligence (C^3I) structures, owing its existence purely to the Soviet iron fist and politico-ideological rhetoric of "self-defense" against NATO.

Warsaw Pact membership varied: Albania (left in 1963), Bulgaria, Czechoslovakia, East Germany, Hungary, Mongolia (since the 1970s), Poland, Romania (left in 1968), Vietnam (since the 1980s), Yugoslavia (left in 1956) and the USSR. Communist China was directly allied with the USSR but, like Cuba, was never an official member. Created principally as a tool of Soviet politico-military occupation of Eastern Europe and ideologico-economic expansion of the Communist system, Warsaw Pact militaries provided mostly backup combat forces and logistics for Soviet offensives against NATO in World War III, while Red Army forces present in all Warsaw Pact members also guarded the "rear-front" as a permanent veiled repressive threat against any rebellious satellites ("Brezhnev Doctrine"). Non-Soviet Warsaw Pact forces were not permanently deployed outside their home countries, except for temporary joint military maneuvers and occasional repressions.[11]

In time the USSR copied NATO's organization to turn the Warsaw Pact into a truly integrated alliance system with joint training and military exercises and full weapons standardization using exclusively Soviet matériel, but this only further rationalized its members' original total military integration within the Red Army structure, rather than as independent equal national allies. Thus, as

Alliance and the Integration of Europe (New York: St Martin's, 1992); Stephen Ambrose and David Brinkley, *Rise to Globalism,* Chapters 4–6; Marco Rimanelli, *NATO and Other International Security Organizations: Historical Dictionary* (Lanham, MD: Scarecrow/ Rowman, Littlefield, 2009), 980.

11 H. Kissinger, *Diplomacy,* 493–593; Adam Ulam, *Expansion and Coexistence: Soviet Foreign Policy, 1917–73* (New York: Praeger, 1974); AAVV, *Le Livre Noir du Communisme: Crimes, Terreur, Répression* (Paris: Laffont, 1997); François Feytö, *Storia delle Democrazie Popolari, 1945–71,* 2 volumes (Milan: Bompiani, 1977); Zbigniev Brzezinski, *The Soviet Bloc: Unity and Conflict* (Cambridge, MA: Harvard University Press, 1967); Karen Dawisha and Philip Hanson (eds), *Soviet-East European Dilemmas: Coercion, Competition and Consent* (London: Heinemann/RIIA, 1981).

a Soviet hegemonic alliance, the Warsaw Pact never had nor fostered any true partnership among its members, while NATO instead always cultivated both equal membership rights (including vetoing some controversial US NATO proposals) and an integrated military command structure paralleled by a politico-diplomatic Secretariat.

It was the USSR which fully used its Warsaw Pact alliance network of bases and logistics to mount any repression of rebellious satellites, like the 1953 East German Revolt, 1956 Hungarian Revolution and threats of invasion of Poland in both 1956 and 1981, while the 1968 invasion of Czechoslovakia to destroy its "Prague Spring" reforms represented the only instance of an official Warsaw Pact-wide military operation against a member state. As a result, Warsaw Pact cohesion was badly shaken in 1956 when first Poland threatened to leave, then Hungary revolted and briefly left (hoping for an illusionary NATO rescue), followed by Yugoslavia's withdrawal to protest the Soviet repression of the Hungarian Revolution, fearing in future a similar intervention against itself. Yugoslavia instead became a leader of the independentist Third World Non-Alignment movement (nominally against both West and East). Likewise, Romania left only the organization's military command in 1968 to protest the repression of Czechoslovakia, although still remaining both Communist and pro-Soviet despite its quirky independentism. Regardless, the Kremlin's principle of "Socialist solidarity" was reestablished officially with the 1956 Hungarian Revolution and again with the "Brezhnev Doctrine" during the 1968 Czechoslovak invasion.[12]

As an organization the Warsaw Pact only once openly opposed the USSR, by rejecting in 1980 Kremlin pressures to join the Soviet invasion of Afghanistan (1979–89). Furthermore, despite public and propaganda declarations of mutual politico-ideological loyalty, the Kremlin always remained secretly distrustful of the actual combat reliability in wartime of most Warsaw Pact members. In any World War III scenario the Kremlin trusted only East German and Czech forces (not Slovaks or other East Europeans), considered the best of the lot and closely trained for front-line combat support of Soviet forces. Thus, in World War III the bulk of the Warsaw Pact's anti-NATO offensives on the Central Front and pro-Western "neutrals" (Austria, Sweden, Finland, Switzerland and Yugoslavia) would be undertaken by Soviet, East German and Czech forces. All other Warsaw Pact forces were assigned by the Soviet Command to support, logistics, reserve and occupation duties in both Eastern and Western Europe, with limited diversionary combat operations supporting the Red Army on secondary fronts: the southwestern Front against northern Yugoslavia and Italy through the Gorizia Gap, or the southern Front against Greece, Turkey, the Turkish Straits, Aegean Sea and Caucasus borders.

US President Eisenhower, who as NATO's first SACEUR had led its conventional preparation for World War III, sought in the mid-1950s a less costly way to defend

12 Vojtech Mastny, "Learning from the Enemy—NATO as a Model for the Warsaw Pact" in G. Schmidt (ed.), *A History of NATO*, Volume 1; Jeffrey Simon (ed.), *NATO-Warsaw Pact Force Mobilization* (Washington, DC: National Defense University, 1988); M. Rimanelli, *NATO and Other International Security Organizations*, LXVII–LXXVIII, 1–666.

Europe through the "New Approach" of relying on tactical nuclear weapons to augment Alliance conventional defenses against any massed Soviet armor attack (Massive Retaliation). Indeed, since the mid-1950s NATO's conventional defense of the Central Front was based on the untenable "Forward Defense" strategy, vital to politically bolster a vulnerable West Germany, relying on the old post-World War II deployments of Allied Occupation troops along the Inter-German border. This left the strongest US and French forces clumped away in safer South Germany facing the Czech and Austrian borders, compared to the thin NATO front line with German and US, BENELUX and Anglo-Canadian forces spread along the Intra-German border from the Baltic Sea in the north to the Fulda Gap abutting Czechoslovakia.[13]

Pre-1988 Soviet military doctrines against the West relied upon the USSR/Warsaw Pact's traditional peacetime conventional superiority over NATO's weaker, but immediately deployable, combat-ready forces, while the USSR's geostrategic contiguous location assured a comparatively short, steady flow of Soviet massive reserve units to the Central Front vs the US-Canadian reserves bottled-up across the Atlantic Ocean. To maintain and exploit the Warsaw Pact's peacetime military advantage, the Kremlin planned World War III around surprise attacks and fast Soviet conventional offensives against NATO with massive firepower and vital armour breakthroughs in the Central Front to conquer West Germany's industrial Rhine-Rhur areas. The USSR/Warsaw Pact's strategic goal was to seize the northern Rhine River and BENELUX countries within a three-week window of superiority, before US air-lifted forces and pre-positioned armour could rescue NATO and Germany from collapse. However, such a limited timeframe would also most likely force any East–West conventional combat into a quick escalation to nuclear warfare. On one hand, limited nuclear war scenarios envisaged a desperate US gambling to halt World War III and NATO's imminent defeat by using its tactical nuclear forces (INFs/SNFs) in Europe for limited nuclear strikes targeting only Warsaw Pact massed conventional armoured breakthroughs inside West Germany. On the other, had the US lost such a desperate theater nuclear gamble vs continued sustained Warsaw Pact military offensives even in a nuclearized environment, or had instead the Kremlin chosen to preempt NATO by relying since the start on a strategy of mixed conventional/chemical/nuclear offensives in Europe (as revealed by Soviet Marshall Ogarkhov in 1984 at the Soviet Frunze War Academy), then in either case both the North American continent and the Soviet heartland would have been exposed to quick escalation to reciprocal full retaliatory US–Soviet nuclear strikes in an apocalyptic global nuclear holocaust (as dramatized in the 1985 ABC film *The Day After*).

13 Edward N. Luttwak, *Strategy: The Logic of War and Peace* (Boston: Harvard/Belknap, 2001); John Ikenberry (ed.), *American Foreign Policy: Theoretical Essays* (Boston: Scott, Foresman, 1996); M. Rimanelli, *NATO and Other International Security Organizations*, LXVII–LXXVIII, 1–666; Wolfram Hanrieder and Graeme Auton, *The Foreign Policies of West Germany, France and Britain* (Engelwood Cliffs, NJ: Prentice Hall, 1980); S. Ambrose and D. Brinkley, *Rise to Globalism*, Chapters 10–12; A. Grosser, *Western Alliance*, 59–178.

During the Cold War the USSR's armed forces were deployed on eight combat TVDs (Continental Operational Theaters) along her borders (Atlantic, Arctic, Northwestern, Western, Southwestern, Southern, Far East, Pacific), supported by Warsaw Pact forces in Eastern Europe and Mongolia. Conventional World War III combat scenarios focused on massive, rapid Soviet/Warsaw Pact armored and mechanized strikes against NATO's Central Front (Western TVD), with secondary strikes on both NATO's Northern Flank (Northwestern TVD) and Southern Flank (Southwestern TVD). As epic armor battles pin down both NATO and Warsaw Pact forces in Germany, a collapse of Italy following the Soviet underside flanking strikes could doom the Central Front: if NATO collapsed militarily in few weeks, then the USSR/Warsaw Pact could round off in few more days the conquest of West Germany's Rhineland, Belgium and Luxemburg to isolate France and West Europe, with auxiliary attacks to finish conquering South Sweden and Norway, devastating both NATO's Central Front and Northern Flank, and close off the Eastern Mediterranean. This would force NATO to retreat from Germany into France to evade encirclement from the strategic pincer onslaught of both Soviet Western and Southwestern TVD forces. If no nuclear option emerged (doubtful as NATO and French defense doctrines planned the use of tactical nuclear weapons if NATO forces lost), then the USSR/Warsaw Pact could encircle in France a collapsing NATO and reach the Atlantic coast, forcing also the political surrender of isolated Italy and Turkey. NATO could only regroup behind Spain's Pyrenees line, Great Britain and North America, while building up military support with Japan, Australia, New Zealand, South Korea and the Organization of American States, plus parallel defense collaboration with Communist China against both the USSR and pro-Soviet North Korea.[14]

In any of these World War III scenarios, NATO could still benefit from the fact that even with its combat-ready reserves in Western Russia, the USSR/Warsaw Pact lacked the classic 3:1 superiority in forces deemed vital for any quick successful offensive against the well-entrenched NATO defenses, while Soviet power would also be sapped by wartime unreliability of both Eastern European and the USSR's own Central Asian Soviet conscripts. Furthermore, 40 percent of the Soviet Red Army remained deployed outside Eastern Europe and Soviet Russia: the majority along the extensive Sino-Soviet border against the lightly armed four million Chinese troops, plus around 100,000 in the First Afghan War (1979–89) and residual large forces in Eastern Siberia and Maritime Province against US forces in Japan and South Korea. Additionally, Moscow's lead in tanks, artillery and aviation (7,240

14 Jan Hoffenaar and Christopher Findlay (eds), *Military Planning for European Theatre Conflict in the Cold War* (Zürich: Centre for Security Studies, 2006), 23–231; William Kintner, *Soviet Global Strategy* (Fairfax, VA: Hero, 1987); David Isby, *Weapons and Tactics of the Soviet Army* (London: Jane's, 1988); J. Simon (ed.), *NATO-Warsaw Pact Force Mobilization*, 14–458; John Hackett, *The Third World War: August 1985* (New York: Macmillan, 1978); John Hackett, *The Third World War: The Untold Story* (New York: Macmillan, 1982); Tom Clancy, *Red Storm Rising* (New York: Berkley, 1983); Ralph Peters, *Red Army* (New York: Pocket, 1989); ABC Films, *The Day After* (1985).

Soviet vs 2,975 NATO planes) was partially offset by several Western advantages: a) 60 percent of Soviet air power was only short-range, with also a high rate of engine breakdowns in the most modern attack planes, giving NATO the lead in long-range air-to-ground attacks; b) more modern NATO tanks and high-tech anti-tank weaponry; c) NATO's command of the seas and air for the resupply of Europe; d) the 1980s Franco-Spanish military coordination with NATO; and e) US rapid airlift capability to bring, within two weeks, several combat-ready divisions equipped with stored pre-positioned armored matériel in Europe.

The Cold War longevity and ultimate success of NATO as a military alliance has relied on three advantages over its enemies (USSR and Warsaw Pact states). First, NATO had a common politico-ideological commitment to democracy, freedom and capitalism, stemming from each ally's own political freedom and popularly supported government action, compared to the Kremlin's *diktat* on all its Warsaw Pact satellites of uniform Soviet-style communist dictatorships and command economies. Secondly, NATO is a self-defense alliance under American leadership (not a hegemony as the USSR) against a common enemy, based on equality among Allies (and today Partners too) who retain full control of their national militaries with minimal peacetime NATO integrated military forces (only in wartime would NATO assume full control of all US/Allied militaries, which in peacetime rely on the US nuclear umbrella as a deterrent and combat option).[15]

Instead, throughout the Cold War the Warsaw Pact remained mostly a rhetorically driven political alliance under the USSR's imperialist domination, with total politico-military hegemonic control of subordinate satellite forces by the local Soviet Red Army Fronts deployed in Eastern Europe against both NATO and potential "local" rebellions. Finally, the Soviet Bloc's ideological demonization of NATO and the West as an ideologico-military threat to justify its Cold War combat superiority over NATO in troops, armor and aviation masked a very weak Warsaw Pact with paper-thin common multinational structures and organization evolving slowly through time to both match and apply NATO's more successful inner transformations and combat-preparedness.

At the geopolitical level the Iron Curtain division of Europe during the Cold War represented both the military confrontation between the superpowers and their equally strong politico-ideological commitments to victory over each other. Soviet-style Marxism-Leninism was the orthodox politico-ideological measure of loyalty and subservience of all satellite Eastern European states to Moscow's will, while all "National Paths to Communism" had been eradicated in 1948 by Stalin's ruthless purges. Such ideological communist orthodoxy continued under

15 David C. Isby and Charles Kamps Jr., *Armies of NATO's Central Front* (London: Jane's, 1985); J. Simon (ed.), *NATO-Warsaw Pact Force Mobilization*, 14–458; Michael Jerchel, *USAREUR: The United States Army in Europe* (Hong Kong: Concord, 1989); Ivan Volgyes, *The Political Reliability of the Warsaw Pact Armies* (Durham, NC: Duke Press, 1982); Colin Gray, *Nuclear Strategy and National Style* (Lanham, MD: Hamilton, 1986); Ashton Carter, John Steinbruner and Charles Zraket (eds), *Managing Nuclear Operations* (Washington, DC: Brookings, 1987).

Soviet leader Khrushchëv, who repressed the anti-communist 1956 Hungarian Revolution, and then under Leonid Brezhnev with triple crises in Czechoslovakia (1968), Afghanistan (1979) and Poland (1981). But especially in 1965–68 both rival alliances were shaken to their core by parallel crises: NATO was nearly scuttled by France's President Charles de Gaulle's independentist "Grand Design" between East and West, while the Warsaw Pact's brutal repression of the "Prague Spring" communist-reformist movement in Czechoslovakia heaped international opprobrium upon the Soviet bloc.[16]

On one hand, French President Charles de Gaulle, unable to pry "equal" leadership of NATO ("Triumvirate" proposal) from America, instead pushed for a "Grand Design" to maximize France's global role as a more independent "Third Force" between the superpowers in a period of relative East–West calm by leading a loose confederal "Europe of Nations." Thus, de Gaulle withdrew French forces from NATO's integrated military command in 1966–67 (but not from its political wing, remaining an Ally), hoping that this would precipitate a US disengagement and the collapse of the Alliance, which in turn would entice the USSR to leave Eastern Europe as well under the "friendly" East–West tutelage of Gaullist France ("Grand Design"), while stalling EC integration. Instead, NATO Central Front forces, command and logistics were quickly rerouted to Belgium, Great Britain, Netherlands and West Germany, while relying on Portugal, Italy and externally Spain too. France found itself isolated once the Soviet threat refused to fade away, belying de Gaulle's "Grand Design" pipe-dreams in 1968, while Red Army forces were increased along the Inter-German border. De Gaulle's independentist policy officially continued after he resigned in late 1969, but French defenses slowly realigned with NATO against a World War III Soviet attack, while the EC also quashed French vetoes in 1973 to admit Great Britain, Denmark and Ireland. However, none of this boosted America's sinking leverage in Europe, given Europe's Leftist popular mass demonstrations and worldwide condemnation in 1967–73 of Atlantic solidarity and US combat in the Vietnam War and aid of Israel during the 1973 Yom Kippur War (US airlift). NATO weakened further, due to its inability to address politico-militarily Middle Eastern crises, the First Oil Shock and 1974 Greek-Turkish clash over Cyprus, with Greece's temporary withdrawal from NATO's military. Leftist anti-Americanism waned reluctantly following East–West relaxation of tensions during Détente (1969–79) and US–Soviet nuclear arms control.[17]

On the other, the Spring 1968 relaxations of East–West tensions had prompted Communist-reformist Czechoslovakia to embark on the popular "Prague Spring" socio-politico-economic liberalizations and opening to the West, while remaining a

16 Edward N. Luttwak, *La Grande Strategia dell'Unione Sovietica* (Milan: Rizzoli, 1983); W.R. Kintner, *Soviet Global Strategy*, 79–266; J. Simon (ed.), *NATO-Warsaw Pact Force Mobilization*, 14–458; I. Volgyes, *The Political Reliability of the Warsaw Pact Armies*, 1–112; Enzo Bettiza, *1956 Budapest* (Milan: Mondadori, 2006); H. Kissinger, *Diplomacy*, 550–90, 703–80.

17 Simon Serfaty, *France, De Gaulle and Europe* (Baltimore, MD: Johns Hopkins Press, 1968); H. Kissinger, *Diplomacy*, 468–620; A. Grosser, *Western Alliance*, 111–13, 139–330.

loyal Communist state under the USSR and Warsaw Pact. Very long secret debates within the Soviet collective bureaucratic leadership reflected generalized fears that any weakening of Soviet politico-military controls and ideological orthodoxy over Eastern Europe could also undermine communist rule in the USSR proper. Thus, in the late summer of 1968, Soviet Premier Leonid Brezhnev forced the Warsaw Pact to invade and crush Communist-reformist Czechoslovakia, reconfirming the Cold War "limited sovereignty" of all Socialist states within a "Socialist commonwealth" and the USSR's "right" to repress in any members' deviations from Marxist-Leninist Communism threatening its ideological alignment (the "Brezhnev Doctrine" of Intervention). Unwavering Soviet-Communist rule of Eastern Europe undermined the entire strategic supposition of de Gaulle's delusional "Grand Design" of French limited hegemony in Europe. Finally, the "Brezhnev Doctrine" also worsened Sino-Soviet tensions (the 1969 "Amur-Ussuri Incidents"), with Communist China afraid of a future Soviet invasion to reimpose Soviet control. These international crises finally led the USSR to accept actual reforms of the Warsaw Pact to turn it into a more effective and balanced alliance, with special partnership and superior combat-readiness focused on front line East German and Czech forces, compared to the politically less-reliable Poles, Slovaks, Romanians, Hungarians and Bulgarians.[18]

At the same time, another growing area of concern to NATO was the Middle East/Gulf, both strategically and given the growing fear that regional wars between Soviet client states and pro-US Israel might escalate into an East–West World War III. During World War II the USSR had fleetingly succeeded in expanding its influence outside its borders also to Western-dominated Asia, following in the footsteps of the 1700s–1900s "Great Game" between Great Britain and Russia for control of both the Middle East/Gulf and Central Asia land- and sea-routes to India and China. In 1942–46 Stalin used Soviet Azerbaijan and Central Asia as important staging areas for the Allies' joint military occupation of neutral, pro-German Iran as a strategic land-bridge between the USSR and the Western-controlled Middle East and Persian Gulf. Only in 1946–48, Allied-UN pressures forced the USSR to evacuate Northern Iran, Iranian Azerbaijan and Iranian Kurdistan. Stalin also occupied and sought to annex Nationalist China's East Turkestan (now the "Xinjiang Uighur Autonomous Region") as Chiang Kai-Shek was losing the 1937–45 Sino-Japanese War, but after World War II he reluctantly returned it to China, while helping Mao Tze-Tung win the Chinese civil war of 1946–49.

Conversely, the United States has been the last Power to be involved in the Middle East since World War II (1939–45), and in Central Asia only since 1980. US wartime goals were only to support Great Britain in defeating the Axis in North

---

18 Stephen S. Kaplan, *Diplomacy of Power: Soviet Armed Forces as a Political Instrument* (Washington, DC: Brookings, 1981); F. Feytö, *Storia delle Democrazie Popolari,* 5–280; H. Kissinger, *Diplomacy,* 550–780; Stanley Hoffmann, *Gulliver's Troubles, or The Setting of American Foreign Policy* (New York: McGraw Hill, 1968); Barry M. Blechman and Stephen S. Kaplan, *Force without War: U.S. Armed Forces as Political Instrument* (Washington, DC: Brookings, 1978); Robert H. Ferrell, *American Diplomacy: A History* (New York: Norton, 1988); Peter Calvocoressi, *World Politics, 1945–2000* (New York: Longman, 2001).

Africa, the Mediterranean and Italy, while secretly undercutting the British oil monopoly through preferential US deals with Saudi Arabia and Iran. During the Cold War, the Middle East remained remote, with US foreign and security policy totally focused on "Containment" of the USSR and communist subversion in Europe (the 1947 Truman Doctrine) and Asia (Korean War, 1950–53; two Vietnam Wars, 1946–54 and 1964–75), despite US–Soviet competition in 1948–49 to support Israel's independence against Arab invasion and terrorism. Thereafter, the USSR became involved in the Middle East/Gulf, supporting anti-Israeli Arab nationalism as a politico-ideological tool to undermine Western control of the region's geostrategic international sea-lanes of communication (SLOCs) and petrol/gas sources.[19]

President Truman's pro-Israeli policy was reversed by President Eisenhower's attempt to draw the wider Arab world into the US global containment system against the USSR and Communist China, while distancing America from regional Anglo-French colonialism. Yet US courting in 1954–55 of Egypt's pan-Arab radicalism under its dictator General Gamal al-Nasser failed to keep him from aligning with Soviet penetration of the Middle East/Gulf, or condemning the whole West (America too) as "colonial" oppressors of the Third World, while seeking to destroy Israel and nationalize the Anglo-French Suez Canal in 1956. It was only US politico-economic interventions that saved Egypt's Premier Nasser from the surprise Israeli-Anglo-French parallel attacks during the 1956 Suez Canal War, but Nasser still rebuffed the US and aligned Egypt even closer with the USSR. In 1956–75, Nasser's radical pan-Arabism and Soviet alignment both threatened pro-Western Arab states (Jordan, Lebanon, Saudi Arabia, N. Yemen), while Soviet arms sales fueled new conflicts against Israel (1967 Six Days War; 1969–71 Attrition War; 1973 Yom Kippur War).

The post-Suez collapse of Anglo-French colonial power in the region and anti-Western Arab nationalism stoked by Nasser and the USSR forced the US also to militarily contain local Nasserite and Soviet subversions with the 1958 Eisenhower Doctrine and Anglo-American interventions in Lebanon and Jordan against Egypt, Syria and Iraq. Then, in 1967–75, in partial response to the second Vietnam War debacle, both Presidents Lyndon B. Johnson and Richard Nixon followed Truman's 1949 lead in locking-in the influential US Jewish vote by fully aligning America with Israel during Arab-Israeli conflicts, even at the risk of US–Soviet showdowns in 1970 (Attrition War) and 1973 (Yom Kippur War) which could have escalated to World War III.

Thereafter, all US Presidents (Nixon, Gerry Ford, Jimmy Carter, Ronald Reagan, George Bush "Sr.", Bill Clinton and George Bush "Jr.") followed an even more ambitious and uncertain regional strategy: a) US military commitment to Israel's survival (including under Reagan a secret US-Israeli alliance against the USSR,

19 P. Calvocoressi, *World Politics, 1945–2000,* Chapters 10–14, 20; H. Kissinger, *Diplomacy,* 423–72; Simon Serfaty, "Europe, the Mediterranean, and the Middle East", *Joint Force Quarterly* (Spring 2000), 56–61; "Quel Avenir pour le Moyen Orient?" *Geostratégiques* 6 (2005); Dennis Ross, "The Middle East Predicament," *Foreign Affairs* (January 2005), 61–74.

later abandoned as bilateral tensions boiled over during Israel's 1982–86 Lebanese Intervention against Arafat's PLO guerrillas and the 1982–84 US-led Western peacekeeping coalition); b) US–EU active courting of Moderate Arabs (Anwar al-Sadat's Egypt, Jordan, Saudi Arabia, Morocco, Gulf states) to achieve regional peace with Israel and its withdrawal from the Sinai (Henry Kissinger's "Step-by-Step" Approach and Carter's Camp David Accord, plus the EU's Euro-Arab and Barcelona Mediterranean initiatives). US-Western peacekeeping of the Sinai (1980–now) sealed the successful 1979 Egyptian-Israeli Camp David Peace despite virulent opposition by the Arab Radical Rejectionist Front (Iraq, Syria, PLO, Libya, Lebanon, Yemen, Islamic Iran and the Egyptian terrorist groups who assassinated Sadat). Soviet regional influence quickly reduced, by 1975 to just Syria, S. Yemen, Libya and Iraq, collapsing in 1989–90 when her client states refused to pay back billions of Soviet arms sales.[20]

With the Middle East in turmoil, NATO and the USSR sought at least to spark East–West trade and stabilize relations in Europe during the 1969–79 Détente, with the superpowers signing unprecedented arms control accords to offset the risk of World War III: the Strategic Arms Limitations Treaties (SALT I, 1972, and II, 1979) and the decade-long, unsuccessful arms control talks to cut NATO/Warsaw Pact conventional forces (Mutual and Balanced Force Reductions—MBFR). East–West Détente and trade came just as the Vietnam War was over, with US forces weakened at home and abroad by their controversial withdrawal from that country. All of these events combined to also weaken NATO's overall combat effectiveness (especially US troops) in Europe, while NATO resisted pressures from several Allies (Canada, Great Britain, US) to reduce their forces in Europe. NATO instead sought to balance East–West arms control with major improvements in Allied conventional and nuclear forces, plus improving integrated command, control and flexibility of Alliance forces in Europe in the 1970s–80s (NATO Strategic Planning during the 1970s and *NATO Long-Term Defence Improvement Programme*). Then, under SACEUR Al Haig in 1975, NATO introduced its most important new annual exercise called REFORGER (Return of Forces to Germany): integrated multinational Allied training to support rapid US-Canadian air- and sea-lifts of reinforcements to NATO's Central Front in Germany and Northern Europe in case of a World War III invasion by the USSR/Warsaw Pact.

At the same time, the USSR secretly modernized in the 1970s all of its numerically superior Eastern European-based forces and Warsaw Pact with "state-of-the-art" weaponry, which undercut NATO's traditional qualitative superiority in forces against the Warsaw Pact. To counter this USSR/Warsaw Pact massive conventional and theater nuclear improvement, with SS-20 missiles (Intermediate-range Nuclear

20 John Stoessinger, *Why Nations go to War* (Belmont, CA: Thompson, 2008), 213–284; Hugh Thomas, *La Crisi di Suez, 1956* (Milan: Rizzoli, 1969); Michael Oren, *Six Days of War* (Oxford: Oxford University Press, 2002); Tom Segev, *1967* (Jerusalem: Keter, 2005); Jacques Derogy and Jean-Noel Gurgand, *La Morte in Faccia, 1973* (Milan: Rizzoli, 1973); Peter Tsouras (ed.), *Cold War Hot: Alternate Decisions of the Cold War* (London: Greenhill, 2003), 12–16, 99–125, 186–207.

Forces—INFs) in Eastern Europe, NATO responded with the 1977–83 controversial, vital modernization of NATO's own Theater Nuclear Forces (INFs) in Europe. NATO's December 1979 "Dual-Track Policy" modernized by 1983 its INF missiles with US Cruise and Pershing II in Europe to supplement NATO's inadequate conventional military forces against any Warsaw Pact attack, while negotiating a new arms control treaty to eliminate both Soviet and NATO INFs. But the USSR's parallel invasion of Afghanistan (1979–89) and hostile rejection of NATO's "Dual-Track Policy" precipitated the collapse of NATO–Warsaw Pact relations and East–West Détente, while the USSR again used the "Brezhnev Doctrine" to justify both the invasion of Afghanistan and to prepare a Soviet/Warsaw Pact invasion of Poland to crush its pro-democracy anti-Communist *Solidarność* union. However, this risk was preempted in 1981 by an unprecedented pro-Soviet Polish military coup, although fears of World War III rose as NATO readied its own military should Poland collapse into civil war and the USSR militarily intervene.[21]

NATO kept modernizing its INFs, despite the USSR support for massive anti-nuclear pacifist demonstrations in key NATO countries (Belgium, Germany, Great Britain, Greece, Italy, US), which died out once the INFs were deployed in 1983. NATO also adopted the revolutionary "Follow-On Forces Attack" (FOFA) Plan to improve conventional defenses against any World War III Warsaw Pact invasion: in combat, NATO would not restrict itself to the old losing "Forward Defense" of the Inter-German border, but would launch conventional attacks deep inside Eastern Europe to destroy Soviet/Warsaw Pact second- and third-echelon forces, plus their C^3I and reserves, well before they could reinforce the Red Army's initial offensives on NATO's fronts.

In 1985–91, East–West tensions between NATO and the Warsaw Pact finally abated, once reformist Soviet leader Mikhail Gorbachëv restarted Détente and new US–Soviet arms control accords radically cut nuclear and conventional weapons: the 1987 Intermediate-range Nuclear Forces (INF) treaty eliminating all "Euromissiles" and the 1991 Short-range Nuclear Forces (SNF) unilateral deals halving SNFs and placing the rest in US and USSR "central storages." On the conventional level, by 1988 deep cuts in Soviet forces had already left the NATO–Warsaw Pact balance in Europe and America roughly equal, with 4,788,000 Warsaw Pact men to 4,771,000 NATO and 493,000 French forces. Still, the Warsaw Pact retained a superiority of 1.5/2:1 in total armor-equivalent combat power, and 202 divisions compared to 121 NATO ones. Only the 1990 CFE Treaty eliminated the risk of a surprise Soviet World War III strike by deeply cutting NATO/Warsaw Pact forces in the "Atlantic to the Urals" (ATTU) region for each alliance. For the first time, on-site inspectors could verify the destruction of excess units, while US/Soviet forces in Europe were further cut to 195,000 each. After the 1990 dissolution of the Warsaw

21 Raymond Garthoff, *Détente & Confrontation: American-Soviet Relations from Nixon to Reagan* (Washington, DC: Brookings, 1985); Jonathan House, *Combined Arms Warfare in the Twentieth Century* (Lawrence, KS: University of Kansas Press, 2001), 189–285; M. Rimanelli, *NATO and Other International Security Organizations*, 282–666; S. Ambrose and D. Brinkley, *Rise to Globalism*, Chapters 12–15.

Pact and 1990–91 reunification of Germany within NATO, the USSR was forced to completely withdraw all its forces by 1994 from its ex-Eastern European satellites. US Presidents Bush Sr. and Clinton reciprocated by further cutting US forces in Western Europe from 336,000 in 1989 to 192/175,000 by 1992, 154,700 in 1993 and 100,000 in 1995, despite the protests of NATO and European Allies that such lower levels undercut NATO's war-fighting deterrent.[22]

In 1988–89 Gorbachëv also denounced the "Brezhnev Doctrine" and allowed *Solidarność* back in a power-sharing deal with the Polish military. However, Gorbachëv's openness and reforms failed to keep Eastern Europe in the Soviet-Communist fold once the threat of repression ("Brezhnev Doctrine") ended and local protests erupted in 1989. Gorbachëv's indecision between supporting reforms in the Communist system or repressions against both rebellious Warsaw Pact satellites and breakaway Soviet states (Baltics and Caucasus) unwittingly precipitated the anti-communist 1989 Revolutions in Eastern Europe and the demise of the Soviet "empire".

By early 1990 the USSR had lost control of all of Eastern Europe, Mongolia and the Warsaw Pact front-line defenses against NATO. These dramatic events also pushed NATO to open to its Warsaw Pact ex-enemies with the June 1990 "Message from Turnberry," and the London Summit then declared the USSR and Warsaw Pact no longer "enemies" and invited them and all "neutral" European states to establish diplomatic and cooperative ties with NATO. Consequently, the new North Atlantic Cooperation Council (NAC-C) comprised NATO Allies, USSR, Warsaw Pact and "neutrals" (including the three ex-Soviet Baltic states and Albania), signing a non-aggression accord on the basis of territorial integrity and political independence on the principles of the UN Charter and Helsinki Final Act of the Conference/Organization on Security and Cooperation in Europe (C/OSCE). In the shadows of the 1990 "2+4 Talks" on Germany's reunification (absorbing East Germany into NATO and the EU), the Warsaw Pact was also disbanded in late 1990 upon request of Hungary and Poland: all its European ex-satellites and three Baltic states sought instead US protection and integration in both NATO and the EU, while the USSR agreed to pull out its forces from most ex-Warsaw Pact states by 1991, with the last units leaving Poland and ex-East Germany in 1995. The Cold War's end finally also precipitated the 1991 collapse of the USSR.[23]

---

22 Kenneth Myers (ed.), *NATO: The Next Thirty Years* (Boulder, CO: CSIS/Westview, 1980); ACDA, *Treaty on Conventional Forces in Europe* (Washington, DC: Arms Control and Disarmament Agency, 1991); Marco Rimanelli, "East-West Arms Control and the Fall of the USSR, 1967–1994," *East European Quarterly* 29: 2 (June 1995), 237–73.

23 Michael Beschloss and Strobe Talbott, *At the Highest Levels: The Inside Story of the End of the Cold War* (Boston: Little, Brown, 1993); David Remnick, *Lenin's Tomb: The Last Days of the Soviet Empire* (New York: Random House, 1993); Richard Staar (ed.), *East-Central Europe and the USSR* (New York: St. Martin's, 1991); H. Kissinger, *Diplomacy*, 733–836; S. Ambrose and D. Brinkley, *Rise to Globalism*, Chapters 15–16; William Barbour (ed.), *The Breakup of the Soviet Union: Opposing Viewpoints* (San Diego, CA: Greenhaven, 1994); George Bush and Brent Scowcroft, *A World Transformed* (New York: Knopf, 1998), 3–205, 518–66.

NATO's Success and Transatlantic Security

NATO's success as the longest alliance in history has allowed its transatlantic partnership between America and Europe to guarantee their mutual security interests during the past bipolar Cold War and World War III threats from the USSR and during the current post-Cold War diffuse "Arc of Crisis."

Cemented politically and ideologically in the Allies' common combat and democratic heritage of World War I and World War II, NATO finally provided to both the isolationist "New World" and the hegemonic-weary "Old World" the perfect multinational military shield and common cause against the USSR during the Cold War. As an integrated military alliance in peacetime, under US Command through SACEUR, NATO provided adequate (although not effective until 1985–90) defenses with a fighting mix of conventional and nuclear forces to fight the threat of World War III launched by the superior USSR and Warsaw Pact. As a military deterrent this contained Soviet-Communist expansionism and subversion since the late 1940s, while the US nuclear umbrella and front-line GIs assured the automatic combat involvement of a now-vulnerable American heartland to the cause of democracy and peace in arms across the oceans in distant Europe (reversing the US Isolationism prevalent prior to and during both World Wars).

As an Alliance of "equals" under US leadership (not hegemony), it also guaranteed that no member state ever fight against each other (Germany vs Western Europe, Greece and Turkey), thus democratizing and reintegrating ex-Nazi West Germany and ex-Fascist Spain, while absorbing in 1990 ex-Communist East Germany, and in 1997–2008 the ex-Communist Eastern European and Baltic states. As a military alliance steeped since the 1967 Harmel Report in internal democratization, NATO has also been the beacon of democracy in the post-Cold War (1995 Principles on NATO Enlargement) for all Partners and aspirant Allies seeking its protection against ex-Soviet Russia and against each other over rival minorities and unjust borders. Further, as a credible war-fighting military alliance, NATO compelled the USSR/Warsaw Pact to enter verifiable and stabilizing international arms control treaties, slashing theater and strategic nuclear weapons, as well as conventional armies in Europe facing each other across the Iron Curtain.

NATO's historic, military and geostrategic roles remain vital also in the post-Cold War: its loss as a permanent peacetime transatlantic alliance would be a catastrophe negating the blood sacrifices of Allied nations in both World Wars when the lack of NATO's integrated regional security, joint training and political identity as a "Western Community" resulted in millions of deaths and belated twin US victorious military interventions. NATO has won the Cold War, but its pro-UN post-Cold War global peacekeeping must also resist insidious inward domestic calls for pacifist politics within all its members, especially during economic crises. Otherwise, as US Defense Secretary Robert Gates criticized the Alliance in June 2011, it now "risks collective military irrelevance" if further domestic cuts in forces and budgets erode both its long-term combat peacekeeping (Afghanistan) and short-term interventions (Libya).[24]

24 M. Rimanelli, *NATO Enlargement after 2002*, pp. 1–57; Marco Rimanelli, *NATO's 2002 Enlargement: U.S.-Allied Views on European Security*, in Hall Gardner, ed., *NATO and the European Union: New World, New Europe, New Threats* (London: Ashgate, 2004), p. 91–106. Quote from speech to NATO by US Defense Secretary Robert Gates that NATO risks "collective military irrelevance" in "Gates: NATO Alliance 'Dim, if not Dismal'," AOL News, 10 June 2011, http://www.huffingtonpost.com/2011/06/10/gates-nato-alliance-dim-i_n_874715.html.

The Cold War and the Media: Lessons from America in Vietnam

Steven Ekovich

The relationship of the media with war may not be as old as war, but it is certainly as old as the media—as old as the first representations of armed conflict between human communities. Representations of war encompass all forms of media: the written word, painting, sculpture, architecture, dress, masks… It is expressed as well in music, song, dance and plays. The earliest works of literature are stories told about war. Epic poems, myths, sacred texts, histories and legends recount grand and violent struggles. All innovations in the means of communication, media, have been used for political purposes. To the extent that war is the continuation of politics by other means it is natural that war has included in its panoply of arms any and all media at its disposition. This is particularly crucial when war is understood in its broadest sense as the confrontation of wills. Any instrument that can bend the will of an adversary to one's purposes is vital in such contests. But these chronicles of armed confrontations are not used simply to glorify war. They are essential cultural vehicles that define for societies their virtues and vices: heroism, vanity, venality, treachery, brutality, loyalty, comradeship, love, hate, forgiveness …

In today's postmodern world there is still the ancestral need to continually analyze, mythologize, reinterpret, culturally frame, legitimize and even to render aesthetic the human endeavor of war. Societies never stop revisiting, reviewing and reinterpreting their wars by using all media available to them. In recent experience the American War in Vietnam has been catalogued as the first war in history when the new medium of television played a crucial and decisive role. Nevertheless, even though Vietnam is called the first television war, the printed press remained powerful. The first television war is an interesting and revealing example of the age-old relationship between media and war. Societies carry their technologies, their institutions and their values onto the battlefield. Furthermore, the media, and in particular television, have made a successful effort to ensure that they will forever play a central role in the lasting narrative of the war in Vietnam.

The Legend and the Reality

The most common narrative about the role of the media during the American war in Vietnam is of a vigilant, independent corps of responsible journalists who uncovered the truth that presidents tried to hide, and stopped the war. As this story goes, by revealing the truth and unmasking official propaganda the insurgent media saved American honor while at the same time reaffirming the vitality of the freedom of expression. For those who were in favor of the war, and who today think it was a noble and necessary endeavor, the media malevolently contributed to the US defeat because it is believed that journalists were generally against the war. Many Vietnam veterans also feel that uncensored and overly negative television coverage helped turn the American public against the war, and even against the veterans themselves. There are, indeed, other narratives. However, the variations on the theme that the democratic media brought the war to a halt, for good or for bad, has taken a firm hold on popular imagination not only in the United States, but all over the world, and continues to dominate.[1] It is, of course, laudable that the freedom of expression, "speaking truth to power," should be celebrated as a fundamental value of a democratic civil society. But as frequently happens with potent commonplaces that take on mythological dimensions, the passing of time and rigorous research bring on a more sober and accurate analysis—or at least provide us with a new persuasive interpretation until another is put together. Of course, it is never easy to take apart a myth. Nevertheless, recent scholarly research has called into question the tenacious orthodox interpretation of a valiant media opposed to government, and bringing it to heel.

The Antiwar Media and the Anti-Antiwar Media

The mainstream media and the prestige press did report on the American antiwar movement, but regularly presented it in a negative light as deviant and even

1 Since this is the hegemonic interpretation, it can be found all over the place. Michael Mandelbaum, "Vietnam: The Television War," *Daedalus* 111: 4 (Fall 1982), 157, provides a useful list. To it should be added a standard and widely read history of the war: Guenter Lewy, *America in Vietnam* (New York: Oxford University Press, 1978), 433–4. Also see Samuel Huntington, *American Politics: The Promise of Disharmony* (Cambridge, MA: Harvard University Press, 1981); Austin Ranney, *Channels of Power: The Impact of Television on American Politics* (New York: Basic Books, 1983); Martin Esslin, *The Age of Television* (San Francisco: W.H. Freeman, 1982); Kathleen J. Turner, *Lyndon Johnson's Dual War: Vietnam and the Press* (Chicago: University of Chicago Press, 1985), 4; Peter Braestrup, *Big Story: How the American Press and Television Reported and Interpreted the Crisis of Tet 1968 in Vietnam and Washington* (Boulder, CO: Westview Press, 1977); and Peter B. Clark, "The Opinion Machine: Intellectuals, the Mass Media and American Government," in Harry Clor (ed.), *Mass Media and American Democracy* (Chicago: Rand-McNally, 1974).

dangerous. The reigning doctrine of balanced reporting committed the media to giving voice to the other side, but modulating that voice so that it would not threaten wartime unity. Todd Gitlin, a former president of the leftist Students for a Democratic Society, has engaged in scholarly self-reflection on the relation of the media to the American left and the antiwar movement. As Gitlin summarizes this, "A news story adopts a certain frame and rejects or downplays material that is discrepant. A story is a choice, a way of seeing an event that also amounts to a way of screening from sight." For the movement leader turned scholar, the methods used to achieve this were "trivialization" (mocking the anti-war movement's language, dress style, accusatory posture and values in general), "polarization" (casting the movement as extreme), highlighting dissension within the movement, implying that it was subversive, underestimating the number of people involved in demonstrations, and in general maintaining a contemptuous attitude toward the protests. The net impact of this ensemble of representations painted the movement as an anarchic and subversive conspiracy that threatened the social fabric.[2]

As the American military presence in Vietnam was reaching its peak, protestors even had their movement consecrated by Norman Mailer, the author of one of the most famous novels of World War II. In his 1968 work *The Armies of the Night* (which was awarded both a Pulitzer Prize and the National Book Award) he straddled the genres of history and fiction to render the antiwar march on the Pentagon in 1967 into a cultural spectacle that raised its participants into celebrities, although questionable ones. Mailer's spotlight thrown onto the event reprised the ambiguity, even the schizophrenia, of antiwar protest, portraying it as opposition while at the same time with a certain disdain for, in the author's words, the "jargon-mired" and "middle-class cancer pushers and drug-gutted flower children."[3]

Media historians Michael Mandelbaum and Adam Garfinkle pick up this theme and suggest that the antiwar movement might have produced a result contrary to its goal: "Because of representations which undermined the articulation of any unified or coherent anti-war discourse, along with images which depicted styles of dress, patterns of behavior and sexual conduct which appeared to challenge dominant values and norms, it has been argued that the anti-war movement may have helped prolong American engagement in Vietnam."[4] In short, the many Americans who were distressed by the war were more distressed by the antiwar movement and what they perceived to be its rowdy and unpatriotic activities.

2 Todd Gitlin, *The Whole World is Watching: Mass Media in the Making and Unmaking of the New Left* (Berkeley: University of California Press, 1980), 7–8, 49–50. See also the summary of Gitlin's analysis in Graham Spencer, *The Media and Peace: From Vietnam to the "War on Terror"* (Basingstoke: Palgrave Macmillan, 2005), especially the section "Reporting the anti-war movement," 62–70.

3 Norman Mailer, *The Armies of the Night* (London: Penguin Books, 1968), 292, 45.

4 Michael Mandelbaum, "Vietnam: The Television War," *Daedalus* 3 (1982), 165. Garfinkle develops this idea at length in his Adam Garfinkle, *Telltale Hearts: The Origins and Impact of the Vietnam Antiwar Movement* (New York: St Martin's Press, 1995).

For sociologist E.M. Schreiber, a more credible interpretation of the impact of the antiwar movement would be that it made little or no difference to the length of time America was engaged in conflict. As Schreiber notes, the impact of antiwar demonstrations on public opinion followed changes in the Vietnam-related views expressed by the mainline news media, rather than directly inflecting public opinion.[5] Viewed from a more scientific angle, media researcher Shanto Inyegar sees the lack of impact as an inevitable result of the way in which antiwar demonstrations were structurally formatted by the mainstream media. In his rigorous study of how television news frames political issues, he found that, for example, coverage of mass-protest movements generally focuses more closely on specific acts of protest than on the issues that gave rise to the protests. This pattern characterized network coverage of the protests against the Vietnam War. The protests were covered in an "episodic news frame" that takes the form of an event-oriented case study, rather than in a "thematic frame" that places public issues in some more general and abstract context. "Since television news is heavily episodic, its effect is generally to induce attributions of responsibility to individual victims or perpetrators rather than to broad societal forces, and hence the ultimate political impact of the most common framing is pro-establishment." A related result of his research is that "Americans are generally tolerant of dissent when survey questions frame dissent as a general democratic right but are significantly less tolerant when questions direct their attention to specific dissenting groups."[6]

It should be recalled that the antiwar movement appeared in the United States at the same time that the nation was undergoing a cultural shift related to a generational friction that saw more young Americans than ever on university campuses. The antiwar movement and the campus counterculture were tied together in the same social setting—a cultural bubble disconnected from every other segment of American society except elite intellectuals on both coasts. Inside this bubble was an alternate media universe. Through much of American history, social movements have launched newspapers to publish news that mainstream newspapers refused to print. Like so much else in the 1960s, the underground newspaper boom was located at the intersection of technological progress and social unrest. For most of those involved, organizing the student movement and publishing the paper were indistinguishable activities. The underground press not only wanted an end to the Vietnam War, but to all war, along with black power and an end to racism; civil rights and students' rights, and then women's rights and gay rights; to bring the troops home; and for the most extreme free universities, free love, free music and free drugs. But the alternate media undermined itself when it adopted images of the Vietnamese enemy as an armed and righteous force of underdogs. "On posters and flyers, in underground newspapers and in film,

5 E.M. Schreiber, "Anti-War Demonstrations and American Public Opinion on the War in Vietnam," *British Journal of Sociology* 27: 2 (June 1976), 225–36. This is also the view of Spencer in *The Media and Peace*.

6 Shanto Iyengar, *Is Anyone Responsible?: How Television Frames Political Issues* (Berkeley: The University of Chicago Press, 1987), 13–16.

heroic and determined peasants in poses made popular by socialist realist art or small armed peasants guarding giant American prisoners came to represent a new iconography of triumph" that Tom Engelhardt sees as the turning on its head of the "American victory culture," which, of course, was repugnant and frightening to middle America.[7]

No account of the alternate media during the war would be complete without considering the underground press that blossomed in the ranks of the military. After the 1968 Tet Offensive the GI antiwar movement exploded. In 1967 there were three underground GI papers. By 1972, the Department of Defense reported that 245 had been published.[8] The underground papers were often a bridge GIs used to cross over from private misgivings to open opposition. The GI paper became a significant conduit for news of what was going on in the barracks and the battlefield, what the major prestige press refused to print. By the 1970s resistance to the war had spread to the ranks of all branches of the military. A two-volume study of soldiers' dissent was prepared for commanders in 1970 and 1971 by the Research Analysis Corporation, now publicly available. The reports reveal the startling dimensions of GI resistance, depicting a movement more widespread than known by the public at large at the time. The activists of the GI underground press were overwhelmingly white "middle-class intellectuals." Stockade riots, race rebellions on bases, and insurrections in the field in Vietnam were generally led by black soldiers.[9] Underground publications of soldiers conflated antiwar and anti-military sentiments. The mocking of commanders and careerists got tied up with questions about the wisdom of the war. As with the antiwar movement in general, the mainstream press reported very little about rebellion and resistance in the ranks. There is no doubt, however, that the GI movement weakened the effectiveness of the military endeavor. It certainly contributed to the ending of the draft and the establishment of all-volunteer armed forces.

---

7 Tom Engelhardt, *The End of Victory Culture: Cold War America and the Disillusioning of a Generation* (Amherst: University of Massachusetts Press, 1995), 40. A thorough history of oppositional media in the United States is Bob Ostertag, *People's Movements, People's Press: The Journalism of Social Justice Movements* (Boston: Beacon Press, 2006). I have drawn from pp. 119–22. On elite opposition see Paul R. Gorman, *Left Intellectuals and Popular Culture in Twentieth-Century America* (Chapel Hill: The University of North Carolina Press, 1996), as well as Adam Garfinkle, *Telltale Hearts*.

8 Terry H. Anderson, "The GI Movement and the Response from the Brass," in Melvin Small and William D. Hoover (eds), *Give Peace a Chance: Exploring the Vietnam Antiwar Movement* (Syracuse, NY: Syracuse University Press, 1992), 133.

9 Howard Olson and R. William Rae, *Determination of the Potential for Dissidents in the U.S. Army* (McLean, VA: Research Analysis Corporation, March 1971); R. William Rae, Stephen B. Forman and Howard Olson, *Future Impact of Dissident Elements within the Army on the Enforcement of Discipline, Law and Order* (McLean, VA: Research Analysis Corporation, January 1972). Cited in David Cortright, "GI Resistance During the Vietnam War," in Small and Hoover *Give Peace a Chance*, 117. See all of the chapters in Part Two, "The Military and the Antiwar Movement," in Small and Hoover.

The Myth Challenged

So, it turns out that the relationships of the media to power during the Vietnam War were significantly more complex and ambiguous than the valorous myth of media resistance. In fact, journalists were mostly complicit with those in power and supported the decisions that led to the American involvement and subsequent escalation in Vietnam. This was particularly the case at the beginning of the American intervention, but a certain measure of complicity endured right up to the moment of the precipitous and ignoble withdrawal of the US military in 1975. Of course, there is always some truth to any popularly accepted orthodoxy. This is the case for the prevalent popular narrative. For example, in the conclusion of her study of what she calls "US Official Propaganda" during the Vietnam war, Caroline Page asserts that when "the differences between official perceptions and reality" had become too great to sustain "they constituted a considerable and continuing weakness in the Administration's propaganda campaign."[10] The media is cast in the role of presenting hard reality against the misrepresentations of the government. Other research has shown that this is not erroneous, it just does not take into consideration the internal fragmentations and internecine conflicts of the administrations of Presidents Kennedy, Johnson and Nixon.[11]

American "propaganda" was not monolithic, and even became contradictory and incoherent. Page admits that the use of the term "propaganda" does not necessarily connote inaccuracy or lying, it can be employed interchangeably with the word "information" or "official information." Nevertheless, she uses the more charged term "propaganda" in the title of her study, developing needed nuances throughout the work.[12] The term "propaganda" should be employed warily in reference to the role of the media during the war in Vietnam. The term generally implies a carefully crafted and skillfully managed strategy to explain and justify an undertaking that would not otherwise be popular or accepted by an accurately informed public. This definition is an oversimplification when applied to the war in Vietnam.

In the case of Vietnam, the incoherence of "official information" came not only from clashes of perceptions and reality between the White House and the media—the story of the dominant narrative—but also in very decisive measure from clashes *within* and *between* the executive and legislative branches of American government. Different factions and powerful individuals within government advanced their interests by mobilizing favorably disposed journalists. Journalists rushed into these breaches to advance their own viewpoints and appeal to the preferences of the "consumers" of their interpretations. In this way reporters could also maintain access to the powerful and the inside information they possessed. Journalistic

10 Caroline Page, *U.S. Official Propaganda During the Vietnam War, 1965–1973: The Limits of Persuasion* (London and New York: Leicester University Press, 1996), 299.

11 The most notable is Daniel C. Hallin, *The "Uncensored War": The Media and Vietnam* (Oxford: Oxford University Press, 1986).

12 Pages 5, 41–49 and throughout Chapter 2.

reputations and careers could be advanced by pitching to their "market" while at the same time cultivating the influential. So, the behavior of the media during the Vietnam War was intimately related to the varying unity and clarity of the American government itself, as well as the corresponding consensus in society at large. When the White House does not manage to control a coherent message, this opens the door to policy dissonance, which then leads to criticism and resistance which can be carried into the public arena.

The Cold War Consensus

American political culture and its legal conventions allow for wide-open free speech with very limited constraints. However, all media exist within the confines of its national culture, and that culture defines both overtly and tacitly what is permissible to say and to do. If we use Clifford Geertz's definition of culture as "an ensemble of stories we tell ourselves about ourselves,"[13] it is evident that war stories are no exception. First of all, it cannot be expected that American journalists would tell the story of the Vietnam War from any side other than that of the United States and its allies. As Jill Lepore says in the introduction to her remarkable book on King Philip's War in New England in 1675, "The words used to describe war have a great deal of work to do: they must make clear who is right and who is wrong, rally support, and recruit allies; and they must document the pain of war, and in so doing, help alleviate it."[14] Words and images about war are what political philosopher Michael Walzer has called a "moral vocabulary of warfare," the language by which all sides justify their own actions while vilifying those of their enemies.[15] The Vietnam War was no exception to this.

Of course, the side fighting with a relatively free press at its side is at an inherent disadvantage when confronting the moral vocabulary of warfare employed by authoritarian and totalitarian regimes. As Guenter Lewy reminds us in his *America in Vietnam*, for the Vietnamese communists ideological mobilization at home never had to bump up against protections of free speech. Mounting their side of the story against the enemy and carrying it to the world was therefore much easier, "and they worked at both objectives relentlessly and with great success." Lewy points out that "The Western observer, essentially unable to check out the claims of the communist camp, was left with the image of a tough and highly effective enemy while at the same time he was daily exposed to the human and bureaucratic errors and shortcomings of his own side."[16]

---

13 Clifford Geertz, *The Interpretation of Cultures* (New York: Basic Books, 1975).

14 Jill Lepore, *The Name of War: King Philip's War and the Origins of American Identity* (New York: Vantage Books, 1998), ix. Lepore's book won the Bancroft Prize.

15 Michael Walzer, *Just and Unjust Wars: A Moral Argument with Historical Illustrations* (New York: Basic Books, 1977), 16.

16 Guenter Lewy, *America in Vietnam* (Oxford: Oxford University Press, 1978), 433.

But the outer boundaries of consensus are not just the result of the natural tendency to root for one's own side. Political consensus grows out of political ideology. What is known in the United States as the Vietnam War is known in Vietnam as the "American War." On the American side the war was waged in the name of the fundamental American values that have shaped its democracy from the foundation of the republic. But all American wars have been ostensibly fought in the defense of these values. During the Vietnam War these values were mobilized in the additional effort to contain and roll back communism, the primary threat to liberal institutions following World War II. This Cold War consensus required that the West defend its interests and institutions not only at national borders. The ideological struggle meant that the threat to liberalism had to be met all around the world. The moral vocabulary of the American side thus overshadowed other dimensions of the conflict in Vietnam: a peasant uprising against a feudal social order steered by French-educated leaders of a nationalist struggle against colonial rule.

Nearly all of the reporting at the start of the American War fit snugly into what Daniel Hallin calls the "Sphere of Consensus." However, since democratic values also demand open debate, he also allowed in his analysis for a "Sphere of Legitimate Controversy"—legitimacy defined by deeply-rooted and unconsciously shared values and attitudes. If controversy and criticism attempted to go beyond legitimate criticism it was excoriated as deviant and even subversive. Only in the later stages of the war did reporting cross over from the sphere of legitimate controversy and bravely foray into the "Sphere of Deviance"—or rather it stretched outward the limits of the domain of legitimate controversy.[17] A testimony to the power of the consensus comes from diplomat Richard Holbrooke. Looking back on his youthful days in Vietnam, he would recall: "When I got to Saigon I was twenty-two and I believed everything I had been told by the United States government. I believed that the commitment was correct – freedom of choice, self-determination, save the country from Communism – and that we were doing the right thing because the US government *did* the right thing. In those days you didn't question it."[18] His experience in Vietnam weakened those unquestioned beliefs, as it would for many Americans.

A Civilians' War

Even though throughout the Vietnam War there was a Cold War consensus that viewed the United States as a bulwark against communist expansion (and this consensus traversed the ideological spectrum from left to right, from liberal to conservative), it should also be noted that the American military had always

17 Hallin's presentation of spheres is in Hallin, *The "Uncensored War,"* 117–18.

18 Quoted in Kim Willenson, *The Bad War: An Oral History of the Vietnam War* (New York: New American Library, 1987), 147.

harbored a resistance to buttressing that bulwark by means of a major land war in Asia. As Robert Buzzanco points out in his *Vietnam and the Transformation of American Life*, "Vietnam was very much a civilian's war. Liberals in the John F. Kennedy and Lyndon B. Johnson administrations made crucial decisions to fight there and to constantly escalate the conflict. But from the 1950s on a significant number of ranking military leaders had argued against a war in Vietnam." More specifically, "They believed that it was not an area of vital importance to US security, that the enemy had the capability to fight a long-term guerilla war on its own terrain, that the allied government and military of the RVN [Republic of Vietnam in the south] was corrupt and weak, and that America's heavy firepower would be ineffective or counterproductive."[19]

This disparity between civilian and military views was apparent as early as the communist siege of Dien Bien Phu during the end of the French war in 1954. John Foster Dulles (the American Secretary of State) and others in the civilian leadership were tempted to accept a French request to use the America air force to bomb around the perimeter of the French base. President Eisenhower, the former general, and other military leaders flatly refused. There was also never serious consideration of the use of atomic weapons, a persistent legend about the battle.[20] The Americans finally opted for a policy of merely financing the remaining French war effort, about which in any case there was great pessimism in Washington.

As Lewis Sorely points out in *A Better War*, his recent widely read and debated revisionist view, "The stew of people involving themselves in the war continued to roil unabated, fueled by internecine warfare within the government." For Sorely, this policy cacophony derived from differing perceptions of the participants' "roles and interests, readings of and responses to domestic political pressures, and outlooks on the state of the war and its possibilities."[21] He quotes Admiral Elmo Zumwalt, who in early 1970 returned from Vietnam to Washington to become Chief

---

19 Robert Buzzanco, *Vietnam and the Transformation of American Life* (Oxford: Blackwell Publishers, 1999), 20, 51–3. Buzzanco draws on his earlier work *Masters of War: Military Dissent and Politics in the Vietnam Era* (New York: Cambridge University Press, 1996).

20 Buzzanco, *Masters of War*. On Dien Bien Phu and the American government the sources are voluminous. The place to begin is the widely read Pulitzer Prize winning book by Stanley Karnow, *Vietnam: A History* (New York: Viking, 1983), 213–14. See a standard treatment in George C. Herring, *America's Longest War* (New York: John Wiley & Sons, 1979), 26–9; or Peter A. Poole, *Eight Presidents and Indochina* (Huntington, New York: Robert E. Krieger Publishing Company, 1978), 26–31. A good account based on released archives that also takes apart the atomic strike myth is George C. Herring and Richard H. Immerman, "Eisenhower, Dulles, and Dienbienphu: 'The Day We Didn't Go to War' Revisited," *The Journal of American History* 71: 2 (September, 1984), 357–8. Only very recently has an objective account of the battle been told from the Vietnamese communist side. The publication was a literary event in Vietnam. Only a French translation exists for the moment, Dao Thanh Huyen, et al., *Dien Bien Phu Vu d'en Face* (Paris: Nouveau Monde Editions, 2010).

21 Lewis Sorely, *A Better War: The Unexamined Victories and Final Tragedy of America's Last Years in Vietnam* (Orlando: Harcourt, Inc., 1999), 173. Hallin develops the same point.

of Naval Operations. In his memoirs Zumwalt considered all this maneuvering as "a species of immorality." He described "deliberate, systematic and, unfortunately, extremely successful efforts of the President [Nixon], Henry Kissinger, and a few subordinate members of their inner circle to conceal, sometimes by simple silence, more often by articulate deceit, their real policies about the most critical matters of national security." And he asserted that "concealment and deceit was practiced against the public, the press, the Congress, the allies, and even most of the officials within the executive branch who had a statutory responsibility to provide advice about matters of national security. The result, which I am sure was the one intended, was to foreclose outside and inside discussion of major policy issues."[22] Sorely points out that the admiral's observations are confirmed by William Bundy's careful study, *A Tangled Web*, published nearly three decades later. Bundy was a foreign affairs advisor to Presidents John F. Kennedy and Lyndon B. Johnson.[23] Both of these insiders reveal that tensions over strategy existed, and were even acrimonious. They were eventually carried by the media to the public.

Insiders to Outside to Insiders

In his remarkable and highly researched study *The "Uncensored War"*, Hallin has shown how these tense differences were played out by powerful actors through their media contacts. In the history of American democracy this was not the first time, and not the last, that the political struggles of insiders were carried to the public arena, often discreetly and cunningly, in order to shore up their moves on the internal chessboard of power. Segments of the American public became critical of the war when policy-makers became critical and divided. In the early stages of the US involvement the criticism was usually directed at tactics and military strategy, rather than the overall goals of the war. Criticism was aimed at demanding a *better* war, not ending it. The questioning of the goals of the war only came toward the end of the long military engagement. But even with critical reporting, the American public remained largely behind the war—at least until the communist Tet Offensive of 1968 when it became clear that with over half a million Americans in Vietnam the enemy came close to winning. (Tet is the traditional Vietnamese holiday celebrated at the start of the New Year.) Even though the Americans and their allies turned back the communist attacks, the Tet Offensive has since become frequently cited as supporting the age-old geopolitical principle that one may win a conflict militarily, but lose politically.

Political infighting and the struggle for influence by reaching out to public opinion is not new in democracies. This is less the case in other types of regimes, but

22 Sorely cites Admiral Zumwalt's memoirs *On Watch: A Memoir* (New York: Quadrangle, 1976).

23 William Bundy, *A Tangled Web: The Making of Foreign Policy in the Nixon Presidency* (New York: Hill and Wang, 1998).

even authoritarian governments find it necessary in today's world to employ the tools of public relations in order to maintain legitimacy or undermine the legitimacy of opponents. Half a century ago, Raymond Aron, the preeminent strategic thinker of his time, wrote in *The Century of Total War* how "Public opinion played hardly any part in the limited warfare of the eighteenth century; the professional soldiers, recruited from the lower classes of society felt no need to know why they were fighting."[24] Today, on the contrary, public opinion in democracies and the electoral cycle exercise an overwhelming influence over the conduct of wars. This was certainly the case with the Vietnam War.

The nature of a political regime shapes and limits the way it engages in wars and diplomacy. Even though military establishments are not democracies, the militaries of democratic regimes possess characteristics that support and do not undermine democracy. The defense establishment in a democracy is imbued with the ambient democratic culture, liberal institutions, a market economy and an open civil society. An essential feature of a democratic civil society is the freedom of speech and the transparency of information. Of course, allowances are made for tactical and even strategic secrecy when it comes to military operations and national security.[25] But in democracies the boundary of necessary secrecy is constantly contested. Media power and military power wrangle to protect and advance their perceived interests and responsibilities.

War and Censorship

The War in Vietnam is an interesting case study because the war was not carried out with a Constitutional declaration of war (the last being World War II), thereby limiting the legal reach of censorship and prior restraint. As the title of Hallin's study signals, Vietnam was an "uncensored" war. This is not to say that the US government and military were completely bereft of the means to legally control sensitive information. In Vietnam, the American culture of a free press was carried onto the battlefield. American journalists and the public expected a decent respect for the freedom of expression, while at the same time acknowledging the confining imperatives of military necessity. In American history the demand for transparent journalism became increasingly persistent with each war fought by Americans. In World War II reporters were inducted into the armed forces. They were "soldiers with typewriters" and their articles were subject to strict military censorship. Their reports of the war were essentially military dispatches. Because they were fully integrated into the military machine they identified with the cause, and of course bonded with the troops and celebrated their exploits and sacrifices. They also

24 Raymond Aron, *The Century of Total War* (Boston: Beacon, 1954), 9.

25 Developed more fully in Steven Ekovich, "The Culture of Democratic Public Administration and Military Culture," *Western Balkans Security Observer* 4: 14 (July–September 2009), 24–38.

considered that their task was to keep up morale on the home front. For example, the American public was not aware that the war against Japan was going very poorly in the early days. They learned this only long after the war had ended. As part of the task of keeping up morale back home the American public of the time also did not receive images of the carnage of battle.[26]

The journalistic practices of World War II were carried over into the Korean War. Even though the reporters in Korea were given a relatively free hand, unlike in World War II their identification with the cause would sometimes break down. When this occurred the US military would make it difficult for them to get their most pessimistic and critical stories out of the country. The last rampart against pessimism consisted of the editors back home, who would veto any stories they felt would aid and comfort the enemy.[27] As the famous television reporter Edward R. Murrow put it when he covered Korea for CBS: "I have never believed that correspondents who move in and out of the battle area, engage in privileged conversations with commanders and with troops and who have access to a public platform, should engage in criticism of command decisions or of commanders while the battle is in progress." Nevertheless, contrary to his experiences in World War II, Murrow did not like what he was seeing in Korea and decided that reporting some of the unpleasant facts would help to better inform the public. But those higher up at CBS back in New York decided not to use the most critical reports of their famous correspondent.[28] This very neatly encapsulates what would later happen in Vietnam until consensus broke down at home as well.

Starting with Vietnam, war reporting took on a more open character. Reports from the battlefield were far more accurate, and they also included for the first time images of American soldiers killed in battle. It was in Vietnam that American war reporters first took the view that they had to be "impartial observers." The adjective "war" attached to reporter must be underlined here. American journalism had already developed a professional ethic of impartiality, of so-called objectivity, in covering other areas of life. This is, of course, the ideal goal to be attempted, but not easy to attain. With the rise of the mass circulation, urban newspaper and the professionalization of the craft of journalism, reporters came to see themselves as "engineers" of the news. As a result, newspapers also became less overtly partisan, loosening their ties to political parties while still maintaining their ideological preferences—especially in editorializing, which was uprooted from the news and moved to a separate page, the editorial page.[29] This ethic of the "engineer" of the news was carried over into the other media relevant during the Vietnam conflict: radio and especially television.

26 Phillip Knightley, *The First Casualty: The War Correspondent as Hero and Myth-Maker from the Crimea to Iraq* (Baltimore: The Johns Hopkins University Press, 1975). See in particular for World War II, Chapters 10–13; for Korea, Chapter 14.

27 See Knightley, *The First Casualty*, Chapter 14.

28 Edward R. Murrow, *In Search of Light* (New York: Knopf, 1967), 166.

29 Steven Ekovich, "Corruption, the Media and American Politics," in Anne Deysine (ed.), *Argent, Politique et Corruption* (Nanterre: Publidix Université X, 1999), 153–172.

The First Television War

News reports of World War II and Korea sometimes came complete with battlefield photographs of troops in action. For film reports in World War II and Korea (the first war where filmed images started showing up on the very few television sets then in the United States), war correspondents would simply collect footage provided by other sources, often the military, and the news anchor would then simply add narration. The footage was often staged because cameras were large and bulky and could not easily follow the combat action with agility.

So, Vietnam was truly the first television war. By the mid-1960s television had become the most important source of news for the American public, and possibly the most powerful influence on public opinion itself. During the Korean War the television audience remained small, even miniscule. In 1950 only nine percent of households owned a television. By 1966, the Vietnam War era, this figure had risen to 93 percent.[30] As households in the United States became saturated with television sets more Americans began to get their news from the new electronic medium than from any other source. A series of surveys conducted by the Roper Organization for the Television Information Office from 1964 to 1972 demonstrated the growing presence and power of television. Those polled were asked from which medium they "got most of their news." In 1964, 58 percent said television; 56 percent, newspapers; 26 percent, radio; and 8 percent, magazines. By 1972, 64 percent said television, while the number of respondents who primarily relied on newspapers dropped to 50 percent.[31] Thus, as the Vietnam War dragged on, more and more Americans turned to television as their primary source for news. The Saigon bureau for years was the third largest the networks maintained, after New York and Washington, with five camera crews on duty most of the time. It should be noticed, however, that the print media (what became known as the "prestige press" because its consumers were more educated and upscale), remained very influential.

The Impact of Television

As television reports showed greater realism, American sentiments were no doubt shaken. However, subsequent studies have shown that this was not necessarily crucial in turning the public away from the war.[32] An American killed or wounded

30 George Comstock, "Television Research: Past Problems and Present Issues," in Joy Keiko Asamen and Gordon L. Berry (eds), *Research Paradigms, Television, and Social Behavior* (Thousand Oaks, CA: SAGE Publications, 1998), 11–36. Other statistics regarding television ubiquity and measurements of viewing are found here. See also Ekovich, "Corruption, the Media and American Politics," 154 and references cited.

31 Hallin, *The "Uncensored War,"* 106.

32 For example, ibid.

in combat can be viewed as a necessary and even heroic sacrifice in pursuit of the patriotic goals of a conflict, or on the contrary as a wasteful tragedy in a spurious cause. It is the perceived legitimacy of the armed conflict that shapes which view is taken and the level of casualties acceptable to the public—and its level of support for the war.[33] Nevertheless, a very high number of casualties will strain even the most steadfast support for an armed conflict. John Mueller has shown that the evolution of public support for the War in Vietnam was highly similar to that in Korea, but support for both wars declined as a logarithmic function of the number of American casualties. While public backing for the war in Vietnam did finally drop below those levels found during Korea, it did so only after the war had gone on considerably longer and only after American casualties had far surpassed those of the earlier war.[34]

This elucidates why presidents since Vietnam attempt to limit casualties and to define war goals in a way that will gather public and Congressional support. President Ronald Reagan's Secretary of Defense Caspar Weinberger, drawing lessons from the Vietnam conflict, laid out a series of conditions that should be met before using military force. Among the vital conditions of what came to be known as the Weinberger Doctrine were that "If we decide to commit forces to combat, we must have clearly defined political and military objectives," and "Before the United States commits combat forces abroad, the US government should have some reasonable assurance of the support of the American people and their elected representatives in the Congress."[35] When President Barack Obama was reviewing the US strategy in Afghanistan he and his advisors were reported to be reading Gordon M. Goldstein's *Lessons in Disaster*, whose thesis is that a president should never deploy military means in pursuit of indeterminate ends or mount stagecraft for decisions already made.[36] It was also revealed that the White House was reading Lewis Sorely's book on the under-reported successes of the US counterinsurgency

33 For a recent example that analyzes varying public and media views of combat casualties, in particular in the context of Iraq, see the editorial by Robert D. Kaplan, "Modern Heroes," *The Wall Street Journal*, October 4, 2007. The debate provoked by the editorial illustrates the importance of public trust in the mission shaping media and public views of casualties. See, for example "Responding to Robert Kaplan, Reach for the Civics, Not Psychology, Text," at "Democracy Arsenal," October 23, 2007, available at: http://www.democracyarsenal.org/2007/10/responding-to-r.html. Accessed October 7, 2010.

34 John E. Mueller, *War, Presidents and Public Opinion* (New York: John Wiley and Sons, 1973).

35 Caspar W. Weinberger, "U.S. Defense Strategy," in *Foreign Affairs* 64: 4 (Spring 1986), 159.

36 Gordon M. Goldstein, *Lessons in Disaster: McGeorge Bundy and the Path to War in Vietnam* (New York: Times Books/Henry Holt and Co., 2008). See Goldstein's own summary of his argument in "Lessons in Disaster: Why is the Obama Administration Reading up on its Vietnam History?" in *Foreign Policy*, available at: http://www.foreignpolicy.com/articles/2009/10/06/lessons_in_disaster?page=0,3.

strategy in the last years of the Vietnam War. *The Wall Street Journal* called this "A Battle of Two Books."[37]

Regardless of the impact of poorly defined war aims in Vietnam, the dominant popular narrative was that through television the horrors of war entered the living rooms of Americans and turned them against it. While ordinary citizens led their lives at school, work, and during dinner, they could watch villages being destroyed, Vietnamese children burning to death, and American body bags being sent home. This recurrent popular narrative has also been challenged by scholarly research. Myths and unchallenged clichés flourish if left unexamined.

We return once again to the exhaustive research of Daniel Hallin, who has conducted perhaps the most thorough study of US media coverage of Vietnam. Undermining the standard rhetoric that Vietnam had been the "living room war"—an "uncensored war" showing its "true horror"—Hallin found a war, especially on TV, that was in fact largely sanitized. He demonstrates that this was the result of media coziness with government and military sources. The TV networks adopted policies against airing footage that might offend soldiers' families or shock the sensitivities of civilians in their living rooms. Pictures of US casualties were actually rare, Vietnamese civilian victims almost nonexistent. Even the infamous massacre by American soldiers of the villagers of My Lai was difficult to get into the mainstream news. The atrocity took place in March 1969, but was not reported until November 1969. Only the persistence of the American journalist Seymour Hersh pushed the story into widespread public view.[38] After the turning point of the 1968 Tet Offensive the story gave added impetus to anti-war sentiment in the United States. But it seems the story took hold mostly because the military position of the United States had become precarious and the atrocity at My Lai fed into the new anxiety and self-doubt. The atrocity contributed to weakening the moral dichotomy television and other media had set up between Americans and the enemy.

Rather than bringing the gruesome "realism" of combat into the living room, the very nature of television technology may well have attenuated its brutality by shrinking it in size. As Michael Arlen intriguingly proposes in his *Living-Room War*: "I can't say I completely agree with people who think that when battle scenes are brought into the living room the hazards of war are necessarily made 'real' to the civilian audience. It seems to me that by the same process they are also made less 'real' – diminished, in part, by the physical size of the television screen, which, for all the industry's advances, still shows a picture of men three inches tall shooting at other men three inches tall, trivialized, or at least tamed, by the enveloping cozy alarums of the household."[39] For media analyst Robert Stam, that very coziness may

37 "A Battle of Two Books Rages," *The Wall Street Journal* (October 7, 2009), A1.

38 A balanced presentation of the incident can be found on Wikipedia: "My Lai Massacre," at http://en.wikipedia.org/wiki/My_Lai_Massacre. A good detailed account of the efforts of Seymour Hersh and the difficulty of getting the story out is found in Knightley, *The First Casualty*, 428–31.

39 Michael Arlen, *Living-Room War* (New York: Penguin, 1982), 8.

adulterate the beastly side of war. "The smaller screen, while preventing immersion in any deep enveloping space, encourages in other ways a kind of narcissistic voyeurism. Larger than the figures on the screen, we quite literally oversee the world from a sheltered position – all the human shapes parading before us in television's insubstantial pageant are scaled down to Lilliputian insignificance, two-dimensional dolls whose height rarely exceeds a foot."[40] This conclusion is related to the research that has undermined the import of the "vividness effect." Media researchers Shanto Iyengar and Donald R. Kinder have shown through rigorous testing that the singular, even powerful and poignant image, has very little impact on a citizen's judgment of the broader issue it attempts to illustrate. "Under this assumption, it is one thing to understand that American boys are fighting and dying in Vietnam; quite another to watch them fight and die," they pointed out. The two researchers conclude that such images are too personal, too intimate and thus not easily transferred to the larger issue. So, while televised scenes of brutal combat are too small and intimate for some, for Iyengar and Kinder they are too close to the viewer to call into question broader national concerns.[41]

Compared to the "prestige" print media, television during the Vietnam War had assumed a different role in the purveyance of information. It performed, and continues to perform, a different political function. As Hallin puts it: "The prestige press provides information to a politically interested audience; it therefore deals with *issues*. Television provides not just 'headlines,' as the television people often say, nor just entertainment, but ideological guidance and reassurance for the mass public."[42] At the beginning of the war there was a sharp contrast between television and the print media. The prestige press practiced the kind of "objective" journalism that sometimes ventured outside the "Sphere of Consensus." Television reports, however, remained solidly beyond the ramparts of the consensus. This would change only later in the war when internal disputes among elites mobilized the electronic media.[43]

Nevertheless, both the prestige press, which appealed to an upscale public, and television (whose audience was more "downscale") drew on the deep structures of American ideology. But television is a medium that dramatizes and is more apt to present news in an "entertainment" and "soap opera" mode. It is a vehicle whose characteristics come closest to expressing social values in ways similar to those of fictional series.[44] And fiction of all kinds is an emanation of its cultural environment. Television, more than the print media, blurs the boundaries between "manifest political socialization" (what we learn directly and explicitly about politics) and

40 Robert Stam, "Television News and Its Spectator," in E. Ann Kaplan (ed.), *Regarding Television* (Frederick, MD: University Publications of America, 1983), 26.

41 Shanto Iyengar and Donald R. Kinder, *News That Matters: Television and American Opinion* (Chicago: The University of Chicago Press, 1987), quote from 35.

42 Hallin, *The "Uncensored War,"* 125.

43 Ibid., 118.

44 Kathleen Hall Jamieson, *Dirty Politics: Deception, Distraction, and Democracy* (Oxford: Oxford University Press, 1992).

"latent socialization" (what we learn about politics when we are learning about power, virtue and authority elsewhere and early in our lives).[45] Hallin notes that "Almost all prime-time TV entertainment sets its characters within a family or ... a substitute family of some sort." This was particularly pervasive in the Vietnam War era. And the nation tends to see itself as a family writ large, so that the American cultural suspicion of power and those who seek it coexists with a deep belief in loyalty to the political family and harmony within it.[46] This may help to explain why one of the traditional findings of research on the effects of mass media is that they often tend to reinforce people's deep-seated attitudes and propensities developed in more intimate, face-to-face environments.[47]

The president as something of a father figure is also necessary to understanding the potency of the American leader's words and actions in reinforcing consensus. Studies demonstrate that attitudes toward the presidency stem from messages received in childhood about the virtues of past presidents. They are made into secular saints. Esteem and respect for the office independent of the occupant is established at a rather early age.[48] Until the American presidents of the Vietnam era suffered from a "credibility gap" their power to shape the public desire for unity was very potent. It is also significant that at the time of Vietnam the most trusted man in America was not a president, but the famous TV anchorman Walter Cronkite, an authoritative father figure. When Cronkite concluded a report on the Tet offensive with the comment that the war had become a "bloody stalemate" and that the time had perhaps come to get out, President Lyndon Johnson is said to have turned to his aides and sighed, "It's all over."[49]

Bonding with the Troops

As the Vietnam War grew in magnitude with the increased presence of American troops (more than half a million at its high point in 1968), this also provided a new and expanding source of information for American journalists—the combatants on the ground. At the beginning of the war, reporters' contacts with the troops were mostly with field-grade officers. Later on they increasingly focused on the ranks,

45 A useful presentation of these concepts developed by Lucian W. Pye and Sydney Verba is found in Gerald J. Bender, "Political Socialization and Political Change," *The Western Political Quarterly* 20: 2, Part 1 (June 1967), 390–407.

46 For a thorough account of the American value system see James Oliver Robertson, *American Myth, American Reality* (New York: Hill & Wang, 1980). Hallin relates this to television: *The "Uncensored War,"* 124–5.

47 See the summary in Ekovich, "Corruption, the Media and American Politics."

48 A brief presentation of the relevant research is found in Robert E. Denton, Jr. and Gary C. Woodward, *Political Communication in America* (Westport, CT: Praeger Publishers, 3rd edition 1998), 191–5.

49 Austin Ranney, *Channels of Power: The Impact of Television on American Politics* (New York: Basic Books, 1983) 5.

the "grunts." And since reporters were given open access to the troops, even being ferried to the battlefield by military transport, dispatches from Vietnam frequently became battle reports. This gave journalists a different perspective on the war, even jarringly different from that peddled in the official briefings in Washington or those held daily at a Saigon hotel. Since there was very little censorship, writers and film crews got a view of the war that contrasted and even conflicted with the high-level briefings. Soldiers tended to give their views in a straightforward and honest manner. As the famous television reporter Dan Rather (who covered Vietnam in the mid-1960s) put it in 2006 at a two-day conference held at the John F. Kennedy Library in Boston: "I never had a captain or a sergeant lie to me ... They levelled with you. They knew what was going on. For example (and excuse my reporter's French if necessary) it was not uncommon for you to crawl into someplace with the captain of the line and say, 'What's happening?' And the captain would say, 'We are getting our ass handed to us.' And you try to reflect that in your reporting. That reporting would hit the wall in Washington because the Washington view was that we were handing them their backsides. They were not handing us ours."[50] At least until the growing loss of public support for the war, bonding with the troops usually led reporters to regard them sympathetically in their dispatches.

So, reporting from the field usually did not go against the limits desired by the military and political hierarchy. A very revealing study published after the first Gulf War in 1990–91 compared the reporting from that war, which was rigorously restricted by the military, and the dispatches from the Vietnam War. Journalists in the Gulf War were required to respect 12 rules on the type of information they could communicate to their news outlets. The types of restrictions included, for instance, a prohibition on information on intelligence collection activities (Rule 5), any information that revealed details of future plans, operations or strikes, including postponed or cancelled operation (Rule 2), and most other types of tactical information. The author of the comparative study wanted to see if the application of the strict Gulf War rules would have made reporting from Vietnam any different. The conclusion is astonishing: "Had the 12 rules been in place during Vietnam, would television coverage of that war have been any different? Probably not." But most astonishing of all, "Had Gulf war rules been in place in Vietnam, they would have been violated, but the violations appear to have been on the part of the command structure, not the reporters."[51] So, without the censorship of the Gulf War, reporters in Vietnam tacitly and no doubt unconsciously respected the

50 From "Vietnam Panel Discussion: Media and the Role of Public Opinion (excerpt)," part of a two-day conference "Vietnam and the Presidency," held at the John F. Kennedy Library in Boston, March 10–11, 2006, sponsored by the presidential libraries and the National Archives and Records Administration. Transcript available at: ABC-CLIO, "History and the Headlines," http://www.historyandtheheadlines.abc-clio.com/ContentPages/ContentPage.aspx?entryId= 1199591&currentSection=1194544&productid=10. Hallin also makes the point.

51 Oscar Peterson III, "If the Vietnam War Had Been Reported Under Gulf War Rules," *Journal of Broadcasting & Electronic Media* 39 (1995), 20–29.

reigning positive consensus on the war, at least until toward the end. Journalists appear to have been in tune with the desires of local commanders and their troops and did not violate their confidences.[52]

On the Ground vs The Big Picture

The consensus on the war shifted after Tet, but there remained division among reporters as well as among officials. "For the most part," concludes Hallin, "journalists seem to have interpreted Tet, without consciously making the distinction, for what it *said* rather than what it *did* [emphasis Hallin] ..." After all, Tet was a military defeat for the communists, but it was given the meaning of a political victory. "It may be one of the many ironies of Tet coverage," continues Hallin, "that it gave the public a more accurate view of the overall course of the war through the *inaccurate* view it gave of the outcome of the particular battle." After Tet, the growing American public malaise with the war was decisively solidified into the view that the war would be too long and too costly. The US mission in Saigon, as well as many in Washington (both in Congress and the administration), was also divided on the meaning of Tet. Hallin quotes a former US official in Saigon as telling him that those in the embassy were split roughly along generational lines, with the younger generally feeling that the official claim of victory was "a brave cover story." Here again, the views of officials and journalists seem to have coincided before mutually reinforcing each other.[53]

The clash between the on-the-ground view and the big picture was also lamented by the military leadership. General Creighton Abrams, the last US commander in Vietnam, complained that reporters in the last stages of the war were not seeing the big picture, especially the positive changes resulting from his new counterinsurgency strategy. He was exasperated that journalists who were accompanying his troops on the ground were also deliberately, and even maliciously, seeking out instances of low morale (such as drug abuse and conflict in the ranks). Indeed, the counterculture back home was also carried by American soldiers in Vietnam. Political and cultural changes in the United States were being reflected in its army and its journalism. Abrams complained that journalists were stuck in a mode of reporting that focused on combat operations and did not see, or even understand, the importance of large-scale trends like population security, pacification, infrastructure development, agricultural production, redistribution of land, training of local governments—and all else that was difficult to dramatize. "Christ," Abrams is quoted exclaiming in irritation, "even these young chaps, they've got to get it into sort of a World War II context. Otherwise you can't report it." As in the earlier stages of the war, he noted that there was a tendency on the part of reporters to fit their dispatches into preconceived notions, with their

52 See, for example, Braestrup, *Big Story*.
53 Hallin, *The "Uncensored War,"* 173 and footnote 30 in Chapter 5.

editors back home reinforcing this, except that now the consensus had shifted to a sympathy for an anti-war stance. As an example, Abrams recounted an anecdote related to him by a reporter who was interviewing an enlisted man about the 1970 incursion into Cambodia. The reporter wanted to know what the soldier thought of it, presumably believing he would condemn it because he was wearing a peace medal over his uniform. "Shit," replies the trooper, "it should have been done two *years* ago!" Then the reporter says, "Well, how can you say things like that and there you are wearing a peace medal?" Abrams then recounts that his guy responds, "I'm for peace, but you've got to *fight* for it!" The general ruefully concluded, "Well there went that, the segment never made it on the air."[54]

The contention that existed during the Vietnam War over the duty of the media to focus on the big picture (including war goals) or the reality of the war on the ground would continue well after armed hostilities were over. The media in the United States has unceasingly revisited the war in order to give it a meaning that fits into the ensemble of deeply-rooted American values. This has been carried out not only in the work of historians, but probably more powerfully in fiction, poetry, television documentaries, movies, political speeches, popular music, war memorials—all forms of media in our postmodern world. If a new and comprehensive interpretation of King Philip's War in 1675 came out only in 1999, there is no doubt that the "war" over the meaning of the Vietnam War will be a conflict without end.[55] Part of that ongoing revisionism will always include a place for the debate on the role of the media.[56]

Conclusion

There are those who blame the media coverage of the US war in Vietnam for the American defeat. This comes from some quarters of the American military, who deplore it, and from American journalists, who vaunt their influence on US decision-making. The Vietnam War was covered by the media as no other American military engagement had been covered before or, for that matter, since. Vietnam was the first television war. Even though there were cameras in World War II and in Korea, the reports that used them were framed differently. This was only partly due to the technology available. More basically it was due to the evolving

54 Abrams' views and his quotes in Sorely, *A Better War*, 225–6.

55 Lepore, *The Name of War*.

56 Treatments of the narratives of the war after the war are in Edwin A. Martini, *Invisible Enemies: The American War on Vietnam, 1975–2000* (Amherst: University of Massachusetts Press, 2007) and Mark Taylor, *The Vietnam War in History, Literature, and Film* (Tuscaloosa: University of Alabama Press, 2003). A good collection of fiction and nonfiction is in Stewart O'Nan (ed.), *The Vietnam Reader* (New York: Anchor Books, 1998). A good, but difficult-to-find collection of poetry is Jan Barry and W.D. Ehrhart (eds), *Demilitarized Zones: Veterans after Vietnam* (Perkasie, PA: East River Anthology, 1976).

relationship of American journalism to decision centers. When there is a high level of consensus among policy-makers and the public, journalism tends to follow the White House and is even nationalistic. When consensus breaks down the media reflects this and may even take the lead in criticism, especially when the military and political leaders bicker among themselves and leak to the media as part of their struggle for control over policy. This happened in Vietnam. The larger an American military engagement becomes, and the higher its risks and costs, the more likely this will be to happen. In Vietnam the struggle over policy, as well as the unclear strategic aims of the war, led to the undermining of the Cold War consensus underpinning its goals. This opened the door to harsh criticism from an ever more politically powerful and vigilant media whose practitioners increasingly considered themselves impartial and professional purveyors of information. But it must be emphasized that it was not the mainstream media alone that turned the public against the war. When consensus became fragile for many other reasons, it laid the groundwork for an ever-growing anti-war movement, fortified by veterans of the conflict who spoke out against the war. This in turn prodded the media toward more skeptical coverage. In the end, the collapse of America's "will" to fight in Vietnam resulted from a political process of which the media were only one part. Of course, the media, especially the new media of television, played a role. However, the previous belief of the importance of its weight, on its own, has been called into question—at least until new research can persuasively demonstrate otherwise. As William M. Hammond concludes in his exhaustive history of the role of the media in Vietnam, "In the end, what happened in Vietnam between the military and the news media was symptomatic of what had occurred in the United States as a whole."[57] The narratives of war carried by all media always go deeper than they know, and never end when the last spear has been thrown or the last shot fired.

57 William M. Hammond, *Reporting Vietnam: Media and Military at War* (Lawrence, KS: University Press of Kansas, 1998), 296.

NATO as a Post-Cold War Humanitarian and Peacekeeping Organization

Marco Rimanelli

The Cold War's end with the 1989–90 peaceful anti-Communist revolts in Eastern Europe, and the collapse of both Communism and the USSR, suddenly released old pent-up ethno-nationalisms in Europe and the ex-USSR, while US leadership of the West and NATO was reconfirmed with its permanent commitment to European security. Russia instead survived only as a semi-democratic "courtesy power" with minimal forces, a collapsing population and massive economic reliance on exports of raw materials and arms. In Russia, the political clash between President Boris Yeltsin's reformist government and a Communist/Nationalist parliament was repressed during the "Reds and Browns" failed Coup (1993), but opposition to NATO's expansion toward vulnerable Russian borders remained, pushing Russia into ultra-nationalist anti-Western policies in Yugoslavia and Iraq to retain a figment of its past international influence.

The post-Cold War (1990–now) early global euphoria for the West's "triumph" over the collapsing Communist Bloc and Warsaw Pact suddenly turned into bitter criticism of NATO as having outlived its missions. Blind inward-looking domestic and media advocates loudly called for NATO's "dismantlement" and dissolution, and the recycling of most military funds into social priorities, willingly drawing erroneous parallels with the Warsaw Pact's demise. Thus, the US, NATO Allies and Partners all scaled down their military budgets and NATO commitments (the "Peace Dividend"), while EU politico-economic integration (the Maastricht and Amsterdam Treaties, 1991–98) renewed debates on an autonomous EU politico-security force parallel to NATO. The US, although uncertain about the extent of its post-Cold War security commitments abroad, strongly refused to dissolve NATO and opposed any creation of a French-led autonomous "European Army" within the Western European Union (WEU) or EU, once the two merged in 1999–2001 as the basis of a new European Security and Defense Initiative/Policy (ESDI/P). Instead, the 2004 Istanbul NATO Summit allowed ESDI/P to relieve NATO peacekeeping in Bosnia by 2005, with EU security and logistics subordination to NATO in missions where "the Alliance chooses not to be involved," not the other way around as cyclically dreamed by France, but opposed by most old and all new NATO Allies.[1]

Since 1990, the chaotic post-Cold War world and erosion of Euro-Atlantic cooperation over divisive perceptions of "Future Threats" has been influenced by ethno-nationalist disintegration and genocide (Yugoslavia and, to a lesser extent, the ex-USSR), two international wars against Iraq (1990–91 and 2003–07), proliferation of Weapons of Mass Destruction (Iraq, Libya, Iran and North Korea), and the globalist onslaught of Islamicist terrorism (linked to Al-Qaeda). In this milieu, the sharp reduction of US/NATO military expenditure made the Alliance's Enlargements part of a concerted drive to unify with the EU both sides of the "Iron Curtain" into a broad pan-European politico-security area encompassing most OSCE and NAC-C Partners, with selective force-building through NATO's Enlargements from 16 to 28 member states in 1997–2010.

Thus, to preserve stability in a wider European area and gradually draw in newly democratic ex-enemies from Eastern Europe and the ex-USSR, NATO and the West changed the old Atlantic security system from its rigid, Cold War East–West collective defense strategy to a new, evolving, interlocking Euro-Atlantic security architecture with selective "Out-of-Area" global peacekeeping. The post-Cold War security architecture combined a reformed NATO (a lean, highly mobile, multinational force for both combat and peacekeeping) with all West European states, ex-"neutrals," WEU, EU and Conference/Organization on Security and Cooperation in Europe (CSCE/OSCE). In the early 1990s, US Presidents Bush Sr. and Clinton postponed NATO Enlargement by only integrating as Partners all 13 Eastern Aspirants, Russia and ex-Soviet successor states in the OSCE, 1990 NAC-C, 1995 Partnership for Peace and 1999 Euro-Atlantic Partnership Council (EAPC), with Russia as a symbolically "equal" Partner.

However, all 13 East European and Baltic Aspirants insisted on full NATO membership because they saw NATO as their only guarantor of pan-European regional stability, collective security, politico-military and democratic integration in the West against any Russian "revanchism" or regional ethno-nationalist crises. Instead, the US and many Allies worried that NATO's Eastward Enlargements would provoke anti-Western resentment in an isolated Russia after the collapse of its empire, while destabilizing NATO too if festering ethno-nationalist controversies persisted among old members (Greece, Turkey) and new ones (13 Aspirants), or

1 John E. Peters, *CFE and Military Stability in Europe* (St. Monica: RAND, 1987); Andrew Pierre (ed.), *The Conventional Defense of Europe: New Technologies and New Strategies* (New York: Council on Foreign Relations, 1986); William Cromwell, *The United States and the European Pillar: The Strained Alliance* (New York: St. Martin's, 1992); Werner Feld, *The Future of European Security and Defense Policy* (Boulder, CO: Rienner, 1993); G. Bush and B. Scowcroft, *A World Transformed* (New York: Knopf, 1998) 205–449; Marco Rimanelli, *Strategic Challenges to U.S. Foreign Policy in the Post-Cold War* (Tampa FL: Center on Inter-American and World Studies, 1998); Karl Kaiser and Pierre Lellouche (eds), *Le Couple Franco-Allemand et la Défense de l'Europe*, IFRI-DGAP (Paris: Economica, 1986); Hall Gardner, *Dangerous Crossroads: Europe, Russia and the Future of NATO* (Westport, CO: Praeger, 1998).

if decision-making bogged down. Once Aspirant-Partners embraced democratic values, conflict resolution and peacekeeping with NATO, during 1990–2008 all ex-Warsaw Pact states and Croatia joined both NATO and the EU first as Partners (1990–95), then as full Allies: the 1997–99 First tranche Enlargement (Czech Republic, Hungary, Poland); the 2002–04 Second Enlargement (Bulgaria, Estonia, Latvia, Lithuania, Romania, Slovakia, Slovenia); the 2008 Third Enlargement (Albania and Croatia); and all also joined the EU as equal "Western" members (2002–10).[2]

The Cold War era's bipolar nuclear and conventional balance of terror had also assured an unprecedented degree of regional stability: East–West deterrence and inter-bloc "policing" had cajoled the respective reluctant European allies into overcoming past nationalist hatreds and cooperating as partners on daily inter-bloc politico-military alliance integration within NATO and the Warsaw Pact. The Cold War's end exposed the unexpected fragility of the superpowers' regional security that kept peace and stability within each bloc, clamping down on local ethno-nationalist rivalries and holding a unified front against the enemy bloc. Once freed of previous bipolar control mechanisms, the 1990s explosion of pent-up nationalist and religious hatreds affected mostly the ex-Yugoslavia and part of the ex-USSR, while threatening to spread out of control to most of Eastern Europe and the ex-Soviet successor states with countless deaths and refugees. Local ethno-nationalist clashes proved impossible to stop by the traditional international conflict resolution of the OSCE, EU and UN, while post-Cold War crises threatened both Europe's security and NATO's existence, either by affecting vital oil supplies or by flooding the West with untold refugees and wartime ravages at the doorstep of that very European home that NATO had vowed to defend and all East Europeans sought to join.

In the post-Cold War era, NATO's success in regional self-defense and arms control reductions has sparked radical changes to avoid stagnation and fend off criticisms. NATO remains the only security organization capable of guaranteeing permanent US commitment to European security and an integrated military alliance able to quickly intervene in regional conflicts once consensus is achieved. Thus, NATO repeatedly updated its Strategic Concept from the Cold War (1950s East–West war-fighting missions under Art.V of the NATO Treaty, plus the 1960s "Three Wise Men" and Harmel Reports to protect a free, democratic Europe within the West) to post-Cold War challenges (1991, 1999 and 2010) to allow "Out-of-Area" peacekeeping, humanitarian and collective security missions within a broader Euro-Atlantic security area, supporting the OSCE and UN. Thus, the Alliance is focused now on four overlapping strategic goals: 1) regional integration and NATO Enlargements; 2) a new Euro-Atlantic collective

2 Jeffrey Simon (ed.), *European Security Policy after the Revolutions of 1989* (Washington, DC: National Defense University, 1991); Marco Rimanelli, *NATO Enlargement after 2002: Opportunities and Strategies for a New Administration* (Washington, DC: National Defense University, 2001); Marco Rimanelli, "NATO's 2002 Enlargement: U.S.-Allied Views on European Security" in Hall Gardner (ed.), *NATO and the European Union: New World, New Europe, New Threats* (London: Ashgate, 2004), 91–106.

security Partnership architecture (NAC-C, Partnership, EAPC, NATO-Russia Permanent Joint Council, NATO-Ukraine Charter, Mediterranean Partners, Gulf Cooperation Partners, and Strategic Partners) cooperating with the OSCE, EU and UN; 3) "Out-of-Area" peacekeeping missions in adjoining theaters—like the 1992–95 NATO air-naval support of UN peacekeeping in Yugoslavia; the 1995–96 International Force (IFOR) and 1996–2004 Stabilization Force (SFOR) in Bosnia; the 1999–current Kosovo Force (KFOR) in Kosovo, Albania and briefly in Macedonia too; and the 2001–current International Security of Afghanistan Force (ISAF); 4) the "War on Terror" against Islamicist terrorism—like Al-Qaeda regionally (Afghanistan, Iraq, Yemen, Saudi Arabia, Pakistan, Somalia), in the Mediterranean (NATO anti-terrorist patrols), and within NATO members (US, Great Britain, Spain, Turkey, Germany, Netherlands, Italy).[3]

NATO's first post-Cold War major reorganization severely cut back military expenditures in 1991–99, reducing US and Allied forces, scaling down its Commands and turning over to Europeans several influential traditional US posts by 25 percent, while its *Right Mix Studies* on future force structures shifted NATO training and exercises away from their Cold War focus to newer mobile forces, like ACE's Rapid Reaction Corps (ARRC) for post-Cold War international operations. By 1995 NATO had created four mobile, integrated, multinational NATO corps on the old Central Front, each 50–75,000 strong, with their national units under NATO command, joint inter-corps planning and communications; multinational corps lessen the defense burden for smaller Allied countries unable to field full armed forces. NATO's full operational integration is effective only in wartime, when all national control of Allied forces automatically falls under the US/NATO integrated command in Europe through SACEUR (always a US general). Instead, French and other Allied forces in the Central Front (Belgium and Britain), plus the Northern Flank (reorganized as Allied Forces Northwest Europe with Great Britain and Norway) and the Southern Flank (Italy, Spain, Portugal, Greece, Turkey and the US Sixth Fleet), remained under their original organization/command structure, although sharply reduced and with some forces "redeployed" back home (Canada and Belgium). Subsequently, many commands were cut and consolidated.

However, in 1990–2003, as during the 1980s, NATO's finely tuned military remained often paralyzed by internal political and public opinion fears of bloody post-Cold War guerrilla warfare. The Alliance repeatedly failed to quickly intervene militarily outside its traditional European border into vital "Out-of-Area" regional crises ignited by the collapsing Cold War system, which threatened Western security and economic interests, and were temporally patched with US-

3 David S. Yost, *NATO Transformed: The Alliance's New Roles in International Society* (Washington, DC: US Institute of Peace, 1998); *NATO Handbook: 50th Anniversary Edition* (Brussels: NATO, 1999); S. Ambrose and D. Brinkley, *Rise to Globalism: American Foreign Policy since 1938* (New York: Penguin, 1999), Chapters 16–19; Salvo Andò, "Preparing the Ground for an Alliance Peacekeeping Role," *NATO Review* (April 1993), 4–6; Kori N. Schake, "NATO Chronicle: New World Disorder," *Joint Forces Quarterly* (Spring 1999), 18–24.

led Western "coalitions-of-the-willing:" Sinai MFO peacekeeping (1980–now); Lebanon MNF I and II peacekeeping (1982–84); Western protection of oil tankers in the Persian Gulf during the Iran–Iraq War (1986–now); First Gulf War against Iraq over control of regional oil assets (1990–91); Yugoslav Civil Wars (1991–99); Southern and Northern Watch Air Patrols against Iraq (1992–2003); Second Gulf War against Iraq (2003) over risks of proliferation of weapons of mass destruction; and Iraq's Occupation and peacekeeping (2003–11).[4]

Iraq's invasion of Kuwait and threat to Middle Eastern oil routes propelled President Bush Sr. to organize a grand US-led coalition under UN mandate to defeat Iraq. Although NATO did not openly fight in this 1990–91 First Gulf War (due to domestic opposition in Germany and Greece to "Out-of-Area" combat), most Allies and NATO assets contributed to the war, with the bulk of US troops in Europe shipped over and REFORGER used to trans-ship, in six months, 600,000 men, heavy weapons and materiel to Saudi Arabia (twice the distance planned in REFORGER). SHAPE also protected Mediterranean Allies from feared Iraqi missile strikes through NATO Airborne Early Warning aircraft, naval protection of Mediterranean shipping, massive logistics and the air-defense of Turkey. Iraq's defeat in a short war was a vindication of both the diplomatic skills of Bush Sr. in crafting a fragile but wide war coalition, and of the US/NATO 1980s Cold War combined arms combat "Follow-On Forces Attack" (FOFA) strategy, which decimated the Soviet-trained and armed Iraqis.

Only by 1991 did the risks of exponential regional destabilization finally force NATO to adapt its missions to this new strategic environment: from anti-Soviet conventional/nuclear deterrence to mobile regional interventions and peacekeeping in many "Out-of-Area" missions against both ethno-nationalist hatred in the Balkans (ex-Yugoslavia; Bosnia, Kosovo, Macedonia and Albania) and regional destabilization near the oil-rich Persian Gulf (Afghanistan). NATO's reorganization brought into fruition the combat lessons learned in the 1990–91 "Desert Storm" operation against Iraq, and contributed to the smooth US/NATO peacekeeping for the UN in Bosnia (1995–2004), Macedonia (1998–2003), Albania (1995–2003), Kosovo (1999–now) and Afghanistan (2001–now).[5]

The collapse of Communism in Eastern Europe (1989–90) and the USSR (1991–92) also led to the break-up of independent Communist Yugoslavia along regional religious lines away from Serb-dominated Yugoslavia under President Slobodan Milošević, while Serb minorities in the new states fought to create a "Greater Serbia"

4 David S. Yost, *NATO Transformed*; *NATO Handbook: 50th Anniversary Edition*; S. Ambrose and D. Brinkley, *Rise to Globalism*, Chapters 16–19; Salvo Andò, "Preparing the Ground," 4–6; Kori N. Schake, "NATO Chronicles," 18–24.

5 G. Bush and B. Scowcroft, *A World Transformed*, 416–517; Bob Woodward, *The Commanders* (New York: Simon & Schuster, 1991); J. Stoessinger, *Why Nations go to War* (Belmont, CA: Thompson, 2008), 293–336; Dilip Hiro, *Desert Shield to Desert Storm: The Second Gulf War* (New York: Routledge, 1992); James F. Dunnigan and Austin Bay, *From Shield to Storm: High-Tech Weapons, Military Strategy and Coalition Warfare in the Persian Gulf* (New York: Morrow, 1992); William Head and Earl Tilford Jr., *The Eagle in the Desert: Looking Back on U.S. Involvement in the Persian Gulf War* (London: Praeger, 1996).

with Yugoslav military aid. The OSCE failed to live up to its inflated expectations as a new regional conflict resolution power-broker during Yugoslavia's 1991–99 Civil Wars, and Serb-led ethnic cleansing and atrocities dashed all EU, UN and NATO diplomatic mediations, while isolating the UN Protection Force in Bosnia (UNPROFOR) peacekeepers in 1992–95. The break-up of Yugoslavia ushered five civil wars in the decade 1991–2000—Slovenia (1991), Croatia (1991–92, 1995), Bosnia (1992–95), Kosovo (1998–99) and Macedonia (2000–2001)—forcing NATO to intervene "Out-of-Area" in this humanitarian catastrophe to preserve the Balkans' fragile post-Communist ethnic balance and Alliance security which were threatened by 250,000 dead, tens of thousands of rapes and hundreds of thousands of refugees fleeing to Europe. However, NATO also had to confront US and Allied divisiveness over possible ground combat operations to restore peace regionally in its first "Out-of-Area" peacekeeping mission on behalf of the UN in the Balkans.

NATO air-naval patrols of the Adriatic Sea imposed a UN naval embargo against arms shipments to the ex-Yugoslavia in 1992–96, with the blockade under direct NATO command through Allied Forces Southern Europe (AFSOUTH). From 1992, NATO also enforced UN no-fly zones over Bosnia to prevent attacks, followed by selective NATO air strikes from April 1993 against the Serbs. These were NATO's first combat actions since its founding in 1949, destroying selected Serb positions, providing NATO humanitarian air drops and protecting from the air UN humanitarian convoys. But persistent Allied divisiveness and US opposition prevented any NATO ground combat operations until the Bosnian-Serbs overran the UN "Safe-Area" of Srebrenica in summer 1995, slaughtering all Bosniak Muslim males, and attacked other UN Safe-Areas. This precipitated NATO's Operation Deliberate Force (August–September 1995), with air strikes destroying all Bosnian-Serb command, control and heavy weapons, while Croatia's entry into the war defeated Serb forces both inside Croatia (the Serb-controlled Krajina area) and in Western Bosnia.[6]

These parallel, independent NATO/Croat actions forced all the warring ethnic factions to sign the UN Dayton Peace Accords (November 1995), with their disarmament and peace enforced on the ground by NATO's heavily armed peacekeepers pouring in from Germany during the winter (Operation Joint Endeavour) to absorb local UN peacekeepers into NATO's Implementation Force in Bosnia (IFOR, 1995–96). IFOR was the largest military operation in Europe since World War II, with 65,000 IFOR peacekeepers—50,000 NATO troops from all Allies and 17 non-NATO Partners with Russia (NATO's historic Cold War enemy), plus 15,000 ex-UN peacekeepers—who quickly separated the three ethnic armies into cantonment and storage sites, while transferring areas between hostile

6 Steven L. Burg, "Why Yugoslavia Fell Apart," *Current History* 92 (November 1993), 291–297; Jasminka Udovički and James Ridgeway (eds), *Burn This House: The Making and Unmaking of Yugoslavia* (Durham, NC: Duke University Press, 1997); Kofi Annan, "U.N. Peacekeeping Operations with NATO," *NATO Review* (October 1993), 3–7; D.S. Yost, *NATO Transformed*, 189–301; Richard Holbrooke, *To End a War* (New York: Random House, 1998).

communities. IFOR's success enabled the High Representative for Bosnia to implement Dayton's civil provisions, while IFOR was replaced by a smaller NATO Stabilization Force (SFOR, 1996–2004) with 32,000 peacekeepers (Operations Joint Guard/Joint Forge). SFOR force levels were gradually drawn down through six-month reviews (NAC+N Meetings) until replaced by the EU Force (EUFOR, 2005–now). NATO's 2006 Riga Summit agreed to turn Bosnia into a Partner, while EU aid promotes the economic integration of all the southwest Balkan states (Bosnia, Serbia, Montenegro and Macedonia) through the EU Stabilisation Pact.

Nevertheless, NATO had to fight again in the Balkans against Yugoslavia/Serbia with the 1999 Kosovo War to stop ethnic cleansing of Albanian Muslims inside Serbia's province of Kosovo. During 1998, tensions among ethnic Albanians and the Serb minority within Kosovo turned into conflict between Serbian military and secessionist insurgents of the Kosovo Liberation Army (KLA), leaving over 1,500 Kosovar-Albanians dead and 400,000 refugees in a major humanitarian crisis as the Serbs forced civilians to flee Kosovo. International diplomatic pressures coupled with NATO's threat of air strikes on Yugoslavia/Bosnia forced President Milošević to withdraw large numbers of Serbian security forces from Kosovo in October 1998. Yet, as fighting resumed in January 1999 between Kosovar-Albanian insurgents and massive Serb reinforcements violating the October 1998 Accord, only renewed threats of NATO air strikes forced the two sides to the Rambouillet talks in France (February–March 1999). In the end the peace talks collapsed, while the Serb forces' scorched-earth strategy evicted 80 percent of Kosovo's Albanians (1.5 million people as 90 percent of the population of Kosovo compared to the Serb minority); by late May 1999, 5,000 were dead, 800,000 had fled abroad and 580,000 more were homeless inside Kosovo.[7]

NATO launched massive air strikes in March–May 1999, while building refugee camps with vast amounts of aid for a catastrophic humanitarian crisis bigger than Bosnia. Pro-Serb Russia and China broke up temporarily with NATO over its UN-backed intervention in a sovereign state (both feared Kosovo foreshadowed future UN interventions against Russian repression in Chechnya and China's in Tibet and Xinjiang), but the start of NATO's ground offensive was met by a total Serb military withdrawal from Kosovo on June 9, 1999, relinquishing it to NATO's peacekeeping Kosovo Force (KFOR), while civilian duties were ran by the UN Mission in Kosovo (UNMIK). KFOR comprised 50,000 peacekeepers, with 40,000 from all 19 NATO members and 20 non-NATO countries, including 16 Partners, Switzerland and an unwelcome Russian contingent of 3,200 men who broke off from their Bosnian base to race with NATO to occupy Pristina Airport in Kosovo (Russia temporarily broke relations with NATO until 2001). Notwithstanding the return of all refugees and

7 David Halberstam, *War in a Time of Peace: Bush, Clinton and the Generals* (New York: Simon & Schuster, 2001); D.S. Yost, *NATO Transformed*, 70–301; Vaughan Lowe, Adam Roberts, Jennifer Welsh and Dominik Zaum (eds), *The United Nations Security Council and War: The Evolution of Thought and Practice Since 1945* (Oxford: Oxford University Press, 2008); M. Rimanelli, *NATO and Other International Security Organizations: Historical Dictionary* (Lanham, MD: Scarecrow/Rowman, Littlefield, 2009), 102–666.

the NATO/international aid to rebuild Kosovo, ethnic tensions now forced KFOR to protect the local Serb minority from revenge, while in 2008 the UN allowed Kosovo to become independent, despite vigorous condemnations from now democratic Serbia (once a popular coup deposed Milošević, he was later prosecuted with the Serb-Bosnian leaders for their war crimes by the UN International Criminal Tribunal on the ex-Yugoslavia—ICTY).

The final Balkan conflict came in summer 2001, with the incipient civil war in the ex-Yugoslav Republic of Macedonia between insurgent local Albanian Muslims and the majority Slav Macedonians. NATO's Task Force Harvest was deployed in just five days, successfully enforcing a ceasefire, collecting all weapons and stabilizing the country with political power-sharing (Operation Amber Fox) until turning over its task to EU peacekeepers (Macedonia remains volatile, given Greece's veto in both NATO and the EU against Macedonian membership).[8]

NATO's continued transatlantic security relevance against the magnitude of new post-Cold War crises and "New Threats" has been questioned cyclically since the USSR's collapse in 1989–92 by critics mindless of History's lessons, who keep portraying it as a military Alliance bereft of its old Cold War common enemy. Instead, according to US Ambassador to NATO Nicholas Burns in 2008, "NATO is at the center of nearly every major security issue facing Europe and America" by forging common transatlantic policies on "building a Europe whole, free and at peace," while facing "New Threats," peacekeeping, Balkan stabilization, expansion to new Allies and Partners, WMD proliferation, and even applying for the first time its Art.V collective defense to the post-9/11 anti-Terrorist War in Afghanistan, despite increased strains on Allies and Partners from additional peacekeeping missions. Also, the USSR/Russian politico-military decline in the post-Cold War period curtails any opposition, while US diplomatic assiduousness towards Russia has, since 2001, co-opted both Chinese and Russian national security needs to the US-led global anti-Terrorism War, through even closer post-9/11 cooperation with the US and NATO on Missile Defense, new nuclear arms cuts (START III in 2001 and START IV in 2010) and a new 2002 NATO-Russia Council.[9]

---

8 Madeleine Albright, *Madam Secretary: A Memoir* (New York: Miramax/Hyperion, 2003); Associated Press, "Kosovo Campaign no Model," *AOL News* (2 July 1999); D.S. Yost, *NATO Transformed*, 70–301; Stuart E. Eizenstat, John Porter & Jeremy Weinstein, "Rebuilding Weak States," *Foreign Affairs* (January 2005).

9 Henry Kissinger, *Does America need a Foreign Policy?: Toward a Diplomacy for the 21st Century* (New York: Simon & Schuster, 2001); Jay Hines, "From Desert One to Southern Watch: The Evolution of U.S. Central Command," *Joint Force Quarterly* (Spring 2000), 32–3, 42–8; Pedro Moya, "Security in the Greater Middle East" (Brussels: NATO Parliamentary Assembly, 1998). Quote: R. Nicholas Burns, Speech "NATO in a New Era" in at The American University of Paris Conference "New World, New Europe, New Threats" (Paris: French Senate, 8 December 2001), cited in Hall Gardner, "Toward New Euro-Atlantic Euro-Mediterranean Security Communities" in Hall Gardner, ed. *NATO and the European Union* (Aldershot: Ashgate, 2004).

America's "first tier" strategy against "New Threats" spearheaded the US/NATO/UN Second Afghan War against terrorism in 2001–02, followed by NATO's new International Security of Afghanistan Force (ISAF) peacekeeping mission and NATO/Russian/EU cooperation against terrorism and WMD through NATO's new Rapid Deployment Force alongside the EU's sluggish ESDI/P. Both NATO's Summits at Prague (2002) and Istanbul (2004) confronted post-Cold War "New Threats" (international terrorism and WMD proliferation) and committed the Alliance from 2001 "to strike at threats anywhere in the world, and harden domestic targets and urban centers against terrorists with WMD and missiles." But NATO's successes in the 1990s and US global leadership as "sole Superpower" were suddenly put in question by the severe transatlantic rift of 2003.

Inter-Allied controversy over US "unilateralist" policies reemerged under President Bush Jr. when he first fought the Second Afghan War as a three-tiered global coalition based first on US/British forces, then an *ad hoc* coalition, and also NATO's ISAF peacekeepers, which by 2008 had absorbed all the other components as a new insurgency war by the Taliban raged. But the severe 2003–04 transatlantic rift was sparked by Bush Jr.'s highly controversial "Preventive Doctrine," which added to his well-lauded anti-Terrorism War ("first tier" strategy) an internationally controversial parallel "second tier" strategy of preventive "Out-of-Area" wars against "rogue states" engaged in WMD proliferation (the "Axis of Evil:" Iraq, North Korea, Iran and Libya), or a lauded "third tier" international strategy of multilateral diplomatic pressure and sanctions against any WMD threat. Finally, a controversial "fourth tier" strategy addressed anemically both the stalled Israeli-Palestinian "Two States" peace initiative and Middle East democratization (which paradoxically helped the Islamicist Palestinian Hamas seize power in the Gaza Strip after winning local elections against a now moderate PLO).

The US remained focused in 2001–08 on quick diplomatic or military strikes at all WMD "rogue states" (especially Iraq militarily, while threatening Iran, North Korea and Libya) well "before" they could mature their individual growing WMD threat against the West. The international consensus among the US, NATO, the UN and all intelligence services, plus high-placed witnesses (like Saddam's sons-in-law, before he had them killed), was that Iraq retained some secret WMD weapons despite losing the 1990–91 First Gulf War, UN sanctions and the 1991–96 UN inspectors' disarmament. Bush Jr. also floated embarrassingly unsubstantiated allegations that anti-Islamist Saddam Hussein's Iraq could even provide WMD technologies to Islamicist terrorists as a way to indirectly strike at the West under the cover of Al-Qaeda's 9/11 success.[10]

10 Bob Woodward, *Bush at War* (New York: Simon & Schuster, 2002); Don Oberdorfer, *The Two Koreas: A Contemporary History* (New York: Basic Books, 2002); Ahmed Rashid, *Taliban* (New Haven, CN: Yale University Press, 2000); Norman Podhoretz, "World War IV: How it Started, What it Means and Why We Have to Win," *Commentary Magazine* (September 2004), 1–56; Hall Gardner, *American Global Strategy and the "War on Terrorism"* (London: Ashgate, 2005); Stephen J. Flanagan and James A. Schear

Thus, the sharp international and Allied divisiveness over a 2003 Second Gulf War against Iraq split the UN, NATO and the EU: on one hand, acrimonious international public opposition and unprecedented eruption of acrimonious anti-Americanism fostered by the unlikely "pacifist" front of a few Allies/Partners (Jacques Chirac's "nationalist" France, Gerhard Schroeder's "leftist" Germany, Greece, Belgium, Turkey, Austria, Sweden and, ambiguously, Turkey), semi-rivals (Russia, China) and Western Leftist public opinion (Great Britain, Italy, Spain), which scuttled US attempts to forge an interventionist consensus with the UN, the EU and NATO military support in a Second Gulf War against Iraq, unless a new UN Resolution declared war; on the other, most "old" and "new" Allies joined a US-led anti-Iraqi Coalition fighting in the name of the UN's previous 12 years of anti-Iraq Resolutions.

But most troubling for Western cohesion was the open Franco-German anti-war vetoes at NATO, EU and UN (backed by Russia and China at the UN) against a US-led Second Gulf War on Iraq, which briefly undermined NATO's 50 years of common security gains and transatlantic solidarity, until the 2006–08 reversals of governments in Germany and France favored pro-US and pro-NATO policies (likewise, Italy, Spain, Australia and Japan flip-flopped electorally twice between governments supporting the US in Iraq with national combat troops and Leftist governments that quickly reversed these international engagements and withdrew their troops from Iraq, followed by new pro-US anti-Leftist governments). At NATO, France remained marginal since 1966, in spite of its "veto," while French influence at the EU rapidly waned from 2002 over many European members' resentment of Chirac's high-handed leadership of the EU, attempts to split ESDI/P from NATO, and France's failure in 2005 to ratify the very EU Constitution France had helped craft (a new EU Constitution was ratified only in 2010). Further, NATO agreed in 2004 to train Iraqi forces battling insurgents, while in 2008 under President Nicolas Sarkozy France finally re-entered NATO's integrated military command.[11]

(eds), *Strategic Challenges: America's Global Security Agenda* (Washington, DC: National Defense University, 2008); Alexander Moens, *The Foreign Policy of George W. Bush: Values, Strategy and Loyalty* (London: Ashgate, 2004); Robert Pauly Jr. & Tom Lansford, *Strategic Preemption: US Foreign Policy and the Second Iraq War* (London: Ashgate, 2005); Kenneth Pollack, *The Threatening Storm: The Case for Invading Iraq* (New York: McGraw-Hill, 2002).

11 Bob Woodward, *Plan of Attack* (New York: Simon & Schuster, 2004); Henri Vernet and Thomas Cantaloube, *Chirac contre Bush: L'autre guerre* (Paris: Lattès, 2004); Rowan Scarborough, *Rumsfeld's War: Remaking the American Military and Winning the War on Terror* (Washington, DC: Regnery, 2004); Josef Joffe, "America and the Euroweenies: The Future of the Transatlantic Relationship," AEI (June 5, 2002); "Europe and America: Sixty Years On," *The Economist* (June 5, 2004), 9; "The Transatlantic Alliance: A Creaking Partnership," *The Economist* (June 5, 2004), 22–4; Kathleen C. Bailey, *Doomsday Weapons in the Hands of Many: The Arms Control Challenge of the '90s* (Urbana, IL: University of Illinois Press, 1991); "Battling Proliferation: Win some, lose some," *The Economist* (June 5, 2004), 40–41.

The global storm of anti-US criticism in 2002–03 reflected mostly a dramatic chasm in style and international authority between Bush Sr.'s 1990–91 "Desert Storm" Coalition against Saddam's Iraq and Bush Jr.'s Second Gulf War coalition of 2003–10 that destroyed Iraq and captured Saddam, but was unable to exercise an effective Occupation over a fractured country in near-civil war between rival Iraqi insurgencies (Baathists and nationalist Sunnis; Al-Qaeda terrorists; Shi'a fundamentalists) fighting the US and each other for power. Although all Allies, Partners, EU, UN, Russia and China agreed that WMD proliferation did put them all equally at risk, they still supported exclusively multilateral diplomacy and sanctions against Iraq and other "Rogue states." Instead, both US presidents pursued energetic wars as the main US strategy to prevent WMD proliferation and imperialism by Saddam's Iraq, but the first had the more compelling "casus belli" (Kuwait's invasion) and diligent assembly of a UN-sponsored international effort, while the other mostly "went through the mechanics" to clothe US unilateralism in the face of nebulous threats posed by Saddam's Iraq. Thus, the US Coalition in 2003 raced out of its Kuwaiti bases and destroyed, in a few weeks, all Iraqi forces, reaching Baghdad and toppling Saddam.[12]

America, albeit deeply divided domestically over Bush Jr.'s leadership, remains the "sole SuperPower" committed to Iraq's post-war democratization and reconstruction, regardless of years of virulent Iraqi anti-Coalition insurgency, while Europe too is targeted by Al-Qaeda mass terrorism (Madrid 2002; Casablanca 2003; London 2005; Heathrow 2006). But the "Iraqi lesson" and Saddam's defeat in 2003 perversely accelerated two other "rogue states" in a WMD arms race as a hoped-for national shield or bargaining chip against other "preventive" US strikes, while only Libya understood the lesson and, in 2004, bargained away its WMDs to US/UN inspections and disarmament in exchange for international reintegration. Thus, the 2002 dramatic public nuclear race by North Korea (violating its 1995 accord with the US and the 1967 Non-Proliferation Treaty—NPT) and Islamic Iran since 2003 (also an NPT transgressor) led only to repeated failures of North Korea to enforce the Six-Party Accords it had twice signed and violated in 2007–08, while Islamic Iran in turn trumped UN sanctions using international condemnation of the US non-negotiating stance compared to the EU-Iranian talks, then dismissed European talks, and when the US started its own talks, all efforts by Presidents Bush Jr. and Barak Obama came to naught as well.[13]

---

12 Bob Woodward, *Plan of Attack* (New York: Simon & Schuster, 2003); Michael R. Gordon and Bernard E. Trainor, *Cobra II: The Inside Story of the Invasion and Occupation of Iraq* (New York: Pantheon, 2006); J. Stoessinger, *Why Nations Go to War*, 293–390; John Keegan, *The Iraq War* (New York: Vintage, 2005); Thomas Ricks, *Fiasco: The American Military Adventure in Iraq* (New York: Penguin, 2007); Bob Woodward, *State of Denial: Bush At War Part III* (New York: Simon & Schuster, 2007); John Lewis Gaddis, "Bush and the World, Take 2: Grand Strategy in the Second Term," *Foreign Affairs* (January 2005), 2–15.

13 Dexter Filkins, *The Forever War* (New York: Knopf, 2009); Bob Woodward, *The War Within: A Secret White House History, 2006–08* (New York: Simon & Schuster, 2008); *Géopolitique: Iran*, 64 (Paris, 1999), 1–100 ; Kenneth Timmerman, *Countdown to Crisis: The*

This post-Cold War "Arc of Crisis" in the 1990s–2000s indicating the global emergence of diffused "New Threats" (WMD proliferation, ethno-nationalist civil wars, Islamicist terrorism, illegal migrations, trafficking, pandemics, ecological blight) contrasted starkly with the West's earlier 1990s optimistic Euro-transatlantic "Arc of Stability" replacement of Cold War divisions with a new spirit of international cooperation, Europe's politico-economic unification, and NATO-EU enlargements. These "New Threats" and a diffused "Arc of Crisis" pushed NATO and the US to implement their third major parallel military reorganization in 12 years to better cope with the entire range of missions. On the one hand, NATO's old Central Front Allied Command Europe (ACE) was changed into Allied Command Operations (ACO), covering all combat operations throughout the entire NATO and Euro-Atlantic area, not just Europe. On the other, Allied Command Atlantic (ACA) is now Allied Command Transformation (ACT), focusing on innovative twenty-first-century technologies and policies. Additionally, the US announced in 2004 that in the period 2006–16 it would massively restructure forces a third time since the end of the Cold War by pulling out 70–100,000 troops from Europe and Asia, plus 100,000 dependants, given the "lack of Cold War strategic justification and the need to face 21st Century world threats and terrorism." Two-thirds of the repatriated US troops will come out of Europe—where in 2004 the US had 100,000 forces, mostly in Germany with 70,000 men of which 50 percent will leave (including the two US armored divisions)—with NATO bases in Germany being cut and thousands of US troops transferring to new rapid-deployment bases in East European Allies (Bulgaria, Hungary, Poland and Romania).[14]

Despite these "New Threats" and the diffused "Arc of Crisis," Bush Jr. secured full NATO-UN backing only on Afghanistan, which had already been a key area for US opposition to the USSR during the Cold War, when the US global containment geostrategic vision extended to remote Central Asia. The 1979–89 First Afghan War was sparked by the USSR's invasion to rescue the local fragile communist state from being overrun by Islamized guerrillas (Mujahedeens), while containing any Islamicist penetration of Soviet Muslim Central Asia/Caucasus. Plagued by the parallel 1979 Teheran Hostage Crisis with Islamic Iran, US President Carter was morally outraged at Moscow's brutal demotion of the decade-old East–West Détente, while Soviet conquest and air power in Afghanistan implicitly threatened nearby Western and world oil routes through the Persian Gulf. Unaware and unwilling to diplomatically exploit the USSR's own fear of Islamicist contagion

Coming Nuclear Showdown with Iran (New York: Crown, 2005); "Guerra Infinita," *Giano*, 38 (Roma: May–August 2001), 3–202; Robert W. Tucker and David C. Hendrickson, *The Imperial Temptation: The New World Order and America's Purpose* (New York: Council Foreign Relations, 1992).

14 Hélène Carrère d'Encausse, "Russie, États-Unis et l'Arc de Crise du 11 Septembre," *Géopolitique: Caucase et Asie Centrale*, 79 (Paris, 2003); M. Rimanelli, *NATO and Other International Security Organizations*, 282–666; Nancy Soderberg, *The Superpower Myth: The Use and Misuse of American Might* (New York: Wiley, 2004); Deepak Lal, *In Praise of Empires: Globalization and Order* (New York: Palgrave/Macmillan, 2004).

in Soviet Central Asia from Iran and Afghanistan, Carter never sought to trade US acquiescence to the USSR's First Afghan War in exchange for Moscow's lifting its veto on crippling UN sanctions against Islamic Iran. America passed economic embargoes against both Iran and the USSR, undertook a massive rearmament (continued by Reagan) and produced the 1980 Carter Doctrine restating the World War II policy of militarily defending the vital Persian Gulf oil routes from any threat by creating a US Rapid Deployment Force (US Central Command since Reagan) with regional bases and operational range from East Africa and the Middle East/ Gulf to Central Asia. The Carter Doctrine was the foundation of Reagan's 1986–2003 Western combat naval patrol of the Gulf straits against Iran in 1986–88 and the USSR in 1986–90, as well as of Bush Sr., Clinton and Bush Jr. against Iraq in 1990–2003. Equally, the 1981 Reagan Doctrine of armed "liberation" of oppressed people under communist yoke fuelled the US/Moderate Arabs/Pakistani arms flows into Afghanistan to the Mujahedeens fighting Soviet forces (although only when anti-air Stinger missiles destroyed Soviet air superiority did the Kremlin withdraw in 1989), while also arming the Contras in Nicaragua and Unita guerrillas in Angola.[15]

However, after Moscow's withdrawal from Afghanistan the US did not stay behind, or help Pakistan stabilize the Mujahedeen warlords' corruption, while the parallel 1991 collapse of the USSR into 15 rival states confirmed to the West the disappearance of any geostrategic threat in the region. Instead, America's swift withdrawal from Afghanistan and the 1990–95 civil war among the warlords led the Pakistani-based Taliban to seize control of most of the country against the besieged Northern Alliance of ethnic Tajik and Uzbeki Afghani-Mujahedeen. The Taliban government also supported fundamentalist terrorist groups in Central Asia and Chechnia, while being the protector of Al-Qaeda.

In the greater Middle East/Gulf region, in the 1980s–90s, the US remained mostly focused on a "Dual Containment" policy of both Saddam's Iraq and Islamic Iran (1980–86 and 1990–2001), becoming embroiled in the 1980–88 Iraq–Iran War, then the Two Gulf Wars against Iraq (1990–91 and 2003), while supporting Israel. Until the 2001 9/11 terrorist attacks, the US and NATO did not perceive any imminent threat in Central Asia or globally from Islamicist terrorism, while both were also mired in Eurocentric post-Cold War security, focused on Russia and NATO's 1999–2004 dual Enlargements to East Europe rather than new post-Cold War global threats. Indeed, despite US economic involvement in opening Central Asia/Caucasus' remote land-locked oil/gas sources through the new East–West pipelines to Georgia (bypassing both Russia's monopoly of ex-Soviet pipelines in the North, and blocked Southern routes via Iran and Afghanistan), the US and NATO remained reluctant to cross Russia's "red lines" on its "Near-Abroad" area of influence. The US instead promoted an ever-wider ambitious strategic oil/gas network of pipelines to outside world markets, in cooperation with most regional players (Caucasus, Turkey, Russia, Central Asia, China, Afghanistan, Pakistan and India), while also fostering increasing interdependence and stabilization

15 Lester Grau, *The Bear Went Over the Mountain: Soviet Combat Tactics in Afghanistan* (London: Cass, 1996); Mark Urban, *War in Afghanistan* (London: Macmillan, 1990).

among ex-rivals and ex-enemies. The goal for the US/NATO and European now is to diversify energy supplies away from their traditional Middle East/Gulf dependency by stabilizing politico-militarily the Caucasus and Central Asia, while developing economically its oil/gas resources (Azerbaijan and Kazakhstan on land, and offshore under the Caspian Sea) for future export worldwide outside the traditional ex-Soviet oil ducts running in the north.[16]

Within this evolving geostrategic post-Cold War context, the ex-Soviet Partners looked for closer security ties with the US/NATO to rebalance Russia's predominance (with an option to even join NATO in the future through its 1999 "Open Door" policy). Yet, since the Second Afghan War these same states have also cooperated closely with Russia and China in the anti-Western "Shanghai Six" Asian Economic and security association. Thus, US/NATO low-key politico-military cooperation with newly independent ex-Soviet Central Asian/Caucasus sought mostly to both stabilize the region against either a remote Russia/Communist resurgence, or more likely extreme Islamicist revolutions. On one hand, Central Asian/Caucasus states worked with the US and NATO to implement the US-Soviet nuclear disarmament accords on ex-Soviet Theater Nuclear Forces (US-USSR 1987 INF Treaty and unilateral 1991 SNF Accords), and all ex-Soviet Strategic Nuclear Forces on their territory (1991 START I Treaty involving also Kazakhstan, Ukraine and Belarus) transferred back to Russia. On the other, NATO's security Partnership was extended to the ex-Soviet Central Asia/Caucasus through the NAC-C and OSCE (1990–95), followed by the Partnership and EAPC (1995–99). From 1992, NATO's Partnership with Central Asia focused on bilateral military and regional security cooperation, joint peacekeeping, democratization and intelligence-sharing, plus the 2000–01 rotation of US special forces in Central Asia (training for possible joint actions) against common Islamicist threats of the Afghan Taliban, Islamic Iran and transnational terrorist groups (Al-Qaeda, Islamic Movement of Uzbekistan (IMU), Chechen rebels) operating in Afghanistan, Uzbekistan, Tajikistan and Russia (Chechnya, Daghestan, Ingushetia and Moscow). In this context, NATO's new annual Partnership exercises since 1995 (CENTRASBAT) involved in 1997 the longest ever airborne/peacekeeping operation, 6,700 miles from North Carolina, USA to Kazakhstan and Uzbekistan, with also US, Russia, Turkey and Central Asian forces (an exercise which paved the way for the 2001 US Special Forces deployments to Afghanistan in the Second Afghan War.

Following Al-Qaeda's 9/11 Terrorist strikes on America in September 2001, remote, overlooked Central Asia and Caucasus were propelled to the center of Euro-Atlantic security concerns through the Second Afghan War (2001–02) and NATO peacekeeping. At first, in October 2001, under UN Mandate, an international coalition of forces led by America and NATO (under Art.V for attacks against its members) swiftly injected inside Taliban Afghanistan thousands of mobile forces

16 See *Géopolitique* 79, especially Brenda Shaffer, "Opportunités et Menaces en Asie Centrale et Caucase," 76–81; Édith Ybert, "Le Caucase: Toile de Fond Historique," 7–18; Charles Zorgbibe, "Frontières et Litiges en Mer Caspienne," 33–7; and Jean Radvanyi, "Vent Américain au Caucase," 25–32.

(US, British and some French) in "Operation Enduring Freedom," to destroy the local Islamist Taliban government allied to Al-Qaeda, while helping the besieged Northern Alliance Afghan rebels into power (December 22, 2001). This US/NATO global operation deep in ex-Soviet Central Asia combined special forces, carrier aviation, long-range bombing, joint combat forces and multinational diplomacy, but not large-scale armies.[17]

Central Asian Partners played key security roles against both Al-Qaeda and Taliban-Afghanistan by providing Central Asian/Russian military support, intelligence and logistics (strategic land routes, plus the ex-Soviet air bases of Khanabad in Uzbekistan, Manas in Kyrgyzstan and refueling facilities at Dushanbé airport in Tajikistan), which paralleled existing local Russian bases temporarily opened to NATO with unrestricted use of Russian and Kazakhi airspace and logistics. Additionally, NATO relied on Turkish military leadership among its ethnic brethren in Central Asia to avoid Muslim reaction to foreign troops, while refusing to deploy Central Asian peacekeepers in Afghanistan.

NATO's International Security Assistance Force, Afghanistan (ISAF) was created in December 2001 to help the new Afghan government democratize, and to provide humanitarian aid, military and police training, plus combat peacekeeping to relieve US/Coalition forces. By July 2002 ISAF had 5,000 peacekeepers from 19 states based only in Kabul, then from 2005 ISAF assumed command from the US/Coalition in northern and western Afghanistan, while by 2006 it expanded also in pro-Taliban southern Afghanistan. By December 2006 ISAF had complete peacekeeping control of Afghanistan, including US/Coalition forces, while having to fight in 2006–11 resurgent Talibani attacks and suicide bombings out of Pakistan (where local Talibani also attacked the government in the border Tribal North-West Frontier until crushed in 2009–10 by the Pakistani Army).

NATO's regional and international role remains fragile, notwithstanding its application of Art.V (self-defensive war) over the 9/11 attacks and Second Afghan War. As NATO grew after the Cold War from a tighter massive military Alliance to a larger, political security organization of 28+ Allies and 20 Partners, its post-Cold War influence has increasingly supported US-led "Coalitions-of-the-Willing" and EU humanitarian peacekeeping globally for joint UN/Western goals. But with casualties rising in Afghanistan and that war also fast becoming targeted by broad Western public opinion ("pacifism," despite the fact that NATO is enforcing UN Mandates for peace), NATO has had to bolster its "Out-of-Area" geostrategic reach since 2004 by also integrating its Mediterranean Dialogue and Gulf Partners

17 Muhammad-Reza Djalili, "Asie Centrale ex-Soviétique, 1991–2002," 54–61, and Catherine Poujol, "L'Asie Centrale dix ans après la chute de l'URSS," 18–24, both in *Géopolitique* 79; AAVV, *Asie Centrale* (Paris: Cahiers de Mars, 2003); Rajan Menon, "The New Great Game in Central Asia," *Survival* 45: 2 (Summer 2003), 187–204; Lyle J. Goldstein, "Making the Most of Central Asian Partnerships," *Joint Force Quarterly* (Summer 2002), 82–90.

(Istanbul Initiative), as well as new Strategic Partnerships in 2006–08 (Australia, Japan, New Zealand, South Korea, India and Pakistan).[18]

Conclusion: NATO and Transatlantic Security from Past to Future

NATO's success as the longest alliance in history has allowed its transatlantic partnership between America and Europe to guarantee their mutual security interests during the past bipolar Cold War and World War III threats from the USSR and the current post-Cold War diffuse "Arc of Crisis".

Cemented politically and ideologically in the Allies' common combat and democratic heritage of World War I and World War II, NATO finally provided to both the isolationist "New World" and the hegemonic-weary "Old World" the perfect multinational military shield and common cause against the USSR during the Cold War. As an integrated military alliance in peacetime, under US Command through SACEUR, NATO provided adequate (although not effective until 1985–90) defenses with a fighting mix of conventional and nuclear forces to fight the threat of World War III launched by the superior USSR and Warsaw Pact. As a military deterrent this contained Soviet-Communist expansionism and subversion from the late 1940s, while the US nuclear umbrella and frontline GIs assured the automatic combat involvement of a now vulnerable American heartland in the cause of democracy and peace across the oceans in distant Europe (reversing the US Isolationism prevalent prior to and during both World Wars). As an Alliance of "equals" under US leadership (not hegemony), it also guaranteed that no member state would ever fight against each other (Germany vs Western Europe, Greece and Turkey), thus democratizing and reintegrating ex-Nazi West Germany and ex-Fascist Spain, while absorbing in 1990 ex-Communist East Germany and in 1997–2008 the ex-Communist Eastern European and Baltic states. As a military alliance steeped since the 1967 Harmel Report in internal democratization, NATO has also been the beacon of democracy in the post-Cold War (1995 Principles on NATO Enlargement) for all Partners and Aspirant Allies seeking its protection against ex-Soviet Russia and against each other over rival minorities and unjust borders. Finally, as a credible war-fighting military alliance, NATO compelled the USSR/Warsaw Pact to enter verifiable and stabilizing international arms control treaties, slashing theater and strategic nuclear weapons, as well as their conventional armies in Europe facing each other across the Iron Curtain.

---

18 Patrick Hayden, Tom Lansford and Robert Watson (eds), *America's War on Terror* (London: Ashgate, 2003); Gilbert Étienne, "Afghanistan: carrefour à hauts risqué," 41–7 and Michel Bozdémir, "Turquie et l'espace Turcophone," 48–53, both in *Géopolitique* 79; Bob Woodward, *Bush at War*; Gary Berntsen, *Jawbreaker: The Attack on Bin Laden and Al-Qaeda* (New York: Crown, 2005).

In the post-Cold War era since the collapse of Communism and the old Soviet threat, NATO has endured both barrages of unwarranted criticism from the domestic Left over its defense budgets and continued existence (seeking a utopian peace at all costs at the "End of History"), as well as from US "Neo-Conservatives" faulting it for being militarily irrelevant to the emerging global "New Threats." Yet US and European political resolve to preserve and strengthen the transatlantic Alliance in the post-Cold War era guarantees America's continued commitment to European security after two global wars and a Cold War. Thereafter, the Alliance enlarged both its membership (through multi-tiered Partners and new Allies) and exclusive security role to the entire unified Europe and Partners, protecting also the parallel EU politico-economic integration. Finally, NATO's triple military restructuring and new "Out-of-Area" missions under UN mandates (from the Mediterranean and Balkans to Central Asia) confounded critics who, in the early 1990s and again after 2001 and in 2008–11, cyclically predicted its demise after 45 years of Cold War service and 20 of post-Cold War peacekeeping.[19]

Thus, NATO has been successfully revamped from a Cold War regional/international defensive Alliance that deterred the Soviet threat of World War III to a post-Cold War global collective security network of expanding Allies and Partners within the broader Euro-Atlantic and Mediterranean areas, with "Out-of-Area" peacekeeping missions for regional stability, arms control, WMD anti-proliferation and anti-terrorism. NATO and EU Enlargements in 1999–2010, reaching 28+ members in each parallel organization, strengthened the democratic unification of all of Europe, although critics also worry that a "NATO at 30" has integrated too fast with too many militarily weak "free riders" leaving it unable to increase military capabilities in line with its rising "Out-of-Area" global security commitments or jointly intervene as an Alliance in more controversial conflicts (like the Iran–Iraq War or the two Gulf Wars against Iraq).

NATO cooperation with Central Asia and especially Russia over Afghanistan and the "War on Terror" (enhancing the NATO/Russia Joint Partnership in May 2002 as "NATO+1" decision-making on common security) also finally enhanced US-Russian ties under President Obama, strengthening also US/EU diplomatic leverage with Russia and China to stop Iran's WMD proliferation through new UN sanctions in 2010. But following the Second Afghan War many international critics, "Neo-Cons" in the Bush Jr. administration and Leftist Liberals in the Obama administration, once again faulted NATO as irrelevant in the US-led global "War on Terrorism," because actual combat was done mostly by US, British and French Allies (the only ones with rapidly deployable forces), while many European Allies were forced by their national parliaments and public opinion to seek only low-profile ISAF peacekeeping in Kabul and other "safe areas."

19 David Calleo, *Rethinking Europe's Future* (Princeton, NJ: Princeton University Press, 2001); Simon Serfaty, *Stay the Course*; Markus Meckel, "Key Issues for Future Transatlantic Relations and European Security" (Brussels: NATO Parliamentary Assembly, 2001).

NATO's historical, military and geostrategic advantages remain as current today in the post-Cold War era as during the Cold War. The loss of NATO as a permanent peacetime transatlantic Alliance would be a catastrophe, flying in the face of the blood sacrifices of Allied nations in World War I and World War II when the very lack of NATO's integrated structures, political purpose and training was the difference in so many millions of deaths and belated US military interventions. But as NATO will inevitably continue to survive, it faces the more insidious domestic challenge of pacifist democratic post-Cold War domestic politics within its own members, especially during global economic crises, with consequent limitations to forces, blood and treasure needed to sustain the Afghan War or future combat peacekeeping. And should NATO not survive a Taliban return to power in Afghanistan or the collapse of Iraq due to US-Allied wavering combat commitments and withdrawals, the Alliance might likely retrench in the future to just transatlantic security at home for the EU, with limited regional peacekeeping and low-cost security cooperation on the "War on Terror."[20]

20 M. Rimanelli, *NATO Enlargement after 2002*, 1–57; Hall Gardner (ed.), *NATO and the European Union*.

NOT a Clash of Civilizations: The Conflict in Kosovo Revisited

Oleg Kobtzeff

Patriotism based upon intolerance against other nationalities, upon growing national hatred, cannot become a basis for progress.

Slobodan Milošević, who pronounced these words in Kosovo, in April 1987, was sent by then president of Serbia Ivan Stambolić to bring calm to the autonomous region, where interethnic relations were rapidly deteriorating. At the time, the Serbian Republic, and the entire Yugoslav Federation, was sinking into deep economic crisis with 1 million people unemployed, 200 percent per annum inflation, and a 21 billion dollar foreign debt.[1] Kosovo was the worst affected, with 57 out of 100 workers unemployed.[2] Under such conditions, the slightest incident sparked by hooliganism, property rights or family feuds (vendetta is an important feature of local sociology) could lead to violence. Should the feuding parties belong to different ethnic communities, their private conflict could lead to a geopolitical explosion.

"When a given world and social structure collapse," writes the Hungarian sociologist Elemer Hankiss, "a substantial proportion, if not the majority, of the population will suffer from an unsettling of, if not an insult to, their sense of identity. Changes of social setting, disruption of networks, crises of confidence, loss of work, the devaluation of skills, the necessity to relocate – all these things can weaken and destroy people's sense of themselves, their self-identity."[3]

---

1 Tatiana Globokar, "L'année économique: sortir de la crise," in Thomas Schreiber and Françoise Barry (eds), *L'URSS et l'Europe de l'Est* (Paris: La Documentation Française, 1988), 225.

2 Ibid., 234.

3 Elemer Hankiss, "Transition or transitions? The transformation of eastern central Europe 1989–2007," *Eurozine* (2000), accessed (November 11, 2010) at: http://www.eurozine.com/articles/2007-07-26-hankiss-en.html. See all other chapters by O. Kobtzeff commenting at length on the relation between social instability, identity and violence in the present volume.

At the time, the population of the Albanian majority had risen dramatically, while the demographic figures of the Serbian minority were proportionally decreasing. Moderate ethnic Albanians feared the worst, as the Serbian minority frequently presented their frustrations to Belgrade as if they were suffering martyrdom. The stage was set for an explosion. So why did no one react? Why did the decision makers at the time allow the most moderate of the Communist regimes to become the stage of what is still the worse bloodbath in Europe since 1945?

The Tipping Point: The 1987 Crisis in Kosovo

Milošević was expected to find peaceful solutions with the local Communist Party of Kosovo (headed at the time by Azem Vllasi). Unexpectedly, he also met with Serbian nationalists and crowds of Serbs who gathered spontaneously, such as those for whom he would soon pronounce the words quoted above. The vocabulary of the Serbian President's relatively unknown subordinate sounded like standard discourse under Tito's regime—neutrality of the state in ethnic affairs and internationalism with a pretense of ethnic pluralism. Yet Milošević had already broken a basic rule of Yugoslav communism by addressing a crowd of nationalists. But in the light of further events, was Milošević speaking against "ethnic intolerance" addressing the Serbian nationalists, in a last plea for calm? Or was it already an accusation directed against any form of separatism, even moderate nationalism, manifested by non-Serbian ethnic groups, such as the local Kosovars?

In any case, the career of Slobodan Milošević, as future leader of a nationalist Serbian regime, began there in Kosovo, in that month of April 1987. An incident was to occur on the 24th which would constitute the first milestone in his ascension to power. A public meeting was organized to allow citizens to express their worries in front of the local Communist leadership, with Milošević playing the role of arbitrator. Outside the building where the meeting was held, a fight broke out between Serbian demonstrators and the local police (composed mostly of ethnic Albanians, reflecting the ethnic distribution of the region). Piles of stones appeared, very conveniently, out of nowhere, and were thrown at the police who fought back. Hearing the noise coming from outside, the Communist authorities interrupted the meeting and exited the building with Milošević at their head, followed by Vllasi.

While a somber Vllasi appears to contemplate the complete loss of control over the situation, Milošević seems worried and hesitant as shouts and screams are addressed to him. "Speak out," he hears as he advances toward the crowd. Emotions are running high in the ranks of the Serbian demonstrators. An elderly man wails: "The police have charged! They have beaten women and children!" Others accuse the Albanians of helping the police. "They are beating us up!"

Suddenly, Milošević is no longer hesitant. He marches through the crowd with determination, listening intently. The moment is historic; it will be seen by millions on Yugoslav national television. He raises his finger and declares: "You will never be beaten up again."

From that moment, whether he had planned it or not, Milošević seized a chance to become more than a second-rate politician in a decrepit Yugoslav Communist government struggling to maintain a difficult patchwork of ethnicities in an atmosphere of economic decline, just when the Communist ideology that had supported the legitimacy of the government was seen as a charade at home and worldwide. Now, he had become the champion of the Serbian cause.

Returning to Belgrade, Milošević was accused of "anti-Communist discourse" and faced severe censorship. But what should have been the sanction of a nationalist deviation turned instead into a confrontation which resulted in the removal from power of the more moderate Serbian president Stambolić. The easy political removal of Stambolić and his political apparatus by the supporters of Milošević illustrates the evolution of the regime.

It was the best way of preserving the power of a communist élite who had neither ideas nor resources to offer its people in order to justify its power. Reforming the economy was not as exciting for Serbia as it was for other communist regimes, nor for Croatia and Slovenia where the major industrial and tourist resources are found. Those Republics would benefit the most from a free market. One of the options to secure an electorate in poorer Serbia was to offer Serbs a forum for venting old frustrations. Since the empowerment of Tito's regime, many expressions of national history and identity were censored—religion, the exact history and figures of at least a half-million Serbs massacred for their ethnic identity by pro-Nazi Croat nationalists. The suppression of many aspects of national history by Marxist historiography in the name of proletarian internationalism was a characteristic of every territory under a Marxist-Leninist regime, including every Yugoslav republic. It only exacerbated frustrations, causing nationalism among all ethnies, aggravating the feeling of domination among peoples of Soviet or Yugoslav peripheries or a feeling of being deprived of their identity among Serbs and Russians. The atmosphere in Belgrade in 1988 or 1989 illustrated the change. The building of a gigantic new Orthodox cathedral in the middle of the federal capital, and the solemnities leading to the celebration of the 600th anniversary of the battle of Kosovo, aroused Serbian memory and their self-image as victims. As the Serbs felt that they were regaining their long-frustrated identity, they forgot, in those crucial late 1980s, to question the nature of the communist regime as other nations did at the same time elsewhere in Central and Eastern Europe. The diversion became the foundation of the Milošević regime. But to continue surviving in the absence of any genuine political or economic program, that regime needed more causes, more struggles.

Trapped in a Logic of Diversionary Politics

Even the most internationalist, socialist or pro-Serbian Croat could fear what could happen to his or her country if the memory of the genocide committed by the fascist *ustaši* was exploited to create revanchist sentiment among Serbs. Eventually it *would* be exploited for that very purpose. Having catalyzed in that manner Slovenian, Croatian, Macedonian and Bosnian separatism at the turn of the 1980s and 1990s, the Milošević version of national-Bolshevism could feed on the anti-Serbian feelings of the growing nationalist movements. This was a self-fulfilling prophecy confirming that Serbs were surrounded by enemies who would not hesitate to repeat the massacres of the 1940s. Rumors were systematically spread by Belgrade's specialists of "parallel strategy." Many observers noted, on the eve of the first armed engagements, the role played by the large diasporas of all ethnic origins in escalating the language of hatred and paranoia. I personally witnessed the frenzy into which the Serbian community of Rome was driven by newly appointed members of the clergy (who used all means at their disposal to eliminate the old moderate church leaders), and the subtle "planting" of false information. Sincerely persuaded, by 1988–89, that a new Holocaust was in preparation, the community was torn apart by internal strife caused by the atmosphere.

The situation in Western Yugoslavia deteriorated at a faster pace than in Kosovo. The first real war began in Slovenia in 1990, continued later in Croatia, and reached the summits of horror in Bosnia, in the mid-1990s. After the Dayton peace agreements, the Belgrade regime, trapped in its logic, could only return to where it first conquered its legitimacy: Kosovo. Meanwhile, everything was in place to create a conflict: local institutions were being dismantled and cultural rights (such as the use of the Albanian majority's language in the school system) were constantly under attack. This renders immaterial questions regarding which side actually initiated the first violent actions, or the arms and other illicit trades in the area, or the origins of the UCK (connected with the élite of Enver Xoxha's former state police). Milošević simply needed more confrontations to justify his power and the constant postponing of full democracy and economic improvements.

Since there was no guarantee that after Kosovo, such logic would not turn Milošević towards other targets, Serbian military operations in Kosovo became a clear and present danger. The differences from the previous Balkan wars of the 1990s were not only the quarter of a million casualties and the effects of ethnic cleansing over international public opinion. Ethnic warfare was now becoming more international than ever. A swarm of refugees was beginning to rewrite the human geography of Albania and Macedonia—already a familiar experience at the turn of the last century. In addition, a rapid victory for Milošević in Kosovo would immediately have created the need for a new battlefield. For those who knew the Yugoslav situation even superficially, the next region on the list was evidently Vojvodina and its Hungarian populations. That would have meant a war on the borders of a NATO country, with the usual flow of refugees spilling over the international boundary and *necessarily* involving Hungary. Would Milošević have dared to take such risks in the future? Or would he have stopped in time before triggering a major international catastrophe? Waiting any longer was an extreme risk, and the line had to be finally drawn.

Reasonable solutions were not ruled out at first. But the Rambouillet conference organized by the French only made them appear weak. Placing a Western power in a position of weakness probably served the immediate interests of the Milošević regime, and that is probably why Belgrade made no efforts to seize France's last chance of an honorable compromise. One of the major French actors of the conference revealed to me in confidence that the delegation sent by Belgrade was essentially composed of low-ranking politicians with no capacity for negotiating at that level and even less capacity for making any reliable commitments. France, where a strong pro-Serbian tradition resisted the turmoil of the Yugoslav wars, was now alienated.

With no obvious intentions of stopping at Kosovo, Milošević could inspire a variety of dangerous potential scenarios. After Vojvodina and Hungary, there existed unlimited possibilities for more conflict over maritime rights, border disputes with every neighbor (above all with Albania and Macedonia), commercial conflicts, more ethnic conflicts with other minorities living on the national territory (Gypsies) and beyond its borders. Any international situation involving Serbia and Macedonia, Hungary, Romania, Croatia, Slovenia and Bosnia, Bulgaria, Greece and, last but not least, Turkey and Russia, would become extremely volatile.

Russia's Ambivalent Situation

Western media in the 1990s frequently underlined the strong cultural and emotional ties existing between Russians and Serbs through their linguistic similarities and religious heritage. During its difficult history, the Serbian state received little support from foreign powers other than Russia. A Montenegrin expression illustrates the popular Serbian perception of its geopolitical meaning: "we, *plus* the Russians, constitute a force of 200 and fifty million *and* a half." Such feelings, rarely manifested towards the West but generously shown by Serbs every time a Russian peacekeeping force entered a former Yugoslav territory (it would happen again in Kosovo), held an important significance. It gave Russians a sense of purpose in a world that perceived Russia either as a decadent prey for carpetbaggers or an eternal menace. Emotions could, apparently, run as high in Moscow as in Belgrade. In a time of near-statelessness and loss of social, economic and cultural foundations, a cool, unemotional analysis of geopolitics is a luxury. Consequently, having to manage its public opinion, the Russian élites from various parties may have seemed caught in the same logic as the Serbian leadership. The temptation was great to let the Russians release steam by rallying around pro-Serbian themes. Moreover, Moscow could not easily dismiss the value of an ally, one of the few in the region with the strategic importance of the Balkans and the Mediterranean. Neglecting Serbia would also have sent a disturbing message to Greece, Bulgaria and other potentially friendly countries in the Balkans who were also struggling with economic problems and identity. Russia could not afford to compromise its "historic responsibilities" towards Orthodox nations of the Balkans and towards an embryonic axis that seemed at the time to tie Athens to Belgrade, Sofia and Moscow (and possibly Bucarest). Was Russia, then, locked in a relationship with Serbia that could be defined as "my ally right or wrong?"

The situation is not that simple. There also exists a history of tensions between Russia (or the USSR) and Yugoslavia. In the beginning of the twentieth century, Russia's support of the Serbian kingdom brought more trouble than advantages. In the Balkan wars preceding the conflagration of 1914, Russia was placed in the difficult situation of choosing sides during the border conflicts that opposed former Orthodox allies. The most painful situation for Russia was Bulgaria. Territorial bickering and conflicting nationalism finally pushed Bulgaria to side with the former Ottoman enemy in the war of 1913. During World War I, having to defend Serbia against the Axis and its allies, Russia found itself at war with Bulgaria, a country bearing as much cultural resemblance (if not more) as Serbia. Other problems limited unconditional pro-Serbian sentiment among Russians looking back to the old empire in search of their roots after the fall of the Soviet regime and ideology. After the "Great War," Serbia achieved all its goals by uniting all the southern Slavs and growing into the larger state of Yugoslavia, while Russia, for its support, inherited a civil war and 70 years of Soviet regime. Those Russians who remained Communists or who retained at least a little nostalgia for the power of the USSR, could have remembered when the ungrateful Tito ridiculed successive Soviet governments by defending his independence. In any case, one of the principal assets of Russia's stature in Central Europe—pan-Slavism—is in shambles because, or so it appears, of Serbian nationalism. Any dream of re-establishing pan-Slavic ties between Russia, Serbia, Croatia, Slovenia or other Central European countries (the flags of which are often variations on the Russian white-blue-and-red banner) vanished in smoke on the battlefields of the Balkans in the 1990s. Of course, such dreams died when Soviet troops crushed Central Europe under their boots in the 1940s. But the dream was buried forever by Serbian fanaticism, and that is an argument that Russian public opinion or the government, still in search of an ideology to replace Marxism-Leninism, could not neglect.

Also, representing Russians as a unanimous pro-Milošević block was one of the myths perpetrated by Western media throughout the Kosovo conflict. That a then strong liberal Russian press opposed Milošević was systematically omitted at the time from Western media coverage that insisted on "ancestral hatreds" and natural alliances between Orthodox nations, according to the clichés of Samuel Huntington that had just become notorious.[4] The Russian *nomenclatura,* no matter how interested it was in maintaining geopolitical leverage and supporting a useful ally, had its own reasons to fear Milošević. In a tense international atmosphere, commercial opportunities are dramatically reduced. Objectively, there was very

4 This ignorance of the West marginalized the liberal Russian press on the international stage making it more fragile to the future attacks that led to its decomposition under Putin.

little to be gained by Russia should Milošević have succeeded in Kosovo. Such success could have forced Russia again and again to support him any time he wanted to create problems elsewhere.

Such could be the explanation for Russia's attitude throughout the conflict: a mixture of faithful pro-Serbian rhetoric (using threats rarely surpassed since the Cuban missile crisis) and great prudence. The West acknowledged Russia's difficult position, which might explain why Russian peacekeeping troops were the first to be allowed to enter parts of Kosovo. There remains a doubt. Was the situation exploited in order to advance NATO hegemony at the expense of Russia's international stature? Unfortunately, the events of the following years, with more and more former Warsaw Pact countries joining NATO and the exacerbation by Western rhetoric of differences between Russia and the Ukraine, legitimizes the impression that such a tendency emerged during the crisis in certain Western political and military circles.

The Aftermath

Once again, Western troops, after what was a rapid and easy military victory for NATO (followed by a difficult occupation that continues more than a decade later), are stuck in a situation where they have to manage an extremely tense interethnic society in a land even poorer than before, with few perspectives for reconciliation. NATO may have prevented a far bloodier conflict (as experienced by Bosnians), but it is still faced with the following dilemma: what political and cultural project can it envision to secure stability in the region? The independence of Kosovo only complicated matters.

The great winners in this conflict were the Cassandras who announced such geopolitical disasters ever since the bipolar world began changing. Thanks to the media, the conflict in Kosovo popularized the growing fashions in political thought such as Kaplan's "coming anarchy" and especially Huntington's "clash of civilizations," which was the magical lens through which all were expected to understand everything happening in Yugoslavia in simple terms, and was also easily quoted in television soundbites. Very little was mentioned about the economic motivations behind the conflict and about the important mineral riches hidden in Kosovo. Even less heard were the sociologists who could have provided more profound explanations of the conflict as Elemer Hankiss had done for Hungary.[5] The former Marxist intellectuals were also more absent than ever; "Whatever happened to class struggle?" one could have been tempted to ask. But how could there have been a Marxist interpretation of Marxism's failure? In the mid-1990s, for raising such disturbing questions, the Bosnian Muslim director of

5 See Elemer Hankiss, *East European Alternatives* (New York: Oxford University Press, 1990), and Elemer Hankiss, *Hongrie: Diagnostiques. Essai en pathologie sociale* (Geneve: Georg Editeur, 1990).

the movie *Underground* (Emir Kusturica) was literally portrayed as a pro-Milošević traitor in a vicious campaign waged by the left-wing French intelligentsia. His personal political evolution in the following years became another self-fulfilling prophecy.

The repeated references to "ethnic hatreds," "religious intolerance" and "century-old conflicts" only diverted Western public opinion from more complicated political and economic interpretations. The clash of civilizations is easier to explain in a few minutes, between the reports on the latest sex scandals and the sport results. The danger is that references to conflicts between civilizations, by creating an impression of inevitability, not only imitate the paranoid discourse of Milošević but even reinforce it.

In the case of Kosovo, to realize the inaccuracies of such theories about cultural confrontation is far more useful to international security. Indeed, presenting the policies of Milošević as an Orthodox crusade against Islam was a lie very useful for his own propaganda. In the Balkans, as in the West, public opinion had been too frequently preserved from the following facts. First of all, an important proportion of Albanians are Orthodox. Of all the Orthodox churches in the world, the Albanian Church has suffered some of the worst persecutions in history, caused by the extremist regime of Enver Xoxha. In Albania, as in Kosovo, Orthodox Albanians have been struggling recently to rebuild their communities in the face of extreme difficulties. Now, their religion makes them the target of non-Orthodox Albanian extremists both in Albania and Kosovo. For the Serbs, their names and language make them fodder for persecution. To view the war in Kosovo (or the other Yugoslav wars) as a religious conflict misses an important aspect of Yugoslav life: there were as few religious people at the time in Yugoslavia as in the secular societies in Western Europe. Viewing twentieth-century Serbs as hordes of religious fanatics is as inaccurate as presenting Slobodan Milošević (whose wife had presided over the League of Young Atheists) as an Orthodox Joan of Arc. As a matter of fact, the Russian religious renaissance has been exaggerated as well. Its photogenic qualities, visibility and inspiring humanitarian actions (when the religious leaders are open-minded) cannot hide the fact that religion in Russia, although far from being marginal, has no more influence on modern life than it does in Los Angeles or Paris.

It is true that anti-Muslim feeling does not need Orthodox fervor if you present Muslims as fanatic fundamentalists ready to massacre anyone who does not accept their faith. "You may not be religious but the 'others' will slaughter you for the religion of your forefather" is a slogan that can easily rally agnostics or atheists. Paranoia has been exploited, but it exists and *could* explain some cultural conflicts. The stature of victim is an essential feature of Serbian identity. It is characteristic that one of the historic events which is the touchstone of the Serbs' vision of their past and of their present identity is (precisely) the battle of Kosovo of 1389. It was not a victory but a bloody defeat (the King and a frightening number of the Serbian élite were executed) followed by centuries of Ottoman oppression. The Alamo was also a defeat, but it was also a sacrifice to obtain a victory for Texas and America in the long term. The Serbian struggle for independence in the nineteenth century needed the cult of the martyrs to justify its existence and the emergence of the new State. But state legitimacy could never rest upon the head of state or the government, since the bloody personal vendetta between the Obrenović or Karagjorgjević royal families constantly created extreme dynastic instability. Serbs were forced to build part of their identity as modern citizens upon the history of their occupation by Muslims. This identity constructed upon the representation of martyrdom was materialized in a spontaneous initiative at the time that spread throughout Serbia and Serbian communities abroad: Serbs posted on the windows of their homes or automobiles a target (in the style similar to the targets used in shooting alleys). It became a popular symbol of Serbian resistance to NATO bombings. It was a provocation not unlike the jihadist call to martyrdom.

* * *

Presenting this as an inevitable consequence of the end of the bipolar world, even on the basis of sincere scholarly research, is as dangerous as Milošević building state legitimacy on collective paranoia. It is, maybe, what blunted Western reflexes at a moment when a more careful analysis of Yugoslav political dynamics could have identified the moderates who could then have received greater and faster support.

Worst of all, the disinformation that dominated academic and political debates at the time blunted the critical skills of entire societies when the war in Iraq was presented as another inevitable conflict that had to be fought out. The collective emotion caused by September 11 was not the only reason why we failed to reflect upon the consequences of launching a colonial adventure that is still not finished.

Child Soldiers: The Pursuit of Peace and Justice for Child Combatants

Susan Hitchcock Perry

The pivotal role played by children in war is both ancient and complex. Children have served as armed combatants and in supporting roles as spies, porters and sex slaves since the beginning of organized conflict. In most instances, children are the victims of war, losing their formative years to a maelstrom of violence and destruction. But in an important minority of cases, both male and female children benefit from the upending of traditional hierarchies and the opportunity to lead their comrades or contribute to what is perceived as a just struggle for the survival of their communities.

Humanitarian organizations vigorously advocate against the use of children in armed conflict, while a growing number of scholars are willing to examine the moral ambiguity surrounding the issue of child soldiers in contemporary warfare.[1] While children suffer from numerous forms of exploitation, child soldiering is an especially complex phenomenon in that it renders most victims unable to fully integrate with society as productive adults; long after the peace has been declared, former child soldiers may remain especially violent and marginal members of their communities. This chapter will present an overview of the problem of child soldiering in law and in practice, examining several salient questions that highlight the multifaceted nature of the issues at stake: Who may be considered, *de jure* and *de facto*, a child soldier? Should child soldiers be perceived as victims or perpetrators of organized warfare? And to what extent can child combatants be successfully reintegrated into their home communities once the war is over? The first section of this chapter will provide a brief, historical summary of the role

1 Sabine Collmer, "Child Soldiers—An Integral Element in New, Irregular Wars?" *The Quarterly Journal* 3: 3 (September 2004); Scott Gates and Simon Reich (eds), *Child Soldiers in the Age of Fractured States* (Pittsburgh: University of Pittsburgh Press, 2009); David M. Rosen, *Armies of the Young: Child Soldiers in War and Terrorism* (New Jersey: Rutgers University Press, 2005).

of youth and children in conflict, positing that the root causes of historical and contemporary child soldiering are, in fact, quite similar. The second section will examine the ever-evolving social perception of what constitutes a child and how that evolution has been translated into law. Within the context of this articulation of children's rights in humanitarian and human rights law, the third section will examine the obligation on the part of the international community to disarm, demobilize, rehabilitate and reintegrate child soldiers after conflict, and comment on the range of initiatives currently underway. This section will present Chad as an example of the tensions inherent in the simultaneous pursuit of peace and justice during the demobilization of children in post-conflict situations. In societies where youth constitute half the population, the successful reintegration of child soldiers may be the single most critical investment that the international community can make to achieving lasting peace in an unstable region, and should be considered the linchpin of restorative justice.

Children in War

International law sets the minimum age for active combat service for men and women at 18, a standard that is problematic at best.[2] The majority of child soldiers cited by humanitarian organizations are between 14 and 18 years old, although there are many reported cases of children as young as 8 actively participating in battle.[3] In addition to carrying a gun, children also serve in a variety of support activities, such as spying, scouting, carrying heavy loads, cooking, cleaning and providing sexual services as wives and as prostitutes of both sexes. Children are desirable recruits in war for three reasons: early training of malleable minds assures unwavering commitment to a cause; due to their small size and assumed innocence, children make excellent scouts, spies or suicide bombers; and finally, if the majority of a given population is young, available recruits for war will most likely range between 14 and 18 years of age, since both national and rebel forces must draw from the population at hand.

History is replete with examples of child soldiers, children and youth who served an armed cause, some to great acclaim by their local community, others forgotten by history. Spartan male citizens were celebrated as soldiers at age 18, after years of rigorous military training and exposure to battle, while Athenian citizens engaged in two years of obligatory military training between the ages 18 and 20. But slaves were not so lucky. The helots of Sparta and the slaves of Athens served as rowers, porters, messengers and providers of sexual services throughout the Peloponnesian War, essential components of combat barely recorded by

2 Optional Protocol on the Involvement of Children in Armed Conflict, 2000. See later in this chapter for a detailed discussion of the minimum age requirement.

3 Coalition to Stop the Use of Child Soldiers, *Child Soldiers Global Report, 2008*: http://www.childsoldiersglobalreport.org.

chroniclers. Thucydides does inform us that those male slaves who assisted Athens during battle were sometimes granted their freedom, while others ran away to the enemy camp. We do not know how many enslaved men and women assisting in battle or at camp were under the age of 18, but since entire urban populations—young and old—were forced into slavery after the defeat of their cities, it is unlikely that rowers or providers of sexual services were expected to meet a minimum age requirement to serve their masters in war.[4]

The American Civil War is rarely viewed as a campaign fought by a greater number of child soldiers than are presently engaged in any one single contemporary conflict. And yet conservative historical analysis places the number of Union and Confederate recruits under the age of 18 at between 250,000 and 420,000 men, or 10–20 percent of combatants, a high figure when we consider that UNICEF estimates the current number of child soldiers to be 300,000 in 30 conflicts worldwide.[5] These nineteenth-century adolescents were considered to have willingly sacrificed their greatest asset, their youth, and were glorified in hagiographic accounts by their contemporaries. Those adolescents who survived the Civil War returned to their communities to live out their lives as honored citizens, leaving behind tens of thousands of fallen teenage comrades who were not so fortunate.

Twentieth-century child combatants often adopted a vigorous response to conflict in their communities. An interesting parallel between the Jewish resistance fought in the forests of Eastern Europe during World War II and the first Palestian *intifada* engaged against the Israeli occupation in the West Bank and Gaza demonstrates that youth between 12 and 25 years old defied insurmountable odds without the benefit of regular access to arms and munitions.[6] In both cases, community elders continue to honor these young men and women for their sacrifices. Scholars have also carefully examined the role that adolescent women have played in nationalist struggles ranging from East Timor to Eritrea, establishing that their contribution as combatants, scouts and spies has had a determining impact both on the outcome of the struggle for independence and, in some cases, on their positive self-image and continuing leadership roles within their communities.[7] But not all women combatants benefit from war. The one reoccurring complaint of these former child soldiers is the inability of post-independence governments to deliver on promises

4 For an analysis of Thucydides' recounting of the Peloponnesian War, see Donald Kagan, *Thucydides: The Reinvention of History* (New York: Viking, Penguin Group, 2009).

5 UNICEF, *Report on Child protection from violence, exploitation and abuse* (2010), available at: http://www.unicef.org/protection/index_3717.html.

6 David M. Rosen, *Armies of the Young*.

7 Examples are numerous. See Heike Becker (2006) "'New Things After Independence': Gender and Traditional Authorities in Postcolonial Namibia," *Journal of Southern African Studies* 32: 1, March; Patricia Campbell (2005) "Gender and Post-Conflict Civil Society, Eritrea," *International Feminist Journal of Politics* 7: 3; Nina Hall (2009) "East Timorese Women Challenge Domestic Violence," *Australian Journal of Political Science* Vol. 44, No. 2 (June); and Kimberly Nettles (2007) "Becoming Red Thread Women: Alternative Visions of Gendered Politics in Post-independence Guyana," *Social Movement Studies* 6: 1.

of gender equality as a reward for young women's contributions during their respective armed nationalist insurgencies; despite the lack of sufficient empirical data, local activists and scholars cite the sinister ramifications of continuing gender-based violence long after the signature of the peace treaty.[8] More recent scholarship has begun to focus on the role children play in urban gang warfare in the city of Kingston, Jamaica.[9]

Many of today's conflicts are protracted civil wars over contested resource exploitation; these exceptionally violent conflicts take place in areas inhabited by traditional societies for whom the coming of age usually takes place in early adolescence. In Sierra Leone, for example, Mende girls and boys participate in the *sande* and the *poro*, coming of age ceremonies that usher them into the world of adult privileges and responsibilities at 13 or 14 years old. When civil war broke out in 1993 for control of the nation's diamond mines, government and rebel forces were recruited from amongst a rural population that considered adolescents to be old enough for marriage or combat. Although experts have pointed out time and again that children are desirable recruits because they are easily manipulated, we must also take into account the fact that the median age in Sierra Leone for the period 1990–95 was 18.7 years old.[10] Thus, United Nations estimates that one quarter of all volunteers and forced conscripts in Sierra Leone's civil war (1993–2002)—boys and girls alike—were under the age of 18 are probably conservative; many children were kidnapped or forced into soldiering through rape and other forms of extreme violence, while others volunteered for service with government or rebel forces to avenge the loss of their families or simply to eat.[11]

The claim that child soldiering is a recent phenomenon, exacerbated by globalization and an increase in the small arms trade, is not borne out by careful analysis.[12] The assault on civilians by adult and child soldiers, although forbidden

8 Myriam Denov and Richard Maclure, "Engaging the Voices of Girls in the Aftermath of Sierra Leone's Conflict: Experiences and Perspectives in a Culture of Violence," *Anthropologica* 48: 1 (2008); Vanessa Farr, Henri Myrttinen and Albrecht Schnabel (eds), *Sexed Pistols: The Gendered Impacts of Small Arms and Light Weapons* (Tokyo: United Nations University Press, 2009); Stijn Smet, "A window of opportunity - improving gender relations in post-conflict societies: the Sierra Leonean experience," *Journal of Gender Studies* 18: 2 (2008).

9 Laurie Gunst, *Born fi' dead: A Journey through the Jamaican Posse Underworld* (New York: H. Holt, 1995); Natalya Kitchin, *Rehabilitation and Penal Reform in Jamaica: An Unresolved Issue* (unpublished undergraduate thesis) (Amherst MA: Amherst College, 2009); Ivelaw Lloyd Griffith, *Drugs and Security in the Caribbean: Sovereignty under Siege* (University Park, PA: Pennsylvania State University Press, 1997).

10 UN Department of Economic and Social Affairs, *World Population Prospects* (2009).

11 UNICEF, *From Conflict to Hope: Children in Sierra Leone's Disarmament, Demobilization and Reintegration Program* (2004). See also Ishmael Beah, *A Long Way Gone: Memoirs of a Boy Soldier* (New York: Sarah Crichton Books, 2006).

12 Scott Gates and Simon Reich, "Think Again: Child Soldiers," *Foreign Policy* (May 2009); Vera Achvarina and Simon F. Reich, "No Place to Hide: Refugees, Displaced Persons, and the Recruitment of Child Soldiers," *International Security* 31: 1 (Summer 2006).

by twentieth-century laws of war, has been a reality for those populations caught in the crossfire of conflicts ranging from the Rape of Nanjing in 1935–36 to the Sabra and Shatila massacre of 1982, both of which involved ruthless combatants under the age of 18. The flood of small firearms on the market hardly made a difference in the 1994 Rwandan genocide; many of Rwandan *interahamwe* armed with machetes and knives were unemployed high-school-aged students eager to participate in the campaign to "cut the tutsi down to size."[13] Thus, it is not the use of children in war but our awareness of the phenomenon that has changed. Although the 300,000 children estimated to be child soldiers as far back as the early 1990s is a figure impossible to verify, the numbers have attracted the attention of the donor community and thrust child soldiering onto the international agenda. The seminal reports by Graça Machel in 1996 and Olara Otunnu, UN Special Representative for Children and Armed Conflict, in 1998 led to important initiatives to bring an end to the contemporary practice of child soldiering. The next section examines the accelerated evolution of the international legal framework and the role of internationally funded tribunals in advancing protection for children in war.

Law and Justice for Child Soldiers

International law has been slow to address the issue of children in combat, in large part due to the fact that children have been active and valued participants in war throughout history. One of the cornerstones of the laws of war, the Lieber Code of 1863, makes no mention of children other than as civilians to be removed along with women before a bombardment campaign (art 19), and only if secrecy is not a priority.[14] Neither the Hague Conventions of 1899 and 1907, nor the first three humanitarian law conventions signed in Geneva provide protection for children in war,[15] lacunae that were only addressed in 1949, when a new series of Geneva Conventions afforded fundamental guarantees for civilians during conflict, including children not taking part in hostilities.[16] But it was not until 1977 that a standard minimum age for combatants was codified in law. In the first

13 *The Prosecutor v. Jean-Paul Akayesu* (Trial Judgement), ICTR-96-4-T, International Criminal Tribunal for Rwanda (ICTR), September 2, 1998.

14 General Orders No. 100—Instructions for the Government of Armies of the United States in the Field by Francis Lieber, April 24, 1863. It is notable that although the American Civil War was fought by a significant minority of soldiers under the age of 18, little attention was paid to the plight of under-age recruits or child civilians. Further instruments, such as the Geneva Declaration of the Rights of the Child of 1924, were inapplicably vague.

15 Cf. the Conventions of 1864, 1928 and 1929.

16 Those children who do take part in hostilities are guaranteed the same rights as adult combatants. See Advisory Service on International Humanitarian Law, *Legal Protection of Children in Armed Conflict* Document 0577/002;03 01 2003 (Geneva: International Committee of the Red Cross, 2003).

Additional Protocol to the Geneva Conventions, State Parties are obliged to take all *feasible* measures to prevent children under the age of 15 from taking direct part in hostilities, and their recruitment into national armed forces is expressly prohibited; article 77 encourages Parties to give *priority* to the oldest, when recruiting soldiers between the ages of 15 and 18.[17] Article 4 of the second Additional Protocol goes one step further in prohibiting both the recruitment and the participation—directly or indirectly—of children under the age of 15 in hostilities.[18] In both cases, the cutoff point of age 15 accommodated the use of young recruits in the armed forces of many of the signatories; this, and the use of terms such as "feasible" or "priority" in the first Additional Protocol, weaken the level of protection afforded to children.

More recently, human rights law has also addressed the issue of a minimum combat age. Article 1 of the 1989 Convention on the Rights of the Child, ratified by all but two UN member states,[19] defines a child as anyone under the age of 18, but article 38 of the same treaty reiterates the accommodating language of the first Additional Protocol to the Geneva Conventions, setting the minimum age for child recruitment at 15.[20] This contradiction embedded within the clauses of a single treaty illustrates the difficulty in setting norms for military recruitment; both the United States and the United Kingdom actively recruit into their armed forces at age 17, while high school military academies worldwide provide an avenue for advancement for tens of thousands of children. The failure of the Convention on the Rights of the Child to provide a maximum standard of protection for child combatants was not addressed until 2000, when the Optional Protocol on the Involvement of Children in Armed Conflict essentially amended the treaty, raising the minimum age of active combat to 18.[21]

---

17 Author's italics. Protocol Additional to the Geneva Conventions of 12 August 1949, and relating to the Protection of Victims of International Armed Conflicts (Protocol I), 8 June 1977.

18 Art. 4, §3c, Protocol Additional to the Geneva Conventions of 12 August 1949, and relating to the Protection of Victims of Non-International Armed Conflicts (Protocol II), 8 June 1977.

19 Neither the United States nor Somalia has ratified the treaty. The Somalian government announced in November 2009 its intention to ratify the Convention on the Rights of the Child, whereas the Obama administration committed to a "thorough and thoughtful" review of ratification of the treaty. *Reuters Africa*, November 20, 2009, online version available at: http://af.reuters.com/article/topNews/idAFJOE5AJ0IT20091120.

20 For a thorough explanation of efforts on the part of the US government to maintain the minimum age for recruitment at 15, see Sharon Detrick (ed.), *The United Nations Convention on the Rights of the Child: A Guide to the 'Travaux Préparatoires'* (Dordrecht: Martinus Nijhoff Publishers, 1992).

21 Articles 1–4, Optional Protocol to the Convention on the Rights of the Child on the Involvement of Children in Armed Conflict (2000). The Capetown Principles (1997), the ILO Worst Forms of Child Labor Convention (1999) and the Paris Principles (2007) have further contributed to the establishment of a straight 18 minimum recruitment standard.

Increased legal protection for children was reinforced by the setting up of mixed and international tribunals to try those individuals responsible for established patterns of child soldier recruitment in civil and international conflicts. The international tribunals for Rwanda (ICTR) and for former Yugoslavia (ICTY), along with the mixed tribunal for Sierra Leone (SCSL) and the cases before the International Criminal Court (ICC) in The Hague, all rely on humanitarian law in the prosecution of child soldiering, particularly the clauses of the four Geneva Conventions of 1949, their Protocols, and the Statute of the International Criminal Court, commonly referred to as the Rome Statute. The Rome Statute is the first international treaty to criminalize the use of child soldiers. And yet, because the Statute was adopted in 1998, two years before the Optional Protocol to the Convention on the Rights of the Child, and the negotiators hoped to bring in the United States as a State Party to the Statute, the clauses on child soldiering fall short of the straight 18 requirement, defining as a war crime only the active involvement of children in hostilities or the recruitment of children under the age of 15.[22] Despite this lower standard of protection for children, the International Criminal Court currently has in custody Congolese rebel commander Thomas Lubanga Dyilo, standing trial for child soldiering as a war crime. The Court will undoubtedly benefit from the seminal jurisprudence provided by the Special Court for Sierra Leone in its landmark 2008 Appeals Chamber decision in *Prosecutor v. Brima, Kamara and Kanu*.[23]

The consolidated indictment against Commanders Brima, Kamara and Kanu, Armed Forces Revolutionary Council (AFRC) officers who, in conjunction with the Revolutionary United Front (RUF), established a military government over part of Sierra Leone from 1994–1998, included the crime of conscripting or enlisting children under the age of 15 into armed forces and using them in hostilities. This charge marked the first time that an international criminal tribunal invoked the offence of recruiting child soldiers. The Trial Chamber convicted Kanu for conscripting, enlisting and using children under age 15 in hostilities. Kanu contested the Trial Chamber decision, demanding that the Appeals Chamber overturn his conviction for recruiting child soldiers because he lacked the requisite *mens rea* or intent and child recruitment was not a war crime at the start of the SCSL's temporal jurisdiction in 1996. The Appeals Chamber responded by finding it "vexatious" that Kanu suggested he did not possess the necessary *mens rea* for child recruitment,[24] and then cited its previous decision in *Prosecutor v. Norman, Fofana and Kondewa* to determine that using children in hostilities was a violation of Article 4(c) of the Special Court's Statute.[25] The SCSL Appeals Chamber decision sets down seminal

22 Rome Statute of the International Criminal Court, Art. 8, §2b(xxvi) and §2e(vii).

23 *Prosecutor v. Brima et al.*, SCSL-2004-16-A, Appeals Chamber, Judgment, February 22, 2008.

24 Ibid., §296.

25 UN Security Council, Statute of the Special Court for Sierra Leone, January 16, 2002.

jurisprudence for both international and national tribunals in the prosecution of commanders responsible for the recruitment of child soldiers.[26]

As a result of the Appeals Chamber decision in *Prosecutor v. Brima, Kamara and Kanu*, legal scholars expect clarification of the crime of child soldiering to facilitate prosecution of Thomas Lubanga at the International Criminal Court, despite a series of procedural difficulties that have slowed down the case.[27] It is the interplay between the legal texts, on the one hand, and the willingness of international prosecutors and judges to apply the clauses on prevention of child soldiering, on the other, that promises redress for child combatants. But is legal redress enough? Nation states, in conjunction with UN agencies, international NGOs and local actors, have played a pivotal role in the disarmament, demobilization and reintegration of child soldiers worldwide, working with and sometimes against the interests of justice.

Is Peace or Justice the Priority for Child Soldiers?

Although the author has advocated pursuit by the International Criminal Court of those individuals and corporate actors who benefit from conflict resources in Central Africa,[28] the current situation in Chad demonstrates the limits of this position. Chad, an oil exporting state, has been a signatory to the Rome Statute since 2007 and, as a party to the treaty, is obliged to cooperate with the Court in its pursuit of those responsible for heinous violations of international humanitarian law. In March 2009, an ICC Pre-trial Chamber issued an international arrest warrant for Sudanese President Omar Al Bashir, suspected of responsibility for the war crimes, crimes against humanity and genocide that occurred in Darfur during his presidency.[29] President Al Bashir visited Chad in July 2010, thereby violating the terms of Chad's engagement as a signatory to the Rome Statute. Under article 59 of the treaty, Chadian President Deby's government should have arrested Sudanese President Al Bashir as soon as he landed on Chadian soil and turned him

26 For further discussion of the impact of this case, see Charles Jalloh and Janewa Osei-Tutu, "*Prosecutor v. Brima, Kamara, and Kanu*: First Judgment from the Appeals Chamber of the Special Court for Sierra Leone," *American Society for International Law Insights* 12: 10 (2008), electronic journal available at: http://www.asil.org/insights080520.cfm#_edn1.

27 Ibid.

28 Paper on "La justice internationale pour les enfants soldats: la responsabilité des commandants commerciaux," presented at the conference on "Ending Recruitment and Use of Children by Armed Forces and Groups: Contributing to Peace, Justice and Development," N'djamena, June 7–9, 2010. I would like to thank US Chargé d'Affaires in Chad, Sue Bremner, along with her colleagues and staff, for their unstinting support.

29 ICC-02/05-01/09 "Warrant of Arrest for Omar Hassan Ahmad Al-Bashir," issued 04.03.2009 for the case *The Prosecutor v. Omar Hassan Ahmad Al Bashir*, situation in Darfur, Sudan, Public Court Records—Pre-Trial Chamber I.

over to the requisite authorities at The Hague. Instead, the Chadian government opted to ignore its treaty obligations, thereby snubbing the United Nations and the ICC, claiming to privilege regional peace over international justice. While this choice led to an outcry amongst international human rights and humanitarian organizations, such a policy is the motor behind the steady demobilization of child combatants in Chad and Sudan over the course of the past year. With overt support from US Special Envoy General Scott Gration, Sudan and Chad are engaged in a rapprochement that may end decades of conflict across their borders and stem the massive flows of refugees generated by the humanitarian crisis in Darfur.[30] Given that famine set in earlier than usual in 2010 due to poor harvests and drought, I would argue that the choice of peace over justice is the right one for Chadian child combatants, but only in the short term.[31]

Demobilization of child soldiers creates a whole host of new problems for a post-conflict society in terms of distribution of resources and capacity for rehabilitation. In the context of a UN-led demobilization campaign that has led to the release of hundreds of child combatants into disarmament, demobilization and reintegration (DDR) programs across Chad, the six Central African states signed the N'djamena Declaration in June 2010. The signatories—Chad, Sudan, Niger, Cameroon, Central African Republic and Nigeria—have pledged that no child under 18 should take part, directly or indirectly, in hostilities. These countries also agreed to prevent any form of recruitment of children and to ratify the Additional Protocol to the United Nations Convention on the Rights of the Child on the involvement of children in armed conflict, if they have not already done so.[32] As a contributor to the drafting of the N'djamena Declaration, I would contend that its primary value is the emphasis on effective DDR, including monitoring procedures; if children are not rehabilitated after conflict in this highly unstable region, they will either turn to crime or return to child soldiering, posing a real threat to the long-term peace and development of the Chadian state.

UNICEF estimates that there are approximately 15,000 child soldiers in Chad, Sudan and the Central African Republic.[33] DDR professionals in the Central Africa region consistently support the "do no harm" principle, and point to agency coordination, staff codes of conduct and community participation as key factors in the successful rehabilitation of child soldiers. They pay special attention to the long-term trajectories of children and the impact of war on their development, acknowledging that the related erosion of local capacities and institutions after

30 General Scott Gration, US Special Envoy to Sudan, Testimony before Senate Foreign Relations Committee, May 12, 2010. Hostilities between Chad and Sudan ended on January 15, 2010 with the signing of a security protocol.

31 FEWSNET, at: http://www.irinnews.org/Report.aspx?ReportId=8837.

32 As of September 2010, Chad and Sudan have signed and ratified the Optional Protocol on the Involvement of Children in Armed Conflict, while Cameroon and Nigeria have signed the protocol, but have not yet ratified it. Niger and the CAR have neither signed nor ratified.

33 UNICEF, *Fact Sheet: Children Associated with Armed Groups and Forces* (2010).

conflict acts is detrimental to their work.[34] These professionals make a distinction between a child's inner resources, such as self-esteem and intellectual ability, and external resources, such as social support networks and relationship with a caregiver; external resources, in fact, receive more attention in framing responses to encourage resilience amongst demobilized children. In addition to resilience, support for the reunification of children with their families or other culturally appropriate caregivers, along with encouragement for the prompt reinstitution of schooling or vocational training and the strengthening of wider community activities and institutions, form the backbone of rehabilitation.[35] Unfortunately, however, there is very little empirical evidence to assist in determining whether DDR strategies have a positive, long-term impact. In Sierra Leone, although UNICEF estimated that 98 percent of the children demobilized were reunited with one or both parents, older siblings, or relatives,[36] some of those who returned home were rejected by their families and subsequently migrated to other areas rather than reintegrate locally.[37] Communities often stigmatized the returning child soldiers, while many families were unable to integrate former child soldiers due to problems ranging from violent behavior and chronic drug use to the murder or rape of family members by the child during the period of conflict.[38]

The care and protection of children in crisis settings pose a challenge to the criminal pursuit of war commanders in post-conflict regions. The cooperation of military commanders is essential for disarmament and demobilization and, as influential figures, they are often in the front lines in terms of the allocation of resources for effective rehabilitation and reintegration. Chad has a history of conflict, rendering the environment for DDR troubled, at best.[39] The Chadian-Sudanese rapprochement holds out hope for a relatively stable environment in which to return child combatants to their families and to rebuild infrastructure

34 Conference proceedings for "Ending Recruitment and Use of Children by Armed Forces and Groups Contributing to Peace, Justice and Development," N'djamena, June 7–9, 2010. These statements by DDR professionals correspond with the empirical evidence gathered by Alastair Ager, Lindsay Stark, Bree Akesson and Neil Boothby for their article on "Defining Best Practice in Care and Protection of Children in Crisis-Affected Settings: A Delphi Study," *Child Development* 81: 4 (July/August 2010).

35 Ibid. I would also like to thank Dr Marzio Barillo, UNICEF Country Officer for Chad, and Sophie da Camara, UNDP (New York), for discussing their work in the Central African region with me in detail.

36 UNICEF, *From Conflict to Hope: Children in Sierra Leone's Disarmament, Demobilization and Reintegration Program* (2004).

37 John Williamson, "The disarmament, demobilization and reintegration of child soldiers: social and psychological transformation in Sierra Leone," *Intervention: The International Journal of Mental Health, Psychosocial Work and Counseling in Areas of Armed Conflict* 4: 3 (2006).

38 Ibid.

39 See Susan H. Perry, "Resolving Conflict in Chad," in Lionnet, Nnaemeka, Perry and Schenck, *Signs: Journal of Women in Culture and Society*, special issue on "Development Cultures: New Environments, New Realities, New Strategies" 29: 2 (Winter 2004).

to assist with reintegration: the traumas of war require clinics and hospitals to treat damaged bodies and minds; new teachers and classrooms to absorb the numbers of students returning to school; and improved roads and markets to facilitate increased economic interaction as a follow-up to vocational training. Despite its burgeoning oil wealth, Chad is at the bottom of the list in terms of human development, ranked 175th out of 182 countries in the 2009 UNDP Human Development Index.[40] In the past six years, the country has hosted approximately 255,000 refugees from Sudan's Darfur region and close to 77,000 refugees from the Central African Republic, and 188,000 Chadians have been displaced by fighting in the east of the country.[41] Moreover, the Chadian judicial system is fragile, the result of decades of failed state-building. As military and rebel commanders continue to demobilize hundreds of children, the Chadian Parliament is unable to promulgate a coherent corpus of laws on children's rights, relying instead on a web of diverse legislation that will make the domestic prosecution of military commanders who have recruited child soldiers extremely difficult.[42] Finally, both the Chadian and Sudanese presidents seem determined to remove all international "interference" from their soil, whether MINURCAT forces or EUFOR troops.[43] The lessening of an international presence on Chadian soil will most likely hinder access by international aid workers to demobilization or interim care centers, and will certainly do nothing to bolster the reputation of the International Criminal Court. Given the current circumstances, international justice may have to wait.

Conclusion

Child soldiering is a contemporary phenomenon with deep historical roots. The phenomenon continues today with more than 70 military organizations in nineteen nations worldwide that recruit and use children in armed hostilities of an international or non-international character.[44] In response to the questions posed at the beginning of this chapter, the ambiguity over what constitutes a child in peace and in war remains a factor in the protection of children. While the Optional

40 UNDP Human Development Index 2009: http://hdr.undp.org/en/reports/global/hdr 2009.

41 United Nations Security Council, *Report of the Secretary-General on the United Nations Mission in the Central African Republic and in Chad (MINURCAT)*. S/2010/217, April 29, §7 (2010).

42 Author's field notes, N'djamena, June 2010.

43 Ibid. The European Union Peacekeeping Mission to Chad and the Central African Republic (EUFOR Chad/CAR) served from January 2008 to March 2009, and was replaced by the UN Mission in the Central African Republic and in Chad from early 2009 to May 2010. See UN Security Council, Resolution 1861. S/RES/1861 (2009), January 14.

44 Coalition to Stop the Use of Child Soldiers, *Child Soldiers Global Report 2008*, http://www.childsoldiersglobalreport.org.

Protocol on children in armed conflict sets the minimum age for recruitment into the armed forces at 18, the Rome Statute follows the precedent set by the Geneva Conventions and allows for recruitment of children as young as 15 years of age. International criminal tribunals thus focus on the youngest child combatants; they have rendered unambiguous jurisprudence for the prosecution of commanders who recruit children under the age of 15 into rebel or national armed forces. More importantly, international tribunals have firmly established that children must continue to be considered the victims of organized warfare, regardless of whether they commit atrocities or benefit from the chaos of war. Finally, United Nations agencies, in partnership with governments, as well as global and local NGOs, have created disarmament, demobilization and reintegration programs for child soldiers. These DDR initiatives have been able to get children off the battlefield, but have often been less successful in rehabilitating them as fully functioning members of society.

The difficulties inherent in DDR campaigns are particularly apparent in Chad. Although a State Party to the Rome Statute, the government has chosen to consolidate peace with neighboring Sudan, while turning its back on its international obligations. But should Chad pursue peace at the expense of international justice? There is a dearth of educational and livelihood opportunities for children in a post-conflict economy, where health and educational infrastructure and market exchange have been destroyed by decades of war. For every child soldier turned pastry chef in the kitchens of an international hotel, there are dozens of former child combatants who cluster together in migrant shanty towns at the edges of the capital, unable to access schools and vocational training once they leave their home villages. And yet the challenges inherent in the pursuit of peace are no excuse to abandon justice. The long-term solution lies in providing both peace and justice for demobilized child combatants. Corporate actors in the Chadian oil industry—be they Chinese or American—have a security interest in financing the schooling and vocational training of former child soldiers, in an effort to discourage a rise in criminality that would negatively impact the oil extraction industry. And local NGOs have an unprecedented opportunity to bolster basic freedoms in a post-conflict state. By vigorously collecting child soldier testimony, while working to provide former combatants with the tools necessary to rebuild their lives, current activists may oversee the emergence of a healthy civil society for the next generation. As integrated adults, former combatants should then be encouraged by the international community to pursue restorative justice in an effort to heal the scars of war once and for all.

The Instrumentalization of Gender in War

Carol Mann

The official discourse on war has stressed armed conflict as a masculine preserve from which women have systematically been excluded, relegated to the, supposedly passive, 'home front'. Since the late eighteenth century, gendered stereotypes have given assigned roles to men and women in wartime that conceal the historical participation and agency of women in warfare, resistance and peacemaking – and still do.

It is on conventional representations of women that war still thrives, obliterating the far more complex and truly interesting reality. In this chapter, I will consider two situations typical of armed conflict where women are not considered as individual persons, but as global categories, quasi-interfaces through which male military forces and political agendas confront each other. The first is about causality, the second about tactics; both are products of a patriarchal society built on sexual inequality that depoliticizes the male–female relation by naturalizing male domination. In a sense, these could be termed gender wars as they aim to install a specifically unbalanced gender relation, based on a supposedly 'natural' order that needs to be preserved, albeit through the most extreme violence against women.[1] In brief, this will be an attempt at deconstructing the instrumentalization of women during armed conflict.

Women remain one of the major excuses to start or pursue a war, often to justify colonialist expansion. They are beings to be saved, protected, at any rate so manipulated as to evacuate any possibility of agency. And as a result they suffer doubly: not only from the horrors of war but also from the repression that their own males subject them to, as a reaction to their own wounded masculinity. Likewise in the case of wartime rape, which is seen as the supreme strategy where women are used against 'their' men in order to contrast the ultra-virile superiority of one male group against the other and bring out the incapacity of the supposedly weaker group to defend their property. Nevertheless, the shame and

---

1 Cynthia Cockburn, *The Space Between Us: Negotiating Gender and National Identities in Conflict* (London: Zed Books, 1999).

dishonour is frequently carried by the women themselves, who find themselves doubly victimized as they are so often excluded from their own society after rape, especially in the patriarchal societies where wars take place today.

The place of women in armed conflict has produced three distinct kinds of writing, as well as a number of sub-genres: one, the inspirational kind, part and parcel of wartime propaganda designed to inspire patriotic emotion – be it tales of Florence Nightingale or the semi-beatification of Soviet resistance heroines after World War II. The other, the critical discourse, is more rarefied and is emerging today with pioneers Cynthia Enloe and Cynthia Cockburn, who have reshaped thinking about war along feminist, antiwar and antimilitaristic lines. Between the two lies the journalistic writing and TV reporting that presents women-and-children exclusively as perennial defenceless victims, fortunately photogenic enough to fire virtuous patriotic/nationalistic anger, and just as important to inspire donors to give to certified charities. These naturally do not question the premises upon which such instrumentalization is built.

This chapter is partly based on my own work, namely the recently published *Femmes dans la guerre 1914–1945* (Paris: Flammarion/Pygmalion) and *Femmes d'Afghanistan* (Paris: Le Croquant), both published in 2010, as well numerous articles, mostly based on extensive fieldwork in wartime Bosnia, Pakistan and Afghanistan since 2001.

Woman as Nation, Nation as Woman

Throughout history, an ongoing process of routinely gendering community and subsequently the nation has taken place, usually to the dubious benefit of the female sex. In pre-national times, that is to say largely before the French Revolution, communities were united by religious practice, language, modes of alliance (endogamous/exogamous) and above all the management of private domestic space that defined cultural continuity. The constructed male/female binary that produces essentialist definitions of maleness and femaleness has been traditionally built round biological explanations, much criticized by feminist authors (from Simone de Beauvoir to Judith Butler and many more between and after). Yet what amounts to a political construction and division of labour dominates patriarchal societies, even today in countries where war takes place.

Private space (the home, in the inside world) was allotted to women (and remains so in traditional societies such as Afghanistan). This supremely female sphere covers areas from birthing, death rituals, accepted signs of modesty signifying male ownership of female bodies (head coverings throughout the Mediterranean), crafts, specific foods, the shape and way of making bread loaves. Land – that of its male proprietors – comes to life symbolically through a series of specific rituals enacted by its female inhabitants. Even though intended to preserve a traditional way of identification (to oppose 'foreign' invasion, takeover or annihilation by force), these were not static, modified by variations in religion, colonialism and elements of the outside world which inevitably seeped into the remotest societies. The globalized market – often in the shape of shoddy-quality Chinese goods – appears in the most distant Afghan and Congolese villages today and is appropriated according to traditional practice, as mediated by the dominating females of the household. Furthermore, post-Cold War nationalist movements have invented traditions in the previously Communist world, especially in the Balkans, that deliberately sought to break with Communist egalitarianism.[2]

However, this spatial/cultural construct has been used as an excuse for war. There is an old Pashtun proverb which claims that men fight over three things: *Zan, Zamin, Zar* – women, land and gold. Women indeed are causes to fight over in that they might be considered as property, personifications of a national or tribal ideal. Bringing these three factors together in this way implies proprietorial rights on behalf of those fighting and performing their political manhood ('doing their duty'), who place these elements on an equal footing within the patriarchal *Weltanschaung* which, after all, continues to rule the classic world order.

It seems logical that the next step, from the late eighteenth century, would be a female construction of the nation, as dominated by the male state. As Cynthia Enloe states: 'If a state is a vertical creature of authority, a nation is a horizontal creature of identity.'[3] The latter, which is at present territorially defined, presumes a more or less unified set of criteria linked to what previously constituted acceptable forms of domestic space, reconfigured as a national tradition. The nation is an all-pervasive simultaneously castrating and empowering mother-figure – France, Britannia, Russia, Italy (invented in the late nineteenth century), America, Felix Austria, Germania (even though it's theoretically a 'Vaterlan'd) – defended by her citizen-sons. This influences post-colonial national constructs: hence the emergence of 'eternal' Mother India, even 'Modar Afghanistan' in this deeply tribalized region. Senegalese academic Malik Ndyaye spoke of l'Afrique mère-patrie, the ultimate motherland for all her sons and daughters, that is to say countries inhabited by descendants of African slaves.[4]

Women are indeed causes to preserve, in that they might be considered as property, and as Deniz Kandiyoti writes 'the symbolic repository of group identity';[5] in post-Cold War wars, the latter creates a sense of illusory community, based on an imagined communion between the male citizens defending it in war,[6] and also includes a reconfiguration of past and future generations.[7] This metaphorical equivalence between women and nation, women as repositories for

2 Rada Ivekovic, *Le Sexe de la nation* (Paris: Leo Scheer, 2003).

3 Cynthia Enloe, *Bananas, Beaches and Bases* (Berkeley: University of California Press, 1990).

4 Ernest Harsch, 'Solidarité africaine avec Haïti, Argent, aide alimentaire et médicaments affluent de la "Mère patrie" africaine', *Afrique Renouveau* 24: 1 (April 2010), 22.

5 Deniz Kaniyoti, 'Identity and its Discontents', *Millennium* 20: 3 (1991), 435.

6 Benedict Anderson, *Imagined Communities, Reflections on the Origin and Spread of Nationalism* (London: Verso, 1983).

7 Nira Yuval-Davis and Marcel Stoelzer, 'Imagined Boundaries and Borders: A Gendered Gaze', *European Journal of Women's Studies* 9: 3 (August 2002), 329.

national culture and wealth, suggests that indeed women's bodies are turned into battlegrounds. Rape becomes the most potent weapon of war, as was shown by the mass rape of Bosnian women by Serbian paramilitaries (1992–1995), whereas the escalating domestic violence in Serbia in those same years was deemed a 'normal' part of everyday life.[8]

The nation-as-woman invites a vocabulary of war based on and legitimized by notions of assault and rape. 'The rape of Belgium' during World War I, and 'the rape of Kuwait'[9] at the start of the Gulf War in 1991, offered passive and helpless stereotyping of nations which entitled foreign intervention, as that of chivalrous superior ultra-virile forces, especially in the case of the USA.

A War to Save Women

In his first State of the Union address on January 2009, the then President George W. Bush triumphantly declared: 'The women of Afghanistan are now free.' This message was reinforced by passionate speeches delivered by Laura Bush, his wife, and Cherie Blair, spouse of British Premier and primary US ally Tony Blair. The two first ladies set themselves out to be champions and arbiters of the Afghan women's cause. The desperate condition of this half of the population of Afghanistan was considered the ultimate expression in a veritable internal gender war which could only be stopped through Western military intervention. Indeed, the future of Afghan women became one of the major modes of legitimation of Operation 'Enduring Freedom', organized in retaliation for the destruction of New York's Twin Towers on September 11, 2001. Was this a case of window dressing for the unsuccessful pursuit of the evanescent Al-Qaeda and the failure to capture Osama Bin Laden? At any rate, gender became and remains the key notion of every kind of aid project – even though, viewed nine years later, the promises have not been kept and the overall condition of women remains abysmal.

The blue burqa, customarily worn by Pashtun women and forced upon the entire female population by the Taliban, became the logo for the operation, taken up by global media in a popular rhetoric which owes much to Samuel Huntington's writings.[10] This sartorial obligation crystallized Huntington's thinking that Western and Eastern thinking were so opposed as to be irreconcilable. This, in itself, was a theorized, more elaborate version of British colonial thinking, as Rudyard Kipling,

8 Unpublished interview of author with representatives of 'Women in Black' in Sarajevo in February 1995.

9 This was even the title of a work by a popular American writer: Jean Sasson, *The Rape of Kuwait: The True Story of Iraqi Atrocities Against a Civilian Population* (New York: Knightsbridge, 1991).

10 Samuel Huntington, 'The Clash of Civilizations', *Foreign Affairs* (Summer 1993).

the bard of the Raj, stated in the first line of what is probably his most famous poem: 'Oh, East is East, and West is West, and never the twain shall meet.'[11]

The reform of women's lives has always been a central part of the colonial process which necessarily aims at destroying the fabric of indigenous societies in order to submit them to its power. The near decade of American military presence in Afghanistan is in every way a colonial one, with the notable exception of a territorial takeover, deemed no longer necessary these days and replaced by the conquest of market share for a supposedly global economy based on mass consumerism. A frequently fuzzy but standardized notion of human rights belongs with this package deal (as it is with humanitarian aid everywhere) because, ultimately, real social change has to be enforced in order to adjust to the new world order, especially in a traditional society where the notion of state and common good is singularly absent. Tackling the 'woman question' in a frontal manner cuts through all social strata.

Frantz Fanon eloquently described this process as applied by the French to colonized Algeria:

> *The colonial administration can then define its doctrine: "If we want to reach Algerian society at its heart, through its powers of resistance, we must first conquer their women, we need to seek them out behind the veil where they conceal themselves and in the houses where the men hide. ... The dominating powers officially set themselves up to defend the humiliated, secluded and concealed woman."*[12]

The challenge of male domination inherent to any attempted reform in the private sphere has always been instantly perceived by its opponents for the past century, be they Algerian or Afghan. This helps explain why the Taliban today are so virulently opposed to women's rights in their country: they are perceived as the ultimate insult to male honour. This needs to be set in its historical background, with a brief consideration of the European colonial process and how this created a set of norms and ideals in Asia and Africa.

Colonial Feminism

The Western model of progress, which acquired the status of paradigm in Afghanistan, was experimented in British India (also known as the British Raj) from the late nineteenth century onwards and has remained influential in the whole area. Progress meant medical and scientific improvement as well as reforms deemed necessary to bring the colonized lands up to Western standards of propriety

11 Rudyard Kipling, *The Ballad of East and West* (1889), available at: http://www.everypoet.com/archive/poetry/Rudyard_Kipling/kipling_the_ballad_of_east_and_west.htm.

12 Frantz Fanon, *L'An V de la révolution algérienne* (Paris: Maspero, 1959).

and morals. Social legislation aimed at improving the life-conditions of women, which included raising the age of marriage and introducing education, became emblematic of Imperial achievement, rather like the priorities of present-day humanitarian agencies.[13] The vigorous support of women's rights (of which there was no parallel back in London) served principally to show up the backwardness of their menfolk and the superiority of British civilization, which can be seen as one form of virility emasculating and triumphing over the other in the conquest of the native female population – which explains much of the opposition such measures encountered. The feminization of representations of the Oriental male is part of this process, as Edward Saïd has famously explained.[14] Leila Ahmed has called this strategy 'colonial feminism or feminism as used against other cultures in the service of colonialism'. Ideals of upper-middle-class Victorian femininity were exported to India, aspects of which had much in common with the local prototype in their dependence on male validation and their primacy of a self-effacing maternal role.[15]

Since the early twentieth century, progressive Afghan monarchs tried to reform the country according to Western criteria, as influenced by the nearby British Raj – which started just over the southern border, and where they themselves were, more often than not, educated. In the 1920s, the most progressive of all Afghan rulers, Amanullah (1892–1960), went further and took up the lead from Mustafa Kemal, known as Ataturk, the Turkish reformer of the newly founded republic, himself much inspired by Lenin.[16] Heralding the later Communist government, he simultaneously attempted reforms concerned with land tenure and marriage customs, all of which remain at the heart of Pashtun identity and self-definition. When, following the lead given by Turkey and Persia, Amanullah ordered the population of Kabul to wear Western clothes (something which the British never dared attempt in India), the enforced unveiling of women (1929) was designed to turn them into equal partners of their freely chosen husbands (1924) in the national quest for modernity, reinforced by co-education. As a result, he had a near revolution on his hands, coming from the staunchly conservative Pashtun southern part of the country, and had to flee into exile.

The educated classes living in the cities – be it in Afghanistan or India during the Raj – had always associated themselves with Western modernity, which permitted access to privilege and power in the capital city, something which is equally true in present-day Kabul, occupied by allied forces. Then, as now, the poorer classes were excluded from this increasingly incomprehensible process. This reconfiguration of Afghan identity with Western style aspirations bringing together British, Turkish, Soviet and (today) globalized American influences did not help unite the Afghan population and indeed achieved the opposite. The result of such policies applied until now may well account for at least 75 percent of the

13 Dagmar Engels, *Beyond Purdah? Women in Bengal 1890–1930* (Delhi: Oxford University Press, 1999).

14 Edward Saïd, *Orientalism* (London: Penguin, 1978).

15 Leila Ahmed, *Women and Gender in Islam* (New Haven, CT: Yale University Press, 1992).

16 Louis Dupree, *Afghanistan* (Princeton, NJ: Princeton University Press, 1973).

population being non-literate which really shows a rejection of modern policies regarding the link between literacy and progress. The absence of communication and mutual incomprehension between the urban ruling class and the vast rural majority were and remain largely caused by the foreign and secular criteria for modernization. This meant cultural alienation for those who felt threatened by these new norms, especially those which threatened to change the status of women, perceived as the guarantors of continuity and tradition, the passive representative of family honourability.

They retaliated by setting themselves up as the champions of ethnic and religious tradition, which they defended with orthodox zeal, as proven by riots against the Raj in Peshawar yesteryear and today against international aid agencies in rural Afghanistan. This widespread anger contributed to the development of Wahabi Muslim fundamentalist reformist movements which, from the mid-nineteenth century onwards (especially active in the frontier region with Afghanistan), assimilated any form of Western progress to Infidel propaganda, especially secular education. This indeed is the historical origin of the Taliban movement, which carries the self-same ideology. Above all, the schooling of girls was (and is) seen as the supreme *casus belli*, the root of all evil for giving women ideas above their station and affording them a measure of autonomy that might cause them to challenge traditional modes of male domination.

The Schooling of Girls as *Casus Belli*

The rural, traditionalist opposition to any form of challenge concerning women's rights in Northern India (also populated by Pashtuns known as 'Pathans' on this side of the frontier) was echoed in Afghanistan from the early twentieth century onwards, and has continued unabated until the present day. The ever-resilient reticence towards government schools, especially for girls, was displayed by the Afghan rural population each time the latest ruler in Kabul, be it an emir, a king or the Communist government, attempted to enforce secular education for girls as well as boys. This problem was also encountered by the UNHCR when they set up schools in the refugee camps during the Soviet intervention in Afghanistan (1979–1989) which caused over three million mainly rural Afghan citizens to flee over the frontier to Pakistan. Unprecedented repression of women was licensed by the fundamentalist brand of Islam which had poured out from the neighbouring conservative mosques into the refugee camps, where the future Taliban were raised in the 1980s. When the Soviet army retreated and ex-combatants fought for several years in an extraordinarily bloody war (1989–1994), the fundamentalist ideology was imported from the deeply rural camps into the capital, reversing the usual trend where change goes from city to countryside.[17] The repression of

17 Carol Mann, *Models and Realities of Afghan Womanhood: A Retrospective and Prospects* (UNESCO, 2005).

women reached its acme in the Taliban years (1994–2001) and came to symbolize the entire regime.

Despite the enforcement of education and the building of thousands of schools, the traditionalist, and especially Taliban, opposition is as virulent as ever, especially since 2005 as it has gained support both on the Pakistani and Afghan sides, its finances bolstered by the vigorous opium trade. It has resurged with a vengeance as witnessed by the frequent torching of girls' schools (to date about one thousand schools), gas attacks, constant threats to pupils and the killing of teachers despite (or perhaps because of) the government commitment to girls' education. Similar events are occurring in Taliban-controlled areas of Pakistan: in Swat alone 162 girls' schools were destroyed in 2009.[18] It has to be pointed out that the new-style Taliban are not identical to the ones who ruled the country in the 1990s. Despite their ideology, they use the most up-to-date technology, media and business methods, especially when it comes to drug trafficking, all of which Mullah Omar's generation opposed.[19] It has to be said that the Taliban, from their own point of view, consider that what they do protects women, as inferior helpless beings, from the evil that may befall them and that they might inadvertently provoke: exposing them to novel options produced by education and knowledge would do them more harm than good, perturbing the God-given order of things and the naturalized male domination and authority.

Women as Victims of War and Peace in Afghanistan Today

Women are the veritable victims both of 30 years of unabated war and the compensatory violence at the hands of their own men who need to offload their frustrations and anger at being incapable of handling their own lives. They are also victims of a symbolic violence, as they have become the official cause for armed confrontation between conflicting approaches to human rights and civilization generally, a now territorialized war between a rural majority and a minority composed of a small urbanized elite, backed by US and Western forces. It is in these terms that the recent slaughtering of a humanitarian team by the Taliban, in August 2010, on the border of Nuristan and Badakhshan provinces, needs to be considered.[20] Although the aid was medical and destined for women and children in areas where maternal mortality is the highest in the world,[21] it was nevertheless

18 'Taliban Destroy Girls' Education, Pakistan Is Powerless', IPS, Peshawar (January 28, 2009).

19 Patrick Porter, *Military Orientalism: Eastern War Through Western Eyes* (Columbia: Columbia University Press, Critical War Studies, 2009).

20 International Assistance Mission (IAM), August 12, 2010: http://www.iam-afghanistan.org/.

21 Carol Mann, 'A death sentence for women', *The Guardian* (November 7, 2008), available at: http://www.guardian.co.uk/commentisfree/2008/nov/07/afghanistan-gender.

perceived as an invasion of private territory, as symbolized by women, by foreign maleness. In a culture where male honour is pivotal, empowerment of women (including access to independent health care) effected by external powers (British or American, colonial or neo-colonial) can only be seen as an attack against collective virility and patriarchal privilege. The peculiar brand of Afghan fundamentalist Islam continues to provide God-given legitimation to the most violent prescriptions of customary law.

In the days of the British Raj, Victorian ideals were to a certain extent more in keeping with the upper-class mores than what has been attempted today. The feminist principles which major aid agencies support today are rooted in first-generation secular feminism, which considers women as individuals with their own personal desire for fulfilment, preferably overriding traditional female roles such as marriage and family. The Communist government ran into the same trouble and found themselves having to deal with a war that was as much a civil war as one directed towards the Soviet troops. Is the United States heading towards a similar situation?

From 2002 to 2005 there seemed to be a real improvement in women's lives and expectations: enrolment in the new schools was high, new avenues to employment were opening, especially for girls who had learned English in Pakistan and other skills in Iran. Through humanitarian aid the younger generation of rural refugees (the vast majority) had been exposed to schooling abroad they never would have had access to, had they remained in Afghanistan. Women's associations have been set up, taking up structures that already existed in the far more progressive 1960s and 1970s: indeed, the Communist era provided more opportunities for women than ever before or since. All kinds of projects have been attempted, with varying efficiency, sponsored by aid agencies. They have allowed the new generation of educated girls in cities to find paid employment and therefore become the financial mainstay for their families. Naturally such examples of success – unfortunately limited to cities – bring out the value of schooling for families otherwise reluctant to send their daughters to school. But it is likely that these attempts will be short-lived, in view of the growing force of reactionary opinions.

The superficial nature of the investments in the future of the women in Afghanistan become evident as one looks at their place in politics. A quarter of parliamentarians today are women, which constitutes indisputable progress. But this does not mean that human rights are a priority for these female MPs. Their solidarity goes first to their political family and the clan to which they belong. One of the problems is that they are totally untrained and are incapable of conceptualizing problems and their possible solutions. Furthermore, female MPs, lawyers and activists who dare to voice any opposition are continually threatened. Murders are frequent, perpetrated by the Taliban and their allies, and are intended to inhibit any female ambition through sheer terror. The number of women in the civil service is steadily declining. Because of the impunity for these crimes, the government is, in fact, amplifying the deterrent effect of this campaign of continuing intimidation. What happens in high places reflects a trend at all levels. According to a survey conducted in 2008, some 87.2 percent of women of all ages have suffered at least one act of brutality, physical, sexual or psychological.[22] Less than 15 percent of the victims dare to complain, because they may be jailed by policemen always sympathetic to the male side of the story.[23]

Today, most projects for female empowerment (a favourite catchword amongst donors) are completely insensitive to the realities of Afghan life. The notion of the individual, so typically Western, fatally collides with the centrality of the family, the institution of marriage and the importance of religion. The main problem is that these ambitious schemes are developed thousands of miles away from Afghanistan. The reaction of the women themselves is a mixed one: the youngest, usually urbanized high school students, can benefit from the advantages, yet depend on the approval of their own families, whose dictates they have to follow. In rural areas, women's solidarity necessarily goes towards their menfolk on whom they depend, and takes priority over individual decisions concerning schooling or health. The lack of even the most basic education ensures that they are kept unaware of other possibilities. However, through the media (TV, wherever there is even a couple of hours of daily electricity) they are increasingly ambitious for their children, a factor which is generating a measure of change.[24]

In the media, the country is inevitably reduced to being a primitive Muslim country thirsting for Western salvation, which has enraged one part of the Muslim-American academic community, as Lila Abu-Lughod's aptly entitled article 'Do Muslim women really need saving?'[25] has demonstrated. This has been the major reason for their failure in Afghanistan: it enraged the most conservative factions as well as the Taliban, who have actively fought against foreign troops as well as aid and the ideology the agencies foster.

If armed conflict shows no sign of abating, and on the contrary the Taliban are taking the country over, their main victims have been women, even though the violence being enacted against them is seen to serve society's higher principles, which in the case of Afghanistan is a unique mix of political Islam and pre-Islamic tribal law.[26] A tug of war is seen to be going on between the US and allied forces and the Taliban, with the issue of women's rights at stake. What of women's voices? A law is being debated to stop women working for foreign aid agencies, which indeed will reverse any of the progress achieved.

---

22 Global Rights, 'Living with Violence: A National Report on Domestic Abuse in Afghanistan', March 2008: http://www.globalrights.org/site/DocServer/final_DVR_JUNE_16.pdf?docID=98.

23 Human Rights Watch, 'We Have the Promises of the World Women's Rights in Afghanistan' (New York, 2009): www.hrw.org/en/reports/2009/12/03/we-have-promises-world.

24 Carol Mann, *Femmes d'Afghanistan* (Paris: Le Croquant, 2010).

25 Lila Abu-Lighod, 'Do Muslim Women Really Need Saving? Anthropological Reflections on Cultural Relativism and Its Others', *American Anthropologist* 104: 3 (2002).

26 Carol Mann, 'The law's the problem in Afghanistan', *The Guardian* (December 26, 2009), available at: http://www.guardian.co.uk/commentisfree/2009/dec/26/afghanistan-women-customary-law?showallcomments=true#start-of-comments.

To make matters worse, the United States is presently backing an alliance between the government, the fundamentalist former warlords already in power and their new neo-fundamentalist Taliban allies, and, amongst others, the arch-criminal Hekmattyar whose true place is at the Hague tribunal. This policy may be the only way to get out of an inextricable war, but the fate of the entire female population of Afghanistan will be sacrificed on the altar of expediency in the process. Naturally this is hardly popular with the public at large in the West for whom the war in Afghanistan is getting increasingly unpopular.[27] Rather than bring up the nature of the compromises the US is willing to make, and what the mission of the 17,000 extra men drafted into the country will be, women's rights are brought to the forefront by the conservative media. In August 2010, Time magazine carried on its cover page the photograph of 18-year-old Bibi Aisha, who had her nose and ears cut off by her husband after she ran away from the continuous violence he and his family inflicted on her. What was not stressed is that Aisha and her sister were 'given' as compensation in a marriage through a customary arrangement called *Baad*, whereby frequently very young brides (under 14) are given to pay debts by opium farmers or as compensation for murder to the injured party: this generally means that the bride (in this case Aisha) is subjected to the most brutal treatment in lieu of the murderer, usually her brother or father. The increase of such marriages has less to do with ancient tradition than modern catastrophes that cannot be resolved by an impotent state.

The caption of the cover of Time magazine simply read 'What happens if we leave Afghanistan.' The trouble is that this is already whilst 'we' are *in* Afghanistan, which goes to show that all these grand empowerment-based projects, inadequately funded and improperly thought out, have not affected the benighted majority of women of Afghanistan who continue to hold world records in the field of illiteracy and maternal and child mortality. They have been used as an excuse by the United States to invade their country and establish a neo-colonial rule that has failed them, and remain the victims of rising reactionary politics in Afghanistan that seeks to confront the West, through its own female population.

Sexual Politics: Mass Rape as a Weapon of War and Ethnic Cleansing

Since the second third of the twentieth century, women have been given the role of personifying the nation, the supreme mother figure of presumably grateful citizens ready to defend her honour. The latter is perceived as sexual – hence the 'rape of Belgium', as the German attack on Belgium during World War I was called by

27 A Gallup poll claims that 43 percent of Americans believe the war in Afghanistan to be a mistake. See http://www.gallup.com/poll/141716/New-High-Call-Afghanistan-War-Mistake.aspx.

the British media. The actual wartime rape of real women does not deserve the same consideration, as it is generally depoliticized and has been until very recently considered an inevitable consequence of armed conflict, along with looting. This explains why the mass rape of German women in Berlin in 1945 by Soviet troops has been ignored for so long. As Cynthia Enloe puts it so eloquently: 'The women who suffer rape ... merge with the pockmarked landscape, they are put on the list of war damage along with gutted houses and mangled rail lines.'[28]

This stereotype reflects deep-rooted gender inequality, typical of patriarchal societies, that places women in a category devoid of personhood, on the level of objects that can be pillaged, understood as 'damaged goods', and at the same time reflects the naturalizing of sexual violence against women. The Afghan proverb previously quoted about women, land and gold as causes for war also reflects the traditional level of consideration afforded to the female half of the world's population.

Rather than being considered as the extreme expression of political violence inflicted by one social group on another, a specific crime against the female enemy, rape, as been thought of, in the best case, as a natural outlet for exhausted soldiers or the cultural practices of some barbaric human groups. This, above all, rules out intentionality and the responsibility of perpetrators, as well as expressing racism and arbitrary class distinctions. The implication is that officers and gentlemen of the civilized nations don't do such things, it is rather the low-life drunken brutes lurching back from the battlefield. However, mass rape, as distinct from more individualized acts, are planned acts of war and as such put in place through a chain of command that put officers and gentlemen, generals and rulers, at the top, and imply their direct responsibility.

Ethnic Cleansing

There are many categories of rape during war, ranging from 'casual' rape by a soldier perpetrated on a civilian (the story of Elsa Morante's most famous novel *La Storia*, set in Italy during World War 2); to the Korean 'comfort women' in the same war; to rape as a form of specific torture in jail during military dictatorships (Chile, Argentina, Uruguay during the 1970s); to mass rape used as a specific weapon of war in recent and ongoing wars, as was the case in ex-Yugoslavia (1991–95), Rwanda (1994) and today in many parts of Africa, especially in the Democratic Republic of Congo (DRC, former Zaire), which has the highest number of rape victims in recorded history.

This is what we shall be looking at specifically here, in particular the notion of 'ethnic cleansing', which first appeared in the war in former Yugoslavia. According to the UN definition, this describes 'the planned deliberate removal from a specific

28 Cynthia Enloe, *Maneuvers: The International Politics of Militarizing Women's Lives* (Berkeley: University of California Press, 2000).

territory, persons of a particular ethnic group, by force or intimidation, in order to render that area ethnically homogenous'.

I wish to stress, however, that the ubiquitous term 'ethnic' remains a persistent misnomer, as it really designates different religious/national communities within the same ethnic group. Men who defined themselves as Serbian and Orthodox nationalists sought to occupy territory through forced migration, mass murder (as in Srebrenica) and the rape of women from Croatian Catholic and especially Bosnian Muslim communities. The further uncritical use of such extraordinary terminology ('ethnic cleansing') does imply, for the outsider, a condoning of the principle: if 'cleansing' is seen as a positive virtue, then its opposite must forcibly be 'dirtying'.[29] In this case the adjunct of any extra ethnic group means that any kind of multicultural society (as Kigali and Sarajevo were before their respective wars) is necessarily unclean and deserves severe purging ...

Thus, in Rwanda, the mass (mainly) Hutu rape of Tutsi women became a mode of terror for entire communities which accompanied the slaughtering and maiming of the most vulnerable, including infants. Organized and planned by the hierarchies, these rapes were systematic, in that they were part and parcel of military tactics against a perceived enemy, which is why it is so important that the UN consider them as such, describing them as veritable crimes, not just a random and inevitable side-effect of war effected by irresponsible brutes. The estimated total number of rapes in the war in Bosnia ranges from 30 to 50 thousand, and those for Rwanda, just during the three-month process of warfare in 1994, are rated around ten times these figures. Yet real numbers are impossible to evaluate, because of the silence and shame that the victims themselves feel and the social stigma they fear by denouncing the crimes.

The Case of DRC

Today, the world's most catastrophic rape 'epidemic', for want of a better word, is happening in the Eastern provinces of DRC. The blame has been put first on disbanded Hutu militia, the Democratic Forces for the Liberation of Rwanda (FDLR), remaining from the Rwandan war and roaming the jungles of the area since late 1994. Several groups are allied to them and operate in the area, as do assorted Congolese armed groups, the most (in)famous being the Mai-mai and other splinter bands that include members of the Congolese army meant to be fighting the rest.[30] Child soldiers have been forcibly recruited into these groups, which are often constructed on ethnic lines, and rapists are in fact often teenagers. On top of rape, genital mutilation of the most sadistic kind takes place, the different

29 My thanks to gender specialist Nermina Trbonja from Sarajevo for drawing my attention to this important point.

30 'DRC: Who's who among armed groups in the east', *IRIN* (July 10, 2010), available at: http://www.irinnews.org/Report.aspx?ReportId=89494.

groups symbolically staking their territory by additionally raping their victims with specific tools (guns, sticks, broken bottles, etc.) that act as a form of signature for the group that is attempting domination through 'ethnic cleansing'.

Apart from these organized mass assaults, civilian rape, especially of children (as in South Africa), is on the increase, fuelled by impunity and the normalizing of extreme brutality as a mode of relating to women and girl children. There are no easy explanations, but part of it is anger and the feeling of social and economic powerlessness of men who compensate through hyper-virilized aggression against the most fragile: the extreme suffering of their victims gives them the illusion of power.[31] Solely in the north and south Kivu provinces, 8,000 rapes are estimated to have been committed in 2009,[32] nearly half of those registered for the whole country in that year. The latter number just represents the total of the cases that have been reported and the reality is a multiple of that estimation.

In a country riddled by corruption, rampant impunity, extremes of wealth and poverty, it follows that violence and crime constitute ideals for youths in violent all-male groups who have lost their links to society and family and the possibility of any autonomous planning of their own destiny. The repeated pain inflicted by their systematic cruelty surely gives them an addictive feeling of collective power and virile authority that numbs any kind of sensitivity.

What Wartime Rape Means

There is much more to mass rape than misogyny, testosterone and nationalism gone mad. It is essential to see what the rapists are aiming at, and how this brutal instrumentalization of women both fits in with military strategy and an overall conceptualization of the intertwined notions of race, class and sex, as applied to the rapists' communities and those they are seeking to annihilate.

What do these women represent, seeing that victims can be from age four to eighty, so that it is not a question of hyperactive libido? Be they Bosnian Muslim, Rwandan or Congolese, they share the following in the eyes of their enemy: as the female members of a designated community, each one represents the backbone of the society they come from, in many cases the promise of future generations, or the living memory of the past. In the areas where Serb militia operated, they also took pains to destroy ancient mosques and change the names of the villages. Furthermore, in all these communities, these women play an active economic role by working in farms and fields: the disappearance of the income they traditionally generate will impoverish their families. As female members of patriarchal families,

31 Véronique Nahoum-Grappe, 'Anthropologie de la cruauté' in Marie Rose Moro and Serge Lebovici, *Psychiatrie humanitaire en ex-Yougoslavie et en Arménie* (Paris: Presses Universitaires de France, 1995).

32 'DRC: Behind bars for rape', *IRIN* (July 7, 2010), available at: http://www.irinnews.org/Report.aspx?ReportId=89761.

clans, each represents the honour of its men and the respectability of the group. We have seen how this works in the Afghan context, where young victims of rape can be killed by their own families because the collective honour has been sullied through this crime. This goes a long way to explaining why women in Kosovo and remote Bosnian hamlets did not want to come forward to indict the rapists, even when they recognized them. These villages were home to both Muslim and Orthodox inhabitants, and the Serbian Bosnian militias recruited men from these places. The same thing happened in villages populated by Hutus and Tutsis. The male rivalries inside these closed communities were played out in a fully-fledged war: rape became an attack on the men defined as enemies, through the instrumentalization and humiliation of 'their' women's bodies. In fact, the latter turned into war zones for confrontation and violent exchange between opposed sides. They become the text on which war propaganda is writ large: during the confrontation between Muslim, Sikh and Hindi populations during the partition of the British Raj into Pakistan and India (1947), extreme violence was exercised against women, including rape, the tattooing of nationalist slogans and the chopping-off of breasts, as the supreme attack against the nurturing mother-figures in each other's communities.[33]

Naturally, the mass-rapists from the countries listed imagine that they are going to get away with their crimes, knowing full well the penalty their victims will have to pay inside their own communities. The men whose honour has been sullied by their incapacity to defend their wives, mothers and daughters often turn their humiliation into anger aimed precisely at those who are the living witnesses of their failure to protect them. Genocide accompanies mass rape. Where women are raped, men are tortured, have their throats slit and are slain. In the case of Bosnia, in places such as eastern DRC, women were raped, boys and men murdered, leaving whole communities of distraught females, many of whom found themselves pregnant, infected with HIV. The status of the baby becomes fraught with uncertainty because she no longer represents continuity, but disruption, the constant reminder of personal and social humiliation and pain. The rapists can pride themselves on having fathered more citizens from their own communities through the abuse of bodies deprived of agency and even personal identity in this mass twinned process of rape and enforced reproduction which has sometimes taken place in rape camps, as a subsection of the whole ethnic cleansing project. During the Spanish Civil War, pro-Franco Fascists would inscribe the following slogan on street walls: 'Perhaps we will perish but your women will give birth to fascist children.'[34]

The whole process is curiously reminiscent of the Nazi Lebensborn, where 'racially appropriate' young German women were made to produce babies

33 See the descriptions of that period in the remarkable novel by Bapsi Sidwah, *Ice Candy Man/Cracking India* (London: Penguin, 1988).

34 Yves Ripa, 'Armes d'hommes contre femmes désarmées: de la dimension sexuée de la violence dans la guerre civile espagnole' in Cécile Dauphin and Arlette Farge (eds), *De la violence et des femmes* (Paris: Albin Michel, 1997).

conceived through what could be called organized and enforced procreation with selected studs from the German army. This was known, through Nazi propaganda, as 'having baby for the Führer'. Naturally, nothing else bears comparison, as the women were well treated and valued as mothers of future citizens of the Reich. What is similar, however, is the nationalist exploitation of their reproductive capacities, women reduced to their wombs. The Nazis put in place the exact reverse tactic for Jewish women and little girls, who were specifically targeted as potential mothers.

Sometimes, rapes fire the men with the desire for vengeance and can be further instrumentalized by governments, as was the case for the Croat leader Franjo Tudjman: the spiral of war becomes endless as individual women become subsumed into a global national cause and excuse to pursue organized armed violence. Croatian rape victims might not have been given the support and attention they desperately needed, but their plight served Tudjman's nationalistic and military ambitions. When, in 1992, I tried to find out if it was possible to adopt babies born from rapes in Croatia by Serb militia, I was told that these children were kept in orphanages (ill-treated as enemy spawn) as pawns for future bargaining over territory: they would be reclassified as (future) Croat citizens, therefore entitling the government to ask for greater *Lebensraum* in the inevitable post-war negotiations, whenever these would come about. In brief, the Croatian government was turning a crime against their otherwise neglected raped female citizens into a profitable enterprise, never mind the cost to the actual victims.

Looking Beyond the Stereotypes

It will take time before tribunals set up explicitly to try wartime rape in Bosnia and Rwanda actually have a long-term deterrent effect, as they alone cannot be effective unless stereotypes concerning women as passive vehicles for male honour and ethnic continuity are truly challenged by the communities of the perpetrators as well as the victims concerned, be it in war or in peace.

Despite over a century of women's struggle for emancipation and the changes that have been effected, armed conflict is still reduced to a confrontation of stereotypes. As Cynthia Cockburn reminds us, before being reduced to the status of 'Muslim', 'refugees', 'victims', the women of Bosnia were educated, secular members of an egalitarian society: when I went there the first time during the war, a woman, an engineer, of Sarajevo asked me: 'Is it true that women in Paris are paid less than men? It sounds incredible.' As we spoke, bombs were raining on the city. Likewise, the Communist regime in Afghanistan created a class of educated, sophisticated women, albeit an urban minority, that was to suffer the most from the brutal rule of illiterate fundamentalists.[35] In Rwanda, RDC and other parts of the African continent, victims are not just remote village women, but articulate,

35 Valentine M. Moghadam, *Modernizing Women: Gender and Social Change in the Middle East* (Boulder, CO: Lynne Rienner, 2003).

educated citizens aware of their rights. The rampant social inequality means that educated women in developing countries are usually privileged and urban. The relative wealth and status that these usually middle-class women enjoy has helped them to find routes (including migration) to escape the fate inflicted on their poorer compatriots, and also to seek justice.

Thinking of society in terms of essentialist stereotypes that lump humanity into active/passive, warriors/victims, necessarily perpetuates male domination and the entire mechanism behind nationalist wars and mass rapes. Men are not born rapists or combatants, just as women are not natural pacifists. The stereotype of the woman as angelic mother-figure just does not fit the likes of Ilse Koch, the sadistic guard at Buchenwald concentration camp in World War II, or more recently Linndie England, the American military policewoman who was convicted for torturing prisoners at the notorious Abu Ghraib jail in Baghdad, in 2005.

What emerges is a kaleidoscope of conflicting images revealing how shallow stereotypes of womanhood really are, however potent they remain, especially in the areas where wars take place today. When examining more closely wars in Africa and Asia today, it becomes obvious that women often have to compensate for areas in which their male counterparts have failed, mainly supporting the bulk of civil society and ensuring a future for their families. The lack of recognition of women's agency is certainly a constant feature, as is the systematic recourse to gender stereotypes to legitimate the absence of women in politics, be it in occupied Gaza, Somalia, DRC, Iraq or Afghanistan.

Wars and Climate: The Effects of Climatic Change on Security

Ben Cramer

Environmental security, a relatively recent concept, has provoked intense debate among theorists of international relations. What are the indicators for environmental security? To what extent is the scarcity of a natural resource likely to cause a "green war"? Are crises, such as Darfur, likely to be more frequent? Taking into account the amount of land soon to be engulfed by rising sea levels, is climatic change then a threat to national or international security? To what extent are these climatic disturbances going to represent the drop that causes the vase to overflow, the vase being already full to the brim with demographic pressure, soil erosion, deforestation and diminishing sources of drinking water and fish stocks. Do we include among these new threats States that are demanding compensation for environmental damage perpetrated by other States, or the attempts of those who seek to delocalize their pollution? Or those who would rely on military solutions in situations of environmental insecurity?

A New Context?

> *Prediction is very difficult, especially about the future.* (Attributed to Niels Bohr, Danish physicist, 1885–1962)

There is much research that attempts to establish links between climate change or disturbances, "human security," migrations and armed conflicts.[1] However, by way of an introduction, we will recall here the road that has been travelled in recent decades. During the 1980s, just prior to the ending of the Cold War, the United Nations Group of Governmental Experts on Disarmament and Development wrote: "It is now clear, without any doubt, that resource shortages and ecological

1 Fabrice Renaud, Institute for the Environment and Human Security, United Nations University.

constraints pose threats, real and imminent, to the future well-being of all peoples and all nations. These problems have a primarily non-military character and it is absolutely necessary that they be treated as such."[2]

Is the military option still the only one imaginable when we speak of security? In 1983, the Danish political scientist Barry Buzan, in his work *People, States, and Fear*, proposed "revisiting" the security field and constructed a five-part typology of security issues. Besides military security, Buzan identified other forms of security: 1) political security, which concerns the institutional stability of the State and of its political regime; 2) economic security, which concerns the conditions for maintaining the well-being and prosperity of the State; 3) environmental security, as protection of the conditions for human life on earth; 4) lastly, social security, which seeks to protect against attacks on the culture and language of a political entity, or its identity.[3]

The formalization of the concept of environmental security in international relations theory occurred at the beginning of the 1990s. The post-Cold War period made it more and more evident that military threats were not the only elements of insecurity in the world. In 2005, the former UN Secretary-General, Kofi Annan, had asked in the framework of the High-Level Panel on Threats (the Challenges and Conflicts report) that the environment be recognized as a source of conflict, which was a cultural revolution insofar as questions of environmental security are not mentioned in the UN Charter. Since then, the International Panel on Climate Change, and its President, Rajendra Patchauri, have been awarded the Nobel Peace Prize.

In the Brundtland Report, generated in 1987 by the World Commission on Environment and Development, later known as the Brundtland Commission and presided over by Gro H. Brundtland, the impact of climate is not absent. Their report, entitled "Our Common Future," notes:

> *Environmental threats to security are now beginning to emerge on a global scale. The most worrisome of these stem from the possible consequences of global warming caused by the atmospheric build-up of carbon dioxide and other gases. Any such climatic change would quite probably be unequal in its effects, disrupting agricultural systems in areas that provide a large proportion of the world's cereal harvests and perhaps triggering mass population movements in areas where hunger is already endemic. Sea levels may rise during the first half of the next century enough to radically change the boundaries between coastal nations and to change the shapes*

2 The United Nations Group of Governmental Experts on the Relationship between Disarmament and Development (1978–1981), chaired by Ms. Inga Thorsson (Sweden); UN General Assembly document A/36/536, reproduced by the Centre for Disarmament Affairs in 1982 as Disarmament Study Number 5.

3 Brundtland Commission, *Our Common Future*. Available at: http://www.un-documents.net/ocf-cf.htm.

and strategic importance of international waterways – effects both likely to increase international tensions.[4]

In June 2009, almost 20 years later, the UN General Assembly adopted a resolution concerning the harmful effects of climate change and their implications for international security.[5] This was the first time that the delegations of different member countries of the concert of nations reached a consensus on the issue and adopted a resolution which established the link between climate change and international security.

In the words of the UN resolution A/RES/63/281, sponsored by more than 90 countries, adopted on June 3, 2009, the General Assembly, "*Deeply concerned* that the adverse impacts of climate change, including sea-level rise, could have possible security implications, … *Invites* the relevant organs of the United Nations, as appropriate and within their respective mandates, to intensify their efforts in considering and addressing climate change, including its possible security implications …"

The Ongoing Debate

The December 2007 report of the German Federal Government's Advisory Council on Global Environmental Change (WBGU—Wissenschaftlicher Beirat der Bundesregierung Globale Umweltveränderungen) is a point of reference. The report, entitled *World in Transition: Climate Change as a Security Risk,* is based on the work of international experts and organizations such as the UN Environment Programme (UNEP). It concluded that if climate change is not brought under control, it is likely to aggravate old tensions and to provoke new ones in certain parts of the world which could then sink into violence, conflict and war. Professor Hans Schellnhuber, the main author of the report, who is Director of the Potsdam Institute for Climate Impact Research and a professor at Oxford University, wrote, "Without the means to combat it, climate change will destroy the capacities of adaptation of a number of societies in the decades to come. This could in turn lead to destabilisation and violence that will endanger national and international security at an entirely new level."[6]

In terms of expertise, we can refer to various schools of thought. According to the "American School" represented by Arthur Westing, environmental changes and resource shortages contribute largely to the emergence of armed conflicts. Other institutes are working on this sensitive and controversial issue; they try to assess the

4 Ibid. See Chapters 4 and 7.

5 Appendix to UN resolution A/63/L.8/rev.1.

6 German Federal Government's Advisory Council on Global Environmental Change (WBGU), *World in Transition: Climate Change as a Security Risk* (December 2007), available at: http://www.wbgu.de/wbgu_jg2007_engl.html.

influence of environmental problems on the emergence of armed conflicts and on the course of events, and to evaluate the influence of the management of resources on peace and security, whether at the national level (civil wars) or the international level, considering that each indirectly touches upon the climate question.

The pioneers in this realm include the Toronto Group around Thomas Homer-Dixon. The Toronto Group acts as a programme of studies, at the University of Toronto, on peace and conflict. The conceptual and theoretical bases of the work were presented in two articles published in the journal *International Security* (1991, 1994). The work of this group puts at our disposal a range of interpretations of phenomena and their consequences by seeking to show how resource shortages and violent conflicts are inseparable, based upon case studies that include Mexico (Chiapas), the Middle East (Gaza), Pakistan, and South Africa.

Among the other authoritative institutions, mention should be made of the group created by the Environment and Conflicts Project (ENCOP) established by Bächler and Spillman of the Polytechnic University (ETH) in Zurich. The ENCOP has focused its empirical studies on trying to establish the correlations between environmental degradation and the escalation of conflicts.

To expand this list, we must take account of the work of the Oslo Group around Nils Petter Gleditsch. This group is composed of researchers from the Peace Research Institute of Oslo (PRIO), and uses statistical methodologies; for these researchers, ecological and sociological variables are connected. Finally, there is also the GECHS-UCI, established in 1999 at the University of California, Irvine, with support from the Center for Global Peace and Conflict Studies (GPACS), by Dr Richard A. Matthew. GECHS-UCI has the following objectives: 1) to undertake original, interdisciplinary and participatory research; 2) to collaborate with academics and policy-makers in developing countries; 3) to develop policy recommendations; and 4) to educate the public on the ways in which environmental change interacts with other transnational forces to affect the lives and welfare of individuals and groups around the world, especially in developing countries. It focuses on the "capacities of adaptation" of human societies,[7] but challenges the neo-Malthusian notions of "carrying capacity" according to which human population growth and per capita rates of consumption will cause more and more demands, shortages and distributional conflicts.[8]

7 See http://www.gechs.org.

8 Ehrlich, P.R., A.H. Ehrlich, and G.C. Daily, *Population, ecosystem services, and the human food supply* (Morrison Institute for Population and Resource Studies Working Paper No. 44, Stanford University, Stanford, CA, 1992). Ehrlich, P.R., A.H. Ehrlich, and J.P. Holdren. Ecoscience: Population, Resources, Environment (W.H. Freeman, San Francisco, 1977). Ehrlich, P.R., and A.H. Ehrlich, "How the rich can save the poor and themselves: lessons from the global warming," in S. Gupta and R. Pachauri (eds), Proceedings of the International Conference on Global Warming and Climate Change: Perspectives from Developing Countries (Tata Energy Research Institute: New Delhi, 21–23 February, 1996), 287–294. Homer-Dixon, Thomas, "Environmental Scarcities and Violent Conflict: Evidence from Cases," *International Security* 19: 1 (Summer 1994), 5–40.

Before reviewing the issues that are subject to debate, we can already identify some key points from the different approaches. To summarize, environmental factors are rarely, perhaps never, the only cause of violent conflicts. It is impossible to isolate environmental factors from their contexts, from the socio-economic, political and cultural factors which are interrelated and interact. In the case of civil wars, intra-state wars, which today represent the majority of conflicts in the world, the factors leading to war, for the most part, are explained through structural contexts: the political exclusion of certain ethnic groups (or "the micro-nations" as they are called by Nobel Peace Prize laureate Wangari Maathaï), the weakening of economies, without forgetting the collapse of the system created by the Cold War. We could say that the structural causes of conflict, such as the withdrawal of the State, the emergence of markets of violence, the exclusion or extermination of certain population groups, find themselves reinforced and accelerated by ecological problems and the loss of resources like water and soil. It is these factors which lead the political scientist Halvard Buhaug (from the PRIO Institute in Oslo) to write "the principal causes of civil war are political, and not environmental."[9] According to his findings, there is virtually no correlation between climate-change indicators such as temperature and rainfall variability and the frequency of civil wars over the past 50 years in sub-Saharan Africa.

So far, there has been no evidence that environmental problems are the direct cause of war—that is to say, there have been no "environmental wars" as manifestations of the most extreme form of interstate conflict. At least, no evidence exists to date to suggest any unambiguous causal links between environmental change and violent interstate conflict. The greatest danger posed by climatic disturbances is not the degradation of ecosystems in itself, but rather the disintegration of human society brought about by widespread famine, mass migrations and recurring conflicts over resources.

The Disintegration of Human Society

In the next twenty years, the world will be adversely affected by famines, and available food stocks are not sufficient when facing these climatic changes. Bohle et al. (1994) make the link between food security and climate change, and Döös (1994) insists on the fact that climate change will lead to exacerbated food shortages caused by soil degradation. Molvaer (1991) had foreseen that soil degradation would become a source of conflict between farmers and herders in the Horn of Africa.[10] Even while being the continent the least responsible for greenhouse

9 Proceedings of the US National Academy of Sciences, 10.1073/pnas.1005739107 (2010).

10 H.G. Bohle, T.E. Downing and M.H. Watts, "Climate Change and Social Vulnerability: Toward a Sociology and Geography of Food Insecurity," *Global Environmental Change* 4: 1 (1994), 37–48; B.R. Döös, "Environmental Degradation, Global Food Production, and Risk for Large-scale Migrations," *Ambio* 23: 2 (1994), 124–30; R. Molvaer,

gas emissions, Africa is considered as running the highest risk of being subject to conflicts generated by climate change; this risk emanates from the economic fact that the continent's economic stability is based on sectors that are themselves dependent on climate (such as rain-fed agriculture). An example is Niger: 75 percent of the population lives on 12 percent of its territory in the South, which creates a heavy pressure on an already fragile environment. Desertification threatens an agricultural economy which employs more than 90 percent of Niger's inhabitants. The precarious nature of the agricultural sector has obliged the country to import 60 percent of its food requirements. From another angle, concerning the vulnerability of Africa, it is necessary to evaluate its past history: ethnic and political conflict, and disputes linked to resources. At the dawn of the twenty-first century, the wars in Africa have caused more casualties than the combined total of all the others taking place in the rest of the world over the same period.

Lastly, the general link between the level of economic development of a given country and its propensity for conflict is recognized.[11] Some studies have been undertaken to evaluate the impact of economic growth, or absence of growth, on conflicts. A recently issued report by the Brookings Institute, generated jointly with researchers from the Wilson Center and other specialized institutes, found that, in the years to come, there would be little correlation between violent conflicts and variables such as political repression, ethnic fragmentation, colonial history or population density. Rather, it would be economic factors which would be determinant: "a 1% drop in the GNP raised the probability of conflict (civil conflict) by more than 2 points."[12] Even if the demand for the available resources becomes more insistent, successful economies were less likely to experience conflicts, and more likely than weaker economies to avoid resorting to war as a remedy.[13]

Migratory Phenomenon

Towards the end of 2007, the UN High Commissioner for Refugees (UNHCR) counted close to 67 million persons displaced through force due to reasons of conflict, persecution and natural disasters. Nearly half of those, approximately 30 million, migrated on account of natural catastrophes, and another 30 million on account of armed conflict; more than 30 percent of the world's refugees are displaced persons in Africa, and are accommodated—because they are neighbors—by other African

---

"Environmentally Induced Conflicts? A Discussion Based on Studies from the Horn of Africa," *Bulletin of Peace Proposals* 22: 2 (1991), 175–88.

11 P. Collier and A. Hoeffler, "Greed and grievance in civil war," *Oxford Economic Papers* 56: 4 (2004), 563–95. Commission Européenne: http://www.consilium.europa.eu/ueDocs/cms_Data/docs/pressdata/FR/reports/99389.pdf.

12 C.S. Hendrix and S.M. Glaser, "Trends and triggers: Climate, climate change and civil conflict in Sub-Saharan Africa," *Political Geography* 26 (2007), 695–715.

13 Ibid.

countries.[14] Twenty million people, according to the UN figures from 2007, had already been displaced following the erosion of arable lands. According to a report emanating from the British humanitarian organization Christian Aid entitled *The Human Tide: The Real Migratory Crisis*, there are presently 163 million people who have had to leave their homes as a result of conflict, natural disasters and massive infrastructure development projects, such as mines and dams.[15]

The estimate of 200 million climate refugees by 2050, forecast by Dr Myers, is now a figure of reference, even if he himself admits that it is based on "major extrapolations." Myers was one of the first (in 1989, and then in 1993) to predict that climate change would lead to vast population movements. These estimates have been appearing in various publications emanating from such organizations as the GIEC (*Groupe d'Experts Intergouvernemental sur l'Evolution du Climat*), in the Stern Report, the International Organization for Migrations (IOM) based in Geneva, and the UN's Institute for the Environment and Human Security. All forecast that migrations will involve 50 million persons in 2010, rising to 200 million in 2050, and towards 700 million in the years beyond. To understand this phenomenon better and put it in context, it means that one out of 45 persons in the world will have been displaced due to climate change between now and 2050. However, no one can know with any certainty which part of the human population will be most affected by climate change. The present estimates run from 25 million to 1 billion displaced persons between now and 2050.

McGregor (1994) has suggested the need to ensure that the movements of population do not affect, in turn, access to food resources in the areas where "climate refugees" seek refuge or relocate.[16] However, this point is rarely emphasized. The importance of the issues linked to migration is in part due to the impending disappearance of certain land masses (islands for example) whose inhabitants will not survive the rises in sea level. It is emphasized that these exiles, who are persecuted by natural elements "aggravated" by abusive usage of the planet's resources, do not have "refugee" status. In fact, there is a fear that the distinction between refugees fleeing from war and those who will flee from their environment, between political refugees and climate refugees, will no longer be relevant insofar as the number of new wars will increase as a result of environmental degradation. Legal experts—such as those at the University of Limoges Research Centre on Environmental, Planning and Urban Law—have written much on this subject.[17] From another perspective, the French frigate captain Jérôme Origny, during a

14 D. Garcia, "The climate security divide: Bridging human and national security in Africa," *The African Security Review* 17: 3 (Institute for Security Studies, 2008), 2–17.

15 WBGU, *World in Transition*.

16 J. McGregor, "Climate Change and Involuntary Migration: Implications for Food Security," *Food Policy* 19: 2 (1994), 120–32.

17 Jean-Marc Lavielle, Michel Prieur, Jean-Pierre Marguénaud, Gérard Monédiaire, Julien Betaille, Bernard Drobenko, Jean-Jacques Gouguet, Séverine Nadaud and Damien Roets, "Projet de Convention relative au statut international des déplacés environnementaux," *Revue européenne de droit de l'environnement* 4 (2008), 381; http://www.cidce.org.

demography seminar directed by the University President G.F. Dumont, noted that "This is to ensure that the 200 million candidates for climate immigration will not be followed by billions."[18]

If one believes the forecasts of researchers who mix with policy-makers, and who serve certain ambitions that are far from academic considerations, mass migrations are flows that will affect everybody. According to the writers Peter Schwartz and Doug Randall (in their Pentagon report), "drought and/or frost will push populations towards the South or towards the interiors of land masses, provoking or amplifying conflicts between States situated in turbulent zones like countries of Europe, the United States and Canada, India and Pakistan, and even in China and Africa."[19] Nonetheless, this literature omits to specify that these migrations will most often be internal to the countries concerned, as shown in studies of population reactions to floods and droughts in Mozambique or Ghana. Of the 200 million people referred to, around one million will be led to seek refuge in a country other than their own, which represents 0.5 percent of the total.

There is a consensus regarding the typology of the conflicts believed to involve an environmental element and the climatic element that emphasized it. They are predominantly intra-state conflicts; even when they can be categorized as cross-border conflicts, they are generally not classical interstate conflicts in the sense of large-scale wars between countries, but rather regionally limited clashes at the sub-national level, such as between States that border on the same rivers and lakes.

Conflicts Over Resources

In stating that humanity will experience mass migrations, can solutions other than violent ones to the problems of the refugees be imagined, given the tensions which surround the issues involved: the right to water and its exploitation? When we speak of resources, we inevitably think of natural elements like water, air and land that are essential for the survival of any human community. Thus they constitute fertile ground for many kinds of environmental antagonism.

If we are to believe UNEP experts, since 1990 at least eighteen conflicts have been fuelled by the exploitation of natural resources. Some recent studies have suggested that over the course of the last 60 years, at least 40 percent of all interstate conflicts involved a link to natural resources. Some civil wars such as those of Liberia, Angola and the Democratic Republic of Congo have been centered around "high value" resources such as wood, diamonds, gold, minerals and oil. Other conflicts, for instance in Darfur (the conflict here is linked to climate, according

18 G.F. Dumont, *Les Territoires face au vieillissement en France et en Europe* (Paris: Ellipses, 2006).

19 Peter Schwartz and Doug Randall, *An Abrupt Climate Change Scenario and Its Implications for United States National Security* (October 2003), available at: http://www.edf.org/documents/3566_AbruptClimateChange.pdf.

to one of the UNEP reports) and in the Middle East with Palestine and Lebanon, are about the control of scarce resources like fertile land and water. Tensions will increase as a result of the high growth rates of emerging economies and their demands for uranium, cobalt, titanium, zinc and other rare earth metals, which are increasing by 10 percent per year.[20] In the same vein, access to certain raw materials and/or supply infrastructures, especially pipelines, constitutes a sensitive area of "globalized insecurity." We can see this with Iraq, Nigeria, and also Afghanistan.[21] The growing importance of oil, natural gas and uranium production is itself a source of instability, acting as a magnet that attracts the arms trade and external interventions.

"Green" Conflicts and the Stakes Linked to Water

In *Africa – Up in Smoke?* Andrew Simms illustrates the vulnerability of African ecosystems to global climate change with the situation of the Nile, noting that "most scenarios estimate a decrease in river flow of up to more than 75 per cent by the year 2100."[22] Simms then cites a 2003 article in the journal *Mitigation and Adaptation Strategies for Global Change* which states that "a reduction in the annual flow of the Nile of above 20 percent would interrupt normal irrigation," and he concludes: "Such a situation could cause conflicts because the present distribution of water, negotiated during a period of high flow, would become untenable."[23]

Sudan has been seeking to irrigate the Sahel, but Ethiopia has made clear that any Sudanese attempt to divert water from the Nile would provoke a military response.[24] Along the same lines, Egypt has threatened to oppose Sudan or Ethiopia if one or the other attempts to manipulate the waters that feed the Nile.

20 Michael Klare, *Rising Powers, Shrinking Planet: The New Geopolitics of Energy* (New York: Henry Holt & Company, 2008).

21 George Monbiot, *Heat: How to Stop the Planet Burning* (London: Allen Lane/Penguin, 2006).

22 Andrew Simms (Policy Director at the New Economics Foundation), *Africa – Up in Smoke* (London: New Economics Foundation, 2005), available at: http://www.unep.org/PDF/Africa_climatechange_report.pdf.

23 R. Dixon, J. Smith and S. Guill, "Life on the edge: vulnerability and adaptation of African ecosystems to global climate change," *Mitigation and Adaptation Strategies for Global Change* 8: 2 (2003), 93–113; cited in A. Nyong, "Impacts of climate change in the tropics: the African experience," keynote presentation at "Avoiding dangerous climate change: a scientific symposium on stabilization of greenhouse gases," Met Office, Exeter, United Kingdom, February 2005. Available at: http://www.stabilisation2005.com/Tony_Nyong.pdf (accessed 1 June 2005).

24 K.M. Campbell, J. Gulledge, J.R. McNeill, J. Podesta, P. Ogden, L. Fuerth, R.J. Woolsey, A.T.J. Lennon, J. Smith, R. Weitz and D. Mix, *The Age of Consequences: The Foreign Policy and National Security Implications of Global Climate Change* (Center for Strategic and International Studies (CSIS) and Center for a New American Security (CNAS), 2007).

Professor Frédéric Lasserre at Laval University (Quebec), a geopolitical expert on "blue gold" (water), presents scenarios of a synergistic correlation between the scarcity of water and what he qualifies as "hydraulic conflicts" or "water wars." He claims that climatic disturbances will exacerbate pre-existing tensions. On the other hand, the willingness to settle disputes between opposing parties is manifest, whether it be over the usage of "blue gold" between Israelis and Palestinians or the cooperation between Israelis and Egyptians in the context of the Mediterranean Action Plan. Since 1953, Israel and Jordan have held secret informal meetings on the management of the Jordan River, even though the two countries were officially at war from 1948 to 1994 (the date when their peace treaty was signed). The Indus River Commission survived two major wars between India and Pakistan, providing further proof of a willingness to settle water disputes.

Another textbook case of regulating water conflicts is the sharing of the Shatt Al-Arab between Arabs and Persians from the sixteenth century. Certainly, there has been, in the recent past, conflict between two nations wishing to establish their regional supremacy, but cooperation has tended to prevail. Admittedly, scenarios exist in which States would be prepared, weapons in hand, to obtain from a neighbor a vital resource of which they are deprived. The UN has identified about 300 potential areas of "hydro-conflict" based on common groundwater and trans-boundary Rivers. However, it cannot be said with certainty that the use of armed conflict will be the *de facto* response to future water supply problems. In reality, it is rather the opposite trend which has emerged from current historical research. Researchers at Oregon State University have created a databank of all reported interactions (be they cooperative or conflictual) between two (or more) countries that have been provoked by water issues over the last 50 years. The results indicate that the rate of cooperation massively exceeded the cases of "serious" conflict. During this period, only 37 disputes involved the use of violence, 30 of which were between Israel and one of its neighbors. Outside the Middle East, the researchers found that only five violent conflicts had erupted, while as many as 157 treaties had been negotiated and signed. Between 1945 and 1999, the cases of cooperation have been two times more numerous than the cases of conflict between States sharing the same water source.[25]

Green Conflicts and Raw Materials

The overexploitation of natural resources is likely to boost global tensions to a degree hitherto unknown as nations seek to satisfy their needs for energy, water, food and raw materials. Global energy consumption could double by 2030. The countries of the European Union today depend on production zones situated in the

25 Aaron T. Wolf, Annika Kramer, Alexander Carius and Geoffrey Dabelko, "Water can be a pathway to peace not war: Global Security Brief #5," *State of the World 2005* (Washington: Worldwatch Institute, 2005).

Middle East, Africa and Russia for more than 75 percent of their oil consumption. The figures are comparable for gas. Countries with high economic growth rates, like India and China, are searching for new sources of supply all over the planet. Competition and perhaps conflicts could result from tensions that become too strong and are not regulated.

The US military has warned that surplus capacity production of oil could lead to severe shortages by 2015, with significant political and economic consequences. This energy crisis, elaborated in the Joint Operating Environment (JOE) 2010 of the American Joint Forces Command, came just as the price of petrol had reached record levels. The report specifies that in 2012 the surplus capacity production could completely disappear, and from 2015 the production deficit would be close to 10 million barrels per day. It adds: "Such an economic slowdown would exacerbate other unresolved tensions, push fragile and failing states further down the path toward collapse, and perhaps have serious economic impact on both China and India."[26]

Has the War for Uranium Begun?

In the context of an international market, ten countries in the world possess 90% of the world's uranium production. Will this production be enough to satisfy the demand of the growing nuclear power sector? The remaining available uranium reserves are becoming scarce, while demand continues to increase since the Kyoto Protocol, alongside the production of new reactors, notably in China. Since 1991, and in every year after that, the global demand for uranium has exceeded mining production. The difference between mining production and demand is approximately one-third, which is being offset by what is called "secondary uranium," or uranium coming from military stockpiles: for example, uranium recuperated from the dismantlement of nuclear weapons, via the Russo-American programme "Megatons to Megawatts." However, the sale of uranium coming from dismantled warheads will end by 2013. With or without the conflict linked to the Imouraren site in Niger (conceded to the French energy company Areva), any restrictions on the exploitation of this deposit site could lead to a "uranium shock" similar to the oil shock of the 1970s, with consequences that are unpredictable.

In terms of rare earth minerals, indispensable for the production of sophisticated military equipment, there is nothing to keep powers such as China, which now holds a near monopoly (97 percent) over the extraction of these materials, from being tempted to use this powerful position to pressure or even strangle certain partners. (See Cindy Hurst, "China's Rare Earth Elements Industry: What Can the West Learn?" Fort Leavenworth, KS: Institute for the Analysis of Global Security (IAGS) March 2010.)

26 US Joint Forces Command, *Joint Operating Environment (2010)*. See http://www.fas.org/man/eprint/joe2010.pdf, 27.

The Debate Over Carrying Capacity

Carrying capacity, a proposition that comes close to the neo-Malthusian theories espoused by Gaston Bouthoul, one of the French fathers (with Louise Weiss) of War Studies or "polemology" and a pioneer in his time, has associated tensions and violence with the scarcity of food resources. Carrying capacity is defined here as the capability of a given natural environment to sustain a given population. It follows that, once the carrying capacity of a given environment is exceeded, there will be a growing imbalance between population growth and dwindling resources, which almost automatically will lead to conflict.

This interpretation, originating from an ecological theory, is questionable. Hence the reservations or criticisms of the determinist and negative aspect of this line of thinking that could be called "cultural ecology."[27] When examining cases such as the conflicts of the last few years in Chiapas, Mexico, for example, or in Rwanda, this theory espoused by Steve Leblanc has been described as too "ecological." "We, the anthropologists," declares Ferguson,

> *discover that if a population suffers from a lack of basic resources, the principle cause of this shortage is an inequitable distribution of resources within the society, a political question, and an economic question, rather than the combination of a too large population and a lack of resources. Among the peoples of the Northwest Pacific Coast, before the depopulation of the 19th century, groups competed for access to basic resources, such as salmon-sheltering estuaries. But the singularity of the situations does not allow them to be transposed. Thus, in numerous places across the globe, as in the case of the Yanomami people (Brazil and Venezuela), war is not linked to a battle for food. Studies of modern conflicts show that very diverse factors can interact, including the need for food, local ecological relations, as well as struggles for power within governments, and cultural characteristics like beliefs and symbols.*[28]

While one might tend to emphasize the inevitability of clashes which "naturally" will hit or are hitting failed states—states being considered as "unsafe, in the process of failing or having already failed"[29]—it appears that the capacity of these states to manage or adapt to conflicts is not uniform.[30] As argued by Oli Brown

27 C. Obi, "Globalised Images of Environmental Security in Africa," *Review of African Political Economy* 27: 83 (2000), 47–62.

28 R.B. Ferguson and N. Whitehead (eds), *War in the Tribal Zone: Expanding States and Indigenous Warfare* (Santa Fe, NM: School of American Research Press, 2000). Interview with Brian Ferguson in 'La Recherche'.

29 Harald Welzer, *Les guerres du climat, Pourquoi on tue au XXIème siècle* (Paris: Gallimard, collection NRF essais, 2009).

30 S.P.J. Batterbury and A.Warren (eds), "The African Sahel 25 years after the Great Drought," *Global Environmental Change* 11: 1 (2001), 1–96 (8 papers).

of the International Institute of Sustainable Development (IISD): "We have seen, across regions and the world, that conditions of stress have provoked conflicts in certain regions but not in others."[31]

The Unknowns Linked to Climate

We can find solid historic links between civil war and temperature in Africa, with warm years resulting in the significant increase of the probability of war, or so it is claimed in the study entitled "Warming Increases the Risk of Civil War in Africa."[32] Based on computer modeling, this study, undertaken by a team of scientists from Stanford University, the University of California (Berkeley), New York University and Harvard University has already calculated the prospective number of victims in 2030: 393,000 additional war victims if the future conflicts are as deadly as recent conflicts; according to the projections, we should expect an increase of approximately 54 percent in the incidence of armed conflict by 2030.

Is the collected data—based on events occurring in Africa between 1981 and 2002—a good indicator? According to the Norwegian political scientist Halvard Buhaug of the Peace Research Institute in Oslo (PRIO), it is fallacious to establish overly simplistic and mechanical links between climate and civil war(s). Buhaug does not deny the reality of global warming: he admits that the African continent, over the last 50 years, has become hotter and dryer (see above). But there is no tangible correlation between the frequency of civil wars in Sub-Saharan Africa over the last 50 years and temperature and rainfall trends. Especially since, although the temperature continued to rise over the last ten years, conflicts on the other hand actually declined. Seen from another perspective, the panorama of 73 conflicts listed between 1980 and 2005, when considered as environmental, shows that they were limited to a certain space and did not in fact present any real threat to international security.

The science of climate change is quite complex. According to the report of Peter Schwartz and Doug Randall (2003–2004), commissioned by Andrew Marshall, strategic adviser to the Pentagon, "We must prepare ourselves for the inevitable effects of abrupt climate change, which will likely occur regardless of human activity."[33] Moreover, the scenario of the authors is modeled on a climatic event, which occurred suddenly (according to ice-core samples from glaciers in Greenland), took place 8,200 years ago, and lasted a century.

31 Oli Brown, "Migration and climate change" International Organisation for Migration. IOM Migration Research Studies No. 31. (2008).

32 Marshall B. Burke, Edward Miguel, Shanker Satyanath, John A. Dykema and David B. Lobell, "Warming Increases the Risk of Civil War in Africa," *Proceedings of the National Academy of Science* 106: 49 (October 2009), 20670–74, available at: http://www.pnas.org/content/early/2009/11/20/0907998106.full.pdf+html.

33 Peter Schwartz and Doug Randall, *An Abrupt Climate Change Scenario*.

The debate remains open. Though not between supporters and adversaries of global warming, or between climate-sceptics and the others, but rather between the interpretations of global warming: controllable or not, progressive or/and discontinuous? For the geopolitical analyst, what consequences will follow an inevitable yet unpredictable process? These unknowns are to be taken into account if one should consider their repercussions on societies endowed with vastly different resources and capabilities of adapting to an array of diverse exterior shocks. One can say that the projected repercussions of climate change on societies are manifestly *even more uncertain* than anticipated climatic changes since they are essentially a projection based on another projection.

Geopolitical Perspectives

Resource scarcity could dictate the terms of international relations in the years to come. This is the line of thinking in the report *The Age of Consequences* produced by the Center for Strategic and International Studies (CSIS) in the United States.[34] Could the countries considered as rich, over the next thirty years, focus on their own survival while neglecting and maintaining a distance from poor countries? Leon Fuerth, one of the authors of the CSIS report and former National Security Advisor to former Vice-President Al Gore, believes that climate change and its consequent effects, including causing wars, could lead to the end of the globalization process as we know it today; different regions of the world would withdraw in on themselves to conserve what they "need" for their own survival, or to avoid sharing resources or having to redistribute resources which could be perceived as global public goods.

This trend is already perceptible in the lack of financial support that the West is willing to offer in the framework of official public development assistance, and in the difficulties faced by partners in finding innovative financing in order to achieve the Millennium Development Goals. In any case, a reconfiguring of nation-state power structures cannot be excluded; this could involve exclusive economic zones, which recently arose out of the Law of the Sea treaty, with the prospect for navigation and exploitation of the Poles. In the scenario concocted for the benefit of the Pentagon,[35] we can look forward to seeing new alliances and alliances of circumstance. The US and Canada could unite and become one country. The two Koreas could re-unite to create an entity with technological knowhow and nuclear weapons. The reconfiguring includes a new international sharing of the threat of nuclear death. Amongst the countries who would develop their enrichment and processing capacities (of nuclear materials) to guarantee their national security, the report lists not only countries that already possess nuclear weapons but also Japan,

34 K.M. Campbell et al., *The Age of Consequences*. Available at: http://csis.org/files/media/csis/pubs/071105_ageofconsequences.pdf.

35 Peter Schwartz and Doug Randall, *An Abrupt Climate Change Scenario*.

South Korea, Germany and Egypt. The subtitle of the 2003 report is "Imagining the Unthinkable." It sounds like an echo of the slogan of Herman Kahn's "Think the Unthinkable," voiced during the darkest hours of the Cold War during which the superpowers relied on MAD (Mutual Assured Destruction).[36]

Even if the West were somehow compelled to "toe the line" after its long period (five centuries) of domination,[37] Europe could act or react as a unified, fortified and bunkered block to limit the problems associated with immigration and organize its protection against "aggressors." Unless Russia, with its abundant mineral resources, oil and natural gas, seeks a rapprochement with Europe, a Europe cut loose from the other side of the Atlantic. As a result of these readjustments, which also concern the emerging powers, the lines of demarcation between the North and the South, between rich countries and the South, will change.

Unabated climate change could thus plunge the "industrialized countries" in particular into crises of legitimacy, and limit their international scope for action. The worst-affected countries are likely to invoke the "polluter pays" principle, so international controversy over a global compensation regime for climate change will probably intensify.

Faced with this complicated future, the thesis developed by Roy Woodbridge in his book *The Next World War* is appealing: it proposes that the international community should literally go to war, in a unified and coordinated manner, against environmental degradation.[38] This proposal comes close to somewhat Utopian ideas of a reconciled humanity facing a common exterior enemy (the arrival of Martians), and to the views of those in the pacifist movement, including Sara Parkin, former director of Forum for the Future, who insists with a dash of wishful thinking that the military are incapable of sewing up the holes in the ozone layer and are powerless against tidal waves and tsunamis.[39]

Climate as a Tool of Strategic Influence

For the United States, the climate question is a means of establishing their leadership and perpetuating their strategy of preeminence over other countries, with Europe and the BRIC (Brazil, Russia, India and China) countries in the front row.[40] While the CIA opened (in September of 2009) a center on climate change and national

36 Ibid.

37 Hervé Kempf, "La crise écologique: une question de justice," *Revue Défense Nationale* (February 2010).

38 Roy Woodbridge, *The Next World War: Tribes, Cities, Nations, and Ecological Decline* (Toronto: University of Toronto Press, 2004).

39 "As shooting the ozone layer or bombing empty water aquifers is not an option ..." Sara Parkin, "Environmental Security: Issues and Agenda for an Incoming Government," *RUSI Journal* (June 1997), 24–8.

40 Romain Lalanne, "Quand la sécurité devient verte," *Revue Défense Nationale* (February 2010).

security, the United States had begun to develop a program to investigate climate as a security risk factor since 1979. With the arrival of President Bill Clinton in the White House, and under Vice-President Al Gore, the intelligence community saw itself assigned an environmental mandate. Since 1994, NATO has launched a series of conferences on the theme of "environment and defense" and has evoked at times the concept of "ecological security." Studies of the impact on national security of phenomena such as desertification, rising sea levels, population movements and the rising competition for natural resources are not, moreover, preoccupations that are exclusive to the Pentagon. But the consequences can affect military installations, such as Diego Garcia for example, endangered by rising sea levels. Furthermore, the consequences of climate change could impose "an additional burden on military forces in the world," as noted in the *Quadrennial Defense Review* published by the American Army.[41]

Certain rival powers could be tempted to appropriate the Arctic region at the risk of leading the world into "another Cold War," in the words of Frank-Walter Steinmeier, the German Minister for Foreign Affairs. Thus, "Additionally, an Arctic with less sea ice could bring more competition for resources, as well as more commercial and military activity, that could further threaten an already fragile ecosystem," as stated in the April 2007, CNA report *National Security and the Threat of Climate Change*, that articulates the concept of climate change acting as a "threat multiplier" for instability in some of the most volatile regions of the world.[42]

For Russia, its new Antarctic strategy has been framed around the preservation of "Russian national interests." The need to develop geophysical knowledge about the Antarctic geological make-up is linked to the presence of ore deposits found on the continent and its continental shelf, and also to the potential presence of hydrocarbons. Although the Antarctica Treaty prohibits any mining activity in the region, States with interests there must prepare themselves to assert their "geopolitical claims."

By Way of a Conclusion

In spite of the controversy about the graphs and data furnished by climatologists, a majority have drawn the conclusion that the potential consequences of climate change on the availability of water, food security, the prevalence of disease, coastal boundaries and population distributions could aggravate already existing tensions and generate new conflicts. It is, at the same time, difficult and risky to establish relationships of cause and effect, especially since the current period has been characterized by a decrease in conflicts in comparison to the Cold War period. Can

41 http://www.defense.gov/qdr.

42 CNA, *National Security and the Threat of Climate Change* (April 2007), 21. Available at: http://securityandclimate.cna.org/report/National%20Security%20and%20the%20Threat%20of%20Climate%20Change.pdf.

we therefore venture to establish a link between natural disasters and conflicts? The data collected by the Research Centre on the Epidemiology of Disaster (2006) seems to disavow this conclusion, stating that "of the 171 disasters caused by storms and flooding since 1950 – each having at least 1,000 victims – a clear link has been established in 12 cases – between natural disasters and the intensification of a conflict or a political crisis."[43] Only 12 cases.

If no one is capable of determining the type of conflicts and their degrees of intensity, then there is no reason to believe that they will be confined to a given area. In a world that is more and more interdependent, the rise in the number of armed conflicts in Africa will have repercussions that go beyond the boundaries of that continent.

From a retrospective analysis of interstate conflicts over the last 60 years, it is possible to conclude that conflicts associated with natural resources are twice as likely to flare up again within five years, which could be termed a "boomerang" effect. And yet, less than a quarter of the peace negotiations aimed at resolving conflicts over resources have seriously sought to address the mechanisms of resource management.[44]

There are other factors than those linked to climate—poverty, governance, conflict management, regional diplomacy and others—which will determine, in large part, if the status of climate change will evolve from being a challenge to sustainable development for the most vulnerable to being a global security threat. The Brundtland Report already stated more than 20 years ago that "In recent years, international relations have been characterised by a marked tendency to resort to the threat or use of military force in response to security threats of a non-military character."[45] It's a safe bet that this prognosis is even more valid today.

---

43 P. Hoyois, R. Below, J.-M. Scheuren and D. Guha-Sapir, *Annual Disaster Statistical Review: Numbers and Trends 2006* (Brussels: Centre For the Research on Epidemiology Disasters, 2007). Available at: http://www.cred.be/sites/default/files/ADSR_2006.pdf.

44 "The death of a large number of people following the hurricane Katrina, the heat wave of the summer of 2003 in Europe, and the Indonesian tsunami of December 2004 have sadly demonstrated that, in the case of natural disasters, countries and communities, even the most developed ones, are badly prepared to protect and help their citizens, the vulnerable ones in particular, i.e. poor, isolated, sick, handicapped people who live in inappropriate lodging conditions. Such disasters are never totally 'natural.' In fact, many experts consider that so-called 'natural' disasters are largely due to human negligence or to inappropriate land use. When causes and consequences of disasters are considered, it is crucial to address the notion of individual and social vulnerability of a person as much as human adaptation to stress." Cited in abstract: Danielle Maltais and Simon Gauthier, *Les Catatastrophe Dites Naturelles : Un Construit Social?* Proceedings of the 4th Canadian Conference on Geohazards: From Causes to Management. Presse de l'Université Laval, Québec (2005) http://www.landslides.ggl.ulaval.ca/geohazard/0_Keynotes/maltais.pdf.

45 Brundtland Commission, *Our Common Future.*

23 ASHGATE RESEARCH COMPANION

Cyberconflict and the Future of Warfare

Athina Karatzogianni

Introduction

Writing a brief history of cyberconflict of the last decade and speculating on the future of warfare is by no means an easy task. The reasons are plenty and it is worth mentioning a few here, as they do tend to get lost in colleagues' specialized debates in the fields of international relations and global politics, global and national security, internet security, new media political communication, international governance, internet governance, information warfare, critical security and the geopolitics of new technologies. Information communication technologies (ICTs) have unsettled in an unprecedented way the majority of academic fields, all of which are currently required to negotiate multi-level conflicts transferring from the real world to cyberspace or being created originally through cyberspace and spilling over to real life. Equally, this is a very fast-moving field. It is also a field which is not solely dominated by states and traditional wars, but by movements, civil society organizations, protest events, insurgencies, network resistances and ad hoc assemblages. These groups and their use of ICTs are the subject of this work. These players are using social media technologies to punch above their weight, to challenge the supremacy of the state as having the monopoly of violence and propaganda, through using ICTs as a weapon or as a tool for mobilization, organization and recruitment, and to provide instant access to the global public sphere to influence the strategy, tactics and justification of wars, and resist the violent oppression of citizens by totalitarian and authoritarian regimes. The relevance of these actors and their use of technological innovation is currently more than critical: the use of social media networking to accelerate regime changes in the Middle East region and elsewhere, and the military interventions the international community had to respond with to protect the citizens of these states following the undeniable publicization of their plight in the virtual public sphere, point to the need to examine the history of the use of ICTs by these actors.

This global media transformation affecting international communications has created theoretical and empirical problems reflecting the multi-disciplinary, interdisciplinary and cross-disciplinary character of cyberconflict,[1] and has resulted in a disparate literature which only rarely comes together as one area of study.[2] In ontological, theoretical and methodological terms, the claims and politics stemming from the various disciplines quite often clash – for instance, information warfare, counterterrorism, cybersecurity and global communications studies are influenced by an inevitably conservative state and status quo bias with a positivist methodology resting on a realist and/or neo-liberal politics, in direct contrast to more sociological, media and political theories resting on the centre or centre-left of the spectrum, engaging with more qualitative methodologies and perhaps focusing on postmodern and poststructuralist explanations. Exceptions to this crude generalization are evident all over the place, but it is indicative of the overall state of the literature in the last decade. An example of this amalgam of debates, academic areas and methodologies is this author's research monograph *The Politics of Cyberconflict*, which theorizes cyberconflict between 2000 and 2005 in terms of elements from three academic areas (media, social movement and conflict theories), looking at earlier ideas on information warfare and security and engaging with sociopolitical cyberconflicts (anti-globalization and anti-war movements, cyberdissidents and internet censorship) and ethnoreligious cyberconflicts (various examples such as Israeli-Palestinian, Indian-Pakistani spilling over into cyberspace), as well as the effect of the internet on the anti-war movement, coverage and cyberattacks.[3]

Since the publication of that work, a proliferation of linked subjects to cyberconflict has emerged. Even if it is impossible to encompass the diversity of issues here, it is worth mentioning briefly the kind of breadth one is faced with when discussing cyberconflict, in a way setting an agenda for future study:

- *The Individual* and individual security in cyberspace (i.e. internet safety in vulnerable groups, for instance underage users);
- *Class, Gender, Minority, Migration* issues, individuals and groups (i.e. the digital gap, digital have-less, digital working class, digital diaspora networks and the digital development of migrants);
- *Private Corporations* in the IT industry and elsewhere and their corporate, social and moral responsibility (i.e. issues coming up in the Google–China cyberconflict, issues of human rights, censorship, cybersecurity professionals hawks vs doves, etc.);

1 Athina Karatzogianni, "Confronting internal and external problems of cross- inter- and multi-disciplinarity: Researching cyber conflict and global politics", in Maria Karanika and Rolf Wiesemans (eds), *Exploring Avenues to Cross-Disciplinary Research* (Nottingham: Nottingham University Press, 2009).

2 For an example of the diversity of contributors and areas of study see Athina Karatzogianni (ed.), *Cyber Conflict and Global Politics* (London and New York: Routledge, 2009).

3 Athina Karatzogianni, *The Politics of Cyberconflict* (London and New York: Routledge, 2006).

- *Civil Society* – non-state actors (i.e. the role of these actors in ensuring digital freedom, the methods used and the ethical debates and cybersecurity issues raised by NGOs, etc.);
- *The State* – the role of the state and the difficulties of the boundary-less character of cyberspace, the inability of the state to embrace ICTs quickly or adequately, if at all, depending on its position in the global system (linked to that, cyberconflicts in unrecognized and small states and cybersecurity effects in the struggle for statehood and survival); also, questions of e-government as the last effort at state relevancy and survival;
- *International Relations* – International Regulation and International Law regarding cyberspace (i.e. the problems related to the non-existence of these for situations such as Estonia, NATO, UN, EU, major INGOs, and serious problems in addressing violations and cyberattacks between states, see for example the accusations made concerning China and Russia by Western governments over the last decade);
- *Global Politics, Political Economy* – wider implications for global politics beyond states and NGOs to include social movement organizations and their demonstrated overuse of ICTs, the transformations due to network forms of organization, mobilization and recruitment;
- *Media Convergence, Digital Economy Regulations* – illegal file-sharing, fandom and brand purity control, transmedia marketing and story telling;
- *Global Media* – the effect of the internet on media ownership and media coverage (for instance the Iraq war), security implications stemming from cybersecurity problems, radicalization media;
- *Global Resistances, Uprisings, Movements* and their organization, mobilization, recruitment and ideological development/framing through cyberspace.

In this chapter, a brief summary of cyberconflict during the first half of the last decade will be provided, followed by a discussion of the major incidents of the second half of the decade, thereby engaging more broadly with the popularization of cyberconflict, cybercrime and cybersecurity as fields of enquiry in the media, government and private sectors. Moreover, the advent of Web 2.0 post-2005 creates unprecedented access to data and social networking, leading to a flourishing of flamewars, national homeland patriotic hacking and mobilization, and the emergence of governance, privacy, safety, piracy issues, for example on Youtube, Facebook, Wikipedia and other platforms and virtual communities, which despite their empirical richness will not be part of our discussion here. Instead, after engaging with the most important empirical cyberconflicts and their linked subjects of the past decade, the remainder of this chapter discusses theories of future warfare and future effects of war and conflict on politics, culture, media and society.

Cyberconflicts 2000–2005

Cyberconflict, defined as conflict in computer-mediated environments, has been witnessed as early as 1994 with the Zapatista guerrilla movement in Mexico transferring their mobilization online and linking with the anti-globalization movement through the internet. In the late 1990s, Arquilla and Ronfeldt[4] expressed the idea that conflicts will increasingly revolve around knowledge and the use of soft power. Additionally, these Rand theorists defined *netwar* as the low, societal type of struggle, while *cyberwar* refers more to the heavy information warfare type. Here the focus is on netwar type cyberconflicts, as historical incidents are explained and their implications for global politics and security are considered.

Cyberconflicts can act as a 'barometer' of real life conflicts and can reveal the natures and the conflicts of the participating groups. Before the advent of Web 2.0, two types of cyberconflict were prevalent: ethnoreligious (between ethnic or religious groups fighting in cyberspace) and sociopolitical (conflicts between a social movement and its antagonistic institution).

In sociopolitical cyberconflicts, such as the anti-globalization and anti-capitalist movements, an alternative programme for the reform of society is evident, asking for democracy and more participation from the 'underdogs', be they in the West or in the developing world. New social movements (NSMs) are not new, but, rather, part and parcel of the dominant modern culture, which makes it difficult to think of movements as flowing either from 'pre-modern' or 'postmodern' subcultures. However, the structure of NSMs – open, decentralized, nonhierarchical – makes them ideal for internetworked communication. The movement is composed of autonomous units that expend an important part of their resources on internal solidarity. A network of communication and exchange keeps the cells in contact with each other. Information and resources circulate in networks, and leadership is not concentrated but diffuse. NSMs advocate direct democracy, self-help groups and cooperative styles of social organization. The fewer and weaker the social ties to alternative networks, the greater the structural availability for movement participation. Sociopolitical movements, such as the political dissidents in China, can test the limits of a system, pushing the system beyond the range of variations that it can tolerate without altering its structure.

In an anti-war movement, which is a single-issue movement, the demand is for a change in power relations in favour of those that believe the war to be unjustified. In new social movements, networking through the internet links diverse communities such as labour, feminist, ecological, peace and anti-capitalist groups, with the aim of challenging public opinion and battling for media access and coverage. Groups are being brought together like a parallelogram of forces, following a swarm logic, indicating a web of horizontal solidarities to which power might be devolved or even dissolved. The internet encourages a version of the commons that is ungoverned and ungovernable, either by corporate interests or by

4 John Arquilla and David Ronfeldt (eds), *Networks and Netwars: The Future of Terror, Crime and Militancy* (California: Rand, 2001).

leaders and parties. An early example of hacktivism (online activism) is the Seattle anti-WTO mobilization at the end of November 1999, which was the first to take full advantage of the alternative network offered by the internet.

Also, dissidents against governments are able to use a variety of internet-based techniques to spread alternative frames for events and a possible alternative online democratic public sphere. Online efforts, such as pro-democracy, activist or anti-government websites point to the fact that people believe in the power of the medium enough to organize and run thousands of these sites. In many cases, they are able to initiate and control events, and mobilize and recruit others for their cause, as in the case of sites in the Islamic world, in China, in Latin America, activist sites for anti-globalization and single-issue protests and mobilizations both on national and international levels.

To continue, ethnoreligious cyberconflicts primarily included hacking enemy sites and creating sites for propaganda and mobilizational purposes. In ethnoreligious cyberconflict, despite the fact that patriotic hackers can network, there is a greater reliance on traditional ideas, such as protecting the nation or fatherland and attacking for nationalist reasons. The Other is portrayed as the enemy, through very closed, old and primordialist ideas of belonging to an imagined community, which 'patriotic' hackers will have to fight for in cyberspace.

For instance, in 2001–03, the Israeli-Palestinian cyberconflict saw the use of national symbols, explicitly drawing attention to issues of national identity, nationalism and ethnicity. Also, the language used by hackers relied on an 'us' and 'them' mentality, where the internet became a battleground and was used as a weapon by both sides, and full-scale action by thousands of Israeli and Palestinian youngsters involved both racist emails and the circulation of instructions on how to crush the enemy's websites. Similarly, in the Indian-Pakistani cyberconflict, the Indian army's website was set up as a propaganda tool and was used as a weapon: in particular discourses religion was mentioned (religious affiliation), the word 'brothers' was used (collective identity and solidarity), as was 'our country', a promised land.

In contrast, the Al-Qaeda network and its ideology relies more on common religious affiliation and kinship networks than strict national identity, which fits well with the borderless and network character of the internet. The internet has been used as a primary mobilizational tool, before 9/11, especially more after the breakdown of cells in Afghanistan, Saudi Arabia and Pakistan. On the internet, Al-Qaeda replicates recruitment and training techniques and evades the security services, because they cannot be physically intercepted due to the virtuality of their networks. The internet is used as a propaganda tool via electronic magazines, training manuals and general recruitment sites, as well as a weapon for financial disruption aimed at financing operations, or stealing data and blueprints.

The internet's role was crucial in the March 2003 Iraq conflict, on the organization and spread of the movement, its impact on war coverage and war-related cyberconflicts. These last involved hacking between anti-war and pro-war hacktivists (sociopolitical cyberconflict), but also between pro-Islamic and anti-Islamic hackers (ethnoreligious cyberconflict). Moreover, mobilization structures were greatly affected by the internet, since the peace groups used the internet to organize demonstrations and events, to mobilize in loose coalitions of small groups that organize very quickly, and to preserve the particularity of distinct groups in network forms of organization. Furthermore, the framing process was affected as well, since email lists and websites were used to mobilize, changing the framing of the message to suit the new medium. The language used to mobilize through the internet differs from traditional political discourse (for instance, speeches or texts in traditional media) in that it can combine various technical media (video, satellite images, file-sharing) in a way that delivers on the one hand a richer message, but on the downside a sometimes hasty and crude, under-analytical political message. The political opportunity structure in this particular case can refer to the rise of alternative media, but also to an opening of political space, and an opening of global politics to people who would not or could not get so involved before. In virtual terms, hacktivism was apparent in anti-war/pro-war hacking, for example a Virtual March on Washington, which impacted the city's communication infrastructure.

Pro-Islamic/anti-Islamic hacking was an example of ethnoreligious cyberconflict during the Iraq war. The link between ethnoreligious affiliation and discourses of exclusion/inclusion is evident when considering the al-Jazeera hack by American hackers and the movement of Islamic hackers united in a common anti-US, UK, Australia, India and Israel agenda. Furthermore, on the sociopolitical side (anti-war/pro-war), the use of the internet as a propaganda and mobilizational tool was common to both sides through the considerable number of websites advocating one view or another, mobilizing, countermobilizing and anti-mobilizing against each other.

On the media front, it is clear that political discourse was constructed in the American mainstream media to mobilize support for the war, since, for example, it is estimated that more than two-thirds of all sources in news programs were pro-war. Also very important was the issue of alternative sources and censorship.[5] Because of the embedded system, journalists having their work jeopardized for not being 'patriotic' enough, and the American media generally following the government line, Americans and the rest of the world went online to find alternative news and first-hand eyewitness accounts via emails, blogging and video logging. The result was the integration of the internet into media coverage and the distribution of online material challenging official sources. Anti-war groups had the ability to initiate and control protest events and to mobilize supporters, but were not as successful in dominating political discourse. The media effects on policy were, above all else, technical. As a result, there was instant 24-hour access to the war, bringing with it the pressure this would inevitably put on any administration. However, no actual debate or impact on policy took place, since the American media failed, at least until 2005, to question any decisions being taken by their government.

5 For more analysis, see S. Rendall and T. Broughel, 'Amplifying officials, squelching dissent: FAIR study finds democracy poorly served by war coverage,' *Extra!* May/June 2003.

In the final analysis, the internet played a distinctive role in the spread of the peace movement, on war coverage and on war-related cyberconflicts, in relation to which the full potential of the new medium in politics was shown. In the months preceding the actual war in Iraq, a plenitude of phenomena both on and off the internet emerged which in previous international conflicts were only embryonic. Anti-war groups used email lists and websites, group text messages and chatrooms to organize protests and, in some cases, to engage in symbolic hacking against the opposite viewpoint. The integration of the internet into mainstream media, the effect of online material challenging official government sources and the mainstream media, and blogging, were indicative of future coverage, and were the first signs of what we are witnessing today.

Lastly, between 2000 and 2005 there was a duality of cyberconflict, where ethnoreligious cyberconflicts were mapped as representing/defending loyalties of hierarchical apparatuses, and sociopolitical cyberconflicts were empowering network forms of organization. Neo-liberal governments and institutions face a counter-hegemonic account of globalization, to which they have responded in a confused and often contradictory manner. One of the interesting sides to the argument is that the information revolution is altering the nature of conflict by strengthening network forms of organization over hierarchical forms. In contrast to the closure of space, the violence and identity divide found in ethnoreligious discourses, sociopolitical movements seem to rely more on networking and rhizomatic structures.

Cyberconflict 2005–Present

Besides the acceleration of the use of the internet for radicalization purposes,[6] and by dissident and social movements around the globe,[7] the first serious incident of cyberconflict, in International Relations terms at least, occurred in 2007 in Estonia.

The real life event that sparked the cyberconflict was the removal of a Soviet war hero statue from Tallinn's square, which caused riots in Estonia for a couple of days around April 26, 2007, resulting in one death and several injuries. But by April 29, although the real-world riots calmed down, the country's digital infrastructure was crumbling from cyberattacks. The statue incident expressed the deeper tensions and the cultural conflict between the ethnic Russians in Estonia (around

6 See Athina Karatzogianni, *Cyber Conflict and Global Politics*.

7 Lincoln Dahlberg and Eugenia Siapera (eds), *Radical Democracy and the Internet* (Basingstoke: Palgrave Macmillan, 2007). For conflicts relating to peer production and the open source, see Athina Karatzogianni and George Michaelides, 'Cyberconflict at the Edge of Chaos: Cryptohierarchies and self-organization in the open source', in Phoebe Moore and Athina Karatzogianni (eds), *Parallel Visions of P2P production: Governance, Organization and the New Economies Capital and Class* Special issue (January 2009), 143–59.

one-quarter of the Baltic republic's population of 1.34 million) and the Estonian state. The country is considered to be a success story due to its e-commerce, which even sees government activities conducted on line.

In this cyberconflict we witnessed denial of service attacks, clogging the country's servers and routers, and the infiltration of the world with botnets (banding computers together and transforming them into 'zombies' hijacked by viruses to take part in such raids without their owners knowing). Multiple sources flowed into the system, the attackers even renting time in botnets. The attacks lasted for three weeks. The plans of the attackers were posted in Russian-language chatrooms with instructions on how to send disruptive messages, and which websites to target. The targets were on all social, political and economic levels: the Estonian presidency and its Parliament, almost all of the country's government ministries, political parties, three of the country's six big news organizations, two of the biggest banks, and firms specializing in communications. The effect was a rapid organization by the Estonians to fight the war, utilizing contacts in several countries and asking NATO and the EU for help, blaming the Russian state for the attacks.

Although Estonia claimed the attacks originated in Russia, and the global press linked the attacks to the Russian government, it was eventually accepted that it was, in fact, nationalist hackers that had done most of the work. Members of Nashi, a private pro-Kremlin youth group, also claimed to have had a hand in launching attacks, and state-controlled media were reported to have helped whip up anti-Estonian fervour that may have helped recruit hackers. An ethnic Russian student, Dmitri Galushkevic, was convicted of attacks against the website of Estonian Prime Minister Andrus Ansip and would pay a fine of roughly 1,100 Euros. The Estonian government, which complained that the attacks were orchestrated by Russia, were also portrayed as panicking, exaggerating the situation, when their networks were attacked in cyberspace.

Several issues emerged because of the attacks in Estonia. One of the main problems is that NATO does not yet define electronic attacks as military action, and therefore cannot intervene even when the origin of an attack can be proven. This issue has been a problem addressed by various authors.[8] What is more interesting is that in June 2010 it was reported in the *Sunday Times*[9] that a team of NATO experts, led by former US Secretary of State Madeleine Albright, prepared a report saying, among other things, that a cyber attack on the critical infrastructure of a NATO

8 Central European University. Participants' reflections on workshop themes: 'Cybersecurity: Europe and the Global Society Revisited: Developing a network of scholars and agenda for social science research on cybersecurity', June 7–8, 2010, Budapest, Hungary: http://ww.cmcs.ceu.hu/cybersecurity/main. Also see Andrew Liaropoulos, 'War and Ethics in Cyberspace: Cyber-Conflict and Just War Theory', *Proceedings of the 9th European Conference on Information Warfare and Security*, University of Macedonia, Thessaloniki, Greece, July 1–2, 2010.

9 Michael Smith and Peter Warren, 'NATO warns of strike against cyberattackers', *The Sunday Times*, June 6, 2010, available at: http://www.timesonline.co.uk/tol/news/world/article7144856.ece.

country could equate to an armed attack, justifying retaliation. The organization's lawyers were reported as saying that because the effect of a cyberattack can be similar to an armed assault, there is no need to redraft existing treaties. If an attack on critical infrastructure resulted in casualties and destruction comparable to a military attack, then the mutual defence clause, Article Five, could be invoked. Still, the level of attack required is not exactly clear.

Also as a result of the Estonian incident, the role of information communication technologies was yet again explored as a very convenient and cost-effective tool for protest – usually related to hacktivism and the ethical debates involved. Linked to the real life protests and their online incarnation is the real spark, the uncertainty about the enemy within and the anxiety about the always incomplete project of national purity as reflected in the ethnic Russians living in Estonia and elsewhere. These cultural struggles are exacerbated by the media and propaganda, with groups defending the purity of their national space using online technologies. The Estonian cyberconflict is also a reflection of the instability of the EU/NATO enlargement project, especially in relation to Russia's hegemonic aspirations, energy disputes and legacy in the region, with Western reports pointing to an emerging second Cold War through asymmetric warfare by Russia, such as the missile dispute with the US and Russia's relentless involvement in the region as a whole (supporting secessionist states, intervening in 'coloured' revolutions, embargoing products, etc.).

Another major incident was in the case of Georgia, although the circumstances were different. It was reported as a 'virtual war' in cyberspace, accompanying the brief actual war in the summer of 2008 between Georgia and Russia. Again, Russia was accused of orchestrating the cyberattacks and again it turned out that although coordination with the military was not deemed impossible, it was largely due to patriotic hacking.

Russia sees armament in Georgia as a serious problem, and it has brought it up, after the war, in NATO meetings in Brussels. In December 2009, NATO and Russia resumed their political dialogue, which NATO had broken off after the war in Georgia. All this discussion regarding global security, espionage and cybersecurity is accompanied in the media with questions over NATO, and Russia seeing the participation of former Soviet-influence countries as threatening. This is especially prevalent in the media debates when the Estonia and Georgia cyberconflicts are covered, as well as NATO's cybersecurity capabilities, doctrine and general regulation of cyberconflict.

The link of cybercrime to cyberconflict is explicit in reports that one of the botnets drafted for the Georgian cyberattack was Black Energy, a Trojan horse-hijacked army of PCs thought to have been used to hit Citibank, while Black Energy 2 was being used to launch distributed denial of service (DDoS) attacks against Russian banks. Political and patriotic hacking is not only linked to cybercrime, but to the Russian intelligence service, the FSB, in the Georgian press.

In November 2009, Russian hackers and Russia were immediately implicated in the Climategate hack, when emails exchanged between key climate change scientists of the Climate Research Unit at the University of East Anglia were posted on a Siberian server, creating a debate over peer review and climate change while the Climate summit in Copenhagen was occurring. The Russians were portrayed in a Cold War propaganda discourse in the media as having the motive to want to discredit the summit, while the use of patriotic hackers by the Russian secret service, the FSB, fitted the narrative (poor, unemployed but talented Russian hackers would have been easy to employ). Although there is no resolution on who was responsible for the hack, it seems that the Russian connection definitely is not as strong as in the original reports, with analysts even talking about computer security failures at the university involved to have been more likely, coupled with the likelihood of American climate-sceptic bloggers having played a role at least in the dissemination of the files.

A few months later, in January 2010, one of the most complex cyberconflicts occurred, when Google reported as originating in China attacks which took place towards the end of 2009, penetrating their network to steal intellectual property (source code) and hacking into gmail accounts held by human rights activists; Google responded with a declaration on changes in their China policy. Revelations were made about similar attacks involving more than thirty companies in a 2009 US-China Economic and Security review, which reported to Congress a steep rise in attempts to infiltrate and disrupt US government sites from all over the world, with China the largest single source.

The effects of this incident are wide-reaching in this field of research, as it brings together, in one discussion, a complex matrix of debates. In the bigger picture, this cyberconflict event adds to the debate on the position of China in the world system, and creates insecurities about the ambitions, capabilities and hidden desires of the 'next hegemon',[10] while it punches more holes in the odd Sino-American relationship. Further, it raises questions about China's information warfare philosophy and military doctrine, and the bizarre and contradictory ways they develop their virtual society, i.e. exploiting the technologies commercially, but using surveillance and censorship in ways contradicting liberal ideals of universal digital rights. On top of these concerns are the transformations the internet has brought in regards to civil society, citizenship and activism;[11] the relationship between business and activism in China and beyond; the relationship between the state and the plethora of 'patriotic' hackers; and the question of the working class digital have-less inside China.[12]

The result of the Estonia cyberconflict was the establishment by NATO of a Cooperative Cyber Defence Centre of Excellence (CCD COE or code name K5) in 2008 in Tallinn.[13] In May 2010, the US Secretary of Defense, Robert Gates,

10 Athina Karatzogianni and Andrew Robinson, *Power, Resistance and Conflict in the Contemporary World* (London and New York: Routledge, 2010), 115.

11 Guobin Yang, *The Power of the Internet in China: Citizen Activism Online* (New York: Columbia University Press, 2009).

12 J.L. Qiu, *Working-Class Network Society: Communication Technology and the Information have-less in Urban China* (Cambridge, MA and London: The MIT Press, 2009).

13 For an enjoyable tour of K5, see Bobbie Johnson, 'No one is ready for this', *The Guardian,*

announced the activation of the Pentagon's first comprehensive, multi-service cyber operation, the US Cyber Command (CYBERCOM), with General Keith Alexander as its commander. Talking about cyberspace as the fifth battlespace, transferring soldiers from communications and electronics to an Army Forces cybercommand, and wondering on how cyberwarriors should be trained, confirms a trend toward militarization of what were previously criminal and commercial matters.[14] With Russia and China frequently the usual suspects, the US and its NATO allies have had to address cyberwarfare questions in its twenty-first-century strategic concept. With 120 countries developing cyber capabilities, NATO's Director of Policy Planning, Jamie Shea, has commented that 'there are people in the strategic community who say cyberattacks now will serve the same role in initiating hostilities as air campaigns played in the 20th century'.[15] NATO will have to create a coherent strategy for cyberwarfare.

These historical incidents of cyberconflict of various types raise questions of cybersecurity as a part of global security in global politics today. Unless the precise level at which a cyber attack becomes part of armed conflict is defined by international law on cyberconflict, any cyber attack could be framed as cybercrime and prosecuted as such. This in turn would mean that any political hacking, even for protest, will be prosecuted as cybercrime. This could potentially mean that electronic disobedience or hacktivism as we have known it, despite having mostly symbolic effects, can be also prosecuted under this logic. An added problem for global politics is the difficulty in understanding where attacks originate from and whether there are state-sponsored or ad hoc assemblages. Not having defined the level at which a cyber attack becomes equivalent to an armed attack, there is no way currently to plan a reaction at an international level. Furthermore, it is not clear whether cyberattacks and cyberespionage will eventually be considered as a kind of war, as information warfare and espionage historically have not been recognized as war or grounds for war.

Cyberconflict and Future Warfare

The most common view on information warfare and the future of conflict, whose best-known exponents are Heidi and Alvin Toffler,[16] extrapolates from the idea that territory, population and natural resources are becoming less important, relative to human capital and the possession of information. Taking this process to its logical

April 16, 2009, available at: http://www.guardian.co.uk/technology/2009/apr/16/internet-hacking-cyber-war-nato.

14 Rick Rozoff, 'U.S. Cyber Command: Waging War in World's Fifth Battlespace', May 26, 2010, available at: http://rickrozoff.wordpress.com/2010/05/26/u-s-cyber-command-waging-war-in-worlds-fifth-battlespace.

15 Ibid.

16 Heidi and Alvin Toffler, *The Third Wave* (New York: Bantam Books, 1980).

conclusion, these theorists believe that information will soon become the key source of wealth and power – equivalent to steel, coal and oil in the industrial age, or fertile land in the agricultural age. This change will eventually amount to a social revolution, whose scope is equivalent to only two previous such transformations: the agricultural and industrial revolutions.

The transition will be from industry to information-based services and this will correlate with the 'informatting' of warfare. Sun Zi is an icon in this pantheon, with his observation that the 'acme of skill' consists in winning without fighting. Advanced technological systems will not only help shape the environment of future conflict, but will also magnify the importance of the psychological battle in the conflict outcome. At the systemic level, information warfare is the organization of information to provide warriors with what has been termed 'dominant battlespace knowledge'. Insofar as the ability to kill what can be seen makes seeing (locating, identifying and tracking) the key to war, seeing is increasingly best done by networking sensors and human observers to create a shared foundational truth that forms the basis of command, control and operations.

Arquilla and Ronfeldt[17] argued at the dawn of the millennium that power seems to be migrating to non-state actors who are able to organize into 'sprawling multi-organizational networks' which are more flexible and responsive than hierarchies in reacting to outside developments, and appear to be better than hierarchies at using information to improve decision-making. The battlespace of information warfare is cyberspace – an ethereal place which does not fit neatly into the land/sea/air space. Taking out all information-transfer media would bring down a country's stock market, banking system, air traffic control, emergency dispatches and more. The rise of networks is likely to reshape terrorism in the Information Age and lead to the adoption of netwar, a kind of Information Age conflict that will be waged principally by non-state actors. The Rand corporation in the US also predicts that cyberterrorists will use new tactics such as 'swarming', which occurs when members of a terrorist group, spread over great distances, electronically converge on a target from multiple directions, a tactic different from the traditional form of attacking in waves, which delivers a knockout blow from a single direction on the internet.

Besides this type of argumentation on the future of conflict stemming from the counterterrorist literature in the US and focusing on the Revolution in Military Affairs (RMA), there have been other contributions theorizing war, media and culture more broadly, for example its relationship to postmodernity,[18] to culture and media as 'militainment',[19] and virtual war in general.[20] Hammond, for instance,

17 J. Arquilla and D. Ronfeldt, *Networks and Netwars.*

18 Phillip Hammond, *Media, War and Postmodernity* (London and New York: Routledge, 2009).

19 Roger Stahl, *Militainment Inc.: War, Media and Popular Culture* (London and New York: Routledge, 2010).

20 James Der Derian, *Virtuous War: Mapping the Military-Industrial-Media-Entertainment Network* (London and New York: Routledge, 2009).

explains that beyond the high-tech weaponry and the RMA discussion, war is becoming postmodern both in the sense of intra-state conflicts where we witness wars about identity politics, in the cosmopolitanism vs exclusivism fashion, but also wars of humanitarian intervention, 'spreading democracy'. Hammond argues that the West's crisis of meaning after the end of the Cold War and the collapse of the grand narratives has caused a shift first to the therapeutic war (salvaging the reality of war in our own eyes – humanitarianism) to the War on Terror (postmodern terror, as the west at war with itself, with Other regarding imperialism and nihilistic terrorism as products of the crisis of meaning). In Hammond's explanation of postmodern politics, he cites Žižek's argument that the elite takes over the language of the left: from identity politics to official multiculturalism as the ideal form of ideology of global capitalism, which does not disturb the circulation of capital. The idea of war as distraction is replaced by war used to engage a disengaged citizenry. The postmodern war becomes an exercise in risk management.

In this kind of logic Stahl[21] talks of the fusion of military and entertainment as *militainment*: the transformation of war aesthetics from the 1991 Gulf war, where we consume a clean surgical sanitized war, a computer game techno-fetishism with the citizen spectator to Iraq 2003, where we have depictions of war as sports coverage, reality television and video games, with similarities to all these entertainment genres. Identity is absorbed into the military-entertainment matrix: a migration of identity to the interactive war. The spectator of 1991 becomes a virtual citizen-soldier, annihilating the viewer's capacity to distinguish between fact and fiction. This is similar to embodying the body in the military machine, as in the movie *Iron Man*: 'An integrated machine of hardware and software interfacing the subject with the military apparatus'.[22] As Stahl explains, conflict becomes a celebratory event, an exercise in recreational violence within a larger sea of fictitious violent entertainment.[23]

In turn, Der Derian in his *Virtuous War* argues that the global media is e-motive: a transient electronic affect conveyed at speed, where it is difficult to maintain the distinction between war and peace: 'In this high tech rehearsal for war, one learns how to kill but not to take responsibility for it, one experiences "death" but not the tragic consequences of it'.[24] In this type of infowar, Der Derian tells us, they did not invent a new game: they made the virtuous war the only game worth playing.

Although it is impossible to predict the future of warfare, this chapter has attempted to show historically how cyberconflict, the role of networks and communication technology infrastructures will be of paramount importance, not only in the way wars are fought, but also the way wars are communicated and justified to the global public. Not only that, but the acceleration of protest, when there is a crack in the global political structure, by ad hoc assemblages, protest networks and other resistant movements (due to the digital virtual enabling the

21 R. Stahl, *Militainment, Inc.*

22 P. Hammond, *Media, War and Postmodernity.*

23 R. Stahl, *Militainment, Inc.*

24 J. Der Derian, *Virtuous War.*

grasping of political opportunity), such as the situation with WikiLeaks and its effect on diplomacy, and spill-over effects currently in the Middle East, points to the critical importance of political communication in the global transformations taking place all over the world. The move to overthrow repression, violence and fear through peaceful means with virtual protest and its real life materialization of revolution seems to be perhaps rendering war an extraordinary response to be used only to protect and not maim life. The politics of justifying war beyond the protection of life will likely be debated for a long time to come, but the importance of ICTs as a factor in the political communication of future wars, protests and resistance is unquestionable.

The Future of Asymmetric Warfare

François Géré

Introduction: War, Defense and Security—What's New?

History will probably look at this period of our common history as a moment of uncertainty between traditional and irregular warfare.

Today the average citizen is living and thinking in an environment made up of multiple hurdles to the exercise of a free understanding of reality. His judgment is framed by prejudices, *set ideas*, repeated for the sake of repetition, of bureaucratic concerns estranged from the real world, of industrial interests indifferent to everything which is not the final account presented to the Board, and finally of deliberate, planned disinformation.

There is little doubt that the 9/11 aggression has generated a new relationship between security and defense, between law and justice operations, and the use of military forces. The push for homeland security created a new mindset in the United States. It has created a new perception: military operations abroad are just a component of the protection of the homeland. They are not only in its national interest, like in the days of the Cold War where operations in Vietnam were portrayed as an essential (not vital) way to counter the communist threat. Fighting in Afghanistan is portrayed as the allegedly best shield for American and European territories and against terrorist aggression. International terrorism has created a linkage, very often politically manipulated, between domestic security and limited military engagement (with the additional problem that the war in Iraq has been presented as a shield against cooperation between Saddam Hussein and Al-Qaeda). The linkage between Weapons of Mass Destruction (WMD) and "rogue states," between nuclear weapons and terrorist organizations, has created a new strategic environment where asymmetry stands as a new and central dimension of the "threat," which has been reflected in the way states deal with their defense and security concerns. In France, the White Paper (2006) deepens the interrelationship between homeland security and military operations abroad, paying particular attention to the asymmetric nature of the new threats and to non-conventional adversaries.

That situation has created a new imbalance of perceptions among politicians, the media and public opinion. Major wars between states regulated by a principle of symmetry have faded. Communism has almost disappeared as an ideological threat to liberalism and democracy. Even if Russia and China are not the ideal friends, the perspective of war seems very questionable and, to be sure, decades ahead. Therefore, as a result of the new balance of threats, Serious Organized Crime (SOC) has reached the top of the list and has been boosted by its alleged connections with terrorism under the shadow of the threat of WMD. (As early as 1992, newly elected President Clinton made similar statements.)

During the last ten years, after the 9/11 attacks and the insurgency in Iraq, the notion of asymmetric warfare has become fashionable. Several authors have made the case that major wars have ended and will never reappear. They argue that defense budgets are kept at a high level without real purpose, just to keep the military-industrial complex alive. These authors argue that wars will be asymmetric on the grounds that wars between states have decreased in number, leaving room for interethnic, interreligious conflicts inside the borders of a single state. However, the magnitude of recent human disasters have led to the conclusion that there was a duty of intervention in order to stop or prevent a conflict that could become a genocide. That view has been very popular since 1990 and at the end of the Cold War, but after the limited success of military interventions in the Balkans, and bad experiences in Somalia, Iraq and possibly Afghanistan, there has been a clear step back.

Iraq and Afghanistan are counter-guerrilla and/or counterinsurgency in nature. War against terrorism from Al-Qaeda belongs to the same form. Therefore it is necessary to establish a security/defense combination that is capable of addressing the present threats and challenges which will persist over the coming decades in the absence of a major power war.

Nowadays, visions of warfare are blinded by the overwhelming superiority of the American military, which has created a general asymmetry. Today, no country would dare to directly confront US forces. That vision, which is unfortunately widespread in Europe, is challenged by analysts, civilian and military, in emerging countries. Why? Not only because they are considering how to avoid US superiority, but also because they are paying attention to three issues:

First, the reasons to go to war are increasingly linked to the demand for global resources, and the freedom to secure their access. In part this is not new, but the dimensions are now global and the number of players are many more than a century ago. Appetites are driven not only by greed but also by real needs.

Second, the use of smart weapons in warfare. In this respect: "The true revolutionary shift of the battlefield stems from the extension of war to non-natural space. The electromagnetic spectrum is a new form of combat space based on technical creativity."[1] These weapons now take the lead in the fighting and then let traditional

1 Colonels supérieurs Qiao and Wang, *La guerre hors limites* (Paris: Payot & Rivages,

weapons finish the job if necessary. (This reminds one of the relationship between the artillery and the infantry: artillery prepare, infantry conquers and then holds the positions.) But the problem remains: Occupying a territory seems to be less and less interesting. It is too much trouble with uncertain benefits.

Third, there is little doubt that the perception of the use of violence has shifted in the Western world, including on the part of the United States, particularly after Iraq and Afghanistan. War is no longer about destroying, killing people or inflicting pain to the population. The period of coercion should be quick, and the quantity of violence limited and proportionate. War is much more designed to permit the restoration of law and order, the rebuilding of the country, and the establishment of conditions of good governance according to the free choice and good will of the population.

On Symmetry and Asymmetry

The myth of symmetry

Same level of forces, same quality, same doctrine, same culture—such a perfect model reflects the wars in Europe in the eighteenth century.[2] Above all, it refers to World War I, which proved to be so symmetrical that instead of the expected quick decision it lasted for four and a half years, slaughtered millions and crushed western economies. Symmetry of equivalent industrial potential facing each other separated by a front line was the recipe for indefinite war and the concurrent massacre. In that case, let asymmetry be blessed!

Incredibly, two major strategic thinkers drew opposite conclusions. The British strategist B.H. Liddel-Hart built his theory of "indirect approach," and considered the speed and firepower of the battle tank as elements of asymmetry. The Italian general Giulio Douhet also considered a new weapon: airplanes. He regarded massive, indiscriminate air bombing as the ultimate weapon. The devastating effect of airpower upon cities—both population and industries—in the rear no longer protected by the soldiers on the front line would quickly put an end to a war. Massive bombing was supposed to break the will to resist. In the end, for both these theorists, the conditions for a *guerre-éclair* in Napoleonic style

2003), 79. The French edition was directly translated from the 1999 Chinese language original.

2 As presented by Maurice de Sax, maréchal of Saxony, in *Mes Rêveries* (1757). Adopted with enthusiasm by Liddle-Hart.

would reduce casualties to a minimum and likewise reduce the related suffering of a protracted conflict.

Indeed, during the Cold War "High Intensity Conflict" was related to the exchange of strategic nuclear strikes, if deterrence were to fail. By contrast, the strategic concepts of "limited wars" and "Low Intensity Conflict" (LIC) were characterized by the exclusive use of conventional weapons (chemical weapons belonged to a disturbing "grey zone") against an enemy which combined a partly irregular, partly regular, army of a modest size. War in Vietnam was initially considered a LIC. Only in 1966–67 did President Johnson pay attention to the risk of an escalation with a nuclear-armed China. Therefore initial air bombings were first limited to Haïphong harbor and its industrial area close to Hanoï (Operation Rolling Thunder). Later on, President Nixon authorized Linebacker I and II bombing campaigns which were more ambitious and appeared to be more effective.[3]

However, the US vision of war remained focused on the deadly strategic nuclear parity and the risks of a sudden break-out. Even in Europe, NATO claimed that the Warsaw Treaty Organization (WTO) enjoyed a conventional superiority which had to be balanced with European land-based nuclear weapons as a deterrent. The nuclear component still stands out as a disturbing factor in the military equation since it modifies the traditional "bean counting" of conventional forces and may create an unexpected equalizing symmetry between two adversaries. Nonetheless, such alleged parity has been labeled as "peer competitors" in the official documents of the US Department of Defense (DOD) since the end of the Cold War. This fact refers to a symmetry between the US and the Soviet Union which will never, ever be repeated. Seldom does reality offer such a situation, which can be regarded as a kind of Weberian strategic "ideal-type." The real world is always made of a permanent dynamic interaction between factors of symmetry and asymmetry. Asymmetry between two regular armies: symmetry in one domain, but asymmetry in another. For centuries the competition between infantry and cavalry has been a major factor of asymmetry and uncertainty at the operational level.

Operational asymmetry can reach a structural level between two great powers: Napoleon, master of the land, against insular England which was dominant over the oceans. The duration of the conflict: 1802–1815, regardless of the fact that the war actually erupted in 1793. The old Clausewitzian model is based on land warfare. It pays no attention to the naval dimension, which played a decisive role in the ultimate British victory and the final success of a coalition which was widely funded by the "English gold." These alliances were aimed at creating imbalance.

In addition, the nature of alliances make the strategic equation even more complex. Such was the calculus of the German HQ which made it necessary to attack preemptively due to the asymmetry in the balance of military potentials. The skill of military strategy is precisely to look for, and to identify, the relevant

3 Whether or not this was a side effect of the ongoing negotiations with China which were ended by the spectacular Shanghaï Communiqué of 1972 remains an interesting point. Matriochka strategy: one major goal with, underneath, apparently less important issues.

elements of asymmetry and to translate these findings into general orientations for the operational level. Operational know-how is about asymmetry, about the Achilles heel of the enemy. The confrontation of regular armies creates some elements of symmetry, but the operational art is aimed at introducing dissymmetry.

Surprise can be considered a provider of asymmetry in the use of the moment and the location, and sometimes a combination of both—a sudden change in the traditional doctrines, a leap out of the established rules of the game. Admiral Nelson adopted the "T" line of battle at Trafalgar (1805) instead of engaging in the traditional parallel line. Lieutenant General Heinz Guderian disobeyed the orders of German Chief of Staff Erich von Manstein and exploited in depth the success of the blitzkrieg in the Ardennes in 1940.

More recently, John Boyd has created some principles based on the search for the actual creation of asymmetry. The Observation-Orientation-Decision-Action (OODA) time cycle or loop is aimed at creating an asymmetry in the operational and tactical decision-making processes of Command, Control, Communications, Computers, Intelligence, Surveillance and Reconnaissance (C4ISR).[4] Through speed and the ability to anticipate the decisions made by the communication systems of the enemy, OODA creates an operational paralysis which generates a major imbalance between two armies which, at the outbreak of the conflict, were supposed to be symmetric. But does OODA prove that it is capable of working against asymmetric enemies, in which the ways of communication are very different in nature and technology?

There are many reasons to explain why symmetry has generally been considered the normal nature of warfare. First, the lazy mindset of large bureaucracies and administrations which replicate their counterparts. In effect, structural "mimetism" leads to mutual intellectual arthrosis which can be considered a form of symmetry. Second, intellectual conformism or "homothetical thinking"—the parallel evolution of French-German military thinking between 1870 and 1940 is a striking example.

Third, an intellectual effort to introduce rationality inside the chaos of war and the uncertainty about the efficiency of operational initiatives—how many brand new solutions have turned into deadly disasters? The search for asymmetry is a risky business, not a quiet and comfortable endeavor. Strict/absolute parity does not exist. Models try to create them in order to inject some order and rationality into the chaos of war. The mathematician and engineer Frederick W. Lanchester elaborated his equations to do so. However, he was smart enough to write: "there will probably never be a close and easily recognizable relationship between real world attrition rates and those produced by the use of (my) equations"; Lanchester considered that his equations were valid only in a situation "in which the opposing forces are of equal quality and are affected equally by the circumstances of the battle."[5] The French artillery's *abaques* were designed to deliver the adequate number of artillery shells per square meter. They were supposed to reduce uncertainty and create a reliable, rational combat environment.

4 John Boyd "Destruction and Creation," 1976 (no formal publication).

5 T.N. Dupuy, *Understanding War* (New York: Paragon House, 1987), Chapter 16, 221.

Fourth, and not least, partly nurtured by the previous factors, stands a common, sustained, very widespread mistake: the confusion between symmetry and asymmetry on the one hand, superiority and inferiority on the other. Traditionally, regular armies consider themselves as superior to an irregular force just because that enemy does not enjoy the same level of technology, the same quality of training, and does not have, for instance, air superiority (sometimes it does not possess any aircraft at all).

Finally, the illusion of asymmetry of stronger/weaker represents an element of deception: France against Vietminh; the US against North Vietnamese forces and the South Vietnamese National Liberation Front (NLF); Israel versus Hizbollah/Hamas.

The Iraqi army has been crushed to pieces by the US. But the real question should be: does the enemy retain the weapons, organization and related tactics needed to achieve its goals? Based on that question, it would be possible to understand who is weak and how asymmetry plays its role. *Superiority is the result of cumulative and combined asymmetries in favor of one side.*

The real world of asymmetry in its various dimensions

In 1896, Charles Callwell wrote "Small Wars," which quickly became a classic.[6] Not so "small" today, because they are much more demanding than those of the end of the nineteenth century, but why? Which factors make them more demanding? Asymmetry of what and between whom?

Callwell used a very simple criterion as a starting point: regular armies on one side, irregular fighters on the other. However, throughout his book he brings together a large number of nuances. At that time, the dominant paradigm was the "great war" in Europe against "real armies," not just bunches of "savages" using light, obsolete weaponry. J.F.C. Fuller mentions the success of "Swiss peasants" against the armies of traditional knights as a consequence of appropriate weapons used inside the framework of a democratization of warfare.[7] Theoretically this is important, in that it is not understood as an asymmetrical situation, but rather as a vision of an "order" which substitutes for another and for a period of time becomes dominant; it is a linear conception as opposed to a structural approach.

As a matter of fact, it appears that modern high-tech armies were badly defeated in their first attempts to dominate Sudan (Khartoum), Ethiopia (Adoua), Madagascar and, of course Afghanistan (Khyber pass). The UK, France and Italy had to reconsider their engagements and make sure that they would benefit from overwhelming superiority. Incidentally, they would acquire—through their first bad experiences—a better knowledge of the ground, the climate and the dangers posed by diseases.

6 C.E. Callwell, *Small Wars* (1896); reprinted by Greenhill Books, 1990. French translation, *Petites Guerres* (Paris: Economica, 1998), Preface by François Géré.

7 J.F.C. Fuller, *Armament and History* (New York: C. Scribner's Sons, 1945).

The situation is reflected in the asymmetry of diplomatic relations. Traditional diplomats dream of a Vienna Conference in the old-fashioned style and the "concert of great powers." First, from this dreamy perspective, we should reach an agreement between European Powers. Then, as partners, we dictate our law to the rest of the World: as Europe did to China and to Africa during the Berlin Conference in 1884–85, for example. The unique world power tried to do that in 1995 by imposing the Dayton Agreement on Bosnia. The French tried to do the same through the Marcoussis Agreement on DRC (Congo) in 2003.

Recently, Dave Johnson from RAND has proposed the notion of "hybrid war" between two different adversaries.[8] A military has to perform well with a relatively low level of capabilities, at the same time, it also has to be prepared to face several different adversaries. Johnson takes the example of Israel. He compares fighting against the Palestinian organizations, particularly Hamas in the Gaza strip and Hezbollah in southern Lebanon. In fact, for Israel the challenge is very demanding. Military planners have to continue to prepare Tsahal for a real high intensity conflict against Syria, possibly supported by an Arab ally. Such a war could trigger action from Hezbollah and Hamas, posing a serious threat to the limited human resources of the Israeli State. How should such a conflict be defined? Of course, it's a combination of symmetric and asymmetric operations. And that is probably the worst case scenario in general.

Napoleon had to fight against the Russian Empire while his troops were subject to an aggressive Spanish guerrilla movement supported by the Wellington expeditionary forces which landed in Portugal. It was the same for Hitler, who had to cope with all the kinds of oppositions one can imagine: "resistance fighters," "partisans," mountain guerrillas and desert special forces in their foxholes. The final conclusion indicates that modern western armies should avoid focusing only on counterinsurgency. Making such a mistake means they could ultimately lose the fundamental knowledge of classic inter-army fighting. Most of the resources, training and equipment should therefore continue to be dedicated to missions of high intensity warfare against powerful regular armies. Nonetheless, it is worthwhile to plan for specific tactics adapted to the operations of stabilization.

Asymmetry of the will: Political goals and moral

Asymmetry of the stakes (Afghanistan, Somalia, Iraq)

There is a fundamental asymmetry between wars of choice ("optional wars") and "wars by necessity" (when your territory is under attack, invaded, or partly occupied). Wars of necessity are when the Armed forces are fighting on their own territories, protecting their people. And the people will spontaneously resist the invader, giving assistance to their soldiers, even though they do not necessarily agree with the government. Optional wars are the basis of the so-called intervention

8 David E. Johnson, *Military Capabilities for Hybrid Warfare* (RAND Corporation, 2010).

or *"droit d'ingérence"* to bring freedom and democracy. But let's remember the famous words of Maximilien Robespierre in October 1792, disapproving of the French decision to go to war: *"on n'apporte pas la liberté à la pointe des baïonnettes."* ("One does not bring freedom at the point of a bayonet.")

The difference between those two categories has a major impact on the psychological quality of the fighting forces. It creates an asymmetry of the stakes, which at the present time is the state of affairs in Afghanistan, just as it has been the case in Iraq since 2004 and in Somalia in 1994. As the former Chancellor of Germany, Otto von Bismarck, used to say, "all these conflicts [in that case it was Bulgaria] are not worth the life of a single good Pomeranian grenadier." If the stakes do not look as great given the human sacrifice and financial resources of the nation that are required when it fights abroad, then there is little chance for success in the long term—the immediate impact is upon the motivation of the soldier himself.

Asymmetry of stakes and motivations are indeed a crucial element since they bear on the asymmetry of fear. Often, fear is supposed to create some kind of shared empathy between the adversaries on the battlefield. That is symmetry. This might be true when they share the same culture, the same values (but in that case, why do they fight each other?). This concept of shared empathy therefore represents a fundamental misinterpretation when the adversaries belong to different cultures, particularly societies where the warrior enjoys social prestige (India, Pakistan, Nepal, most of Africa, ...). Nonetheless, it remains partly a mystery.

It is possible that a suicide bomber may "break down" and not engage in an act of "terrorism" not because he is afraid of dying but because he "feels" the life of the "enemy" in viewing children, or women like his mother, or sister. Beyond his initial determination, he finds inside himself a sense of humanity which makes him identify with those whom he is supposed to blow up with him. That feeling, however, resembles the ambiguity of contact in hand-to-hand combat. As a paradox, the more a suicide bomber is motivated by a sense of vengeance (his father killed or brother tortured), the more he keeps a sense of humanity. That is the reason why those who are indoctrinating him must insist on their ideology and the supreme value and goals of the cause; they indoctrinate the suicide bomber through spiritual exercises, in order to reduce as much as possible the personal motivations.

The asymmetry of the stakes refers to widespread "clichés." First is the problem of "time asymmetry"—guerrillas and insurgents both have time on their side. They merely have to continue to fight, inflict casualties, be active on several portions of the territory, and erode the morale of the "invader" and its public opinion. Eventually, they will win. Sometimes the opposite comes true. After a period of surprise and irrelevant responses, the regular armed forces understand the nature of the enemy and its organization. They then adapt and tailor a new response in kind. In the long run, they take the upper hand and ultimately create the military conditions for a political victory.

That point is related to a second one, political defeat versus military "victory": you win militarily but you lose politically. This argument has been made for Algeria and Vietnam. We can expect that it will appear soon in the case of Afghanistan. It is a way for the military to save honor and prestige. It is true that the US has not been defeated or suffered a *"Dien Bien Phu."* But the political decision to accept defeat comes from the incapacity of the armed forces to destroy the enemy and to bring convincing evidence that they were, in fact, winning. It is an artificial construct to disconnect one from the other. To be sure, the military does its best to achieve the strategy, sometimes dragging their feet, sometimes arguing about the best military strategy to serve the goals. But this is not the real issue.

Asymmetry on the traditional info-com battlefield

The nineteenth and twentieth centuries, until the Kuwait war of 1990, have been characterized by the absence or the scarcity of information because of poor and slow communications. No one was reporting about General Bugeaud's brutal punishment raids in Algeria. Limited, but significant, beginnings came with the reports on the War in Crimea in 1854 by the British press, which had decided to send war correspondents to the theater. Disinformation then followed rapidly. For instance, in the case of the journalistic manipulation of American public opinion by Randolph Hearst against Spain, the leadership in Madrid or Havana was not able to counter the lies about Cuba in asymmetrical mode.

Attacking the morality of the invaders is on the top of the communication agenda of the "resistance fighters" or insurgents. But they need to work hard on the population. First, they need to counter the "propaganda" of the enemy. Second, to promote their own values. Third, to instill fear among those who could envisage cooperation with the enemy. Fourth, to spread the information to neighboring countries. These are "basic" goals. Fifth, and finally, one can question whether they are really interested in reaching the public opinion of the enemy. (Let's say the NATO countries and the US.) It depends. In the case of Somalia, it is very unlikely. In Afghanistan, certainly. It is according to the degree of maturity of the groups, both in terms of internal organization and the ambition of their political goals. It depends, for example, upon their sense of, and inclination to, globalism, the nature of the main adversary and finally the communication strategy which results from all those factors.

Large organizations are not best suited for efficient modern communication. Efficient means active, reactive, flexible and fast. A good, although rather dull, example of symmetry is the NATO communication struggle with the Soviet Union and the Warsaw Pact. Both were huge bureaucracies exchanging hostile communiqués at slow speed inside a shared general context of propaganda and disinformation. (In the last part of this paper we will consider how much the information age affects asymmetry in the present and in the future.)

The combination of asymmetric motivations and the growing role of info-com transforms the ethics of warfare.

Ethical asymmetry and the "rules of engagement"

Regular forces have to respect the laws of war. Yet peacekeepers under UN leadership faced militias (in Somalia, former Yugoslavia and, to some extent today, Afghanistan) which did not pay attention to those laws. Moreover, militias can profit from the limitations of the "Might" of their adversary by the "Right." However, some rules can be imposed by force and the prospect of retaliation, but *peacemaking* is not exactly what *peacekeepers* are supposed to do.

Therefore, militias will use all the dirty tactics that they can conceive of. For instance, consider Human Shields. After all, it is an old, ordinary tactic used by many military leaders in ancient times. Prisoners of all kinds, women and children included, were often put in front of the line of attack as a shield against the offense. Prisoners were used in order to clear minefields. At that time, a psy-ops dimension was already there, and it included a strong disinformation component, in effect asserting: "We do not kill your people, it is you who have created minefields, therefore you are responsible for putting them at risk…" and so on. Regrettably, on that specific issue the international community has been pretty much unable to do better than referring to the Geneva Conventions.

A modern, if not more sophisticated version, was initiated by Saddam Hussein in 1990 in order to try to prevent military action against his invasion of Kuwait. More important was the use of Human Shields by the Serbian Army in order to complicate air operations by NATO and nurture a reasonably efficient propaganda and disinformation air campaign against NATO bombing.

The location and protection of Command and Control (C2) has become more and more crucial because it has become decisive for the conduct of operations. The groups described above have used their own population as shields against long distance strikes—Hezbollah tactics during the Lebanon war of 2006 demonstrated the same disregard for the respect of the laws of war. C2 were installed in the basements of hospitals, or close to schools inside Beirut. Furthermore, the destruction of such assets is welcomed as a tool for propaganda. It allowed Al Manhar, the Hezbollah radio and TV company, to broadcast an exhibition of dead or bleeding harmless civilians who were just looking for cover in shelters. Finally, it is possible to counter the efficiency of a well-equipped army using smart technology by creating media campaigns against the vicious effects of some specific weapons: phosphorous bombs, high penetration shells using enriched uranium heads producing radioactive contamination; the dispatch of submunitions create a dangerous environment that is lethal for the civilians for years. In the latter case, the use of such weapons by the Israeli Air Force in Lebanon has proven to be counterproductive. It has created the image of a barbarian aggressor, damaging the traditional image of Israel as a small country fighting for its survival. Political leaders have to weigh the military benefit of the use of such weapons against the political disadvantages in public opinion both at home and in the international community. However, there are limits to "dirty tactics." They come from the targeted people themselves and they can be countered by a strong counter-disinformation campaign.

Economic asymmetry: what price glory?
Irregular forces use cheap civilian technology: a combination of civilian devices obtained either on the open market or by invention—like Improvised Exploding Devices (IEADs), the value of which has been created by the US forces themselves in a major communication error. One should never overvalue the power of enemy devices, even if you want to justify your failures or obtain better protection. If the enemy can use a "magic weapon" such as the Stinger in Afghanistan in 1982, he may retain superiority and the initiative, at least for a limited period of time, but, above all, he obtains prestige by using that weapon. The shock effect on public opinion has been produced. This is true even if in the following days or weeks indisputable evidence arrives that contradicts that version of events. A reputation has been established which will affect the morality on both sides.

Let's now consider high technology, such as the American B-2 bomber and the French Rafale. These costly sophisticated weapons have to be protected or engaged in environments where there is almost no risk. Initially conceived for a major battle on the wide flat plains of the Ukraine, the Leclerc main battle tanks appeared to be useless in an environment of narrow segmented valleys like the Balkans, Lebanon or Afghanistan. Sophisticated equipment can tolerate mild climate, some dust, but not sand storms and excessive heat. Nonetheless, in order to appear relevant in some theaters they have been sent abroad. But they are too heavy for air transportation, and therefore have to be partly disassembled and reassembled upon arrival, and it takes a lot of time assuming that these heavy weapons need to reach a safe base or a usable airport. The deployment of a naval task-force requires an important mobilization of devices in order to protect the precious air carrier. The engagement of sophisticated air fighters requires electronic measures and countermeasures in order to provide the precious weapon with a reliable shield. Emerging gradually in the West is a "military economy of resource saving," in personnel, materiel and the related technologies—all have become too expensive. Europeans can no longer afford the price, moral and economic, of large casualties. So far, in the West, only the US has been able to do so, but for how long? Gradually, the cost of technology has exhausted the capacity to wage war. This dilemma has not, however, been the case for the "rustic" enemies with limited means whom the developed countries have been fighting since the end of the Cold War.

Serious Organized Crime

Who are the real enemies? Who is covertly destabilizing and ruining the global village? To many after the Cold War the immediate answer has been: Serious Organized Crime (SOC). Piracy is far from being new. All the Navies are well aware that piracy around the Strait of Malacca and in the South China Sea was prevalent during the years of 1985–2005. It took a number of very controversial international conferences to reach an agreement to establish efficient cooperation between the navies of those countries which had a major interest in protecting their interests in those areas.

The problem today in Somalia is widely linked to jurisdiction procedures and the respect by each country of the rule of law against criminals whose legal status, identity and citizenship are hard to establish. Piracy can be linked to traditional drug smuggling and slavery (which has been a traditional form of behavior for centuries in the Red Sea area). The bad news is the linkage between piracy and extremist Islamist groups (Somali *sheba* militias) which are taking control of parts of the disrupted country.

Drugs are by far the only domain which reaches a strategic level. Even though the public name of the "war on drugs" is much oversold, a large number of factors make drugs a threat to national and international security. This is true, first, because the drug industry is directly linked to the structure of states in both political and bureaucratic terms. It is also related to the economy and the ordinary social life of hundreds of thousands of peaceful villagers who need to live in one way or another. It is likewise related to powerful organizations which have a vested interest in penetrating the political structure of the state. Those organizations retain a wide spectrum of illegal activities (gambling, prostitution, arms, but above all drugs). They demonstrate very little interest in nuclear smuggling. They are looking for quick profit, not protracted trouble from authorities with whom they are not familiar, such as the FBI's counter-proliferation section.

Drug trafficking is more important than the traffic of human beings (slave manpower, prostitution), and even more important than arms trafficking. Why? Because of the magnitude of the market, the nature and dimension of the demand. Drugs have become a real strategic issue, since it is related to conflicts, to the stability of societies and economies in many countries. It has become much more than an illicit activity; it is an instrument of power and sometimes of statecraft. Very often, governments do not want to recognize the leverage of that instrument of power within their own country: Morocco, Turkey, Afghanistan, Thailand, Colombia, just to mention a few of them. Central and South America deserve particular attention. Since 2005, 28,000 people have been killed in Mexico due to rivalries between drug gangs. Massacres have been indiscriminate in bars, supermarkets or any other location: All the innocent people who, by accident, are close to the main target are exposed to firepower. Traditional security forces are frequently infiltrated by the gangs and fall into an inferior position, incapable of effective action. Therefore armed forces are required. Only they have the firepower and, above all, the communications required to counter the quality and sophistication of the devices used by the gangs.

An interesting point is that SOC does not care about a communication strategy. While political or ideological organizations need communication and look to gain superiority over the communication strategy of their adversary, gangs, mafias and other cartels have no ideology. They do not intend to gather popular support for their activities. But they do practice disinformation in order to create hurdles to the action of the security forces, the government and the allies who support their operations. It is useful to launch rumors about the massacre of villagers by the US military to present the government as if it were a puppet in the hands of the "gringos." Finally, by penetrating deep into the state administration they can influence governmental communication in a direction supportive to their profit goals.

However, glory does matter for warlords and other "capi" who, in the macho tradition, pay a lot of attention to their reputation. That fact makes them vulnerable to well-crafted psychological operations.

Counter-terrorism: How much is enough?

In the span of less than five years the world has been shaken by a major wave of internationalist terrorism inspired by a violent *Salafist* ideology, the importance and impact of which had been neglected since the end of the Soviet war in Afghanistan in 1988. The United States, Indonesia, the UK, Spain and other Middle East countries have suffered attacks through suicide bombings. The story of counter-terrorism since 2001, will appear in the future as a blend of successful actions and correct decisions, and, on the other hand, of wrong strategic interventions (Iraq invasion) which have aggravated the problem rather than bringing a solution. One thing can be considered for sure: fighting terrorism in its new form and its multifaceted threats must rely on an integrated strategy where the use of armed forces is only one element in support of the traditional use of the most efficient tools: police and justice acting through efficient intelligence cooperation and exchange.

Major mistakes came from the pollution of a prudent strategy against terrorism by other political goals which were NOT related to the fight against terrorism but were rather linked to imperialist visions, such as the transformation of the Middle East according to a neo-conservative vision of the world.

Linkages between Serious Organized Crime and terror

At present, terrorism works like cement, because of a powerful ideology which has acquired for a limited period of time the ability to connect with, or take control of, traditional forms of organized crime.

Terrorism seeks financial resources through the illegal traffic of goods which can be sold through covert channels. Therefore terrorist groups may connect with corrupted governments, with corrupted administrations, in order to penetrate the black market of diamonds, gold and other precious substances, drugs being one of them. In addition, traditional forms of transaction in the Muslim world, such as the *hawala* system, allow the discreet and reliable transfer of funds. Over time, contemporary terrorism has created a powerful strategy of flows—migrant fighters, clandestine traffic, information flow, use of internet and swarming—which is able to challenge the power and efficiency of heavy and hierarchical bureaucracies. Terrorists, jihadists, militants, no matter what the name, in many cases prove to be experienced warriors who travel from one theater to another.

That flow began in 1980, when *mujahidin* came to Afghanistan and the flow of terrorist actions continued after the claimed victory against the Soviet Union, and then spread around the world. First in Bosnia and later on they came back to their home countries or began to develop their activities in Peshawar, creating Al-Qaeda.

One can rightly argue that the correct strategy against terrorism should concentrate on the connections, the "hubs" which are the centers of gravity of non-hierarchical organizations. Some areas like Somalia can be considered hubs—"instability distribution centers." They are not "safe havens" where terrorists can establish fixed bases. In that case they would be vulnerable. It is rather they support a loose combination of traditional regional criminality[9] and more international activities, which are politically and religiously oriented. Drug dealers and pirates belong to the former, and jihadists to the latter. However, there is room for connections, particularly through extended family ties which play a very important role in cultures where polygamy remains a common practice. It is possible, from time to time, to counter those random activities through fresh intelligence, but it is generally too chaotic, too uncertain to elaborate and implement a real enduring strategy. All these criminal activities are too fragmented and do not obey a clear hierarchy. There are no clear political goals. This is the true nature of the cultural gap.

On Men and Robots the Technology Solutions: California Dreaming and Vacuum Cleaners

The importance of imagination

"*Vous en avez rêvé, Sony l'a fait*" ("You dreamed it, Sony made it"). One's dreams also become true when one considers the spin-off of civilian advanced technologies upon the military. Vacuum cleaner technology, very often made in Japan (thanks to rare raw materials exported by China), is often used by US armament companies for the production of demining robots.

A robot is man's creation. If it is a machine, the problem remains technical. If the robot is the image of man, it turns into metaphysics. Traditionally, the robot is considered an offense against God, a challenge from man's *hubris* against his own creator. (See the Jewish Golem, Dr Frankenstein's creature.) For centuries, our imagination has been vibrant with that fantasy.[10] However, two issues are

9 For the pleasure of reading, see Henry de Monfreid, *Les Secrets de la mer rouge* (*Secrets of the Red Sea*) (Grasset, 1931), who describes a situation extraordinarily similar to the present one. At that time there was the traffic of slaves in addition. But today, smuggling of manpower seems to be very comparable.

10 In recent times, consider TRON, RoboCop, The Matrix, Terminator, Star Wars, Alien and many others.

not fictional. First, education and culture—young children are no longer using tin soldiers. They "work hard" on play stations crowded with images of human beings and ambiguous monsters. Who is human and who is not remains uncertain. Ultimately, the choice stands between the "good" and the "bad." However, the core issue is combat, regardless of any motivation. Agressivity, aggression, speed and destruction are central. The killer is the winner. That's the education of the future warrior.

Second, there is a deep cooperation between civilian and military companies in order to produce special effects, and real "things" which work, using advanced artificial imagery (Titanic). In Hollywood, Disney Studios and Lucas Company share advanced technology with the Defense Advanced Research Projects Agency (DARPA). Just as in the case of the Reagan-era, the Strategic Defense Initiative (SDI) of 1983, a Californian dream coincides with a human dream.

The central problem: The phobia of contact

Combat and battle (*not war*) are about the human body. The most important military inventions have been the weapons which avoid full contact by increasing the distance between combatants. Not only do they protect the soldier but, more importantly, they allow him to have no knowledge, no vision and no smell of the persons he kills. Certainly, the bow was an important step forward, but clearly artillery has introduced a new dimension: perfect anonymity. Still, the problem was how infantry or cavalry could reach artillery positions in order to destroy them (the last hundred meters).[11] Using fixed bayonets for contact has always been of major concern. Today no conscript likes that perspective, just like their predecessors.

Professional soldiers have struggled to meet that challenge. The "drill" in the Old Prussian tradition of infantry training was designed to transform the soldier into an intelligent machine. That is the perfect robot—indifferent to fear, executing all together, at the same time engaging in only movements necessary for efficient combat. Elite infantry units such as the Napoleonic guards or the Russian "black guard" were trained on that model and traditionally kept as reserves in battle. In that perspective, we must consider suicide bombers. It is an asymmetric weapon, mainly psychological, that is capable of disrupting a small society but not an army, not even in Iraq or in Afghanistan.

However, the suicide bomber represents a new form of smart yet irregular fighter. He receives a special education and training. This is a combination of ideological brainwashing and serious techniques aimed at creating a familiarity with the tactics and use of the device which becomes part of the body in a mission of no return. Moreover, physical fear is related to the perception of the body of "the other" as a human being. "The other" becomes "the same," as demonstrated by so many wars, particularly in the case of trench warfare during World War I. That

11 Ardant du Picq, *Etudes sur le combat* (1866).

unexpected consciousness creates highly disturbing problems as continuously illustrated in various movies.[12]

The traditional drill is gradually replaced or, to be more precise, complemented by the vaunted "robot warrior." The French army has adopted the concept of the *combattant FELIN*. This notion, which has many similarities in the US Army, is based on the idea of a soldier who is equipped with a lot of devices which provide him with the ability to capture the key elements of the combat environment in which he is acting: vision, hearing, even smell, but not touch.

The fighter is protected against the hidden threats because he can anticipate, analyze and react. Consequently, he retains the ability to fire first, preempting the enemy. This is particularly important in urban warfare. To be accurate, robot fighters, as smart machines operating on their own and replacing human fighters, are not coming soon, if ever. Nonetheless, two evolutions should be seriously considered: first, a modification of the human body; second, the relationship between the warrior who guides the robot and the warrior himself.[13]

In the first case, we are seeing an evolution from prosthesis *on* the body to implants *inside* the body, creating some form of "hybrid" human who still retains the quality of the human brain and reactivity but is enhanced by the adjunction of implants. It is a very dangerous domain which is already filled with rumors, fears and disinformation, notably concerning genetic manipulations related to the progress of research on genetics and the ability of surgery to penetrate the human body (to cure, of course, but there is always a possible dark side). These developments raise a large number of issues which are not directly related to that study but which remain very close to the use of robots from an ethical perspective.

The "solution": More technology push; ground and air robots

Robots can be aggressors and protectors. The German Bundeswehr, although on a modest scale, began to experiment with the remote control of small vehicles aimed at exploding under the attacking enemy tanks. It was one example of a "techno-guerrilla," coinciding with the utmost concern about protecting German soldiers.[14] Such weapons can be used in combat to penetrate the enemy defenses just like the "bangalores"—tubes for the penetration of mined defenses, barbed wire and other obstacles used during World War II. The notion of "clearing" is, once again, the same coin with two faces depending on the aim and the circumstances (in war or after war), in order to open the way for offensive forces or to make the ground safer for those who are living there. Deliberately mediatized tests, supported by

12 Samuel Fuller, *The Big Red One*; Steven Spielberg, *Saving Private Ryan*.

13 Peter W. Singer, *Wired for War: The Robotics Revolution and Conflict in the 21st Century* (Brookings, 2010).

14 Horst Afheldt, "Alternative defence postures for NATO and the WTO," in Hylke Tromp (ed.), *War in Europe: Nuclear and Conventional Perspectives* (Aldershot: Avebury, 1989).

Princess Diana of the UK, were useful in the context of the promotion of the Ottawa Interdiction Convention on land mines.

Robots fit perfectly in the American vision of the evolution of warfare and the search for overwhelming superiority in order to create a permanent asymmetry. The notion of a "system of systems" introduced by many strategists in the mid-1990s, among them Admirals Owen and Cebrowski,[15] has created a new dynamic in the elaboration of future weapons systems. Infantrymen of the future, robots and stand-off weapons are integrated in the gigantic web of C4/ISR (Command, Control, Communication, Computing/Intelligence, Surveillance, Reconnaissance). That situation is particularly obvious in the case of air robots.

Initially, Unmanned Aerial Vehicles (UAVs) were used, and still are, for reconnaissance missions and targeting. Gradually, UAVs have been equipped with precision missiles. They perform destruction missions in Iraq (such as the elimination of Abu Musab al-Zarqawi, leader of Al-Qaeda), as well as in Afghanistan, Pakistan, Yemen, and Somalia since 2002. Decapitation operations or "selective targeting" have also been used by Israel against Hamas leaders (Sheikh Yacine, Abdel Aziz Al-Rantissi), and in Iraq (Zarqawi), raising many criticisms.

Introducing ethics: Good and bad robots, relevant or not

There are laws of war. They have been consistently violated, but they have also been partly respected. That imperfect system has saved the lives of millions of people throughout history. Asymmetry has always been suspected as an element of treacherous behavior. The introduction of new weapons has been regarded as a potential break away from the codes and limitations that had been painfully and gradually obtained throughout history. At the beginning of Modern history, the introduction of the *harquebus*, for example, was stigmatized as an illegitimate (evil) weapon. When captured, the special units using *harquebuses* were hanged as war criminals not soldiers.

Is it ethically acceptable to use a robot against a human being? If robots can take on the mission, if computers can do the job at stand-off distances, there is no more fear of contact. If soldiers can fire and forget, one would argue that they feel relieved from moral responsibility and the sense of guilt has been erased. As demonstrated by neurophysiologists, ethics are very much related to sense experience.[16] Others would argue that the answer is related to a different question. In combat, what do you value the most? The lives of your soldiers? The civilian population? In the case of the Israeli Defense Force (IDF) there is a clear concept: Israeli lives come first, civilians and soldiers as well. Civilians on the "other" side stand below on

15 An insightful presentation of the theories and practice is provided in David S. Alberts, John J. Garstka and Frederick P. Stein, *Network Centric Warfare: Developing and Leveraging Information Superiority* (CCRP, 1999), available at: http://www.carlisle.army.mil/DIME/documents/Alberts_NCW.pdf.

16 Nayef Al-Rodhan 2006.

that scale (Palestinian, Lebanese, Turkish, etc.). To Tsahal's soldiers, there is no clear distinction between Palestinian "soldiers," since there is officially no army, only "terrorists." This is also a factor of asymmetry related to the laws of war. The Palestinian authority has only security forces.

Finally, let's consider the ratio between technology and ethics from an efficiency standpoint using ethics/efficiency as a paradigm. Relevance and efficiency are two different issues depending on the nature and the phase of the war and the characteristics of the combat situation.

In the use of robots, some ill-defined ratio has to be respected between the efficiency and the psycho-political impact of their effects. In the Federally Administered Tribal Areas (FATA), US operations are quite successful from an operational point of view. But they trigger rage from both Afghan and Pakistani allies who feel compelled to condemn the operations. They have not been informed of the operation. They have no power to refuse and therefore they cannot endorse them. They appear incompetent and powerless. This creates a perfect environment for damaging disinformation and political propaganda. The challenge thus is in the ratio between the impact of successful military operations and the resulting political benefit.

If we consider counter-insurgency, even counter-terrorism, or the fight against drug cartels, two intertwined elements are crucial. First, one must win the battle for the "hearts and minds" of the people. Second, one must gather as much Human Intelligence (HUMINT) as possible. That is the kind of job robots are not made for. Just like the much vaunted and criticized U-2 high speed airplanes, UAVs offer valuable data on the nature of the movements of the enemy (if the enemy has not taken cover). However, this author remains convinced that the goal is to maximize the quality of Signals Intelligence (SIGINT) and other data collection by robots by integrating that data with HUMINT data.

To make it clear, the information provided by the average countryman has to be compared with the information obtained by drones on a permanent basis. If the two coincide, then it creates reliable information for an efficient mission. If the two don't coincide, it becomes a wasted asset or, worse, a recipe for major damaging mistakes both in terms of ethics and psychological efficiency. No one can pretend to kill people by robots and conquer "the hearts and minds" of the locals. It has become a major problem today in the area between Afghanistan and Pakistan (FATA) related to the uncertainty about the right strategy: counter-terrorism or counter-insurgency. The error factor related to "collateral damage" induces political problems because they are used by hostile propaganda. At the same time an additional question is evoked: Who is responsible (ethically as well) for what, according to which rules and principles?

When it comes to robots, a crucial element is that they are less reliable than the human brain. The risk in the forthcoming war will stem from the incorrect belief that a robot is more intelligent than a human being. This will not create a futuristic domination of man by robots; it will only result in more confusion, and more mistakes. Robots are not autonomous and ethically responsible for their acts. Robot "mistakes" refer to those human beings who are guiding them with difficulty.

There is no such a thing as "the system did not work properly." In addition, one's interpretation of robot data might be wrong. This then becomes a human failure—"robot originated data" should not be considered more valuable than human data. To conclude, one risk is to develop, sell and produce new expensive systems for technologies which are non-usable (unfriendly to the user) or needless because they are irrelevant in a combat situation.

The major mistake is to ask the situation to adapt to our capabilities, rather than ask for capabilities that actually fit the real situation.

Conclusion: War Transformed Symmetry and Asymmetry in the Information Age

Asymmetry in info-com and psy-ops have always been an important element of warfare. In 1960, during the war in Algeria, the new transistor radio helped in countering the military coup. In Vietnam, US privates could hardly phone their relatives. Now they have cell phones, and use Twitter and Facebook. Tomorrow, other more sophisticated devices will be at hand. The traditional notions of propaganda, psy-ops and other activities based on communication have become partly obsolete for two reasons.

First, humanity has entered the information age. We are already different people. That evolution means that we are now living, competing and fighting in the information age: everything becomes different. Above all, new perceptions, new relations to the information are developing at a rapid pace. A new citizen is emerging. His requirements and awareness have become different.

Second, the last twenty to thirty years have created a new information communication environment (satellites, cell phones, the internet). A new domain of confrontation has emerged: cyberspace. New vectors (the internet and Facebook, Twitter, and Bluetooth) create new opportunities. With them come new capabilities for aggression, along with new rules—or rather the absence of them, since no international law rules the use of cyberspace. There is no symmetry in the use of those vectors. Therefore, the issue is not about upgrading or adapting. It is about transforming the psy-info-com system and giving it the correct importance, according to the real stakes, reconsidered in depth on the basis of the potential of new technologies and of the lessons learned from the tactics of different kinds of enemies. 1990–1995 can be considered a nexus. CNN provided for the first time "almost" real time coverage, although very limited, of the Kuwait war. That performance improved rapidly as demonstrated by the lessons from Somalia and the Balkans. The Department of Defense (DOD) created "media pools" in Kuwait (1990), then the embedding of journalists in Iraq (2003), the success of which is arguable in terms of objectivity and critical distance.

Western militaries are still relying on concepts and doctrines which belong to the Cold War, at best. The enemy, terrorists or guerrillas have no problems with doctrine. As irregular forces they are flexible and opportunist. They use technologies when they need to and are innovative according to the situation and their specific needs. They just practice info/psycom/ops in a spontaneous way, which is at best coordinated at the tactical level. They do not need a doctrine. In the case of Al-Qaeda, it is enough to have a general ideology established and regularly dispatched by a few respected *imams*. From time to time broad strategic orientations are indicated. Therefore it becomes urgent to review the traditional conception of psy-ops and their role in war fighting as well as peacekeeping operations in the information age.

The basic principles are not new. Since World War II, plugging in to the media of the enemy to introduce information and disinformation that will be communicated through their channels are classic operations. Terrorism does the same. New tools of communication allow the creation of new strategies, new tactics aimed at destabilizing the enemy, and at penetrating inside his own country, creating troubles with his allies and friends. Disinformation is certainly not new as well. However, its power has increased in an amazing way because of the sky-rocketing importance of communication during a conflict.

The enemy learns at the same time ways to benefit from the same new technologies, and discovers new vulnerabilities in our communication info-systems, the way we use it or consume it. In Africa and the Middle East, "warriors" are being paid in photographs. The opposite happens as well, the enemy can even pay reporters to exhibit those photographs that will create shock at home. We should recognize that the traffic of photographs and videos plague the most important and respected magazines. Media are not only in the loop of war fighting, they are creating a new mode of confrontation on a new battlefield. Info-com actions of the enemy have to be countered permanently. First, by occupying the communication battlefield. Second, by anticipating its operations—keeping the initiative by creating surprise in order to destabilize him. Boyd's model mentioned above should apply to information-communication. In other words, the benefits of asymmetry should be on our side and not that of the enemy. As demonstrated by Al-Qaeda, Hizbollah, Hamas and the Taliban, communication strategy and tactics become as important as operations of destruction. Some would argue that they have become even more important. That is, of course, a never-ending debate. However, there is wide room for accepting that they are more complementary than ever.

It is equally crucial to recognize that these traditional practices take place in the information age and that technology has been able to create a new immaterial dimension, cyberspace, which can produce physical effects bearing on the life, prosperity and security of individuals, organizations and states. Cyber-warfare is at an experimental level, as demonstrated in Estonia, Georgia and Moldova. In a few years, cyberspace will become one of the main battlefields, contributing to the dangerous confusion between peace time and war time. Some think in terms of symmetry, others envisage the benefits of asymmetry. In reality, so far, we have little idea of what could be a High Intensity Information Conflict in the future.

Symmetry and asymmetry appear to be the result of complex situations combining many elements, coming from different levels and different areas (public opinion, the relevance of armed forces, the "morale" and the quality of the political and military leaders). Technology by itself does not create asymmetry or symmetry. Finally, once a war has begun, all these factors are affected by the variation of events, unpredictable by definition.

PART IV
Long Cycles and Major Power Conflict

Metastrategy: Sociostrategic Systems in the West

Jean-Paul Charnay

Introduction

In 1990, I published a book with Economica entitled *Métastratégie: Systèmes, Formes et Principes de la guerre féodale à la dissuasion nucléaire*. The book attempted to articulate and synthesize the doctrines and practices of war and revolution in the West: How does one order the historical proliferation of wars and revolutions at length without outlining a simple chronology? Can one bring together geopolitical concepts, operational strategies, the sociological factors and tactics?

The third section, "Principles," analyzed the reversibility which occurred between the classical "principles of war" (concentration and economy of force, leadership and security, freedom of action), and the new "principles of deterrence" (destructive capacity, postponement, credibility). This section sought to reveal the interchangeability of classical principles of war and new principles of deterrence. The second part describes the morphological changes of fighting: the order of battle, geostrategic maneuvering, "major" or "great power" war, and revolutionary struggles.

The first part presents the sociostrategic systems through empirical sociology by arranging large *macro historical periods*, the succession of conflicts and overlaps according to two criteria: the *intensity of reciprocal negations* between opponents (its total destruction or a simple re-balancing of power), and *a scale of constraints* (modes of force, military or not, and geographical and social theaters where this power is exerted). It lists the Western "sociostrategic systems" according to philosophical definitions, tactics such as "material reification of the adversary," and strategy such as "a more or less extreme negation of the Other."

Twenty-three systems were accordingly defined, from the fall of Rome to the current intra-continental projections, basing itself not on the history of wars, *but on the mutations of the art of war through the major periods of Western civilization*. And so discussed are the barbarian invasions, overseas settlements, feudal contractions, the Reformation, the major monarchies and nation states, the Enlightenment and revolutions, total wars, colonization and decolonization, and deterrence and globalization, all of which raise questions about the impact of civilization, and the acceleration or the complexity of the story.

The definition of socio-strategic systems results from a historical statement: it must be noted that these systems coincide neither with the great epochs of civilization (Late Antiquity, Medieval, Renaissance, Baroque and Classicism, Enlightenment, Revolutions, Nationalism, Globalization), nor with the industrial changes in arms, nor with the chronology of wars *per se*. Because throughout the same conflict, socio-strategic systems replace themselves, can overlap, and are of great variability, over several decades or centuries. They are thus constructed through empirical sociology by combining three plans.

The macro strategic stages

Starting with conflicts occurring in the West itself, or in confrontation with other civilizations, we set off in history various stages induced by demographic transfers, economic upheaval, religious or cultural changes, and ideologies of violent protest (revolutionary, subversive, and anti-revolutionary doctrines).

Thus we have divided the historical axis extending from the end of the Roman Empire to the present according to 23 stages—16 having since elapsed, and seven which are in progress—for which we are unable to attach outermost dates, each stage extending itself due to recurring events. They were therefore structured on known historical facts, made up of events either initial or final and on symbolic dates. These events describe a temporality of open duration and not a chronology.

The intensity of negations

In every phase, we must evaluate the "intensity of reciprocal negations:" identification, coexistence or the struggle with the "Other." Often these reciprocal negations do not coincide with each other. An opponent might want to just take over a province or to subjugate a population, while his opponent wants his death or absolute defeat. Such was the unconditional surrender of Nazi Germany or Japan in 1945. The will to destroy the other cannot be objectively determined. The extent of destruction (on a human scale) depends on the overall nature of intolerance and the extent to which each opponent is willing or able to subdue his enemy. The nature of destruction will differ in intensity depending on whether or not one or both of the antagonists accept the prevailing sociopolitical and normative frameworks.

Impact inflicted in conflicts varies in intensity; on a scale from 1 to 7, the variation ranges from the development of all communities in a comprehensive peace (0) to the risk of de-humanization (7). The intermediary levels are: the knockout of the organized opposing forces (the destruction of the weapons of war) (1); the annihilation of the enemy's will to fight leading to "decisive" victory in the envisioned strategy of the conflict (2); partial conquest: territorial, economic,

... (3); the disintegration of social structure and opposing values, resulting in the formation of a new ethical and institutional order (4); the destruction of the society, physical liquidation, or else the "conversion" or dispersion of survivors (5); reciprocal attacks against the potential for growth—that is to say the risk of decline or change in the state currently achieved by humanity if the main poles creating progress suffer irreparable impact (6); and finally, this ends in damaging the very chances of survival for humanity (7).

Clarification is needed. Variations in the intensity of the negation of the Other do not necessarily reside only in his destabilization or in his radical destruction (genocide in extreme cases), but conversely in a search for an alliance, or a union which looks forward to a symbiosis with the Other (-1). This poses the problem of the harmonization of the values and social structures of the Other with one's own values and social structures. However, in logic, if not in practice, any effort to persuade the Other, meant to improve friendship or forge a union, also represents a desire to change a prior situation; so it represents a form of constraint exercised upon the Other. Thus non-violent strategies are located on the fourth level of the previously mentioned scale: the construction of new structures and values.

On the other hand, these conflicts can be regrouped into two broad categories: those that define external socio-political groups opposing one another (tribes, nations, empires and religious confessions) can be called *vertical struggles*, while those defining confrontation between differentiated social strata within the same society (clans, orders, classes, ...) can be called *horizontal struggles*. In practice, the possibilities of interference in societies are numerous.

Scale of constraints

In what areas, upon which political or military entity, social class or economic structure, will changes be successful? The stages can be determined in relationship with strategic phenomena and not in relationship with political, economic, and social history.

At the foundation lies the classical instrument of strategy: organized force (the army) (a), in which the more or less military character and importance varies according to different societies. In (b) appears the most basic display against all attacks: a habitat-based defense, more broadly, urban planning: strong houses, fortified towns, fortified camps. The consistent association of fortification systems in geographic space shapes the organization of this space (c), in connection with the natural features of the region and the scale of the theater of war that varies in accord with extensions permitted by technology. This organization of geographical space has, for a long time, covered the ocean surface, then extended to undersea and air, and now space. These three elements constitute the normal "environment" of "great war."

The next three constitute the field of indirect strategy. In (d) appears the nature of individual protest, in which propaganda, subversion, terrorism, etc. bring forth a wide range of disobedience which will eventually institutionalize itself. Hence, in (e), struggles at the political structural level: their remodeling increases the combat capability (dictatorship of the proletariat, revolutionary committees, "Motherland in danger," systems of political mobilization, etc.). Then the economic organization which is rationalized in order to support the war effort or just the arms race, or competition for markets, raw materials or prestige in science and technology. Then comes the pressure from social struggles (f).

Appearing above the two stages that reach the extremes of human life: the individual, and the international order. Reification of the individual (g), which is about the phenomenon of mass psychology or population movements in which most of the potentialities of the person, multiplied by the number, are collected together and organized as an instrument of struggle. Any closed society—the city, sectarian religion, the single party—contributes to a relative reification. This is also developed by regulating the distribution of the population: from the suppression of individual freedom (hostages), to expulsion, mass exodus or encirclement of populations.

Finally, the last stage, the recasting of the international order (h) combining the various human communities in accord with a hierarchy and rules predetermined by one of the adversaries, or else by multilateral acceptance.

Bygone Systems

Romano-barbarian kingdoms of the feudal pyramid: From the Catalaunian Fields to the beginning of the Hundred Years War (451–1337)

Germanic and Roman civilizations and methods of warfare opposed each other in the Late Empire. But they also merged, and at the battle of the Catalaunian Fields both pushed out the Huns of Attila. The Germanic chiefdoms were romanized and christianized but constituted a dissociated and fragmented geopolitical entity where wars and raids continued on waste land punctuated by stone fortification as multiple points of resistance (castles, cities hosting fairs, monasteries), "capitals" of fiefdoms nested within the feudal pyramid between feudal lords and vassals marching to war on their call. The old distinction between peasant-citizen and slave was fused into the institution of servitude, but it was humanized by the Church which restricted the use of force (knighthood) between the lordships of which the most powerful became kingdoms, as was Edward III's claim of the crown of France in order to remove his French fiefs from under the suzerainty of Philip VI of Valois.

Renovatio imperii and temporal kingdoms: From the deposition of Romulus Augustus to the Outrage of Anagni (476–1303)

The dissolution of the Western Empire did not abolish the hope of a restoration of the empire. Three renaissances were attempted: from Byzantium, the Justinians who took over Rome in the sixth century. Then the Romanized Germanic kingdoms achieved their own imperial renaissance: the Carolingian experience of the ninth century, which defended Christianity against the Moors and converted Upper Germany but exploded in the Lotharingian partition (Treaty of Verdun, 843); the Ottonian Saxon experience in the tenth century gave birth to the Holy Roman Empire. In the struggle between the Papacy and the Empire, the papal authority wanted to change the *pax romana* into the *pax evangelica,* and declare itself *rex regnum*. This project triumphed at Canossa (1077), but having to rely on weaker kingdoms it sanctified their kings, who eventually rejected its secular significance (except in Italy: Papal States), as exemplified by the Outrage of Anagni, committed by Philip the Fair against Pope Boniface VIII. The idea of the empire fell apart in the West during the Great Interregnum (thirteenth century), as in the East, where Byzantium was faced with local empires and the Turkish advance.

The implementation of overseas incursions into Normandy in the West and the Franks' principalities in the East: From the siege of Paris to the fall of Saint Jean d'Acre (845–1291)

From the eighth century, Christianity lost both the eastern and southern fronts of the Mediterranean as well as the western islands to Islam. The Arab-Berbers settled for centuries in the Iberian Peninsula. The Italian maritime cities and the East continued to trade or capture each others' vessels. Then, coming from the north, the Scandinavian populations followed the inlets and river systems to carry out raids that resulted in remodeling a military topography now focused on fortifying the cities, meant to become safe havens, and protection of transportation (fortified bridges). The conversion and implementation of the Vikings in the territory and in the feudal pyramid led to a demographic renaissance that led to dreams about the "purification" of the Holy Land and of the riches of the East. The sanctity of holy war, the *Crux Transmarina* and the departures of the Crusades led to the establishment of Frankish principalities in Palestine, a new art of war, the formation of terrestrial and maritime networks and enculturation. But despite the "sacred militia," the great religious orders and transnational militaries, the demographic weakness of the Latin principalities of the Orient and the European *Realpolitik* of Western kingdoms led to the victory of the Counter-Crusades. Meanwhile, in the eleventh century began the Iberian *Reconquista,* with its alternate periods—as during the Hundred Years War—of battle and of truce, and with its shifting alliances.

Feudal decline and communal movement: From the last Crusades to the end of the Hundred Years War (1250–1475)

At the death of the "last" crusader, St. Louis, a pluralist Europe dispersed its centers of power between the kingdoms and the great feudal fiefdoms that competed using armored cavalry riding through flat country, or marriage strategies. But it remained unitary through its intellectual networks, universities and Latinized monasteries, and the commercial networks that connected cities who bought their town charters from their lords. Their militias stayed ineffective in open country, but fortification made great progress. Through this geographic space major pilgrimage routes also unfolded, and benefited a popular mysticism propagated by wandering populations soon to be repressed by the *crux cesmarina*, the internal repressive crusades (such as the one against the Cathars) carried out by the Inquisition and the knighthood. In contrast, Du Guesclin, Constable of France, fought against the English a war of attrition based on stratagems, while the realist policy of the great rulers subordinated the remains of the religious and military orders. The end of the Hundred Years War was signed at Picquigny between Louis XI and Edward IV, still involved in feudal wars (Burgundy, Deux-Roses). The papacy failed in its two dreams of re-conquest: theological over Eastern Orthodoxy, military over the Byzantium Ottoman.

Urban turmoil and peasant terror: From the Great Jacquerie to the Great Fear (1358–1789)

When the major structures of the late Middle Ages, the Church and kingship, lordship and county, could not provide sustenance for the "scrawny," the populace, misery provoked explosions and mystagogic exaltation: the revolt of the Shepherds' Crusade, of the Cabochien revolt, Lollardy, and the revolt of the Ciompi—peasant revolts in Germany in the sixteenth century, provincial ones in the seventeenth century as far as Russia, revolutions in England. When the disassembled "common" took arms in community settings, some similar points appeared: the revolt of hopelessness, the aspiration to achieve justice with a more humane order against the High Church and the princes (Joaquim de Flore Savonarola, Wyclef and Huss) by mass struggle, organized by parishes or corporations under the military leadership of a few nobles or clerics leading to the capture of towns and castles and the burning of archives. This was a long period of underpaid craftsmen or oppressed peasants, of people living in abject poverty or as outcasts, and a period of repression after the powers in place, surprised at first by the social movements, regrouped their soldiers. Later, a few survivors received letters of remission, sometimes with a raised status.

Appearance of the "Great War" in the temporal kingdoms: From the Treaty of Troyes to the treaties of Westphalia (1420–1648)

Humiliated by the Treaty of Troyes which recognized King Henry V of England as King of France, divided between Armagnacs and Burgundians, then having had to face the Spanish Habsburgs and the Empire, the kingdom of France built itself into a structure based on its institutions (regular army: Company Ordinance, 1439) and, in a series of wars in Italy with the papal states, the city-republics and their *condottieri,* and with the Imperials (Thirty Years War), undertook a transformation of the art of war to the art of the "Great War:" a regulated and salaried infantry once again the main weapon on the battlefield, fortifications buried because of the power of the artillery—a weapon reserved, due to its price, to the great rulers repressing the last of the feudal revolts (the Fronde, 1650–1653)—organization of geographical space through networks of castles and fortresses or through the drive towards natural boundaries. Gold in the Americas, and powder as energy, gave rise to modern armies and projects to establish peace through the balance of power. The Habsburgs continued over the centuries in the Danube basin and on the steps of the Empire a type of war where a few great sieges, battles, and "limited war" actions were waged by light militia troops.

The dissociation of the Reformation: From the Peasants' War to the War of Camisards (1524–1705)

A complex system endorsed the second fracture of Christianity in favor of temporal princes *(cujus regio, ejus religio*). The wars of religion involved implacable negations of the Other, the largest geopolitical space (the Invincible Armada, the Swedish dream of a Protestant Baltic empire, the independence of the Netherlands, the conquest of Scotland and Ireland by England), the dogmatic controversy (the theologian becomes statesman: Calvin in Geneva, Luther an adviser to princes), setbacks in political alliances and emigration (the *Mayflower*, the Huguenots) following massacres that got out of the control of those who had inspired them and following the radicalization of political acts (tyrannicide), but also the secularization of political and humanist views (edicts of tolerance) and the strengthening of national military institutions by means of dismantling religious strongholds. The dissociations were gradual, ranging from extremist heresies (Baptists, Presbyterians) to canonical schism (Anglicanism). The papacy responded with the Counter-Reformation (Council of Trent).

The classic wars between great monarchies: From the Thirty Years War to the partition of Poland (1618–1795)

The secularization of European monarchies continued through their centralization, their internal use of vernacular languages instead of Latin, the establishment of a legal system for the people, a concern for the protection of civilian populations through a new legal system regulating war (Spanish theologians, Protestant publicists), and the economic organization of the nation (Colbert, Dutch Mercantilism). Wars became conducted primarily by "entrepreneurs"—Wallenstein rented his army to the Emperor, Richelieu "purchased" that of the Protestant Bernard of Saxony-Weimar. The infantry moved away from the Swiss model of the *tertio* column battle arrangement, adopting the Spanish combined use of pike-men and musketeers, and alignments of riflemen: these were the times of controversies over thin or deep, parallel or oblique layouts. The soldiers were given uniforms and subjected to drills, the artillery became lighter. War was aimed at the conquest of a province and the annexation of the small principalities that were still scattered between the great nations that were in search of historical or natural borders. Their cost led to the organization of theaters of operation (networks of roads and forts, food depots, and a consequent controversy between the geographical and geometric schools). Two poles of power appeared in the east—Prussia and Russia, which absorbed Poland—and two disappeared in the Mediterranean: Venice and Malta, which had been arming against the Barbary piracy.

Enlightenment and liberal revolution: From the Encyclopedists to the Libertadores (1751–1826)

The two English revolutions of the seventeenth century (the power of Parliament against royal absolutism, Catholicism and Anglicanism against Puritanism), then the French Revolution, brought deep inner turmoil. The American revolutions were first wars of liberation. 1751: between the Wars of Succession of Austria and the Seven Years War, that opposed great dynasties, Diderot and d'Alembert launched the encyclopedic movement, measured the progress of the human mind, assumed guidance by reason as a method, taking into account the evolution of technology. In their chaotic competitions, absolute monarchy and enlightened despotism they tried to organize the manufacturing industry with the early colonial capitalism and banks. But this degraded the society of orders (clergy, nobility, third estate) in favor of a fraction of the latter: the enlightened middle class, some of which would err into revolutionary terrorism. After the intellectual failures of the Reformation and the revolutions of the seventeenth century—the English political one and the agricultural technological one—the late seventeenth century provided the precursors of the first revolution in manufacturing and a (controversial) Atlantic revolution encompassing on the one hand American independence, and on the other hand, the various stages of the French Revolution with its internal and external "days and battles" "great war" which escalated death, rendering it massive (citizen-soldiers and peasants granted *seigneurial* rights), at the expense of "small war" (ambushes, skirmishes, harassment) that would still be carried out during the revolutions of independence by the Latin American *Libertadores*. After fighting small wars and raiding forts across an immense geographical expanse, the insurgents, backed by both the French government and the advocates of

Enlightenment, prevailed, while at the Congress of Panama (1826) Bolivar could not obtain Latin American unity.

The conventional wars between nation states: From the first Coalition to the outbreak of the Triple Alliance (1793–1914)

After the declaration by the Girondins of the war "against the king of Bohemia and Hungary" bringing about seven Coalitions against France for the ensuing years until Waterloo, European wars were primarily national wars. With the Emperor Napoleon, the art of the "Great War" conducted with regular armies, articulating strategic maneuvering and decisive battles with great fire power, reached its classical age. This art combined an industrial production of weapons, an expansion of conscription, a preliminary planning of mobilization, with transport by land, river and rail, a new representation of geographical areas on the vast European plain from northern France to the Niemen. A certain "humanization" of war was organized (Red Cross, The Hague) for the wounded and prisoners. All this was facilitated by capitalist manufacturing, which increased economic competition (the Continental Blockade against England) and the colonies. But beneath the diplomatic game, later played at the League of Nations, the great states were now faced with dual rivalries on two sides of their borders—old (France/Habsburg, France/England) and new (Russia/Japan, Germany/England). After Greece freed itself, Austria and Russia chased the Ottomans away from the Balkans. In 1914, Italy withdrew from the Triple Alliance (1888) and abandoned the Central Powers to the *Entente Cordiale* and the Russian "steamroller." The use of artillery spread, and chauvinistic doctrinal controversies arose on who was the best strategist-captain, Frederick or Napoleon, or on Clausewitz: which should have primacy, the offensive strategy or the defensive?

The British peripheral strategy: From the Invincible Armada to the first battle of the Marre (1585–1914)

Without a land border defense system, but equipped with iron, wood and coal that would build the *capital ship* of naval power, the *ship of the line* and the *Dreadnought*, the British would affirm their ambition to trade with the whole world (economic liberalism) based on the principle of the freedom of the seas. The command of a naval force was the key. This guaranteed the import of what it did not produce and the export of what it did produce. England was thus laying out a planetary system: banking power, contingent alliances against any power that tended to challenge its supremacy (notably France and Germany, considered disruptive forces in the world), limited use of military force, from the installation of harbors that were naval bases and counters at the same time (island and peninsular strategy) acting on the interior of continents to the projection of genuine expeditionary corps. But in 1914, England was obliged to engage real armies in the ground battles of total industrialized warfare.

The projection of Europe on the planet. The colonial conquests: From Christopher Columbus to Mussolini (1492–1936)

The Iberian Re-conquest did not continue in North Africa. The Ottoman thrust into central Europe deterred possible resumption of the Crusades. Then, disunited Christian kingdoms launched competing discovery expeditions and conquests of new worlds, first based on mining and commerce, then shifting to a mainland strategy using two basic tactics: control waterways by means of nautical science (with wind, then steam, power) and the use of gunpowder and firearms. This "coverage" of the planet (excluding Japan) by European powers was accomplished over four and a half centuries in several phases: establishment of lines of competing domestic outlets for local trade; military penetration into demographically dense continental territories by mobile columns supported by strong points (which occasionally experienced harsh military defeats);[1] centripetal organization of colonial empires in their relations with the colonizing mainland (mining predations, export agriculture, linguistic and fiscal constraints, military constraints such as the presence of or conscription of natives into the colonial armies); links between the various colonies (route to India, Euro-African networks, march towards the Pacific in the American West and Russia, ...); diplomatic agreements between the major colonizers over their areas of influence. Colonization effected the largest transfer of civilization ever carried out on the planet: values and techniques asserted against "colored people," discovery of non-Western civilizations, triumphant science.

Empires, nationalities and minorities. The circle arc of North Cape/ Bosphorus: From the Congress of Vienna to the Anschluss (1814–1938)

After two decades of revolutionary and Napoleonic Wars, the Congress of Vienna attempted to establish a community of nations which is in fact the balance in *realpolitik* of the great powers. Intermediate states were established as more or less neutral: Belgium, the Netherlands, Swiss cantons. In fact, two types of tension and military conflict overlapped: the revolutionary principle of nationality and the right of peoples to self-determination. The first results were centripetal movements of large populations identifying themselves with a strong state, its army and its language: Germany, Italy, France (with Savoy and Nice), Poland in 1918. On the contrary, the second result was the centrifugal aspirations in the ethnically and culturally diverse empires, the temptations of autonomy or even independence for minorities supporting their identity with cultural romanticism, language, literature, history: as during the "Spring of Peoples" in 1848. In the nineteenth century

1 Jean Paul Charnay, *Colonial Disasters* (Paris: Les Editions d'En Face, 2007).

regular armies alternated some "great wars" with bloody reprisals against liberal or anarchist claims that exploded in urban riots with barricades. The disintegration of the Austro-Hungarian and Ottoman empires after 1918 redefined the "Eastern Question" ("*la Question d'Orient*") by releasing small ethnically complex countries: the Baltics, Central and Eastern Europe, the Balkans, along a line once named the "Tragic Diagonal of Europe" running from the North Cape to the Bosphorus.[2] This disintegration endorsed by the Treaty of Versailles was used by Nazi Germany to reclaim its ultimate irredentist targets (Sudetenland) into the Third Reich (believed to "last one thousand years").

Socialism and the proletarian revolution: From the Montagnard Constitution to Stalin's purges (1793–1936)

With the first Industrial Revolution (coal, steam, steel and oil, electricity, aviation), the nineteenth century witnessed the beginning of the transfer of peasants to manufacturing cities, changing the system of ethics and beliefs of societies structured into orders and transformed into societies of classes and "class struggle." Against the great conventional war doctrines, a multiplicity of doctrines of revolutionary struggle asserted themselves. They were in search of a new social organization bringing happiness to the wretched of the earth—hence the proliferation of diverse philosophies and social movements. Some constants appeared in the debates about strategies: the leading role of an organized party (dictatorship of the proletariat) or anarchic impulse; denial or use of the legal system; seizure of power through coup or armed insurrection; the general strike as myth or reality; revolutionary days or permanent revolutionary agitation; intellectual propaganda or individual terrorist attacks; the threat of the impoverishment of workers and middle classes countered by state socialism; controversies between the revolution in one country or permanent revolution; a vanguard class (workers and intellectuals) versus collectivized peasants against the Russian bourgeoisie. In contrast, revolutionary workers' revolts were repressed after several attempts by the police and the military, the archetypes being the revolt of the Canuts of Lyon and the Paris Commune of 1871. The theory of the party giving the masses a class consciousness and strategic direction is developed in the conflict between the advocates of revolutionary force versus the advocates of social reform to be opposed against conquering capitalism. The remaining stumbling block was that internationalism, in its anti-imperialist struggles, could degenerate into social conflicts driven by chauvinistic centralist party bureaucracies. The century of proletarian revolutions blended with total wars.

---

2 Jean Paul Charnay, "La Diagonale Tragique de L'Europe," *Geostrategique* 8 (July 2005), 33.

Total manufacturing war: From the American Civil War to Hiroshima (1861–1945)

The French Revolution had ignited the war by the mass requisition of mobilized males. This war of the masses began its ascension at the height of the first Industrial Revolution: the clash became mechanical. A new type of war became established in time (through anticipation, the production of weapons is planned to make them most efficient just when they need to be used), and space (troops are deployed in the field by the new vehicles of transport and combat: rail, truck, tank and airplane). The battlefield coincided with the theater of war, which reached sub-continental dimensions, sub-continental theaters being connected by transoceanic fleets and naval air forces. These wars became total wars because of four exponential factors: increased firepower reaching civilian populations, geopolitical remodeling of the vanquished to be reset on the chessboard of balance of power, psycho-ideological excitement by propaganda leading to the negation of the opponent, which renders necessary measures ranging from demanding unconditional surrender to ethnic genocide. Fierce economic competition was legitimized by the nobility of political philosophy: the abolition of slavery during the Civil War, the awakening of the peoples of the East during the Russo-Japanese war, German nationalism turning into congested discriminatory racism during World War II, capitalist and democratic nations fighting for world freedom. After the failure of the democratic pacifism of the League of Nations, UN multilateralism entailed the sharing of the world into two blocs, but also the shelving of total war because of the total manufacturing war being rendered *instantaneous*, assuring the mutual destruction that would result from the use of the atomic weapon.

Decolonization: From the insurrection of St Domingo to the end of the Portuguese empire and the fall of Saigon (1801–1975)

The expansion of the peoples of Western Europe over the whole planet had extended over several centuries. Heterogeneous by their ethnic groups, languages, legal systems, and civilizations, colonial empires exploded after World War II when their weakened colonizing mainlands could no longer retain demographic, ideological and economic control. Their reflux was accomplished in less than half a century. In fact, the process of decolonization from the outset was concomitant with the conquest of phases specific to each specific country, but that followed a similar pattern—from the early nineteenth century: revolt and brutal repression and progressive abolition of slavery; from the early twentieth century: nationalist struggle using the values of the colonizer (human rights and the right of people to self-determine); coexistence of armed struggle, partisan activism and diplomatic action supported by internationalism and anti-imperialism, followed by positive neutralism; systematization of a renewed type of warfare: guerrilla, extending into a people's war and revolution, prompting a reply and psychological and counter-guerrilla warfare, with very diverse types of violence ranging from individual terrorist attacks sometimes prompting torture as a preventive measure, the urban riot, or real war (Vietnam, Algeria). The term decolonization has several meanings: restoration of self, reinstatement of an internationally recognized sovereignty and, in a narrower sense, the historical process by which the reinstatement was made and the intellectual movements rationalizing the ex-colonized desire for Western techniques, while affirming the native culture, language and religion. Hence the controversy about the need for the decolonization of the historiography of colonial times, supposed to have legitimized colonization, which leads, in turn to the demand—as a counter-offensive tactic—for the repentance of the former colonizer, and the accusation, uttered by local oppositions against their governments, that former colonies remain subject to economic and cultural neo-colonialism.

Contemporary Systems

Ideological and nuclear deterrents: since Hiroshima, 1945 ...

The Cold War, transposed into a peaceful coexistence, contained the counter-weights for the balance of the Blocs. The two pillars of this balance were technological equivalency (Mutual Assured Destruction) and an ideological confrontation (political and economic liberalism versus anti-imperialist communism). Nuclear deterrence was a praxeological break in strategy: the doctrine of non-use of ballistic nuclear weapons strengthened the stabilization, ensured the sanctuary of the country and its vital interests, the use of limited conventional war being driven back to gray areas, the periphery. A paradox that had to be accepted. Hence the need for a pedagogy that led to extensive casuistry: first or second strike, a warning shot or graduated response, anti-arms or anti-city targeting, intercontinental missiles or weapons meant only for the fields of operations, and more. But America had pursued such a sophisticated technological escalation that the Soviet Union, facing its own economic crisis, pressures from popular movements and the war in Afghanistan, could not continue to compete, demonstrating the weakness of its economic ideology. One may also doubt that non-war between the superpowers since 1945 resulted from the nuclear deterrent and whether "classic" deterrence still stands today. Conversely, under the system of general deterrence between the superpowers, it may seem that a system of deterrence by regional weapons, real or virtual (Israel/Iran, India/Pakistan, North/South Korea, ...) has been established. In fact, despite the environmentalist and anti-globalization denunciations, public opinion in Muslim countries thrives on the denunciation of the nuclear *apartheid* maintained by the major powers against nuclear and missile proliferation in the name of space security and the geopolitical balance. Nationalist fervor, religious faith, atomic blackmail, a fear of the "mad" or "rogue" leader, challenges the indemonstrable assumption that stabilization is guaranteed by nuclear deterrence and ideology. The United States has hesitated between the installation of a "missile shield" that would strengthen the capacity of its rockets and a partial nuclear disarmament with Russia (START, 2010).

Para-manufacturing wars in the developing world: Since the Arab-Israeli War, 1948 ...

General deterrence and peaceful coexistence drove conflict between "little forts" back into the areas of internal warfare where ancestral fractures are revived or where new breaches appear in the borders that emerged from decolonization. These wars were fought with weapons imported from industrialized countries and remain wars of attrition: after the initial fighting, the parties must restore their weakened armed forces (Morocco/Algeria, China/India, Iran/Iraq, India/Pakistan, Congo/Rwanda, or possibly as during the explosion of former Yugoslavia). But since 1990, these wars have partly changed: emerging countries are increasingly producing their own armament and conflicts oppose local armies against expeditionary forces projected by industrialized countries toward the end of their lines of communication: USSR/Afghanistan, Serbia defeated by NATO bombing, the two Iraq wars, the Coalition invading Kuwait in 1990 and again in 2004; Coalition in Afghanistan. This is asymmetric warfare where, after the victory over the enemy's military strength, restructuring for peace through *high technology* and *management* fails to curb the guerrillas and self-sacrificial terrorism. The so-called "attribution wars" were sometimes financed by drugs and carried out by means of computer communication, but exalted the ideological denials that are not contained by anti-insurgency tactics.

Areas of civilization and para-economic wars: Since the conference in The Hague, 1899 ...

From the conferences in The Hague to the Geneva Conventions of 1949, the international concert tried to humanize the law of war because warfare, made more and more deadly by the arms industry, hacked away at military as well as civilians. That legislative effort confirmed the system of values and laws in the West, but it was imperfect in regulating competition between capitalist economies, anti-imperialist and liberal, fascist or communist regimes. Hence the regional regroupings trying, below the international flows of common trade interests, not so much to achieve self-sufficiency as to establish a favorable balance between imports and exports, and to ensure a stable currency. Also, a country could be coerced more easily through its debt or essential aid than by arms. The attempts at global economic regulation were conducted in conjunction with international democratization by the League of Nations, then the United Nations, and with negotiations on disarmament and arms trade. The result: measures were taken against the illicit brokering of small firearms (for guerrillas) and technologies of weapons of mass destruction (proliferation agreements), and the organization of an international legal regime against great leaders accused of genocide, crimes against humanity or war crimes. At the same time, the distribution between rich and emerging countries, developing countries and fragile or failing states, between countries accumulating riches and countries subject to the international division of labor and production, leads to the emergence of "a Fourth World" with a dual facet: the proletarian nations and the countries that are still rich, where outsourcing and unemployment has created a new class of underprivileged coexisting with illegal immigrants. In addition, crises are being caused by the volatility of global financing flows and the environmental degradation of the planet. Meanwhile, a controversy is swelling over the anthropological specificity or universality of values from the humanism of the Enlightenment and of the nineteenth and twentieth century European Socialist traditions being applied to other civilizations.

The revolution infrastructure in industrialized countries: Since the Spanish War, 1936 ...

In this system, the concept of revolution is ambiguous: in its strategic sense, the seizure of power is accomplished through violence by a "hard" left; in the sociological sense, it is the overthrow of a regime and of a social order through a large popular movement; but in its ideological sense, it means changing institutional structures to achieve greater happiness. Surrounding World War II, the Spanish and Greek civil wars had not established a republican revolution of workers and peasants. The fascists took power through a mixture of formal democracy and military coups (besides the case of Vichy's National Revolution). After World War II destroyed Nazism, state communism was imposed in Central and Eastern Europe by the Soviet Union. Communism "by the tanks" continued to impose itself against national uprisings (East Berlin, Budapest, the Prague Spring, Poland), until the implosion of the USSR. The fall of the Berlin Wall could be described as an act of popular revolution, a non-violent revolution, just as could the resistance against the failed attempt to restore the old regime in August 1991, and the passages to formal democracy after the velvet or colored revolutions in the former communist republics. In Western Europe, Spain and Portugal came out of fascism, the communist parties of institutionalized liberal democracies abandoned the principle of dictatorship of the proletariat and turned to reformism. Public opinion condemned minority organizations using anarchist terrorism, individual commandeering (robberies) and symbolic murders (of CEOs, of general secretaries of organizations, ...). However, they become sensitive to the environmentalists' concerns and to the railings of anti-globalization activists at leaders during summits. But modern equipment (tanks, helicopters, drones, mace, ...) has reduced fighting between police forces and protesters to small skirmishes. Did technical progress begin a new phase of this struggle? Police repression and optical and electronic surveillance by the established order are countermanded by the "sub-veillance" of demonstrators' mobile phones broadcasting police brutality to global public scrutiny.

The ultra-revolution in the developing world: Since the Baku Congress, 1920 ...

Was the idea of revolution, with its double meaning—a social and internal institutional change as well as explosions in various parts of an established geopolitical entity—a new idea imported from the West for non-Western societies? Lenin, who opposed the kind of revolution that had "cooled off" in parliamentary democracies, called for the revolution of all the peoples of the East: the colonized. In fact it was the second total manufacturing war that drew the eastern half of Europe and central Asia, from the Baltic to the Pacific across Soviet Siberia, Mongolia and China into a communist revolution. So, in the spirit of Bandung (1955), Nehru, Nasser, Nkrumah, Sukarno, Tito, by anti-colonial wars of liberation, flourished the revolutionary doctrines from the Orient: the Viet Minh War, Algerian-style workers' self-management (*autogestion*), the Palestinian revolution, Maoism and revolutionary war according to Ho Chi Minh and Giap (until the degeneration of the Khmer Rouge). The peninsulas of the progressivist world island were Korea, Indochina, Vietnam. Then came more popular explosions in Africa: Congo, Rwanda, Liberia, Sierra Leone. Once independence had been obtained, the revolution was confronted with ultra-conservative governments—military juntas involved in the capitalist economy using torture against terrorist movements. After the Libyan and Iranian revolutions, Islamic extremism increased self-sacrificial attacks while demanding ethical, economic and institutional transformations. In Latin America, ultra-nationalist military regimes faced progressive movements (Allende in Chile), sometimes Christian (liberation theology), and outbreaks of communist guerrillas using ultra-violent strategies (Shining Path, Tupamaros, and Che Guevara), some of which were maintained in Mexico, Colombia (FARC) or Guatemala, triggering the terrorist/hostage/repression through torture or exile dialectic. There remains the question of Cuba's future after Fidel Castro. Then a more moderate model appears: more credible elections, the fight against drug trafficking and Amerindian revival (Venezuela, Bolivia, Peru, Brazil, ...). More generally, the ultra-revolution in the Third World, sometimes fueled by food riots, exacerbates the massacres/refugees binomial scenario under more or less "humanitarian" care, then oscillates between erratic violence extinguished by bloody repressions and the denunciation of presidential elections by more or less muzzled oppositions. In emerging countries, does the ultra-revolution transform itself into an ultra-mutation?

Ideo-racial instincts, ethnocultural antagonisms: Since the Nazi concentration camps, 1933 ...

Human history is punctuated by massacres. In the twentieth century, technology and bureaucratic racial *statolatry* determined a qualitative change in the elimination of the Other. Dachau was created in 1933, and two types of prison camp began to be distinguished: concentration camps (death by the extreme exploitation of the workforce) and extermination camps (racial ethnocide). Crimes against humanity, war crimes, are surpassed by genocide (defined as a crime with no statute of limitation). But in the newly independent, often heterogeneous, states, old disparities crisscross with economic strife and cultural incompatibilities. The major opposition between the white ex-colonizer and those who were colonized, North and South, developed countries and developing countries, are glorified by the representation of certain historical hatreds resulting from forced population shifts: Serbs and Albanians in Kosovo, Tutsis and Hutus in Rwanda, Hindu Tamils and Sinhalese Buddhists in Sri Lanka. The fate of the remaining white minorities in the former settlements—Algeria, Zimbabwe, South Africa at the end of apartheid—is also uncertain. Hence the attempts at appeasement, either through lengthy and haphazard judiciary procedures, or reconciling through asking for forgiveness after confessing faults and crimes, or laws on historic memory by which, not without anachronism, the *present* judges the *past* to appease the *future* (on the issue of slavery). There remains the problem of a mass immigration of people from South to North: a threat of rupture, the coexistence of diversity, a hope for a rewarding synthesis? Is Barak Obama the first Black American president or the first president of a multicolored humanity?

Intracontinental projections: Since the Islamic-American War, 1979 ...

Was it a symbol? In 1979, Europeans go to war against Muslims, Russians in Afghanistan and Americans, indirectly, in Iran (despite their support of the Taliban against the USSR). Hence an immense geostrategic extrapolation. In 1945, the United States made use of peripheral strategy—insular (surrender of Japan, the Philippines, Indonesia,) and peninsular (Korea, Vietnam, Malaysia) to stem the Communist tide. They settled sixty years later in Central Asia, fighting Taliban extremists. The implosion of the USSR led to the transition to NATO of its advanced territories, partly internal and external, and comforted the hope of Western expansion of democratization throughout the world. The projected operations reached cyberspace: destruction of observation and telecommunications satellites, electromagnetic pulse disrupting the regulation of electricity networks, water, financial flows and air traffic control.

In fact, two reversed projections opposed one another all the way to the center of their respective continents. On the one hand, military and humanitarian action with its geo-ideological messianism and its collateral damage, where the right of interference becomes a duty of protection and swells in long-term occupation fighting local guerrillas. On the other hand, a terrorism of mass destruction, if not nuclear, striking inside Western societies. Projection of expeditionary forces of trained soldiers with hyper-tech equipment coupled with mercenaries and Non-Governmental Organizations against geopolitical moralism and self-sacrifice. Asymmetric warfare where Western legal supervision of the terrorism/torture couple is facing a religious, nationalistic and transcendental fervor. Everyone professes to seek a victory that would establish a better distribution of natural resources, and sustainable ecological and ethical development. All of this being underscored by vast population transfers from South to North, challenging the very notion of West, as Europe is becoming post-Caucasian with the influx of Turko-Arabic-African migrants and the United States moves beyond the anti-segregationist White/Black period of history because of the Latino-Asian influx.

Prospects

Socio-strategic systems combine two levels. One assembles a rebalancing of geopolitical, social structures, religious and philosophical doctrines, trade flows, demographics. They are part of the long time-span ("*longue durée*") and often remain misunderstood by contemporaries who treat them according to short term/incidental ("*évènementiel*") political and economic analysis. Consequently, the second level is more aware of and often punctuated by disasters, strategic doctrines, tactics of war and revolutionary movements. The art of the strategist consists then in defining military or revolutionary action on the first level, and in the modulating mutations on the second level.

Hence some dissociation in the major disruptions in the West: Christianization, the Reformation, humanism, rationalism, the Enlightenment, nationalism, Proletarian Marxism, decolonization, human rights and socialism. The distribution is made to defend the established order at the second tier: annihilation of the will of the opponent. For the challenger, it is made at the fourth tier: destruction of established structures.

It may be noted, however, that the periods of *vertical struggle* are more frequent than the periods of *horizontal struggle*. Chronologically, and taken separately, they are also longer. Among these are the emergence of nation states, monarchic wars, colonial imperialism and continental strategies (Pan-Slavism, Pan-Turanism, Pan-Germanism mutating into Nazism), peripheral thalassocratic strategies, the wars of decolonization, the nuclear duopoly, or intra-continental nuclear projections. The periods of horizontal struggles include the wars of religion, the rationalism of the emancipatory Enlightenment of France and the Americas, the proletarian revolutions, the socialist and anarchist movements. Four periods are mixed, effecting the transition from horizontal conflict to vertical conflict: feudalism, national wars of the nineteenth century, the principle of nationalities, decolonization. They all tend to organize, in a fairly coherent way, what is, even now, the basic strategic unit: the central state with a homogeneous population and civilization without cultural, racial or religious differences, the excess of which would be likely to ravage it through internal ideological or racist impulses. Finally, the Cold War phase which in Europe and Asia conjugated horizontal conflict (more psycho-political in Europe, despite East Berlin, Hungary, Czechoslovakia or Poland, and more violently revolutionary in Asia) and the vertical conflict between the two nuclear powers, the real intensity of which, if it wasn't purely philosophical, was highly variable. The system collapsed after 1990 due to the fragmentation of states and the proliferation of financial and economic networks which emerged after the fracture of international communism, and globalization in part managed by the United Nations diplomatic democratization and supremacy of the United States.

The emergence of nuclear weapons had caused a mutation. The distinction between the objective capacity to destroy using that weapon and the goal subjectively desired by its owners was attenuated by the occultation of what would really happen if it was detonated. This goal is now integrated within the strategy of non-use. Under the American atomic monopoly, which represented the established order, it objectively ranks on the fifth level—which was never a factor—and subjectively on the second level—a mixed success, since the challengers used methods of indirect strategy, less to prevent the rise of the adversary to the fifth floor as to prevent him from using the presence of the objective capacity of destruction against one's own advance, whether physically or to weaken its principles. Hence the anti-Soviet *containment*: Korea, Vietnam. A distinction that became even more clear during the nuclear duopoly (Cold War), the objective capacity rising from one level to reach the elements of industrial growth, as the confrontation took place on the two levels.

Among contemporary systems two are vertical struggles: nuclear-ideological deterrence—real or virtual—between certain emerging countries (India/Pakistan, Israel/Iran, North/South Korea), and para-manufacturing wars in the developing world. But both contain some elements of horizontal conflict: the pacifist or anti-nuclear environmentalist alternative opposition in post-industrial cities, and the opposition against technological and economic dependency—expressed or implied—throughout the geopolitical and cultural areas of the developing world. Two are horizontal struggles: *infra* and *ultra* revolutions in industrial countries or in developing countries; three are mixed: areas of civilization, intra-continental projections, or racism, which, through nations and social strata, bring together or oppose ethnicities, creeds, clans or classes.

Six systems have thus effected "clashes of civilization" or, more accurately, have exported doctrines and values, institutions and technology: the barbarian invasions and conversions, overseas implantations during the Middle Ages, colonial conquests/decolonization, ideological racial movements, intra-continental projections. Five systems illustrate the failures of the Western internal ideological systems: Reform, liberal revolutions (Enlightenment) and proletarian revolutions (socialism), totalitarianism (Nazism and Communism), and manufacturing wars exalted by racial genocidal impulses. The other systems internal to the West expressed nationalist and cultural clashes or economic competition. All have interfered in the passage of time.

Should it be considered that the *acceleration of history*, that is to say changes that are a more or less visible player in the consistency and the functioning of societies, causes a faster rotation and deskilling of socio-strategic systems? The acceleration phenomena giving the impression of abrupt and irreversible changes have already occurred in the past, such as the one experienced by French society during the Regency or the Revolution. This feeling of acceleration is often the result of a break (a "catastrophe" as understood in theory) involved in the political and ideological suprastructure of society: the end of the absolutist reign of Louis XIV, the Enlightenment materializing after 1789, and socialist struggles after 1917. But the in-depth evolution of societies, their tensions and conflicts, will remain less visible and will have a slower transition, as happened during the evolution from paganism to Christianity, the heresiological schismatic breaks with Christianity, the confrontation between bourgeois and proletarian revolutions, decolonization, or immigration from South to North. The big United States/Russia confrontation around the Pacific was foreseen by the best minds at the end of the Napoleonic Wars: de Pradt, de Tocqueville, Herzen and Napoleon who, by predicting the rise of China, has foretold the emergence of the developing world which is perhaps, with the United States, the double act of the twenty-first century.

A new element appeared to be capital since the Napoleonic era: the exponential and invasive growth of technological opportunities—weapons of power, means of communication and of standardization of lifestyles. It is necessary to find a balanced view between time and space.

The technology causes the globalization of socio-strategic systems in both respects of the term: the interdependence of all international policies with many conflicts becoming transnational, the extension of theaters of operation to places that, until now, eluded them in principle (the civilian population) or in practice (oceanic depths, air then space). But the lengthening of conflict becomes an equalizer, whether through similarity (nuclear deterrence) or by heterogeneous asymmetrical complementarity (revolutionary war, subversion against heavy firepower) which leads to extending the *sociostrategic* system in the long duration and (more or less) empower one system as compared to the others. Again, the acceleration of history seems relative.

All systems, within a few years, succeed each other: after the development of the art of the great war in the secular kingdoms follows the classical age of the wars between monarchies and nation-states, then it is extrapolated by the total manufacturing wars before the transformation operated by the ideological-nuclear empires, the para-manufacturing wars in the developing world, and the current intra-continental projections against revolutionary wars. But after the Reformation, the liberal and proletarian revolutions, nationalities and minorities in the contemporary world split between the *infra*-revolution in industrial countries and *ultra*-revolution in developing countries. In this case, therefore, it does not seem that there is acceleration. On the contrary, even for the longest-lasting systems: four centuries for the urban riots and peasant revolts, five for the colonial conquests, more than three for the British peripheral strategy. The introduction of naval feudal principalities lasted four and a half centuries. How long will the current intra-continental projections—that could be retroactive to 1917—last for US involvement in Europe?

The hardness of a conflict sometimes leads to a reduction in the number of systems that are exalted in interlocking into one another. Again the *complexity of the story* does not seem to be characteristic of the contemporary world, where it is commonplace to oppose the rich countries to emerging economies, in a world economy in which the urban universe grows by migration from the countryside opening up the secondary centers on the outskirts. This disturbed dichotomy is, however, one major factor in the merging of contemporary strategies into two major systems of opposition.

One assembles the possessors of nuclear forces (and their immediate dependents) and has given rise to sophisticated strategic concepts based on deterrence. The other divides the various parts of the developing world (decision poles or poles of causality) between them, and from the main proponents of the nuclear game quota. It therefore contains multiple subsystems, real or virtual, each led by disrupting specific classic character (nationalism) or developed character (socio-economic ideologically argued confrontation). Analysis of the amplitude and duration of these disturbances compared with those of closer events, which, in a more fragmented way, describes the true story of the peoples and human communities, reveals the real morphology of conflict. The second set of oppositions, much less conceptualized, is characterized by the length of heated but limited conflicts, and by the uncertain segmented aspects, unevenly nested in the total social life of the protagonists.

Consequently, the transition from system to system is effected less by "rupture" during a "catastrophe" than by undetectable gradual transition. The disaster is accomplished in the short term, but does not fatally impact a long time system perceptible on the geo-historic and meta-strategic scale proposed here.

Islamic Warfare

Jean-Paul Charnay

A warrior society without a military institution, Arab tribes practiced the art of war by alternating commerce by caravans with pillaging expeditions. Quranic revelation gives them a geopolitically expansionist objective, as well as an ethical motivation: the *jihâd,* which evokes notions of effort and self-perfection. The *jihâd* also instills a theocentric view of the world, the center of which revolves around the Ka'aba in Mecca, rippling out to the war-torn territories of the infidels.

Historically, macrostrategic phases of the Muslim world are quite visible: the offensive period of the Arab caliphs (seventh–tenth centuries); the defense of Islamic borderlands (eleventh–fourteenth centuries); the establishment of non-Arab empires (fourteenth–eighteenth centuries); colonization (nineteenth century); de-colonization (twentieth century); the social, political and extra-governmental fragmentations of contemporary wars (twenty-first century). Certain intellectuals and martyr-terrorists, sometimes self-sacrificing, have attempted to articulate the ideas of *jihâd* and of revolution. The controversy rests upon the question: Is the *jihâd* only a defensive strategy? Or is it foremost a spiritual exercise?

The Concept of *Jihâd*

Originating from Arab inter-tribal warfare, the Islamic art of war was transcended by the *jihâd* consciousness. *Jihâd* can be taken to mean a means of combat or spiritual asceticism, holy war or legitimate defense, the controversies abound …

The rationale of *jihâd* is absolute: it is a defense of the laws of God over the "friends of the devil" (Quran IV, 76). This defense constitutes an exercise of justice, and assures the renewal of the initial pacts through which God bestowed upon Adam his office as lieutenant on Earth.

From an etymological viewpoint, the root *"j h d"* connotes, beyond the notion of effort, the idea of constant resistance: from non-surrender to despair. It is not only a (strategic) principle of initiative, but also an (ethical) consciousness of the possibilities offered by the human condition. It appears in the Quran with three principal meanings: 1) the dynamic surpassing of the being of one's self; 2) warlike undertaking *stricto sensu*; and 3) a spiritual asceticism. It brings together the two opposing poles of human action: the mystical and the political.

In Islamic history, and still today, each change of dynasty or of regime, every instance of political challenge, tends to be legitimized by the use of an ideological weapon: the canonical and moral return to Quranic revelation, the accusation that the authority currently in place has let itself be corrupted in its application of the *chari'a* (Muslim law) or that it has subjected it to impure innovations. From the first important Muslim leaders to the statesmen and captains of the next generations (Muhammad, the four legitimate caliphs, notably Omar and Ali), and over the course of internal wars that led to the establishment of the Omeyyade and Abbasidian empires, one remarks that the implementation of the *chari'a* follows the rhythm of conquest and that the majority of successive dynasties were created by a complementary and antithetical duo: the *jawad* (nobleman of the sword), the chief of the warring tribe who rallied the thousands of sabers—or guns—and the *chaykh*, the theologian who gathered together an army of intellectuals. Thus is incarnated the double reality of Muslim power: the logocratical principle of revelation and the materialization of force. So it was for the Fatimides, the Almoravides, the Almohades, the Sunni madrasa, for the Turkish Seljuks b. 'Abd al-Whhab and Muhaamad b. Sa'id for the Wahhabites, and even so today for the reformist theologians of moderate governments, while the "Free Officers" and the Ayatollah Khomeyni join in doctrinal and nationalist reformulation and revolutionary political action.

As a result, one can distinguish a *gradation* in the nature of wars, in which the inferior classes experience a diminished, but not total disappearance, of the quintessential sacred character of war: the *jihâd fi sabil Allah*, the combat in the way of God, which is essentially aimed against infidel territories. Next to appear are the various political or internal wars (*hurub*, singular *harb*) demanded by the common interest, such as wars against apostates, and those against schismatic groups, which are in part military. Such an intellectual and effective schism in Islam was caused by the division between Sunnism and Shi'ism after the death of 'Ali, the death of his son H'usayn a Kerbala' and the victory of the Umayyads. Finally, there are the expeditions and police measures directed against lawbreakers, leaders of disorder, and bandits who rebel against God and his Prophet and who violate the Islamic order of urban society and public peace.

There is opposition and complementarity between the two important virtues of the combatant: the *hilm*, bravery and generosity of the task, and the *çabr*, consistency in the face of adversity. These two then transcend themselves in the supreme image of the Muslim combatant, the martyr (*shahid*) who is immediately admitted into paradise.

Historically

Each victory can be nothing other than a step towards a much wider extension of Islamic influence. In this light, *jihâd* is only a particular application of *'adala* (justice) and a regulatory means of the "rights of people." Inside the *umma* (Muslim community), judicial inferiority, social humiliation and theological controversy are used simultaneously to accelerate movements of true religious conversion. On the outside of the Muslim community, the *jihâd* will be pursued throughout the world—on the assumption that it could be implemented on the day of the Resurrection, a point on which opinion is divided since, in practice, the Muslim community is not obliged to take on a total, universal offensive, if there was no need to use up all of its strength. The Muslim community can organize a state of specific and prolonged co-existence with non-Muslim populations.

*Jihâd*ist doctrines thus describe several perspectives of action.

Ethically

The figure of the warrior possesses ambivalent connotations in Islam. From its Arab origins, this image is at the summit of several "chains" between which it has been divided, according to the functions, roles and activities, and social types of the *Jahiliya* (pre-Islamic period) and primitive Islam. Thus was constituted, from the negative to the positive, a hierarchy of "military" types. The first is the opponent of the Muslim faith who promotes crimes of blood, such as the murder (*istirad*) of orthodox believers, and who is esteemed by true believers as an apostate and who undertakes an intense, secretive or poetic propaganda that exalts the glory, bravery and violence of the anti-Muslim combatants; the next is the rebel against Muslim public order who advocates disobedience against the established sovereign; the third is the soldier who is entrusted to make sure that the established public order is respected; the fourth is the warrior aristocrat who is either "tribal" or "corporatist" in his social context and who is either a nobleman of the sword (*djuwad*) or else a privateer (*raïs*) including their clients throughout their expeditions; and finally, the *ghazi* who arm themselves for the *jihâd*.

As such, the ideal figure of the *mujahid*, the combatant for the faith according to Muslim ethics, is not that of a vanquishing warrior so much as that of the warrior who, with reason and stoicism, sacrifices himself for the common well-being and the glory of the laws of God.

But what of he who sacrifices himself in the shadows, the terrorist, *fida'ï*, who engages in individual attacks against the innocent? He is condemnable when the attack is against Muslims. Hence, since the origins of Islam, there has been the political and police repression of the Kharedjites, who were the assassins of the last legitimate caliph, 'Ali (661), and who were purist partisans of an egalitarian democracy, even to the benefit of the *Malawi* (new converts), by the established power: the Umayyad caliph. Hence the ambiguous renown, after the tenth century, of the Qarmates in the ninth century; of the Ismaliens (shi'ites of Alamut) who, being influenced by Hellenic philosophers, battled against the Sunnism of established dynasties (the *Nizārī*, nicknamed Hashashiyyin[1] of

1 Editor's note: *Nizārī*, nicknamed Hashashiyyin, means hashish smokers, but the origins

the eleventh to thirteenth centuries, establishing a transnational terrorism—before this term was used—against the established authorities, both Muslim and Crusaders, in the Middle East and sometimes in Europe); of the *'ayyarun*, a type of urban bandits made up of "justice" seekers in Iraq from the eleventh to the thirteenth centuries; of contemporary movements, such as the Muslim Brothers, and other organizations of ideological Islamists. The murderers of Anwar El Sadat invoked in their defense an illustrious Hanabali author from the fourteenth century, Ibn Taïmiya, and used his justification for tyrannicide.

Since the term *fida'ï* was given more valor during the wars of anti-colonial independence, the lonely, isolated soldier had his place in these battles. Re-appropriating the old term, Egyptians, Algerians and Palestinians purged it sociologically and semantically of its anarchistic terrorism connotations, so as to reinvigorate it in revolutionary tactics: infiltration into the social framework of the enemy in order to sabotage their material infrastructure or to liquidate a particular leader, who is seen as harmful due to his achievements and prestige. Refusing romanticism and nihilism, the new *fida'ï* is no longer sacrificed, but integrated into the process of a collective battle. Calculating the risks, he accomplishes his mission in solitude, deprivation, intellectual asceticism. He does not act on the basis of fanaticism, but on a total engagement, with sublime effort and endurance—which is divinely aided by the *niya*, a pious intention. From that point, according to the professed legitimacy, whether one is a bearded countryman armed with a rocket, a *kalachnikov*, or else a computer scientist capable of piloting a Boeing, one would either be a justified *mujtahid*, seeker of the faith, or else an *irhab*, a reprobate terrorist, an isolated young person, a cyber-warrior intoxicated by Islamist sites on the Internet.

Sociologically

The *jihâd* organizes the use of violence in society as a non-differential military function; according to *h'adith*, a year should be divided into three parts: one quarter consecrated to one's profession, one half to learning, and one quarter to the armed defense of the community.

Ritualistically

In a religious society that perpetually cleanses the impure with sacred prescriptions (prayer, absolution, fasting), the *jihâd* takes the forefront in certain privileged times: in the morning (the inverse of the silent mystical ascension of the night, for which

of the term have been contested. The Nizari are presently the second largest branch of Shi'a Islam and form the majority of the Ismā'īlī denomination. There are presently an estimated 12 to 15 million Nizari Ismā'īlī residing in more than 25 countries and territories.

the model remains the *mi'raj* of the Prophet in Jerusalem), Mondays and Thursdays. It evokes with even greater power the effort of supererogatory fasts.

Eschatologically

Jihâd recalls the struggle of the faith against infidelity or rebellion; it will evoke combat, if necessary, between the *umma* and the populations of infidels (or rather between good and bad angels). In this sense, the purpose of the *jihâd* for the warrior is not the more or less immediate conversion of individuals, since the effort to force other individual consciousnesses would be against divine volition (Quran X, 99 at 105, and V, 54). The eventual execution of armed enemies will be carried out only as a necessity, to "liquidate" that which is irreconcilable with the state of peace—not to kill non-believers.

Teleologically

Jihâd describes a separation. The collective Muslim order set up on the Earth constitutes the best manner through which a person might accede to his final, celestial reward: paradise, or a *rapprochement* to God. It presents itself as a method: spreading an absolute truth, the divine *logos* of Revelation, it forces the individual to transcend his earthly destiny. The pedagogy involved in the *jihâd* engages him in the *musabara* (the fight against the enemy), the *qital* (combat), the *harb* (war): simple activities of re-equilibrating politics, economy and society. It institutes a geographical rearrangement of the earth.

Geopolitically

The spiritual and spatial unipolarity of Islam, expressed in the convergence of its sacral "lines of force" toward one single point, is reinforced by a series of concentric rings which cut back and forth through these lines of force.

At the center, in the heart of the ellipse representing Muslim territories on the planet from the Atlantic to Indonesia, from Central Asia to the Gulf of Guinea, can be found Mecca, the *Ka'aba*: the Black Rock. Around it, Haram is the sacred territory which encircles the two holy cities of Mecca and Medina (currently a radius of about 25 kilometers). It is forbidden to non-Muslims. There, every non-native must consider himself in a state of pilgrimage and must avoid staying too long, for fear of no longer feeling an ardent desire to return, or else, for fear of falling into sin. There, each infidel risks death, or at least time in prison. Medina, "the noble," also constitutes a privileged zone: polytheists cannot live there, nor can they be buried there. A deeply established belief thus classifies the three principal holy places of orthodox Islam: one prayer in the mosque of Medina is worth ten thousand prayers; one prayer in the mosque *al-Aqça* (of the Rock) of Jerusalem is worth a

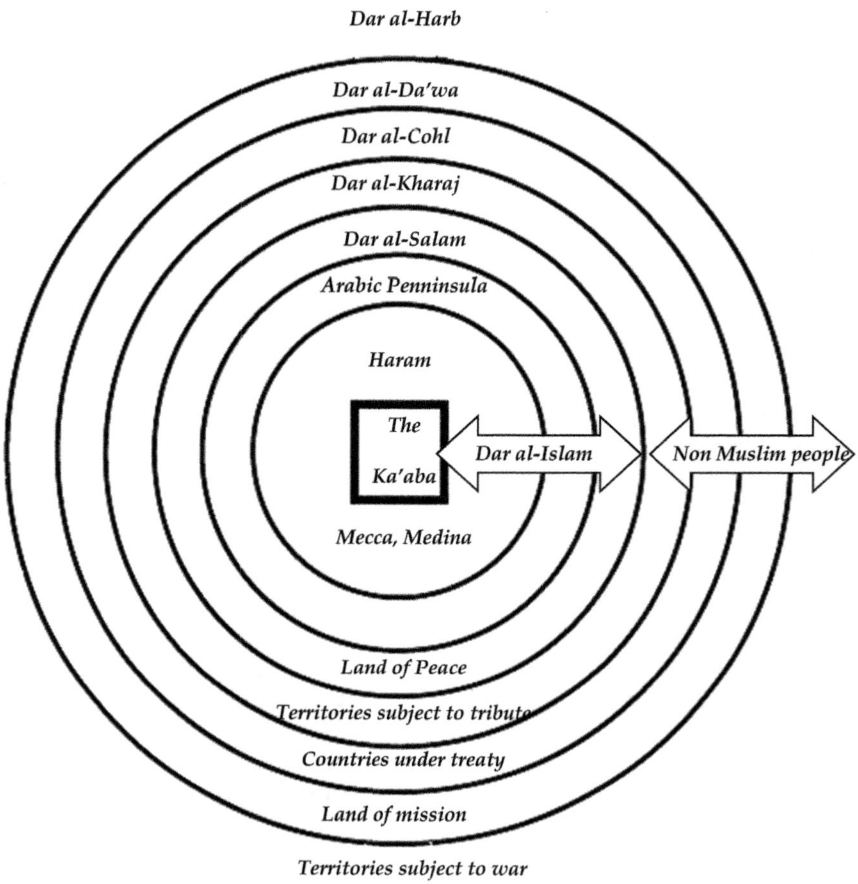

Figure 26.1 Islamic representation of the world

thousand; one prayer in the mosque of Mecca is worth one hundred thousand. Around the two holy cities, the Hedjaz also makes up a sacred zone where infidels are not allowed to settle down or be buried. In the Arabian Peninsula, according to Islamic purists, no church can be built, nor any crosses erected, since the peninsula constitutes a vast, open-sky mosque. Beyond this lies the territory of Dar al-Islam or Dar as Salam (land of peace).

Beyond that limit, the sacred nature of the land diminishes according to dogmatic-judicial categories on a geopolitical level. The soil conquered by the Prophet or his followers makes up the Dar al-Islam, which includes, on the one hand, the lands subjected to canonic tithe (*zakat*): the lands given to Muslims, conquered lands in which reside converted habitants, lands that have been "invigorated" by a Muslim, i.e. not seized from an enemy but conquered over nature, abandoned lands or desert. On the other hand, lay the lands of *kharâj*: tribute is paid there by non-converted inhabitants maintaining their property on Muslim territory, most notably, the *kitabiyun* ("people of the Book:" Christians, Jews and Mazdeians, and more generally all of the *dhimmi,* the protected). Finally, in Dar al-Islam infidel pagans may be admitted and allowed safe conduct for a limited time. Beyond Dar al-Islam appears a controversial category: the *dar al-çolh*. Those territories are subjugated not by conquest, but by treaty recognizing the suzerainty of Islam. Still further beyond lies the *dar al-h'arb*: land of war, because it is there where rules the infidel with whom temporary treaties may be made, if a Holy War is not to be continued. According to the political circumstances and the different interpretations of the Quran, the non-Muslim countries ruled by moderate governments are *Dar al-Cohl. Dar al-Da'wa* are open to movements engaged in missionary activities. Fundamentalist movements justify their terrorist actions in *Dar al-harb.*

In fact, this circular geocentrist representation of Islam is undermined by the notion of *Umma Al islamyia,* a community of believers around the world, a demographic and not a geographic entity destined to expand until it covers the planet. The Arab wars have often proceeded on to population transfers and efforts of religious conversion. Contemporary Muslim immigrants could represent a vector.

The Strategic Phases

Invoking more or less canonical *jihâd* has not always maintained the same functions or raised the same resonances. In six phases, it has undergone significant changes.

Strategy and tactical offensives (the Arab caliphate, seventh–eleventh centuries)

A warrior and caravan society without military institutions, pre-Islamic Arabia remained tribal. From the beginning, the virtue of enthusiasm, representing a psychological offensive, had pushed the Prophet's Companions and their descendants to assemble a world from the Pyrenees to the Indus. The offensive then inspired the creation of a policy-oriented conquest and turned toward assimilation of foreign social structures. It then tended towards the building of an Arab Empire, Umayyad and Abbasid. But in the eleventh century, with the arrival of the Turks and the Crusaders, came the dissociation of Arab power for generally economic reasons: dilution of the stock of gold taken from the Persian palaces and Byzantine churches, which had allowed the establishment of a marine and metal-working industry; weakening of trade flows because Europeans circumvented the Cape of Good Hope; end of the maritime control sought after by the Umayyads and which would be taken over by the Turks, crossing the Bosphorus and finally capturing Constantinople.

Defensive strategy on the borderlands and offensive tactics (eleventh–fourteenth centuries)

The fragmentation of multiple more or less dissenting religious sects, the slow decline of the Abbasid Caliphate in the thirteenth century, trapped between the Mongols and the Crusaders, the emergence of increasingly important non-Arab elements in political management and in the armies are both causes and effects of the stagnation of the Muslim offensive. On the Arab side, *jihâd* underwent two radical changes.

The lack of volunteers in sufficient numbers to continue Muslim expansion by demographic force compelled Islam to develop a defensive strategy. From the Atlantic to Transoxiana (central Asia), the Muslim *limes* (borders and coasts) is covered by *ribadt'*(s), convent-like barracks that welcome those who guard against the infidel. They are the purest and most fervent. Their zeal is heated by numerous treaties of holy war which, unlike those written during the previous period, are less institutional and strategic (organization of the conquests made by a non-stop offensive) than apologetic (exhortations, for example, even inventing *h'adîth*(s) to warm the religious zeal) and belligerent (theory of military combat, exalting both the mission of sacrifice and earthly victory); these are counter-Crusades.

At the same time came the idea that *jihâd* is maintained by actions instigated by the common man, by individual initiative in pursuit of profit, as well as sovereigns: Arab and Malay piracy in the Indian Ocean, Turkish and Barbary pirates in the Mediterranean; struggles against the irredentist Berber households (which eventually descend into tribal feuds) and conflicts between military and financial aristocracies in the Maghreb; attacks on caravans (transfer of goods by booty) or their protection (security ensured through royalties); half-commercial, half-apostolic expeditions to Africa, which also fueled the African slave trade. From Central Asia, the Qaznavids and then the Ghourides invaded northern India.

Imperial, non-Arab strategies and the politicization of war (fifteenth–eighteenth centuries)

The Middle East Kurdish dynasties (Nurides and Ayyubid) ensured the success of the Counter-Crusades and the Mamluks resisted the Mongols. Then the Seljuk Turks took over Anatolia and pushed towards the Balkans.

Every sovereign maintained troops who were frequently mercenaries, foreign slaves undergoing a process of conversion to Islam who would then constitute real political and military parties, such as Schiavoni in Andalusia, the Janissaries in Turkey, the Mamluks in Egypt. They were primarily centered around a mystical brotherhood (Bektashi for the Janissaries), and occasionally founded dynasties whose ambitions were terrestrial. Then followed the resumption of the Muslim offensive (Ottoman) toward the Indian Ocean, Persia, once reaching from the Balkans (following the failure of the Crusade, led by Emperor Sigismund at Nicopolis, in 1396) to the plains by the Danube river.

In the sixteenth century the Ottomans controlled the Arab world, except for Morocco. India's Great Mogul dynasty established its supremacy on the subcontinent by attacking Hindu principalities, as well as independent Muslim sultanates.

In the seventeenth century, the halting of the Ottoman advance begins, first by the Habsburgs in the west, and by the Persian Safavid sovereigns in the east. In the late eighteenth century, the Austro-Turkish wars gave way to Russian-Turkish wars, culminating in the nineteenth (war of 1828–29, Crimean War, 1856), and exerted a thrust which still continues (Chechnya). In reality, the imperial terrestrial fortune of the Ottomans, Safavids or Great Mogul, and the alliances with infidels, corrupted the purity of *jihâd*.

Strategy and defensive tactics: Colonization (nineteenth century)

Despite efforts in the Muslim world, which tended to combine religious exaltation, modern warfare and nationalism, European colonial conquests often met only dispersed socio-political movements, fighting by means of traditional ways of tribal warfare rather than with religious and revolutionary impulses, and who favored compromises as opposed to the original notion of *jihâd*, since they institutionalized a foreign, infidel, domination over Muslim lands.

Thus, over the course of the decades, one can see some troubling of the consciences: was it necessary to accept, provisionally, infidel domination in order to strive towards perfection, considering that Islam continued to be recognized and practiced? In essence, European sovereignty did not damage Muslim lands, therefore the European efforts to colonize Muslim regions may not require an immediate obligation to resort to violence or to emigrate to a land under Muslim sovereignty (the Mashreq, Ottoman). Or else, did it affirm the immediate need to fight? In fact dynasties subjected to the colonizers (Morocco, Egypt, Iraq, Tunisia) tended to benefit from their physical power and their administrative construction, and were able to maintain themselves against Islamic fighters not succeeding by their own strength to put the old canon law of the City Muslim and Islamic just war in the service of building a modern state.

The loss of political sovereignty over the majority of the *Dar al-Islam* was perceived as a trial sent by God. Despite the theory of the Sultan-Caliph, the Ottoman Empire failed to awaken a general revolt of Muslim populations and troops against the Allies through the proclamation of 1914.

Jihâd was invoked against colonization in the many rebellions that punctuated this period, particularly those moved by the ideas of the *murābiṭ* or of the *Mahdîeh* (Sudan, 1881–98), or those close to the peasantry (Palestine, 1936). But overall, after defeats in the colonial period, the canonical concept of *jihâd* was shelved because of the misfortune of the times, through nationalist literature, from the tribal warrior, or the peasant, to the nationalist activist or to the politician reasoning in terms of parliamentary and administrative support, local and international public opinion, and revolutionary and subversive war. In short: the secularization of national struggle had begun.

Parallel counter-offensive strategy and direct offensive tactics: Decolonization (twentieth century)

The idea of renewed attacks or, more accurately, the physical liberation and spiritual purification of Arab land, was still dormant. The premonitory symptoms manifested themselves immediately after World War I, apparently by urban riots (Cairo, Damascus, Constantine, Fes), and by the action carried out by the militants, rather than by the peasant war (Abd el Krim in Morocco, Jebel Druze in Syria and Palestine, 1936). It was not until the end of World War II, as part of a global rebalancing, that revolutionary war demonstrated the inefficiency of technocratic military tactics (the war in Algeria, the Suez crisis), and that the popular revolt led to victory and independence of the Muslim peoples from the Atlantic to Indonesia. During the wars of independence, *jihâd* was not officially proclaimed. But Palestinian combat poetry sanctified violence by comparing the passion of the Palestinian people with the passion of Christ, while theologians evoked *jihâd*. Some authors argue that although Muslims could not canonically invoke *jihâd* in the liberation struggles, they nevertheless, in a way, constitute its political and sociological extension. On the other hand, even more than the notion of apostate, political propaganda and the fight against atheism and materialism evoked the spirit of *jihâd*.

It was not taken into account in this way by the Arab states, but since the end of World War II, the call to *jihâd* has been launched in a number of cases. The Muslim Brothers have never ceased to invoke *jihad*. Indonesia's appeal was launched during the Civil War in 1965 that eliminated the communists and the Chinese. The *jihâd* was proclaimed in Damascus in favor of the Algerian liberation in 1960, and again in Damascus, and in Cairo, during the wars of 1967 and 1973, and was launched by King Faisal of Saudi in 1969, after the burning of the Al Aqsa mosque in Jerusalem. Then it was invoked during the storming of the mosque of the Ka'ba, and in Afghanistan after the Soviet invasion in 1979, by the Islamic Summit in Taif in 1981 in favor of Palestine and again in 1982 upon the Israeli invasion of Lebanon.

During the war between Iraq and Iran (1981–90) Khomeini claimed not *jihâd* (in Shi'ism only the *1'imâm* could do so upon his return), but *harb al Qods* (holy war), and the combatants, the Revolutionary Guards, who died were declared martyrs. The more "secular" Saddam Hussein initially reminisced about the struggle against the heretical Iranians, then once in the quagmire of the war against the Islamic Republic, he proclaimed the *jihâd*, but in a non-canonical war of *fatwas*. This was also the case during the two Gulf Wars (1990 and 2003).

In fact, nationalist impulse and religious passion exalted one another, while most Islamic states kept their distance from the invocation of bellicose *jihâd* in the context of UN international law (Dakar Declaration of 1992). But there now there appears to be a new phase.

Non-governmental offensive and sacrificial violence (twenty-first century)

The large factions within the Muslim world have tended to bring the invocation of *jihad* back onto the agenda. Extreme Islamist ideology exalts a bloody struggle between the *jihâd*ists of "international terrorism" and government repression.

These calls for a rigorously warring canonical *jihâd* made in the heat and pain of dramatic events, have become part of a discourse both spiritual and political, referring to individual ethical development, the sociological depth of mode of operation and defense, and political and revolutionary strategies. Because of the collusion and schemes established with the "corrupt" West, anti-governmental extremist organizations take up the idea of *jihâd* and make it into an individual obligation and a supreme sacrifice, to the destruction of the "human bombs" of the second intifada or Al-Qaeda.

Osama bin Laden declared the illegitimacy of the presence of armed "infidels" in the Arabian Peninsula and proclaimed the necessity of a contestant *jihâd*—now an individual obligation because of the "betrayal" of the governments of Muslim countries. He proclaimed the pursuit of *jihâd* as a counter-offensive strategy and as an offensive tactic for the defense and expansion of Islam throughout a planet subjected to American globalization. Thus the joint fight in *jihâd*, against the "far enemy," the infidel Judeo-Crusader, and the "near enemy," the bad Muslim.

Contemporary Wars

In the historic view of Arab strategies, the memory of the *futūh*, the first conquests and victories, is tempered by the memory of the fifteenth-century Ottomans, and obscured by the nineteenth and early twentieth centuries, when colonization captured the entire Arab world, except the center of the Arabian Peninsula. Despite many internal and external wars since the fifteenth century, Arab armies achieved no decisive victories despite the great success of Mehmet Ali in Egypt against the Sublime Porte, and the Mahdî of Sudan against the Anglo-Egyptians, and later those of the Hashemite Mecca (Sharif Hussein and his son Faisal) allied to the British during World War I: in the first case, the Powers would save the Ottoman Empire, in the third, they took the opportunity to establish their direct or indirect domination of the Middle East, endorsing only the conquest of the holy cities by Ibn Saud over the Hashemites. During the second half of the twentieth century, the Arabs seized independence in metropolitan areas exhausted by two world wars, against a background of demographic shifts, in a process of decolonization profiting from the anti-imperialist complicity of the two superpowers. But neither revolutionary wars (Algeria) nor the guerrillas and the popular urban riots (Egypt and Morocco) resulted in military victories as in Vietnam.

After the World War II, the colonies became the Third World. The *Umma al Islamiyya*, the Islamic community ,and the *al Watan al Arabiyya*, the Arab nation, seemed in need of being rebuilt. The military regimes of newly independent states consisted of "Free Officers" or sovereign-soldiers, except in Tunisia and Morocco.

Since 1945, a dramatic chronology developed in the Middle East:

- 1946–48: Indonesia is free; Pakistan is born of the bloody partition of India, and Israel is established by the UN.
- 1948–56: Iran fails in its attempt to nationalize oil; the Arab League asserts itself as a league of states, not of peoples.
- 1956–67: the Ten Glorious Years of Arab history. Nasser's Egypt regained Arab dignity; the Maghreb gained its liberation; nationalized oil gushed abundantly from the soils of Arabia, Iraq, the Sahara; the Palestinian revolution, positive neutralism, Muslim or Arab socialism, Algerian workers' self-management (*autogestion*) and industrialization enchanted progressive and anti-imperialist Orientalism.
- 1967, destiny changed: Six Day War, loss of East Jerusalem, the Lebanese civil war, "Black September" (1970 conflict between Jordanians and Palestinians), explosion of Pakistan (Bangladesh).
- 1973: rising oil prices and October war, but oil as a weapon remains inefficient despite the second oil crisis of 1985.
- 1977: Sadat in Jerusalem.
- 1979: fighting in the mosque of the Ka'aba; second fight in 1987; the Iranian revolution seeking a Muslim renaissance is contained by three wars: Soviet-Afghan, Lebano-Lebanese, Iraqi-Iranian; Egypt recognizes Israel.
- Since the 1989 implosion of communism and of the Soviet Union: Central Asia rids itself more of communism than of Russian influence; the Gulf War heightens the ideological and economic rivalries as well as the manipulation by the great powers; Bosnia ignites, Kosovo suffers "ethnic cleansing;" Algeria is torn apart; the Palestinians falter between violence and negotiation; the Arabian Peninsula, Libya, Sudan, Pakistan, Iran and Afghanistan demand *Sharia*; Chechnya enters a phase of open rebellion, then "kamikaze" terrorism; in the West, worried immigrants have to deal with the question of their identity.
- September 11, 2001: The destruction of the World Trade Center towers in New York exalts the US-Islamic war, the hatred of the radical Islamist fundamentalist and of Salafi Jihâdists, and the projection of US military force into Afghanistan (training bases) and Iraq (suspected of possessing weapons of mass destruction).
- And now: a system of asymmetrical warfare between one strategy—an operational art using military macro-technology executed through lengthy preparation at long distance and based on enormous firepower, and another strategy—an art of precise yet unpredictable attacks using micro-technology and manipulating the emotions of the adversary.

Historically, moreover, the military seemed (yet) unable to master the art of the "great war." Except in Egypt and Iraq, there is no armaments industry; hence it was believed that the "Islamic bomb" would solve the problem. The expression belongs to the Prime Minister of Pakistan, Ali Bhutto, who called for one to counter the "Christian, Communist Chinese, and [as suspected at the time] Israeli and Indian" bombs. Hence the almost unanimous approval of public opinion welcoming the Pakistani nuclear tests in response to India's tests (in 1998). Moreover, these views express resentment at the fact that both Western and Russian strategists believe Arab leaders are unable to conceive of the notion of *deterrence*, consisting in the possession of a weapon the employment of which is forbidden by that doctrine (of deterrence). The Arabs are concerned as to why they could not be credited with the influence they had in the process that brought down the Soviets, and why they are believed to be run by "mad dictators." Hence, given concerns similar to the Arab states, the Iranian leadership has resisted international control of its production of uranium.

Moreover, Arab bitterness is exacerbated by the nightmares experienced by the people of Algeria, Lebanon, Palestine and Iraq (oil embargo). Some wars in particular have been fueling Arab and Muslim resentment for years: the Six Day War (1967) and the Gulf Wars (1990, 2003). Public opinion accuses the West of "state terrorism" by practicing bombing of factories and exercising its economic might, and of demonizing the leaders of oil producing countries (Iran, Libya, Iraq) which results in a preventive war against a "rogue state" allegedly possessing weapons of mass destruction.

In fact the dividing lines interweave. The Sunni–Shia divide opposed old clan and tribal rivalries; religious differentiation was then one of the clear and present elements of identity. But the distinction between Salafi Jihâdist and Muslim liberal reformist splits across all Sunni and Shi'ite interpretation—literal or contextualized—of Quranic prescriptions, depending on historical and social developments. The government police forces repress extremist movements who by their attacks or hostage-taking disturb local populations. These were divided between the official mosques and politically-oriented private mosques. Al-Qaeda (The Base), even if it appears in the backstage, remains a model that inspires, by its preaching and Internet networks, vocations to martyrdom. The governments of some countries economically aligned with Western countries do not want to engage in a decisive struggle with insurgent organizations such as Al-Qaeda in Maghreb, the Egyptian Muslim Brotherhood, the Palestinian Hamas, the Lebanese Hezbollah, the Madrasas in Pakistan, the Afghan Taliban or the Indonesian irredentists. Mobilized by the struggle for independence, the revolutionary spirit of Muslim societies has been in a state of super-cooling since the end of World War II. The League of Arab States was for the states, not for the Arab people.

Jihâd and Revolution

Jihâd is a rich concept; in recent decades, certain interpretations have been privileged over others. It is frequently treated as one of the urges that perpetuates in the minds of Muslims the sense of duty, the sense of effort, the *Ittihad*, which, in varied fields (theological, intellectual and legal development, steady population growth, wealth creation, empowerment in future history, devotion to the ideal of original purity, desire for personal salvation), ensures that the ideas (that are not monolithic) of Islamic political philosophy take shape in the best possible way.

Historically, the argument that arises very early of revolt justified by the injustice of power, was likened by some to the *jihâd* by certain movements: *kharédjisme* (literally leave, revolt—murder of the last legitimate Umayyad Caliphe, 'Ali in 661); Shi'ism (literally party, sect, fraction); sometimes Mahdîsm. In fact, the eschatological constructs of the hidden imam (who must return one day to establish a state of justice in Shi'ism) and the doctrine (non-Sunni) of the *Mahdî* ("the rightly guided one" who, by the letter and the sword, stands in the name of God against an impure system) provide a dual function: to transpose into future history the exhausted cycles of prophecies announcing the arrival of a righteous one from the House of the Prophet (for the *Mahdî* against the Antichrist), and to legitimize revolutionary impulses. All this offers a vision surpassing the humble scale of social functions, participates in the maintenance and ethical purification of Muslim societies, and provides a model of an eschatological Muslim revolution, but through violence, at least in Mahdîsm.

The conjunction of the tradition of revolutionary struggle against imperialism and colonialism and of the hope for rapid economic development, the concept of *jihâd* has been transposed, ideologically and strategically, into that of revolution (*thawra*). The root *th wr* originally conveyed a sense of turmoil and excitement—swarming locusts, cornered lion—but also an impetuous call to vengeance. It is opposed to the plain will of resistance, of disobedience. Through historical thinking, the intelligentsia sought in *thawra* a model of a political philosophy and a rationale for the mission of the Arab nation located in the perspective of Anglo-Saxon revolutions of the seventeenth and eighteenth centuries, of the French revolutions in the eighteenth and nineteenth centuries, and of the Russian and Chinese revolutions in the twentieth century.

Most of these revolutionary constructions, influenced by European models, were originally located outside the theological framework, and referred to the sovereign power and decisive action of the people. The idea of having the people on one's side is associated with brilliant intellectual sources, while the old ideas of solidarity and advice (*chur*) in the Arab-Muslim community are associated with struggles aiming at a revolutionary, more or less socialist, horizon: the most disadvantaged, the rural or urban masses become aware of themselves, become masters of their destiny. The wars of liberation had combined, in political philosophy and strategy, two discourses and two practices. The first type of discourse, focusing on revolution, acted on behalf of regionally interpreted Arab-Muslim identity (Algerian, Egyptian identity), but using, against the "bourgeois" revolutions of 1688, 1789 and 1848, the vocabulary and methods of analysis and action of the great Western Socialist and especially Marxist socialist revolutions (role of the masses and of the party in revolutionary consciousness) with the addition, in ideology and tactics, of anti-imperialist discourse and methods of revolutionary warfare modeled after the Far

East (China and Vietnam) and then conceptualized by the progressive Western left: the Palestinian left was its extreme formulation. The second type of discourse, focusing on the formulation and practices of ideology and of Arab-warrior Muslim social history concerned symbolic and political recovery, but not in the canonical way, of classical terms referring to the cultural depth of the history and the teleology of strategy and sacrifice: *Mujahid fida'i, chaheid.*

These exaltations result in the security policies of the Israeli right which is unable to respond to the strategy of offensive *jihad*-fighting organizations (Hezbollah, Hamas and Palestinian Al-Aqsa Martyrs' Brigades) in any other way than by tactical counter-offensives dismantling the human (through "targeted attacks") and institutional resources of the Palestinian force, by placing collective responsibility on communities, by reoccupying Palestinian areas, and by destroying the homes of the combatants. It is a war strategy.

The Arab discourse on the revolution has taken two directions: one more or less Marxist or socialist-nationalist, a commonplace discourse on revolution in the contemporary world (mass mobilization, collective leadership of the organized party); and the other, a transposition of the old religious ideas of the solidarity of the whole community, and aid for the weak. In its transformation into a liberating action of the oppressed, the *jihâd* should lose its defensive essence, and be directed against any aggressor to save anyone oppressed (Quran XXVIII, 5 and 6, also see , 193, 217; IV, 75). The *jihâd* becomes an immediate and general struggle against all injustice or, more accurately, it recovers the beauty and the integrity of its function: to establish an immanent justice on Earth by the pressure of a world revolutionary party (true Muslims as forces or as a state), of negotiation or international mediation, or of punitive war (Quran XLIX, 9–11).

Believers have therefore combined the concepts of *jihâd* and revolution to compensate for the delays and uncertainties of Arab or Muslim socialism through the call to an Islamic revolution. A revolution that focuses essentially on the total liberation of a people or a community, rather than their division into antagonistic groups—the social classes. If populations, "nations or tribes, are made to live in peace" (Quran XLIX, 13), internal divisions should be seen all the more as nefarious: (Quran XXVIII, 4). In fact, the contradictions within each country, the Islamic renewal, the reorganization of lifestyles, class struggle, the Israeli presence, and the Palestinian question resulted in an overestimation of the revolution as a strategy, mainly because of the action of terrorist groups in the Middle East. A practice opposed to the assertion of a non-violent *jihâd*.

Therefore, to counterbalance the loss of prestige of Arab revolutionary movements and impulses of any Islamic revolution in world public opinion, some scholars emphasize another transmutation of *jihâd*, conjoining media resonance and a phase opening the *jihâd*: missionary effort, a reasonable call to repentance, an effort in rhetoric, in persuasiveness, if not to convert, at least to raise awareness. The Christian–Muslim dialogues, seminars at UNESCO, many publications and several academic projects insist at will on the irenicism of Islam and how the vision of it was distorted by Westerners. The missionary effort (*da'wa*), is indeed strongly invoked by the Quran (XVI, 125, also 11, 256, 111, 61). This *Da'wa al jihad* presents itself, in total contrast, as an exhortation to peace, as a form of mutual tolerance. It often attempts to integrate Islam in the formulation of Western values (the principles of Christian humanism, of Enlightenment philosophy and of idealist socialism, human rights, freedom, equality, fraternity) in claiming their pre-existence in Islam. In fact, it is a complex mixture of acculturation originating in the West and in a defense and illustration—actually, a system of legitimization and affirmation—of Islam in relation to the non-Muslim world, particularly the West. It transforms the individual call for conversion to a confrontation between moral codes (like a denunciation of the "corruption" of the West) and a reformulation of concepts of peace and international law.

But it also revived the exaltation of *jihâd* as the "great *jihâd" or jihâd al akbar*, that is to say, in the process of action not against the Other but in action upon one's own nature, through deeper spiritual enrichment, through self-improvement. This sublimation of the *military jihâd* has already appeared to transform it into its complete opposite: the personal struggle conducted by each individual, in one's own actions and in one's own heart, to constitute, according to an Islamic order, a better, a nobler type of humanity. A hope which, for the combatant, joins the warrior's effort with spiritual asceticism, according to an exacerbated nationalism and strict Quranic interpretations, against materialism and moral degradation, for the spread of Islam.

Long Cycle Theory and Concentrations/Deconcentration of Economic and Political-Military Resources

William R. Thompson

Leadership long cycle theory first began to appear in the 1970s. Among other things that it had to offer, it seemed to fit a United States that appeared to be in relative decline in the 1970s and 1980s, and about to be overtaken economically by a Japanese challenger. A number of changes have taken place since then. The United States rebounded in the 1990s. Not only did the Japanese challenge turn out to be hollow, the United States' principal rival, the Soviet Union, collapsed and disintegrated. For some observers, the post-Cold War United States was the world system's new and unrivaled Colossus. Is it possible that leadership long cycle theory has been rendered obsolete by changes in the real world? Not surprisingly, my answer is no. But to explain why will require some elaboration both about how to interpret the last 20 years or so and about some of the features of leadership long cycle theory. With that foundation, it will be possible to take on the question of the continued applicability of leadership long cycle theory in the twenty-first century and some of its implications for future world politics.

The End of the Cold War

The end of the Cold War that so preoccupied the United States and the Soviet Union for nearly a half-century was beneficial to most of the world's population. The elimination of threats that are linked to the possible extinction of the human race are, not surprisingly, welcome occurrences. Nevertheless, there were casualties. Governments fell. Civil wars and rebellions withered. Some asymmetrical interstate rivalries were resolved. Most of these activities will not be missed. One casualty, however, was an innocent and unfortunate victim. The casualty that is referred to is the systemic explanation of warfare. The wounds incurred were not terminal but they were serious nonetheless.

Systemic explanations of warfare were wounded by the end of the Cold War in multiple ways. Unipolar triumphalism convinced observers that a new, unprecedented era had set in. One state, unrivalled, was the surviving superpower. The old concerns with the implications of bipolarity and multipolarity seemed obsolete. Unipolarity, whether it applied or not, is a systemic attribute. Yet it was not an attribute with which most international relations scholars were accustomed to dealing. If the older structural concern with bipolarity versus multipolarity was moot, why even bother with the systemic level of analysis?

A second perceived implication of the end of the Cold War was the related belief in an end of history and the end of major power warfare. Liberalism had defeated its last ideological rival and there were no more ideas to fight over. If there was only one genuine major power left intact, warfare between major powers was inconceivable by definition.

Probably not coincidentally, the end of the Cold War was synchronized with the elevation of dyadic analyses in quantitative international relations. The main carrier of dyadic analyses was the discovery of the "democratic peace," an argument that is almost always couched in dyadic behavior. Two democratic states are unlikely to go to war with each other. Putting aside the problems encountered in explaining this observation, the overwhelming empirical support for this generalization left little need for systemic arguments. If one can explain most things dyadically, why bother with other levels of analysis?[1]

Yet we still need systemic explanations. The world is not exclusively arranged on a bilateral or dyadic basis. That is, all international relations do not reduce to what state A does to state B and how state B responds. Nor are we likely to be able to explain conflict between states A and B solely on the basis of their interactions. States A and B operate in environments with attributes (for instance, hierarchy, scarcity, history), processes (governance, economic development, war making), and other actors whose behavior may also be relevant. The classic example is the outset of World War I, the global war that no one wanted or anticipated. Assuming one could make sense of Austro-Hungarian-Serbian relations, World War I is inexplicable without also considering a number of other interacting dyads, including minimally Russia–Japan, Germany–Austria-Hungary, Germany–Russia, Britain–Germany, and Britain–France. As soon as one acknowledges the need to juggle a number of different dyadic "balls" simultaneously, the limitations of dyadic analyses become apparent. Behavior takes place in the context of complexities to which systemic analyses can contribute some explanatory power.[2]

1 The ascendance of dyadic explanations in international relations after the Cold War was presaged by a tendency to explain everything in terms of the world system's most predominant dyad, the US–USSR during the Cold War.

2 Let me be clear that the argument here is not that systemic analyses trump all other levels of analysis—only that no single level of analysis is sufficient. Monads

There is no single systemic explanation of warfare. Indeed, most systemic research programs are not even oriented to explaining warfare *per se*. But most of them do offer specific insights into how at least some warfare comes about. The focus in this chapter will rest entirely on one systemic version—leadership long cycle's interpretation of global war. But to show how the leadership long cycle research program accounts for global war, some attention must be paid to more general features of this approach.

Leadership Long Cycle Analysis[3]

The first place to begin is an all-important assumption. International relations struggles with a number of ontological assumptions about how the system is structured. Some think there is no such thing as an international system. States are real; systems are not.[4] Others think that the interactions of largely European great powers constitute the main system. The history of international relations is thus equated with the history of the European region, at least to 1945, and then Europe became less central to international relations. The system then switched to the two non-European superpowers as its main organizing device.[5] Another approach is to argue that the world is composed of a number of regions. Some states are powerful enough to operate in other regions besides their own, but at some capability and

and dyads operate in contexts that are easily lost without also figuring in systemic considerations.

3 Some of the main statements of leadership long cycle theory can be found in George Modelski, *Long Cycles in World Politics* (London: Macmillan, 1987); George Modelski, "An Evolutionary Paradigm for Global Politics," *International Studies Quarterly* 40: 3 (1996), 321–53; William R. Thompson, *On Global War: Historical-Structural Approaches to World Politics* (Columbia: University of South Carolina Press, 1988); William R. Thompson, *The Emergence of the Global Political Economy* (London: University College London Press/Routledge, 2000); William R. Thompson (ed.), *Evolutionary Interpretations of World Politics* (New York: Routledge, 2001); George Modelski and William R. Thompson, *Sea Power in Global Politics, 1494–1993* (London: Macmillan, 1988); George Modelski and William R. Thompson, "Long Cycle Critiques and Deja Vu All Over Again: A Rejoinder to Houweling and Siccama," *International Interactions* 20 (1994), 209–22; George Modelski and William R. Thompson, *Leading Sectors and World Powers: The Coevolution of Global Politics and Economics* (Columbia: University of South Carolina Press, 1996); Karen Rasler and William R. Thompson, *The Great Powers and Global Struggle, 1490–1990* (Lexington: University Press of Kentucky, 1994); and George Modelski, Tessaleno Devezas and William R. Thompson (eds), *Globalization as Evolutionary Process: Modeling Global Change* (London: Routledge, 2009).

4 Lake's useful emphasis on hierarchy over anarchy seems to proceed on this basis since there is no attention to systemic hierarchy: David A. Lake, *Hierarchy in International Relations* (Ithaca, NY: Cornell University Press, 2009).

5 Waltz is the classical example of this approach: Kenneth N. Waltz, *Theory of International Politics* (New York: McGraw Hill, 1979).

influence discount because they usually must project their power across water to reach the other regions.[6]

Leadership long cycle analysis rejects these initial assumptions as starting points. Instead, a two-level system is envisioned. The world is fragmented into a number of regions and has been for millennia. A fair amount of international relations is thus region-centric. A good deal of the history of European international relations is just that—one region's history of international relations. For a variety of reasons, European international relations became increasingly significant for the rest of the world between 1494 and 1945, but it was never the case that it was only European international relations that mattered.

A second, global systemic level emerged in world politics increasingly after 1494.[7] Global politics focuses on inter-regional transactions. These transactions are not restricted to trade, but commerce looms large as a principal consideration. There had been intermittent attempts at inter-regional governance prior to the sixteenth century CE, but they had been more limited efforts tending to span two adjacent regions. Alexander the Great's attempt to conquer his known world, or the creation of the Mongol Empire, represent trans-regional imperial enterprises. From the sixteenth century on, however, attempts to manage policy problems emerging from the control of trans-regional trade (as in Europe–Africa, Asia–Europe, Europe–Americas, and Americas–Asia) generated a new type of activity in which some states specialized as global powers. The most specialized of these global powers sought to avoid territorial expansion in their home region so that they could better focus on the development of their inter-regional networks of profit and power. Yet not all global powers were pure trading states. Nor has it been possible for global powers to remain entirely aloof from regional politics in their own backyards. It is the occasional fusion and interweaving of regional and global politics that confuses the issue for most observers. Still, we need to differentiate between the two levels of action in international relations if we want to make sense of how the two levels work differently (and how they occasionally fuse).

6 Mearsheimer provides a good example: John J. Mearsheimer, *The Tragedy of Great Power Politics* (New York: W.W. Norton, 2003).

7 The roots of the emergence of the modern global system are traced back to economic developments in Sung China in the eleventh–twelfth centuries CE and transmitted to Europe via the Pax Mongolica and Eurasian commerce (see George Modelski and William R. Thompson, *Leading Sectors and World Powers*). Genovese and Venetian activities in the thirteenth through fifteenth centuries served as prototypes for the subsequent serial development of lead economies in Portugal, the Netherlands, Britain and the United States. The year 1494 is used as a marker instead of the more Eurocentric 1648 because French intervention in Italian politics accelerated the development of European regional activities, while Portuguese activities in the Indian Ocean were initiating a more explicit phase of trans-regional economic activity that shortly thereafter became truly planetary wide in scope. Modern global politics did not spring forth in full form in 1494, but developments after 1494 are more alike than developments before 1494.

One of the implications of this initial assumption is that analysts are sensitive to different hierarchies. Some states are important to regional hierarchies and others to global hierarchies. Some are important to both. But the distinction allows us to avoid exaggerating the significance of actors that are only important at the regional level.[8] It also allows us to appreciate the salience of global predominance while not also requiring these global system leaders to be equally predominant at the regional level.[9]

A third implication is that the analyst must also be sensitive to the need to focus on different capabilities at different levels. Regional political-military interactions require land power. Global political-military interactions require the manipulation of different capabilities. To operate in trans-regional networks, sea power, more recently supplemented by air and space power, has historically been critical. Usually, the states that are good at developing land power are not the same states that have been good at projecting power at long distances. There is a reason and we cannot assume that land and sea powers should be evaluated on the same capability scale. Land and sea powers (elephants and whales) are not necessarily playing the same game.

Another theoretical-empirical contribution, and certainly a principal one, is that modern world politics (that is, since 1494) have been characterized by long cycles of concentration and deconcentration. Although some observers have misread the theoretical motivation, one of the primary foci of leadership long cycle analysis in the early years was developing long series in, first, naval capabilities and then, subsequently, in leading sector production.[10] One of the central premises of leadership long cycle interpretations is that capability concentration has oscillated, as opposed to being constant or random, and that a distinctive set of actors have succeeded one another in providing a leadership sequence. The claim was that first Portugal and then the Netherlands were the first two system leaders of the past half-millennium. They were followed by two British terms and at least one US turn at the helm.

Many professional observers might accept the US claim to systemic leadership in the post-1945 era. Many of these analysts would also accept a significant leadership role for Britain in the nineteenth century. Rather few scholars seem comfortable with or care about the eighteenth century British claim. A few more would acknowledge a significant Dutch role in the seventeenth century. Of the five, the sixteenth-century Portuguese claim seems the most outlandish to most people.

8 Some of the classical European great powers, such as Austria-Hungary, were never all that important as global actors.

9 Too many observers assume that global hegemony must look exactly like regional hegemony when, in fact, they are predicated on much different power principles. The United States is the one exception to the historical "rule" that global hegemons have not also been regional hegemons. In the US case, regional hegemony was acquired before the US became very prominent in global activities. But it was also acquired in a regional system that was less competitive than has been typical.

10 It would make for an interesting historical experiment if we had reversed these priorities and stressed the leading sector series first, rather than second.

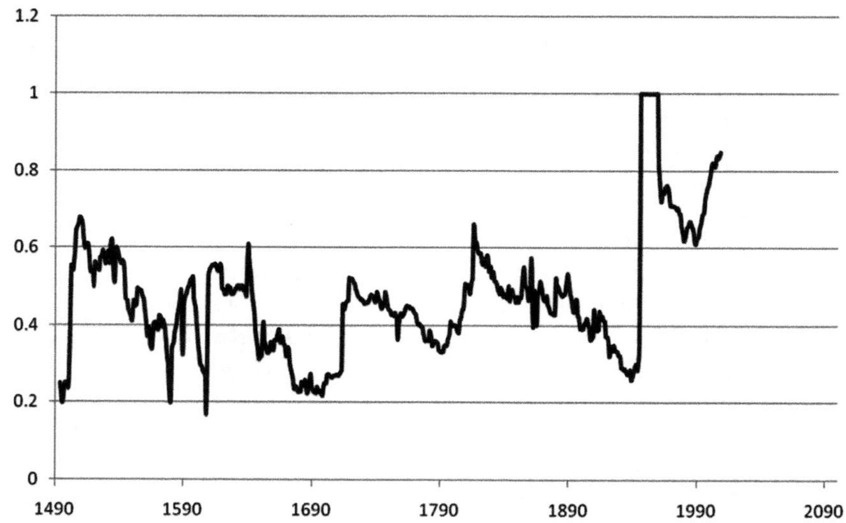

Figure 27.1 System leader naval share

Since support for a basic list of system leaders is a rather central pillar for any theory focusing on systemic leadership, considerable effort was put into creating a 500-year series of fluctuations in global reach capabilities, illustrated in Figure 27.1, to test the proposed sequence.

Given the technological changes experienced by navies over the last half-millennium, it was impossible to find any single and consistent indicator capable of spanning the whole time period. Instead, a more complex schedule or set of indicators was constructed that changed in keeping with real-world changes. Beginning with armed sailing ships owned by the state, through ships of the line with an escalating minimal number of guns, to the mix of battleships, heavy aircraft carriers, and nuclear attack and ballistic missile submarines of more recent years, it proved possible to operationalize the distribution of sea power over a fairly long period of time. The empirical outcome,[11] as updated in Figure 27.1, supports the hypothesized sequence and timing of leadership between 1494 and the current period.

One of the frequent criticisms of an emphasis on sea power is that it is both anachronistic and incomplete. Sea powers are no longer only sea powers; they are also predominant in air and space. We have always been sensitive to this assertion and have responded that we acknowledge air and space as critical domains for what Posen calls the global commons,[12] but that our sea power indicators captured some

11 George Modelski and William R. Thompson, *Sea Power in Global Politics*.

12 Barry R. Posen, "Command of the Commons: The Military Foundation of U.S. Hegemony," *International Security* 28: 1 (Summer 2003), 5–46.

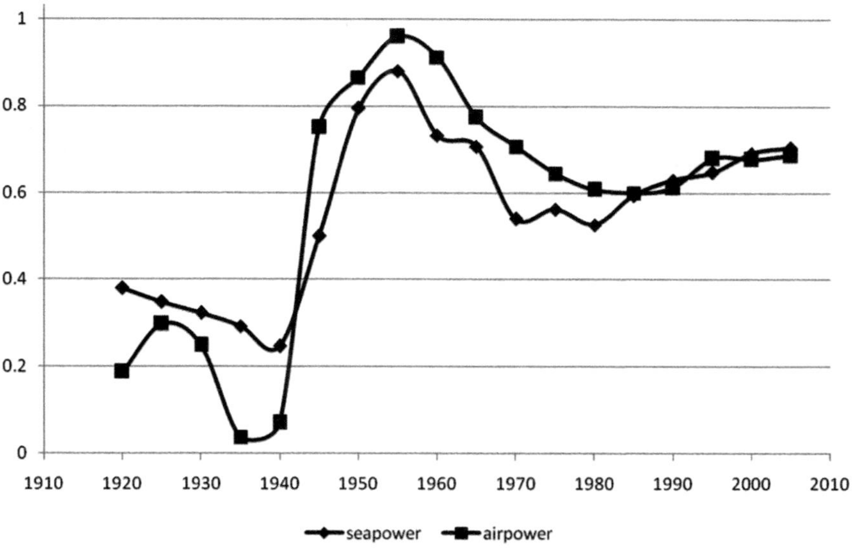

Figure 27.2 US sea and air power shares, 1920–2005

of this more recent capability and that, in any event, it was unlikely that expanding the nature of the indicators to air and space would alter the nature of concentration and deconcentration in world politics. We are now able to demonstrate this with a new index of air and space capability concentration going back to 1916.[13] Figure 27.2 shows that introducing air and space capabilities does little to change our ability to chart fluctuations in global reach capabilities.

Yet it is not just sea power and political-military systemic leadership that oscillate in long cycles. Sea (and aerospace) power is expensive. Decision-makers also need ample incentive to construct blue water fleets. The funding, and the basic motivation, for constructing sea power, is found in patterns of economic innovation. Very much fundamental to leadership long cycle theorizing is the idea that long-term economic change is stimulated by radical innovations in commerce and industry. These innovations are spatially and temporally concentrated in one state for a finite period of time, as delineated in Table 27.1 and Figure 27.3. After they are introduced, they bring about major changes in the way economies function as their techniques and implications diffuse throughout the pioneering economy

13 1916 marks the emergence of the employment of strategic bombers in global warfare. This new index combines indicators on the distribution of strategic bombers, land-based ICBMs and military satellites: Michael Lee and William R. Thompson, *Measuring Command of the Commons and Global Capability Reach* (Bloomington: Department of Political Science, Indiana University, 2010).

and then to other advanced economies that are in a position to adopt or adapt the new ways of doing business.[14]

Table 27.1 Leading sector timing and indicators, fifteenth to twenty-first centuries

Lead Economy	Leading sector indicators	Start-up phase	High growth phase
Portugal	Guinea Gold	1430–1460	1460–1494
	Indian Pepper	1494–1516	1516–1540
Netherlands	Baltic and Atlantic Trade	1540–1560	1560–1580
	Eastern Trade	1580–1609	1609–1640
Britain I	Amerasian Trade (especially sugar)	1640–1660	1660–1688
	Amerasian Trade	1688–1713	1713–1740
Britain II	Cotton, Iron	1740–1763	1763–1792
	Railroads, Steam	1792–1815	1815–1850
United States I	Steel, Chemicals, Electronics	1850–1873	1873–1914
	Motor Vehicles, Aviation, Electronics	1914–1945	1945–1973
United States II?	Information Industries	1973–2000	2000–2030
	?	2030–2050	2050–2080

As pioneers, the initial source of new best-practice technologies reaps major profits and economic development. They need sea power to protect from potential predators the affluent home base and the sea routes via which its products are distributed around the world. In the early leaders, major advances in ship construction were critical to the packages of innovations being introduced to the world economy. More generally, though, the gains from pioneering new commercial networks and industrial production financed the leading arsenals of global reach capabilities developed by system leaders. Those same gains later led to system leaders becoming a, if not, the principal source of credit for the world economy.

Thus, at the heart of leadership long cycle theorizing is a model of long-term economic growth.[15] There is no denying the importance of population size, resource endowment wealth, mass and elite consumption, savings and other standard foci

14 Thompson and Reuveny argue that the unevenness of the technological diffusion process is the main source of North–South economic inequality: William R. Thompson and Rafael Reuveny, *Limits to Globalization and North-South Divergence* (London: Routledge, 2010).

15 George Modelski, "An Evolutionary Paradigm for Global Politics;" Joachim Rennstich, *The Evolution of the Digital Global System* (New York: Palgrave-Macmillan, 2008).

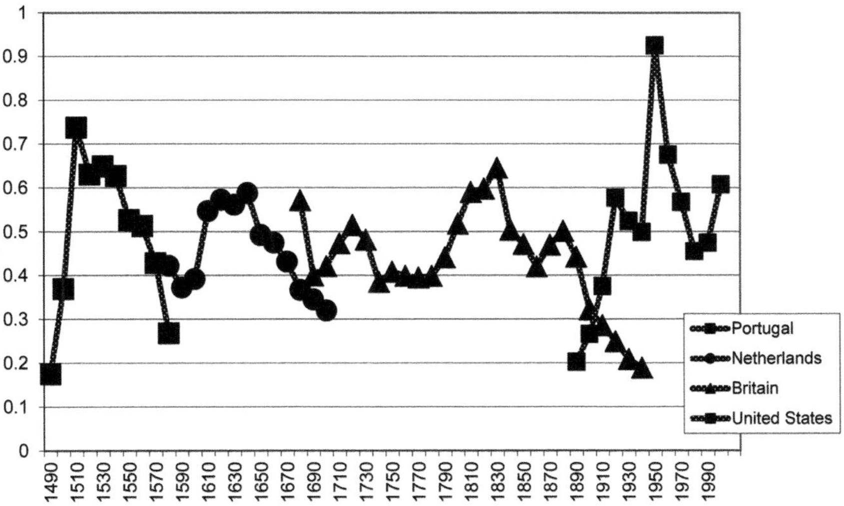

Figure 27.3 Leading sector concentration

of economic growth models. But these are primarily short-term considerations. Over the long haul, development is driven by radical technological revolutions that occur roughly every half-century or so. These are the long waves of economic growth that are also referred to as Kondratieffs or k-waves. They are more irregular waves than neat cycles coming and going with precise periodicity. They also can best be visualized as S-shaped, logistic growth. New technology enters the lead economy slowly at first, then accelerates and ultimately peaks at some point before decaying in impact as new technology becomes increasingly routine.

By focusing on the leading sectors that are at the heart of these technological breakthroughs, it is also possible to measure them, thereby providing important empirical support for the claims that these phenomena exist. It has also been possible to demonstrate that their main carriers, the leading sectors, stimulate the economic growth of the system leader's national economy and the world economy.[16]

Combining the emphases on systemic leadership and long waves of economic growth leads to a third principal contribution, a distinctive explanation of global warfare, that involves a number of related models that address when, why and by whom global wars are fought. Global wars, identified in Table 27.2, represent systemic crises for the global system in the sense that deconcentration has undermined the

16 Rafael Reuveny and William R. Thompson, "Leading Sectors, Lead Economies and Their Impact on Economic Growth," *Review of International Political Economy* 8 (2001), 689–719; Rafael Reuveny and William R. Thompson, *Growth, Trade and Systemic Leadership* (Ann Arbor: University of Michigan Press, 2004); Karen Rasler and William R. Thompson, *Puzzles of the Democratic Peace: Theory, Geopolitics and the Transformation of World Politics* (New York: Palgrave-Macmillan, 2005).

earlier hierarchy and created an opportunity for leadership succession. Global wars last about 30 years on average, mobilize all major powers as participants, and usher in a new era of re-concentration in global politics. Obviously, most wars are not global wars. Yet while they have been few in number, they claim a disproportional share of war deaths and war impacts.[17] They have also been a central switch in global processes of power concentration and deconcentration.

Table 27.2 Global wars

Global war	Timing	Issues
Italian/Indian Ocean Wars	1490s–1510s	Franco-Spanish contest over Italian states; Portuguese breaking of Venetian/Mamluk eastern trade monopoly.
Dutch Independence War	1580s–1600s	Opposition to Philip II's expansion; Dutch breaking of Spanish/Portuguese eastern trade monopoly.
Louis XIV Wars	1680s–1710s	Opposition to Louis XIV's expansion; French attempt to break Dutch trading monopoly in Europe and elsewhere.
French Revolutionary/ Napoleonic Wars	1790s–1810s	Opposition to French expansion; French attempt to resist British industrial lead and systemic leadership.
World Wars I and II	1910s–1940s	Opposition to German expansion; German attempt to succeed Britain as system leader.

When?

The first model divides each long cycle into four sequential phases, specified in Table 27.3.[18] Each phase encompasses distinctive behavior. As leaders of victorious coalitions, system leaders complete their ascension in global wars. With their opponents (and often their allies) exhausted, the system leader possesses its strongest relative capability position. Thus, the post-global war phase maximizes the opportunity for making new rules and institutions for global affairs. The next two phases are periods of relative decline. Challenges of the leadership established in the preceding phase begin to appear in weak form in one phase. These early challenges are followed by a phase described by maneuvering among the major

17 Karen Rasler and William R. Thompson, *War and State Making: The Shaping of the Global Powers* (Boston: Unwin Hyman, 1989).

18 Precisely where in the cycle one starts is somewhat arbitrary. An alternative perspective, for instance, involves starting as a new world power begins its ascent to systemic leadership two phases before the onset of global war.

powers to form alliances for a future war about regional hegemony and leadership succession issues.

Table 27.3 Systemic leadership phases

Global system leader	World power	Delegitimation	Deconcentration / coalition building	Global war
Portugal	1516–1540	1540–1560	1560–1580	1580–1609
Netherlands	1609–1640	1640–1660	1660–1688	1688–1714
Britain I	1714–1740	1740–1763	1763–1792	1792–1815
Britain II	1815–1850	1850–1873	1873–1914	1914–1945
United States	1945–1973	1973–2000	2000–2030	2030–2050

Table 27.3 offers a structural calendar of "seasons" of political behavior. One period is devoted to intensive fighting (1494–1516, 1580–1609, 1688–1714, 1792–1815 and 1914–1945) and the development of new or renewed leadership. The next period is characterized by optimal conditions for engaging in systemic leadership, especially in comparison to the next two periods of relative decline on the part of the system leader. Preparing coalitions for the next round of struggle tends to dominate the second of these phases (1560–1580, 1660–1688, 1763–1792 and 1873–1914).

To the extent that the periodization captures the last half-millennium reasonably well and the pattern persists into the future, one can project the phase sequencing into the future. The last global war (1914–1945) was followed by a period of optimal systemic leadership (1945–1973). Various types of challenge to US system leadership were exhibited in the 1973–2000 phase. Coalition maneuvering should be observed in the present period (2000–2030). Some type of intensive conflict might then be anticipated, on the basis of the past 500 years, in the 2030–2060 phase.

Why?

The first model establishes when but does not say why global wars are likely to occur. A second model, referred to as the Twin Peaks model, advances an answer to the why question. The Twin Peaks model[19] emphasizes technological change. Radical innovation tends to come in intermittent clusters and is concentrated geographically in the system's lead economy. The twin peaks reference refers to the tendency for each lead economy to experience at least one pair of clustered surges in innovations. The first surge peaks in the deconcentration/coalition building phase (see Table 27.3). It pushes one state ahead of the others in terms of technological/

19 George Modelski and William R. Thompson, *Leading Sectors and World Powers.*

commercial advantage and wealth that contributes to the destabilization of the world economy's status quo. Yet the wealth generated in this phase also helps defray the costs of the coalition building and war fighting of the global war in the next phase.

At the end of the global war, the system's lead economy is set up to undergo its second surge of economic innovation (in the world power phase). The "set up" is partially due to the end of a long period of warfare creating an environment to facilitate improving on the innovations of the earlier surge. War-induced necessities may also facilitate further innovation if new technologies are harnessed to war purposes and then later released for civilian applications. Winning the past war is also better than losing it when it comes to creating or re-creating new sources of affluence.

Innovation surges are propelled by identifiable commercial and industrial sectors that lead the way to extending the technological frontier. Representatives of leading sectors in history are subject to a two-phased sequence of preparation and actualization that is summarized in Table 27.4. The asserted pattern of innovation surge/global war/innovation surge seems well documented in the table.

Table 27.4 The timing of leading sector growth surges and global war

First high growth surge	Global war	Second high growth surge
Portugal		
1460–1494	1494–1516	1516–1540
Netherlands		
1560–1580	1580–1609	1609–1640
Britain		
1660–1688	1688–1714	1713–1740
1763–1792	1792–1815	1815–1850
United States		
1873–1914	1914–1945	1945–1973

Who?

A third model, focused on commercial rivalry, helps to explain the destabilization of the world economy wrought by new spikes of innovation and, as well, which states/economies are most affected. Contrary to economic models of divisions of labor, the main economic competitors find themselves producing similar products and wrestling over the same markets for raw materials and consumers. This propensity is dependent on production efficiencies encouraging expansion and exports. Transportation infrastructures, not unlike the economic equivalent of strategic bases, must also be constructed, and it is difficult to avoid frictions as competitors collide in space.

Latecomers aggravate the probability of friction by choosing more predatory commercial practices to catch up to the pioneering leader. Protectionism, dumping and exclusive monopolies are seen as unfair by trading states espousing free trade principles. Authoritarian political structures are more likely given catch-up goals, which makes conflict with early democratizers more likely. Economic depressions make both the competitions and appearance of predatory practices more acute. Inject some increased economic interdependence which can both constrain and promote greater conflict and one has a recipe for economic conflict enhancing the likelihood of political-military conflict between defensive economic pioneers and challenging late comers.

A fourth model, regional–global dissynchronization,[20] relies on a co-opted Dehioan premise[21] that aspiring regional hegemons in Europe were thwarted by counterweights that were able to win by introducing extra-regional resources into the balancing contest. One authoritarian counterweight supplied land power from the east. The other entered from a western position based on its sea power and commercial mediation focused on Europe/Asia/America. Operating simultaneously, the two counterweights were able to defeat a sequence of hegemonic attempts by forcing states attempting to take over the region to fight on two fronts, and usually without much access to extra-regional resources.

A 1494 French bid for Italy was stopped primarily by Spain. France and the Ottoman Empire combined to block a sixteenth-century Habsburg hegemonic effort. A true western counterweight emerged later in the sixteenth century in the form of Dutch/English resistance to Philip II and Spanish ambitions. The same western counterweight was mobilized to help defeat Louis XIV in the late seventeenth and early eighteenth centuries. By the end of the eighteenth century, Britain alone served as the western counterweight to the regional threat posed by the French Revolutionary and Napoleonic Wars. Against the German challenges of World Wars I and II, Britain and the United States cooperated in the west while the earlier Austrian role in the east had been taken over by Russia.

What Dehio contributed was an appreciation for the pattern of regional–global structural change. At the regional level, peaks in concentrated land power alternated with troughs in regional concentration. New regional aspirants tended to emerge during the regional trough periods. It was during these same low regional concentration periods that western counterweight strength was greatest. Thus, the principal dynamic was one of alternating relative strength. The leading land power

20 William R. Thompson, "Systemic Leadership and Growth Waves in the Long Run," *International Studies Quarterly* 36 (March, 1992), 25–48; Karen Rasler and William R. Thompson, *The Great Powers and Global Struggle*; Karen Rasler and William R. Thompson, "Malign Autocracies and Major Power Warfare: Evil, Tragedy and International Relations Theory," *Security Studies* 10 (Spring 2001), 46–79.

21 Ludwig Dehio, *The Precarious Balance: Four Centuries of Balance of Power in Europe* (New York: Knopf, 1962).

in the region tended to be strong when the western sea power was relatively weak and vice versa.

Moreover, the two states pursued dramatically different agendas. The leading land power utilized absolute royal powers and large armies/bureaucracies to expand its territorial control in the home region. The leading sea power sought to evade territorial control by specializing in monopolizing markets at home and abroad. Yet success for the leading land power implied a direct and indirect threat to the leading sea power. Regional hegemony either led to the occupation of the reigning or former sea power and/or created a very strong base for a subsequent challenge for extra-regional colonies and markets. Western sea powers, therefore, never lacked structural incentives to coordinate the suppression of regional hegemonic aspirations.

Western sea power specialized in maritime containment of regional hegemony. Such a strategy could limit the expansion of a regional hegemon beyond the confines of the region but it would not suffice to defeat the hegemon on land. Someone needed to supply substantial land power for that purpose. A coalition combining the requisite naval and army capabilities could then squeeze the aspiring hegemon into over-extending itself fighting on eastern and western fronts at the same time, ultimately unsuccessfully.

The key argument is that declining global powers are likely to be threatened by ascending regional powers. Accordingly, maritime/commercial powers will seek to build coalitions to contain and defeat leading land powers before they become genuinely global threats. In the process, sea power capabilities are renewed and some land power capability is exhausted. This fourth model directs our attention to situations involving declining leading sea powers that are threatened by ascending or leading land powers focused primarily on regional territorial control. It is not a matter of the presence or absence of one or the other factor but their interaction that is most critical.[22] Whether a second counterweight is present or potentially recruitable is more a question of who wins and loses regional wars than it is a concern for anticipating systemic transitions.

Thompson argues that World War I was due at least in part to multiple interstate rivalries that were "ripe" for escalation.[23] He generalizes about the circumstances that increase the probability that rivalry ripeness is translated into nonlinear war escalation: three conditions are advanced in this fifth model. The first concerns serial conflicts within rivalry relationships. An increasing number of rivalries experiencing multiple clashes increases the probability of rivals going to war.

The degree to which antagonists are bound together or "coupled" is another factor increasing the probability for expanded conflict. For this reason, the extent to which rivals are bipolarized into two mutually exclusive camps is highly suggestive for conflict escalation. Allies are not only closely linked to allies in a bipolarized

22 Karen Rasler and William R. Thompson, "Malign Autocracies and Major Power Warfare."

23 William R. Thompson, "A Streetcar named Sarajevo: Catalysts, Multiple Causation Chains and Rivalry Structures," *International Studies Quarterly* 47 (2003), 453–74.

situation, the sheer simplicity of the conflict pattern means that enemies are also closely coupled even if their only shared attributes are mistrust and conflict.

Power transitions represent a third type of conducive factor. A leader being overtaken is apprehensive about future losses. The overtaking challenger may be overly confident of its imminent success. As they approximate capability parity, the probability of one or the other type of actor initiating conflict with the other antagonist increases. The more central these power transitions are, moreover, the more likely the resulting conflict will lead to nonlinear conflict expansion. Examples of more central power transitions would encompass global–regional ones and ones involving central regional leaderships.

This fifth model, therefore, has us look at the frequency of serial clashes within rivalries, the extent to which groups of rivals are bipolarized, and whether multiple power transitions, especially more central ones, are ongoing. In general, we are looking for situations that might be expected to make rivalries more ripe for bloodshed in nonlinear ways.

The Prospects for Global War

Sensitivity to the historical pattern of systemic changes prompts one to ask what "system time"[24] is it now? As noted earlier, we are currently in the decline trajectory of the US systemic leadership era established in 1945. The end of the Cold War confused many observers into thinking something novel had taken place. But that reaction was in part a function of bestowing too much significance on the Cold War in the first place. It reflected an early challenge of American systemic leadership that ran out of steam because the challenger was unable to develop an appropriate economic foundation for the challenge. The Soviet Union's economy was not developed to extend the frontiers of technology. It could barely service the demands of a traditional, if very large, land power (albeit one with an untraditional ideological mission) attempting to expand its influence throughout Eurasia and, to a lesser extent, beyond.

When the Soviet challenge collapsed, it appeared as if 1945 had taken place all over again. The threat to the incumbent systemic leadership had been defeated. The old system leader was somehow re-invigorated. A new world order was about to be established. Instead, the incumbent system leader relatively quickly became bogged down in further delegitimizing activity (delegitimizing for system leadership purposes) in Afghanistan and Iraq. The system leader retained its monopoly on global reach capabilities (much of which it had maintained throughout the Cold War) but it had not regained the kind of technological lead it had possessed in the 1940s. The Cold War was not a global war and the post-Cold War system leader no longer had the foundation for systemic leadership it had once possessed.[25] By 2010,

24 William R. Thompson, *On Global War*.

25 The foundations for systemic leadership are high growth rates in leading sectors, high

it had become more apparent that US relative decline had not been interrupted by the end of the Cold War. It had only fooled a number of observers into thinking otherwise.

The relative decline of US systemic leadership is manifested clearly in a world economy that is increasingly multipolar. Not only is the US economy dependent on Chinese loans and finding it increasingly difficult to pay for guns and butter (see the political turmoil over domestic health insurance or the bemoaning of crumbling infrastructure), its standard of living is matched and sometimes exceeded in many parts of the global North. Another useful indicator is the expansion of the number of elite states that meet each year to discuss world problems. More states have significance to the world economy, so what was once seven major economies has been expanded to as many as 20.

Nothing is more emblematic of world order decline and decay than the resurgence of maritime piracy and the reliance on non-US navies to police the Indian Ocean. The decay of its world order is also evidenced by the many problems encountered by the global institutions created at the end of World War II. They are under-funded, over-worked, and little heeded. Many of the activities that they now attempt to address were not foreseen in 1945. It has proven difficult to re-orient them to new problems in a context of world order decay in which even the system leader has given up on their ability to provide institutional solutions to global problems.

If the decline of systemic leadership appears to be on track, despite some earlier miscues, that could suggest that we are still on track with the pattern sketched in Table 27.3. In an era of deconcentration and coalition-building, the transitioning of the system leader towards appearing more like a "normal," if still powerful, state is to be expected. So, too, is the maneuvering, increased belligerence, and flirting with alliance-making on the part of potential challengers such as China and Russia. Fundamental Islam's rejection of US-led globalization and liberalism is another sign of increasing attacks on the world status quo and the order established in 1945.

Does that mean we should expect the outbreak of a global war some time after 2030? The answer is "not necessarily." Leadership long cycle theory is evolutionary in nature and the world system that it studies continues to evolve.[26] The twenty-first century does not look exactly like the sixteenth century did, any more than the sixteenth century resembled the twelfth century. Things change and the question is whether or to what extent what changes alters how things work in some fundamental fashion.

---

shares in the production of leading sectors, and a monopolistic lead in the possession of global reach capabilities. The United States possessed only the third leg of the foundation after the Cold War ended. It possessed a great deal of economic wealth, but affluence is not the same thing as pioneering radical new technologies that revolutionize industrial/commercial practices.

26 George Modelski, "An Evolutionary Paradigm for Global Politics"; William R. Thompson (ed.), *Evolutionary Interpretations of World Politics*.

One problem with an evolutionary perspective is that the complexity of interacting processes in the future makes forecasting hazardous. We can say that in the short term there seems little reason to expect anything on the order of a global war in the next 15–20 years. Going beyond 20 years is trickier. Here are some possibilities that seem compatible with past tendencies, even if they are not all equally probable:

- Perhaps the key to the future lies with what happens to the generation and location of radical economic innovation. If we are due for a new spurt of revolutionary technology, it will probably take the form of nanotechnology, vehicles that no longer depend on petroleum-run engines, and biotechnology applied to health, water and perhaps even computing. We can see these transformations beginning to appear more or less on schedule. Who will pioneer these new innovations?[27] Will they be as concentrated in time and space as their predecessors have been? One possibility is that they will be concentrated in time and space within the US economy. Britain held the lead through four spurts of technological innovation. The United States has led through two and could conceivably lead the way through two more. Whether this happens or not, it does not necessarily reduce the possibility of global war. But it might lead to a change in the pattern if the US technological lead is so great that challengers hesitate to take it on.
- A second possibility is that the radical innovations that are coming will continue to be concentrated in time and space but will appear primarily elsewhere, as in, say, China. A Chinese lead in technology may not be apparent in 2010 but it is certainly not inconceivable that the US economy could stumble and not capitalize on its head start. There should also be no question as to whether Chinese science and industry is attempting to take the lead in technology. They are. Let us assume that they are successful. Does that increase or decrease the probability of a global war? Other things being equal, it might increase the probability of a global war because the uncertainties associated with the nature of a Sino-centric world system would presumably be more threatening than the perpetuation of the more familiar American-centric system.[28]
- It is possible that the next wave of new technologies will not be as concentrated in time and space as earlier. Should this come to pass, it could reduce the chances for a global war according to Table 27.3's schedule. The new technology, if more equitably distributed, might be less destabilizing to the system's hierarchy. Less destabilization should mean a lessened probability

27 Zakaria is convinced that the United States will lead in the development of new technology.

28 This statement is not meant to suggest that some will find a Sino-centric system more desirable and less threatening than the perpetuation and reinvigoration of the US-shaped world order; Fareed Zakaria, *The Post-American World* (New York: W.W. Norton, 2008).

of fighting about whose preferences will lead the system. On the other hand, it might also mean less systemic hierarchy which might increase the chances of fighting over whose preferences will shape the world economy, especially in an environment characterized by deterioration and increased scarcity.
- Leadership long cycle arguments about global war have been based on the intermittent fusion of European regional dynamics with global dynamics. The European region seems to have played out its potential for making trouble in the world system. Old aspirants to regional hegemony have exhausted themselves. New aspirants are not even on the horizon. If we remove Europe from the global war equation, does that mean the probability of global war moved to zero after 1945? Maybe, but it is also possible that other regions will supplant the historical role played by European affairs. East Asia has considerable potential to become the new cockpit of the world system. It could become the richest and most powerful region in the system. It certainly has potential for continued struggle among the most powerful Asian actors for regional hegemony—as was witnessed in the first half of the twentieth century. Many of the same elite actors remain in place and new ones such as India have joined the contestants. East Asian international relations need not resemble identically the past 500 years of ultra-competitive, European international relations, but it cannot be assumed that Europe's past is not Asia's future.
- Whatever happens in terms of technological concentration, other evolving processes might be expected to reduce the probability of global war. A simple way[29] of framing factors thought to be associated with interstate violence is to array drivers that are conducive to increased violence against factors thought to constrain conflict. The question then is which variables are stronger—factors that encourage or discourage conflict and violence?

In the models reviewed above, the drivers are capability deconcentration, rivalry and struggles over primacy. Not much has been said about possible constraining factors. They certainly exist and seem to be growing stronger, at least in some parts of the world. There are three main sets of constraints: the rising costs of warfare, the increased benefits of peace, and the Kantian cluster of variables (democratization, international organizations and economic interdependence).

- The rising costs of warfare are due to the increased lethality of weaponry and the ability of strong states to mobilize ever greater resources for interstate combat purposes. The deterrence effect of nuclear weapons may suffice to deter global war. We do not really know yet how strong the deterrent effect of nuclear weapons is, or has been. The rising costs of war may be able to dissuade

29 Bruce Russett, "Violence and Disease: Trade as Suppressor to Conflict When Suppressors Matter," in Edward D. Mansfield and Brian M. Pollins (eds), *Economic Interdependence and International Conflict: New Perspectives on an Enduring Debate* (Ann Arbor: University of Michigan Press, 2003).

ascending challengers and declining leaders from attacking one another. It may at least give them some pause. But we also think that the possession of nuclear weapons can encourage aggressive behavior up to a point, with some expectation that an opponent's reaction will be less than maximal. Then the question becomes how well decision makers can be expected to manage "tit for tat" exchanges without losing control of the escalation process.

- The increased benefits of peace refer primarily to affluence, economic development and trade.[30] The question is whether or to what extent populations are willing to give up their comfortable living styles in order to fight. One problem here is that the benefits of peace are unequally distributed. Where they are strongest is also where one would expect their constraint to work most intensely. The problem here is that the most likely challenger (China) is neither affluent nor developed. Do we know what happens when the benefits of peace are asymmetrical? Are states with more benefits more hesitant to go to war? Are states with fewer benefits less hesitant to go to war? Peace benefits can be very attractive, but what happens when affluence, development and trade are all threatened by a challenger? Presumably, benefits that are seen as slipping away are less likely to constrain actors from fighting than are benefits that remain constant. Still, this is an assertion based on logic, not empirical evidence.
- The Kantian cluster[31] also has impressive claims to work as constraints, but they also possess some important liabilities.[32] Democratization remains incomplete. If global war confrontations pit democratic states against authoritarian states, it may not matter that we continue to have problems nailing down just what it is about regime type that constrains conflict in some instances.[33] International organizations are less effective when system leaders are in decline than when they are strongest. Organizational membership density is also greatest where global war is least likely to break out (Europe) and weakest where global war is somewhat more probable (Asia). Something similar applies to economic interdependence. Although it overlaps

30 Richard Rosecrance, *The Rise of the Trading State: Commerce and Conquest in the Modern World* (New York: Basic Books, 1987); Azar Gat, "The Democratic Peace Theory Reframed: The Impact of Modernity," *World Politics* 58 (2005), 73–100.

31 Bruce Russett and John Oneal, *Triangulating Peace: Democracy, Interdependence, and International Organizations* (New York: W.W. Norton, 2001).

32 David P. Rapkin and William R. Thompson, "Kantian Dynamics and Systemic Transitions: Can International Organizations Influence U.S.-China Conflict?" in William R. Thompson (ed.), *Systemic Transitions: Past, Present and Future* (New York: Palgrave-Macmillan, 2008).

33 For different long cycle interpretations of the democratic peace, compare Modelski and Gardner with Rasler and Thompson: George Modelski and Perry Gardner III, "Democratization in Long Perspective," *Technological Forecasting and Social Change* 39 (1991), 23–34; George Modelski and Perry Gardner III, "'Democratization in Long Perspective' Revisited," *Technological Forecasting and Social Change* 69 (2002), 359–76; Karen Rasler and William R. Thompson, *Puzzles of the Democratic Peace*.

> considerably with the benefits of peace, greater economic interdependence might be expected to give decision-makers pause when contemplating interstate violence. But it may also prove to work best twenty years down the road where it is least needed. Economic interdependence also has a down side in the sense that interdependent economies producing similar products and competing for the same markets can be conducive to increased conflict, not less.[34]

So, is a global war probable twenty or more years down the road? It is hard to say. There is no strong reason to assume that leadership long cycle theory has been rendered obsolete by a changing world. The world does not seem to have changed that much just yet. Yet there are a range of factors that are difficult to predict with any degree of accuracy. At least we know what to look for: What happens to new technology? Will it be concentrated in time and space? If so, where? Will strong challengers emerge? Will rivalries become more "ripe" for nonlinear escalation? Will Asian states fight for regional hegemony? Will possible constraints on interstate violence evolve fast enough to restrain urges to fight over contested primacy in a region or in the world economy?[35] We will have to wait and see how things evolve over the next 20 to 50 years. While it would be nice to have definitive answers to these questions, the next best thing is to know which questions to ask.

34 David P. Rapkin and William R. Thompson, "Economic Interdependence and the Emergence of China and India in the 21st Century," in Ashley Tellis and Michael Wills (eds), *Strategic Asia 2006–07: Trade, Interdependence and Security* (Seattle, WA: National Bureau of Asian Research, 2006).

35 At the very least, it is difficult to imagine a global war lasting for 30 years in the middle of the twenty-first century. Should one occur, it will likely be much shorter than in the past. Yet one could imagine a long-running, constrained feud, perhaps along the lines of the US–Soviet Cold War, in the middle of the twenty-first century. A constrained China–US rivalry already appears to be underway.

Preventing Global War

George Modelski

The theory of long cycles in global politics predicts that the global polity, an emerging property of the world system, is moving, over the long run, toward a higher degree of organization, and that the approaching macro-decision (succession to global leadership and new global agendas) may probably avoid the turmoil of global wars that has marked the previous five cycles. However, some chance remains that in the process of competition for global power, major conflict (or conflicts) of worldwide impact might actually erupt in the next generation. In the light of the theory, this chapter examines conditions and scenarios that might reduce such chances, thus favoring the emergence of a non-global-war-like trajectory for the political process that is part of world system evolution.

Long Cycle Theory

The title of this chapter, "Preventing Global War," signals the review of an important global problem, that of understanding and mastering the possibility that global politics might, some day in the near future, break out into violent conflicts of horrifying proportions. Some might argue that such a possibly catastrophic outcome should not be debated, and should be tabooed, on the grounds that the mere discussion of it might help to bring it about. Others believe that such a topic needs to be relegated to the sub-rational realm, such that even the possibility of it might never emerge in rational decision-making.

This presentation adopts the view that the subject needs to aired in public discussion so as to improve the chances that global war will not in fact recur, perhaps banned forever from the experience of the world system. The review of this problem will be carried out from one special vantage point, that of a research program that has become known to scholars as "long cycle theory." What are long cycles of global politics about?

Long cycles and global wars

The research program may be said to have been launched with my paper presented to the Tenth World Congress of the International Political Association in Edinburgh in 1976 and published as "The Long Cycle of Global Politics and the Nation-state" in *Comparative Studies in Society and History* 20(2) April 1978. The paper introduced to the literature the concept of a (phased) political long cycle of some one hundred years in length, each cycle moved forward by a "world power" exercising "global leadership"[1] (since 1500: Portugal, the Dutch Republic, Britain—twice, the United States). The ensuing succession was marked by repeated confrontations with "challengers," culminating in "global wars," defined as (generation-long) "conflicts that determine the constitution of the global political system." That paper was, of course, building on the work of earlier scholars, in particular Quincy Wright and Arnold Toynbee, but it was only the beginning of a long adventure.

By the end of the 1980s the long-cycle research program had yielded a number of publications on these themes, including theoretical and collaborative,[2] documentary[3] and empirical-statistical studies[4] supporting the existence of regularities in global politics and placing it in the context of other historical-structural approaches.[5] Nuclear deterrence theory was subjected to a critique from a long cycle perspective.[6] A paper on "Long Cycles and Global War" appeared four years later.[7] Most important perhaps was the insight that long cycle competition was not about "world domination" but about the emergence and the constitution of the global system.

In the course of further developing the theory, it became clear that each long cycle is a four-phased process, and that the same process might be viewed from two perspectives, either systemic, or learning. In the *systemic* perspective, the generation-long phases begin with world power (say US 1945–75), and go on to delegitimation, followed by deconcentration (2000–) and global war. In the

1 Sometimes also referred to as "global primacy" or "hegemony" (Greek for leadership). "Global leadership" is a term specific to long cycle theory; primacy or hegemony do not readily resonate with the experience of Portugal and/or the Dutch Republic that may nevertheless be described as a leadership exercise in global system-building; T. Devezas and G. Modelski, "The Portuguese as system-builders," in G. Modelski, T. Devezas and W.R. Thompson (eds), *Globalization as Evolutionary Process* (London: Routledge, 2008).

2 George Modelski and William R. Thompson, *Sea Power in Global Politics 1494–1993* (London: Macmillan, 1988).

3 George Modelski and Sylvia Modelski, *Documenting Global Leadership* (London: Macmillan, 1988).

4 G. Modelski and W.R. Thompson, *Sea Power in Global Politics*.

5 William R. Thompson, *On Global War: Historical-Structural Approaches to World Politics* (Columbia: University of South Carolina Press, 1988).

6 G. Modelski and P. Morgan, "Understanding Global War," *Journal of Conflict Resolution* 29: 3 (September, 1985), 391–418.

7 George Modelski and W.R. Thompson, "Long Cycles and Global War," in M.I. Midlarsky (ed.), *Handbook of War Studies* (Boston: Unwin, 1989), 23–54.

(evolutionary) *learning* sequence, the focus is on the process of learning and selection, starting with agenda-setting (1975–2000), going on to coalition-building (2000 to ~2030) and macro-decision (that selects new global leadership and reforms the political structure with a new agenda), to be implemented in the phase to follow. The learning perspective clarifies a key point about global wars: viewed as a learning process, the long cycle does not "need" a global war but makes it possible to envisage that the "selective," macro-decision, phase assumes a non-violent form, more akin, for example, to an electoral process.

In the 1990s, the scope of the research grew to cover at first the entire millennium of the modern era (from 1000 onward), and then the past five thousand years of world system development. Political long cycles were shown to co-evolve with the rise and decline of leading economic sectors.[8] It also became clear that global wars were a feature not of world politics in general but of a rather distinct segment of that experience, from the late fifteenth century onward (see Table 28.6 for some basic data on the five global wars 1492–1945), and that they need to be seen in the context of an evolutionary process: the emergence of a global political system (of which each long cycle is one phase).

In that context, the long cycle is a mechanism that has driven global political evolution for the past millennium and continues to do so, but has passed through two periods, those of (1) preconditions (and failure of world empire), and (2) formation of global nucleus, before entering the third one, that of (3) global organization, since about 1850. The characteristic institution of period (2) was that of global leadership, and it continues into the preparatory phases of period (3). Successive iterations of global leadership have produced increasing increments of global order that have been the products chiefly of global war settlements (see the "Outcomes" column of Table 28.6) but were weakly institutionalized. The global political system is now in the second of the preparatory phases, but not as yet the third, decisive, phase of the "formation of global organization." On this broader canvas, global wars appear as a time-bound form, an aspect of a long transition "from leadership to organization."[9]

Since 2000, the long cycle has attracted attention in China.[10] Work has continued on firming up the understanding of the "cascade of evolutionary processes" that make up world system evolution, and of which the long cycle of global politics is an essential element. Analysis has shown that such processes co-evolve (as long cycles do with waves of economic innovation, and also democratization), that shorter-range processes nest within longer-term ones (as long cycles nest within global

8 George Modelski and William R. Thompson, *Leading Sectors and World Powers: The Coevolution of Global Politics and Economics* (Columbia: University of South Carolina Press, 1996).

9 George Modelski, "From Leadership to Organization: The Evolution of Global Politics," in V. Bornschier and Ch. Chase-Dunn (eds), *The Future of Global Conflict* (London: Sage, 1999), 11–39.

10 Jian-shu Cui, "Cyclical Logic in the Transition of Hegemony," available at: http://faculty.washington.edu/modelski/Cyclicallogic.htm.

political evolution), and that all these processes are self-similar, each replaying the four-phased evolutionary learning algorithm, albeit at different scales. The interrelationship of these processes is governed by a power law, a signature of self-organization.[11] All this may appear rather abstract, but the theoretical structure, subject of a number of empirical tests, does lend conviction to the forecasts that may transpire in relation to the topic of this paper.

Five global wars

"Global war" is a theoretical term of long cycle theory.[12] It connotes a macro-decision for the global system taking the violent form of a conflict for political ascendancy to a position of global leadership that, in a spectrum of global institutions, stands mid-way between world empire and global organization. Table 28.6 at the end of this chapter briefly catalogs the salient characteristics of these conflicts, of which five have been identified in the historical record between 1492 and 1945. The listing reveals the commonalities, but also shows developmental trends that deserve emphasis. Here are the noteworthy characteristics of these event-sequences:

- *Global power alignments*: In each case, all global powers (as defined by their investment in naval capabilities) were engaged in the conflict. The challengers were continental powers, with imperial—and some naval—aspirations; the winning side was that of maritime-oriented nation-states and their allies. England/Britain was a key participant in all five cases, most of the time; it was also able to hold on to its global leadership position for two "terms of office."
- *Triggers* were various, against a background of tension around the question of ascendancy. Imperial aspirations, or the liquidation of empires, was a cause of friction (as in the case of Spain, France and Germany, also the Ottomans or Austria-Hungary in the Balkans). Territorial issues mattered. Regional issues tended to escalate to global levels.
- *War theater*: In each case, warfare occurred on more than one continent; starting with the Italian wars that, via the role of Venice and Egypt, were linked with events in the Indian Ocean, and leading up to World Wars I and II that were clearly multi-continental. While Europe has been the main theater, with the Low Countries at the very center of it, most recently emphasis has been shifting to Asia.
- *Outcomes*: Each conflict produced a leading power for the global political system, with a primary role in the settlement of the war and in fashioning the

11 T. Devezas and G. Modelski, "Power Law Behavior and World System Evolution: A Millennial Learning Process," *Technological Forecasting and Social Change* (2003).

12 To be distinguished from the term "Global War on Terror" used by United States' Department of Defense since 2003 to describe the wars in Afghanistan, Iraq, Horn of Africa, etc., but discontinued by the Obama Administration that has focused on Al-Qaeda. The global war on terror was not a substitute for global war.

rules for intercontinental trade and relations, i.e. global leadership, but not world empire. Each such "world power," part of the "democratic lineage" of republican, parliamentary, liberal states tending toward democracy, made its own contribution to fashioning the emerging global polity by creating international institutions for it. In turn, the increased institutionalization of the global system made it likely that the role of global leadership would be more narrowly circumscribed.

- *Global war duration*: The global wars of the past half-millennium each lasted for about the length of a generation (the replacement interval, of some 25 to 30 years). That does not mean that warfare was continuous. In the most recent case, the "inter-war period" (1919–39) was marked by general peace, though also regional wars—as in China, Spain, etc.). But not until the questions at issue, the composition and the agenda for new, or renewed, global leadership was settled, was the conflict ended.
- *The interval between global wars* might thus be expected to extend over three generations, or three phases of the long cycle (world power, delegitimation, deconcentration in the systemic mode). But it has been getting longer. While the average, for four cases, is 80 years (see Table 28.1 below), the most recent interval, between 1815 and 1914, was one of 99 years.

Table 28.1 Intervals between five global wars

Interval	1515–1580	1609–1688	1713–1792	1815–1914	1515–1914
No. of years	65	79	79	99	80 (average)

Long cycle risk assessments

A macro-decision has two elements: procedural, and substantive. Major war is one such procedural characteristic, an election is another. Examples of substantive (structural) change are installing new global leadership, or choosing new agendas for global problem solving. The five cases just reviewed illustrate both of these elements. In a contribution to a "Handbook of War Studies," Modelski and Thompson wrote nearly a generation ago that "the theory of long cycles does not view another global war as inevitable," but they affirmed "that an evolutionary learning process does, from time to time, require macrodecisions and ... these macrodecisions can be either violent or non-violent."[13]

Not much has changed since. At the time of writing this chapter (2010), 65 years since the end of the last global war, the world system still has not had the experience of another such great conflagration. But in the long-cycle "calendar" the onset of another macro-decision is approaching fast. It might be as close as 15

13 Modelski and Thompson, "Long Cycles and Global War," 48–9.

years (2025—that would make it the average interval, at 80 years) or somewhat longer, more nearly that of the last time elapsed, 99 years. That pre-World War I experience of an extended interval could have been an outlier, created by a sense of rising tensions that, for a time, made it possible to de-escalate mounting crises (over Morocco, German naval build-up, the Balkans, the Near East) and led far-sighted observers to advocate a search for alternatives, such as William James and Norman Angell[14] and German historians, mindful of historical precedents, warning against provoking Britain. The financial crisis of 2008–9 looks like a herald of approaching change. The imminence of the onset of macro-decision may also be a function of the rising quality and capacity of global institutions that might facilitate changes without violence. All in all, an interval closer to 15 than to 39 years seems more likely.

In the context of long cycle theory predictions, the possibility of averting a global war that may be looming on the horizon resides in changing the procedural element in a non-violent direction while preserving the opportunity for systemic adjustment and structural change in the form of reviewing performance, changing leadership and adjusting agendas. The procedural element in turn affects substantial issues (as a fall in the probability of war reduces, without eliminating, the salience and character of the forces of global reach). The matrix of opportunities for the next phase of long cycle (LC) 10 (long cycles may be numbered for the modern era, since about 1000) is laid out in Table 28.2.

Table 28.2 Matrix of opportunities for macro-decision (in Long Cycle 10)

Procedural substantive	Global war	No global war
Structural change	1. Like 1914–1945 (LC 9)	4. Preferred outcome
Status quo	2. Disaster	3. Frozen order

The spectrum of opportunities ranges from, as an example, (1) the World Wars of the twentieth century that, while costly and destructive, nevertheless ushered in substantial and needed changes in world politics and economics. But a repetition of that combination of global war with hopes for systemic transformation, as in (2), appears to hold nothing less than prospects for total disaster. Scenario (3), that of no global war and no structural change, is one of a world of "frozen order" refusing to adapt, possibly an empire; an unlikely recipe for a stable future. The preferred alternative is (4), combining the possibility of a world that allows for competition and for new initiatives, but has distanced itself from memories of persistent and widespread global conflict (while possibly even tolerating continuing—non-

14 William James, "The Moral Equivalent of War" (1906), http://www.constitution.org/wj/meow.htm (accessed May 7, 2011); Norman Angell, *The Great Illusion* (1910).

nuclear and diminished—violence at the regional or local levels), while managing competition on a non-violent basis.

Long cycle risk assessment suggests that a higher probability[15] should attach to scenario (4). That trajectory, while clearly competitive and therefore also conflictual, may prevail if only because all the other alternatives are so unattractive. Such an up-beat prediction relies on the knowledge that the long cycle is an evolutionary process that not only allows for adaptation but also organizes it in tandem with those that co-evolve with it. That process is currently (2010) in the phase of coalition-building, in conditions of a democratic transition (to majority status), an information age founded upon an array of new technology, and a movement of world opinion away from viewing major warfare or nuclear armaments as "normal" features of international coexistence. The glue of global solidarity may be congealing from the existential threat to human survival stemming from the possible use of nuclear weapons, or from climate change. But these influences might also be tempered by the fact that global institutions fully and uniquely capable of responding to these threats have yet to take shape, and may not do so for some time.

Conversely, long cycle theory assigns lower risk to scenarios (1), (2) or (3). The first of these, a simple replay of the world wars, seems unlikely because such crucial but also obviously flawed events as global wars in conditions of nuclear arms are now coming to be regarded as unthinkable. The second, too, raises the specter of nuclear devastation and is so awful that it generates little serious discussion. The third offers little but stagnation, possibly in the shadow of empire(s). All in all, the risk factors for global war have receded in the past two decades, but they cannot be regarded as non-existent.

As far as one can tell, neither the United States nor the Chinese defense establishments envisage the possibility of a global war in the foreseeable future. The United States' Quadrennial Defense Review of 2006 made no mention of such a contingency, being all about the "long war" on terror.[16] The 2010 review endorsed nuclear deterrence, but no more than hinted at possible future conflicts, seen to depend on whether "rising powers fully integrate into the global system."[17] According to the 2010 Department of Defense report on China's "military and strategic developments": "while remaining focused on Taiwan, China will, by 2020, lay the foundation for a force capable to accomplish broader regional and global objectives."[18]

15 Can probabilities be assigned to the occurrence or non-occurrence of global war in the next few decades? The sequence of five global wars just discussed might yield a sixth case. Furthermore, estimates may be made of the relative strength of trends and/or processes that may foster or hinder such an event, or event-sequence. Such probability rises as the postulated transition to the macro-decision phase draws nearer. The following section of this chapter gives examples of such trends.

16 "Quadrienial Defense Review Report." *US Department of Defense,* February 6, 2006, pp. 9–16

17 "Quadrienial Defense Review Report." *US Department of Defense,* February, 2010. p. 30

18 "Military and Security Developments Involving the People's Republic of China 2010." *Office of the Secretary of Defense,* 2010, p. 38

China's long-range posture has been, so far, in line with the "Twenty-four character strategy" laid down by Deng Xiaoping c. 1991 (apparently in response to the collapse of the Soviet Union): "observe coolly; secure our position; cope with affairs calmly; hide our capabilities; bide our time; maintain a low profile; never claim leadership." This last phrase, *never claim leadership*, may in long cycle perspective be the most interesting: it seeks to avoid placing China in the position of *challenger* that has been the hallmark of earlier global wars. (A later addition to the "24," "make some contribution," also points to the awareness of the necessity of seeking engagement for a "harmonious" world.)

That is not the only instance of Chinese leaders formulating a coherent global strategy on the basis of historical experience. In November 2003 the Politbureau held special sessions to study the "rise of great powers," and these were followed by similar discussions of "China's peaceful rise" at lower levels of the party organization. In the discussions that ensued, critics observed that the historical record of the rise of great powers is fraught with wars and wondered if China was following the same route. "Peaceful rise" soon morphed into President Hu's "harmonious world."

In late 2006 Chinese Central Television broadcast a 12-part documentary program prepared by the team of historians that had earlier briefed the Politbureau. The series depicted "the experience of nations and empires [China] had once condemned"—Portugal, Netherlands, Britain, the United States, etc. (all long cycle powers)—putting much stress on economic development. In 2007, a critique of long cycle theory authored by member of a PLA (Peoples Liberation Army) think tank, affirmed that China's rise in sea power is not incompatible with the "perpetual peace of humanity" that is "irreversible."[19]

As of 2010, the full implications of China's "peaceful rise" are yet to be spelled out. Some observers wonder, though, if Deng's "24" "hide and bide" strategy is still in force. The 2006 White Paper laid down a three-step strategy for modernizing national defense: "the first step is to lay a solid foundation by 2010; the second is to make major progress around 2020, and the third is to basically reach the goal of building informatized armed forces and being capable of winning informatized wars by the mid-twenty-first century." While maintaining only a minimal nuclear deterrent, China has been widely noted to be placing emphasis on building up sea power, not only launching a large submarine force and developing missiles capable of striking carriers at long distances, but also apparently planning, for the next decade, a force of multiple operational aircraft carriers with support ships. It is also fielding an independent (apparently dual-use) space program maintaining numerous satellites and aimed at a manned moon mission by 2017 and the completion of a Mir-class space station by 2020. All this would indicate that its military power is yet to peak, apparently in the second quarter of this century.

That is why, in the light of all considerations, long cycle theory affirms a residual possibility that, in the phase of deconcentration (multipolarity) and because of weak institutionalization at the global level, systemic inertia might yet break the

19 Jian-shu Cui, "Cyclical Logic in the Transition of Hegemony."

thin thread holding up the sword of Damocles suspended over humanity. A bid for regional security that others view as hegemony might, unwittingly, and as in the past, set off a global confrontation. As noted, in the long cycle timetable the formative phase of robust global organization is expected basically in the next cycle (LC 11), sometime early in the twenty-second century. In the meantime, nation-states remain the weightiest part of multi-level (global, regional, national and local) governance, and accident, miscalculation or madness, or reliance on faulty memories or irrelevant precedents, might unleash unanticipated, and disastrous, consequences.

Students of world politics, such as John Mueller, have presented a strong case for the "obsolescence of major war."[20] They argue that major war (or is it all war? that is not always clear) might disappear from human practice and become abnormal, just as slavery, or dueling, that have become abhorrent, are now unthinkable and have faded away. War that before 1914 was thought to be virtuous and ennobling is no longer so regarded, and prestige and status accrue to economic performance. If major war is unthinkable, then maybe scholars should avoid discussing it, and decision-makers let it slip from conscious thought and never consider embarking on such?

These are powerful ideas, but they do not cover all of the ground. Mueller dismisses the thought that war needs to be replaced, in the manner of William James, "by some sort of moral or practical equivalent." But he refuses to recognize that past global wars have had formative consequences for global politics, and that such a function must continue to be performed, albeit in new forms. In any event, so long as some states retain their nuclear arsenals, and others try to emulate them, the possibility of major war is not entirely unthinkable. The accession to nuclear power status of India, Pakistan, North Korea, and such a prospect for Iran, has been received by wide popular acclaim in their respective countries.

Some Possible Substitutions

The question is: what might replace global war? What aspects of recent world politics might be so labeled? What emergent process might be recognized as trending in that direction, offsetting systemic inertia and the thought that "war has always been part of human existence?"

This section features brief discussions of three such lines of thought, and actual or possible action: the concept of "democratic peace," the "Global Zero" action plan that aims at a "world without nuclear weapons," and the implications of plans for "planetary defense."

20 J. Mueller, *Retreat from Doomsday: the Obsolescence of Major War* (New York: Basic Books, 1989).

"Democratic peace"

In recent years, the proposition that "democracies rarely if ever fight each other" has gained wide acceptance among scholars, and has even had an impact on policy-making. It is supported by empirical research on wars of the last two centuries that has shown that "democratic dyads" are markedly less violent. This is not to be confused with the claim that democracies are pacifist or do not fight wars well, because when they do fight, as in recent global wars, they tend to prevail.

An original source for this line of argument are propositions first advanced by Immanuel Kant in an essay penned in 1795 in which he declared that a conjunction of three conditions, republican regimes, a federal structure, and commerce, would tend to bring about a condition of "Perpetual Peace." It is noteworthy that these conditions, also now known as the "Kantian peace," are quite close to the long cycle and its co-evolutionary processes that make up the emergence of the global polity by a process of self-organization.[21]

For estimating the strength of "democratic peace" it is necessary to look not just at "democratic dyads," or even at the formation of a community of democracies, but at the status of democracy at the systemic level. For a condition in which, say, one half of the world population lives in democracies, will tend toward maintaining peace for that segment of humanity but not necessarily for the other half. To act as a factor decisively influencing the coming macro-decision, it would be necessary for the world system to reach a condition of overwhelming majority for democracies, accounting for, say, 90 percent of the world population. On current estimates, represented in Table 28.3 below, such a condition is unlikely to be attained until later, after the middle of this century. That would suggest that "democratic peace" cannot be counted on for averting global war in the current cycle.

Table 28.3 Predictions for a democratic world

Author	90% democratic (population) (year)	100% democratic (population) (year)	Method
G. Modelski and G. Perry, "Democratization from a Long-term Perspective"	By 2075		Logistic innovation-diffusion model, POLITY data for 1800–1986 http://www.systemicpeace.org/polity/polity4.htm

21 G. Modelski and G. Perry, "Democratization from a Long-term Perspective," in N. Nakicenovic and A. Gruebler (eds), *Diffusion of Technologies and Social Behavior* (Berlin: Springer, 1991), 19–36; G. Springerski, "Emergence of a global polity," *Encyclopedia of Life Support Systems* (2010).

Author	90% democratic (population) (year)	100% democratic (population) (year)	Method
M.W. Doyle, *Ways of Peace and War* (New York: Norton, 1997)		By 2050–2100 (Liberal republics)	Arithmetic, data for 1800–1990
R. Rummel, "The Democratic Peace Clock" (2006), available at: http://www.hawaii.edu/powerkills/DP.Clock.HTM	By 2071	By 2096	Polynomial regression, Freedom House data, 1900–2000, cited in R. Rummel (2006)
R. Rummel, "The Democratic Peace Clock"	Expect it by the middle of the century, but due to speed-up (critical mass) might make it in second quarter of the century		Personal judgment

Global zero

Almost as soon as the two nuclear bombs exploded over Japan in the closing act of the last global war, plans began to be drawn up for taming, and even abolishing, this new weapon. The Acheson-Lilienthal committee (1946) drew up a proposal (that became the substance of the Baruch Plan) under which "no nation would make nuclear bombs" and all "dangerous" activities would be managed by an international authority. In 1961, President Kennedy launched a design for "general and complete disarmament" that called (in the McCloy-Zorin joint statement) for the elimination of all nuclear stockpiles and of all means of delivery of weapons of mass destruction. The Non-Proliferation Treaty (NPT) of 1968 that was the main outcome of that initiative continues to bind all its parties "to pursue negotiations in good faith on effective measures relating to cessation of the nuclear arms race at an early date and to nuclear disarmament ..." The NPT set the stage for the strategic limitation (START) talks that brought, *inter alia*, the US-USSR "Agreement on the Prevention of Nuclear War" (1972). Soon after attaining power, Mikhail Gorbachev, in a letter to President Reagan, broached a proposal "for the complete liquidation of nuclear weapons throughout the world ... before the end of the present century," a proposal that was seriously discussed but failed to be adopted at the Reykjavik summit (1986).

The end of the Cold War reduced tensions, and brought such a significant decline in both the Russian and the American stockpiles of nuclear weapons (roughly halving them) that the urgency of the problem appeared to recede. But, in the new century, continuing critiques of nuclear deterrence, the fear of terrorists acquiring such weapons, and rising problems of proliferation (North Korea, Iran, etc.) gave new life to plans for abolishing nuclear weapons.[22] That is why President Obama's Prague speech (April 2009) declaring that the United States seeks "the peace and security of a world without nuclear weapons" (and carefully qualified: "perhaps not in my lifetime") was not really such a radical departure but a continuation of a long line of practical statecraft. It was followed by a meeting with President Medvedev, the signing of the New Start treaty with Russia, and the launching of the Global Zero initiative: "an international, non-partisan effort ... dedicated to achieving the phased, verified elimination of all nuclear weapons." Part of the initiative is the Global Zero Commission of 23 members[23] that drew up a plan for a step-by-step process to achieve the goals of "global zero." Key elements of that action plan, adopted at the Global Zero Summit in Paris (February 2010), feature in Table 28.4 below.

At this time, Global Zero appears to be carefully thought out, well staffed, and in tune with important segments of the interested public. In contrast to grandiose plans for "general and complete" disarmament, it seems to have assimilated lessons from decades of arms control experience by choosing a narrower target: nuclear arms only. In contrast to earlier insularity, when schemes (e.g. the Baruch Plan) were launched without any external input, or arms talks were confined to Soviet-American contexts, Global Zero has from the outset established a multilateral support system founded upon an appeal to world opinion.

Table 28.4 Outline of the Global Zero Process 2010–2030

Phase 1 2010–2013	Following the conclusion of a START replacement accord, negotiate a bilateral US-Russian accord to reduce warhead totals to 1,000 each (by 2018); by ratification of the US-Russia accord, all other nuclear weapons countries freeze the total of their warheads, and commit to multilateral negotiations.
Phase 2 2014–2018	In a multilateral framework, the US and Russia cut warheads down to 500 (by 2021), while others freeze until 2018, and proportionately cut by 2021. Establish comprehensive verification and enforcement (no notice, on-site). Strengthen safeguards on civilian nuclear fuel cycle.
Phase 3 2019–2023	Negotiate legally binding Global Zero accord, for all nuclear capable countries, for phased, verified, and proportionate reductions of all nuclear arsenals to zero warheads by 2030.

22 Sparking the contemporary debate was a January 2007 *Wall Street Journal* article by W. Perry, G. Schultz, H. Kissinger and S. Nunn, "A world free from nuclear weapons." For a review of the issues, see G. Perkovich and J.M. Acton (eds), *Abolishing Nuclear Weapons: A Debate* (Washington, DC: Carnegie Endowment for International Peace, 2009).

23 Made up of American (5), Russian (5), Chinese (3), Indian (3), Pakistani (2), Japanese (2), French (1), UK (1) and German (1) political and military figures.

Phase 4 2024–2030	Complete the phased, verified, proportionate dismantlement of all nuclear arsenals to zero total warheads by 2030, and continue the comprehensive verification and enforcement system prohibiting the development and possession of nuclear weapons.

Source: "Global Zero Action Plan," February 2010, available at: http://static.globalzero.org/files/docs/GZAP_6.0.pdf.

The United Nations Security Council endorsed the goal of "a world without nuclear weapons" in September 2009. The Nuclear Posture Review of the United States Department of Defense (released April 2010) is guided by that same idea (that "will not be achieved quickly"). An unprecedented Nuclear Security Summit convened in Washington in May of that year. NATO's Strategic Concept for the coming decade (adopted November 2010 in Lisbon) committed the alliance to the goal of "creating the conditions for a world without nuclear weapons," but also reconfirmed the centrality of nuclear deterrence "as long as there are nuclear weapons in the world."

This is an ambitious start to an ambitious project, but also one that is only just starting in earnest and has yet to show an effective track record. The New Start treaty, which may be thought of as a foundation stone for Global Zero, is relatively modest, and provides for a cut in each side's warheads from 2,200 to 1550. It reduces the number of deployed delivery vehicles (strategic missiles, submarines, bombers) to 700, with another 100 in reserve, from a current level of about 850 for the United States and 565 for Russia. It also reestablishes a system of mutual inspections. It was ratified by the US Senate (December 2010) by a vote of 71:26, but only after a prolonged and arduous debate.

The debate revealed a growing divide between Democrats and Republicans on what the nuclear agenda should be about. What was thought to be a bipartisan issue turned out not to be not quite so. Leading Republican senators launched a serious of criticisms of perceived flaws of the treaty in respect of missiles defenses, tactical nuclear weapons, verification, etc. Some even began to question the legitimacy of the goal of a "nuclear-free world" itself. Looking forward, what one headline summarized as an "uphill climb for Obama" was likely to get "steeper."

Will motivated leadership persist for two more decades? Years of tough politicking, deals and negotiations lie ahead, not only domestically among the nuclear powers—some of which are of only recent vintage—but also with maybe 50 nuclear-capable states, all of whom must be brought into the system of inspection and enforcement. And if enforcement will be needed, then wars might also be in prospect.

In a long cycle perspective, Global Zero seems as if timed toward completion at the exact moment when the global political system will be entering upon the time of macro-decision. But might it in fact make the world safe for conventional warfare? It would certainly dismantle the nuclear deterrence that has been in place for more than half a century and, some have claimed, has helped to secure systemic peace. In the view of others that claim is doubtful and unverified, and the world would be better off without it. In long cycle terms, the general (global-level) peace so far among the major powers since 1945 has been entirely unsurprising, as first argued in 1985.

The critics of Global Zero have found their voices in the Senate debate, calling the project "a fantasy," a "utopian dream," if not outright "dangerous." Its advocates claim, on the other hand, that it stands for "an idea whose time has come." For all we know, it might come just in time to avert a devastating conflict. All things considered, it might yet succeed. And if it does, it would be an achievement of major proportions, easily the greatest and most complex project of political reconstruction and global cooperation in humanity's experience. But surprises are likely to be in store, and even if it did succeed in its proclaimed goals it would have to be accompanied (as it might be) by innovative structural change that has yet to be spelled out.

Planetary defense[24]

The parting shots of World War II were heard not only over Japan, but also in missiles fired against London. Just like nuclear questions, those concerning space became prominent only since the end of the last global war in 1945 and, building on the development of rockets, they immediately assumed a competitive character. The Soviet Union launched Sputnik, the first man-made satellite, into earth orbit in 1957, claiming it was evidence of the superiority of the Soviet system. The United States' Apollo mission landed humans on the Moon in 1969, and evened out the score. The Soviets were unable to match that success, but they did proceed to build the first space station, Mir 1 (with a ten-year life cycle). Meanwhile, satellites became a powerful aid to telecommunications worldwide, becoming a vital industry.

With the end of the Cold War, space questions assumed more of a cooperative character. By 1994, the space agencies of the United States, Russia, Europe and Japan, each of which was making plans for a space station, joined forces to construct an international installation to serve as a research laboratory, and possibly to be used as a test bed for missions to Mars and the Moon. By 2000, the ISS (International Space Station) had become operational. Its cost, estimated at up to $150 billion over 30 years, may make it one of the greatest public projects ever undertaken. With a permanent crew of about six, it has hosted astronauts from 15 countries, and the participation of Brazil, Italy, South Korea and India (in addition to Canada) was being discussed in 2010 (but China's joining was, as of 2009, reportedly meeting with US objections, China announcing plans for an independent station by 2020).

Imaginative writing has long suggested that invasions of "aliens from space" might do wonders for human solidarity. More recently, it has been proposed that the dangers from outer space come not from "little green men" but from Nature:

24 Acknowledgement: Professor J. Longston and Dr W. Ailor have helped clarify the discussion of "planetary defense" (personal communications, August 2010).

from "near-earth objects" (NEOs), the asteroids and comets that are the products of cosmic evolution, some of which might impact our planet.

An asteroid initially thought to have a diameter of 350 meters was first observed in July 2004. Its discoverers gave it the name Apophis, a Greek term for an ancient Egyptian enemy of Ra, Apep, a serpent that dwells in darkness. First calculations of its orbit suggested a small probability that it might impact earth in 2029. More refined observations the following year determined that that was unlikely but they did open another window, for 2036, when Apophis would pass through a "gravitational keyhole" that could (with a small probability) lead to an impact. Further observations foresaw another encounter, in 2068.[25]

Earth scientists and others who follow these matters have been aware for some time of the dangers looming in the skies from extraterrestial objects, and in response to these concerns a conference for "Protecting the Earth from Asteroids" (see Table 28.5) convened in the Los Angeles area early in 2004 under the sponsorship of the American Institute of Aeronautics and Astronautics and the Aerospace Corporation (a federally funded research and development center). The conference report declared the risks of a collision to be "small but very real," considering that several hundred NEOs of a size greater than one kilometer are now known and that anyone would be likely to wreak havoc. It was only a few months later that the asteroid Apophis was observed, and initial calculations suggested a close approach. The second meeting, in 2005, in Washington DC, carried on the theme of "Planetary Defense" with a broader backing from several major space agencies. Attention focused on conceptual problems of soft deflection of a threatening object, and the necessity for international cooperation. The third meeting, in 2009, transferred the responsibility for hosting the series to an international body, the International Academy of Astronautics. A fourth one is scheduled for 2011, in Bucharest, Romania. From an initially chiefly American enterprise, the project has gradually morphed into a wider international undertaking, marked by an approach that is hard-headed and non-alarmist.

Table 28.5 Planetary defense conferences 2004–2011

	Sponsors	Committees, participants	Focus/outcome
2004, February Los Angeles, CA "Protecting Earth from Asteroids"	American Institute of Aeronautics and Astronautics (AIAA), Aerospace Corp.	Organizing Com.: 16 (inc. ESA 2, India 1). 140 participants	Issued white paper: "risk small but very real"; raise NASA funding from $4m

25 "99942 Apophis," accessed on Wikipedia, August 16, 2010.

	Sponsors	Committees, participants	Focus/outcome
2007, March Washington, DC	AIAA, NASA, ESA, ISRO (India), Japan Space Agency, Aerospace Corp. Space Studies Institute, GWU, etc.	Steering Com.: 31, inc. US 23, UK 4, Italy 2, India 1, Russia 1	"Future impacts by asteroids and comets are a certainty" and threaten "even ending civilization and humanity's existence"
2009, April Granada, Spain	International Academy of Astronautics (IAA), ESA (European Space Agency), ISRO, Aerospace Corp.	Planning Com.: 39 inc. US ~20 Co-chairs: W. Ailor, Aerospace C.R. Tremaine-Smith, UK	Emphasis on Apophis; student participation
2011, May 9–12 Bucharest, Romania "From Threat to Action"	IAA, ESA, Romanian Space Agency, Aerospace Corp.	Co-Chairs: W. Ailor, R. Tremaine-Smith. Com.: 37, inc. Romania 2, India 1, China 1	Emphasis: Apophis, inc.: "organizing, coordinating, and managing an international effort"

Source: http://www.aero.org/conferences/planetarydefense, http://www.esa.int/SPECIALS; accessed August 16, 2010.

The concept of "Planetary Defense," in conjunction with the Apophis asteroid, deserves interest in the present context on two grounds. For one, in defense of easily understood common interests it could lead to the establishment of a basis for long-term cooperation in space on a broad and gradually worldwide foundation. US legislation passed in 2010 that authorizes an unmanned government mission to an asteroid by 2025 is justified in part by planetary defense considerations.

Second, the timing of the asteroid's approaches (2029, 2036, 2068) makes them largely coincide with the expected macro-decision of the long cycle (~2030 to ~2060). The perception of a common interest in averting a looming threat and the practical necessities for cooperation in space would likely contribute to smoothing the path toward a non-violent decision for political change. It might also divert attention and resources from planning for a war in space and would also make inconceivable a threat to the existing, and increasingly dense, network of satellites of various kinds that would become one of the first casualties of a major military conflict, creating a vast ocean of debris in space, and havoc on earth.

Two "No Global War" Scenarios

This brief discussion of three possible "substitutes" for global war makes it plain that these are, at best, only partial solutions to the problem. Democratization may yet, by itself, produce general peace over the course of the century, but may not weather easily the vicissitudes of the nearer future, for instance if the Chinese model of economic growth without democracy appears more compelling. Global Zero is an ambitious project that is a well-targeted but demanding single-minded pursuit over a period of two or more decades, and that is hard work. And planetary defense could be more the dream of science fiction writers than an expensive project of uncertain returns.

Long cycle theory is open to two scenarios for averting global war: (1) fostering "engagement" and the continued expansion of the multiple networks of cooperation that now extend worldwide, such that the possibility of global war is seen as unthinkable; (2) bringing about, and consolidating, a condition wherein united democracies carry such enormous weight that war against them becomes unthinkable. These two scenarios correspond to the risk assessment discussed earlier in this chapter: the first echoes the higher probabilities attached to the no-global-war trajectory, the second reflects the lower probabilities suggested by the possibility of systemic war.

Engagement in global problems

Global war may be averted if, in coming decades, global political structures are duly "refreshed" and looming global problems are competently managed, with the assent of major powers and world opinion. The problems must be efficiently diagnosed, prioritized and placed into public agendas in order to garner the necessary support. Priority global problems of the current long cycle (LC 10) dovetail with global political evolution that has now reached the phase of consensus-building. That is, a consensus for the formation of a democratic community is forming as the support base for an institution- and rule-based global political system.

Long cycle theory makes it plain that global politics is not just about selection for global leadership (or, more loosely, about who's Number One), but also about policy agendas and solutions for global problems. Obviously, these are related questions, in as much as some candidates for global leadership (those with imperial ambitions) also stand for some well-known agendas. In principle, each candidacy should be evaluated first of all in terms of its agendas and the problems most likely to be attempted by it. Agendas need to be seen to serve the public interest if they are to command wide approval.

The summation of the chief characteristics of the five global wars of the past half-millennium found in Table 28.6 includes, in its last column, a brief reference to the outcomes and agendas of those major conflicts. In three cases, the post-global war global system was "refreshed" with a new occupant in a leadership position, while in the case of Britain a second "term of office" materialized, albeit on a new basis after shedding what historians have called its "first empire" in America and executing an "industrial revolution." It is quite possible that a similar "renewal" might occur in the current cycle (LC 10), given the United States' current leadership in the information sector, its modernized and "informatized" military forces, and its role in advancing democratization.

Aside from questions of succession, the outcomes of the five global wars also bore on matters of agendas, and the "platforms" on which the parties to these wars were "running." In the case of Portugal the agenda was clear: the construction of an oceanic pathway to Asia and building a new trade route, thus laying the foundations of a global system. That was the "platform" for which support was sought from the crowned heads of Europe in 1499,[26] and it was implemented by 1515. For the most recent global wars, those of 1914–45, the principal statements of war aims were President Wilson's "Fourteen Principles" (1918) and Roosevelt/Churchill's "Atlantic Charter" of 1941. Such declarations of broad aims and general principles were important because they embodied a claim that the goals being pursued went beyond national interests, narrowly defined, and had more general appeal and universal interest, something that served as rallying grounds for wider coalitions.

What might be a "platform" of a candidate for global leadership in the coming decades? One part of such a platform might be the nuclear question. It stands for an existential problem of a high order and has implications not only for the United States and Russia—known to hold the great bulk of nuclear arsenals—but also for near-nuclear states such as Japan, Germany or Brazil, that at this time have the capacity but not the political will to go nuclear, but might change policies in the future, and finally for the many other countries, e.g. in Europe or in the Middle East or East Asia, that might become the victims of nuclear hostilities. To the last in particular, Global Zero might hold an especial appeal. It is a project of broad significance, but one that cannot succeed without at least the acquiescence of all nuclear-weapon and nuclear-capable states.

On an equally serious note, there is also that well-publicized issue of climate change, likewise bearing the imprint of a potentially existential threat. Like the nuclear issue it affects numerous parties all over the globe, even if some areas may be less exposed to it than others. Coordination problems are immense, as shown by the difficulties encountered at the Copenhagen meeting in 2009, but quite a few states are making the necessary investments and scoring advances in alternative energy technologies.

An even more interesting, if hypothetical and less likely, plank for a global platform could be drawn from planetary defense. If a large enough asteroid were found hurtling toward earth, and a space power (otherwise qualified, and perhaps jointly with non-state actors), were to devise a successful scheme for averting catastrophe, that might go some way toward propelling it toward leadership.[27]

26 G. Modelski and S. Modelski, *Documenting Global Leadership*, 56–7.

27 The path of risk where Apophis (diameter now estimated at 270m) might impact in 2036 leads across large parts of southern Russia, and might also produce a large tsunami off the West Coast of the United States. Roscosmos (the Russian Federal Space Agency) announced in 2009 that it will study designs for deflection methods. These include gravitational tractor, kinetic effects and nuclear bomb. NASA maintains a "Near Earth

On all previous occasions global wars were fought on the issue of imperial ambition and territorial aggrandizement (in defense of power balances). At such times, small but independent states usually sought protection, and often found it, at first in Europe and more recently elsewhere, among powers of the oceanic-liberal persuasion. There are no empires at the turn of the twenty-first century, but there are imperial memories (or legacies) that continue to influence the policies and public attitudes of, for example, China, or of Russia. Territorial issues have almost vanished from the European area but they do linger elsewhere, where, for example, ocean space could be appropriated for a variety of national purposes. Climate change is exposing the polar regions to potentially fierce competition.

Engagement is the key to global leadership, at least in certain defined spheres. It may take such forms as participation in UN peacekeeping, earthquake, flood or tsunami disaster relief, contributions to international development projects or counter-piracy operations. Apparently heeding the injunction "make some contribution," China's armed forces have enhanced their capabilities "for the delivery of international public goods."[28] China's Defense White Paper for 2006 listed participation in 21 UN peacekeeping operations, 13 of which were still continuing, as well as numerous exchanges. These were positive signs, even if observers noted problems with transparency.

The method of selection (or re-selection) is at this stage uncertain but must occur either within or without the UN framework. A recent example of a problem-solving mechanism at the global level has been the employment of an existing (non-UN) forum for discussing financial issues (formed in 1999 at the level of finance ministers and central bankers) to respond to the financial crisis of 2007–08. A series of meetings was set in motion at the Summit (heads of state/government) level, convening first at the invitation of the United States in Washington in late 2008, followed by semi-annual meetings in London, Pittsburgh, Toronto and Seoul. From 2011 the summits will continue annually, supplanting the G-8 system, as the principal forum for the review of financial and economic issues.

The G-20 is, of course, broader than the G-8 that was seen as an assembly of rich Western countries; it importantly includes, *inter alia*, China, India and Brazil, as well as the European Union (EU),[29] and is said to represent 85 percent of the world's GDP, 80 percent of world trade and two-thirds of the world population. It is, effectively, weighted in a majority-democratic direction. Its work so far has helped to calm the situation, but the group has been criticized for being self-

---

Object Program" web site at http://neo.jpl.nasa.gov/.

28 US Department of Defense, *Military and Security Developments Involving the People's Republic of China* (2010), available at: http://www.defense.gov/pubs/pdfs/2010_CMPR_Final.pdf.

29 That means that Europe is represented twice: having Britain, France, Germany and Italy individually at the table, as well as the EU as such. The EU has a total of 27 member states. The 2010 meetings included "outreach sessions" with the chairs of the African Union, ASEAN and NEPAD (New Partnership for African Development).

appointed, and failing to represent the views of the nearly 150 other UN members that have not been included (directly and/or indirectly).

Democracies united?

Global war may be averted if the world's democracies are so successful in their endeavors and so powerful in their unity that no other power would dare to confront them or allow a regional conflict to escalate to the global level. The conditions of such unity might now be within sight.

As many observers maintain, in a number of dimensions, the world of democracies has been significant as a "majority party" for quite some time. Democratic countries now comprise over one half of the world's population (with the United States and the European Union (EU) accounting for about 12 percent, and India close to 20 percent of the world total), and they represent over one-half of the members of the United Nations. They have been the sources of innovation in leading industrial sectors and make up over two-thirds of the annual output of the world economy (with the US and the EU each contributing about 20 percent). NATO is a powerful military alliance, such that Russian strategic doctrine now includes reliance on tactical nuclear weapons to offset an inferiority in conventional forces in the European theater, a neat reversal from the post-1950 situation when NATO planning drew on US nuclear strength to neutralize the Soviet tank armies in East Germany. Democracy is the preferred source of legitimacy for social structures today, even if the democratic label is often appropriated on false pretenses. Democracies, finally, are the major sources of debates that define world problems. Only three members of the new G-20 Group just mentioned do not currently rate as democracies: China, Russia and Saudi Arabia. The rise of China might qualify that picture but does not change it in the medium term.

That being the case, in the past decade a number of commentators have raised questions about the possibility of a union of democracies. Among others, Robert Jervis has revived Karl Deutsch's concept of "security community" originally meant for the Atlantic area, and has argued that the United States, West Europe and Japan, what he calls the "leading powers," now form such a community that "war is literally unthinkable" among its members and "will not occur in the future."[30]

In the 2008 US Presidential campaign, Senator McCain declared that, if elected, he would sponsor the creation of a "League of Democracies," linking more than 100 democratic nations, to work together for "peace and liberty" as a supplement to the United Nations. Advocates of a "concert of democracies" argued that it would "ratify and institutionalize the democratic peace."

These proposals have not been implemented. To assess their feasibility, Theodore Piccone reviewed the record of an organization of a more modest scope, the "Community of Democracies," an inter-governmental body founded in 2000

30 Robert Jervis, "Theories of war in an era of Leading-Power peace," *American Political Science Review* 96: 1 (March, 2002).

in Warsaw at a conference, attended by some 120 delegations, that was convened, on the initiative of US Secretary of State Albright and Poland's Foreign Minister Geremek, to "consolidate and strengthen democratic institutions ... jointly to cooperate to discourage threats" to democracy, and to support "emerging democratic societies."[31]

The Community of Democracies has since met fairly regularly (in Seoul 2002, Santiago 2005, Bamako 2007, Lisbon 2009, Vilnius 2011), mostly issuing general declarations. But in the overall judgment of Piccone, the organization has "little to show for the effort." At Bamako, it decided to establish a permanent secretariat, in Warsaw, but with few resources. Between general meetings it is led by a Convening Group that meets monthly in Washington. That group was originally composed of seven members: the United States, Poland, India, Chile, the Czech Republic, Mali and South Korea, and has now expanded to number 17, including, *inter alia*, Italy, Mexico and South Africa, but what is most notable about this "clumsy and non-transparent" body is the absence of Britain, France, Germany and/or Japan.

A new development has been the Democracy Caucus established at the United Nations that backed the creation, in 2005, of a voluntary UN Democracy Fund with over $95m in pledges; by 2009, some $58.7m for 204 projects "to support democratization throughout the world" had been channeled through it. But, overall, Piccone calls the Caucus "largely moribund." Another novel development has been the institution, for meetings from 2007 onward, of an "Invitation Process" by an independent International Advisory Board that would vet the democratic credentials of states to be invited to the meetings. For the 2007 meeting, the Board recommended that 54 countries, including Pakistan, Russia and Singapore, not be invited, but the Convening Group declined 28 of these decisions, and, *inter alia*, brought Russia in with "observer" status, but did disinvite Pakistan and Singapore.

What are the reasons for this mixed performance? According to Piccone, most European governments either participated reluctantly or were outright opposed to the initiative because they feared it represented "the start of a US campaign to undermine the UN, to isolate China and/or block the ascension of Europe." "The White House's cloaking the war of Iraq as a crusade for democracy in the Middle East ... was probably the main culprit for the failure of the CD to get off the ground." Fears were aroused that this might be a campaign of "the West" for "invading sovereignty in the name of promoting democracy." And another major concern was that this move might spark a new Cold War.

The experience of the Community of Democracies so far suggests that even if more than one-half of the world might be thought to be democratic, that part cannot be treated as a monolithic bloc. Two conceptual issues help to clarify this problem: (1) a community is not an alliance, and democracy means more than a functioning electoral system; (2) global political leadership does not translate smoothly into community guidance, or support.

---

31 T.J. Piccone, "Democracies in a League of their own? Lessons Learned from the Community of Democracies," *Brookings Foreign Policy Paper* 8 (October, 2008).

The Community of Democracies is not an alliance with a clear and specific purpose; it is an intergovernmental, hence political, organization operating in the context of a diffuse-character, general-purpose, social association with membership of multiple loyalties. It brings together a pool of possible cooperators but does not guarantee cooperation on any one aim. The glue binding communities are common memories, membership, ceremonies and language. Moreover, global leadership, that is a political office, does not automatically translate into a position of community guide. When weighty questions arise, such as participation in warfare that might be urged by political leadership, they might not reflect the legitimacy of the issues involved, and the merits and justice of rival claimants. Support is hard to garner for a preventive war, but is more likely to come forth in case of unprovoked aggression (other things being equal).

In short, the unity of democracies is not to be taken for granted, and needs to be earned, even by beacons of democracy, by the quality of their societies and the justice of their policies. Democracies hold formidable assets when entering international competition: they inhabit a zone of peace, they are stable societies open to new ideas, their economies sponsor innovation-based global lead industries, and their political systems work. But they cannot be expected to march in lockstep, and they might be subject to divisive campaigns. A community of democracies might be a powerful engine of cooperation—possibly up to averting a global shootout—but it also calls for great care, consumes a lot of energy and requires regular maintenance if it is to serve the cause of global "democratic peace." On the other hand, a strong alliance would likely comprise both democracies and non-democracies.

Table 28.6 Five global wars

Global wars	Global power alignments*	Other participants	Trigger	War theaters	Agendas/ outcome
World Wars I and II 1914–1945 LC 9	Germany#, (Japan) v. Britain, France, Russia/USSR, USA	Austria-Hungary, Italy	Imperial decay (Ottoman, Hapsburg); Belgian neutrality; German claims	Europe: East and West, West Asia and Africa, China, Japan, Southeast Asia	US global leadership; United Nations System; Atlantic Charter
French Revolutionary and Napoleonic 1792–1815 LC 8	France#, (Spain) v. Britain, Russia	USA, Prussia, Austria	French revolution; French attack on Dutch Republic	Europe: West and East, North Africa, North America	Britain's global leadership renewed; Concert of Europe

Global wars	Global power alignments*	Other participants	Trigger	War theaters	Agendas/ outcome
Wars of Grand Alliance, Spanish Succession 1688–1713 LC 7	France#, (Spain) v. Britain, Dutch Republic	Austria, Sweden (Great Northern War, 1700–21)	French moves in Germany; England's "Glorious Revolution"	Low Countries, Germany, North America, Baltic Sea	Britain's global leadership; Balance of power in Europe
Dutch-Spanish 1581–1609 LC 6	Spain#, v. Dutch Republic, England, France		Dutch indepen-dence; Spain's aggrandize-ment	Low Countries, Atlantic, East Indies	Dutch global leadership; International Law; Freedom of the seas
Italian and Indian Ocean wars 1492–1515 LC 5	Portugal, Spain (England) v. France	Hapsburgs, Venice, Gujerat	French thrust into Italy; Portugal's entry into Asian trade	Italy, Low Countries, Indian Ocean	Portugal's global leadership (network of (fleets, bases, alliances); Treaty of Tordesillas partitions ocean space

Note: * = global powers as listed in G. Modelski and W.R. Thompson, *Sea Power in Global Politics 1494–1993*, 98, Table 5.1; bracketed names = participating for part of the time; LC 5–9 = long cycle numbers; # = challenger.

Reflections on Polemology: Breaking the Long Cycles of Wars of *Initial Challenge* and Wars of *Revanche*

Hall Gardner

As *Homo geopoliticus* enters into a period characterized by a highly uneven polycentrism, in which centers of power possess highly uneven force capabilities and political and economic influence, and differing degrees of irredentist and revisionist claims, a reassessment of global security and development concerns is imperative. A new security consensus and geopolitical framework involving more concerted or multilateral policies and the restructuring of the global system must be forged in order to meet the needs of the twenty-first century, in order to prevent the possibility of the continued widening of regional wars, if not the possibility of wars among the major powers. In effect, the apparent cycles of previous global wars in which a core power begins to decline from its position of leadership or "hegemony" in the global system, and is challenged by new and differing kinds of upstarts or ascendant rivals, must be broken.

On the one hand, a number of regional states have been challenging the former US/Soviet-dominated global order by seeking core and amphibious status through the development of a blue water navy and/or advanced military technological capabilities, including advanced missiles and nuclear weaponry. On the other hand, a number of anti-state actors and surrogates of foreign powers appear to be playing a "negative" role in seeking to weaken or break up pre-existing states through wars of attrition. Both these developments place additional strains on the United States and its allies, as well as other major and regional powers under attack. The rise of regional powers, combined with the international presence of a number of nomadic anti-state organizations willing to use violence to assert their cause, has resulted in the widening spread of conflicts to a number of collapsed and collapsing states, raising the possibility that major or regional powers may opt for multilateral or unilateral military interventions.

The question remains whether the dynamics of the contemporary global system are, in fact, leading to the possibility of global conflict, and if so, what form might that conflict take? Could such a conflict look more like the wars of the eighteenth century, in which states fight it out with professional armies and high-tech weaponry, relying on speed and stealth? Or could such a conflict look more like World War I, a conflict between alliance systems set off by terrorist sparks, in which the wars taking place in the "Greater Middle East" parallel the series of Balkan wars that helped set off the "war to end all wars"? Or could it take the form of a revanchist movement, much like the National Socialism movement led by Adolf Hitler or Mussolini's fascism? Or could it take place as a new form of Bonapartist, yet multilateral, "democratic nationalism" in which the United States, as the still-leading hegemonic power, along with its partners, engages in "regime change," and then seeks to "democratize," potential challenges to its rule? Or, by contrast, could it take place as a result of multiple state collapse, resulting in permanent wars of attrition and intermittent interventions by major and regional powers and anti-state "terrorist" groups that ultimately draw a number of states into the crossfire—perhaps more like the Thirty Years War, with the region of the former Ottoman empire largely playing the shatterbelt role of the former Holy Roman empire? Or, more positively, will the very threat of such conflicts lead major and regional powers to engage in more concerted policies in an effort to establish internationalized regional security and development communities, among other means to reduce or eliminate violence and conflict, through arms control, perhaps ultimately leading to the abolition of weapons of mass destruction?

The authors of this book have provided a number of theories and perspectives that impact upon the origins, immediate causes, processes and consequences of different wars, while concurrently attempting to look at wars in a long term and multidisciplinary *polemological* perspective, as pioneered by the work of Gaston Bouthoul, among others. No one can say precisely whether such a systemic "war" might or might not take place, or what form it might take—or whether we might already be just at the beginning of its outbreak—but we can *try* to ask the right questions as to how such a war might break out. Whether or not a systemic war could break out in the near future, the key point is that state leaderships should work to prevent such a reoccurrence by means of developing irenic precautionary measures wherever possible and implementing concerted interstate and inter-societal policies primarily through a step-by-step restructuring and re-equilibration of the "anarchical" global system.

In presuming the relative decline of contemporary American power, William Thompson expects that a period of what he calls "coalition maneuvering" will take place in the period 2000–30. (See Thompson, Chapter 27.) In this period, the possibilities of a global war could augment, due to shifts in alliance networks, military technological innovations and the nature of state and socio-political rivalries with respect to "contending force capabilities, strategic intent and international norm." (See Gardner, Chapter 1.) A series of interstate wars could, for example, be fought for control over energy or resource-rich territories. At the same time, however, as both Thompson and Modelski have pointed out, "the theory of long cycles does not view another global war as inevitable," but "that an evolutionary learning process does, from time to time, require macrodecisions and ... these macrodecisions can be either violent or non-violent." (See Modelski, Chapter 28.)

From this perspective, international leaderships should begin to engage in major systemic adjustment and structural change *sooner rather than later* (i.e. engaging in "macro-decisions") in order to avert the very real possibility of systemic conflict. This is because the nature and structure of the geohistorical inter-state system makes it more difficult for individual states to make radical, irenic alterations *without long term and concerted efforts*. The so-called "anarchical" nature of the international system and structure is only, to a certain extent, what states make of it, in Alexander Wendt's views (see Gardner, Introduction and Chapter 1). The nature and structure of inter-state systems in differing geohistorical epochs, combined with the nature of domestic socio-political restraints *within* individual states, works to precondition and restrict the ability of leaderships to make significant alterations in both domestic and international relationships and inter-state actions. Averting global war accordingly requires more than a mere adjustment in the foreign policy of the leading power or powers toward their rivals: global peace *necessitates significant systemic and structural transformations and re-equilibration at the local, regional and global levels.*

From Wars of *Initial Challenge* to Wars of *Global Revanche*

At least since the eighteenth century, hegemonic wars of *initial challenge* for control over continental focal points and overseas possessions have been followed by the collapse and disaggregation of at least one of the challenging powers, leading to a war of *global revanche*. Such a war is fueled by the collective memories of elites and socio-political interest groups that seek the return of territory, lost possessions, as well as power and global influence (even if that collective memory might be self-serving and biased). In geohistorical terms, what largely distinguished the Thirty Years War from the later Seven Years War, World War I and the Cold War, is that the *indecisively defeated* amphibious-core states of the latter three conflicts (France, Imperial Germany, Soviet Russia) all lost their overseas possessions/ colonies. Unlike Spain, which only gradually lost its control and influence on both the continent and overseas, and which generally sustained its domestic political-economic stability, monarchist France, Imperial Germany and the Soviet Union all lost their relative position of power and influence on the European or Eurasian continent quite rapidly *without gaining some form of significant compensation in return*. (For differing dimensions of interstate and military-technological conflict in the long term from a geo-sociological perspective, see Charnay, Chapter 25; for a geo-economic perspective emphasizing technological innovation, but also recognizing the land–sea schism, see Thompson, Chapter 27.)

In addition, both France of the *ancien régime* and Imperial Germany lost their capacity to maintain domestic political-economic stability, ultimately leading to the very different French *national democratic* and German *national socialist* revolutions. Despite the extreme ideological differences of these two "revolutions," the geostrategic goals of both countries can be compared and contrasted as the wars progressed. The actions of a highly unstable France and Germany and counter-actions (or non-actions) of states that sought to contain or limit French and German behavior then helped set off wars of global *revanche* from a position of geostrategic and political-economic collapse, at the same time that such conflicts were promoted and fueled by state elites and socio-political collectives who sought to regain territories, possessions, power and influence that had generally been lost in past conflicts—or which were seen as necessary in order to seize a geostrategic or political economic advantage over state rivals.

One of the major *systemic* reasons for the latter globally *revanchist* phenomena lies in the fact that the "balance of power" concept was largely confined to Europe alone at the 1648 Treaty of Westphalia at the end of the "time of troubles" (see Kobtzeff, Chapter 10); the balance of power concept was then formalized at the Treaty of Utrecht in 1713 after socio-political and religious conflict over Germany and the central and east European "shatterbelt" was mediated.[1] But this essentially Eurocentric and predominantly geopolitical concept of the "balance of power" (which is better seen as a dynamic equilibrium/ disequilibrium; see Gardner, Chapter 1) was largely thrown out the window with the rise of global overseas rivalries among the European powers themselves. In addition, the rise of powers *within* Europe as well as *outside* of Europe likewise began to challenge the initial European status quo or "power balance." (This was the case for the rise of Prussia/ Germany *within* Europe and the rise of the USA, Japan and then China from *outside* of Europe.) The expanding nature of overlapping strategic theaters and the rise of overseas rivalries consequently led to a highly uneven "imbalance" or *disequilibrium* of power, force and influence (including financial capabilities) among the European states themselves and increasingly overseas, among extra-European powers.

In the Seven Years War, a heavily indebted amphibious France lost its hegemonic position in Europe, plus most of its overseas possessions.[2] This defeat was eventually followed by the French Revolutionary Wars and Napoleonic *revanche* after the general failure of domestic political economic reforms. Somewhat similarly, Imperial Germany (on the rise since the defeat of Prussia by Napoleon at Jena in 1806 and the later humiliation at Olmültz in 1850) collapsed in World

1 Prior to the Treaty of Westphalia, the Italian city states first established the principles of non-aggression and "balance of power" among themselves in northern Italy at the 1454 Peace of Lodi, following the 1453 conquest of Constantinople by the Ottoman empire.

2 William Thompson (see Chapter 27) calls the period 1740–63 one of British de-legitimization and 1763–92 as a period of coalition building; yet he appears to overlook British victory in one of the primary global conflicts, the 1756-63 Seven Years war in which France lost control over North America and India (with Britain using the private army of the East India Company). The US Revolution can be seen as an offshoot of this latter conflict. France's support of the US Revolution was largely a pyrrhic victory, in that Britain was able to retain "hegemony" or "leadership" over the US until the war of 1812–14. Modelski (Chapter 28) does not list the Seven Years War as a "global" war.

War I—an indecisive defeat to be followed by Hitler's *revanchist* national socialist movement, which repudiated German debt. The Cold War then resulted in the collapse of Soviet overseas holdings and Soviet disaggregation into 15 separate states by August 1991, with an unstable Russian Federation as the major actor.

The post-Cold War situation has not yet been followed by the rise of a Russian *revanchist* movement, despite the expression of *revanchist* claims by certain pan-Slav, Eur-Asian and xenophobic factions within the Russian Federation, and the attempts of these groups to link with Slavic or Russian nationals (or socio-political identity groups that seek Russian support) in former Soviet states, plus links to extreme nationalist groups in Europe and abroad. At the same time, Soviet collapse has, to a large extent, permitted the formation of a polycentric world system involving the emergence of states and socio-political movements with highly uneven capabilities of force and influence. This includes the rise of China as a regional, and *potentially* global, hegemon—given its financial influence and industrial capacity. Soviet collapse has also opened up geopolitical and economic rivalries over its former spheres of influence and security, and has resulted in what Moscow sees as a provocative and largely uncoordinated NATO and European Union enlargement into what Moscow now considers its "near abroad."

Unlike the case for eighteenth century France and early twentieth century Imperial Germany, Soviet collapse may or may not set off a Russian *revanchist* movement; yet a number of emergent powers, including China, India, Turkey, Brazil and Iran, among others, which possess differing degrees of irredentist claims, do seek to challenge global inequities in power and wealth, while concurrently sparking largely regional state and socio-political disputes. Russian elites could find themselves increasingly estranged from hopes to sustain Russia's major power status, particularly with respect to the rise of China as a potential military and economic rival. The point is that a clash of major power and rising regional power interests (combined with the actions of surrogate actors and nomadic anti-state organizations) could provoke systemic conflict—if the global system is ultimately unable to re-equilibrate itself after Soviet collapse. Another scenario is the continual *widening* of regional conflicts into "no man's lands" that do not possess clearly designated spheres of influence and security and that increasingly draw in major and regional powers, whether or not the latter act unilaterally or in "coalitions of the willing" and under a UN mandate or not. The danger is that major and rising regional powers could possibly intervene on opposite sides of a specific conflict or conflicts—if multilateral accords or concerted policies cannot be reached and implemented prior to such interventions.

Geohistorical Roots of the Global Crisis

The roots of the modern global crisis, and of contemporary disputes between insular-core and amphibious-core and continental states, can be traced to the break-up of the global Habsburg empire, where "the sun never set" under

Charles V.[3] The Thirty Years War from 1618 to 1648 was initially sparked by religio-political confrontation in Prague following the 1617 election of the Catholic Duke of Styria as king of Bohemia, a Protestant–Catholic dispute set off largely over a question of land rights and the freedom of Protestant churches to worship, resulting in the second defenestration of Prague. The Thirty Years War, originating within the Bohemian/German shatterbelt, represented the culmination of an overlapping series of wars involving Dutch independence, all-out war between the French Valois and the Spanish Habsburgs, in addition to Swedish, Danish, Polish and Russian strife over the Baltic littoral and eastern Europe. The war thus tended to intensify major power rivalry with Russia for control over key regions in the northeast as well as in central and eastern Europe—along with the later sequels of the 1701–13/14 War of Spanish Succession and of the 1700–21 Great Northern Wars. The 1648 Peace of Westphalia (which represented a series of pan-European treaties that ended the 1618-48 Thirty Years War largely fought within the Holy Roman Empire and the Eighty Years War between Spain and the Dutch republic) largely put an end to the Habsburg dream of world empire, limited the power of the Vatican, and established Sweden as a major power in northern Europe; it likewise provided the quasi-insular Netherlands and continental Switzerland with sovereign recognition, while likewise recognizing Austria as a separate dynasty and territorial state. Westphalia consequently helped to establish a new land-sea schism between Britain, Holland and France by breaking up the Spanish-led Habsburg empire in Europe and to formalize a still contentious "balance of power" or "equilibrium" that was not quite as "religiously" charged.

Westphalia's version of the principle of *cuius regio, eius religio* (whoever controls the territory determines the religion), which had previously been established at the 1557 Treaty of Augsburg, extended international recognition from Catholics and Lutherans to only Calvinists (with other sects cautiously tolerated under the 1648 Treaty of Osnabruck). While limiting papal authority, Westphalia did not, in all cases, establish full state sovereignty over territory, but likewise recognized multinational empires (the Holy Roman Empire did not collapse overnight), principalities, in addition to emerging "nation-states" (which generally did not consist of a single ethnic or "national" group). The treaty did indirectly seek to restrict, but not eliminate, warfare through the application of international law (as opposed to canonical law), and thus *attempted to set legal limits on state and imperial sovereignty* (under the influence of Grotius, de Victoria, Gentili, among others). The Treaty of Westphalia thus helped put an end to the age of "planetary upheaval" (See Kobtzeff, Chapter 10)—at least in Europe itself.

At the same time, however, the geostrategic dimensions of "balance of power" were increasingly becoming global and based on *extra-European* geostrategic factors, while the political-economic dimension of the overall "equilibrium" depended, to a large extent, upon the question of national finance and debt. On the continent, the Thirty Years War did not put an end to France's efforts to assert

3 George Liska, *Ways of Power: Pattern and Meaning in World Politics* (Oxford: Basil Blackwell, 1990).

effective control over all "gates" leading to France and to force Spain to surrender the Burgundian Circle, as well as Spanish possessions in Italy and Catalonia, in particular. Despite assistance from England in 1657, France was unable to settle its continuing dispute with Spain at the 1659 Peace of the Pyrenees, after the Eleven Years War. Louis XIV consequently invaded the Spanish Netherlands in 1667 and the Dutch Republic in 1672, largely bringing about a self-fulfilled prophecy. Louis XIV's efforts to break a predictable Spanish-Austrian "encirclement" in 1688 and to block William of Orange from ousting James II from Great Britain by means of a limited "preemptive" engagement in Cologne and Philippsburg in the Palatinate, led to the more general Nine Years War (1688–97).

The latter conflict involved the formation of a "Grand Alliance" against amphibious France and its continental claims to the Palatinate (a territory that then included the Rhenish Palatinate, Heidelberg, Mannheim and parts of French Alsace), once insular England and quasi-insular Holland joined the League of Augsburg in 1689. Following the 1688 Glorious Revolution and the overthrow of James II, limiting the authority of the monarchy in England, and the ascension of William III of Orange and Mary II to the throne (who had been backed by Holland), England opted to declare war on France once the latter decided to assist James II in Ireland.[4] France was consequently unable to prevent an alliance between Britain and Holland, despite previous Anglo-Dutch conflict over the slave trade and the American colonies (New York). (During the Second Anglo-Dutch war from 1665 to 1667, the Dutch fleet had sailed up the Thames in 1667; France and England then combined against the Netherlands in 1672 until the 1674 Treaty of Westminster with Britain, and the 1679 Treaty of Nijmegen with France following the Dutch flooding of the dykes.)

The Nine Years War (or War of the Grand Alliance, 1688–97) represented the first war which decisively used sea power, and corresponded with Anglo-French colonial conflicts and King William's War in the American colonies from 1689 to 1697.[5] The Nine Years War involved a Grand Alliance of the League of Augsburg states (Austria, Sweden, Spain, plus German shatterbelt states of Bavaria, Saxony and the Palatinate) joined by England and Holland. France supported the Ottoman

4 The Williamite/Jacobite war in Ireland (in which Irish Jacobites had been backed by France) was largely at the roots of the conflict in Northern Ireland. William III, following in the footsteps of Charles I, imposed Protestant rule over the long term. Despite initial promises of toleration of Catholicism at the 1691 Treaty of Limerick, Pope Innocent's decision to support France in 1693 toughened Protestant opposition to both Catholicism and France.

5 Paul Kennedy, *The Rise and Fall of the Great Powers, Economic Change and Military Conflict from 1500 to 2000* (New York: Random House, 1987), called the Nine Years War the first "world war," but it is more of an extra-European war fought among European powers *outside of Europe*. The first truly world war *involving major powers outside of Europe* was World War II, in that Japanese-Chinese conflict in Asia linked up with conflict in Europe through the alliance of the Axis powers, while the United States was drawn into both Asia and Europe (and northern Africa) as well. The rise of non-European powers really began to expand geostrategic interconnections across the globe or globalization.

Empire against Austria; Poland, Russia and Venice also engaged. After the 1697 Peace of Ryswick, France relinquished its claims to Cologne, and recognized William III of Orange as king of England, but retained Alsace and Strasburg, leaving Franco-German conflict to fester. Negotiations in the period 1697–1702 likewise failed to settle disputes over the Spanish succession.

The 1701–13/14 War of the Spanish Succession began once a childless Charles II named the Bourbon Philip, duke of Anjou, as his successor. Then, once Philip took the Spanish throne as Philip V, his grandfather, Louis XIV, invaded the Spanish Netherlands, resulting in a Grand Alliance of England, Holland and Austria against France—and against the threat of a dynastic union with Spain. Like the Nine Years War, the War of the Spanish Succession represented an extra-European war as it linked overseas rivalries (Queen Anne's war in the colonies from 1702–13) with conflicts in the Dutch littoral and the German/central and eastern European shatterbelt (in addition to interlinking with the Great Northern War). The Peace of Utrecht partially partitioned Spanish holdings in Europe, but did not totally uproot quasi-insular Spain's overseas possessions and trade. The Treaty of Utrecht gave Britain control of the Mediterranean at Gibraltar and Minorca, for example, plus a monopoly of the slave trade with Latin America, but England's intent to seize large parts of Spanish America had largely been frustrated during the war.[6]

The 1713 Treaty of Utrecht (which more formally established the principle of "balance of power" than did the 1648 Treaty of Westphalia) was followed by the 1716 Anglo-French entente after the death of Louis XIV in 1715. The Anglo-French entente was formulated to counter efforts by Philip V of Spain to overturn the Utrecht (1713) and Rastatt (1714) peace settlements; it was also forged, in part, in response to the Russian occupation of Mecklenburg, which bordered Hanover, which was of major geostrategic concern to the English Crown due to the personal union of the monarchs. France and England were then joined in 1717 by the United Provinces and then Austria. (The United Provinces, however, did not sign the Quadruple Alliance for fear of being cut off from Spanish "most favored nation" status.) The War of the Quadruple Alliance (1718–20) then represented an effort to counter Spanish King Philip V's *revanchist* effort to retake territories in Italy and to claim the French throne. The Quadruple Alliance of Great Britain, France, Austria and the Dutch Republic, plus Savoy, ultimately defeated a revisionist Spain and, by 1720, compelled Madrid to accept the provisions of the peace treaties at Utrecht and Rastatt.

The Anglo-French entente, which lasted roughly fifteen years, represented an uneasy effort to reintegrate France into the European "balance of power" system after the War of Spanish Succession and thus represented a historical break from the traditional antagonism between insular-core and amphibious-core states. The

6 See Derek McKay and H.M. Scott, *The Rise of the Great Powers, 1648–1815* (London and New York: Longman, 1983), 95 and passim. Spain would give up its claims to Brazil to the Portuguese in the Treaty of Madrid, 1750, while the long struggle against Spanish and Portuguese colonialism in Latin America would begin during the Napoleonic wars.

Anglo-French entente was additionally designed to "contain" the ambitions of Austria and, primarily, Spain, but also the ambitions of Russia in the Baltic. Once the Quadruple Alliance (England, France, Austria and the Dutch) began to break down in the period 1720–24, however, the Anglo-French entente subsequently began to break down in the period 1728–31. The breakdown of Anglo-French relations occurred as Britain secretly reached for an alliance with Austria, forging the Second Treaty of Vienna in 1731 after the Anglo-Spanish war of 1727–29 in which France failed to support Britain's interests. (See Rimanelli, Chapter 11.)

The Second Treaty of Vienna in 1731 was also signed by Spain and the Netherlands, once again isolating and encircling France with a new "grand alliance." In addition, the death of Augustus II destabilized Poland in 1733; France, Spain and most of the Polish nobility backed Stanislaus Leszczynski, who had previously ruled Poland in 1704–09. Russia, Austria and the Lithuanian nobility backed Augustus III, son of Augustus II. The War of the Polish Succession then ended with the fourth treaty of Vienna in 1738. But by 1739, however, the Anglo-Spanish War of Jenkins' Ear led to a general overseas conflict involving both Spain and France. Overseas conflict then began to interlock more tightly with the Austro-Prussian war over Silesia and the German shatterbelt in the 1740–48 War of Austrian Succession in which England and Austria aligned. But Austria then complained that Britain did not support Austrian claims to Silesia at the 1748 Treaty of Aix-la-Chapelle.

In many ways, the Seven Years War (1756–63) doomed the "balance of power" concept *as it concerned interstate relations within Europe—but not overseas*—and represented France's penultimate effort to counter British overseas supremacy in India and North America. The Seven Years War was, in part, sparked by a "diplomatic revolution" once Austrian Foreign Minister Count Kaunitz took steps to align Austria and France (joined by the nascent *pivotal* state, Russia, as well as Saxony and Sweden) against Prussia. Austria no longer saw England as supporting its revanchist claims to Silesia; Austrian actions were then countered by a 1756 British-Prussian convention to protect Hannover. France's Duke Choiseul's efforts to play "two games" together, whereby he sought negotiations with England on the one hand, and threatened an alliance with Spain on the other, helped to exacerbate conflict.[7] While supporting Hanover, England increasingly looked to Prussia to counterbalance France—instead of supporting its more traditional ally, Austria.

Ongoing extra-European wars overseas (French and Indian Wars in North America from 1754–63 and the Third Carnatic War over Bengal and South India from 1757–63) concurrently interlinked with wars in Europe (the Pomeranian War in Sweden from 1757–62 and the Third Silesian War between Prussia and Austria from 1756–63), not to overlook Anglo-French conflict over Hanover. French defeat resulted in the loss of French colonies in North America and subservience to Britain in India, plus loss of French influence in Europe following the 1763 Peace

7 J.S. Corbett, *England in the Seven Years War* Volume II (London: Longman, Green and Co., 1907), 185.

of Paris.[8] The 1763 Peace of Hubertusberg guaranteed that Prussian King Frederick II the Great would retain Silesia. Despite protest by those opposed to making any concessions to France and Spain at the 1763 Treaty of Paris, French defeat and inability to reform likewise led to increased domestic instability and mounting debt, plus the loss of French hegemonic influence in Poland and Central Europe (resulting in the further rise of Russian influence in this region). The decline of French support for an independent Poland led the latter to be partitioned in 1772 by Austria, Prussia and Russia; Russian victories over the Ottoman empire also weakened Poland's position. Poland would again be partitioned in 1793 and 1795—actions taken in part to take advantage of a temporarily weak France and to contain French revolutionary support for Polish independence, as well as to assert Russian, Prussian and Austrian geostrategic and political-economic interests in establishing a condominium over Poland.

After its defeat in the Seven Years War, France immediately looked for ways to revenge itself upon Britain; the American Revolution presented itself as the first opportunity. Following Duke Choiseul's fall from power in 1770, France under Foreign Affairs Minister Count de Vergennes began to implement Choiseul's call for *revanche* in 1774 once revolution in the American colonies appeared imminent. Vergennes accordingly sought to maintain the Family Compact with Spain as a means to sustain a rough semblance of naval parity with England. He also sought a defensive alliance with Austria as a means to counterbalance Prussia, so as to prevent England from using the latter to pressure the Rhine. Shielded by close relations with Spain and Austria, it was argued that France could then build up its naval and military capabilities, but without coming into direct conflict with England—unless the chances for success appeared strong.[9]

Britain's efforts to impose taxes on the American colonies to pay for its efforts during the Seven Year War (what was called by the American colonists "the French and Indian wars" in North America and which used "unconventional" tactics for that era) ultimately resulted in the American Revolution based on the creed "no taxation without representation." France's support for the American colonies represented France's first war of *initial challenge* after defeat in the Seven Years War. Yet French naval support for the American colonies following the 1777 victory at Saratoga represented a case of failed *revanche* in that the English ("perfidious Albion") were not substantially weakened by the American Revolution. Contrary to its hopes, France (which had been supported by Spain through the Bourbon Family Compact and 1779 Treaty of Aranjuez) was not able to capitalize on US independence, in that insular Britain, not France, was largely able to monopolize American trade up at least until the War of 1812. The Seven Years War, coupled with French support for the American Revolution, along with domestic corruption and mismanagement, consequently helped to send France into deep bankruptcy. French bankruptcy indirectly sparked the French

8 Treaty of Paris, 1763. See http://avalon.law.yale.edu/18th_century/paris763.asp.

9 Samuel Flagg Bemis, *The Diplomacy of the American Revolution* (Bloomington: Indiana University Press, 1935/1957), 17–18.

Revolution and French Revolutionary/Napoleonic wars, undermining political-economic efforts to sustain a semblance of "balance of power."

By 1789, France could no longer sustain a relationship of global parity with Britain, nor could it find the means to reform itself: Revolution in France upset de Vergennes' system of protective alliances, and opened France's eastern flank to both Prussian and Austrian pressures. The revolutionary abolition of feudal privileges challenged monarchist and aristocratic legitimacy in general; yet the main causes of the war can be seen in the efforts of Austria and Prussia to threaten a new "Grand Alliance" against France, and French efforts to counter both Prussia and Austria through the launching of a war that was at least initially intended to be limited. In effect, the Girondin faction regarded the 1756 alliance with Austria as leading to the downfall of the French empire, resulting in French defeat in the Seven Years War. One of the major leaders of the Girondin factions, Jean Paul Brissot de Warville, accordingly demanded his compatriots avenge themselves against the crimes of foreign powers and to fight for "universal liberty."

As tensions mounted, the Austrian and Prussian rivals signed a formal military convention in July 1791. Austro-Prussian goals were, on the surface, limited; they were aimed in part at limiting Russian influence in Poland, in addition at advocating support for monarchist legitimacy. As revolutionary nationalism had worked to undermine overlapping aristocratic Franco-German controls in Alsace and Avignon, as well as in Prussian-controlled territories, Prussian King Frederick Wilhelm was willing to intervene if necessary to obtain territorial compensation. The largely vacuous July 1791 Padua circular and August 1791 Declaration of Pillnitz, in which Austria threatened to intervene militarily in concert with other major European powers if Louis XVI was not soon restored to the throne (a new Grand Alliance), were largely intended to pressure radicals and to support the moderate Feuillant faction and the monarchy.

The Girondins believed Russia, Sweden and England would remain neutral (despite their own propaganda denouncing an aristocratic conspiracy against France) in a war directed against their Austrian enemy, while Prussia would consequently break its alliance with Austria, ultimately isolating the latter. The war-prone Girondins backing Jean Paul Brissot believed a limited war would test the loyalty of Louis XVI and help consolidate the gains of the revolution. Lafayette, by contrast, believed that a limited victorious war against Austria and Prussia might strengthen a new constitution limiting the power of the monarchy. Louis XVI believed that defeat in war might help him to restore his throne.[10] *With the notable exceptions of Marat and Robespierre, the major political factions believed that some form of war might help their specific cause.*

Despite the efforts by the Girondist French foreign minister Charles-Francois Dumouriez to obtain assurances of Russian, Swedish, Dutch, Spanish and Swiss neutrality, and to seek an alliance with England while breaking ties with Austria, the failed French effort to attack the Austrian Netherlands in April 1792 led to the

10 For a brief overview, see Stephen M. Walt, *Revolution and War* (Ithaca, NY: Cornell University Press, 1996), 62–74.

July 25, 1792 Brunswick Manifesto, just prior to the joint Austro-Prussian July 30, 1792 invasion of France—and caused a militant backlash among the French revolutionaries. (See Kobtzeff, Chapter 4.) The November 1792 French attack on the Austrian Netherlands and Belgium (the "buffer" state between England and the continental powers) then worked to convince the English parliament to vote for war preparations against France in December 1792. Ironically, in February 1792, British Prime Minister William Pitt had stated his conviction that Europe could expect 15 years of peace; yet two months later, the European continent was in flames and, less than one year later, Great Britain was engaged.[11] The ultimate result was that actions by revolutionary France that had been intended as a "limited" conflict instead sparked a general war in Europe by 1793—just 29 years after the collapse of France's overseas empire and hegemony in Europe. As Perfidious Albion had blocked the French navy in the North Atlantic, and as Britain controlled the Cape of Good Hope, Napoleon's failed invasion of Egypt had been intended as *revanche*: to make the Mediterranean a "French lake" while the opening of a route between the Suez, the Red Sea and India was aimed at recovering Pondicherry and other French possessions on the Coromandel and Malabar coasts.[12]

Following the Napoleonic wars (which were settled by the Treaty of Vienna which permitted a rather rapid reintegration of France into the Concert of Europe), Britain then became a truly core hegemonic power until World War I, primarily challenged by France and Tsarist Russia up until 1894–1904/07, when Britain finalized the Triple Entente with France and Russia. The Triple Entente resulted in the "encirclement" of the Imperial Germany and the Triple Alliance—once Berlin was perceived as the primary amphibious-core challenger to London in the period 1901–14. (See Gardner, Chapter 13.)

In the background, the continental United States began to move up from continental semi-peripheral to quasi-insular core status after the American civil war (war of secession), and just before and after World War I (the war to end all wars)—when the US moved from net debtor to net creditor status with the influx of European capital into the US. Likewise, Japan began to claim ascendency in Asia after its defeat of China in 1894–95; this defeat then led to the Chinese revolution under Sun Yat Sen, and ultimately to Maoist *revanche* against Japanese imperialism. The Chinese revolutionary May 4th movement of 1919 was then set off by the 1919 Treaty of Versailles, which shifted Germany's control of Shandong province to Japan. (See further discussion, this chapter.)

By contrast with the 1815 Treaty of Vienna which had worked to reintegrate France into the European concert (once Russian troops left Paris in 1818), the 1919 Treaty of Versailles alienated Germany through the "war guilt" clause and long term debt reparations. Woodrow Wilson's promises of "open covenants openly arrived at" had nevertheless excluded German interests. Moreover, the 1925 Locarno Treaty may have ostensibly guaranteed Germany's western borders with

11 See Hans Morgenthau, *Politics Among Nations: The Struggle for Power and Peace* (New York: McGraw-Hill, 1993), 23.

12 Juan Cole, *Napoleon's Egypt* (New York: Palgrave MacMillan, 2007), 14.

France (until 1936), but did not address Germany's eastern borders. An "Eastern Locarno" failed to materialize following the formation of "The Little Entente System" in 1933 which represented a regional system of defense for the Little Entente states of Czechoslovakia, Romania and Yugoslavia, a system initiated by Czechoslovak leader Edvard Beneš in order to check Hungarian irredentism and to block efforts of the Habsburg monarchy to return to power. The belated plan of French Foreign Minister Jean Louis Barthou to forge an Eastern Locarno that would bring Poland and Germany to guarantee each other's borders was not given strong support by either Britain or Poland; and it was denounced by Nazi Germany as "encirclement." In 1936 Hitler renounced the Locarno treaty and remilitarized the Rhineland—after the 1935 Franco-Soviet mutual defense pact.

The assassination of the King of Yugoslavia (Kingdom of Serbs, Croats and Slovenes) in 1934, who had been accused of repressing Bulgarian minorities in Macedonia, by a member of the Internal Macadonian Revolutionary Organization (IMRO), linked to the Croatian Ustaše, and possibly assisted by Italian or German intelligence, resulted in French Foreign Minister Barthou's death as well. This terrorist attack helped to undermine France's support for the Little Entente. The Little Entente had increasingly been supported by France from 1924 as a means to counter German influence—until the system finally broke down in 1936–38.

The Versailles "war guilt" clause, along with a massive financial depression in Germany, served Hitler's propaganda machine, ultimately enabling him to come to power (with the Nazi's winning 107 seats in the German parliament in 1930 and the majority of seats in 1933, with the Communist Party a close second) along with other pan-nationalist parties throughout eastern Europe (in Hungary and Bulgaria) which likewise opposed the Versailles treaties. Invited by Weimar President Paul von Hindenburg into the Chancellery in January 1933 to counter the Communists, Hitler manipulated the burning of the Reichstag (ostensibly an act of "terrorism") as means to eliminate all opposition parties by 1934, and likewise assassinated his rivals (Ernst Roehm). British Prime Minister Neville Chamberlain's effort to turn Hitler against Stalin at Munich in September 1938 failed miserably, once the August 1939 Soviet-German Non-Aggression Treaty was signed, in effect breaking the 1935 Franco-Soviet mutual defense pact. The Munich agreement had permitted Germany to annex the Sudetenland following Hitler's appeal that Germany should ostensibly have the right to "protect" national minorities (rather than demand that ethnic Germans obtain greater autonomy *within* Czechoslovakia).[13] Poland then seized Teschen in Czech Silesia in September 1938 and Hungary annexed southern portions of Slovakia with Nazi German backing.

Convinced that neither Britain, France, nor the United States, would intervene despite Anglo-French pledges to defend Poland, and having forged the "Pact of Steel" with Mussolini, Hitler marched into Prague in March 1939 and then into Poland on 1 September 1939, setting off World War II in Europe. The Soviet Union

13 Hitler reversed Woodrow Wilson's concept of "national self-determination," from a tool to break up empires toward a tool to re-build empires—in asserting the right to German self-determination.

then thrust into Poland on September 27 in accord with the secret protocol of the Hitler-Stalin Non-Aggression pact. The latter represented an even more grotesque action than the Austrian, Prussian and Tsarist partitions of Poland in the late 18th century. World War II would thus explode roughly twenty one years after Imperial German defeat and collapse.

Unlike the period after World War I, when the US Senate refused to consider security guarantees to Britain and France, and refused to back the formation of the League of Nations despite President Wilson's efforts, the US took a leading role in the post-World War II reconstruction through the Marshall Plan, and supported the creation of the United Nations. The US and Britain, plus France, occupied a divided Germany along with the Soviet Union, forming a power sharing arrangement under Four Power controls. Yet the Marshall Plan represented one of the major factors that began to alienate Moscow after World War II, in that Marshall Plan—without Soviet inclusion—was regarded as potentially undermining Soviet political-economic controls over its newly acquired east European buffer. Washington then became the paramount or hegemonic insular core state in rivalry with the increasingly amphibious Soviet Union. American ascendance to global power status was largely due to the fact that these intra-European Anglo-French-German-Russian wars (in which Imperial Germany was only *indecisively* defeated while Nazi Germany was decisively conquered and divided) ironically undermined British insular core paramountcy—permitting the Americans to predominate in the post-World War II era.

The Question of Long Term Russian Alienation

Not only did the Thirty Years War help to open a dichotomy between insular and amphibious powers by breaking up the global Habsburg empire, but it also helped to further intensify major power rivalry with Russia for control over key regions in the northeast as well as in central and eastern Europe—along with the later sequels of the 1701–13/14 War of Spanish Succession and of the 1700–21 Great Northern Wars, which initially represented an alliance of Russia (under Peter the Great), Denmark-Norway (under Frederick IV) and Saxe-Poland-Lithuania (under August the Strong) against Swedish hegemony (under Charles II) in northern and eastern Europe. From this perspective, the Thirty Years War not-so-inadvertently re-sparked Polish-Ukrainian conflict (the 1648 Ukrainian uprising) and ultimately permitted Russia to establish hegemony over Ukraine following the 1654 Pereiaslav Agreement. Russia was then able to seize much of Polish-held Belarus, as well as eastern Ukraine (Smolensk and Kiev) following the 1654–67 Thirteen Years War.

Swedish efforts to liberate Ukraine and march on Russia failed disastrously in the 1709 battle of Poltava. In addition, following Russian occupation of Poland in 1709, Russian troops forced the return of Augustus II to the throne of Poland; the latter absentee king (he was also elector of Saxony) had been deposed by Sweden in 1704 and then replaced by Stanislas Leszczynski until 1709. With the defeat of

Sweden in the Great Northern War (1700–21), Russia (in alliance with Poland and Denmark) occupied Mecklenberg (the neighbor of Hanover) and made a decisive step to the Baltic, opening the door to Europe by 1721.

This latter act by Peter the Great consequently initiated a long cycle of geopolitical struggle to remove Russia from the Baltic littoral (as well as from central and eastern Europe) in the Napoleonic Wars, World War I, World War II, as well as during the Cold War (in that American containment policy was intended to loosen Russian grips over eastern Europe). Offensives by Napoleon I had ultimately failed to push Russia out either the Baltics or central and eastern Europe. Often overlooked, the British engaged in minor battles with Russia in the greater Baltic region during the Crimean war (1853–56), but the major focus of conflict remained the region around the Crimea, Sea of Azov and the Caucasus.[14] The 1876–78 Russian-Ottoman wars which, in effect, represented the culmination of a series of wars at least since the sixteenth century and prior to World War I (in which Turkey was supported by Imperial Germany) sought to advance Russian controls to the Ottoman straits, in a Slavic "crusade" to break apart the Ottoman Empire, open up the Balkans, the Caucasus and central Asia.[15] In March 1915, Britain secretly promised to hand Constantinople and the Dardanelles over to Russia, in an effort to keep Russia in World War I, but the 1917 Leninist revolution overturned Tsarist hopes for imperial expansion.

During World War I, Imperial Germany stunned Tsarist Russia at Tannenberg in 1914, and then at Brest-Litovsk in 1915, knocking Russia out of central and eastern Europe following the 1917 Treaty of Brest-Litovsk. Berlin had provided secret diplomatic and financial support for Lenin, in which Lenin was permitted to take a sealed train car from Switzerland through Germany to Finland Station in what was then Petrograd, when Lenin wrote his *April Theses*. At the same time, however, the effort to remove Russia from central and eastern Europe (including the undeclared US, French and British war against Russia from 1918–20) was ultimately countered by Russian efforts to reestablish its presence in the region. The Bolsheviks regained much of Ukraine (but not Polish Lvov) in the Russian civil war. Following the breakdown of the 1939 Molotov–Ribbentrop pact, and the failure of Operation Barbarossa, Stalin moved into eastern Europe to counter Hitler's forces, but also to counter the American forward presence.

One can only speculate what might have happened had George Kennan, as director of the US State Department's Policy Planning Staff, been able to negotiate "Plan A" with Moscow in 1949 which was intended to establish a "disarmed and

14 The mid-nineteenth century Crimean War raises parallels to unresolved NATO-Russian-Ukrainian tensions in the Black Sea region today, coupled with NATO-Russian tensions in the Baltic state area, not to overlook Polish-Belarusian disputes.

15 The Russian-Ottoman wars likewise possess troubling parallels to attempts to revitalize Russian "spirituality" and influence in the near abroad since the second Chechen war from 1999 to 2000 and Russian reaction to NATO's war "over" Kosovo in 1999. See Platov, chapter 12. See also footnote 20 on Kosovo.

neutralized" Germany in the midst of Europe, under modified four-power control.[16] The failure to resolve the German question in the aftermath of World War II was the primary factor setting off the massive conventional and nuclear arms race that did risk the real possibility of World War III. (See Rimanelli, Chapter 16.) As Cold War rivalry over Germany, eastern Europe, the Korean peninsula and Indochina (as well as the Middle East) intensified, West Germany would be integrated into the European Coal and Steel Community in the Treaty of Paris by 1951–52, in an effort to permanently prevent war with France. Having divided Germany under Four Power controls, the Soviet Union eventually forged the Warsaw Pact in May 1955 as a military buffer in direct response to NATO's incorporation of West Germany in 1955 a week earlier. (The Warsaw Pact was officially to disband only once NATO had dissolved.) In addition to challenging US controls over Berlin, Moscow began to support pro-Soviet revolutionary forces throughout the world. Not only did it back the Cuban revolution (but only after Castro came to power) within the American sphere of influence, Moscow deployed intermediate range nuclear weapons in Cuba. John F. Kennedy's diplomacy and concessions to Moscow over US nuclear weapons deployed in Turkey helped to avert the real possibility of nuclear war.

US withdrawal from Vietnam in 1975 (after a war deemed necessary to check Communist expansionism) was largely impelled by anti-war groups, including Vietnam Veterans who had turned against the war (see Ekovich, Chapter 17). There was also a general unwillingness to confront China, which had restrained US actions against North Vietnam: Washington feared that Beijing could intervene militarily in Vietnam as it did in Korea in October-December 1950.[17] The American defeat in Vietnam appeared to herald a number of Communist victories around the world through the "domino" effect; yet forging a US-Chinese rapprochement in the early 1970s, Washington generally sought to play pro-Chinese movements against pro-Soviet movements in the belief that Moscow represented the more immediate "threat." The Carter/Reagan administrations then sought to overturn Soviet gains throughout the developing world, through support of "freedom fighters" against pro-Soviet regimes, most crucially in Afghanistan, where the Carter administration "purposely augmented the chances that the Soviet Union would intervene."[18] Once Moscow reluctantly and massively intervened, the US forged a Saudi, Pakistani and Chinese alliance in support of the pan-Islamic Afghan resistance against the

16 George F. Kennan, "Letter on Germany" (to Gordon Craig, July 28, 1998), *New York Review of Books*, available at: http://www.nybooks.com/articles/archives/1998/dec/03/a-letter-on-germany/. Portions of Kennan's Plan A were leaked to the press, and US-Soviet negotiations were suddenly cut off.

17 Not only did the US strike Chinese airbases, but the US and Soviets came close to direct confrontation in Korea, when the US bombed a Soviet air force base in Vladivostok on October 8, 1950, as Soviet pilots were flying jets with North Korean markings. See Jon Halliday, "A Secret War" *Far Eastern Economic Review* (22 April 1993), 32–36. This was confirmed to me by Andrei Grachev, Mikhail's Gorbachev's spokesperson.

18 Zbigniew Brzezinski, *Le Nouvel Observateur* (Paris, 15–21 January 1998).

Soviet Union, while concurrently opting to permit Islamabad direct that resistance and ignore Pakistan's quest for nuclear weaponry.[19]

Often not mentioned in the ideological battle between capitalism vs. communism, and democracy vs. totalitarianism, is the fact that the US also manipulated religion against Soviet atheism by supporting Jewish emigration from the Soviet Union (1974 Jackson-Vanik Amendment), Polish Catholicism (recognition of the Vatican in 1983 plus support for the *Solidarnosc* movement), not to overlook pan-Islamic movements, among other means, such as promoting dissidents, to undermine perceived Soviet *legitimacy*. At a time when the US had backed the Afghan resistance against the Soviet intervention and had initiated a massive arms build-up involving the deployment of Cruise and Pershing missiles in Europe to counter Soviet intermediate range SS-20s, Washington and Moscow came close to nuclear war in late 1983 when the Soviets shot down a South Korean airliner on September 1st, when the Soviet satellites falsely detected a US nuclear missile attack on September 26th, and when NATO was engaging in provocative military maneuvers, Able Archer 83, beginning November 2nd.

By the end of the Cold War, Mikhail Gorbachev's reforms from 1985 to 1989 led to the demise of the Warsaw Pact, while Germany seized the moment to unify under a federal model within NATO—in opposition to Soviet proposals for a unified, yet neutral Germany or else a proposed re-association between West and East Germany under a confederal model. More specifically, no NATO troops or nuclear weaponry were to be deployed in eastern Germany once Germany unified. The failed August 1991 *pronunciamento* by Soviet bureaucrats against Gorbachev was intended, in part, to reestablish the Brezhnev-era system, but resulted in the break-up of the Soviet Union altogether, the "loss" of the three Baltic states and a significant retraction of Russian overseas influence. Combined with the first wave of NATO enlargement (eastern Germany, Poland, the Czech Republic and Hungary), NATO's engagement in the war "over" Kosovo in 1999 and defeat of Serbia (see Rimanelli, Chapter 18; Kobtzeff, Chapter 19) helped to forge a "national security consensus" that gave rise to Vladimir Putin's authoritarian government in Russia,[20] coupled with Putin's promises to put an end to the Chechen insurgency in 1999–2000. Putin's actions were ostensibly taken to counter the internal threat of secessionism as well as the alleged external threat from NATO encroachment upon the Russian "near abroad." At present, Russia has appeared to be attempting to secure its continental position in the Black Sea/Caucasus (the August 2008 Georgia–Russia war) against NATO

19 Hall Gardner, *American Global Strategy and the 'War on Terrorism'* (Aldershot: Ashgate, 2007), 112–13.

20 See Charles Krauthammer, *Congressional Record—Senate* (April 12, 2000), 5323. One can raise the question as to whether a NATO-Russia interpositionary force deployed in Kosovo as a preventive war force could have helped to ameliorate the conflict. On proposals for a joint NATO-Russia inter-positionary force to be deployed in Kosovo in 1999, see "Update: June 4, 1999 – World Policy Institute – Research Project": http://www.worldpolicy.org/projects/arms/updates/june4.html. See also Introduction, this book.

and European Union enlargement, while also seeking to rebuild its influence in the Eurasian "near abroad" as well as its former amphibious-core overseas influence where possible.

Soviet Collapse and the Relevance of the Interwar Period

Following Soviet disaggregation, the Russian Federation has become a largely landlocked continental power somewhat like Weimar Germany after Imperial German collapse.[21] Much as was the case for Weimar German pan-nationalism even prior to the rise of Hitler, Russian pan-nationalists have claimed the right to "protect" Russian nationals in the "near abroad." With other apparent parallels to the interwar period, Kaliningrad, as a military enclave, appears to play a role similar to east Prussia as a discontiguous region. Belarus appears to play a role somewhat similar to interwar Austria in terms of recurrent proposals to "unify" Russia and Belarus, somewhat similar to demands for *Anschluss* between Nazi Germany and Austria—a unification that had been deemed illegal by the Versailles Treaty. Ukraine plays a strategic role as a key pivot state somewhat similar to interwar Poland, in that Kiev could shift toward the US and Europeans or else move closer toward Moscow (and/or China)—or else break apart, possibly resulting in a tacit or overt partition. Here, Russia has been trying to pressure Ukraine into closer geostrategic and political economic relations by extending the lease of the Black Sea fleet at Sevastopol and by seeking to push Ukraine into a customs union with Belarus and Kazakhstan, while likewise extending controls over South Ossetia and Abkhazia, claimed by Georgia. India appears to play a role in the Indian Ocean parallel to that of Italy in the Mediterranean, with Indian nuclear weaponry countering that of China (with disputes over the Trans-Karakoram Tract as part of Kashmir) and Pakistan (with disputes over Jammu/Kashmir, Baluchistan and Afghanistan). Since the end of the Cold War, China and Russia appear to have forged a neo-Rapallo pact that has been intended to prevent Russia from aligning with the US against China and to prevent China from aligning with the US against Russia—with both Moscow and Beijing seeking hegemony over central Asia (where they can agree) through cooperation in the Shanghai Cooperation Organization.

Yet a key difference between the present and the interwar period is that the insular-core United States has *thus far* taken a far more active role in asserting its interests overseas than did insular core Britain prior to World War II. In many ways, US actions appear more like those of late nineteenth-century Great Britain, which had intervened in Afghanistan, Egypt, the Sudan, among other strategic areas, before seeking to forge a condominium agreement with Russia over Tibet as well as Afghanistan, Baluchistan, and most crucially, Persia (despite the democratic

21 One could argue that Russia has even returned to a landlocked position much like Tsarist Russia *before* Peter the Great—in that Moscow no longer exercises total control over either the Baltic states or central Asia.

reforms taking place in the Persian empire since 1905)—in the effort to confront the rise of Imperial Germany. Somewhat similarly, the United States has led military interventions in Afghanistan and Iraq, and continues to place military pressure and sanctions on Iran, but also appears to have targeted a number of Russian overseas arms clients, including Serbia, Iraq and Libya, as well as Syria. While the US has sought also Russian support in checking Iranian influence, the latter actions nevertheless appear to indicate that the US policy of "containment" of Russia has not entirely subsided in the aftermath of the Cold War. Whether or not radical political changes through the "greater Middle East" from 2001 to 2011 will counter Russian, Chinese and Iranian influence, or whether these latter states can still find ways to re-assert their interests, remains to be seen.

China and the Rise of New Powers

During the Cold War, the US and the Soviet Union often tacitly "double restrained" the rise of states and socio-political movements which were generally opposed to both US and Soviet interests. To a certain extent, this involved tacit US/Soviet collaboration against China, to prevent the latter from directly challenging either the US or Soviet Union. Ironically, however, Soviet collapse permitted the very rapid rise of China. Beijing has generally sustained a number of territorial claims against Russia in response to the "unequal treaties"—not to overlook claims to retake Taiwan and other regions, at least since China's destabilizing defeat in the two Opium Wars with Great Britain (which helped to provoke the Taiping Insurrection, for example), plus China's loss of Taiwan to Japan in the 1894–95 Sino-Japanese war. The 1899 Open Door policy that was intended to assert American influence in Asia was followed by the military intervention of the Eight Nation Alliance (of Austria-Hungary, France, Germany, Italy, Japan, Russia, the United Kingdom and the United States) in an effort to repress the rebellion of the xenophobic Righteous Fists of Harmony in 1901. Yet it was the Versailles Treaty that alienated Chinese revolutionaries with respect to international interference in domestic Chinese affairs and that largely provoked the May 4th 1919 movement.[22] While the Europeans were engaging in fratricide in World War I, Japan insisted upon the "21 Demands" that would permit Tokyo to control former German-held territory in Shandong among other territories in southern Manchuria and Inner Mongolia. Britain, France and Italy secretly supported Japanese claims, while the US had no interest in confrontation with Tokyo. The Versailles Treaty then permitted Japan

22 Mao himself had described Wilson in Versailles: "like an ant on a hot skillet. He didn't know what to do. He was surrounded by thieves like Clemenceau, Lloyd George, Makino and Orlando. He heard nothing except accounts of receiving certain amounts of territory and of reparations worth so much in gold. He did nothing except to attend various kinds of meetings where he could not speak his mind." Stuart R. Schram, *Mao's Road to Power: Revolutionary Writings 1912–1949*, Volume 1 (East Gate, 2002).

to reassert its hegemony over China by shifting Germany's control of Shandong province to Japan, despite Chinese protest. As the Chinese civil war between Mao Zedong and Chiang Kai-chek accelerated, the later 1931 "Manchurian incident" then set the stage for the 1937 Japanese invasion of Manchuria and China. This was followed by Japanese demands to impose a Greater East Asian Co-Prosperity Sphere in 1941. (See Kobtzeff, Chapter 15.)

After the defeat of Japan in World War II, the US military occupation led to the formation of the US-Japanese military alliance in 1952 (replaced by the 1960 Treaty of Mutual Cooperation and Security). On the one hand, the US-Japanese alliance has served to restrain the possible resurgence of Japanese *revanchism*. On the other hand, the treaty has likewise defended Japan against potential threats in Asia during the Cold War that manifest themselves most strongly during the Korean war from 1950–53, after the formation of 1950 Sino-Soviet alliance.[23] In the meantime, the US has tacitly backed Japanese interests in Taiwan and sought to check Chinese attempts to reclaim the island.

In the post-Cold War period, following the repression (followed by co-optation) of the Chinese democracy movement in 1989, Beijing has become a rapidly ascending center of power that has increasingly moved from continental and semi-peripheral status toward amphibious naval and core status. On the one hand, Soviet collapse has meant diminished restraints on China's growth; on the other hand, the break-up of the Soviet Union into separate states has raised Beijing's fears that China could likewise break apart following the rise of Tibetan, Uighur and Mongolian secessionist movements within China itself. Yet despite threats of secessionism and "democratic" opposition to Communist Party rule, and despite its relatively weaker GNP, Beijing increasingly appears capable of challenging US insular-core hegemony given its high savings and significant financial and economic capabilities. Beijing continues to pursue the development of a blue water navy, with aircraft carrier and missile capabilities that could eventually challenge American and Japanese military/ naval predominance. The question remains whether China will maintain a low profile and avoid playing the role of a "challenger" that might provoke global war—despite its claims to a "peaceful rise" and support for a "harmonious world." (See Thompson, Chapter 27; Modelski, Chapter 28.)

Continuing Games of "Encirclement" and "Counter-Encirclement"

Russian and American elites have continued to play games of "encirclement" and "counter-encirclement" despite American promises to "reset" US-Russian relations after the 2008 Georgia-Russian war. On the one hand, in addition to an attempt

23 Emma Chanlett-Avery, "The U.S.-Japan Alliance" Congressional Research Service (January 18, 2011): http://www.fas.org/sgp/crs/row/RL33740.pdf.

to check Georgian military capabilities (and Tbilisi's alleged support for Chechen secessionism), Russian efforts to check NATO enlargement to Georgia and Ukraine helped provoke the August 2008 Georgia–Russia war (see Rimanelli, Chapter 18; Karatzogianni, Chapter 23). The "threat" of NATO enlargement likewise led Russia to press for the extension of the lease of the Russian Black Sea fleet based in Sevastopol from 2017 to 2042 at least. The 2008 Georgia-Russia war, coupled with provocative Russian military maneuvers, sent shivers down the spines of all states along the NATO-Russia/CSTO border. This resulted in a tightening of NATO defense commitments to the Baltic states and Poland, matched by a build-up of Russian/CSTO defenses in the Baltic region, as well as the extension of Russian influence over South Ossetia and Abkhazia (claimed by Georgia) while concurrently tightening Russian naval and military positions in the Black Sea and Caucasus.

On the other hand, despite opposition to NATO enlargement and Russia's previous opposition to the US and NATO military presence in central Asia, Moscow has begun to fear that a US/NATO pull-out from Afghanistan could give rise to pan-Islamic groups that could undermine Russian influence throughout central Asia, if not within Islamic areas of Russia itself. Moreover, while Moscow continues to flirt with the possibility of an alliance with China, due to China's needs for Russian energy and mineral resources, Russia also needs to diversify and develop its economy (and reduce its dependence upon exports in arms, energy and raw materials) by working more closely with the US and Europeans, through the World Trade Organization (WTO) and the G-20, for example. The fact that China obtained WTO membership in 2001, but not Russia, continues to vex Moscow.

While the US and Europe should continue to take steps to "reset" US, European and Russian relations, the US and Europe could simultaneously seek out better relations with China, in part as a means to pressure Russia into accepting better terms. Yet this approach, by pressing for a US-Chinese "G-2" framework so as to better coordinate US-Chinese relations, could potentially alienate Russia, if not Taiwan and Japan, among other states in Asia—if Moscow eventually begins to see itself as becoming a junior partner to China. Concurrently, Taiwan and Japan would believe that the US is "appeasing" Beijing. While the US and Europe should engage in joint development projects with China, Moscow is concerned that Beijing will take priority over relations with Russia in terms of US and European Union interests, possibly resulting in an "encircling" US-European-Chinese alliance that could support Chinese claims and influence in Asia, including Russian Siberia.

In addition to significant US and European investment in China, which tends to draw Washington closer to Beijing's interests, the flexibility of US policy is further handicapped by the fact that China has become the major holder of US treasury securities (holding more than 1 trillion dollars in February 2011). Given the more than $14 trillion of American public debt, coupled with China's expanding military program, this fact gives Beijing significant bargaining leverage, making it more difficult for Washington to confront China on exchange rate and trade disputes, human rights issues, pollution and climate change, if not policy toward Japan,

North Korea, Taiwan, among other concerns.[24] As of this writing, the US-Chinese relationship appears tense: Beijing appears to be the beneficiary, taking advantage of US, European and Russian rivalry.

The rise of China as a military and economic power, plus concerns with North Korea's missile and nuclear weapons program, have been pressing the US to strengthen its defense strategy with Japan and other states in the Asia Pacific region. On the one hand, the US has hoped that China will play a stronger role in pressuring North Korea to give up its nuclear weapons program. (Here, however, China's priority has been to prevent a potentially destabilizing collapse of the North Korean regime.) On the other hand, Washington has feared the so-called "Finlandization" of states surrounding a militarily powerful China—in that Beijing may soon possess a significant regional, if not, ultimately, intercontinental missile and nuclear weapons capability that can strike Taiwan, Japan, as well as the United States (if not Russia).

The rise of China—in addition to the threat posed to both American and Russian interests by a number of differing pan-Islamist movements in central and south Asia—is consequently of concern for both the United States and Russia, which fears Chinese demographic and economic pressures and influence on the Russian Far East. This latter assessment makes the assumption that Chinese growth is not a "paper tiger" fuelled by significant internal debt, overvalued companies, and that China will not become bogged down in significant domestic class conflict and secessionist movements in the near future if it begins to move away from essentially export-led growth. These factors, involving the rise of China and the spread of pan-Islamic movements throughout much of the former Ottoman empire or "greater Middle East," if not Africa, should increasingly require closer US-European-Russian foreign policy and defense coordination, and could potentially help press the US, Russia and Europe into an entente, if not alliance, relationship. A stronger US-NATO-Russian-European relationship could likewise strengthen NATO-UN relations, and help bring the UN Security Council into closer coordination, and thereby open the door to far reaching UN reforms, assuming China, Britain or France are not, in turn, alienated. If, however, the US, NATO and Russia can not establish a new entente or alliance relationship, then the possibilities of wider regional wars, if not major power war, will become more likely. The fact of the matter is that so-called "post-historical" Europe has not yet transcended its historical relationships with its own colonies, nor with Russia, and could be drawn into more wars on its peripheries (after ex-Yugoslavia and Libya). How the US, Europe and Russia re-construct or "reset" their geostrategic relationship will be one of the keys to question as to whether or not an uneasy post-September 11, 2001 "global peace" can be sustained well into the future.

---

24 On the G-2 and the G-20, see, for example, Geoffrey Garrett, "G-2 in G-20: China, the United States and the World after the Global Financial Crisis," *Global Policy* 1: 1 (January 2010). "… The stubborn persistence of … massive Sino-American imbalances will be a constant source of tension between the world's two most important countries."

Consequences of the "Global War on Terrorism"

The essential thesis presented in this final chapter is that alliance threats of "encirclement" and "counter-encirclement" (resulting in an "insecurity–security dialectic") have continued to play themselves out in the US-European-Russian-Chinese-Japanese-Indian power constellation. Contrary to the hopes for a "peace dividend" raised by the end of the Cold War, the threat of major power war has not entirely receded in the aftermath of NATO enlargement and the subsequent (but unrelated) September 11, 2001 Al-Qaeda attacks—which worked to spark the "global war on terrorism."[25] In addition to opposing the state of Israel, pan-Islamic movements have sought to overthrow "corrupt" Islamic regimes in areas historically influenced by Islamic culture and thus to reunify the Islamic *ummah*—in effect, seeking "revenge" for the collapse of the Ottoman empire after World War I, as Bin Laden has argued. The American-led "Global War on Terrorism" has consequently been played out in much of the former Ottoman Empire or "Greater Middle East" and has led to at least three major military interventions (prior to the French and UK-led military intervention in Libya in 2011).

Yet the consequences of the US-led interventions appear to be in extreme disproportion to the original crime. In ostensible reaction to the roughly 3,000 killed in the September 11 attacks, resulting in tens of billions of dollars lost in property damages, insurance costs, hazardous clean-up, as well as job loss, these post-September 11, 2001 wars have nevertheless represented wars of strategic choice—as opposed to wars of strategic necessity—in Afghanistan, Iraq and Pakistan which have largely overstretched American military and financial capabilities.[26] Differing cost estimates range from 3.2 to 4 trillion dollars. As many as 225,000 people have died altogether in these three conflicts—including over 6,000 US soldiers with as many as 550,000 veterans making disability claims up to the Fall 2010, and with at least 137,000 civilians dying thus far in Afghanistan, Iraq, and Pakistan. The number of war refugees and displaced persons is estimated to be 7,800,000. These wars have furthermore been accompanied by significant erosions in civil liberties in the US and Europe and human rights violations abroad.[27]

While post-September 11, 2001 US-led interventions have largely focused on Afghanistan, Pakistan, and Iraq (with military pressure placed on Iran), geopolitical rivalry—primarily between Israel, Saudi Arabia (plus other energy-rich Gulf states) and Iran—has continued to impact the 'greater Middle East.' This conflict

25 See Hall Gardner, *Surviving the Millennium* (Westport, CT: Praeger, 1994); Hall Gardner, *Dangerous Crossroads* (Westport, CT: Praeger, 1997); Hall Gardner, *American Global Strategy and the "War on Terrorism"* (Burlington, CT: Ashgate, 2007).

26 Chester A. Crocker, "The Place of Grand Strategy, Statecraft and Power in Conflict Management," in *Leashing the Dogs of War*, ed. Chester A. Crocker, Fen Osler Hampson, and Pamela Aall (Washington, DC: US Institute of Peace Press, 2007)

27 For overall human and financial costs of post-September 11, 2001 conflicts, see Costs of War: http://costs ofwar.org (website checked September 11, 2011). For military costs, see Amy Belasco, "The Costs of Iraq, Afghanistan and Other Global War on Terror Operations Since 9/11," Congressional Research Service (March 29, 2011).

has largely been the result of the 1979 Iranian revolution, the Soviet invasion of Afghanistan in December 1979 (with the Afghan resistance supported by Saudi Arabia, Pakistan and the US), the 1980–88 Iran-Iraq war, the 1990 Iraqi invasion of Kuwait, leading up to the US-led military intervention in Afghanistan in 2001 and then in Iraq in 2003 (not to overlook Israeli interventions in Lebanon in 2006 and Gaza in 2008–09). Here conflict throughout the greater Middle East has tended to intersect with conflict in South and Southwest Asia, including India, Afghanistan and Pakistan (not to overlook US intervention by drone attacks in tribal areas of Pakistan). Concurrently, Iran and Saudi Arabia (with Turkey largely in the background) have been seeking to influence or to manipulate, if possible, popular demands for socio-political change and greater "democratization" throughout much of the Arab-Islamic world.

US-led intervention in Afghanistan may have overthrown the Taliban, but it has not generated a stable democratic and popularly supported government in Kabul. And despite US and NATO efforts to suppress the Taliban in the south, the latter appears to be resurgent, infiltrating the north; this has raised questions as to whether or not Afghan national forces (whether Kabul can reform itself or not) can be built up fast enough and effective enough to prevent a Taliban victory, once US and NATO forces are expected to withdraw in the period 2014–16. Options include: a negotiated settlement drawing the Taliban into a federal or confederal system of governance; a 'partition' between Pashtun regions to the east and south of Kabul and the essentially anti-Pashtun north; or the seizure of Kabul by the Taliban, aligned with other jihadists.[28] The future may well depend upon whether a proposed Southwest Asian 'Contact Group' and a peace *jirgah* among all Afghan factions can ultimately help bring a political settlement to Afghanistan.

The US-led intervention into Afghanistan in 2001, and then across the border into Pakistan by 2008, has tended to further destabilize Islamabad. In part in an effort to stave off both Baluch and Pashtun secessionist movements within Pakistan itself, which the latter sees as supported by India,[29] Islamabad's support for the Afghan Taliban and for other pan-Islamic movements throughout Central Asia has represented a means to establish "strategic depth" versus India. As an integral aspect of this quest for "strategic depth," Islamabad has additionally hoped to establish a greater trade association with predominantly Islamic Central Asian states as a means to counter-balance Indian pressures, simultaneously angering Moscow. Here, however, despite its claims to support pan-Islamic movements, Islamabad has increasingly come under attack by the predominantly Pashtun anti-

28 On US/NATO policy toward Afghanistan after a potential US withdrawal, see Kenneth Katzman, "Afghanistan: Post-Taliban Governance, Security, and U.S. Policy" *Congressional Research Service Report for Congress* (April 15, 2011).

29 "Baluchistan's strategic significance and natural endowment makes it a critical province for Pakistan." See Tarique Niazi, "The Geostrategic Implications of the Baluch Insurgency," *Jamestown Foundation Terrorism Monitor* 4: 22 (November 16, 2006), available at: http://www.jamestown.org/programs/gta/single/?tx_ttnews%5Btt_news%5D=973&tx_ttnews%5BbackPid%5D=181&no_cache=1.

Pakistani Taliban for playing a double game, given both US and Pakistani efforts to suppress a number of Islamicist organizations.

After US Navy SEALs killed Osama Bin Laden in Pakistan by evading Pakistani defenses, Washington may no longer be able to play the role of "honest broker" between India, Afghanistan and Pakistan, despite the fact that the latter had been declared a "major non-NATO ally" in 2004. Much will depend upon whether the US tilts closer to Indian and Russian interests and how Pakistan reacts (the possibility of a radical pro-Islamist coup should not be ruled out). Peace also depends upon whether negotiations between Kabul and Islamabad can find a compromise over the Durand line (imposed by Great Britain in 1893), and find ways to engage in border trade cooperation through the potential formation of an India-Afghan-Pakistan free trade and transit zone. Likewise, a settlement over Kashmir, perhaps involving a limited autonomy, might help to cool Indian-Pakistani tensions. At least since Bin Laden's assassination, Islamabad has appeared to be distancing itself from the US and Europe and has threatened to look closer to China for geostrategic and political economic supports, while India has not yet appeared ready to accept an overall geopolitical settlement with Pakistan.

American intervention in Iraq in 2003 may have overthrown the dictator, Saddam Hussein, consequently leading to the collapse of the next-to-last bastion of secular pan-Arabism (except for Syria), but the establishment of a true "democratic federalism" in Iraq, as promised by the Bush administration, appears a long way off. Iraq has subsequently become a battleground between a number of differing Arab-backed Sunni, Iranian-backed Shi'a and Kurdish socio-political groups and factions. In addition to sectarian tensions and recurrent acts of terrorism between minority Sunnis and majority Shi'a, the relative autonomy of Iraqi Kurds has led Turkey, Iran, Armenia and Syria to oppose the rise of pan-Kurdish movements within their respective countries—with Turkey intermittently intervening militarily against the Kurdish Workers' Party (PKK) across state lines into northern Iraq. Concurrently, both Turkey and the central Iraqi government have opposed Kurdish demands for control over energy resources in Kirkuk and Mosul—a potential *casus belli.* With the United States opting to withdraw its military (but not its civilian security) presence in Iraq by 2011, it is not clear that the withdrawal of US military forces will lead to more or less conflict. It appears dubious, however, that that violence will end altogether until each there is a clear political settlement among the rival Sunni, Shi'a and Kurdish communities and their differing political factions.

While the Iraqi dictatorship fell by the massive use of American force, the rapid and largely unexpected collapse of dictatorships in Tunisia and Egypt in 2011 (in which the militaries of both countries backed the demands of popular protest) appears to indicate that a general loss of *legitimacy* by means of a new media *pronunciamento* (using Facebook and Twitter) can work to transform regimes. (See Emerson, Chapter 6 and Davis, Chapter 7 for proposals prior to these events.) At the same time, however, the direction of the Arab transformation is not entirely certain due to the continuing role of the military and the rise of Islamic parties in these states as well as the conflicting international geostrategic and political-economic interests involved (particularly those of Sunni Saudi Arabia versus Shi'a Iran).

There is a continual danger is that the ongoing conflict in the greater Middle East risks drawing in major and regional powers in the not-so-long term. Colonel Qaddafi's violent efforts to crack down on dissident tribal opponents in Libya in 2011 helped to provide the rationalization for international military intervention ("responsibility to protect") led by France and Great Britain. These actions were supported by the Gulf Cooperation Council, the Arab League and the UN Security Council, and subsequently backed by the US and then NATO. While the initial goal shifted from the "responsibility to protect" to "regime change," Qaddafi's control over Tripoli was finally overthrown by August 2011[30] by significant force involving NATO air supports and popular resistance (to the regret of Russia, South Africa and the African Union which had hoped to mediate a diplomatic settlement).

Assuming it does obtain effective control over the whole country, the new Libyan National Transition Council will need to mediate between differing tribal and ethnic groups in each of the three major regions, including those clans loyal to Qaddafi, through a Truth and Reconciliation process; it will also need to establish trade and revenue sharing accords between Libyan domestic tribal interests and international economic and energy companies. The government will likewise need to control arms trafficking including Qaddafi's leftover shoulder-fired missiles (with excess weaponry purportedly being exported to Egypt, Mali and Chad, if not elsewhere) and to check Al-Qaeda or Salafist influence, as well as the Tuareg insurgency, in the southern Sahara and Sahel regions. As Libya borders Tunisia and Egypt (as well as Algeria, Niger, Chad, and Sudan) continual political and economic instability could block the potential formation of a regional trade accords and formation of a wider regional security and development community.

The question remains whether or not the "democratic" movements in the greater Middle East will be hijacked by military *coup d'état* or by extremist movements of differing forms, or else manipulated by outside powers. The Moslem Brotherhood (though divided between older and younger members) could gain strength in Tunisia, Egypt, Libya and Syria, as well as Algeria, in rivalry with other Islamicist parties as well as with secular socio-political movements. Egypt, for example, could become divided between parties backed by the Muslim Brotherhood, the Army, and remnants of the old regime as they attempt to cling onto power, cutting out less-organized democratic parties. These movements in the "greater Middle East" (particularly with Egypt as the *pivotal* Arab country) are in desperate need

30 On October 20, 2011, Qaddafi was killed. Contrary to expectations that the war would be over by September 2011, Qaddafi supporters engaged in strong resistance in Sirte, Qaddafi's stronghold. Qaddafi's fall raises questions as to what might happen to other socially and politically divided African countries, also ruled by dictatorship (including neighboring Algeria which initially supported Qaddafi), as France, Britain, the US, the Arab Gulf states, as well as China and India, all vie for access to Africa's mineral and energy resources. Whether the new Libyan regime will implement a form of "democracy" blending US, European and Libyan traditions remains to be seen.

of appropriate and well placed regional/international trade and international development assistance if they are to ultimately establish *consensual* democracies (see Emerson, Chapter 6) and not fall back into new forms of dictatorship. The dilemma, however, is that political-economic turmoil generally scares away investors (and tourists), and can open the doors to mafia-type black and gray market activities, permitting the rise of dictatorship—as well as "terrorist" partisans.

In addition to US and Saudi military intervention against Shi'a secessionist movements and Al-Qaeda partisans in the potentially energy rich littoral state of Yemen (which guards the Bab el-Mandab shipping choke point), Saudi Arabia has engaged troops in Bahrain (invited by the latter's Sunni minority leadership) in order to repress what it has regarded as a regional pan-Shi'a movement backed by Iran. While Teheran has denounced political repression in Palestine, Yemen and Bahrain, it has in turn repressed its own domestic Green Movement and the demands of many of its own population. Riyadh has argued that a burgeoning pan-Shi'a movements, ostensibly supported by Iran, could destabilize the predominantly Shi'a regions of the oil-rich province of eastern Saudi Arabia (Shi'a make up roughly 15 percent of the Saudi population), as well as Shi'a regions in Iraq, Lebanon, oil rich Bahrain and Yemen (with conflict once again brewing on the Horn of Africa in Kenya, Somalia and Ethiopia), while Teheran itself has feared Arab claims to Iranian Khuzestan.

Multi-confessional Syria (led by Bashar al-Assad of the minoritarian Alawite community) has appeared to play a role as a dictatorial pivot state that has generally sought to balance Iran and the Arab world, at least since the 1979 Iranian revolution and Israeli interventions into Lebanon. But with the rise of Arab "democracy" movements, Syria's position (caught between Turkey, Iran and Saudi Arabia) appears to be collapsing due to political-economic instability leading to protest and violent repression. The crisis in Syria could isolate Iran (and weaken its links to Hizb'allah in Lebanon and to Hamas in Gaza). Yet a full scale civil war in Syria could also impact upon Turkey, Lebanon, Jordan, and Iraq—and could possibly provoke a wider regional conflict with Israel. A wider regional conflict appears plausible given the degeneration of relations between Israel, Turkey and Egypt—assuming Tel Aviv cannot reach out for an entente with Saudi Arabia and other Arab states.

With the Palestinians essentially divided between Fatah in the West Bank and Hamas in Gaza, tensions between Israel and post-Mubarak Egypt have come to surface following the new Egyptian government's reluctance/ inability to control the Sinai, raising questions as to ability of the two states to sustain the 1978 Camp David accords. At the same time, however, the new Egyptian government may have greater influence over Hamas. Concurrently, Israeli-Turkish relations, which were much more positive than negative at least up until the 2003 Iraq war, have begun to degenerate significantly particularly after Israel's military intervention in Gaza in December 2008–January 2009, if not earlier. The Obama administration's apparent backtracking over a "two state" solution to the Israeli-Palestinian question in 2011 has threatened to undermine American relations with its primary ally, Saudi Arabia and much of the Arab/Islamic world.[31] A rupture in US-Saudi relations would potentially intensify Saudi-Iranian-Israeli rivalries in the region, perhaps leading to a regional nuclear arms race—if Iran should finally opt to deploy nuclear weaponry. By contrast, an Israeli-Saudi entente based on the 2002 Arab Peace initiative (re-confirmed in 2007) could help to isolate Iran and work to stabilize the Middle East by weakening Syria, Hizb'allah and Hamas. Potentially an Arab-Israeli entente could provide greater security for Israel in close cooperation with a new Palestine, possibly in a form of loosely interdependent "confederation." (See continuing discussion in this Chapter.)

Here, the question remains: As Arab and Iranian "democracy" movements attempt to overthrow differing forms of dictatorship and, in effect, take destiny into their own hands, at the same time that Saudi Arabia (plus the Gulf countries) and Iran, not to overlook Turkey, battle it out for influence behind the scenes, who will win out? And how might domestic pressures for democratization change Arab and Iranian policies toward Israel? And how will Israel respond? Can Israel and Saudi Arabia (and the other Arab states) engage in strategic cooperation? Will these domestic socio-political conflicts ultimately simmer down, establishing new forms of governments, democratic or not? Or will these domestic conflicts result in foreign military intervention, whether multilateral or unilateral? And if unilateral, could such interventions draw major and regional powers into direct confrontation?

Polemological Questions

The extent of the contemporary global crisis (and of specific regional crises) introduces a number of fundamental *polemological* questions as to the very possibility of a peaceful resolution of conflicts, or at least a transformation of those conflicts toward a more "positive" and *disalienating* direction.

From a gender perspective, for example, the fact that a handful of key women elites in the United States pressed for military intervention ostensibly to stop Qaddafi's repression of civilians against the advice of a generally more circumspect American male leadership raises questions with respect to the relationship between gender and war—although a number of male neo-conservatives and democratic liberals urged direct military intervention as well.[32] Women's membership in various

31 Turki Al Faisal, "Veto a state, Lose an Ally" *New York Times* (September 11, 2011).

32 Secretary of Defense, Robert Gates, National Security Adviser Thomas E. Donilon and the counter-terrorism chief John O. Brennan had argued against American military action in Libya, yet Secretary of State Hillary Clinton, who obtained some Arab League support, overrode their objections. Samantha Power of the National Security Council and UN ambassador Susan Rice, who pressed the UN Security Council to obtain a 10–5 vote, likewise argued for the deployment of military force. "Obama Takes Hard Line With Libya After Shift by Clinton," *New York Times* (March 18, 2011), available at: http://www.nytimes.com/2011/03/19/world/africa/19policy.html?_r=3&ref=us. Republican John McCain, among other neo-conservatives—as well as liberals such as John Kerry—

violent anti-state organizations of differing political-ideological perspectives and cultural backgrounds[33]—and the use of women suicide bombers in a number of devastating attacks and assassinations—likewise raises a number of profound questions with respect to the ostensible feminine predilection for non-violence. In addition to support for ideological causes and expressions of group solidarity, suicide bombings and assassinations, such as that engaged in by the Chechen "Black Widows," often represent revenge for the murder of family members, for rape, or for other forms of humiliation or crimes perpetrated by state militaries or security forces.

Much as Mann (Chapter 21) and DeLaet (Chapter 3) have pointed out with respect to Afghanistan, the essentially humanitarian demand to "protect" women (and civilians in general) has become a cause and justification for military intervention—without necessarily thinking through the immediate and long term consequences of such wars and consideration for issues of *jus post-bellum*. (See also Kobtzeff, Chapter 4; Robinson, Chapter 5; Charnay, Chapter 25.) While such wars may be fought in the name of "human rights" and "democracy," the actual result may be the collapse of the state and the fostering of wider areas of socio-political instability, coupled with black and gray market activities, if appropriate aid and assistance are not eventually forthcoming, in part due to the global financial crisis, bureaucratic ineffectiveness and bungling, among other factors.

Concurrently, women's leadership (and the "feminist movement") appears to be splintering as a still limited number of women claim the right to *decision-making power* and to take positions of authority in both public and private sectors, but nevertheless divide ideologically.[34] The recent feminist impact might, to a certain extent, help to reduce (but not resolve) issues surrounding male/female alienation by providing a greater number of women as "role models" for future generations, but the nature of *strategic choices*—whether actually decided by men or women—nevertheless appears to be exacerbating other socio-political divisions, rather than *transcending* what is essentially a male-constructed and male-dominated global system altogether.

On the military-technological front, the US appears to have moved from the "gun boat diplomacy" of the late nineteenth century to "cruise (and drone) missile diplomacy" in the late twentieth and early twenty-first centuries. In addition, new computer technologies, stealth weaponry and robotics are increasingly becoming weapons of war that could help open the door to potential conflict *in that they*

supported military intervention in Libya.

33 Woman "terrorists" have engaged in Lebanese, Palestinian (including Islamic Jihad and Hamas), Chechen, Tamil, Kurdish, Al-Qaeda movements. This is not to overlook the German Ulrike Meinhof of the Bader-Meinhof gang and the Japanese "Red Queen of Terror," Fusako Shigenobu, of the Japanese Red Army, among others, during the Cold War.

34 Ideological splits within the women's "movement" included Hilary Clinton vs. Sarah Palin; US Homeland Security Secretary and former Governor of Arizona Janet Napolitano vs. Arizona Governor Jan Brewer, to name a few; in France, Socialists Ségolène Royal or Martine Aubrey vs. National Front leader Marine Le Pen.

provide optimism that these conflicts will result in a minimal loss of life on one's own side. As pointed out by Karatzogianni (Chapter 23), the Russian-Estonian cyber war in 2007 caused significant financial damage through a denial of service attack that disrupted the country's entire Internet service. In addition, Stuxnet computer worm attacks (ostensibly set off by either the US or Israel) on Iranian nuclear facilities in 2010, which sought to disrupt the uranium centrifuge process, represent a new innovation.[35] While it is not altogether clear the actual extent of damage caused to Iranian nuclear facilities (and whether there were risks of a nuclear meltdown), the Stuxnet worm can apparently act on its own, attack its target, and hide itself in the process. While the Stuxnet virus is ostensibly designed to strike only certain targets and to self-destruct after a certain time period, other computer viruses of the future might not be programmed with such self-restraint and could attack multiple targets, causing human "collateral damage."[36] The Stuxnet computer virus could represent the beginning of a *qualitative* leap in innovation toward more dangerous forms of cyber sabotage, possibly leading to more overt warfare.

On the one hand, major and emerging powers continue to develop advanced weaponry for potential warfare against rival states, while also preparing advanced defenses against the new threats posed by domestic and international anti-state organizations. (See Géré, Chapter 24.) Concurrently, on the other hand, armed partisans in a number of collapsed and collapsing states have been engaging in wars of attrition involving a mix of weaponry left over from the Cold War, combined with more "primitive methods" used in new circumstances, such as AK-47s (dumped at bargain prices in the aftermath of the Cold War), machetes, child soldiers and mass rape, which has been interpreted as a "perk of war." (On rape, see Gat, Chapter 2; DeLaet, Chapter 3; Mann, Chapter 21.) These conflicts (in which drug gangs and mafias tend to thrive on black and gray market activities, and in which corporations require private security services, and in which nomadic "terrorist" organizations can migrate from state to state) have indicated a fanatical willingness to engage in extreme measures to make certain one's own side does not "lose" in its struggles against rival factions and groups—even if that side cannot win altogether. This has raised the question as to when (and if) some of these wars may end, and whether it is at all possible to achieve either peace or justice—or any form of *modus vivendi* whatsoever. (See Perry, Chapter 20; Mann, Chapter 21; Robinson, Chapter 5.)

Such conflicts have become all the more problematic as majoritarian "winner take all" systems of governance and binary "for or against" plebiscites (often seen as "imposed" by the US and Europe) may have exacerbated social and political tensions—in that such approaches may turn socio-political collectives (of different sized populations with differing identities) against one another and have thus appeared to rule out different forms of *power sharing*. Whether these differing

35 "Israeli Test on Worm Called Crucial in Iran Nuclear Delay," *New York Times* (January 15, 2011). For a scenario of an Israeli strike on Iran, see Karim Sadjadpou, "Ayatollah for a Day," *Foreign Policy* (November 10, 2011).

36 For information on the Stuxnet virus, see http://www.stuxnet.net/.

identities are actually based on residual genetic or biological grounds (see Gat, Chapter 2) or are, in effect, socially-constructed and reconstructed over time, remains a moot point as long as *alienated* groups and individuals *perceive* distinct differences between themselves and others and are unable to come to terms in some form of *modus vivendi* (see Gardner, Chapter 1).

In a number of countries, including Afghanistan, Belgium, Bosnia, Honduras, Iraq, Kenya, Lebanon, Northern Ireland and Zimbabwe, it appears that power sharing has been turned to only as a last, rather than a first, resort—and only after majoritarian forms of democracy have failed. (See Emerson, Chapter 6.) This fact appears to challenge democratic peace theory: In some situations, the failure of majoritarian democracy and use of binary plebiscites have helped to foster violent secessionist socio-political movements, which are then transformed into violent interstate conflicts, as was the case for Bosnia during the break-up of ex-Yugoslavia in 1990–95 or for the wars of genocide in Rwanda and the Democratic Republic of the Congo from 1996 to 2003.

It is also important not to overlook the burgeoning drug war right on the US-Mexican border, which has begun to infiltrate and subvert much of Central and South America, if not drug trafficking regions in western Africa. The expansion of the drug war has largely been due to the fact that that US "war on drugs" in the region has generally diverted the drug trade away from its original source, Colombia. The drug war is becoming a significant geostrategic issue due to the ways that drug mafias intertwine with governments and societies by helping to foster black and gray market economies, and corrupt leaderships, judicial systems and the police (see Géré, Chapter 24). In Mexico alone (not counting Colombia), this burgeoning conflict has resulted in more than 30,000 deaths from mafia-style executions, 3,000 deaths in gang clashes plus more than 500 people in collateral damage following gang battles with security forces.

In addition to corrupting officials on the American side, this dangerous situation could, in the near future, result in direct US military intervention (in addition to CIA activities, military or police advising), but in very different circumstances than that of US intervention in the Mexican revolution in 1914 and 1917. Such a scenario could become plausible—in that it appears that Mexico cannot handle the drug war alone, let alone the question of illegal immigration, plus the smuggling of arms and hundreds of millions of drug dollars in cash (primarily from the American side into Mexico). Here, in addition to finding a resolution to the drug question, the US and Mexico need to work jointly to find a reasonable and humane compromise to the question of "illegal" immigration.

The deeper dilemma is that it is not at all at certain that states can "win" wars waged against social phenomena (such as drugs and "terrorism," or even illegal immigration) without concurrently abandoning proclaimed values of democracy and justice. (See Robinson, Chapter 5.) At the same time, as long as the US cannot find ways to significantly reduce domestic drug demand through effective educational programs and/or legalize certain softer drugs (while concurrently cracking down decisively on harder, relatively more dangerous, drugs), it is not clear that police or military action will "resolve" these complex and increasingly interwoven issues—

in which drugs can finance *both* state-supported *and* anti-state "terrorist" activities, in Colombia, Mexico and Afghanistan, among other states. In effect, the US (and Europe) need to engage in a new policy towards drugs (as well as firearms and other black market activities), much as Washington opted to change its policy toward alcohol in ending the Prohibition in 1933. Such a new policy could likewise address the burgeoning debt crisis by bringing in significant revenues through taxation on a newly legalized (and effectively regulated) industry.[37]

In the long term, it is not certain how climate change (coupled with poor management of the natural environment) might impact upon the political economy and societies of a number of countries (see Cramer, Chapter 22), but it is already clear that a number of natural and manmade disasters in various countries (massive flooding, hurricanes, tornados, tsunamis, earthquakes, as well as oil spills and nuclear power breakdowns) are causing more death and destruction than most "terrorist" actions and some wars. One dilemma is that some states and societies might benefit in the not-so-long term from the misfortune of others: global warming could flood some areas, while making others more fertile, for example. Whether states can find common ways to address global warming and work concertedly to address ways to prevent the further poisoning of the natural environment through the development of "green" technologies, and provide direct and appropriate development assistance and capital investment to areas that really need it, in addition to purchasing local products, remains to be seen. What is certain is that leaderships cannot deal with environmental questions from a global basis without close *local* attention to the natural environment. One issue is the presently wasteful nature of demand created by mass consumer society, and the question as to whether or not new technologies that can minimize resource depletion and pollution, and provide alternative energy sources and recycle much needed raw materials, can be developed soon enough. Alternative technological options need to be spread fast (and inexpensively) enough throughout the global economic and energy infrastructure so as to minimize social and political tensions over both actual—and *artificially induced*—scarcity.

One danger is that a new scramble for the Arctic (with conflicting claims among the US, Russia, Norway, Denmark and Canada) and for Antarctic resources appears to be in the making, while a number of energy- and resource-rich regions in the world represent focal points of potential wars. These include: the Spratly Islands (claimed by China, Brunei, Malaysia, Philippines, Taiwan and Vietnam); the Kuril Islands/ northern territories (claimed by Japan and Russia); Falklands/Malvinas (claimed by the UK and Argentina); Diaoyu/ Senkaku Islands (claimed by Japan and China); Dokdo Islands (claimed by South Korea and Japan); Mosul and Kirkuk between

37 See *War on Drugs*, Report of the Global Commission on Drug Policy (June 2011), available at: http://www.globalcommissionondrugs.org/Report; Jeffrey A. Miron, *The Budgetary Implications of Drug Prohibition* (Cambridge, MA: Department of Economics, Harvard University, February 2010). See also Hall Gardner, *Averting Global War: Regional Challenges, Overextension and Options for American Strategy* (New York: Palgrave, 2010), Chapter 9.

Iraqi Kurds, the Iraqi central government, if not Turkey; Saudi Arabia and Iran over Yemen, Bahrain, eastern Saudi Arabia; Abu Musa claimed by Iran and the United Arab Emirates; Greek Cyprus, Turkey, Lebanon and Israel over offshore oil; and possibly Siberia or Mongolia between Russia and China, among other possibilities.

If diplomacy through ASEAN or the UN, leading to confidence and security building measures and the sharing of resources, cannot bring common sense to the issue, political-military tensions in 2011 have threatened to the draw both Vietnam and the Philippines, which seek to internationalize claims in the South China Sea, into conflict with China, which wants to deal with these issues bilaterally. Manila has sought to invoke the 1951 US-Philippine Mutual Defense Accord so that the US will back its extended maritime claims (based upon the March 2009 Philippine Archipelagic Baseline Law) against those claims of China.[38] Other conflicts could be exacerbated by disputes over the water, among other resources, between India and Pakistan over Kashmir, between Israel and Palestinians, between Turkey, Iraq and Syria, among many others. Conflict in Sudan, displacing two million people, is in part a result of environmental degradation. There have been continuing signs of potential conflict over the oil-rich region of Abyei on the border between the North and South Sudan despite the fact that the January 2011 South Sudan referendum initially took place peacefully. (See Perry, Chapter 20 and Emerson, Chapter 6.)It furthermore does not appear that the US (or European states) will necessarily be able to "put their money where their mouth is" in seeking to support democracy movements in the greater Middle East (or elsewhere) and support appropriate development projects where needed (and generally open their domestic markets to products produced abroad)—in attempting to resolve a number of seemingly intractable conflicts. This could mean that states with large financial reserves such as China, Turkey, Saudi Arabia and the Gulf states may obtain greater influence in regions in need of investment. In January 2011, US public debt hit a record $14 trillion—raising political calls for cuts in both defense spending and domestic entitlements, and forcing a reassessment of domestic and foreign policy based largely on cost and financial considerations. European Union states have likewise been confronted with major financial crises, involving the so-called PIIGS (Portugal, Ireland, Italy, Greece and Spain), with France and Germany at odds over how to deal with the crisis. Will these financial crises result in even more significant social-political protest and instability both within Europe (across borders) and along the European periphery, resulting in civil strife, if not acts of terrorism, further undermining European unity? In November 2010, the UK and France forged a new defense sharing accord, in part to reduce defense costs. These steps were also intended to strengthen European military cooperation, given prospects that the US will most likely begin to step back from its expenditure on NATO and European defense, in diverting its attention from a defense of Europe toward the greater

38 http://www.atimes.com/atimes/Southeast_Asia/KC27Ae02.html; http://www.manilatimes.net/index.php/news/top-stories/7928-little-resistance-for-new-us-invasion; http://www.thejakartapost.com/news/2011/07/23/military-bases-pose-a-threat-peace-south-china-sea.html.

Middle East, south and central Asia, and the Far East.[39] (See Rimanelli, Chapter 18; Géré, Chapter 24.) Here it appears that so-called "post-historical" Europe could soon find itself dragged back into conflict given socio-political instability along European Union borders, fears of immigration which tend to test the tolerance of 'liberal' societies and provoke nationalist and xenophobic movements, despite some forecasts for the need of greater numbers of workers in Germany and Europe in the long term. These security concerns have been accompanied by continued conflicts within former European colonies (Libya, Syria, Somalia, Ivory Coast, Sudan, among others), coupled with historical European suspicions of Turkey (refusing the latter's membership in the European Union) and of Russia (due to the latter's claims to its "near abroad")—at the same time that states with growing economies, such as Turkey, the Arab Gulf states, India and China, continue to expand their influence into former European spheres of influence and security.[40]

Due to the global nature of the financial crisis, one can accordingly expect considerable social and political unrest as politicians in the US and Europe alike debate the precise form that budget cuts and austerity measures will take.[41] On the positive side, one possibility for more effective financial and economic management is a more engaged role for the G-20, but this requires that the G-20 develop a stronger institutional capacity. Here, George Modelski has critiqued various "substitutions" for global warfare, including democratic peace theory, abolition of nuclear weaponry, planetary defense against asteroids—in addition to multilateral state engagement through the G-20 and global civil society movements through "unions for democracy" which could influence state and international organizations. (See Modelski, Chapter 28.)

On the negative side, one cannot necessarily count on the new media (internet, Facebook, Twitter, etc.) as a tool to put an end to dictatorship and prevent war. It was not, for example, the mass media itself that worked to end the Vietnam war, the "first TV war." (See Ekovich, Chapter 17.) And while the new media can publicize governmental abuse and human rights atrocities and help organize mass demonstrations by speeding communications, if not help to *delegitimize*

39 Ian Davis, "The UK-France Defence Pact and Nuclear Modernization," *NATOWatch Briefing Paper* 16 (January 6, 2011), available at: http://www.natowatch.org/sites/default/files/NATO_Watch_Briefing_Paper_No.16.pdf

40 Extending Oleg Kobtzeff argument in Chapter 9, most wars today are hybrid conflicts and possess internal domestic, regional *and* external international dimensions. The line between internal disputes within socio-political collectives and disputes among differing state leaderships is not always very clear.

41 There may be a trend to cutting the diplomatic corps and reducing costs of international organizations, as well as international aid, as ways to cut the budget, without realizing that such cuts could result in increased military expenditure, more conflicts overseas, or other unexpected costs, if irenic diplomacy is not permitted to function properly. For a study of the close relationship between austerity and social conflict, see Ponticelli, Jacopo and Voth, Hans-Joachim, "Austerity and Anarchy: Budget Cuts and Social Unrest in Europe, 1919–2009" CEPR Discussion Paper No. DP8513 (August 2011). See also, "Unrest in peace," *The Economist*, 00130613, 10/22/2011, Vol. 400, Issue 8756.

oppressive regimes, the new media *by itself* does not necessarily provide effective policy guidance or help determine strategy. The new media can promote the causes of peace, justice and human rights, but these latter goals actually represent very different policy objectives that can lead elites and populations to clash over precisely which policy priorities and strategic options to choose. By contrast with its presumed ability to expose injustice and enhance transparency, the new media can also be used to spread mis- and dis-information; it can likewise be used as recruitment for states to engage in war by means of glorifying military technological achievements (generally to impress young men, for example). The new media can appeal to alienated Muslim youth (of the demographic "youth surge") to engage in the cause of *jihad*.[42] (See Chapters 3, 23, 24, 25, 26 by DeLaet, Karatzogianni, Géré, and Charnay.) In many ways, the ghost of Napoleonic war propaganda still hangs over the contemporary usage of the mass media. (See Rimanelli, Chapter 11.) The key difference between the post-Cold War world and the propaganda of the past, however, is that the new information and media technology is now very well integrated into the preparation of war (war simulation) and is heavily involved in actual "real time" war fighting—in addition to engaging in the propaganda effort before, during and after the conflict.

Moreover, on an even deeper level—and what represents a key question that long cycle theory should address—is the question as to how the new technological innovations, produced by the information age and computer revolution (among others), will impact upon American and European "leadership"? Do such innovations represent a negative form of "creative destruction" which tends to undermine the overall global economic recovery through highly uneven development in which only certain core, semi-peripheral states and regional "nodes" (centers of trade and communication) will benefit? Or will such innovations result in a more positive form of "destructive creation" which sets the groundwork for a more equitable global development in such a way as to minimize disparities between states and those regions that might not otherwise benefit from economic recovery?

Yet it is not entirely clear that the new innovations in Information Technology (IT) will necessarily help bring about a global economic recovery (as compared to previous technological revolutions in history) in the near future.[43] In addition to the difficulties of managing macroeconomic policy, in which global sales, currency flows and exchange rates have been significantly impacted (and potentially destabilized) by IT technologies, the major dilemma is the length of the "lag time"

42 In post-September 11, 2001 circumstances, Moslems have been accused of disloyalty to US or European policies and values for not supporting wars in Afghanistan and Iraq. One is reminded of disputes between Protestants and Catholics before the Thirty Years War or accusations of socialist sedition during the Cold War.

43 See Hall Gardner "War and the Media Paradox" in Athina Karatzogianni, editor, *Cyberconflict and Global Politics* (Routledge, 2009). See, for example, study by Robert J. Gordon, "Does the New Economy Measure up to the Great Inventions of the Past?" *Journal of Economic Perspectives* Vol. 14, No. 4–5 (Fall 2000).

in which new innovations begin to directly impact the world economy on a greater scale. The longer the "lag time" without global economic recovery, the greater the difficulty in managing macroeconomic policies, and the greater the chance for geopolitical and socio-economic conflict to erupt.

A very long "lag time" without significant economic recovery looks increasingly likely. This appears true given the general rise of world energy prices coupled with the scramble for scarce raw materials, and as a consequence of belated efforts to develop viable alternative energy sources, and energy saving technologies—as forewarned as needed at least since the 1970s. These interrelated geostrategic, political economic and technological factors themselves forewarn of major social and geopolitical upheaval ahead—if more *concerted* geostrategic, technological and political economic policies in which alternative energy and IT technology could play a more positive role—cannot soon be adopted.

Will major systemic differences (including new technologies and the quality of leadership) between historical epochs trump similarities, thus averting either wider wars or wars between major powers? Or will similarities reign, resulting in a "repetition" of major power conflict although in very different geohistorical circumstances? More specifically, will the US be able to retain or restore its "leadership" much as Great Britain did in the eighteenth and nineteenth centuries? (See Thompson, Chapter 27; Modelski, Chapter 28.) Or will China, India or other new core and former semi-peripheral "leaders" arise to replace the US? But with or without global or systemic conflict?

Scenarios

The question as to whether new geostrategic, political-economic, military technological or socio-cultural "innovations" (including cyber warfare) might generate sufficient fear—or even *risk taking*—among rival leaderships to provoke major power war, and how such a conflict (or series of conflicts) might play out in the post-September 11, 2001 era remains open. But it is possible to outline a number of plausible scenarios based on the long cycles of previous conflicts. This approach is to thoroughly compare and contrast today's constellation of power interrelationships with those of previous geohistorical periods, when the paramount or leading hegemonic power or powers began to be seriously challenged by emerging powers—a process which can result in the delegitimization of the paramount power(s), the deconcentration of the international state and socio-political system and structure under that power or powers, and new efforts of coalition building by rival states (and/or by political factions of socio-political movements) who generally seek to obtain either parity with, or else supremacy over, or in some way undermine, the paramount power or powers on either regional or global levels. (See Chapters 1, 4, 8, 13, 23, 24 and Part IV.)

The collapse of the French empire after the Seven Years War (1756–63) ultimately resulted in socio-political revolution in 1789, followed by an all-European war in

1792—a roughly 30-year interval between imperial collapse and a *revanchist* major power conflict. The collapse of Imperial Germany after World War I (1914–18) led to the rise of the *revanchist* Nazi German movement in 1933 and then major power war in 1939, for an interval of 21 years. The collapse of Tsarist Russia led to civil war and then regional *revanche* as Lenin and Stalin sought to re-concentrate Soviet power on the Eurasian continent in rivalry with Nazi Germany, likewise resulting in war between the Soviet Union and Nazi Germany by 1941.

By contrast with both the German and Russian past, Soviet reforms followed by collapse in the period 1985–91 led to numerous conflicts along the fringes of the Soviet empire, particularly in the north and south Caucasus and Central Asia. Although the transition from Soviet rule was largely peaceful in most of eastern Europe, Soviet disaggregation indirectly led to international disputes and conflicts over former Cold War "buffer" zones and Soviet spheres of influence, including ex-Yugoslavia and Serbia, Afghanistan and Iraq, as well as Georgia in August 2008, if not Libya in 2011. Mikhail Gorbachev's decision to reduce Soviet support for Pyongyang and open ties with South Korea likewise, for example, helped to further alienate the North Korean dictatorship. At the same time, both a rising China and a collapsing North Korea, despite their significant differences, generally possess a common cause in opposing historical Japanese efforts to sustain regional hegemony, now backed by the United States.

In addition to Soviet collapse, a number of Cold War dictatorships supported by the US and the Europeans also collapsed (with or without violence), including the Shah of Iran, the Apartheid regime in South Africa, Ferdinand Marcos in the Philippines, Mobutu Sese Seko in Zaire, to mention a few. Of these, the collapse of the Shah of Iran (and rise of the Islamic Republic of Iran) destabilized the Gulf region, while the collapse of the Mobutu regime eventually set off the Second Congo War from 1998 to 2003—or what was called "World War III in Africa." In the latter war, eight states clashed, along with at least two dozen armed partisan organizations, in an inter-state and inter-communal conflict that killed up to four million (or more) people directly or indirectly through disease and starvation, with millions becoming refugees within their own homelands. More recently, the collapse (and feared collapse) of authoritarian regimes of the "greater Middle East"—a shatterbelt region largely created after the dissolution of the Ottoman Caliphate which had ruled Sunni Islam (as argued by Bin Laden)—has continued to exacerbate the possibilities of wider regional conflict.

In geohistorical terms, a number of global or systemic conflicts have broken out in response to the failure of the major rivals to establish long term power-sharing relations on global or regional levels. This failure has often led to defense build-ups and formation of encircling and counter-encircling alliances that not only appear intended to counter military capabilities in a strict realist sense, and weaken the allies of the rival, but which also appear designed to undermine state authority over domestic populations through support for irredentist claims, the backing of dissident or revolutionary movements, or *even through the promulgation of "innovative" practices in governance that appear to undermine the legitimacy of rival state leaderships.*

What would happen if two or more outside powers intervened militarily in support of opposing factions in the "greater Middle East" or elsewhere, much as was the case for World War I, or, going further back in time, in the Peloponnesian conflict between Athens and Sparta over Megara and Corinth? What if the US/NATO, despite efforts to "reset" US and Russian relations, clash with Moscow over Georgia, Ukraine or other states in the Black Sea region (as did *not* take place in August 2008)? What if a resurgent Russia and/or China should fully support Iran or another state in a future conflict with the United States and/or Israel? What will be the impact in Asia of a war between the US and Japan versus North Korea, among other possible conflicts? What if Russia and/or China are seized by a *revanchist* type movement, in which former communist "reds" turn "brown?" What if the US, despite efforts to establish a special G-2 relationship with China in an effort to "appease" the latter, still clash over Taiwan? There is still a danger is that third parties, North Korea, Taiwan, Japan, Pakistan, or even the Philippines, could draw the US and China into confrontation despite US-Chinese efforts to cooperate on other levels.

Which kind of conflicts might draw in the United States, as the still leading hegemonic power, along with "coalitions of the willing," in a new form of Bonapartist, yet multilateral, "democratic nationalism"—as has appeared to be the case with US-led interventions in Bosnia/Kosovo/Serbia, Afghanistan, Iraq, and Libya? Or would multiple state and social collapse result in a new form of "Thirty Years War" within the space of the former Ottoman empire involving permanent internal socio-political revolutions and interstate wars among regional, if not major, powers? Such conflict could involve Saudi-Iranian conflict over Bahrain, Yemen, Iraq, Lebanon, as well as within the largely Shi'a inhabited oil-rich provinces of Saudi Arabia itself, not to overlook Arab claims to Iranian Khuzestan (sought by Saddam Hussein in the Iran-Iraq war in the 1980s). Or, another possibility is that an Israeli preemptive strike on presumed Iranian nuclear capabilities could spark a series of Balkan-like conflicts that could potentially drag in the major powers, as was the case before World War I. Moreover, as the global demand for energy and resources has represented one of the major factors leading to conflict in the greater Middle East, how will burgeoning resource demand and search for markets impact the African continent in which many countries of the region have already found themselves caught up in violent civil and interstate strife (the Great Lakes region, the west Africa conflict zone, north Africa, the Horn of Africa, the countries surrounding Sudan and south Sudan)—as the US, Europe, India and China, among other states, continue their quest for energy and resources?

It is also possible that nomadic anti-state "terrorist" organizations (Al-Qaeda or groups supported by Iran) could provoke hostilities between rival states, seeking to plunge Iran and Saudi Arabia into direct military conflict, for example. Even following the death of Osama Bin Laden, Al-Qaeda affiliates (or other anti-state nomadic groups), could play the role of the pan-Serbian "terrorist" group, the Black Hand, by attempting a *coup d'état* in Pakistan, or in Saudi Arabia or other states. Such groups could foment a major terrorist attack in India (such as the attacks on Mumbai in November 2008 with alleged Pakistani backing)—possibly setting off war between the nuclear states India and Pakistan. Here, it is not inconceivable that a Taliban return to power in Afghanistan could result in yet another Indo-Pakistani war in which Beijing could back Pakistan, while India could be backed by Moscow. On the one hand, Russia would fear that a Pakistani-Taliban alliance would support pan-Islamic movements throughout Central Asia, if not within Russia itself. By contrast, the break-up of Pakistan, or more likely, a pan-Islamic *coup d'état*, could result in the creation of an even more hostile regime with nuclear weapons.

Averting Global War

Although the US, Russian and Chinese defense establishments do not openly predict the scenario of major power war in the near future, the problem is that a number of emerging states, as well as militant anti-state partisan movements, do not necessarily interpret American strategic intent as "benevolent." Even NATO allies, such as Germany and Turkey, appear to be distancing themselves from certain aspects of American policy, making it more difficult to coordinate defense strategy and foreign policy.

There are a number of reasons why states may choose war or forceful imperialist policies as an option, even if there are more rational and diplomatic approaches to conflict resolution. The call for war is: (1) increasingly linked to the burgeoning demand for global resources and the freedom to secure access to those resources, while the number of significant actors in search of vital raw materials and energy resources has augmented significantly; (2) the rivalry to develop and use high tech and smart weapons is being countered by the innovative use of both conventional and unconventional technologies as weapons of war; (3) the US and Europeans no longer see as one of their goals in warfare to hold territory, thus less powerful states and nomadic anti-state partisans see insurgency and asymmetrical warfare as a viable tool to destabilize whole regions. (See Géré, Chapter 24. See also Robinson, Chapter 5 on the anarchist concept of "decelerating" modes of warfare; plus Charnay on Islamic warfare, Chapter 26.)Other technocratic or neo-realist efforts to deter war may not be entirely sufficient, and could actually provoke major power conflict. Fears that Missile Defense systems as a technological innovation could be used to launch a first strike could lead states to seek a first strike capability or engage in secret and preclusive "September 11, 2001-type" attacks. At the same time, the development of tactical nuclear weaponry (still with significant explosive power) could lower the threshold in which nuclear weaponry could be used.[44] A

44 The dilemma is that the development of tactical nuclear weaponry, plus the possibility that BMD systems could be used to shield an offensive nuclear first strike, make the spread of nuclear weapons even more dangerous—not to overlook real concerns that such weaponry could fall into the hands of rogue military elements or anti-state "terrorist" groups intent on using nuclear weaponry as "blackmail." See also discussion of nuclear weaponry in Thompson, Chapter 27 and Modelski, Chapter 28.

danger could also arise whereby a rising power, such as North Korea, decides to call the bluff of the United States, and thus take the risk to engage in conflict with South Korea on the assumption that the US will not risk a counter-attack. Another danger is that asymmetrical attacks by a third "power" or anti-state group could be used to spark conflict between two rivals who would not know who actually attacked, possibly drawing major or regional powers on opposing sides.

One irenic, peace-oriented option is for the US and NATO to work with Russia to build *limited* but collectively managed, Missile Defense systems (assuming such systems are truly functional, accurate and cannot be fooled by decoys). A system of cooperative Missile Defenses, involving Moscow, and possibly Beijing, could help thwart the perception that, like ancient Athens, Washington intends to build an isolationist wall around itself, possibly without NATO or European input, thereby undermining UK, French, Russian and Chinese missile deterrents—while simultaneously strengthening both American defensive and offensive capabilities unilaterally.Moreover, the fact that the US and NATO member states must continue to prepare for contingencies of more or less symmetrical warfare against states such as Russia and China—in addition to countering the possibility of asymmetrical and cyber attacks at unexpected times—stretches available resources and manpower to the maximum in an age of significant budgetary restraints while concurrently perpetuating "dialectics of insecurity and security" (see Gardner, Chapter 1). It is accordingly to prevent these latter contingencies and to prevent *all* states from being caught up in alliance games of encirclement and counter-encirclement that a new more concerted approach to major and regional power relationships should be pursued—*one that in effect seeks to restructure the interstate system through the formation of internationalized, yet interlinking, regional security and development communities*.

Global peace will to a large extent depend upon how the US and European Union ultimately respond to President Medvedev's calls (following the August 2008 Georgia–Russia war) for a new Euro-Atlantic security order, and whether Russia itself is prepared for an entente or alliance relationship with the US and Europe, and does not continue to flirt with an alliance with China. To prevent conflicts from either spreading into wider regions or bringing major powers into military intervention at cross purposes will require the US and Europeans to forge a new concerted relationship with Russia, but without totally alienating China, among other significant states, by providing China and the latter countries with degrees of global or regional power sharing, while secondarily seeking to nudge them toward domestic reforms accepted by those societies, where possible.[45]

45 Neo-realist John Mearsheimer put it this way: "If the People's Republic grows economically over the next thirty years the way it has in recent decades ... We can expect the United States to lead a balancing coalition against China that includes India, Japan, Russia, Singapore, South Korea and Vietnam, among others." John J. Mearsheimer, "Imperial by Design," *The National Interest* (January–February 2011), http://nationalinterest.org/article/imperial-by-design-4576?page=show. The questions raised here are whether: a) such a coalition can be achieved without alienating China; b) whether Russia and other states, such as democratic India, will necessarily join such

At the same time, however, given calls for an "alliance of democracies" based on the rise of democratic peace theory, coupled with the doctrine of the "responsibility to protect," there is a major risk, if not a real danger, that an ostensibly neo-Kantian alliance of US, Europe, Japan, Australia, along with other democratic states, in support of republican majoritarian democracy and free trade, could be countered by the rise of an ostensibly neo-Hegelian alliance of authoritarian and protectionist states such as Russia, China, Iran, India, Turkey and Brazil, which would seek to defend rights of state sovereignty, even if India, Turkey and Brazil are considered democracies. Another negative scenario might involve the complete isolation of either Russia or China causing a *revanchist* backlash. While *deep* "democracy" in Russia and China *involving more truly consensual decision-making* might ultimately help moderate the behavior of these authoritarian states, domestic and international efforts to press both Russia and China toward full democratic and social reforms may not go far enough—and soon enough—to prevent potential socio-political instability from leading to either greater degrees of civil conflict within these countries or else to a *revanchist* backlash. Here, from an alternative, yet realistic standpoint, it seems necessary to work with Moscow in order to better *channel* the aspirations of China, rather than wait for a miraculous transformation of either society—while nevertheless pressing for step-by-step reforms within both countries where possible. Can the US and Europeans forge compromises with Russia over questions involving NATO and EU enlargement to the Black Sea/Caucasus region (while likewise seeking a Turkish accord with Greece and the European Union over Cyprus)? Could a compromise with Russia work to reduce conventional weaponry, if not eliminate tactical nuclear weapons from all of Europe altogether? Can the US, a reformed NATO, and the Europeans cooperate with Russia, India, China and central Asian states over the Afghan/Pakistan conflict in a regional Contact Group—a crisis which has been complicated by virulent anti-Americanism and fear of secessionist movements in Pakistan?[46] Can the US, Europeans and Russia (along with the Arab Gulf states) find a way to pressure and contain Iran, ultimately impelling regime transformation, but without embroiling the region in an even wider war—in that Teheran appears increasingly isolated given the near collapse of the Syrian regime?

Could a more concerted US-European-Russian-Saudi policy work to provide security and economic assistance for an Israeli-Palestinian compromise? Much as the 2003 Geneva Initiative proposed,[47] such an option could involve land swaps

---

a balancing coalition; and c) whether Russia might attempt to forge a counter-alliance with China, India, Iran, among others ... The problem is that neo-realism has done little to help explain how to move beyond deterrence through engaged diplomacy and seek reconciliation between the US, NATO, Europe and Russia, among other actors.

46 Hall Gardner, "Toward a New Strategic Vision for the Euro-Atlantic," *NATOwatch Briefing Paper* 15 (December 13, 2010), available at: http://www.natowatch.org/sites/default/files/NATO_Watch_Briefing_Paper_No.15.pdf.

47 The Geneva Initiative http://www.geneva-accord.org; For issues involving land swaps, see David Makovsky, http://www.nytimes.com/interactive/2011/09/12/opinion/mapping-mideast-peace.html?ref=opinion#nytg-optionsBox.

over the 1967 borderline; it would also bring together Saudi Arabia and many of the Arab states into recognition of Israel, perhaps leading to a *confederated* "two-state" solution negotiated bilaterally between Israel and the Palestinians as well as establishing a closer entente between Arab states and Israel. Peacekeepers (such as NATO's Partnership for Peace, for example) could be deployed under a general UN mandate and could help mediate between Israeli and Palestinian security forces, as well as Syria. This approach could, at least in part, represent a geostrategic step toward more effectively containing, but not entirely alienating, Iran, and checking external Iranian and Syrian support for Hamas and Hizb'allah given the near collapse of the Syrian regime. Or, if the Palestinian question cannot be resolved, will that conflict continue to generate tensions throughout the region, if not much of the world?

A close US-Russian-European-Japanese-Indian entente or alliance could consequently work to bring Russia and a democratic India into a closer relationship with the US and Europe while seeking to dampen China's rise to major power status without fundamentally alienating the Chinese, by permitting Beijing varying degrees of regional and global power sharing, for example. This approach appears plausible on the assumption that Russia perceives China to represent a potential "threat" to its international status as a major center of power and influence, a fear which is coupled by potential Chinese claims to Siberia. Moscow furthermore remains fearful of pan-Islamic movements that could destabilize its soft Islamic underbelly. Despite Moscow's more evident suspicions of NATO and EU encroachment into eastern Europe, if not into the Black Sea and Caucasus, a Euro-Atlantic entente or alliance incorporating Russia would work to formulate a step-by-step, region-by-region system of interlinking "regional security and development communities" throughout the Black Sea, Caucasus and Central Asia, in the Middle East, as well as throughout other key areas around the world. These would be backed initially by major and regional power security guarantees—until the point that such communities can become relatively autonomous and stand on their own.[48] Concurrently, the major and regional powers could seek to stabilize the global economy through closer international economic cooperation in the G-20, and through the strengthening of societal participation and *power sharing* in local, national, regional and global governance.

48 See Hall Gardner, *Averting Global War*; Hall Gardner, *American Global Strategy and the 'War on Terrorism.'* See General Introduction, fn 14, this book. In November 2011, in reaction to US Missile Defense plans, Russian president Dmitri Medvedev threatened to deploy nuclear capable missiles in Kaliningrad. See James Joyner, *Outside the Beltway* (November 23, 2011); General Nikolai Makarov warned that NATO enlargement enhanced risks of Russia being drawn into local conflicts that could "under certain conditions" develop "into a full-scale war involving nuclear weapons." *New York Times* (November 17, 2011). As both states enter elections in 2012, future leaderships need to "reset" the "reset": US and European failure or refusal to work with Moscow in forging "regional security and development communities" will result in a serious deterioration of global relations.

One dilemma to overcome in the process of reaching out to Russia or other non-democratic regimes, including China, however, is that democratic peace theory has generally downgraded the possibility of strong cooperation between democratic and non-democratic states whose governance is not based upon independent legal authority and which may engage in the extreme violations of human rights. There has been considerable recent ideological opposition to alignments with non-democratic states despite historical examples of the Anglo-Russian entente of 1907 before World War I or else the US-Soviet alliance during World War II.[49] Prior to World War I, at a time when Britain was confronted by the rise of amphibious Imperial Germany, insular-democratic Great Britain looked to an alliance with continental-authoritarian Russia in the period 1898–1907—in seeking compromise over Tibet, Afghanistan, Baluchistan, as well as Persia (ironically at the expense of the Persian Constitutional Revolution). From this perspective, much as an insular democratic Britain was able to forge an entente with an authoritarian Russia in 1907, the United States and Europe need to reach out for a Euro-Atlantic alliance with the Russian Federation in finding compromise over the Black Sea and Caucasus, Central Asia, Afghanistan, Pakistan, and now, theocratic Iran, among other countries and regions, while at the same time, working with regional powers such as India and the Gulf states, in the effort to *channel* the rise of China, but once again, without alienating the latter.

The efforts to achieve an Anglo-Russian entente in the period 1898 to 1907 could accordingly serve as a possible model for efforts to establish a substantial transformation of US, European, Russian and Japanese relations—but in new, early twenty-first-century circumstances that demand a more concerted relationship between major and regional powers and greater cooperation with divergent socio-political movements and factions. Likewise, another possible geohistorical model involving an insular core state reaching out to an amphibious state includes Great Britain's effort to forge an entente with France that sought to incorporate Austria and the United Dutch Provinces versus a revisionist Spain (from 1718–1731), while concurrently attempting to reintegrate Madrid into a European concert. Such a model of a Quadruple Alliance would parallel a US-European-Russian-Japanese entente intended to draw a reforming China into a global concert, while simultaneously seeking to resolve tensions within the Euro-Atlantic area.

By contrast, the possible failure to bring Russia into a closer entente or alliance relationship with the US, Europe and Japan could lead Russia and China to eventually

49 An Anglo-French-German alliance that might have prevented World War I, or else deflected that war toward Russia, and not against Imperial Germany, was never really attempted. See Gardner, Chapter 13. But here, to take D'Agostino's critique into consideration (D'Agostino, Chapter 14), one should examine different dimensions of the World War I analogy, not just the Anglo-German relationship. Without systematic contrast and comparison, historical analogies appear flat, one-dimensional and potentially misleading, as D'Agostino argues. If, however, the key differences and similarities are depicted systematically, then the historical analogy can possess greater analytical and heuristic relevance.

forge a tighter alliance relationship, along with states such as India, Turkey, Brazil and Iran—despite the apparent obstacles to such a scenario. An alienated Russia and/or China could likewise seek to support pan-Islamic movements against US and European interests—much as was the case for Imperial German support for the pan-Islamic cause before World War I as illustrated in the 1905 and 1911 Moroccan crises. Another concern revolves around the fact that while Great Britain had succeeded in reaching out to both France and Tsarist Russia, forging geostrategic and colonial compromises prior to World War I, London failed miserably in its efforts forge an entente or alliance with Imperial Germany, which in turn tightened its relations with Austria-Hungary—a failure which culminated in major power warfare, sparked in the Balkans. From this perspective, there is still a real danger that US and European efforts to reach out to Russia even in sincere diplomatic engagement, and in likewise seeking constructive cooperation with China as well, could still fail on one or both counts. Given the depth of the global economic crisis in the core financial centers of the US and Europe as contrasted with the over 3 trillion dollars of foreign exchange reserves of China in particular (raising the question as to which country really "won" the Cold War!), plus the apparent inability of the US and Europeans to sort out their strategic-economic priorities with respect to countries like Russia, Turkey, India, Saudi Arabia, Israel, Iran as well as China (states which individually or collectively could refuse to be pressed into cooperation), there is a real danger that half-hearted efforts to averting global war could fail.

The prospects for wider regional conflicts, if not global war, appear to be burgeoning, not waning, given the global finance crisis and significant policy disputes within democratic countries themselves, opposition to authoritarian regimes within societies that are often polarized between contending socio-political groups and factions, and ongoing conflicts/wars in resource-rich regions across the globe. The range of global disputes and conflicts are furthermore taking place in an era in which differing forms of weapons of mass destruction, conventional and not-so-conventional weaponry and IT technologies continue to proliferate. The task of sustaining global peace will consequently require dedicated leaderships—pressed by an active and informed citizenry—to engage in new diplomatic approaches, concerted policies and actions with both democratic and non-democratic states alike, where possible. It may prove necessary, if not obligatory, to renovate the interstate system by transforming it step-by-step toward a global confederation of interlinked "regional security and development communities." Long cycles of major power war may appear to militate against peace, but there is nothing that prevents engaged socio-political movements and leaderships from seeking to defy the millennia in the effort to resolve, or at least ameliorate, disputes and conditions that could once again lead to seemingly repetitive local, regional and international violence and conflict.

Index

For Product Safety Concerns and Information please contact our EU representative GPSR@taylorandfrancis.com Taylor & Francis Verlag GmbH, Kaufingerstraße 24, 80331 München, Germany

Batch number: 10726296

Printed by Printforce, the Netherlands